Modeling of Complex Dynamic Systems

Modeling of Complex Dynamic Systems

Fundamentals and Applications

Vladimir Stojanović
Faculty of Mechanical Engineering, University of Niš, Niš, Serbia; Department of Civil Engineering, Faculty of Engineering, Lakehead University, Thunder Bay, ON, Canada

Jian Deng
Department of Civil Engineering, Faculty of Engineering, Lakehead University, Thunder Bay, ON, Canada

Marko D. Petković
Department of Computer Science, Faculty of Sciences and Mathematics, University of Niš, Niš, Serbia

Marko A. Ristić
Faculty of Mechanical Engineering, University of Niš, Niš, Serbia

ELSEVIER

Elsevier
Radarweg 29, PO Box 211, 1000 AE Amsterdam, Netherlands
125 London Wall, London EC2Y 5AS, United Kingdom
50 Hampshire Street, 5th Floor, Cambridge, MA 02139, United States

ISBN: 978-0-443-23942-7

For Information on all Elsevier publications
visit our website at https://www.elsevier.com/books-and-journals

Publisher: Matthew Deans
Acquisitions Editor: Chiara Giglio
Editorial Project Manager: Toni Louise Jackson
Production Project Manager: Maria Bernard
Cover Designer: Greg Harris

Typeset by MPS Limited, Chennai, India

Dedication

Dedication by Vladimir Stojanović:
In loving memory of my mother.

Contents

Part II Stability of vibrations of complex moving objects: Modeling and applications

About the authors

Vladimir Stojanović, PhD

Dr. Stojanović currently holds an Associate Professor position at the Faculty of Mechanical Engineering, University of Niš, Serbia, and a Visiting Professor position at Lakehead University, Ontario, Canada. Since 2011, he has published numerous papers in prestigious scientific journals, where he also serves as a reviewer. As an internationally recognized scientist, he has been invited to lecture at universities worldwide and has participated in the most significant conferences in the field of theoretical and applied mechanics. His research interests primarily focus on the modeling of complex linear and nonlinear continuous and discrete dynamical systems, analytical and numerical methods for solving MDOF-based models, and the application of dynamic and stochastic stability principles to engineering problems in vibration.

Jian Deng, PhD, P.Eng

Dr. Deng is currently Chair and Associate Professor at the Department of Civil Engineering at Lakehead University, Canada. He is a licensed professional engineer in Ontario and has over 20 years of research, education, and industry experience. Dr. Deng's research group is developing theories and techniques to better understand dynamic instability, geohazards, and the reliability of structures in civil and mining engineering, especially in geotechnical engineering. His excellence in research has been recognized through several awards, including the Best Paper Award, the Research Excellence Award, and the Best Presentation Award. These accolades underscore the impact of Dr. Deng's guidance and contributions to both research excellence and student development.

Marko D. Petković, PhD

Dr. Petković received a PhD in Computer Science in 2008 from the University of Niš, Serbia. He is a prolific author, the editor-in-chief of the *Facta Universitatis, Series: Mathematics and Informatics*, and section editor for three other well-known journals in the field. He is the author of more than 15 software solutions, both commercial and noncommercial scientific software. His research interests include numerical linear algebra, neural networks, source and channel coding, and optimization methods.

Marko A. Ristić

Marko A. Ristić received his MSc degree in Mechanical Design and Mechanization in 2009 from the University of Niš, Serbia. Currently, he works as a Researcher at the Faculty of Mechanical Engineering, University of Niš, to complete his doctoral thesis. His research interests include machine component design, finite element analysis, computational fluid dynamics, gear design, planetary gear trains, and high-conformal gearing.

Preface

Advances in science are closely related to the progress of society as a whole. The basis of any progress is the previous research used to analyze newly observed natural phenomena or laws and thus enable further development and formation of patterns of application in various fields of science. As in other fields of science, as well as in the field of engineering sciences, the observation of repetitive physical phenomena, which have some specific characteristics, encourages researchers to use mathematical formulations in general to enable their detailed analysis. The integration of scientific research and technical practice leads to the development, understanding, and ability to make the best use of natural processes and resources in the future. Specifically, in engineering, researchers have focused on dynamic processes for decades and have tried to understand them fully by describing them. One of the basic undesirable scenarios in engineering practice is the disruption of dynamic stability and unpredictability of events resulting due to their complexity as well as various external influences that can lead to structure breakage or damage, for example, in frequent repetitive cyclic operations. This book addresses the aforementioned problem and is the result of many years of scientific research conducted by the authors, who invested significant effort in developing the manuscript. The departure of the stable and controlled zone is conditioned by unpredictable natural processes and external disturbances. Fortunately, global technological development has made it possible for computational mechanics to more quickly and easily validate a large number of results obtained in theoretical research in the description of very complex mechanical models. This enables both quantitative and qualitative development of the analysis as well as shortens the time required for its implementation. The development of computational mechanics depends on the type and complexity of mechanical models that are described because their different types in nature lead to the need for modifications and adaptations of previously used both numerical and other research methods.

The entire scientific research covered by this book describes primarily complex dynamic systems, formulated within the phenomenon of oscillations related to the analysis of stabilization characteristics integrated into three areas: stability of vibrations of complex moving objects, nonlinear vibrations, and stochastic stability of mechanical systems. This book contains a considerable number of analyzed complex dynamic models in which the possibilities of new types of stabilization as well as their quantitative and qualitative effects are presented. The analysis of mechanical systems with stabilization parametric physical characteristics includes a comparison with classical dynamic models whose motion and stability are already known to see the advantages of the proposed changes. The presented analysis of mechanical systems contains original results and newly proposed technical solutions in which the benefits of different stabilization characteristics are presented. It is necessary to point out that the analysis of certain mechanical systems required improvements and adjustments of numerical methods. Integrally, within the entire research, a large number of different analytical and numerical methods were applied, both for analysis and verification of the obtained results: modal expansion method, trigonometric method, Newmark method, Richardson extrapolation method, harmonic balance method, continuation method, perturbation method, D-decomposition method, argument principle, contour integration method, Laplace integral transform, p-version of the FEM, Fourier integral transform, Galerkin method, the method of stochastic averaging, Monte Carlo method, and others.

This book is designed for researchers, doctoral candidates, master's and postgraduate students, as well as engineers specializing in mechanical, civil and aerospace engineering, vibrations, dynamic stability, elastodynamics, structural stochastic stability, and vibration control.

Part I

Fundamental mathematical background of dynamics and vibrations: Overview of numerical methods and recent improvements

Mathematical methods and procedures in the analysis of stability of vibrations of complex moving objects

1

1.1 Introduction to complex numbers

The set of complex numbers $\mathbb{C}$ is an extension of the set of real numbers $\mathbb{R}$, such that all quadratic equations of the form $x^2 + px + q = 0$, where $p, q \in \mathbb{R}$ have solutions. The basic building block is the solution of the equation $x^2 + 1 = 0$, which we denote by $\boldsymbol{i}$. In other words, we introduce an element $\boldsymbol{i} \notin \mathbb{R}$ such that $\boldsymbol{i}^2 = -1$ and extend the set $\mathbb{R}$ with the element $\boldsymbol{i}$. To maintain the algebraic operations, one must also include all elements of the form $z = x + \boldsymbol{i}y$ where $x, y \in \mathbb{R}$. Therefore the new set $\mathbb{C}$ is given by $\mathbb{C} = \{x + \boldsymbol{i}y | x, y \in \mathbb{R}\}$. The addition and multiplication of two elements $z_k = x_k + \boldsymbol{i}y_k$, $k = 1, 2$ from the set $\mathbb{C}$ are defined by

$$z_1 + z_2 = (x_1 + x_2) + \boldsymbol{i}(y_1 + y_2),$$

$$z_1 \cdot z_2 = (x_1 + \boldsymbol{i}y_1)(x_2 + \boldsymbol{i}y_2) = x_1x_2 - y_1y_2 + \boldsymbol{i}(x_1y_2 + x_2y_1). \tag{1.1}$$

More formally, we define the set $\mathbb{C}$ to be the set of pairs $(x, y) \in \mathbb{R}^2$ and the operations $+$ and $\cdot$ are formally defined on $\mathbb{C}$, according to Eq. (1.1).

Definition 1.1.1. *The set of complex numbers* $\mathbb{C}$ *is defined by* $\mathbb{C} = \mathbb{R}^2 = \{(x, y) \mid x, y \in \mathbb{R}\}$. *Operations* $+ ,\bullet$: $\mathbb{C}^2 \rightarrow \mathbb{C}$ *are defined as follows:*

$$(x_1, y_1) + (x_2, y_2) = (x_1 + x_2, y_1 + y_2),$$

$$(x_1, y_1) \cdot (x_2, y_2) = (x_1x_2 - y_1y_2, x_1y_2 + x_2y_1). \tag{1.2}$$

Denote the special element $\mathbf{i} = (0, 1)$ *and also denote every element* $(x, 0)$ *where* $x \in \mathbb{R}$, *simply by* x.

It can be directly shown that, for every $z = x + \boldsymbol{i}y \neq 0$, the element

$$z^{-1} = \frac{x}{x^2 + y^2} - \boldsymbol{i}\frac{y}{x^2 + y^2}, \tag{1.3}$$

satisfies $z^{-1} \cdot z = z \cdot z^{-1} = 1$. Similarly, for every $z = x + \boldsymbol{i}y \in \mathbb{C}$, the element $-z = -x + \boldsymbol{i}(-y)$ satisfies $z + (-z) = (-z) + z = 0$. The set of complex numbers $\mathbb{C}$, equipped with the operations $+$ and $\cdot$, unary operations $^{-1}$ and $-$, as well as constants 0 and 1, in the field.

Theorem 1.1.1. *The structure* $(\mathbb{C}, + ,\bullet, - ,\_^{-1}, 0, 1)$ *is the field. Set* $\{(x, 0) \mid x \in \mathbb{R}\}$ *forms the subfield which is isomorphic to the field of real numbers* $\mathbb{R}$.

Modeling of Complex Dynamic Systems. DOI: https://doi.org/10.1016/B978-0-443-23942-7.00001-2

Complex numbers are naturally represented in the *complex plane* (Fig. 1.1), where the point (x, y) in the standard 2D plane corresponds to the complex number $z = x + \boldsymbol{i}y$. Denote by (ρ, ϕ) the corresponding polar representation of the point (x, y), where $\phi \in [-\pi, \pi]$.

The following definition corresponds to the additional fundamental terms regarding complex numbers.

Definition 1.1.2. *If $z = x + \boldsymbol{i}y \in \mathbb{C}$, then we denote real and imaginary parts by* $\mathrm{Re}\, z = x$ *and* $\mathrm{Im}\, z = y$, *respectively, modulus* $|z| = \rho = \sqrt{x^2 + y^2}$, *a conjugate complex number of z by* $\bar{z} = x - \boldsymbol{i}y$ *and the argument* $\arg z = \phi$. *Also, define* $e^{i\phi} = \cos\phi + \boldsymbol{i}\sin\phi$. *It is known as Euler's formula.*

The terms introduced in the Definition 1.1.2 are shown in Fig. 1.1.

According to the definition, one holds $z = x + \boldsymbol{i}y = \rho(\cos\phi + \boldsymbol{i}\sin\phi) = \rho e^{i\phi}$. The general exponent a^{iy} is defined by $e^{iy \log a}$ and $a^z = a^x \cdot a^{iy}$, for arbitrary $z = x + \boldsymbol{i}y \in \mathbb{C}$ and $a \in \mathbb{R} \setminus \{0\}$.

Trigonometric functions of the complex argument are defined by

$$\sin z = \frac{e^{iz} - e^{-iz}}{2\boldsymbol{i}}, \quad \cos z = \frac{e^{iz} + e^{-iz}}{2},$$

$$\tan z = \frac{\sin z}{\cos z}, \quad \cot z = \frac{\cos z}{\sin z}.$$

Theorem 1.1.2. *The following statements are valid for arbitrary* $z_1, z_2 \in \mathbb{C}$ *and* $\phi_1, \phi_2 \in \mathbb{R}$:

1. $\overline{z_1 z_2} = \overline{z_1}\,\overline{z_2}$, $\overline{z_1 + z_2} = \overline{z_1} + \overline{z_2}$
2. $\mathrm{Re}(z_1 + z_2) = \mathrm{Re}(z_1) + \mathrm{Re}(z_2)$, $\mathrm{Im}(z_1 + z_2) = \mathrm{Im}(z_1) + \mathrm{Im}(z_2)$
3. $\mathrm{Re}(z_1 \cdot z_2) = \mathrm{Re}(z_1)\mathrm{Re}(z_2) - \mathrm{Im}(z_1)\mathrm{Im}(z_2)$, $\mathrm{Im}(z_1 \cdot z_2) = \mathrm{Re}(z_1)\mathrm{Im}(z_2) + \mathrm{Im}(z_1)\mathrm{Re}(z_2)$
4. $|z_1 z_2| = |z_1||z_2|$, $|z_1 + z_2| \le |z_1| + |z_2|$
5. $e^{i\phi_1} \cdot e^{i\phi_2} = e^{i(\phi_1 + \phi_2)}$

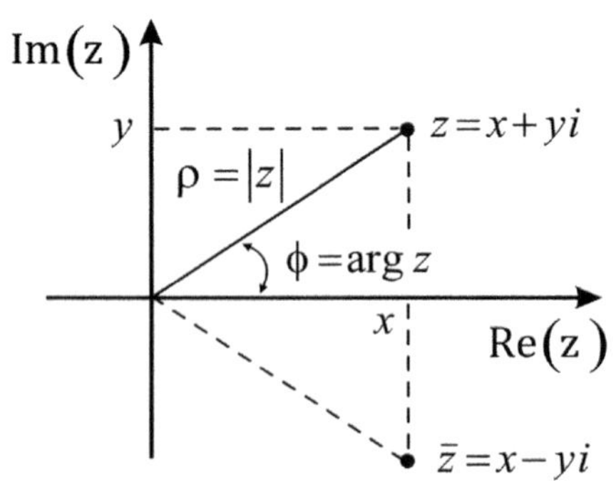

Figure 1.1 The complex plane. Formed by the complex numbers, with a Cartesian coordinate system.

6. $\arg(z_1 z_2) = -\pi + (\arg(z_1) + \arg(z_2)) mod 2\pi$

7. $\overline{z_1}^{-1} = \overline{z_1^{-1}}$, $|z_1^{-1}| = |z_1|^{-1}$, $\arg(z_1^{-1}) = -\arg(z_1)$

Recall that if $b > 0$, we define a mod $b = c$ such that $a = kb + c$ where $0 < c < b$ and $k \in \mathbb{Z}$.

Also, since $d(z_1, z_2) = |z_1 - z_2|$ is the metric and $(\mathbb{C}, d)$ *is the complete metric space*. Using that, one can define convergence $z_n \to z \in \mathbb{C} \cup \{\infty\}$ when $n \to +\infty$, for a given sequence $(z_n)_{n \in \mathbb{N}}$ of the complex numbers. In other words, $z_n \to z$ when $n \to +\infty$, (or $\lim_{n \to +\infty} z_n = z$), if $|z_n - z| \to 0$.

1.2 Complex functions and their continuity and differentiability

The function $f: V \to \mathbb{C}$, $V \subseteq \mathbb{R}$ is known as a *complex-valued function of the real argument*, while $f: U \to \mathbb{C}$, $U \subset \mathbb{C}$ is known as *complex function*. Writing $f(t) = u(t) + iv(t)$, we get two real functions $u(t) = \mathrm{Re}(f(t))$ and $v(t) = \mathrm{Im}(f(t))$. Therefore all definitions of the function limit, as well as the continuous, differentiable, and integrable function f, reduce to the same definitions for both $u(t)$ and $v(t)$. Similarly, one naturally defines the derivative by $f'(t) = u'(t) + iv'(t)$. The case with the complex function $f(z)$ is more advanced. Similarly, to the previous case, one can write $f(x + iy) = u(x, y) + iv(x, y)$ where u, v are two-variable real functions.

By *ϵ-neighborhood* of the complex number $z_0 \in \mathbb{C}$, we assume the set $U_{z_0,\epsilon} = \{z \in \mathbb{C} \mid |z - z_0| < \epsilon\}$, for every $\epsilon > 0$. Similarly, by the *M-neighborhood* of the point, ∞ we assume the set $U_{\infty,M} = \{z \in \mathbb{C} \mid |z| > M\}$, for any $M > 0$. In what follows, we sometimes omit ϵ and M and refer to both neighborhoods simply by *neighborhood* and denote by U_{z_0} (or U_∞) as well.

Definition 1.2.1. *Assume that the function $f(z)$ is defined on some neighborhood D of $z_0 \in \mathbb{C} \cup \{\infty\}$ and let $A \in \mathbb{C} \cup \{\infty\}$. Then $f(z)$ converges to A when $z \to z_0$ (when the notation $f(z) \to A$, or $\lim_{z \to z_0} f(z) = A$) when for arbitrary neighborhood U_{z_0}, there exists a neighborhood U_A such that $z \in U_{z_0} \setminus \{z_0\}$ implies $f(z) \in U_A$.*

The previous definition also covers the cases when z_0 or A is infinite. In this case, $z_0, A \in \mathbb{C}$, it is equivalent to $u(x, y) \to \mathrm{Re}(A)$ and $v(x, y) \to \mathrm{Im}(A)$ when $(x, y) \to (\mathrm{Re}(z_0), \mathrm{Im}(z_0))$.

Definition 1.2.2. *For $z_0 \in D$, function f is continuous in z_0 if every neighborhood $U_{f(z_0)}$, there exists a neighborhood U_{z_0} such that $f(U_{z_0} \cap D) \subseteq U_{f(z_0)}$.*

Similarly, function $f(z)$ is continuous at $z_0 = x_0 + iy_0$ if and only if, $u(x, y)$ and $v(x, y)$ are continuous at (x_0, y_0).

Definition 1.2.3. *For $z_0 \in D$, function f is differentiable (has a derivative) at z_0 if there exists $A \in \mathbb{C}$ such that*

$$\frac{f(z) - f(z_0)}{z - z_0} \to A,$$

when $z \to z_0$. In that case, we denote $f'(z_0) = A$. If it is differentiable at every $z_0 \in D$, then it is differentiable.

The n-th derivative $f^{(n)}$ is defined inductively by $f^{(n)} = [f^{(n-1)}]'$.

It can be shown that f is differentiable at $z_0 = x_0 + iy_0$ if and only $u(x, y)$ and $v(x, y)$ are differentiable at (x_0, y_0). The differentiable functions satisfy the following well-known property, similar to the case of the real functions.

Theorem 1.2.1. *Let f_1 and f_2 be two differentiable functions at $z_0 \in \mathbb{C}$. Then*

1. $f(z) = f_1(z) + f_2(z)$ *is differentiable and* $f'(z_0) = f'_1(z_0) + f'_2(z_0)$;
2. $f(z) = f_1(z) f_2(z)$ *is differentiable and* $f'(z_0) = f'_1(z_0) f_2(z_0) + f_1(z_0) f'_2(z_0)$;
3. *if* $f_1(z_0) \neq 0$ *then* $q(z) = [f_1(z)]^{-1}$ *is differentiable at* z_0 *and holds* $q'(z_0) = -f'_1(z_0)/f_1^2(z_0)$;

Theorem 1.2.2. *Let $f: D \to \mathbb{C}$, $D \subseteq \mathbb{C}$ by differentiable function. Then*

1. *if* $\gamma: [a, b] \to D$ *is differentiable and* $h(t) = f(\gamma(t))$ *then* $h'(t) = f'(\gamma(t))\gamma'(t)$.
2. *if* $g: D' \to D$, $D' \subseteq \mathbb{C}$ *is differentiable and* $h(z) = f(g(z))$ *then* $h'(z) = f'(g(z))g'(z)$.

Table 1.1 shows the derivatives of some common elementary functions (assuming that $a, b, c \in \mathbb{C}$, and $k \in \mathbb{Z}$).

Definition 1.2.4. *Function $f: D \to \mathbb{C}$ is analytic (holomorphic) in $D \subseteq \mathbb{C}$ if it is differentiable at any point $z_0 \in D$. It is entire if it is analytic and $D = \mathbb{C}$.*

For example, functions z^k, e^z, $\sin z$, $\cos z\,(k \in \mathbb{N})$ are entire. On the other hand, the example of a function which is not differentiable at any point $z_0 \in \mathbb{C}$ is $f(z) = \bar{z}$.

Table 1.1 Derivatives of some common elementary functions.

$f(z)$	$f'(z)$
c	0
$az+b$	a
z^k	kz^{k-1}
e^z	e^z
$\sin z$	$\cos z$
$\cos z$	$-\sin z$
$\tan z$	$(\cos z)^{-2}$

1.3 Integrals of the complex functions

Similarly, as in the previous section, we introduce the integral of the function $f: V \to \mathbb{C}$, $V \subseteq \mathbb{R}$ by

$$\int_{t_0}^{t_1} f(t)\mathrm{d}t = \left[\int_{t_0}^{t_1} u(t)\mathrm{d}t\right] + \boldsymbol{i}\left[\int_{t_0}^{t_1} v(t)\mathrm{d}t\right],$$

where $f(t) = u(t) + \boldsymbol{i}v(t)$, and under the assumption that both $u(t)$ and $v(t)$ are integrable.

Assume now that $f: D \to \mathbb{C}$, $D \subseteq \mathbb{C}$ is the complex function, and $z_1, z_2 \in \mathbb{C}$ are given complex numbers. To define an integral of $f(z)$ from z_1 to z_2, we first need to specify the *path of integration* which connects z_1 and z_2. This is the path (in the complex plane) along which we perform the integration.

Definition 1.3.1. *A path γ: $[a, b] \to \mathbb{C}$ is the continuous function, where $a, b \in \mathbb{R}$, and $a < b$. If $\gamma(a) = z_1$* and *$\gamma(b) = z_2$, for given points $z_1, z_2 \in \mathbb{C}$, then γ connects z_1 and z_2. If* $\mathrm{Re}\,\gamma(t)$ *and* $\mathrm{Im}\gamma(t)$ *are smooth (continuously-differentiable) functions on $[a, b]$, then path γ is smooth.*

Usually, the set of values $\{\gamma(t)|t \in [a, b]\}$ is referred to as the path itself, but (as we will see later) different paths can have the same set of values, but also have different integrals of the same function.

Definition 1.3.2. *Assume that f: $D \to \mathbb{C}$ is a continuous function at every point $z_0 \in D$, and γ is the smooth path. Then $\int_\gamma f(z)\mathrm{d}z = \int_a^b f(\gamma(t))\gamma'(t)\mathrm{d}t$.*

If we denote $f(x + \boldsymbol{i}y) = u(x, y) + \boldsymbol{i}v(x, y)$ and $\gamma(t) = x(t) + \boldsymbol{i}y(t)$, expanding the integral on the right-hand side of the expression from Definition 1.3.2, we get

$$\begin{aligned}\int_\gamma f(z)\mathrm{d}z &= \int_a^b [u(x(t), y(t)) + \boldsymbol{i}v(x(t), y(t))][x'(t) + \boldsymbol{i}y'(t)]\mathrm{d}t \\ &= \int_a^b [u(x(t), y(t))x'(t) - v(x(t), y(t))y'(t)]\mathrm{d}t \\ &\quad + \boldsymbol{i}\int_a^b [u(x(t), y(t))y'(t) + v(x(t), y(t))x'(t)]\mathrm{d}t, \end{aligned} \tag{1.4}$$

where both integrals are now integrals of the real functions.

Assume that γ: $[a, b] \to \mathbb{C}$ is the smooth path and $\overline{\gamma}$: $[\overline{a}, \overline{b}] \to \mathbb{C}$ is another path defined by $\overline{\gamma}(t) = \gamma(\psi(t))$ where ψ: $[\overline{a}, \overline{b}] \to [a, b]$ is a continuously-differentiable function such that $\psi(a) = \overline{a}$ and $\psi(b) = \overline{b}$.

Lemma 1.3.1. *Integrals of the continuous function f: $D \to \mathbb{C}$, $D \subseteq \mathbb{C}$ over the paths $\overline{\gamma}$ and γ are equal.*

Proof. Writing $\gamma(t) = x(t) + iy(t)$ and $\overline{\gamma}(t) = \overline{x}(t) + i\overline{y}(t)$, since $\overline{x}(t) = x(\psi(t))$ we get $\overline{x}'(t) = x'(\psi(t))\psi'(t)$. Similarly, we get $\overline{y}'(t) = y'(\psi(t))\psi'(t)$ and after some calculations, using integration by change of variables, one obtains that

$$\begin{aligned}\int_a^b u(\overline{x}(t), \overline{y}(t))\overline{x}'(t)\mathrm{d}t &= \int_a^b u(x(\psi(t)), y(\psi(t)))x(\psi(t))\psi'(t)\mathrm{d}t \\ &= \int_c^d u(x(s), y(s))x(s)\mathrm{d}s.\end{aligned}$$

In the same way, we prove similar statements for the other three integrals in Eq. (1.4), so one can conclude that $\int_{\overline{\gamma}} f(z)\mathrm{d}z = \int_{\gamma} f(z)\mathrm{d}z$.

Therefore we can always assume that the domain of the path is [0, 1] (introducing $\overline{\gamma}$: $[0, 1] \to \mathbb{C}$ by $\overline{\gamma}(t) = \gamma(a + (b - a)t)$) or any other segment. The following theorem can be proved by direct application of the Eq. (1.4).

Theorem 1.3.2. *Let f_1, f_2 be continuous functions in the domain $D \subseteq \mathbb{C}$ and let $a \in \mathbb{C}$ be an arbitrary complex number. Let γ: $[a, b] \to \mathbb{C}$ be the smooth path in D. Then*

1. $\int_{\gamma} (f_1(z) + f_2(z))\mathrm{d}z = \int_{\gamma} f_1(z)\mathrm{d}z + \int_{\gamma} f_2(z)\mathrm{d}z$,
2. $\int_{\gamma} af_1(z)\mathrm{d}z = a\int_{\gamma} f_1(z)\mathrm{d}z$,
3. $\int_{-\gamma} f_1(z)\mathrm{d}z = -\int_{\gamma} f_1(z)\mathrm{d}z$,

where $-\gamma$ is the path defined by $(-\gamma)(t) = \gamma(a + b - t)$ for every $t \in [a, b]$.

The length of the path γ, denoted by $l(\gamma)$, can be computed via the complex integral, that is,

$$l(\gamma) = \int_a^b \gamma'(t)\mathrm{d}t = \int_{\gamma} \mathrm{d}s. \tag{1.5}$$

The following lemma shows the majorization property of the complex integrals. It can be used to show that the limit value of the integral over some class of paths (usually semicircles and circles) is zero, under certain conditions.

Lemma 1.3.3. *If $|f(z)| \le M$ for some $M > 0$ and every $z \in D$, then*

$$\left|\int_{\gamma} f(z)\mathrm{d}z\right| \le M \cdot l(\gamma) \tag{1.6}$$

One of the most important complex integrals is the integral of the function $(z - z_0)^n$ where $n \in \mathbb{Z}$ and $z_0 \in \mathbb{C}$, over the circle centered in z_0 and with radius $r > 0$.

Lemma 1.3.4. *Let γ: $[0, 2\pi] \to \mathbb{C}$ be the path defined by $\gamma(t) = z_0 + re^{it}$ where $z_0 \in \mathbb{C}$ and $r > 0$. Then*

$$\int_{\gamma} (z - z_0)^n \mathrm{d}z = \begin{cases} 2\pi i, & n = -1, \\ 0, & \textit{otherwise}. \end{cases}$$

for arbitrary $n \in \mathbb{Z}$.

Proof. Let $n = -1$. Then from Definition 1.3.2, we get

$$\int_\gamma (z - z_0)^{-1}dz = \int_0^{2\pi} r^{-1}e^{-it}ire^{it}dt = i\int_0^{2\pi} dt = 2\pi i.$$

Similarly, assuming that $n \neq -1$ one gets

$$\int_\gamma (z - z_0)^n dz = \int_0^{2\pi} r^n e^{int} ire^{it}dt = \int_0^{2\pi} ir^{n+1}e^{i(n+1)t}dt$$

$$= \int_0^{2\pi} ir^{n+1}(\cos((n+1)t) + i\sin((n+1)t))dt$$

$$= i\int_0^{2\pi} r^{n+1}\cos((n+1)t)dt - \int_0^{2\pi} \sin((n+1)t)dt = 0,$$

which completes the proof.

Definition 1.3.3. *A precise smooth path γ is the function γ: $[a, b] \to \mathbb{C}$ such that there exists $n + 1$ points $t_0, t_1, \ldots, t_n \in \mathbb{R}$ satisfying $a = t_0 < t_1 < \ldots < t_{n-1} < t_n = b$ and each restriction $\gamma|_{[t_k, t_{k-1}]}$ is the smooth path ($k = 0, 1, \ldots, n - 1$). By restriction $\gamma|_S$ of the function γ: $A \to \mathbb{C}$ on the set $S \subseteq A$, we mean the same function on the domain S, that is, the function $\bar{\gamma}$: $S \to \mathbb{C}$ such that $\bar{\gamma}(z) = \gamma(z)$ for all $z \in S$. The Integral $\int_\gamma f(z)dz$ over the piecewise smooth path is simply defined as*

$$\sum_{k=0}^{n-1} \int_{\gamma|_{[t_k, t_{k+1}]}} f(z)dz. \tag{1.7}$$

Using the basic properties of real integrals and Eq. (1.4), one can show that Eq. (1.7) is valid also in the case of the smooth path γ and arbitrary $a = t_0 < t_1 < \ldots < t_{n-1} < t_n = b$.

1.4 Cauchy integral theorem and formula

The two most important and extensively used results regarding the complex integrals are the formulas for computing these integrals of holomorphic functions over closed paths.

Definition 1.4.1. *Path γ: $[0, 1] \to D$, $D \subseteq \mathbb{C}$ is closed if $\gamma(0) = \gamma(1)$. Closed path γ_1: $[0, 1] \to D$ is homotopic to the path γ_2: $[0, 1] \to D$ if there exists a continuous function H: $[0, 1]^2 \to D$ such that $H(t, 0) = \gamma_1(t)$, $H(t, 1) = \gamma_2(t)$, $H(0, t) = H(1, t)$ for every $t \in [0, 1]$.*

Informally speaking, the closed paths γ_1 and γ_2 are homotopic if one can be obtained from another by (continuous) deformation of the rubber band. In fact, $H(t, 0)$ represents the initial shape of the band, $H(t, 1)$ the final shape, while $H(0, t) = H(1, t)$ shows that the band must always represent the closed path. Also, the condition $H(t, s) \in D$ shows that every intermediate shape cannot leave the set D.

For example, take $D = \mathbb{C}\{0\}$ and paths γ_0, γ_1, and γ_2 as in Fig. 1.2. Then γ_1 is homotopic to γ_2 (and vice versa), but γ_0 is not homotopic to both γ_1 or γ_2 (since it surrounds the obstacle at 0).

One of the main results in the complex analysis is the following Cauchy-Goursat theorem, which characterizes the integral of the holomorphic function over the homotopic paths.

Theorem 1.4.1. *(Cauchy Integral Theorem). Let $f\colon D \to \mathbb{C}$, $D \subset \mathbb{C}$ be the holomorphic function and γ_1, γ_2 be two piecewise smooth, closed, and homotopic paths. Then*

$$\int_{\gamma_1} f(z)\mathrm{d}z = \int_{\gamma_2} f(z)\mathrm{d}z.$$

The most important corollary of the previous theorem is the case when the set D is simply connected, that is, if every closed path γ is homotopic to the constant curve ($\gamma(t) = c$ for some $c \in D$ and every $t \in [0, 1]$).

Corollary 1.4.2. *If $D \subseteq \mathbb{C}$ is simply connected and $f\colon D \to \mathbb{C}$ is holomorphic, then*

$$\int_{\gamma} f(z)\mathrm{d}z = 0.$$

for every piecewise smooth closed path $\gamma\colon [0, 1] \to D$.

Denote by $\gamma_{z_0,r}(t) = z_0 + re^{frm-e\pi it}$ the circular path, centered in z_0 and with radius r.

Another important and widely used result is given by Theorem 1.4.3.

Theorem 1.4.3. *(Cauchy Integral Formula). Let $f\colon D \to \mathbb{C}$, $D \subset \mathbb{C}$ be the holomorphic function, $z_0 \in D$ such that the r-neighborhood $U_{z_0,r} \subseteq D$ for some $r > 0$, and $\gamma\colon [0, 1] \to D \setminus \{z_0\}$ be a piecewise smooth, closed path that is homotopic to the circular path $\gamma_{z_0,r}$. Then*

$$f(z_0) = \frac{1}{2\pi i}\int_{\gamma} \frac{f(z)}{z - z_0}\mathrm{d}z.$$

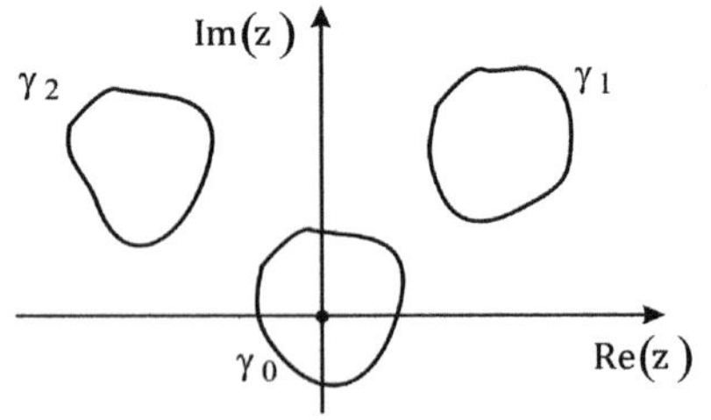

Figure 1.2 Example of the homotopy in $D = \mathbb{C}\backslash\{0\}$. Path γ_1 is homotopic to γ_2, while not homotopic to γ_0.

In a similar way, one can provide the integral formula for the n-th derivative of the holomorphic function f. At the same time, it shows that the holomorphic function f is infinitely times differentiable.

Theorem 1.4.4. *Under the assumptions of the previous theorem, there exists the n-th derivative $f^{(n)}(z_0)$ for every $n \in \mathbb{N}$ and is equal to*

$$f^{(n)}(z_0) = \frac{n!}{2\pi i}\int_\gamma \frac{f(z)}{(z - z_0)^{n+1}}\mathrm{d}z.$$

1.5 Taylor and Laurent series

Every holomorphic function f in the domain D which contains some neighborhood of the point z_0, can be expressed in the form of the Taylor series. It is shown in the following theorem. Recall that $U_{z_0,R} = \{z \in \mathbb{C}\ |z - z_0| < R\}$ is R-neighborhood of the point $z_0 \in \mathbb{C}$, while $\gamma_{z_0,r}(t) = z_0 + re^{2\pi i t}$ is the circle with the center in z_0 and radius r.

Theorem 1.5.1. *(Taylor series). Assume that f: $D \to \mathbb{C}$, $U_{z_0,R} \subseteq D \subseteq \mathbb{C}$ is holomorphic function and let r, $R \in \mathbb{R}$ be arbitrary real numbers such that $0 < r < R$. Then*

$$f(z) = \sum_{n=0}^{+\infty} a_n(z - z_0)^n,\ a_n = \frac{f^{(n)}(z_0)}{n!} = \frac{1}{2\pi i}\int_{\gamma_{z_0,r}} \frac{f(z)}{(z - z_0)^{n+1}}\mathrm{d}z, \tag{1.8}$$

for every $n \in \mathbb{N} \cup \{0\}$.

Taylor series representation directly implies the following convenient classification of the zeros of the holomorphic functions.

Theorem 1.5.2. *(Zeros). Under the assumption of the previous theorem, let $f(z_0) = 0$. Then either $f(z) = 0$ for all $z \in D$ or there exists $m \in \mathbb{N}$ and holomorphic function g: $D \to \mathbb{C}$, such that $g(z_0) \neq 0$ and $f(z) = (z - z_0)^m g(z)$ for every $z \in D$.*

The number m from the previous theorem is known as the order of the zero z_0 of $f(z)$.

Now consider the holomorphic function f on the segment (annulus) $A_{z_0,r,R} = \{z \in \mathbb{C} | r < |z - z_0| < R\}$, for a given point $z_0 \in \mathbb{C}$ and real numbers $0 < r < R$.

Theorem 1.5.3. *(Laurent series). Let f: $D \to \mathbb{C}$ be the holomorphic function, where $A_{z_0,r,R} \subseteq D \subseteq \mathbb{C}$ for some $z_0 \in D$ and $0 < r < R$. Then*

$$f(z) = \sum_{n=-\infty}^{+\infty} a_n(z - z_0)^n,\ a_n = \frac{1}{2\pi i}\int_{\gamma_{z_0,r}} \frac{f(z)}{(z - z_0)^{n+1}}\mathrm{d}z \tag{1.9}$$

for every $n \in \mathbb{Z}$.

For every holomorphic function f: $D\backslash\{z_0\} \to \mathbb{C}$, such that $U_{z_0,R} \subseteq D \subseteq \mathbb{C}$, we say that f has *an isolated singularity in* z_0. The previous theorem is fundamental in the characterization of the types of isolated singularities.

Definition 1.5.1. *Isolated singularity z_0 of the function f is:*

- ***removable****, if there exists the holomorphic function* $\hat{f}: D \to \mathbb{C}$ such *that* $\hat{f}(z) = f(z)$ *for every* $z \in D\backslash\{z_0\}$;
- ***pole****, if* $f(z) \to \infty$ *when* $z \to z_0$;
- ***essential****, if it is neither removable nor pole.*

The characterization is obtained using the coefficients in the Laurent series (Theorem 1.5.3) and given as follows.

Theorem 1.5.4. *(Singularities). Assume that z_0 is an isolated singularity of the holomorphic function f: $D\backslash\{z_0\} \to \mathbb{C}$. Then*

- *z_0 is removable if and only if $(z - z_0)f(z) \to 0$ when $z \to z_0$. It is equivalent to $a_n = 0$ for $n < 0$;*
- *z_0 is pole if and only if it is not removable and $(z - z_0)^{m+1}f(z) \to 0$, when $z \to z_0$ for some $m \in \mathbb{N}$. It is equivalent to $a_{-m} \neq 0$ and $a_n = 0$ $(n < -m)$ for some $m \in \mathbb{N}$;*
- *z_0 is essential if and only if $(z - z_0)^m f(z) \nrightarrow 0$, when $z \to z_0$, for every $m \in \mathbb{N}$. It is equivalent to an $a_n \neq 0$ for infinitely many $n < 0$.*

If z_0 is the pole, minimal index $m \in \mathbb{N}$ such that $a_n = 0$ for $n < -m$ is the order of the pole z_0. It means that

$$f(z) = \frac{a_{-m}}{(z - z_0)^m} + \frac{a_{-m+1}}{(z - z_0)^{m-1}} + \ldots + a_0 + a_1(z - z_0) + a_2(z - z_0)^2 + \ldots \tag{1.10}$$

1.6 Residues

The residue theorem is a powerful tool for evaluating many real and complex integrals of the (holomorphic) functions with the finite number of poles.

Definition 1.6.1. *Assume that f: $D\backslash\{z_0\} \to \mathbb{C}$ is the holomorphic function with the pole z_0. The residue of the function f at z_0 is the coefficient a_{-1} in the Laurent series,* Eq. (1.9). *It is denoted by* $\mathrm{RES}\, f(z)\,|_{z=z0} = a_{-1}$.

Before we formulate the main theorem, we give the following effective formula for the computation of the residue of the function f at the pole z_0 of order m.

Lemma 1.6.1. *Assume that z_0 is the pole of f: $D\backslash\{z_0\} \to \mathbb{C}$ of order m. Then*

$$\frac{1}{(m-1)!}[(z - z_0)^m f(z)]^{(m-1)} \to \mathrm{RES}\, f(z)\,|_{z=z0}.$$

when $z \to z_0$.

As a direct corollary, we get that if $m = 1$, the residue $\mathrm{RES}(|_{f,z0})$ is obtained as the limit value of $(z - z_0)f(z)$ when $z \to z_0$. Now we can formulate the main theorem.

Theorem 1.6.2. *(Residue Theorem). Assume that f: $D \setminus \{p_1, p_2, \ldots, p_m\} \to \mathbb{C}$ is a holomorphic function where $p_1, p_2, \ldots, p_m$ are poles. Assume that γ is a piecewise smooth, closed path which is $D \setminus \{p_j\}$-homotopic to the circle γ_{p_j, r_j} for some $r_j > 0$ and $j = 1, 2, \ldots, k$. Then*

$$\int_\gamma f(z)\mathrm{dz} = 2\pi i \sum_{j=1}^{k} \mathrm{RES}\, f(z)\,|_{z=p_j}. \tag{1.11}$$

The Residue Theorem (Theorem 1.6.2) is effectively used in the computation of the various real function integrals.

Example 1.6.1. Consider the integral

$$\int_{-\infty}^{+\infty} \frac{x^2}{x^4 + 1}\mathrm{dx}.$$

Denote by $f(z) = z^2/(z^4 + 1)$. It is defined for every $z \in \mathbb{C}$ except the points

$$p_j = e^{i\pi(2j+1)/4},\ j = 0, 1, 2, 3.$$

Consider the path γ like in Fig. 1.3, for some $R > 0$. Denote by γ_0 the part on the real axis, and by γ_1 the semicircle path. If $|z| = R$ then $|f(z)| < |z|^{-2} = R^{-2}$ for every z along the path γ_1 so Lemma 1.3.3 implies

$$\left|\int_{\gamma_1} f(z)\mathrm{dz}\right| \le R^{-2} l(\gamma_1) = \pi/R \to 0$$

when $R \to +\infty$. Also

$$\int_{\gamma_0} f(z)\mathrm{dz} = \int_{-R}^{+R} f(x)\mathrm{dx} \to \int_{-\infty}^{+\infty} f(x)\mathrm{dx}.$$

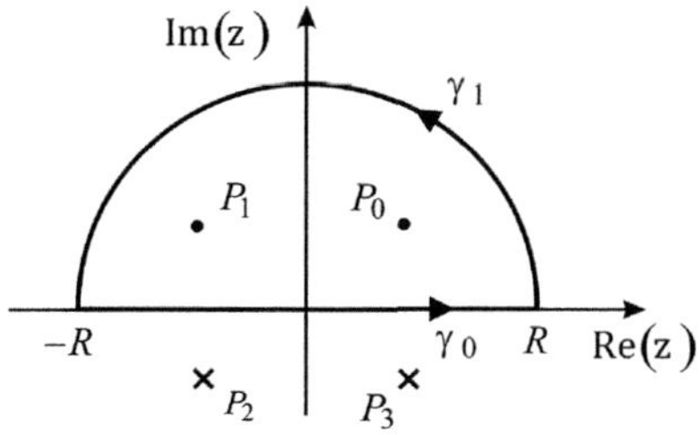

Figure 1.3 Contour of integration. The poles of *f(z)* are according to the Example 1.6.1.

On the other hand, Theorem 1.6.2 implies

$$\int_\gamma f(z)\mathrm{d}z = 2\pi\boldsymbol{i}\,(\mathrm{RES}\,f(z)\mid_{z=p_0} + \mathrm{RES}\,f(z)\mid_{z=p_1}).$$

since only p_0 and p_1 are inside the path γ. From $z^4 + 1 = (z - p_0)(z - p_1)(z - p_2)(z - p_3)$, we get that both p_0 and p_1 has order 1, so $\mathrm{RES}\,f(z)\mid_{z=p_0} = \lim_{z\to p_j}(z - p_1)f(z)$. So, we compute

$$\mathrm{RES}\,f(z)\mid_{z=p_0} = \lim_{z\to p_0}\frac{z^2}{(z-p_1)(z-p_2)(z-p_3)}$$

$$= \frac{z^2}{(p_0-p_1)(p_0-p_2)(p_0-p_3)} = -\frac{1}{4}e^{(3\pi/4)\boldsymbol{i}},$$

$$\mathrm{RES}\,f(z)\mid_{z=p_1} = \lim_{z\to p_0}\frac{z^2}{(z-p_0)(z-p_2)(z-p_3)}$$

$$= \frac{z^2}{(p_1-p_0)(p_1-p_2)(p_1-p_3)} = -\frac{1}{4}e^{(\pi/4)\boldsymbol{i}}.$$

and therefore

$$\int_{-\infty}^{+\infty} f(x)\mathrm{d}x = \int_\gamma f(z)\mathrm{d}z = -\frac{\pi}{4}\boldsymbol{i}\left[e^{\left(\frac{\pi}{4}\right)i} + e^{\left(\frac{3\pi}{4}\right)i}\right] = \pi\frac{\sqrt{2}}{2}.$$

By a direct generalization of the procedure shown in the previous example, we can obtain the following corollary of the Residue Theorem (Theorem 1.6.2) which is effectively used for the integral computation in Chapters 4–9.

Corollary 1.6.3. *Assume that f is a holomorphic function defined on the set $D\setminus\{p_1, p_2, \ldots, p_k\}$, where $D = \{z \in \mathbb{C} \mid \mathrm{Im}\,z \geq 0\}$ (or $D = \{z \in \mathbb{C} \mid \mathrm{Im}\,z \leq 0\}$) and $p_1, p_2, \ldots, p_k \in D$ are poles of f. Assume that $zf(z) \to 0$ when $|z| \to +\infty$ and $z \in D$. Then*

$$\text{v. p.} \int_{-\infty}^{+\infty} f(x)\mathrm{d}x = 2\pi\boldsymbol{i}\sum_{j=1}^{k}\mathrm{RES}\,f(z)\mid_{z=p_j}. \tag{1.12}$$

Recall that v. p. $\int_{-\infty}^{+\infty} f(x)\mathrm{d}x = \lim_{R\to+\infty}\int_{-R}^{+R} f(x)\mathrm{d}x$.

Note that, if the domain of f is a larger set than the upper half-plane D as in Corollary 1.6.3, only poles in D are taken into account in the right-hand side sum of Eq. (1.12). That is the case, for example, for the function $f(z)$ from Example 1.6.1.

1.7 The principle of argument

Another important application of the complex integrals is counting the number of zeros and/or poles of a complex function $f(z)$ in a certain region. This problem is directly related to the stability of the mechanical systems.

Theorem 1.7.1. *(Argument principle). Under the assumptions of the* Theorem 1.6.2, *if* $f(\gamma(t)) \neq 0$ *for every* $t \in [0, 1]$, *then*

$$\frac{1}{2\pi i}\int_{\gamma}\frac{f'(z)}{f(z)}\mathrm{d}z = n - k, \tag{1.13}$$

where n *is the number of zeros in the interior of the path* γ.

If one can write $f(\gamma(t)) = g(t)e^{i\varphi(t)}$, where g and φ are continuously-differentiable real functions. The key point is to define the argument function $\phi(t)$ to be continuously-differentiable. Defining it simply by $\phi(t) = \arg f(\gamma(t))$ does not ensure this, due to the discontinuity at values 0 and 2π. Then

$$\frac{f'(\gamma(t))}{f(\gamma(t))}\gamma'(t) = \frac{[f(\gamma(t))]'}{f(\gamma(t))}$$

$$= \frac{g'(t)e^{i\varphi(t)} + g(t)i\varphi'(t)e^{i\varphi(t)}}{g(t)e^{i\varphi(t)}} = \frac{g'(t)}{g(t)} + i\varphi'(t).$$

Now Theorem 1.7.1 implies

$$n - k = \frac{1}{2\pi i}\int_{\gamma}\frac{f'(z)}{f(z)}\mathrm{d}z = \frac{1}{2\pi i}\int_0^1\left[\frac{g'(t)}{g(t)} + i\varphi'(t)\right]\mathrm{d}t$$

$$= \frac{1}{2\pi i}[\log g(1) - \log g(0) + i(\varphi(1) - \varphi(0))].$$

Now since γ is closed, one has $\gamma(0) = \gamma(1)$ so $f(\gamma(0)) = f(\gamma(1))$ and therefore $g(0) = g(1)$. The previous expression implies

$$n - k = \frac{1}{2\pi}(\varphi(1) - \varphi(0)).$$

Therefore $n - k$ is equal to the ***total*** *change of the argument* of $f(z)$ *along the path* γ, divided by 2π. The problem is how to (numerically) compute this quantity in practice.

The idea is to divide the segment $[0, 1]$ into k segments $[t_j, t_{j+1}]$, $j = 0, 1, \ldots, k-1$, where $0 < t_0 < t_1 < \ldots < t_k = 1$, and compute

$$n - k = \frac{1}{2\pi}\sum_{j=0}^{k-1}(\varphi(t_{j+1}) - \varphi(t_j)).$$

Denote the difference $\Delta\phi_j = \varphi(t_{j+1}) - \varphi(t_j)$. Then it is equal to $\Delta\phi_j = \arg(f(\gamma(t_{j+1}))/f(\gamma(t_j)))$ if $\Delta\phi_j \in [-\pi, \pi]$. This condition is satisfied if the difference

$t_{j+1} - t_j$ is sufficiently small for every $j = 0, 1, \ldots, k + 1$. The step size $\Delta t_j = t_{j+1} - t_j$ can be dynamically adopted, depending on the argument change $\Delta\phi_j$.

The effective procedure for generating the points $0 < t_0 < t_1 < \ldots < t_k = 1$ and computing $Arg = \varphi(1) - \varphi(0)$ can be summarized as follows.

ArgChange$(f, \gamma, \Delta\phi_{\min}, \Delta\phi_{\max}, \Delta t_0)$:

1. Set $j = 0$, $t_0 = 0$, $Arg = 0$ and $\Delta t = \Delta t_0$.
2. Compute $t_{j+1} = t_j + \Delta t$.
3. If $t_{j+1} > 1$, set $\Delta t = 1 - t_j$ and go to the step 2.
4. Compute $\Delta\phi = \arg(f(\gamma(t_{j+1}))/f(\gamma(t_j)))$.
5. If $|\Delta\phi| > \Delta\phi_{\max}$, set $\Delta t = \Delta t/2$ and go to the step 2.
6. Set $Arg = Arg + \Delta\phi$.
7. If $|\Delta\phi| < \Delta\phi_{\min}$, set $\Delta t = \Delta t \cdot 2$.
8. If $t_{j+1} = 1$, stop and return Arg.
9. Set $j = j + 1$ and go to the step 2.

The parameters of the procedure ArgChange are the minimal and maximal values of the argument difference $\Delta\phi_{\min}$ and $\Delta\phi_{\max}$, as well as the initial step size Δt_0. In step 5, if the argument change is too large, we discard it and retry with halved the step size. In step 7, if the change is too small, we double the step size for the next step. The actual values of parameters are adjusted for the concrete function f (see Chapters 4–9).

1.8 Laplace transform

Definition 1.8.1. *Let f: $\mathbb{R} \to \mathbb{R}$ be a piecewise continuous function, such that $f(t) = 0$ for $t < 0$* and *$|f(t)| < Me^{at}$ for some real numbers a, $M > 0$, and every $t > 0$. Function F: $\mathbb{C} \to \mathbb{C}$ defined by*

$$F(s) = \int_{-\infty}^{+\infty} f(t)e^{-st}\mathrm{d}t$$

*is **Laplace (integral) transform** of $f(t)$ and denoted by $F(s) = \mathscr{L}(f(t))$.* Formally, it should be written $F = \mathscr{L}(f(t))$, but for the sake of simplicity, the arguments of the function are kept.

The variable s is known as the *Laplace variable*. Function $f(t)$ is usually referred to as *original*, while $F(s)$ is referred to as an image under the Laplace transform.

The previous definition can be extended to the generalized functions (distributions), including the well-known *Dirac delta function* $\delta(t)$. Recall that Dirac delta function $\delta(t)$ is the "object" such that $\int_{-a}^{b} \delta(t)\mathrm{d}t$ for every $a, b > 0$ and $\delta(t) = 0$ for every $t \neq 0$. It belongs to the class of generalized functions (also known as distributions). More about generalized functions reader can find, for example, in Refs. [1–5].

Theorem 1.8.1. *Assume that f, f1, and f2 are functions satisfying the conditions of the*

Definition 1.8.1 and let $F_1(s) = \mathscr{L}(f_1(t))$, $F_2(s) = \mathscr{L}(f_2(t))$, *and* $F(s) = \mathscr{L}(f(t))$. *Then*

1. $\mathscr{L}(\alpha f_1(t) + \beta f_2(t)) = \alpha F_1(s) + \beta F_2(s)$, *for every* $\alpha, \beta \in \mathbb{R}$;
2. $\mathscr{L}(f(at)) = a^{-1}F(a^{-1}s)$, *for* $a > 0$;
3. $\mathscr{L}(f(t-a)) = e^{-as}F(s)$, *for* $a > 0$;
4. *If f is k-times differentiable, then* $\mathscr{L}(f^{(k)}(t)) = s^kF(s) - \sum_{j=0}^{k-1} s^{k-j-1}f^{(j)}(0)$;
5. $\mathscr{L}\left(\int_0^t f(w)\mathrm{dw}\right) = F(s)/s$.

Table 1.2 shows the Laplace transform of several elementary functions, as well as the Dirac delta function. Note that each entry corresponds to the function multiplied by the function

$$h(t) = \begin{cases} 1, & t > 0, \\ 0, & otherwise. \end{cases}$$

known as the Heaviside function.

One important operation, which has a nice connection with the Laplace transform is convolution.

Definition 1.8.2. *If* $f, g: \mathbb{R} \to \mathbb{R}$ *are continuous functions, the convolution* $f*g$ *is the function defined by*

$$(f*g)(t) = \int_0^t f(t-w)g(w)\mathrm{dw}.$$

Properties of the convolution, as well as the connection with the Laplace transform, are given in the following theorem.

Theorem 1.8.2. *Let f1, f2, and f3 be two continuous functions satisfying the condition of Definition 1.8.1. Then*

1. $f_1*f_2 = f_2*f_1$, $f_1*(f_2*f_3) = (f_1*f_2)*f_3$;
2. $f_1*(af_2*bf_3) = a(f_1*f_3) + b(f_2*f_3)$, *for any* $a, b \in \mathbb{R}$;
3. $\mathscr{L}((f_1*f_2)(t)) = F_1(s)F_2(s)$ *where* $F_j(s) = \mathscr{L}(f_j(t))$, $j = 1, 2$.

Table 1.2 Laplace transform of some common elementary functions.

f(t)	*F(s)*
e^{-at}	$1/(s+a)$
1	$1/s$
t^k, $k \in N$	$s^{-(k+1)}\ k!$
$\cos t$	$s/(s^2+1)$
$\sin t$	$1/(s^2+1)$
$\delta(t)$	1

One can describe the inverse of the Laplace transform $\mathscr{L}$, under certain conditions for the function $f(t)$.

Theorem 1.8.3. *(Mellin inverse formula). Assume that f satisfies the condition of* Definition 1.8.1. *Moreover, assume that f has finitely many extremal points. Let the integral* $\int_0^{+\infty} f(t)e^{-(a+iw)t}dt$ *be uniformly convergent for some* $a > 0$ *and* $w \in \mathbb{R}$. *Then*

$$f(t) = \frac{1}{2\pi}\int_{-\infty}^{+\infty} F(s+iw)e^{(s+iw)t}\mathrm{dt}. \tag{1.14}$$

The following example shows how the Fourier transform can be used to solve the differential equation with the Dirac delta function on the right-hand side.

Example 1.8.1. Consider the differential equation given by

$$\ddot{x}(t) + 3\dot{x}(t) + 2x(t) = h(t) - h(t - t_0),$$

with the initial conditions $\dot{x}(0) = x(0) = 0$. It corresponds to the stability problem of the mechanical system, under the impulse external force, modeled by the finite width impulse term $h(t) - h(t - t_0)$ on the right-hand side. Applying the Laplace transform, and using Theorem 1.8.1, we get that $\mathscr{L}(\ddot{x}(t)) = s^2X(s)$, $\mathscr{L}(\dot{x}(t)) = s^2X(s)$ and

$$(s^2 + 3s + 2)X(s) = \frac{1 - e^{t_0 s}}{s},$$

implying that

$$X(s) = \frac{1 - e^{t_0 s}}{s^2 + 3s + 2} = \frac{1 - e^{t_0 s}}{s + 1} - \frac{1 - e^{t_0 s}}{s + 2}.$$

Now from Table 1.2 and Theorem 1.8.1, one finds $\mathscr{L}(e^{-t} - e^{-(t-t_0)}) = (1 - e^{-t_0 s})/(s + 1)$ and $\mathscr{L}(e^{-2t} - e^{-2(t-t_0)}) = (1 - e^{-t_0 s})/(s + 2)$, so

$$x(t) = \mathscr{L}^{-1}(X(s)) = (e^{-t} - e^{-2t})h(t) - (e^{-(t-t_0)} - e^{-2(t-t_0)})h(t - t_0).$$

Since both coefficients in the exponential function terms have negative real parts (−1 and −2, respectively), we conclude that the solution decays to 0 when $t \to +\infty$, that is, that the corresponding system is stable.

1.9 Fourier transform

Another important integral transformation similar to the Laplace transform is the Fourier transform. It is essential in any sort of harmonic analysis, signal processing, and many more similar topics.

Definition 1.9.1. *Assume that* $f\colon \mathbb{R} \to \mathbb{R}$ *is infinitely differentiable function which satisfies* $\lim_{x\to\pm\infty} f^{(n)}(x) = 0$ *for every* $n \geq 0$. *Then* $F\colon \mathbb{R} \to \mathbb{C}$ *defined by*

$$F(k) = \int_{-\infty}^{+\infty} f(t) e^{-ikx} \mathrm{dx},$$

*is the **Fourier transform** of the function f(x), and is denoted by* $F(k) = \mathscr{F}(f(t))$.

As in the case of the Laplace transform, the Fourier transform can be extended to the generalized functions, including the Dirac delta function $\delta(t)$. The main properties of the Fourier transform are given in the following theorem and are similar to the corresponding properties of the Laplace transform (Theorem 1.8.1).

Theorem 1.9.1. *Assume that f, f1, and f2 are functions satisfying the conditions of the Definition 1.9.1 and let* $F_1(k) = \mathscr{F}(f_1(t))$, $F_2(k) = \mathscr{F}(f_2(t))$, *and* $F(k) = \mathscr{F}(f(t))$. *Then*

1. $\mathscr{F}(\alpha f_1(t) + \beta f_2(t)) = \alpha F_1(k) + \beta F_2(k)$, *for every* $\alpha, \beta \in \mathbb{R}$;
2. $\mathscr{F}(f(at)) = a^{-1} F(a^{-1}k)$, *for* $a > 0$;
3. $\mathscr{F}(f(t-a)) = e^{iak} F(k)$, *for* $a \in \mathbb{R}$;
4. $\mathscr{F}(f^{(n)}(t)) = (ik)^n F(k)$;
5. $\mathscr{F}((f_1 * f_2)(t)) = F_1(k) F_2(k)$;
6. $\mathscr{F}(f_1(t) f_2(t)) = (2\pi)^{-1} F_1(k) F_2(k)$.

The inverse transformation is computed in a similar way to the original, as shown in the next theorem.

Theorem 1.9.2. *(Inverse Fourier transform). Assume that the function f satisfies the conditions of Definition 1.9.1. Then*

$$f(x) = \frac{1}{2\pi} \int_{-\infty}^{+\infty} F(k) e^{ikx} \mathrm{dk}. \tag{1.15}$$

Example 1.9.1. Consider the same differential equation as in Example 1.8.1 but now with Dirac delta impulse on the right-hand side:

$$\ddot{x}(t) + 3\dot{x}(t) + x(t) = \delta(t).$$

Applying the Fourier transform, and using Theorem 1.9.1, we get

$$X(k)((ik)^2 + 3ik + 2) = 1,$$

so

$$X(k) = \frac{1}{-k^2 + 3ik + 2}.$$

Now applying the inverse Fourier transform (Theorem 1.9.2) one obtains the solution

$$x(t) = \frac{1}{2\pi}\int_{-\infty}^{+\infty} X(k)e^{ikt}\mathrm{dt} = \frac{1}{2\pi}\int_{-\infty}^{+\infty} \frac{e^{ikt}}{-k^2 + 3ik + 2}\mathrm{dt}.$$

We use the complex integration and Residue theorem to solve the corresponding integral. The poles of the function $f(k) = e^{ikt}/(-k^2 + 3ik + 2)$ are $p_1 = \boldsymbol{i}$ and $p_2 = 2\boldsymbol{i}$. In order to select the proper integration contour, we distinguish two cases:

1. $t > 0$. The integration contour is given in the Fig. 1.4 (left). Let $k = Re^{i\tau} = R(\cos(\tau) + \boldsymbol{i}\sin(\tau))$ for $\tau \in (0, \pi)$. Then

$$e^{ikt} = e^{itR(\cos(\tau)+\boldsymbol{i}\sin(\tau))} = e^{-tR\sin(\tau)+itR\cos(\tau)} = e^{-tR\sin(\tau)} \cdot e^{itR\cos(\tau)}.$$

Now since $t > 0$ and $\sin(\tau) > 0$, we have $|e^{ikt}| = e^{-tR\sin(\tau)} \leq 1$. That further implies

$$|kf(k)| \leq \frac{|k|}{|-k^2 + 3ik + 2|} \leq \frac{R}{R^2 - 3R - 2} \to 0.$$

When $R \to +\infty$. According to Corollary 1.6.3, we get

$$x(t) = \frac{1}{2\pi}\int_{-\infty}^{+\infty} X(k)e^{ikt}\mathrm{dt}$$

$$= \boldsymbol{i}\left[\mathrm{RES}X(k)e^{ikt}|_{k=i} + \mathrm{RES}X(k)e^{ikt}|_{k=2i}\right]$$

$$= \boldsymbol{i}\left[-\frac{e^{ikt}}{z - 2\boldsymbol{i}}\bigg|_{k=i} - \frac{e^{ikt}}{z - \boldsymbol{i}}\bigg|_{k=2i}\right] = e^{-t} - e^{-2t}.$$

2. $t > 0$. The integration contour is given in the Fig. 1.4 (right). Performing a similar analysis, if $k = Re^{i\tau}$ for $\tau \in (\pi, 2\pi)$, we get $|e^{ikt}| = e^{-tR\sin(\tau)} \leq 1$ since $t < 0$ and $\sin(\tau) < 0$.

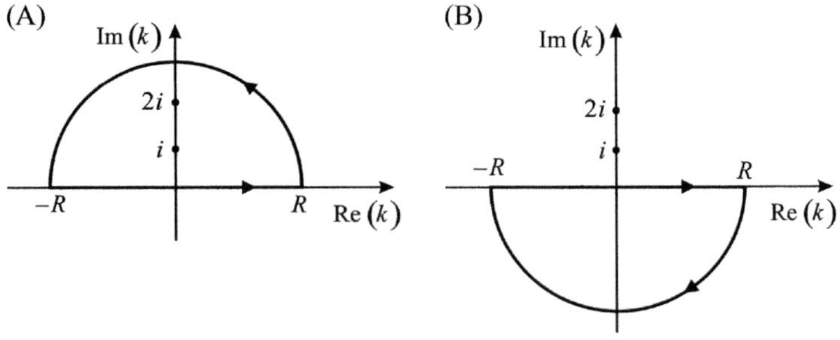

Figure 1.4 Contours of integration for the inverse Fourier transform in Example 1.9.1. (A) $t > 0$, (B) $t < 0$.

According to Corollary 1.6.3, we get

$$x(t) = \frac{1}{2\pi}\int_{-\infty}^{+\infty} X(k)e^{ikt}\mathrm{dt} = 0,$$

since there are no poles in the lower half-plane.

Combining both cases, one gets

$$x(t) = (e^{-t} - e^{-2t})h(t).$$

References

[1] S.M. Sasane, A. Sasane, A Friendly Approach to Complex Analysis, 2nd ed., World Scientific, 2024.

[2] W. Rudin, Real and Complex Analysis, 3rd ed., McGraw-Hill, 1986.

[3] S.G. Krantz, Handbook of Complex Variables, Birkhuser, Boston, MA, 1999, pp. 212–214.

[4] M. Rahman, Applications of Fourier Transforms to Generalized Functions, WIT Press, 2011 ISBN 978-1-84564-564-9.

[5] D. Kammler, A First Course in Fourier Analysis, Prentice Hall, 2000 ISBN 978-0-13-578782-3.

Mathematical methods and applications in the analysis of nonlinear vibrations

2

2.1 Problem formulation

Consider the dynamical system (equations of motion) of the form

$$m\ddot{q}(t) + c\dot{q}(t) + k(q(t))q(t) = f(t), \tag{2.1}$$

where $q: \mathbb{R}^+ \to \mathbb{R}$ is the unknown function, $f: \mathbb{R}^+ \to \mathbb{R}$, $k: \mathbb{R} \to \mathbb{R}^+$ are given functions, and $m, c \in \mathbb{R}^+$ are given constants. This equation models the single mechanical oscillator with the external force and nonlinear spring. The following are mechanical equivalents of the components of the Eq. (2.1):

1. Unknown function q is the displacement;
2. Constants m and c are mass and damping coefficient, respectively;
3. Function f represents the external force;
4. Function k represents the nonlinear stiffness of the spring.

We assume that $k(x)$ is decomposed into the linear and nonlinear parts as $k(x) = k_l + k_{nl}(x)$, where $k_l = k(0) \in \mathbb{R}^+$ is linear stiffness, while $k_{nl}: \mathbb{R} \to \mathbb{R}^+$ is a nonlinear stiffness function, satisfying $k_{nl}(0) = 0$.

The similar dynamical system of the multidimensional mechanical oscillator is given in the matrix form by

$$\mathbf{M}\ddot{\mathbf{q}}(t) + \mathbf{C}\dot{\mathbf{q}}(t) + \mathbf{K}(q(t)) = \mathbf{f}(t), \tag{2.2}$$

where $\mathbf{q}: \mathbb{R}^+ \to \mathbb{R}^n$ is the unknown displacement vector, $\mathbf{M}, \mathbf{C} \in \mathbb{R}^{n\times n}$ are symmetric matrices, $\mathbf{K}: \mathbb{R}^n \to \mathbb{R}^{n\times n}$ is the given matrix function such that $\mathbf{K}(\mathbf{x})$ is symmetric for every $\mathbf{x} \in \mathbb{R}^n$ and $\boldsymbol{f}: \mathbb{R}^+ \to \mathbb{R}^n$ is the given function. Following are the mechanical representations:

1. Unknown function $\boldsymbol{q}$ is the vector of displacements;
2. Constant matrices $\boldsymbol{M}$ and $\boldsymbol{C}$ are inertia and damping matrices, respectively;
3. Function $\boldsymbol{f}$ represents the vector of external forces;
4. Matrix function $\boldsymbol{K}$ represents the nonlinear stiffness coefficients matrix.

Again, we decompose $\boldsymbol{K}$ into linear and nonlinear parts by $\mathbf{K}(\mathbf{x}) = \mathbf{K}_l(\mathbf{x}) + \mathbf{K}_{nl}(\mathbf{x})$ such that $\mathbf{K}_l = \mathbf{K}(0)$ and $\mathbf{K}_{nl}(0) = 0$.

The first step in the analysis of the dynamical systems (Eqs. 2.1 and 2.2) is *time-domain analysis*. It is used to compute the response generated by suddenly applied force $f(t)$, which can be both periodic and nonperiodic. These forces can be applied

Modeling of Complex Dynamic Systems. DOI: https://doi.org/10.1016/B978-0-443-23942-7.00002-4

either for a short period (*transient response*, after which the system remains in the free vibrations mode) or continually for an indefinite period.

2.2 Numerical methods for solving general dynamical systems

One can use and apply an arbitrary method for numerical solving of ordinary differential equations (ODEs) to the dynamical systems (Eqs. 2.1 and 2.2). Although several specialized methods for this type of system are developed and extensively used. In this section, we give some basic theory and introduce some general-purpose numerical methods for solving the (system of) ODEs, while in the next section, we deal with the specialized method for the dynamical system of type (Eqs. 2.1 and 2.2).

Consider the general form of the dynamical system

$$\dot{\mathbf{q}}(t) = \mathbf{f}(t, \mathbf{q}(t)),$$

$$\mathbf{q}(0) = \mathbf{q}_0, \tag{2.3}$$

where $\mathbf{q}$ is the unknown (vector) function, $\mathbf{f} : \mathbb{R} \times \mathbb{R}^d \to \mathbb{R}^d$ is known function ($d \in \mathbb{N}$), while $\mathbf{q}_0 \in \mathbb{R}^d$ is known initial vector. Note that the system (Eq. 2.2) is reduced to the form (Eq. 2.3) by introducing $\mathbf{q}_{10}(t) = \mathbf{q}(t)$, $\mathbf{q}_{11}(t) = \dot{\mathbf{q}}(t)$, and

$$\underbrace{\begin{bmatrix} \dot{\mathbf{q}}_{10}(t) \\ \dot{\mathbf{q}}_{11}(t) \end{bmatrix}}_{=\dot{\mathbf{q}}_1(t)} = \underbrace{\begin{bmatrix} \mathbf{q}_{11}(t) \\ \mathbf{f}(t) - \mathbf{M}^{-1}\mathbf{C}\mathbf{q}_{11}(t) - \mathbf{M}^{-1}\mathbf{K}(\mathbf{q}_{10}(t)) \end{bmatrix}}_{=\mathbf{f}_1(t, \mathbf{q}_1(t))}.$$

The existence–uniqueness theorem (also known as Picard–Lindelöff theorem) states that, under certain conditions on f, the dynamical system (Eq. 2.3) has a unique solution.

Theorem 2.2.1. *(Picard–Lindelöff). Assume that* $D = [a, b] \times \mathbb{R}^d$ *for some* $a < 0 < b$. *Let* f *be continuous in the first argument for every* $(t, \mathbf{x}) \in \mathbb{D}$ *and that exists* $C > 0$ *such that* $\|\mathbf{f}(t, \mathbf{x}_1) - \mathbf{f}(t, \mathbf{x}_2)\| \le C\|\mathbf{x}_1 - \mathbf{x}_2\|$ *for every* $(t, \mathbf{x}_1), (t, \mathbf{x}_2) \in D$. *Then the dynamical system (Eq. 2.3) has a unique solution* $\boldsymbol{q}$ *defined on* $[a, b]$.

Note that in the previous theorem, one might take $b = +\infty$ as well. In what follows, we assume that the value of b is either $+\infty$ or large enough.[1]

The numerical solution of the dynamical system (Eq. 2.3) is the sequence of discrete vectors $\overline{\mathbf{q}}_j$, $j = 0, 1, \ldots$ such that they are approximations to the values of $\boldsymbol{q}$ at discrete time points t_j, for $j = 0, 1, \ldots$ It is usually taken that the time points t_j are equidistant, that is, that $t_j = j\Delta t$, where $\Delta t > 0$ is *time step*. If we denote $\mathbf{q}_j = \mathbf{q}(t_j)$, it means that $\mathbf{q}_j \approx \overline{\mathbf{q}}_j$ for $j = 0, 1, \ldots$.

[1] That assumption makes sense since we explore the period before our system reaches the steady state behavior.

There exist a vast of different techniques to compute the approximation sequence $\overline{\mathbf{q}}_j$, for the general dynamical system (Eq. 2.3). The most commonly used are the following ones:

1. ***Euler method***. It is based on the approximation $\dot{\mathbf{q}}(t) = (\mathbf{q}(t + \Delta t) - \mathbf{q}(t))/\Delta t$, so the next vector $\overline{\mathbf{q}}_{j+1}$ is computed by

$$\overline{\mathbf{q}}_{j+1} = \overline{\mathbf{q}}_j + \mathbf{f}(t_j, \overline{\mathbf{q}}_j)\Delta t. \tag{2.4}$$

2. ***Implicit Euler method***. It is very similar to the Euler method, but the derivative value is taken at the next point $\overline{\mathbf{q}}_{j+1}$, rather than $\overline{\mathbf{q}}_j$:

$$\overline{\mathbf{q}}_{j+1} = \overline{\mathbf{q}}_j + \mathbf{f}(t_{j+1}, \overline{\mathbf{q}}_{j+1})\Delta t. \tag{2.5}$$

   Note that, in order to compute $\overline{\mathbf{q}}_{j+1}$ from the previous equation, one needs to use some method for solving a system of nonlinear equations (like the Newton–Kantorovich method [1]).
3. ***Runge-Kutta-4 (RK-4) method***. It is the improvement of the Euler method when the derivative $\dot{\mathbf{q}}(t)$ is approximated using 4 different evaluations of the function f:

$$\begin{aligned}
\mathbf{w}_{1,j} &= \mathbf{f}(t_j, \overline{\mathbf{q}}_j),\\
\mathbf{w}_{2,j} &= \mathbf{f}\left(t_j + \frac{\Delta t}{2}, \overline{\mathbf{q}}_j + \frac{\Delta t}{2}\mathbf{w}_{1,j}\right),\\
\mathbf{w}_{3,j} &= \mathbf{f}\left(t_j + \frac{\Delta t}{2}, \overline{\mathbf{q}}_j + \frac{\Delta t}{2}\mathbf{w}_{2,j}\right),\\
\mathbf{w}_{4,j} &= \mathbf{f}(t_j + \Delta t, \overline{\mathbf{q}}_j + \mathbf{w}_{3,j}\Delta t),\\
\overline{\mathbf{q}}_{j+1} &= \overline{\mathbf{q}}_j + \frac{\Delta t}{6}(\mathbf{w}_{1,j} + 2\mathbf{w}_{2,j} + 2\mathbf{w}_{3,j} + \mathbf{w}_{4,j}).
\end{aligned} \tag{2.6}$$

The important properties of all numerical methods are the *order of convergence* and *stability*. The first term shows how fast the approximate solution $\overline{\mathbf{q}}_j$ converges to the exact $\mathbf{q}_j$, when time step $\Delta t \to 0$.

Definition 2.2.1. *(Order of convergence). The numerical method has the* ***order of convergence*** *p, if* $\|\mathbf{q}_j - \overline{\mathbf{q}}_j\| = \mathcal{O}(\Delta t^p)$, *for* $j = t/\Delta t$, *when* $\Delta t \to 0$, *for arbitrary* $t > 0$.

The term stability is related to the behavior of the (approximate) solution $\overline{\mathbf{q}}_j$ *under the small perturbation of the initial value* $\mathbf{q}_0 = \overline{\mathbf{q}}_0$.

Definition 2.2.2. *(Stability). The solution* $\mathbf{q}$ *of the dynamical system (Eq. 2.3) is* ***stable*** *if, for every* $\epsilon > 0$, *there exists* $\delta > 0$ *such that for every solution* $\mathbf{q}'(t)$ *satisfying* $\|\mathbf{q}_0 - \mathbf{q}'_0\| < \delta$, *it holds* $\|\mathbf{q}(t) - \mathbf{q}'(t)\| < \epsilon$ *for every t.*

Similarly, numerical solution $\overline{\mathbf{q}}_j$ *is stable if, for every* $\epsilon > 0$, *there exists* $\delta > 0$ *such that for every* $\mathbf{q}'_0$ *satisfying* $\|\mathbf{q}'_0 - \mathbf{q}_0\| < \delta$, *we have* $\|\overline{\mathbf{q}}_j - \overline{\mathbf{q}}'_j\| < \epsilon$ *for every* $j = 0, 1, \ldots$.

For example, the Euler and Implicit Euler methods have the order of convergence $p = 1$, while the RK-4 method has order $p = 4$.

Regarding the stability, solutions obtained by Euler and RK-4 methods are stable (assuming that the exact solution $\mathbf{q}(t)$ is stable) only for some values Δt satisfying the certain conditions. It is known as *conditional stability* solutions [2]. On the other hand, the Implicit Euler method is stable for arbitrary Δt, (under the same assumption of stability of $\mathbf{q}(t)$). These methods are known as *unconditionally stable.*

2.3 Newmark method

In this section, we describe the Newmark method [3] (also known as the Newmark-β method), which is used in the subsequent chapters (i.e., Chapters 10–14). Some of the other similar methods are the Central difference method, Houbolt method, Wilson method, etc. (see for example [4]).

It will be first shown for the scalar case (Eq. 2.1) and then generalized to the vector case (Eq. 2.2). Assume that Δt is the time step and $t_j = j\Delta t$. Denote as usual $q_j = q(t_j)$, $\dot{q}_j = \dot{q}(t_j)$, $\ddot{q}_j = \ddot{q}(t_j)$, and $f_j = f(t_j)$, for $j = 0, 1, \ldots$ Using the Lagrange theorem, one gets

$$\dot{q}_{j+1} = \dot{q}_j + \ddot{q}(t_j + \theta'\Delta t)\Delta t,$$

for some $\theta' \in (0, 1)$. Under the assumption $\ddot{u}_{j+1} \neq \ddot{u}_j$, we can write

$$\dot{q}_{j+1} = \dot{q}_j + \Delta t[(1 - \gamma)\ddot{q}_j + \gamma\ddot{q}_{j+1}], \tag{2.7}$$

for some real number γ. Similarly, using the Taylor expansion, one gets

$$q_{j+1} = q_j + \dot{q}_j\Delta t + \ddot{q}(t_j + \theta''\Delta t)\frac{\Delta t^2}{2},$$

for some $\theta' \in (0, 1)$. So again, under the assumption $\ddot{u}_{j+1} \neq \ddot{u}_j$, one can write

$$q_{j+1} = q_j + \dot{q}_j\Delta t + \frac{\Delta t^2}{2}[(1 - 2\beta)\ddot{q}_j + 2\beta\ddot{q}_{j+1}], \tag{2.8}$$

for some real number β. Although in general case, numbers γ and β depend on j, we assume that they are constant and that $\gamma \in [0, 1]$ while $\beta \in (0, 1/2]$ (such that $2\beta \in (0, 1]$). Their values are considered to be the parameters of the method.

Now $\ddot{q}_{j+1}$ can be computed from Eq. (2.8):

$$\ddot{q}_{j+1} = \frac{q_{j+1} - q_j}{\beta(\Delta t)^2} - \frac{\dot{q}_j}{\beta\Delta t} - \left(\frac{1}{2\beta} - 1\right)\ddot{q}_j, \tag{2.9}$$

while substituting Eq. (2.9) into Eq. (2.7), one obtains

$$\dot{q}_{j+1} = \dot{q}_j + \Delta t(1 - \gamma)\ddot{q}_j + \Delta t\gamma\left[\frac{q_{j+1} - q_j}{\beta(\Delta t)^2} - \frac{\dot{q}_j}{\beta\Delta t} - \left(\frac{1}{2\beta} - 1\right)\ddot{q}_j\right],$$

that is,

$$\dot{q}_{j+1} = \frac{\gamma}{\beta \Delta t}(q_{j+1} - q_j) + \left(1 - \frac{\gamma}{\beta}\right)\dot{q}_j + \Delta t\left(1 - \frac{\gamma}{2\beta}\right)\ddot{q}_j. \tag{2.10}$$

Putting $t = t_{j+1}$ into the dynamic (Eq. 2.1), one gets

$$m\ddot{q}_{j+1} + c\dot{q}_{j+1} + k(q_{j+1})q_{j+1} = f_{j+1}. \tag{2.11}$$

So, by substituting $\ddot{q}_{j+1}$ and $\dot{q}_{j+1}$ from Eqs. (2.9) and (2.7), into Eq. (2.11), we obtain:

$$\frac{m(q_{j+1} - q_j)}{\beta(\Delta t)^2} - \frac{m\dot{q}_j}{\beta \Delta t} - m\left(\frac{1}{2\beta} - 1\right)\ddot{q}_j$$

$$+ c\left[\frac{\gamma}{\beta \Delta t}(q_{j+1} - q_j) + \left(1 - \frac{\gamma}{\beta}\right)\dot{q}_j + \Delta t\left(1 - \frac{\gamma}{2\beta}\right)\ddot{q}_j\right] + k(q_{j+1})q_{j+1} = f_{j+1}$$

Grouping factors in the previous equation with q_{j+1}, yields to

$$\left[\frac{m}{\beta(\Delta t)^2} + \frac{\gamma c}{\beta \Delta t} + k(q_{j+1})\right]q_{j+1} = f_{j+1} + \left[\frac{m}{\beta \Delta t^2} + \frac{\gamma c}{\beta \Delta t}\right]q_j + \left[\frac{m}{\beta \Delta t} - c\left(1 - \frac{\gamma}{\beta}\right)\right]\dot{q}_j$$

$$+ \left[m\left(\frac{1}{2\beta} - 1\right) - c\left(1 - \frac{\gamma}{2\beta}\right)\Delta t\right]\ddot{q}_j. \tag{2.12}$$

From the previous equation, one computes q_{j+1}, and then $\dot{q}_{j+1}$ and $\ddot{q}_{j+1}$ using Eqs. (2.10) and (2.9) respectively. Assuming that $k(\cdot)$ is constant, q_{j+1} can be directly computed. Otherwise, one may use an arbitrary numerical method for solving nonlinear equations (like the Newton method, Secant method, etc.).

Also note that, when q_0 and $\dot{q}_0$ are given initial conditions, one can compute $\ddot{q}_0$ directly from Eq. (2.1) taking $t = t_0 = 0$, that is,

$$\ddot{q}_0 = \frac{f_0 - c\dot{q}_0 - k(q_0)q_0}{m}. \tag{2.13}$$

The usual choices of the coefficients γ and β are:

- $\gamma = 1/2$ and $\beta = 1/4$, where the method is also known as the *average constant acceleration method* or midpoint method. The name comes from the fact that the approximation of $\ddot{q}(t)$ by Eq. (2.7) is reduced to $\ddot{q}(t) = (\ddot{q}_j + \ddot{q}_{j+1})/2$ for every $t \in [t_j, \ t_{j+1}]$.
- $\gamma = 1/2$ and $\beta \to 0$ when the method is also known as the *central difference method*. Eq. (2.12) becomes

$$q_{j+1} = q_j + \dot{q}_j \Delta t + \ddot{q}_j \frac{\Delta t^2}{2}, \tag{2.14}$$

while Eq. (2.10) becomes

$$\dot{q}_{j+1} = \dot{q}_j + \left(\frac{1}{2}\ddot{q}_{j+1} + \frac{1}{2}\ddot{q}_j\right)\Delta t. \tag{2.15}$$

- In this case, $\ddot{q}_{j+1}$ is directly computed from Eq. (2.11). The name of the method comes from the fact that it can also be derived by replacing the central difference approximation formula for $\ddot{q}(t)$ and $\dot{q}(t)$ for $t = t_j$, in Eq. (2.1).
- $\gamma = 1/2$ and $\beta = 1/6$, where the method is also known as the linear acceleration method. It can be derived in a different way, by approximating $\ddot{q}(t)$ with the linear function in the segment $[t_j, t_{j+1}]$, that is,

$$\ddot{q}(t) \approx \ddot{q}_j + (\ddot{q}_{j+1} - \ddot{q}_j)\frac{t - t_j}{\Delta t}. \tag{2.16}$$

Performing the integration over t twice, one gets

$$\dot{q}(t) \approx \dot{q}_j + \ddot{q}_j(t - t_j) + (\ddot{q}_{j+1} - \ddot{q}_j)\frac{(t - t_j)^2}{2\Delta t}, \tag{2.17}$$

$$q(t) \approx q_j + \dot{q}_j(t - t_j) + \ddot{q}_j\frac{(t - t_j)^2}{2} + (\ddot{q}_{j+1} - \ddot{q}_j)\frac{(t - t_j)^3}{6\Delta t}. \tag{2.18}$$

Now taking $t = t_{j+1}$ in Eqs. (2.17) and (2.18), one gets Eq. (2.15) and

$$q(t) = q_j + \dot{q}_j\Delta t + \ddot{q}_j\frac{\Delta t^2}{2} + (\ddot{q}_{j+1} - \ddot{q}_j)\frac{\Delta t^2}{6}, \tag{2.19}$$

$$= q_j + \dot{q}_j\Delta t + \left(\frac{2}{3}\ddot{q}_{j+1} + \frac{1}{3}\ddot{q}_j\right)\frac{\Delta t^2}{2}, \tag{2.20}$$

which is the same formula as Eq. (2.8) for $\beta = 1/6$.

It is worth noting that for $c = 0$, the method is conditionally stable for $\gamma \geq 1/2$, while unconditionally stable if additionally, $\beta \geq (\gamma + 1/2)^2/4$ [4,5]. The order of the method is $p = 2$ for $\gamma = 1/2$ and $p = 1$ otherwise.[2]

2.4 Newmark method in the multidimensional case

The Newmark method described by Eqs. (2.12), (2.10), and (2.9), together with the initial computation of $\ddot{q}_0$ by Eq. (2.24), can be directly generalized to the multidimensional case, that is, Eq. (2.2). Denote by $\mathbf{q}_j = \mathbf{q}(t_j)$, $\dot{\mathbf{q}}_j = \dot{\mathbf{q}}(t_j)$, $\ddot{\mathbf{q}}_j = \ddot{\mathbf{q}}(t_j)$, and $\mathbf{f}_j = \mathbf{f}(t_j)$ for all $j = 0, 1, \ldots$ (recall that $t_j = j\Delta t$). The main Eq. (2.12) is given as follows:

$$\left[\frac{1}{\beta(\Delta t)^2}\mathbf{M} + \frac{\gamma}{\beta\Delta t}\mathbf{C} + \mathbf{K}(\mathbf{q}_{j+1})\right]\mathbf{q}_{j+1} = \mathbf{f}_{j+1} + \left[\frac{\beta\Delta t^2}{\mathbf{M}} + \frac{\gamma}{\beta\Delta t}\mathbf{C}\right]\mathbf{q}_j$$
$$+ \left[\frac{1}{\beta\Delta t}\mathbf{M} - \left(1 - \frac{\gamma}{\beta}\right)\mathbf{C}\right]\dot{\mathbf{q}}_j + \left[\left(\frac{1}{2\beta} - 1\right)\mathbf{M} - \left(1 - \frac{\gamma}{2\beta}\right)\Delta t\cdot\mathbf{C}\right]\ddot{\mathbf{q}}_j, \tag{2.21}$$

[2] See Section 2.2 for the definitions of stability and order of convergence.

while the other two are given by

$$\dot{\mathbf{q}}_{j+1} = \frac{\gamma}{\beta \Delta t}(\mathbf{q}_{j+1} - \mathbf{q}_j) + \left(1 - \frac{\gamma}{\beta}\right)\dot{\mathbf{q}}_j + \Delta t\left(1 - \frac{\gamma}{2\beta}\right)\ddot{\mathbf{q}}_j, \tag{2.22}$$

$$\ddot{\mathbf{q}}_{j+1} = \frac{\gamma}{\beta (\Delta t)^2}(\mathbf{q}_{j+1} - \mathbf{q}_j) - \frac{\gamma}{\beta \Delta t}\dot{\mathbf{q}}_j - \left(\frac{\gamma}{2\beta} - 1\right)\ddot{\mathbf{q}}_j. \tag{2.23}$$

The initial values are $\mathbf{q}_0$ and $\dot{\mathbf{q}}_0$, as well as $\ddot{\mathbf{q}}_0$ computed by solving the linear system

$$\mathbf{M}\ddot{\mathbf{q}}_0 = \mathbf{f}_0 - \mathbf{C}\dot{\mathbf{q}}_0 - \mathbf{K}(\mathbf{q}_0)\mathbf{q}_0. \tag{2.24}$$

Note that in the general case, the first Eq. (2.21) is the system of nonlinear equations. It can be solved by some nonlinear solver (including Newton–Kantorovich method), where the initial value is taken by $\mathbf{q}^0_{j+1} = \mathbf{q}_j$. One can also use the initial value $\mathbf{q}^0_{j+1} = \mathbf{q}_{j-1} + 2(\mathbf{q}_j - \mathbf{q}_{j-1}) = 2\mathbf{q}_j - \mathbf{q}_{j-1}$ computed by linear interpolation.

On the other hand, if $\mathbf{K}(\cdot)$ is constant, then Eq. (2.21) is the system of linear equations of the form $\mathbf{A}\mathbf{q}_{j+1} = \mathbf{b}_{j+1}$, where the system matrix $\mathbf{A}$ is given by

$$\mathbf{A} = \frac{1}{\beta (\Delta t)^2}\mathbf{M} + \frac{\gamma}{\beta \Delta t}\mathbf{C} + \mathbf{K}. \tag{2.25}$$

In that case, one can precompute some suitable matrix factorization (like LU or QR decomposition) of matrix $\mathbf{A}$ and use it for each $j = 0, 1, \ldots$. It considerably reduces the time complexity of every time step, and also improves the accuracy of computation. Also, in the general case if $\mathbf{K}(\mathrm{x}) = \mathbf{K}_l + \mathbf{K}_{nl}(\mathrm{x})$, one may obtain the initial value $\mathbf{q}^0_{j+1}$ by solving the linear system $\mathbf{A}_0\mathbf{q}^0_{j+1} = \mathbf{b}_{j+1}$ where

$$\mathbf{A}_0 = \frac{1}{\beta (\Delta t)^2}\mathbf{M} + \frac{\gamma}{\beta \Delta t}\mathbf{C} + \mathbf{K}_l. \tag{2.26}$$

The factorization technique now can be applied to the matrix $\mathbf{A}_0$. Our experiments show that using this initial value, just needed just a couple of iterations of a nonlinear solver to obtain $\mathbf{q}_{j+1}$ to the satisfactory accuracy.

2.5 Improvement of the accuracy by Richardson extrapolation

Richardson extrapolation is the technique used to increase the accuracy of the numerical method, by repeated evaluation for the different step sizes. The application of the Richardson method has been used to obtain better approximations of the results, which are thoroughly presented through model analysis in Chapter 12. Denote by $F(h)$ the value obtained using some numerical method for the step size $h > 0$, for computing the quantity Q. If the method is convergent, we have that $F(h) \to Q$ when $h \to 0+$. In other words, decreasing the step size h to 0, results in a decrease in the error $|F(h) - Q|$ to 0. Recall that the numerical method has order p if $F(h) = Q + \mathcal{O}(h^p)$ when $h \to 0+$.

Now assume that $F(h) = Q + \mathcal{O}(h^p) + o(h^p)$ when $h \to 0+$, for some (unknown) constant $a \neq 0$. The goal is to eliminate the term ah^p in order to increase the method order, and as well the accuracy. It can be done by evaluating the method for the smaller step size qh, where $0 < q < 1$. Therefore, one gets

$$F(h) = Q + ah^p + o(h^p), \tag{2.27}$$

$$F(qh) = Q + a(qh)^p + o((qh)^p). \tag{2.28}$$

Now since $o((qh)^p) = o(h^p)$, we can compute

$$Q = F(h) + \frac{F(h) - F(qh)}{q^p - 1} + o(h^p). \tag{2.29}$$

The expression $F_1(h) = F(h) + (F(h) - F(qh))/(q^p - 1)$ is a better approximation of the target value Q than both $F(h)$ and $F(qh)$, and has an order of at least $p + 1$. The term $(F(h) - F(qh))/(q^p - 1)$ is known in the literature as **Richardson correction**. This value can be as well used for the determination of the time step for the adaptive-step numerical methods [6].

Assume now that the approximate solution $\mathbf{q}_j$ of the system (Eq. 2.2) is obtained by applying the Newmark method for $\gamma = 1/2$ (so it has order $p = 2$) for the step size Δt, while $\mathbf{q}'_j$ is the approximate solution for the halved step size $\Delta t/2$. Then $\mathbf{q}_j$ and $\mathbf{q}'_{2j}$ both approximate the value of the exact solution $\mathbf{q}(t)$ at time point $t = j\Delta t$, for $j = 0, 1, \ldots$. Now according to the expression (Eq. 2.29) (for $q = 1/2$, $h = \Delta t$), one can construct the improved sequence

$$\bar{\mathbf{q}}_j = \mathbf{q}_j + \left[\left(\frac{1}{2}\right)^2 - 1\right]^{-1} \cdot (\mathbf{q}'_{2j} - \mathbf{q}_j) = \frac{7}{3}\mathbf{q}_j - \frac{4}{3}\mathbf{q}'_{2j}, \tag{2.30}$$

having the order at least $p + 1 = 3$.

The shown procedure can be easily extended to the much more general case when

$$F(h) = Q + a_1 h^{p_1} + a_2 h^{p_2} + \ldots + a_k h^{p_k} + o(h^{p_k}),$$

where $p_1 < p_2 < \cdots < p_k$ are terms in the Taylor expansion of the function $F(h)$. It computes the series of approximations by

$$F_{l+1}(h) = F_l(h) + \frac{F_l(h) - F_l(qh)}{q^{p_l} - 1},$$

for all $l = 0, 1, \ldots, k - 1$, where $F_0(h) = F(h)$. It requires total k evaluations of the function $F(\cdot)$ (i.e., total k running of the model), for arguments $h, qh, q^2h, \ldots, q^{k-1}h$.

2.6 Harmonic balance method

Consider the dynamical system in the vector form (Eq. 2.2). Assume that the external force $\mathbf{f}(\cdot)$ is a periodic function with period $T = 2\pi/\omega$. We are interested in finding the approximate periodic solution $\mathbf{q}(t)$ which can be approximated by the following finite Fourier series

$$\mathbf{q}(t) = \mathbf{Q}_0 + \sum_{i=1}^{k} \mathbf{Q}_{c,i} \cos(i\omega t) + \sum_{i=1}^{k} \mathbf{Q}_{s,i} \sin(i\omega t), \tag{2.31}$$

where $k \in \mathbb{N}$ is the approximation order, while $\mathbf{Q}_0$, $\mathbf{Q}_{c,i}$, and $\mathbf{Q}_{s,i}$ are unknown coefficient vectors ($i = 1, 2, \ldots, k$). Functions

$$f_0(t) = 1, \quad f_{c,i} = \cos(i\omega t), \quad f_{s,i} = \sin(i\omega t), \tag{2.32}$$

for $i = 1, 2, \ldots, k$ are *basis functions*. In this section, we show the method to obtain the approximate solution of the form (Eq. 2.36). The method is known as the **harmonic balance method** and is based on replacing the approximate solution into the dynamical system (Eq. 2.2) and balancing the terms with respect to the same basis function. This method is applied in Chapters 10–12.

Differentiating Eq. (2.36), one obtains

$$\dot{\mathbf{q}}(t) = \sum_{i=1}^{k} i\omega \mathbf{Q}_{c,i} \cos(i\omega t) - \sum_{i=1}^{k} i\omega \mathbf{Q}_{s,i} \sin(i\omega t), \tag{2.33}$$

$$\ddot{\mathbf{q}}(t) = -\sum_{i=1}^{k} i^2\omega^2 \mathbf{Q}_{c,i} \sin(i\omega t) - \sum_{i=1}^{k} i^2\omega^2 \mathbf{Q}_{s,i} \cos(i\omega t). \tag{2.34}$$

Consider now the unknown vector $\overline{\mathbf{Q}}$, which is obtained by stacking the unknown coefficient vectors:

$$\overline{\mathbf{Q}} = \begin{bmatrix} \mathbf{Q}_0^T & \mathbf{Q}_{c,1}^T & \cdots & \mathbf{Q}_{c,k}^T & \mathbf{Q}_{s,1}^T & \cdots & \mathbf{Q}_{s,k}^T \end{bmatrix}^T. \tag{2.35}$$

Since all coefficient vectors are n-dimensional, we have that $\overline{\mathbf{Q}} \in \mathbb{R}^{n(2k+1)}$. We also assume that $\mathbf{f}(t)$ can be approximated by the finite Fourier series in an analogous way

$$\mathbf{f}(t) = \mathbf{F}_0 + \sum_{i=1}^{k} \mathbf{F}_{c,i} \sin(i\omega t) + \sum_{i=1}^{k} \mathbf{F}_{s,i} \sin(i\omega t), \tag{2.36}$$

and that

$$\overline{\mathbf{F}} = \begin{bmatrix} \mathbf{F}_0^T & \mathbf{F}_{c,1}^T & \cdots & \mathbf{F}_{c,k}^T & \mathbf{F}_{s,1}^T & \cdots & \mathbf{F}_{s,k}^T \end{bmatrix}^T. \tag{2.37}$$

is the equivalent vector with stacked force coefficients vectors.

Now replacing Eqs. (2.36), (2.33), and (2.34) into Eq. (2.2) and grouping factors according to the basis functions (Eq. 2.32), one obtains the following system

$$(-\omega^2\overline{\mathbf{M}} + \omega\overline{\mathbf{C}} + \overline{\mathbf{K}}_l + \overline{\mathbf{K}}_{nl}(\overline{\mathbf{Q}}))\overline{\mathbf{Q}} = \overline{\mathbf{F}}. \tag{2.38}$$

The coefficient matrices are given by

$$\overline{\mathbf{M}} = \mathrm{diag}(\mathbf{M},\ 1^2 \cdot \mathbf{M},\ \ldots,\ k^2 \cdot \mathbf{M},\ 1^2 \cdot \mathbf{M},\ \ldots,\ k^2 \cdot \mathbf{M}), \tag{2.39}$$

$$\overline{\mathbf{C}} = \mathrm{diag}(\mathbf{C},\ 1 \cdot \mathbf{C},\ \ldots,\ k \cdot \mathbf{C},\ 1 \cdot \mathbf{C},\ \ldots,\ k \cdot \mathbf{C}), \tag{2.40}$$

$$\overline{\mathbf{K}}_l = \text{diag}\underbrace{(\mathbf{K}_l,\ \mathbf{K}_l,\ \ldots,\ \mathbf{K}_l)}_{2k+1}. \tag{2.41}$$

The term $\overline{\mathbf{K}}_{nl}(\overline{\mathbf{Q}})\overline{\mathbf{Q}}$ is obtained from the nonlinear term $\mathbf{K}_{nl}(\mathbf{q}(t))\mathbf{q}(t)$, by expanding and grouping the terms by the basis functions, and neglecting all other (higher-order) terms.

Note that Eq. (2.38) is the system of $n(2k+1)$ nonlinear equations of the form

$$\mathbf{G}(\overline{\mathbf{Q}},\ \omega) = 0, \tag{2.42}$$

with the unknown vector $\overline{\mathbf{Q}} \in \mathbb{R}^{n(2k+1)}$, parameter $\omega > 0$ and the nonlinear function $\mathbf{G}$: $\mathbb{R}^{n(2k+1)} \times \mathbb{R} \rightarrow \mathbb{R}^{n(2k+1)}$.

The goal is to obtain a solution $\overline{\mathbf{Q}}$ of the system (Eq. 2.42) as the function ω. The classical approach introduces the series of values of ω_i, $i = 0, 1, \ldots$ and solves the system $\mathbf{G}(\overline{\mathbf{Q}},\ \omega_i) = 0$ with respect to $\overline{\mathbf{Q}}$ using arbitrary nonlinear equation solver. The most commonly used nonlinear solver is the Newton method. It constructs the series of approximations of the form

$$\overline{\mathbf{Q}}_i^{(j+1)} = \overline{\mathbf{Q}}_i^{(j)} + \Delta\overline{\mathbf{Q}}_i^{(j+1)},$$

$$\mathbf{J}(\overline{\mathbf{Q}}_i^{(j)},\ \omega_i)\Delta\overline{\mathbf{Q}}_i^{(j+1)} = -\mathbf{G}(\overline{\mathbf{Q}}_i^{(j+1)},\ \omega_i), \tag{2.43}$$

for $j = 0, 1, \ldots$ Here $\overline{\mathbf{Q}}_i^{(0)}$ must be predefined starting value, while $\mathbf{J}(\cdot,\cdot)$ is the **Jacobian matrix** defined by

$$\mathbf{J}(\mathbf{x},\ \omega) = \left[\frac{\partial G_p(\mathbf{x},\ \omega)}{\partial x_q}\right]_{1 \le p,\ q \le n(2k+1)}. \tag{2.44}$$

The approximations (Eq. 2.43) are performed until the stopping criterion $\|\overline{\mathbf{Q}}_i^{(j+1)} - \overline{\mathbf{Q}}_i^{(j)}\| < \epsilon$ is satisfied, and the last computed approximation is taken as the approximate solution, that is, $\overline{\mathbf{Q}}_i := \overline{\mathbf{Q}}_i^{(j+1)}$. For the initial vector $\overline{\mathbf{Q}}_i^{(0)}$, one can take, for example, previously computed approximation (i.e., for $\omega = \omega_i$), by $\overline{\mathbf{Q}}_i^{(0)} := \overline{\mathbf{Q}}_{i-1}$.

The downside of the Newton method is that the double iteration (over both i and j) is required, which can be time-consuming. At the same time, it can compute only one solution of the equation $\mathbf{G}(\overline{\mathbf{Q}},\ \omega_i) = 0$, for every value of ω_i. In the case when the bifurcation occurs (see for example [7]), there is more than one solution for the single value ω_i. In that case, it is convenient to use the method which is specialized for that type of problem.

One such method is the **continuation method** [8,9], which we briefly describe in the rest of this section. To be consistent with that notation, consider the main system of equations in the form $\mathbf{G}(\overline{\mathbf{Q}},\ \omega^2) = 0$, that is, either that $\overline{\mathbf{C}} = 0$ or that there is no term ω in front of $\overline{\mathbf{C}}$. The continuation method consists of the following two steps:

1. **Predictor step**. Having two consecutive pairs of values $(\omega_{i-1}^2,\ \overline{\mathbf{Q}}_{i-1})$ and $(\omega_i^2,\ \overline{\mathbf{Q}}_i)$ construct the initial values for the next step by

$$\overline{\mathbf{Q}}_{i+1}^{(0)} = \overline{\mathbf{Q}}_i + \Delta\overline{\mathbf{Q}}_i,$$

$$\Delta\overline{\mathbf{Q}}_i = (\overline{\mathbf{Q}}_i - \overline{\mathbf{Q}}_{i-1})\frac{\Delta Q_{aux}}{\|\overline{\mathbf{Q}}_i - \overline{\mathbf{Q}}_{i-1}\|}, \tag{2.45}$$

where ΔQ_{aux} is given parameter, while

$$[\omega_{i+1}^2]^{(0)} = \omega_i^2 \pm \frac{s}{\|\delta\overline{\mathbf{Q}}_i\|}, \tag{2.46}$$

where $\delta\overline{\mathbf{Q}}_i$ is the solution of the linear system

$$\mathbf{J}(\overline{\mathbf{Q}}_i,\ \omega_i^2)\delta\overline{\mathbf{Q}}_i = \overline{\mathbf{M}\mathbf{Q}}_{i+1}^{(0)}. \tag{2.47}$$

and $s > 0$ is the given parameter. The sign $\pm$ is chosen depending on the sign of the determinant of the Jacobian $\mathbf{J}(\overline{\mathbf{Q}}_i,\ \omega_i^2)$ (it is changed, if the sign of the determinant is changed from the previous iteration).

2. **Corrector step**. In the corrector step, we obtain $(\overline{\mathbf{Q}}_{i+1},\ \omega_{i+1}^2)$ by solving the system of equations

$$\mathbf{G}(\overline{\mathbf{Q}},\ \omega^2) = 0,$$
$$\|\overline{\mathbf{Q}} - \overline{\mathbf{Q}}_i\|^2 = s^2, \tag{2.48}$$

over $(\overline{\mathbf{Q}},\ \omega^2)$, using Newton's method and $(\overline{\mathbf{Q}}_{i+1}^{(0)},\ [\omega_{i+1}^2]^{(0)})$ as the initial values.

For a more detailed and comprehensive study of the continuation method, the reader is referred to Refs. [8,9].

References

[1] R.L. Burden, J.D. Faires, Numerical Analysis, ninth ed., Brooks/Cole, 2011.

[2] J.C. Butcher, Numerical Methods for Ordinary Differential Equations, Wiley, 2003.

[3] N.M. Newmark, A method of computation for structural dynamics, Journal of the Engineering Mechanics Division 85 (3) (1959) 6794. https://doi.org/10.1061/JMCEA3.0000098 (EM3).

[4] M. Petyt, Introduction to Finite Element Vibration Analysis, second ed., Cambridge Unversity Press, 2010.

[5] G. Lindfield, J. Penny, Numerical Methods Using Matlab, fourth ed., Elsevier, 2019.

[6] Z. Zlatev, I. Dimov, I. Farago, A. Havasi, Practical aspects and applications, Richardson Extrapolation, De Gruyter, 2018.

[7] R. Seydel, Practical Bifurcation and Stability Analysis, third ed., Springer, 2010.

[8] P. Ribeiro, M. Petyt, Non-linear vibration of beams with internal resonance by the hieratical finite-element method, Journal of Sound and Vibration 224 (4) (1999) 591–624.

[9] P. Ribeiro, M. Petyt, Nonlinear vibration of plates by the hierarchical finite element and continuation methods, International Journal of Mechanical Sciences 41 (1999) 437–459.

Mathematical methods in stochastic stability of mechanical systems

3

3.1 Introduction to probability space and random variables

This section introduces the basic concepts of probability and random variables. The introduction is very brief, and for a more detailed and systematic approach, the reader is referred, for example, to Ref. [1].

For the formal definition of probability and the random variable, we first need to define the probability space.

Definition 3.1.1. *(Probability space). The triplet* $(\Omega, \mathbb{F}, P)$ *is the* ***probability space****, where* Ω *is the set known as* ***sample space*** *or* ***space of all possible outcomes****,* $\mathbb{F}$ *is the set of subsets of* Ω *known as* ***event space****, and* $P\colon \mathbb{F} \to [0, 1]$ *is the function satisfying:*

1. $\Omega \in \mathbb{F}$ *and if* $A \in \mathbb{F}$*, then also* $\Omega \setminus A \in \mathbb{F}$;
2. $A_n \in \mathbb{F}$ *for* $n = 1, 2, \ldots$*, then* $\bigcup_{n=1}^{+\infty} A_n \in \mathbb{F}$;
3. $P(\Omega) = 1$;
4. *If* $A_n \in \mathbb{F}$ *for* $n = 1, 2, \ldots$ *are pairwise disjoint sets, then* $P\left(\bigcup_{n=1}^{+\infty} A_n\right) = \sum_{n=1}^{+\infty} P(A_n)$.

Function P is known as the ***probability*** *function or just* ***probability****.*

The sample space Ω can be either finite (e.g., $\Omega = \{1, 2, 3, 4, 5, 6\}$ for a model of the fair-rolled dice), countable or uncountable. The set of events $\mathbb{F}$ does NOT always have to contain all possible subsets of Ω.

The events $A, B \in \mathbb{F}$ are **independent** if $P(A \cap B) = P(A)P(B)$. The independence can be the natural consequence of what is modeled by the probability space $(\Omega, \mathbb{F}, P)$, but can be purely mathematical property as well.

Also, if $B \in \mathbb{F}$ is the event such that $P(B) \neq 0$, we define the **conditional probability** $P(A|B)$ by

$$P(A|B) = \frac{P(A \cap B)}{P(B)}, \tag{3.1}$$

for every $A \in \mathbb{F}$. Note that if $P(A) \neq 0$ as well, it satisfies the well-known **Bayes** formula $P(A|B) = P(B|A)P(A)/P(B)$.

Definition 3.1.2. *If we have the probability space* $(\Omega, \mathbb{F}, P)$*, the* ***random variable*** *X is an arbitrary function* $X\colon \Omega \to \mathbb{R}$ *such that* $X^{-1}((a, b)) \in \mathbb{F}$ *for every* $-\infty \leq a < b \leq +\infty$.

Modeling of Complex Dynamic Systems. DOI: https://doi.org/10.1016/B978-0-443-23942-7.00003-6

The function $F_X: \mathbb{R} \to [0, 1]$ defined by $F_X(x) = P(\{\omega \in \Omega \mid X(\omega) < x\})$ is known as the **cumulative distribution function (CDF)** of the random variable. If there exists a nonnegative function $p_X: \mathbb{R} \to \mathbb{R}^+$ such that $F(x) = \int_{-\infty}^{x} p_x(t)\mathrm{d}t$, then X is known as an **absolutely continuous** random variable, and p_x is the **probability density function (PDF)** of X.

The value $X(\omega)$ of the random variable, X should represent some numerical property of the state $\omega \in \Omega$. So usually, we omit state ω and will denote the evaluation $X(\omega)$ simply by X. Also, for example, the event $\{\omega \in \Omega \mid X(\omega) < x\}$ will be simply denoted by $\{X < x\}$. Also, we use the shorter notation $P(X < x)$ for $P(\{X < x\})$, in this and other similar cases.

The concept of the random variable is similarly defined in the multidimensional case, as shown in the following definition.

Definition 3.1.3. *For a given probability space (Ω, $\mathbb{F}$, P), the* ***k-dimensional random variable*** *is the function X: $\Omega \to \mathbb{R}^k$ satisfying $\mathbf{X}^{-1}((-\infty, \mathbf{x})) \in \mathbb{F}$ for any $\mathbf{x} \in \mathbb{R}^k$.*

Similarly, function $F_{\mathbf{X}} : \mathbb{R}^k \to [0, 1]$ defined by $F_{\mathbf{X}}(\mathbf{x}) = P(\{\omega \in \Omega \mid X_i(\omega) < x_i\})$ is CDF of the random variable $\mathbf{X}$. We denote $\mathbf{x} = (x_1, x_2, \ldots, x_k)$, $\mathbf{t} = (t_1, t_2, \ldots, t_k)$, and $\mathbf{X}(\omega) = (X_1(\omega), X_2(\omega), \ldots, X_k(\omega))$. Also, the function (if exists) $p_{\mathbf{X}}$ satisfying

$$F_{\mathbf{X}}(\mathbf{x}) = \int_{-\infty}^{x_1} \cdots \int_{-\infty}^{x_k} p_{\mathbf{X}}(\mathbf{t})\mathrm{d}\mathbf{t}, \tag{3.2}$$

is known as the probability density function (PDF) of $\mathbf{X}$.

The random variables X and Y are **independent** if the events $\{X \in (a, b)\}$ and $\{Y \in (c, d)\}$ are independent for every interval (a, b) and (c, d). As a consequence, we have that $F_{(X,Y)}(x, y) = F_X(x)F_Y(y)$ for every $x, y \in \mathbb{R}$. The independence is similarly defined in the case of multidimensional random variables.

Assume that X and Y are absolutely continuous random variables and $p_Y(y) \neq 0$ for $y \in (a, b)$, where (a, b) is given interval. The **conditional PDF** $p_{X\mid Y}(x\mid y)$ is defined by $p_{X\mid Y}(x\mid y) = p_{(X,Y)}(x, y)/p_Y(y)$. Similarly, the **conditional CDF** $F_{X\mid Y}(x\mid y)$ is defined by $F_{X\mid Y}(x\mid y) = F_{(X,Y)}(x, y)/F_Y(y)$.

The subscripts for F_X, $F_{(X,Y)}$, and $F_{X\mid Y}$ can be omitted when it is evident which random variables we are dealing with. So, one can write simply $F(x)$, $F(x, y)$, and $F(x\mid y)$ respectively. A similar convention will be used for the PDF functions (i.e., we will write $p(x)$, $p(x, y)$, and $p(x\mid y)$) and in the multidimensional case.

By saying that the random variable X or $\mathbf{X}$ has a certain **distribution**, we mean that it has a corresponding CDF (or in the absolutely continuous case PDF) function.

Example 3.1.1. The most important distribution of the absolutely continuous random variable is **Gaussian** or **normal** distribution. The PDF is defined by

$$p_X(x) = \frac{1}{\sqrt{2\pi}\,\sigma} e^{-(x-\mu)^2/(2\sigma^2)}. \tag{3.3}$$

where $\mu \in \mathbb{R}$ and $\sigma > 0$ are parameters. This distribution is usually denoted by $\mathcal{N}(\mu, \sigma^2)$ and the fact that X has this distribution is denoted by X: $\mathcal{N}(\mu, \sigma^2)$.

It is generalized in the k-dimensional case by

$$p_{\mathbf{X}}(\mathbf{x}) = \frac{1}{(2\pi)^{k/2}|\boldsymbol{\Sigma}|^{1/2}} e^{-(\mathbf{x}-\boldsymbol{\mu})^T\boldsymbol{\Sigma}^{-1}(\mathbf{x}-\boldsymbol{\mu})/2}. \tag{3.4}$$

where $\boldsymbol{\mu} \in \mathbb{R}^k$ is vector and $\boldsymbol{\Sigma} \in \mathbb{R}^{k\times k}$ is a symmetric, positive definite matrix. By $|\boldsymbol{\Sigma}|$ we denoted the determinant of the matrix $\boldsymbol{\Sigma}$. The fact that $\mathbf{X}$ has a ***k*-dimensional Gaussian (normal) distribution** is denoted by $\mathbf{X} : \mathcal{N}(\boldsymbol{\mu}, \boldsymbol{\Sigma})$.

It is evident that if $f : \mathbb{R} \to \mathbb{R}$ is the function and X is a random variable, then $Y = f(X)$ (defined by $Y(\omega) = f(X(\omega))$ for $\omega \in \Omega$ is also the random variable). Similarly, if $\mathbf{f}: \mathbb{R}^k \to \mathbb{R}^l$ and $\mathbf{X}$ is a k-dimensional random variable, then $\mathbf{Y} = \mathbf{f}(\mathbf{X})$ is l-dimensional random variable. The following theorem shows the connection between PDFs of X and Y, and $\mathbf{X}$ and $\mathbf{Y}$ respectively.

Theorem 3.1.1. *If $f : [a, b] \to \mathbb{R}$ is a monotonic and differentiable function, and X is an absolutely continuous random variable such that $X \in [a, b]$, then $Y = f(X)$ has PDF equal to*

$$p_Y(y) = p_X(f^{-1}(y))|(f^{-1}(y))'|, \tag{3.5}$$

for every $y = f(x)$ and $x \in [a, b]$. Also, if $\mathbf{f}: D' \to D''$ is bijective and differentiable function, while $D' \subset \mathbb{R}^k$, $D'' \subset \mathbb{R}^l$ and $\mathbf{X}$ is the k-dimensional random variable with the set of values D', then $\mathbf{Y} = \mathbf{f}(\mathbf{X})$ has PDF equal to

$$p_{\mathbf{Y}}(\mathbf{y}) = p_{\mathbf{X}}(\mathbf{f}^{-1}(\mathbf{y})) \left| \left. \frac{\partial \mathbf{f}^{-1}(\mathbf{z})}{\partial \mathbf{z}} \right|_{\mathbf{z}=\mathbf{y}} \right|, \tag{3.6}$$

where

$$\frac{\partial \mathbf{g}(\mathbf{z})}{\partial \mathbf{z}} = \left[\frac{\partial g_i(\mathbf{z})}{\partial z_j} \right],$$

is the matrix of partial derivatives of the function $\mathbf{g}$, *known as the* **Jacobi matrix**. *Its determinant is known as* **Jacobian** *of the transformation* $\mathbf{X} \to \mathbf{f}(\mathbf{X})$.

The two most important properties of the random variables are given by the following definition.

Definition 3.1.4. *The* ***expectation (mean)*** *and* ***dispersion (variance)*** *of the random variable X are defined by*

$$\mathbb{E}X = \int_{-\infty}^{+\infty} t\cdot \mathrm{d}F_X(t), \tag{3.7}$$

$$\mathbb{D}X = \mathbb{E}(X - \mathbb{E}X)^2, \tag{3.8}$$

The corresponding integral in the definition of $\mathbb{E}X$ is known as the Lebesgue–Stieltjes integral. It is more common to define as $\mathbb{E}X = \int_\Omega X \mathrm{d}P$, that is, the integral of X with respect to the measure P. For more information about the measures and integrals, the reader is referred to any book about measure theory, for example [2]:

In the case of k-dimensional random variable $\mathbf{X}$*, the expectation is similarly defined by*

$$\mathbb{E}X = \int_{-\infty}^{+\infty} \mathbf{x}\cdot \mathrm{d}F_{\mathbf{X}}(\mathbf{x}), \tag{3.9}$$

while instead of dispersion, we define the ***covariance matrix*** *by*

$$\mathrm{cov}\mathbf{X} = [\mathrm{cov}(X_i, X_j)]_{1\le i,\, j\le k}, \tag{3.10}$$

where

$$\mathrm{cov}(X, Y) = \mathbb{E}[(X - \mathbb{E}X)(Y - \mathbb{E}Y)], \tag{3.11}$$

is the ***covariance*** *between random variables X and Y.*

In the case of absolutely continuous function, the expectation formulas (3.7) and (3.9) reduce to

$$\mathbb{E}X = \int_{-\infty}^{+\infty} x p_X(x)\mathrm{d}x, \tag{3.12}$$

$$\mathbb{E}\mathbf{X} = \int_{\mathbb{R}^k} \mathbf{x}\cdot p_{\mathbf{X}}(\mathbf{x})\mathrm{d}\mathbf{x}. \tag{3.13}$$

The expectation and dispersion possess the following linearity properties and similar statements are also valid for the k-dimensional random variable (where dispersion $\mathbb{D}$ is replaced by the covariance matrix cov).

Theorem 3.1.2. *If X and Y are random variables and* $\alpha, \beta \in \mathbb{R}$*, then*

1. $\mathbb{E}(\alpha X + \beta Y) = \alpha\mathbb{E}X + \beta\mathbb{E}Y$;
2. $\mathbb{D}(\alpha X) = |\alpha|^2\mathbb{D}X$;
3. $\mathbb{E}(XY) = \mathbb{E}X\cdot\mathbb{E}Y$ *if X and Y are independent;*
4. $\mathbb{D}(X + Y) = \mathbb{D}X + \mathbb{D}Y$ *if X and Y are independent.*

Example 3.1.2. Consider the Gaussian random variable $X : \mathcal{N}(\mu, \sigma^2)$. Then it holds $\mathbb{E}X = \mu$ and $\mathbb{D}X = \sigma^2$. Similarly, in the case of k-dimensional Gaussian random variable $\mathbf{X}: \mathcal{N}(\boldsymbol{\mu}, \boldsymbol{\Sigma})$, we have $\mathbb{E}\mathbf{X} = \boldsymbol{\mu}$ and $\mathrm{cov}\mathbf{X} = \boldsymbol{\Sigma}$.

Expectation and dispersion are analogously defined in the case of the conditional distribution. In other words, we have

$$\mathbb{E}[X|Y = y] = \int_{-\infty}^{+\infty} t\cdot \mathrm{d}F_{X|Y}(t|y). \tag{3.14}$$

for the (absolutely continuous) random variables X and Y.

Another important numerical property of the random variable is the k-th moment, for any positive integer k.

Definition 3.1.5. *If X is a random variable, then* $m_k = \mathbb{E}X^k$ *is known as the* ***k-th moment*** *of the random variable X.*

Finally, before proceeding to random processes, we introduce the convergence of the sequences of random variables. Note that, instead of the single definition we have in the classical case of the sequence of numbers, we have more than one concept of convergence, in the case of random variables, which need independent definitions.

Definition 3.1.6. *(Convergence of random variables). Sequence of random variables X_n for $n = 1, 2, \ldots$, converges to the random variable X:*

1. ***in probability**, if for every $\epsilon > 0$, it holds $P(|X_n - X| \geq \epsilon) \to 0$ when $n \to +\infty$, which is denoted by $X_n \xrightarrow{p} X$.*
2. ***in the mean square**, if $\mathbb{E}\,|X_n - X|^2 \to 0$ when $n \to +\infty$, which is denoted $X_n \xrightarrow{m} X$;*
3. ***with probability 1 (almost surely)** if $P(X_n \to X) = 1$, which is denoted by $X_n \xrightarrow{a.s.} X$.*

It can be proved that convergence in mean and almost surely convergence implies the convergence in probability, but not vice versa. Also, these two convergence definitions do not imply each other.

3.2 Random processes

The random variable is suitable for modeling the single numerical value corresponding to some phenomenon involving randomness. But if we want to record a series of values, or, for example, the time evolution of the corresponding value, then we need the definition of the random process.

Definition 3.2.1. *For a given probability space $(\Omega, \mathbb{F}, P)$, **the random process** is the function $X: T \times \Omega \to \mathbb{R}$ where $T \subset \mathbb{R}$, such that $X(t, \cdot)$ is the random variable for every $t \in T$. Similarly, the function $\mathbf{X}: T \times \Omega \to \mathbb{R}^k$ is **k-dimensional random process** if $\mathbf{X}(t, \cdot)$ is a k-dimensional random variable for every $t \in T$. A random process is usually denoted by $X(t)$ or $\mathbf{X}(t)$ respectively.*

In simple words, the random process is the collection of random variables indexed by the elements from the set T. It is usually taken to be either finite or infinite segment (mostly with 0 as one of the endpoints) and it represents the time component. If it is not mentioned otherwise, we assume that $T = \mathbb{R}$.

Having the random process, we can consider the (joint) CDF function of the n-dimensional random variable $(X(t_1), X(t_2), \ldots, X(t_n))$ for $t_1 < t_2 < \cdots < t_n$, and denote it by $F_n(x_1, t_1; x_2, t_2; \cdots; x_n, t_n)$. They are referred to as the ***n*-th dimensional CDFs** of $X(t)$. Also, we denote by $m(t) = \mathbb{E}X(t)$ the mean and by $\sigma^2(t) = \mathbb{D}X(t)$ the variance of $X(t)$. Similarly, by $m_k(t) = \mathbb{E}[X(t)]^k$ we denote the k-th moment of $X(t)$. In the same way, we define the n-dimensional PDF functions p_n.

An important class of random processes are **Gaussian processes**, where $X(t)$ has Gaussian distribution, that is, $X(t)$: $\mathcal{N}(m(t), \sigma^2(t))$.

An important class of random processes are stationary processes.

Definition 3.2.2. *The random process $X(t)$ is stationary if*

$$F_n(x_1, t_1; x_2, t_2; \cdots; x_n, t_n) = F_n(x_1, t_1 + \Delta t; x_2, t_2 + \Delta t; \cdots; x_n, t_n + \Delta t) \quad (3.15)$$

for every $n \in \mathbb{N}$, $x_1, x_2, \ldots, x_n, \Delta t \in \mathbb{R}$ and $t_1 < t_2 < \cdots < t_n$.

In other words, the process is stationary, if the n-dimensional CDFs are prone to arbitrary time shifts. In such case, we have only one 1-dimensional CDFs given by $F_1(x) = F_1(x, t)$ and $F_2(x_1, t_1; x_2, t_2) = F_2(x_1, x_2; t_1 - t_2)$, for some functions $F_1(\cdot, \cdot)$ and $F_2(\cdot, \cdot; \cdot)$. Also, it is evident that n-dimensional PDFs of the stationary processes satisfy the same Eq. (3.15).

Examples of the phenomena modeled by a stationary random process are the variations of the pressure in a turbulent fluid when flow is steady or bit-rate when transferring large files of the internet (excluding period over start and end of transfer). On the other hand, the numerical parameters corresponding to the earthquake have no stationary character.

It is worth noting that for the stationary process, we have that the expectation and variance do not depend on time, that is $m = \mathbb{E}X(t)$ and $\sigma^2(t) = \mathbb{D}X(t)$.

Definition 3.2.3. *Let $X(t)$ be a random process. The **autocorrelation function** $R(t_1, t_2)$ of $X(t)$ is defined by $R(t_1, t_2) = \mathbb{E}X(t_1)X(t_2)$.*

If the process is stationary, then the autocorrelation function $R(t, t + \Delta t)$ depends only on Δt and is written by $R(\Delta t) = \mathbb{E}X(t)X(t + \Delta t)$.

We define two more properties of random processes, which turn out to be very important in the following sections.

Definition 3.2.4. *The random process $X(t)$ has **orthogonal increments** if for any $t_1 < t_2 \leq t_3 < t_n$ it is valid $\mathbb{E}[X(t_2) - X(t_1)][X(t_4) - X(t_3)] = 0$. It has **independent increments** for any $t_1 < t_2 < \cdots < t_n$, random variables $X(t_i) - X(t_{i-1})$ for $i = 2, 3, \ldots, n$ are independent.*

The derivative and the integral of the stochastic process are commonly defined in two different ways.

Definition 3.2.5. *Random process $X(t)$ is differentiable*

1. ***with probability 1 (almost surely)*** *if a set of $\omega \in \Omega$ such that $X(t, \omega)$ is a differentiable random event from $\mathscr{F}$ and has probability 1. We denote the derivative by $\dot{X}(t)$;*
2. ***in the mean square***, *if*

$$\mathbb{E}\left[\frac{X(t + \Delta t) - X(t)}{\Delta t} - \dot{X}(t)\right] \to 0, \quad (3.16)$$

when $\Delta t \to 0$, for some random process $\dot{X}(t)$.

The definition of the (Riemann) integral is carried out in a similar way. Recall that the definition of the classical Riemann integral of the function $f\colon [a, b] \to \mathbb{R}$ by

$$\int_a^b f(x)\mathrm{d}x = \lim_{n\to+\infty} \sum_{i=1}^{n} f(\tau_i^{(n)})(t_i^{(n)} - t_{i-1}^{(n)}), \tag{3.17}$$

where $a = t_0^{(n)} < t_1^{(n)} < \cdots < t_n^{(n)} = b$ and $\tau_i^{(n)} \in [t_{i-1}^{(n)}, t_i^{(n)}]$ for $i = 1, 2, \ldots, n$ are arbitrary sequences such that $\max_{1\le i\le n} t_i^{(n)} - t_{i-1}^{(n)} \to 0$ when $n \to +\infty$.

Definition 3.2.6. *Random process $X(t)$ is integrable on $[a, b]$:*

1. ***with probability 1**, if the set of $\omega \in \Omega$ such that $Y(\omega) = \int_a^b X(t, \omega)\mathrm{d}t$ has probability 1, for some random variable Y;*
2. ***in the mean square**, if*

$$\mathbb{E}\left[\sum_{i=1}^{n} X(\tau_i^{(n)})(t_i^{(n)} - t_{i-1}^{(n)}) - Y\right]^2 \to 0, \tag{3.18}$$

for arbitrary sequences t and τ, as in the definition of the Riemann integral.

In both cases, we denote $Y = \int_a^b X(t)\mathrm{d}t$.

These two definitions of derivative and integral are mutually independent, that is, do not imply each other. In what follows, we assume that all derivatives and integrals are taken in the mean square sense.

3.3 Wiener and white noise process

One of the most important random processes both from the theoretical and practical point of view is the Wiener process.

Definition 3.3.1. *Random process $W(t)$ satisfying the following properties:*

1. *$W(t)$ is defined for every $t \in \mathbb{R}$ and is Gaussian random variable;*
2. *$W(t)$ has zero expectation, that is, $\mathbb{E}[W(t)] = 0$ for every t;*
3. *$\mathbb{E}[W(t)W(s)] = R(t, s) = \sigma^2 \cdot \min\{t, s\}$, for some $\sigma > 0$.*

*is known as the **Wiener process**. If $\sigma = 1$ then $W(t)$ is a **standard Wiener process**.*

Putting $t = s$ in the part 3, one gets $\mathbb{D}W(t) = \mathbb{E}W(t) = t\cdot\sigma^2$. There are many important properties of the Wiener process. A few of them are summarized by the following theorem.

Theorem 3.3.1. (Properties of the Wiener process). *Let $W(t)$ be a Wiener process. Then:*

1. *$W(t)$ has independent increments;*
2. *$W(t)$ is continuous and nondifferentiable with probability 1;*

3. *If* $t_0 < t_1 < \cdots < t_n$ *and* $X_i = W(t_i) - W(t_{i-1})$ *for* $i = 1, 2, \ldots, n$, *then* $\mathbf{X} : \mathcal{N}(\mathbf{0}, \mathbf{D})$ *where* $\mathbf{D} = \sigma^2 \cdot diag(t_1 - t_0, t_2 - t_1, \ldots, t_n - t_{n-1})$.
4. *Let* $[a, b]$ *be finite segment and* $a = t_0^{(n)} < t_1^{(n)} < \cdots < t_n^{(n)} = b$ *is the sequence of points such that* $\max_{0 \le i < n} t_{i+1} - t_i \to 0$ when $n \to +\infty$.

Then

$$\sum_{i=1}^{n} [W(t_{i+1}^{(n)}) - W(t_i^{(n)})]^2 \overset{m.}{\to} \sigma^2 (b - a),$$

when $n \to +\infty$. *This is known as the* ***Levy oscillation property***.

Another important random process is the white noise process given by the following definition.

Definition 3.3.2. *The* ***white noise process*** $\xi(t)$ *is the stationary process defined for every* $t \in \mathbb{R}$ *satisfying* $\mathbb{E}[\xi(t)] = 0$ *and* $\mathbb{E}[\xi(t)\xi(t + \Delta t)] = R(\Delta t) = S_0 \delta(t)$, *where* $\delta(t)$ *is Dirac delta function and* $S_0 > 0$ *is the real constant (known as the power of the process* $\xi(t)$*). Additionally, if* $\xi(t)$ *is a Gaussian random variable for every* $t \in \mathbb{R}$, *then* $\xi(t)$ *is the* ***Gaussian white noise process***.

Note that if we assume that $\xi(t)$ is Gaussian random variable, then

$$X(t) = \int_0^t \xi(\tau)\mathrm{d}\tau, \tag{3.19}$$

satisfies the conditions of the Wiener process (the formal proof can be found for example in Ref. [3]). Additionally, since

$$\mathbb{E}[X(t)^2] = \int_0^t \int_0^t \mathbb{E}[\xi(\tau_1)\xi(\tau_2)]\mathrm{d}\tau_1\mathrm{d}\tau_2 = S_0 \int_0^t \int_0^t \delta(\tau_1 - \tau_2)\mathrm{d}\tau_1\mathrm{d}\tau_2 = S_0 t,$$

we conclude that the parameter σ^2 of the Wiener process is $\sigma^2 = S_0$. Having this in mind, one can formally write $\dot{W}(t) = \xi(t)$, that is, that $\xi(t)$ is the formal derivative of the Wiener process. Taking $S_0 = \sigma = 1$, we get the standard Wiener and unit Gaussian white noise processes.

3.4 Markov and diffusion processes

The Markov property is one of the most important properties of stochastic processes, which usually occur in different natural phenomena.

Definition 3.4.1. *Random process* $X(t)$ *is* ***Markov's process*** *if*

$$F_{X(t_n)|X(t_{n-1}),\ldots,X(t_1)}(x_n|x_{n-1}, \ldots, x_1) = F_{X(t_n)|X(t_{n-1})}(x_n|x_{n-1}), \tag{3.20}$$

for any $t_1 < t_2 < \cdots < t_n$ *and* $x_1, x_2, \ldots, x_n \in \mathbb{R}$.

In fact, the process is Markov if the distribution of the future instance $X(t_n)$ depends explicitly only on the current instance $X(t_{n-1})$ and not on the previous ones

$(X(t_{n-2}), \ldots, X(t_1))$. The same equation as Eq. (3.20) is satisfied by the corresponding PDF functions.

Definition 3.4.2. *Random process $X(t)$ is diffusion process if it is the Markov process and $X(t)$ is a continuous function with probability 1.*

The fundamental properties of the diffusion process are given by the following lemma and are crucial in the theory of stochastic differential equations.

Lemma 3.4.1. *Let $X(t)$ be the diffusion process. Then the following limits exist*

$$m(x,t) = \lim_{\Delta t \to 0+} \frac{1}{\Delta t} \mathbb{E}[\Delta X(t) | X(t) = x], \tag{3.21}$$

$$\sigma(x,t) = \lim_{\Delta t \to 0+} \frac{1}{\Delta t} \mathbb{E}[\Delta X(t)^2 | X(t) = x], \tag{3.22}$$

for every $x, t \in \mathbb{R}$, where $\Delta X(t) = X(t + \Delta t) - X(t)$.

Functions $m(x,t)$ and $\sigma(x,t)$ are known as **drift** and **diffusion** coefficient and measure displacement and dispersion of the random process $X(t)$.

3.5 Stochastic integrals and differential equations

Assume that $X(t)$ is a random process with orthogonal increments (Definition 3.2.4). The one attempt to define stochastic integral of the random process $Z(t)$. might be as follows

$$Y^{(n)} = \sum_{i=1}^{n} Z(\tau_i^{(n)})(X(t_i^{(n)}) - X(t_{i-1}^{(n)})) \xrightarrow{m.} Y,$$

when $n \to +\infty$ and Y is a random variable (sequences t and τ satisfy the same conditions as for the Riemann integral). However, as we can see in the rest, different choices of $\tau_i^{(n)}$ lead to the different values of the integral. The following example illustrates this behavior.

Example 3.5.1. Take $Z(t) = X(t) = W(t)$ where $W(t)$ is Wiener process and consider two cases:

1. $\tau_i^{(n)} = t_{i-1}^{(n)}$; Then

$$\begin{aligned} Y^{(n)} &= \sum_{i=1}^{n} W(t_{i-1}^{(n)})(W(t_i^{(n)}) - W(t_{i-1}^{(n)})) \\ &= \sum_{i=1}^{n} \left[\frac{1}{2} W(t_i^{(n)})^2 - \frac{1}{2} W(t_{i-1}^{(n)})^2 - \frac{1}{2}(W(t_i^{(n)}) - W(t_{i-1}^{(n)}))^2 \right] \\ &= \frac{1}{2} \sum_{i=1}^{n} [W(t_i^{(n)})^2 - W(t_{i-1}^{(n)})^2] - \frac{1}{2} \sum_{i=1}^{n} \left(W(t_i^{(n)}) - \frac{1}{2} W(t_{i-1}^{(n)}) \right)^2. \end{aligned}$$

The first sum can be expanded to $(W(t_i^{(n)})^2 - W(t_{i-1}^{(n)})^2)/2 = (W(b)^2 - W(a)^2)/2$. The second sum converges in mean (Levy oscillation property, Theorem 3.3.1) to $(b-a)/2$. Therefore,

$$Y^{(n)} \xrightarrow{m.} \frac{1}{2}[W(b)^2 - W(a)^2 - b + a] \tag{3.23}$$

2. $\tau_i^{(n)} = (1-\lambda)t_{i-1}^{(n)} + \lambda t_i^{(n)}$; by using similar techniques (the full proof is given in Ref. [3]), one can show that

$$Y^{(n)} \xrightarrow{m.} \frac{1}{2}[W(b)^2 - W(a)^2] + \left(\lambda - \frac{1}{2}\right)(b-a). \tag{3.24}$$

Taking $\lambda = 0$ we reobtain case 1.

Two choices shown in the previous example are the motivation to define two fundamental types of stochastic integrals and differential equations.

3.5.1 Ito's integral and differential equation

Definition 3.5.1. *Assume that $X(t)$ is a random process and $W(t)$ is Wiener process such that $X(t)$ is independent of $W(s)$ for $s > t$. Let*

$$Y^{(n)} = \sum_{i=1}^{n} X(t_{i-1}^{(n)})(W(t_i^{(n)}) - W(t_{i-1}^{(n)})), \tag{3.25}$$

where $0 = t_0^{(n)} < t_1^{(n)} < \ldots < t_n^{(n)} = t$ is the sequence of points such that $\max_{1 \le i \le n} t_i^{(n)} - t_{i-1}^{(n)} \to 0$ when $n \to +\infty$. If there exists a random variable Y such that $Y^{(n)} \xrightarrow{m.} Y$ when $n \to +\infty$, then it is known as the ***stochastic integral in the Ito sense*** *and denoted by*

$$Y = \int_0^t X(t)\mathrm{d}W(t). \tag{3.26}$$

The previous definition remains the same if we take the k-dimensional random process $\mathbf{X}(t)$.

We define the random process $X(t)$ to be the solution of the **stochastic differential** equation (SDE)

$$\mathrm{d}X(t) = m(X(t), t)\mathrm{d}t + \sigma(X(t), t)\mathrm{d}W(t), \tag{3.27}$$

if it is the solution of the following integral equation

$$X(t) = X(0) + \int_0^t m(X(t), t)\mathrm{d}t + \int_0^t \sigma(X(t), t)\mathrm{d}W(t). \tag{3.28}$$

The integral in Eq. (3.28) is considered in the Ito sense (Definition 3.5.1). Since both $m(X(t), t)$ and $\sigma(X(t), t)$ are random variables, The first integral is the classic Riemann integral (like we introduced in Definition 3.2.6), while the second is the stochastic integral in the Ito sense. Functions $m(\cdot,\cdot)$ and $\sigma(\cdot,\cdot)$ are the real continuous functions, but we might require additional conditions as we will see later.

In a similar way, if we have independent Wiener processes $W_1(t), W_2(t), \ldots, W_d(t)$, then the solution of the stochastic differential equation

$$\mathrm{d}X(t) = m(X(t), t)\mathrm{d}t + \sum_{j=1}^{d} \sigma_j(X(t), t)\mathrm{d}W_j(t), \tag{3.29}$$

is in fact the solution of the following integral equation

$$X(t) = X(0) + \int_0^t m(X(t), t)\mathrm{d}t + \sum_{j=1}^{d} \int_0^t \sigma_j(X(t), t)\mathrm{d}W_j(t). \tag{3.30}$$

where $\sigma_j(\cdot, \cdot)(j = 1, 2, \ldots, d)$ are corresponding functions.

One can define the system of stochastic differential equations in a similar way as Eq. (3.29). Given the k-dimensional random process $\mathbf{X}(t)$, it is the solution of the system of stochastic differential equations

$$\mathrm{d}X_i(t) = m_i(\mathbf{X}(t), t)\mathrm{d}t + \sum_{j=1}^{d} \sigma_{ij}(\mathbf{X}(t), t)\mathrm{d}W_j(t), \tag{3.31}$$

if it satisfies the following system of integral equations

$$X_i(t) = X_i(0) + \int_0^t m_i(\mathbf{X}(t), t)\mathrm{d}t + \sum_{j=1}^{d} \int_0^t \sigma_{ij}(\mathbf{X}(t), t)\mathrm{d}W_j(t), \tag{3.32}$$

where $m_i(\cdot, \cdot)$ and $\sigma_{ij}(\cdot, \cdot)$ are corresponding functions for $i = 1, 2, \ldots, k$ and $j = 1, 2, \ldots, d$. We use the following vectorized notation for Eq. (3.31):

$$\mathrm{d}\mathbf{X}(t) = \mathrm{m}(\mathbf{X}(t), t)\mathrm{d}t + \sum_{j=1}^{d} \boldsymbol{\sigma}_j(\mathbf{X}(t), t)\mathrm{d}W_j(t), \tag{3.33}$$

where

$$\mathrm{m}(\mathrm{x}, t) = \begin{bmatrix} m_1(\mathrm{x}, t) & m_2(\mathrm{x}, t) & \cdots & m_k(\mathrm{x}, t) \end{bmatrix}, \tag{3.34}$$

$$\boldsymbol{\sigma}_j(\mathrm{x}, t) = \begin{bmatrix} \sigma_{1j}(\mathrm{x}, t) & \sigma_{2j}(\mathrm{x}, t) & \cdots & \sigma_{kj}(\mathrm{x}, t) \end{bmatrix}. \tag{3.35}$$

For the end of this section, we provide the change-of-variable rules for the stochastic differential equations, in both scalar and vector cases. It is given by the following lemmas.

Lemma 3.5.1. (Ito differential rule). *Let $X(t)$ be random process and $Y(t) = \phi(X(t), t)$, where $\phi: \mathbb{R}^2 \to \mathbb{R}$ is the function differentiable in the second and twice differentiable in the first argument. Assume that $X(t)$ satisfies the stochastic differential* Eq. (3.29). *Then $Y(t)$ satisfies the equation*

$$\mathrm{d}Y(t) = \widetilde{m}(Y(t), t)\mathrm{d}t + \widetilde{\sigma}(Y(t), t)\mathrm{d}W(t), \tag{3.36}$$

where

$$\widetilde{m}(x, t) = \frac{\partial \phi}{\partial t}(x, t) + m(x, t)\frac{\partial \phi}{\partial x}(x, t) + \frac{1}{2}\frac{\partial^2 \phi}{\partial x^2}(x, t)\sigma(x, t), \tag{3.37}$$

$$\widetilde{\sigma}(x, t) = \sigma(x, t)\frac{\partial \phi}{\partial x}(x, t). \tag{3.38}$$

Lemma 3.5.2. (Ito differential rule in the vector case). *Let* $\mathbf{X}(t)$ *be k-dimensional random process and* $\mathbf{Y}(t) = \phi(\mathbf{X}(t), t)$, *where* $\phi: \mathbb{R}^k \times \mathbb{R} \rightarrow \mathbb{R}^k$ *is the function differentiable in the second and twice differentiable in the first argument. Assume that* $X(t)$ *satisfies the system of stochastic differential* Eq. (3.33). *Then* $\mathbf{Y}(t)$ *satisfies the system of equations*

$$\mathrm{d}\mathbf{Y}(t) = \widetilde{\mathbf{m}}(\mathbf{Y}(t), t)\mathrm{d}t + \sum_{j=1}^{d} \widetilde{\boldsymbol{\sigma}}_j(\mathbf{Y}(t), t)\mathrm{d}W_j(t), \tag{3.39}$$

where

$$\widetilde{\mathbf{m}}(\mathrm{x}, t) = \frac{\partial \phi}{\partial t}(\mathrm{x}, t) + \sum_{i=1}^{k} m_j(\mathrm{x}, t)\frac{\partial \phi}{\partial x_j}(\mathrm{x}, t)$$
$$+ \frac{1}{2}\sum_{i=1}^{k}\sum_{j=1}^{d} \frac{\partial^2 \phi}{\partial x_i \partial x_j}(\mathrm{x}, t) \sum_{l=1}^{k} \sigma_{ij}(\mathrm{x}, t)\sigma_{lj}(\mathrm{x}, t), \tag{3.40}$$

and

$$\widetilde{\boldsymbol{\sigma}}_j(\mathrm{x}, t) = \sum_{l=1}^{k} \sigma_{ij}(\mathrm{x}, t)\frac{\partial \phi}{\partial x_i}(\mathrm{x}, t). \tag{3.41}$$

Note that

$$\sum_{j=1}^{d} \sigma_{ij}(\mathrm{x}, t)\sigma_{lj}(\mathrm{x}, t) = b_{il}(\mathrm{x}, t), \tag{3.42}$$

where

$$\mathbf{B}(\mathrm{x}, t) = \boldsymbol{\sigma}(\mathrm{x}, t)\boldsymbol{\sigma}(\mathrm{x}, t)^T, \quad \boldsymbol{\sigma}(\mathrm{x}, t) = [\sigma_{ij}(\mathrm{x}, t)]_{1\le i\le k,\ 1\le j\le d}. \tag{3.43}$$

The proofs of these two lemmas can be found in Ref. [3] or [4].

3.5.2 Stratonovich's integral and differential equation

The other way to define the stochastic integral is to use the second case of Example 3.5.1 for $\lambda = 1/2$. That is done in the following definition.

Definition 3.5.2. *Assume that $X(t)$ is a random process and $W(t)$ is Wiener process such that $X(t)$ is independent of $W(s)$ for $s > t$. Let*

$$Y^{(n)} = \sum_{i=1}^{n} X\left(\frac{t_{i-1}^{(n)} + t_i^{(n)}}{2}\right)\left(W\left(t_i^{(n)}\right) - W\left(t_{i-1}^{(n)}\right)\right), \tag{3.44}$$

were $0 < t_0^{(n)} < t_1^{(n)} < \ldots < t_n^{(n)} = t$ is the sequence of points such that $\max_{1\le i\le n} t_i^{(n)} - t_{i-1}^{(n)} \to 0$ when $n \to \infty$. If there exists a random variable Y such that $Y^{(n)} \xrightarrow{m.} Y$ when $n \to \infty$, then it is known as the ***stochastic integral in the Stratonovich sense*** *and denoted by*

$$Y = \int_0^t X(t) \circ \mathrm{d}W(t). \tag{3.45}$$

Similarly, the random process $X(t)$ is the solution of the following SDE in the Stratonovich sense:

$$\mathrm{d}^*X(t) = m^*(X(t), t)\mathrm{d}t + \sigma^*(X(t), t)\mathrm{d}W(t), \tag{3.46}$$

if it is the solution of the integral equation

$$X(t) = X(0) + \int_0^t m^*(X(t), t)\mathrm{d}t + \int_0^t \sigma^*(X(t), t) \circ \mathrm{d}W(t). \tag{3.47}$$

Analogous expressions to Eqs. (3.31) and (3.32) can be also defined in the Stratonovich sense,that is, for the solution $X(t)$ of the following system of equations

$$\mathrm{d}^*X_i(t) = m_i^*(\mathbf{X}(t), t)\mathrm{d}t + \sum_{j=1}^{d} \sigma_{ij}^*(\mathbf{X}(t), t)\mathrm{d}W_j(t). \tag{3.48}$$

for $i = 1, 2, \ldots, k$ and $j = 1, 2, \ldots, d$.

The following lemma shows the connection between SDEs in the Ito and Stratonovich sense. Since the expressions are not complicated, we show them only for the most general case of the system of SDEs (Eq. 3.48).

Lemma 3.5.3. *Assume that $X(t)$ is the random process that satisfies the SDE* (Eq. 3.48). *Then it is the solution of the SDE in the Ito sense* (Eq. 3.31) *where*

$$m_i(\mathbf{x}, t) = m_i^*(\mathbf{x}, t) + \frac{1}{2}\sum_{l=1}^{k}\sum_{j=1}^{d} \frac{\partial \sigma_{ij}^*}{\partial x_l}(\mathbf{x}, t)\sigma_{lj}^*(\mathbf{x}, t), \tag{3.49}$$

$$\sigma_{ij}(\mathbf{x}, t) = \sigma^*(\mathbf{x}, t), \tag{3.50}$$

for $i = 1, 2, \ldots, k$ and $j = 1, 2, \ldots, d$.

The Stratonovich sense SDE is often used for the mathematical modeling of the process where one of the components is determined by the white noise process. However, prior to introducing the variable change, the corresponding SDE is first converted to the Stratonovich sense using Lemma 3.5.3, and then the new variable is introduced using Lemma 3.5.2.

3.6 Orstein–Uhlenbeck and bounded noise process

Besides the white noise process $\xi(t)$, these are the two most important random processes from the application point of view. The **Orstein-Uhlenbeck process** is the solution to the following SDE in the Ito sense:

$$dX(t) = -\alpha X(t)dt + \sigma dW(t), \tag{3.51}$$

where α and σ are real constants. The bounded noise process is defined by $Z(t) = \cos X(t)$, where $X(t)$ is the solution of Eq. (3.51). These two processes have vast properties and are studied in more detail for example in Ref. [3].

3.7 Moment Lyapunov exponent and stability

Consider the SDE in the Ito sense (Eq. 3.33). The p-th moment $\mathbb{E}\|\mathbf{X}(t)\|^p$ is **stable** if, for every $\epsilon > 0$, there exists $\delta > 0$ such that $\mathbb{E}\|\mathbf{X}(t)\|^p < \epsilon$ when $\|X(0)\| < \delta$. In order to characterize the stability of the p-th moment, one introduces the **moment Lyapunov exponent** defined by

$$\Lambda(p) = \lim_{t\to+\infty} \frac{1}{t} \log \mathbb{E}\|\mathbf{X}(t)\|^p. \tag{3.52}$$

It is known that (see Ref. [5]) that the p-th moment is stable if $\Lambda(p) < 0$. The moment Lyapunov exponent $\Lambda(p)$ has the following properties:

1. Value $\Lambda(p)$ does not depend on the initial value $X(0)$;
2. $\Lambda(0) = 0$;
3. Function $\Lambda(p)$ is analytic and convex;
4. Function $\Lambda(p)/p$ is increasing for $p > 0$ and $\Lambda(p)/p \to \lambda$ when $p \to 0+$, where $\lambda > 0$ is the Lyapunov exponent.

In order to efficiently compute the approximation of the function Λ(p), one can use either of two commonly used approaches:

1. Perturbation method;
2. Numerical evaluation using the Monte–Carlo method.

The perturbation method is introduced in the next section and is our main tool in the analyses in Chapters 11–15. Numerical evaluation using the Monte–Carlo method was used for the verification. Various numerical methods can be used for solving SDEs (Eq. 3.33), some of which are shown in Ref. [3] or [6].

3.8 Perturbation method

This method is first introduced by Wedig in Ref. [7] and shown in more detail in Ref. [3]. Consider the following system of SDEs in the Ito sense (Eq. 3.31)

$$dX_i(t) = \sum_{j=1}^{k} a_{ij}X_j(t)dt + \sum_{l=1}^{d}\sum_{j=1}^{k} \sigma_{ij}^{l} X_j(t)dW_l(t) \tag{3.53}$$

where a_{ij} and σ_{ij}^{l} are real constants, for $i,\ j = 1, 2, \ldots, k$ and $l = 1, 2, \ldots, d$. This type of system commonly arises in all applications in the Chapters 11–15.

The first step is to introduce the variable change $\mathbf{X} \to (P, \mathbf{S})$, where $P = \|\mathbf{X}\|^p$ and $\mathbf{S} = \mathbf{X}/\|\mathbf{X}\|$. The transformation is performed using the Ito differential rule (Lemma 3.5.2). These transformations are known as Khasminskii's transformations [8]. The equivalent system now has the form

$$\mathrm{d}\mathbf{S}(t) = \hat{\mathbf{m}}(\mathbf{S}(t))\mathrm{d}t + \sum_{j=1}^{d} \hat{\boldsymbol{\sigma}}_j(\mathbf{S}(t))\mathrm{d}W_j(t) \tag{3.54}$$

$$\mathrm{d}P(t) = P\widetilde{m}(\mathbf{S}(t))\mathrm{d}t + \sum_{j=1}^{d} \widetilde{\sigma}_j(\mathbf{S}(t))\mathrm{d}W_j(t) \tag{3.55}$$

where $\hat{\mathbf{m}}(\cdot)$, $\hat{\boldsymbol{\sigma}}_j(\cdot)$, $\widetilde{m}(\cdot)$, and $\widetilde{\sigma}_j(\cdot)$ are corresponding functions. The next step is to apply the additional transformation of the form $S = T(\mathbf{S})P$ (i.e., transformation $(P, \mathbf{S}) \to (S, \mathbf{S})$ such that the first equation has the form)

$$\mathrm{d}S(t) = \Lambda(p)S(t)\mathrm{d}t + \sum_{j=1}^{d} \sigma_j(S(t), \mathbf{S}(t))\mathrm{d}W_j(t). \tag{3.56}$$

In other words, we want the drift term to be independent of the vector $\mathbf{S}(t)$. Applying again the Ito differential rule (Lemma 3.5.2), one leads to the following equation:

$$\mathscr{L}(T(\mathbf{s})) = \Lambda(p)T(\mathbf{s}), \tag{3.57}$$

where $\mathscr{L}$ is the corresponding operator. In fact, the transformation T is the eigenvector of the operator $\mathscr{L}$ corresponding to the eigenvalue $\Lambda(p)$. The eigenvalue problem (3.57) is often difficult to solve directly. However, the idea of the perturbation method is to expand $\mathscr{L}$, T, and $\Lambda(p)$ into the finite series over the "small" parameter ϵ. If one can write

$$\begin{aligned} \mathscr{L}(\cdot) &= \sum_{l=1}^{m} \mathscr{L}_l(\cdot)\epsilon^l, \\ T(\cdot) &\approx \sum_{l=1}^{n} T_l(\cdot)\epsilon^n, \\ \Lambda(p) &\approx \sum_{l=1}^{n} \Lambda_l(p)\epsilon^n. \end{aligned} \tag{3.58}$$

The eigenvalue equation now can be solved by iteratively solving

$$\sum_{j=0}^{m} \sum_{i=0}^{\min(n,r-j)} \mathscr{L}_j(T_i(\mathbf{s})) = \sum_{l=0}^{n} \sum_{i=0}^{\min(n,r-l)} \Lambda_j(p)T_i(\mathbf{s}), \tag{3.59}$$

or $r = 0, 1, 2, \ldots$. The corresponding equations are much easier than the original Eq. (3.57) and solvable in the closed form.

3.9 New generalized transformations for multiple degrees of freedom mechanical systems

Within the broader scope of stochastic mechanical modeling, both nongyroscopic and gyroscopic systems can be derived through appropriate discretization methods. These systems, defined by n degrees of freedom (DOF), must be converted into $2n$-DOF first-order Stratonovich stochastic differential equations to facilitate further stochastic stability analysis. The differential equations governing a stochastic system with nongyroscopic parametric excitation and n degrees of freedom can be represented as follows:

$$\ddot{q}_i(t) + 2\varepsilon \sum_{j=1}^{n} \beta_{ij}\dot{q}_j(t) + \omega_i^2 q_i(t) + \sqrt{\varepsilon}\xi(t)\omega_i \sum_{j=1}^{n} k_{ij}q_j(t) = 0, \tag{3.60}$$

where $\xi(t)$ denotes an ergodic stochastic process with a mean of zero, ε represents a small parameter, ω_i represents the natural frequencies, and β_{ij} represents the viscous damping coefficients, where i ranges from 1 to n. By introducing a new variable in place of $\dot{q}_i$, Eq. (3.60) can be reformulated into an equivalent system comprising $2n$ first-order Stratonovich stochastic differential equations in the following form

$$\mathrm{d}\{\mathbf{x}\} = \mathbf{A}\{\mathbf{x}\}\mathrm{d}t + \varepsilon\mathbf{B}\{\mathbf{x}\}\mathrm{d}t + \sqrt{\varepsilon}\mathbf{C}\{\mathbf{x}\}\mathrm{d}\xi(t). \tag{3.61}$$

The stochastic component $\sqrt{\varepsilon}\xi(t)$ in Eq. (3.60) represents a low-intensity white-noise process. Subsequently, both the nongyroscopic and gyroscopic systems are converted into a set of $2n$ first-order Stratonovich differential equations. From this stage forward, an identical methodology is applied for the subsequent dynamic analysis of both multiple degrees of freedom (MDOF) systems. Within this context, the dynamic stability of an oscillatory system is determined by evaluating the largest Lyapunov exponent and the p-th moment Lyapunov exponent in the form

$$\lambda_T = \lim_{t\to\infty}\frac{1}{t}\log \|\boldsymbol{q}(t;\, q_0)\|. \tag{3.62}$$

In the context where $\boldsymbol{q}(t;\, q_0)$ is the solution process of a linear dynamic system, the almost-sure stability is dependent on the sign of the maximal Lyapunov exponent, which characterizes the exponential growth rate of the solution in a perturbed dynamic system. A negative maximal Lyapunov exponent indicates almost-sure stability, whereas a nonnegative value suggests instability. The exponential growth of $\mathrm{E}[\|\boldsymbol{q}(t;\, q_0, \dot{q}_0)\|^p]$ is characterized by the moment Lyapunov exponent, defined as $\Lambda_T(p) = \lim_{t\to\infty}\frac{1}{t}\log \mathrm{E}[\|\boldsymbol{q}(t;\, q_0)\|^p]$, where $\mathrm{E}[\cdot]$ represents the expectation. If $\Lambda_T(p) < 0$, then $\mathrm{E}[\|\boldsymbol{q}(t;\, q_0, \dot{q}_0)\|^p] \to 0$ as $t \to 0$, which is referred to as p-th moment stability. Despite the significance of Lyapunov exponents in analyzing dynamic stability in stochastic systems, their practical evaluation poses considerable difficulties. Moreover, ensuring almost-sure and moment stability of the equilibrium state, where $q = \dot{q} = 0$, remains a complex challenge. The stochastic component $\xi(t)$, representing the white-noise process with zero mean and autocorrelation function in the following form

$$R_{\xi\xi}(t_1, t_2) = \mathrm{E}[\xi(t_1)\xi(t_2)] = \sigma^2\delta(t_2 - t_1), \tag{3.63}$$

where σ is the intensity of the random process $\xi(t)$, $\delta(*)$ is the Dirac delta function. By utilizing the recently developed generalized transformations, applied specifically to a mechanical system with 4 DOF as demonstrated in Chapter 19 and expressed as

$$\begin{Bmatrix} x_1 \\ x_2 \end{Bmatrix} = a \begin{Bmatrix} \cos\theta_1 \\ -\sin\theta_1 \end{Bmatrix} \prod_{i=1}^{n-1} \cos\varphi_i,$$

$$\vdots$$

$$\begin{Bmatrix} x_{2k-1} \\ x_{2k} \end{Bmatrix} = a \sin\varphi_k \begin{Bmatrix} \cos\theta_k \\ -\sin\theta_k \end{Bmatrix} \frac{\prod_{j=1}^{n-1} \cos\varphi_j}{\prod_{i=2}^{k} \cos\varphi_i}, \quad (k = 2,\ldots,n-2)$$

$$\vdots$$

$$\begin{Bmatrix} x_{2n-3} \\ x_{2n-2} \end{Bmatrix} = a \cos\varphi_1 \sin\varphi_{n-1} \begin{Bmatrix} \cos\theta_{n-1} \\ -\sin\theta_{n-1} \end{Bmatrix},$$

$$\begin{Bmatrix} x_{2n-1} \\ x_{2n} \end{Bmatrix} = a \sin\varphi_1 \begin{Bmatrix} \cos\theta_n \\ -\sin\theta_n \end{Bmatrix},$$

$$-\infty < p < \infty, \quad 0 \le \theta_i \le 2\pi, \quad 0 \le \varphi_j \le \frac{\pi}{2},$$

$$(i = 1,\ldots,n), \quad (j = 1,\ldots,n-1),$$

$$P = a^p = \sum_{i=1}^{2n} (x_i^2)^{\frac{p}{2}}, \tag{3.64}$$

it becomes feasible to derive the Itô equations for any MDOF stochastic system directly. This achieves the primary objective of the presented transformations.

References

[1] R. Meester, A Natural Introduction to Probability Theory, Birkhauser, Basel, 2008.

[2] K.B. Athreya, S.N. Lahiri, Measure Theory and Probability Theory, Springer, 2006.

[3] W.C. Xie, Dynamic Stability of Structures, Cambridge University Press, 2006.

[4] X. Mao, Stochastic Differential Equations and Applications, second ed., Woodhead Publishing Limited, 2011.

[5] S.A. Molchanov, The structure of the eigenfunctions of one-dimensional unorder structure, Mathematics of the USSR-Izvestiya 12 (1978) 69–101.

[6] P. Kloeden, E. Platen, Numerical solution of SDE Through Computer Experiments, Springer, 1994.

[7] W. Wedig, Lyapunov exponent of stochastic systems and related bifurcation problems, in: T.S. Ariaratnam, G.I. Schueller, I. Elishakoff (Eds.), Stochastic Structural Dynamics - Progress in Theory and Applications, Elsevier, 1988.

[8] R. Khasminskii, N. Moshchuk, Moment Lyapunov exponent and stability index for linear conservative system with small random perturbation, SIAM Journal of Applied Mathematics 58 (1998) 245–256.

Part II

Stability of vibrations of complex moving objects: Modeling and applications

Stabilization and critical velocity of a moving mass

4

4.1 Introduction

The presence of loose soil where the construction of a roadway or railway is in process complicates the conditions related to stability and safety. Geosynthetic materials can be introduced into the soil to improve its characteristics. Fernandes et al. [1] presented a notable application of geosynthetic materials; when geosynthetic materials are used in the construction of trackbeds, they provide one or more of the four functions. These functions are: separation, which prevents the intermixing between the adjoining soil types; filtering, which hinders the seeping of soil particles; drainage, which enables an uninterrupted passage of water; and reinforcement. One of the significant physical parameters that should be considered when analyzing the stability and vibrations is the stiffness of the layers.

In developed areas, the development of road and rail traffic has led to high-speed transportation. In particular, that can be seen in high-speed train traffic. Thus, Ju and Lin [2] studied the vibrations of a high-speed train 'Japan SKS-700' for constant train speeds in the range between 300 and 900 km/h. This kind of traffic modernization has been made possible due to theoretical research that encompasses a variety of scientific fields, from civil engineering to mechanical engineering to automation, control, and electrical engineering. On the subject of mechanics, many researchers have provided substantial initiative and contributions to this field. First, Timoshenko [3] showed that a moving vertical force (of constant intensity) has a critical velocity at which deflections become severe. This finding has led to further studies on the influence of axial forces on critical velocity. Kerr [4] studied the response of an axially compressed, continuously supported infinite beam subjected to a moving vertical force that has constant intensity. In real-world rail structures (like a railway track), normal stresses in the axial direction occur because of thermal expansion [5]. For that reason, studying vibrating beams subjected to axial forces is of interest. The mechanical model of a beam that vibrates and is supported by an elastic foundation, subjected to a moving vertical force of constant intensity, is the basis for further research involving the model of a beam subjected to a moving mass. Thus, Denisov et al. [6] have identified the region of stability for a mass that moves uniformly on a continuously supported beam by applying the D-decomposition method. Metrikine and Dieterman [7] have studied the vibrations of a mass that moves on an axially compressed beam that is supported by a viscoelastic foundation. For that problem, they have determined the zones of the instability of vibrations and have found that there is a critical mass (for the moving mass) for which the axial forces have an influence on the emergence of the zone of instability. Chang et al. [8], Muscolino and Palmeri [9] studied the reaction of a single beam under a moving oscillator to evaluate the vehicle-track and vehicle-bridge interactions. Anyhow, Kozić et al. [10]

Modeling of Complex Dynamic Systems. DOI: https://doi.org/10.1016/B978-0-443-23942-7.00004-8

were the first who, through the use of the regular perturbation method, determined the explicit asymptotic expressions for the moment Lyapunov exponents, for a double-beam system, in the presence of small intensity noises. The critical buckling force of two elastically connected beams is determined by means of an analytical procedure, which can be found in the works by Stojanović et al. [11] and Stojanović and Kozić [12].

This chapter is composed of two separate parts. The first part is about the reaction of a double-beam system where the top beam is axially compressed and subjected to a moving force. The second part is about the regions of the instability of vibrations for the complex model that is analyzed. This model comprises a deformable beam on which a mass is moving. The mass is subjected to a vertical force of constant intensity, while the beam is subjected to compressive axial forces. The analysis in the second part was done using the analytical results from the first part. The regions of stability (in the second part) are determined based on the value of the critical mass. This critical mass is greater when the continuous support has a stabilizer (the double-supported system). This means that this kind of support produces a larger region of stability. The advantages of the double-supported system are shown through the dynamic analysis of stability. In the second part, the values of the critical velocities between the models are also compared. The focus of this comparison was on the influence of compressive axial forces and the influence of the vibration frequency of the moving mass during constant contact between the mass and the support. Unlike the study of the classical model, concluded by Verichev and Metrikine [13], the presented study considers the influence of axial forces on the natural frequency of the system. Analytical works [14–16] in the area of vibrations of solids and structures have brought forth a model in which two dynamic elements are elastically connected [17]. This type of model was used in a study by Stojanović et al. [18] and is also the basis for the study presented in this chapter.

4.2 Model and solution of free waves propagation

To improve the performance of roads and railways, engineering practice implements geosynthetics that provide the four basic functions, as described in the introduction [1]. Ling et al. [19] found the application of the complex beam model. In their model, the embedded rail track was treated as a two-layer system that comprises two rails, slabs, filling material, and an adjustment layer, which is underneath the slabs. The rails (described by the Timoshenko beam assumption) were supported by continuous elastic supports. To reduce the order of the partial differential equations (PDEs) that mathematically describe the beams, the modal superposition method was used. The above work has motivated the study presented in this chapter and has served as its basis. Therefore, the present study is a mathematical analysis of a double-supported system, made of two different parallel beams of infinite length, continuously joined by an elastic layer of Winkler type; the bottom beam rests on an elastic foundation (Fig. 4.1). The analytical analysis

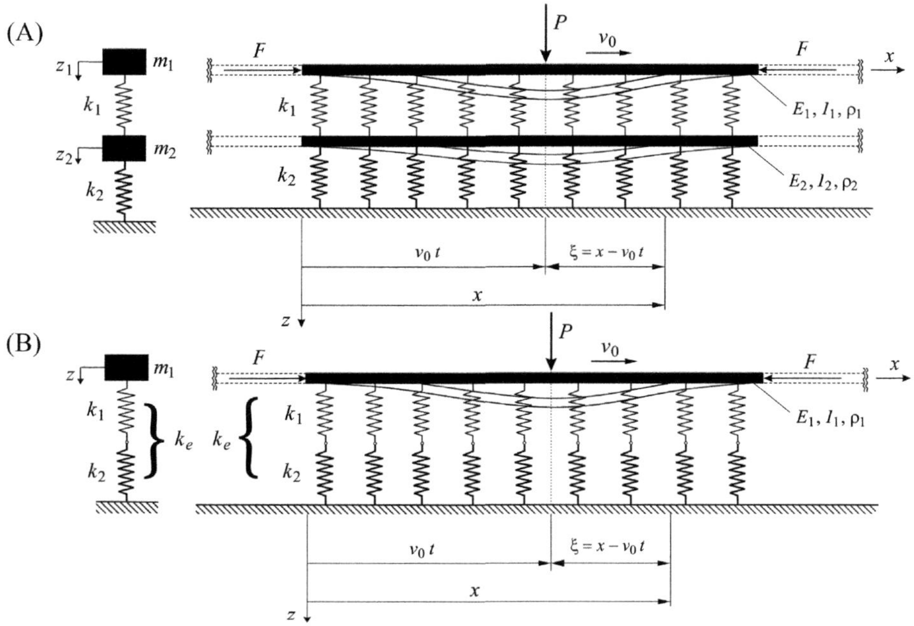

Figure 4.1 Infinite beam systems. (A) Double-supported system with a moving vertical force and (B) Classical beam with a moving vertical force.

begins with the system of two differential equations that describe the continuous beam system and the forces between and acting on them.

$$E_1 I_1 \frac{\partial^4 w_1}{\partial x^4} + F \frac{\partial^2 w_1}{\partial x^2} + \rho_1 \frac{\partial^2 w_1}{\partial t^2} + k_1 (w_1 - w_2) = P\delta(x - v_0 t), \tag{4.1}$$

$$E_2 I_2 \frac{\partial^4 w_2}{\partial x^4} + \rho_2 \frac{\partial^2 w_2}{\partial t^2} + k_2 w_2 - k_1 (w_1 - w_2) = 0, \tag{4.2}$$

in which $w_i = w_i(x, t)$, $i = 1, 2$ are the transverse displacements of the centerlines of the beams; F is the axial compressive force that acts on the top beam; k_i are the stiffness parameters of the elastic layers; ρ_i are the masses of the beams, per unit length; and the products $E_i I_i$ represent the bending stiffness of the beams. The intensity of the moving vertical force is denoted by P, while the velocity of the force, which is assumed to be constant, is denoted by v_0. Lastly, $\delta(\cdot)$ denotes the Dirac delta function. The physical model in Fig. 4.1(A) represents a dynamic system with a stabilizer. The physical model in Fig. 4.1(B) [4] can be derived from the model in Fig. 4.1(A) when the stabilizer is discarded ($w_2 = 0$) and the stiffness moduli k_1 and k_2 are replaced with an equivalent stiffness modulus $k_e = \frac{k_1 k_2}{k_1 + k_2}$. This procedure confirms the results presented by Kerr [4].

By substituting the wave type expression

$$w_i(x,\ t) = W_i \sin\left[\frac{2\pi}{\lambda}(x - ct)\right],\quad i = 1,\ 2; \tag{4.3}$$

into Eqs. (4.1 and 4.2), and putting $P = 0$, the determinant of the system coefficients $(w_1,\ w_2)$ is obtained. This yields a fourth-order (biquadratic) equation in c with two roots:

$$\begin{aligned} c_1^2 = \frac{1}{2k_\lambda^4\rho_1\rho_2}\Bigg\{ & k_\lambda^2\left[\rho_2(E_1I_1k_\lambda^4 + k_1 - Fk_\lambda^2) + \rho_1(E_2I_2k_\lambda^4 + k_1k_2)\right] \\ & -\Bigg[k_\lambda^4\Bigg[2k_1(\rho_1 - \rho_2)(Fk_\lambda^2\rho_2 - E_1I_1k_\lambda^4\rho_2 + E_2I_2k_\lambda^4\rho_1 + k_2\rho_1) \\ & +(Fk_\lambda^2\rho_2 - E_1I_1k_\lambda^4\rho_2 + E_2I_2k_\lambda^4\rho_1 + k_2\rho_1)^2 + k_1^2(\rho_1 + \rho_2)^2\Bigg]\Bigg]^{1/2}\Bigg\}, \end{aligned} \tag{4.4}$$

$$\begin{aligned} c_2^2 = \frac{1}{2k_\lambda^4\rho_1\rho_2}\Bigg\{ & k_\lambda^2\left[\rho_2(E_1I_1k_\lambda^4 + k_1 - Fk_\lambda^2) + \rho_1(E_2I_2k_\lambda^4 + k_1k_2)\right] \\ & +\Bigg[k_\lambda^4\Bigg[2k_1(\rho_1 - \rho_2)(Fk_\lambda^2\rho_2 - E_1I_1k_\lambda^4\rho_2 + E_2I_2k_\lambda^4\rho_1 + k_2\rho_1) \\ & +(Fk_\lambda^2\rho_2 - E_1I_1k_\lambda^4\rho_2 + E_2I_2k_\lambda^4\rho_1 + k_2\rho_1)^2 + k_1^2(\rho_1 + \rho_2)^2\Bigg]\Bigg]^{1/2}\Bigg\}, \end{aligned} \tag{4.5}$$

in which λ represents the wavelength, $k_\lambda = 2\pi/\lambda$ represents the wave number, and c_1 and c_2 are the velocities of wave propagation. Velocity c_1 is the minimal wave velocity in the scenario when the influence of the axial force is considered. With the condition $c_1 = 0$, the critical axial force can be determined based on the existence of its minimal value for a unique real positive value of the wave number. The formulation for that is

$$F = \frac{E_1E_2I_1I_2k_\lambda{}^8 + [E_1I_1(k_1 + k_2) + E_2I_2k_1]k_\lambda{}^4 + k_1k_2}{k_\lambda{}^2(E_2I_2k_\lambda{}^4 + k_1 + k_2)}. \tag{4.6}$$

The minimal value of the axial force can be ascertained from the condition $\frac{dF}{dk_\lambda} = 0$ for a singular real positive value $k_\lambda = k_{\lambda F}$. In this case, the critical axial force of the double-supported system has the following form

$$F_{c,d} = \frac{E_1E_2I_1I_2k_{\lambda F}{}^8 + [E_1I_1(k_1 + k_2) + E_2I_2k_1]k_{\lambda F}{}^4 + k_1k_2}{k_\lambda{}^2(E_2I_2k_{\lambda F}{}^4 + k_1 + k_2)}. \tag{4.7}$$

With an analogous procedure, it is possible to determine the maximally allowed (i.e., the critical) value of the system velocity (with the axial force) from the condition

$\frac{dc_1}{dk_\lambda} = 0$ for the unique real positive value $k_\lambda = k_{\lambda \upsilon}$. In this case, the critical velocity of the double-supported system has the following form:

$$v_{cr}^2 = c_{1min}^2 = \frac{1}{2k_{\lambda\upsilon}^4 \rho_1 \rho_2} k_{\lambda\upsilon}^2 \left(\rho_2 \left(E_1 I_1 k_{\lambda\upsilon}^4 + k_1 - F k_{\lambda\upsilon}^2 \right) + \rho_1 \left(E_2 I_2 k_{\lambda\upsilon}^4 + k_1 + k_2 \right) \right)$$
$$- \left[k_{\lambda\upsilon}^4 \left(2k_1 (\rho_1 - \rho_2) \left(F k_{\lambda\upsilon}^2 \rho_2 - E_1 I_1 k_{\lambda\upsilon}^4 \rho_2 + E_2 I_2 k_{\lambda\upsilon}^4 \rho_1 + k_2 \rho_1 \right) \right.\right.$$
$$\left.\left. + \left(F k_{\lambda\upsilon}^2 \rho_2 - E_1 I_1 k_{\lambda\upsilon}^4 \rho_2 + E_2 I_2 k_{\lambda\upsilon}^4 \rho_1 + k_2 \rho_1 \right)^2 + k_1^2 \left(\rho_1 + \rho_2 \right)^2 \right) \right]^{1/2}. \tag{4.8}$$

For the analyzed systems, shown in Fig. 4.1(A and B), the diagrams in Figs. 4.2 and 4.3 show the critical velocity of the system as a function of the parameter $\zeta = k_2/k_1$; two diagrams are for different axial forces. This and other diagrams in this analysis were delineated using the following set of system parameters [13]: $E_1 = 2 \times 10^{11}$ Pa, $I_1 = 3.055 \times 10^{-5}$ m^4, $\rho_1 = 7849$ kg/m, $k_1 = 10^8$ N/m^2, $E_2 = 2.73 \times 10^{10}$ Pa, $I_2 = 9.165 \times 10^{-5}$ m^4, $\rho_2 = 2400$ kg/m. The diagrams (Figs. 4.2 and 4.3) illustrate that as the stiffness of the bottom elastic layer (k_2) increases, the critical velocity of the double-supported system, Fig. 4.1(A) begins to gravitate toward the critical velocity of the system without a stabilizer—Fig. 4.1(B). For the model with different beams (E_1, I_1, ρ_1 —top beam, E_2, I_2, ρ_2 —bottom beam), the stability

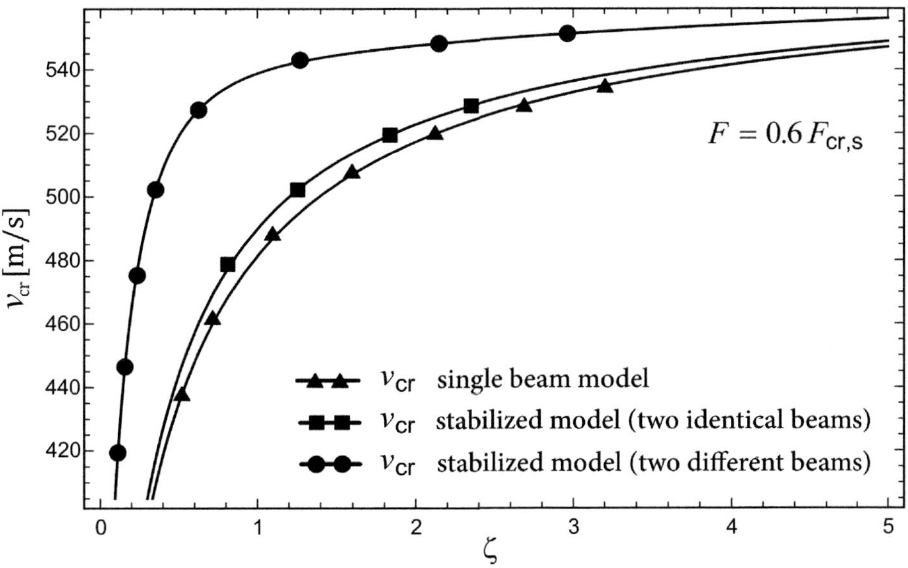

Figure 4.2 Critical velocities for the compared models. For single-beam model, and stabilized two-beam models, when the parameter ζ is the varying parameter.

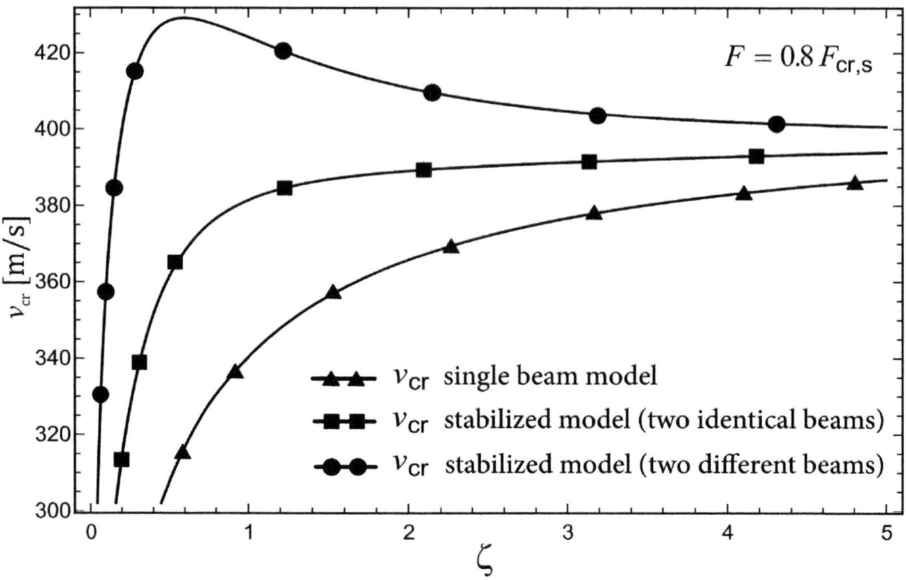

Figure 4.3 Critical velocities for the compared models. For single-beam model, and stabilized two-beam models, when the parameter ζ is the varying parameter.

regions are greater, that is, the system can withstand greater speeds in comparison to the model constructed with the same beams. This effect increases with an increase in the axial force (Fig. 4.3).

For the analyzed systems, Figs. 4.4–4.6 show how the critical value of the velocity changes depending on the applied axial force F for various values of parameter ζ. Looking at the results shown in Figs. 4.4–4.6, it can be concluded that both systems (single-beam and stabilized model) can have the same value of the critical velocity for the same intensity of the axial force $0 \leq F \leq F_c$, which is designated with F_{switch}. This analytically determined intensity of the force F_{switch} is noteworthy, for it shows that the double-supported model, in comparison with the single-beam model, owns a vaster stability region for the same values of velocity, although only when subjected to axial forces larger than F_{switch} and smaller than the critical system force. The double-supported system reveals a significant advantage (in terms of stability) when the stiffness k_2 is reduced; in those situations, the stability region is larger in comparison with the classical (single-beam) model. Fig. 4.7 shows the comparison between the critical forces of the models studied in the function of parameter ζ; the critical force of the double-supported system is designated with $F_{c,d}$, while $F_{c,s}$ is the critical force of the single-beam (classical) system [4].

As can be seen, the decrease in stiffness k_2 leads to the increase of the $F_{cr,d}/F_{cr,s}$ ratio, and as this ratio increases, so does the stability of the system. This is expected because the stiffness of the added support layer takes precedence in providing the system's overall equivalent stiffness. In the case when $k_2 \to \infty$, both of the studied models are reduced to a

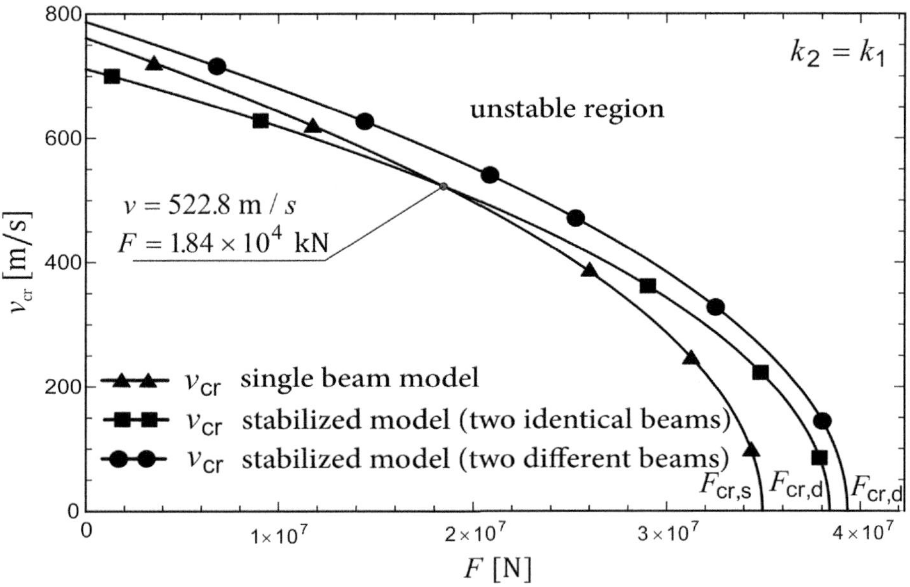

Figure 4.4 Critical velocities for the compared models. For single-beam model, and stabilized two-beam models, when the axial force is the varying parameter, while ζ=1.

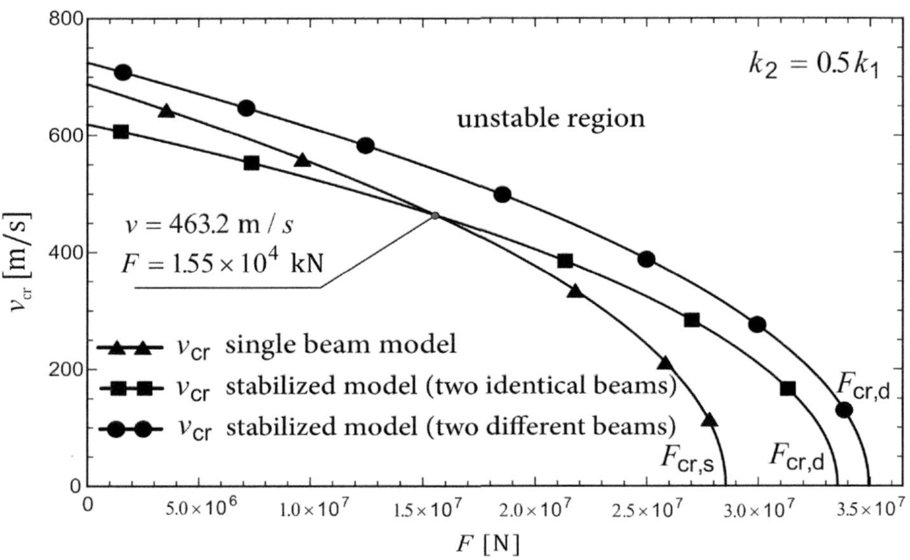

Figure 4.5 Critical velocities for the compared models. For single-beam model, and stabilized two-beam models, when the axial force is the varying parameter, while ζ =0.5.

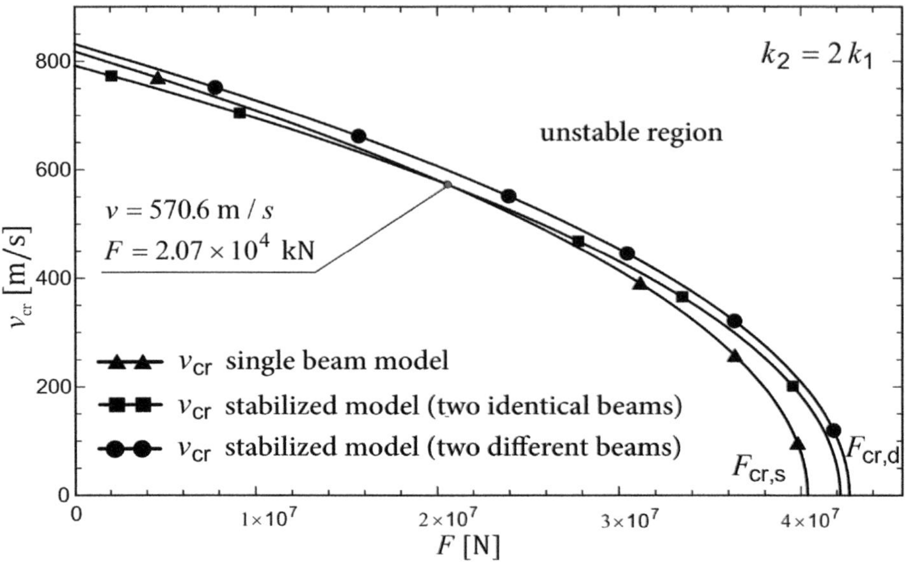

Figure 4.6 Critical velocities for the compared models. For single-beam model, and stabilized two-beam models, when the axial force is the varying parameter, while ζ=2.

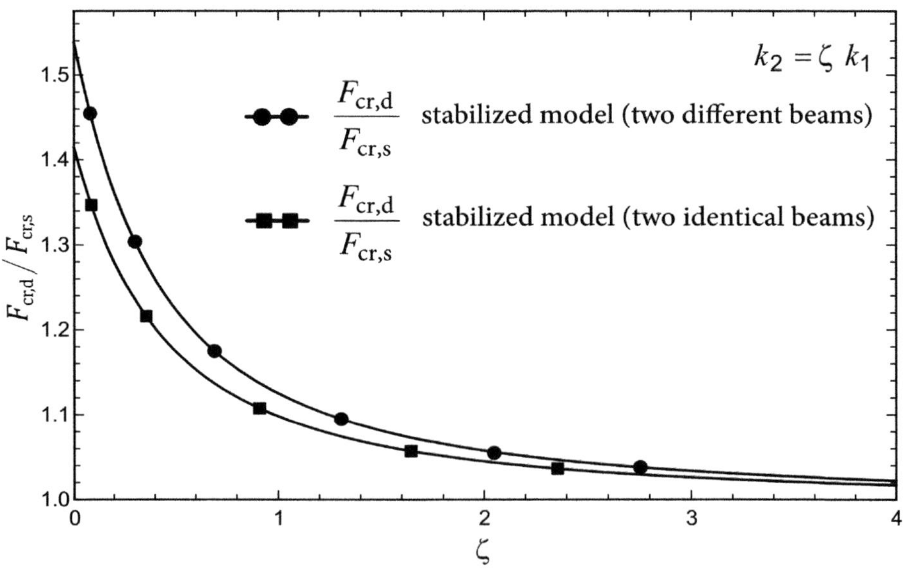

Figure 4.7 Ratio between the critical forces. For the observed stabilized models, when the parameter ζ is the varying parameter.

single-beam model with only one elastic layer, whose stiffness modulus is k_1. The analytical form of the force F_{switch}, can be obtained from the following condition:

$$c_{1\,min}(F)|\text{Present model} = c_{1\,min}(F)|\text{Classical model} \tag{4.9}$$

in its implicit form, Eq. (4.9), in which $F_{switch} = F$ will be

$$\begin{aligned}\frac{2\sqrt{k_e E_1 I_1 - F}}{\rho_1} = & \frac{1}{2k_{\lambda v}^4\rho_1\rho_2} k_{\lambda v}^2(\rho_2 E_1 I_1 k_{\lambda v}^4 + k_1 - F k_{\lambda v}^2) \\ & + (\rho_1 E_2 I_2 k_{\lambda v}^4 + k_1 + k_2)) \\ & - \Big[k_{\lambda v}^4 (2k_1(\rho_1 - \rho_2))(F\, k_{\lambda v}^2\rho_2 - E_1 I_1 k_{\lambda v}^4 \rho_2 + E_2 I_2 k_{\lambda v}^4 \rho_1 + k_2\rho_1) \\ & + (F\, k_{\lambda v}^2\rho_2 - E_1 I_1 k_{\lambda v}^4\rho_2 + E_2 I_2 k_{\lambda v}^4\rho_1 + k_2\rho_1)^2 + k_1^2(\rho_1\rho_2)^2 \Big]^{1/2}.\end{aligned} \tag{4.10}$$

4.3 Response of the double-supported beam system subjected to a moving constant vertical force *P*

After a certain space of time, because of a minute transient motion, the infinite length of the beam, and a constant velocity of the vertical force, steady-state vibrations will come about. By transforming the coordinates into new ones; $\xi = x - v_0 t$, $\eta = y$, $\chi = z$, the system of two PDEs (Eqs. 4.1 and 4.2) can be transformed following the rules: $\frac{d}{dx} = \frac{d}{d\xi}\frac{\partial\xi}{\partial x}$; $\frac{d}{dt} = \frac{d}{d\xi}\frac{\partial\xi}{\partial t}$, where $\frac{\partial\xi}{\partial x} = 1$; $\frac{\partial\xi}{\partial t} = -v$. The fourth derivative by x becomes: $\frac{\partial^4 w_1}{\partial x^4} = \frac{d}{dx}\left[\frac{d}{dx}\left[\frac{d}{dx}\left[\frac{dw_1}{d\xi}\frac{\partial\xi}{\partial x}\right]\right]\right] = \frac{d}{dx}\left[\frac{d}{dx}\left[\frac{d}{dx}\left(\frac{dw_1}{d\xi}\right)\frac{\partial\xi}{\partial x}\right]\right] = \frac{d}{dx}\left[\frac{d}{dx}\left[\left(\frac{d^2w_1}{d\xi^2}\right)\frac{\partial\xi}{\partial x}\right]\right]$ $= \frac{d}{dx}\left[\left(\frac{d^3w_1}{d\xi^3}\right)\frac{\partial\xi}{\partial x}\right] = \frac{d^4w_1}{d\xi^4}$

Following the same procedure, the second derivative by x becomes $\frac{\partial^2 w_1}{\partial x^2} = \frac{d^2w_1}{d\xi^2}$. The second derivative by t is $\frac{\partial^2 w_1}{\partial t^2} = \frac{d}{dt}\left[\frac{dw_1}{d\xi}\frac{\partial\xi}{\partial t}\right] = -v\frac{d}{dt}\left[\frac{dw_1}{d\xi}\right] = -v\frac{d}{d\xi}\left(\frac{dw_1}{d\xi}\right)\frac{\partial\xi}{\partial t} = v^2\frac{d^2w_1}{d\xi^2}$. Now the mentioned system of PDEs (Eqs. 4.1 and 4.2) will turn into

$$\begin{aligned}& E_1 I_1\frac{d^4w_1}{d\xi^4} + \left(F + \rho_1 v_0^2\right)\frac{d^2w_1}{d\xi^2} + k_1(w_1 - w_2) = P\delta(\xi), \\ & E_2 I_2\frac{d^4w_2}{d\xi^4} + \rho_2 v_0^2\frac{d^2w_2}{d\xi^2} + k_2 w_2 - k_1(w_1 - w_2) = 0.\end{aligned} \tag{4.11}$$

The vertical force P moves only horizontally; accordingly, the acceleration in the vertical z-direction does not exist. The force P can be included through the conditions at $\xi = 0$. Assuming the values $w_i = A_i e^{s\xi}$, $i = 1, 2$ and substituting them into the system of equations (Eq. 4.11), the equations that have nontrivial solutions are obtained for the case when the determinant of the system coefficients (A_1, A_2) matrix is equal to zero. This produces the equation that is a reduced fourth-order polynomial in

$$p^4 + ap^3 + bp^2 + cp + d = 0, \quad p = s^2, \tag{4.12}$$

where

$$a = \frac{F + \rho_1 {v_0}^2}{E_1 I_1} + \frac{\rho_2 {v_0}^2}{E_2 I_2},$$

$$b = \frac{\rho_2 {v_0}^2 \left(F + \rho_1 {v_0}^2\right) + E_1 I_1 \left(k_1 + k_2\right) + E_2 I_2 k_1}{E_1 E_2 I_1 I_2},$$

$$c = \frac{\left(k_1 + k_2\right)\left(F + \rho_1 {v_0}^2\right) + k_1 \rho_2 {v_0}^2}{E_1 E_2 I_1 I_2},$$

$$d = \frac{k_1 k_2}{E_1 E_2 I_1 I_2}.$$

In the upcoming, the case when $F \le F_{c,d}$ is analyzed with the purpose to determine the response of the beam system when F gets close to the critical force F_c. In such a case, the roots of Eq. (4.12) are complex numbers.

$$s_{1,2,3,4} = \pm\lambda_1 \pm i\omega_1; \quad s_{5,6,7,8} = \pm\lambda_2 \pm i\omega_2. \tag{4.13}$$

After some algebra, the roots of Eq. (4.12) can be obtained using De Moivre's formula where

$$\lambda_1 = \sqrt{\rho_1}\cos\frac{\varphi_1}{2}, \quad \omega_1 = \sqrt{\rho_1}\sin\frac{\varphi_1}{2},$$

$$\lambda_2 = \sqrt{\rho_2}\cos\frac{\varphi_2}{2}, \quad \omega_2 = \sqrt{\rho_2}\sin\frac{\varphi_2}{2},$$

$$\rho_1 = \sqrt{x_1^2 + y_1^2}, \quad \rho_2 = \sqrt{x_1^2 + y_2^2},$$

$$\varphi_1 = \tan^{-1}\left(\pi + \frac{y_1}{x_1}\right), \quad \varphi_2 = \tan^{-1}\left(\pi + \frac{y_2}{x_1}\right),$$

$$x_1 = -\frac{a}{4}, \quad y_1 = \frac{1}{4}\sqrt{\frac{16c}{a} - a^2 - \frac{16\sqrt{c^2 - a^2 d}}{a}}, \quad y_2 = \sqrt{\frac{c}{a} - \frac{a^2}{16} - \frac{\sqrt{c^2 - a^2 d}}{a}}.$$

The corresponding solution for the top beam for the sector ahead of the force P is

$$w_{1a}(\xi) = C_1 e^{s_1\xi} + C_2 e^{s_2\xi} + C_3 e^{s_3\xi} + C_4 e^{s_4\xi} + C_5 e^{s_5\xi} + C_6 e^{s_6\xi} + C_7 e^{s_7\xi} + C_8 e^{s_8\xi}, \tag{4.14}$$

and for the sector behind the force

$$w_{1b}(\xi) = C_9 e^{s_1\xi} + C_{10} e^{s_2\xi} + C_{11} e^{s_3\xi} + C_{12} e^{s_4\xi} + C_{13} e^{s_5\xi} + C_{14} e^{s_6\xi} + C_{15} e^{s_7\xi} + C_{16} e^{s_8\xi}, \tag{4.15}$$

The corresponding solution for the bottom beam for the sector ahead of the force P obtained from the system of Eq. (4.11) is

$$w_{2a}(\xi) = w_{1a}(\xi) + \frac{1}{k_1}\left(\frac{d^2 w_{1a}(\xi)}{d\xi^2} + \rho_1 v_0{}^2 \frac{d^2 w_{1a}(\xi)}{d\xi^2} + E_1 I_1 \frac{d^4 w_{1a}(\xi)}{d\xi^4}\right), \tag{4.16}$$

and for the sector behind the force

$$w_{2b}(\xi) = w_{1b}(\xi) + \frac{1}{k_1}\left(\frac{d^2 w_{1a}(\xi)}{d\xi^2} + \rho_1 v_0{}^2 \frac{d^2 w_{1b}(\xi)}{d\xi^2} + E_1 I_1 \frac{d^4 w_{1b}(\xi)}{d\xi^4}\right). \tag{4.17}$$

At coordinate $\xi \to \infty$, the deflections of the beams are $w_{1,2a} = 0$; also on the opposite side, at coordinate $\xi \to -\infty$, the deflections are $w_{1,2b} = 0$; from this, the values of eight constants are determined $C_1 = C_2 = C_5 = C_6 = C_{11} = C_{12} = C_{15} = C_{16} = 0$. The other eight constants are determined from the conditions at $\xi = 0$.

$$w_{1a}(0) = w_{1b}(0), \quad \left.\frac{dw_{1a}}{d\xi}\right|_0 = \left.\frac{dw_{1b}}{d\xi}\right|_0, \quad \left.\frac{d^2 w_{1a}}{d\xi^2}\right|_0 = \left.\frac{d^2 w_{1b}}{d\xi^2}\right|_0,$$

$$\left(\frac{d^3 w_{1a}}{d\xi^3} - \frac{d^3 w_{1b}}{d\xi^3}\right) = \frac{P}{E_1 I_1},$$

$$w_{2a}(0) = w_{2b}(0), \quad \left.\frac{dw_{2a}}{d\xi}\right|_0 = \left.\frac{dw_{2b}}{d\xi}\right|_0, \quad \left.\frac{d^2 w_{2a}}{d\xi^2}\right|_0 = \left.\frac{d^2 w_{2b}}{d\xi^2}\right|_0,$$

$$\left.\frac{d^3 w_{2a}}{d\xi^3}\right|_0 = \left.\frac{d^3 w_{2b}}{d\xi^3}\right|_0, \quad \left.\frac{d^4 w_{2a}}{d\xi^4}\right|_0 = \left.\frac{d^4 w_{2b}}{d\xi^4}\right|_0. \tag{4.18}$$

Functions $w_{ia}(x, t)$, $w_{ib}(x, t)$, $i = 1, 2$ are obtained when the constants are substituted into Eqs. (4.14–4.17). For the system parameters given in the paper by Kerr [4], the results are displayed in Figs. 4.8 and 4.9.

According to the results shown in Figs. 4.8 and 4.9, it can be observed that the double-supported system has smaller deflections. The stabilization effect present in the double-supported system is more pronounced when the stiffness modulus k_2 is decreased. A more pronounced stabilization effect leads to a decrease in the amplitudes of vibrations of the double-supported system along which the vertical force is moving. This means that the stabilizing properties of the double-supported

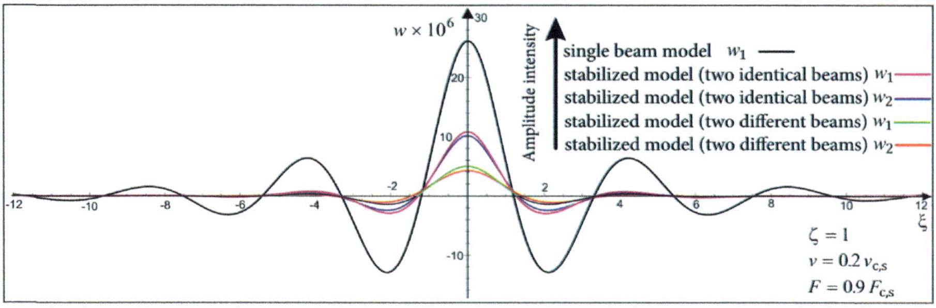

Figure 4.8 Beam deflections. For the parameters ζ=1, v=0.2$v_{c,s}$, and F=0.9$F_{cr,s}$.

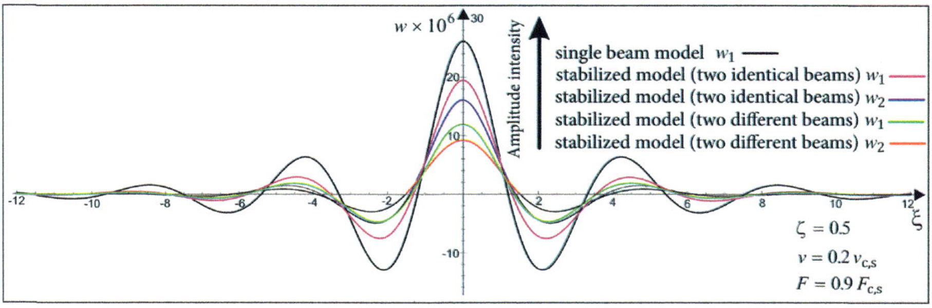

Figure 4.9 Beam deflections. For the parameters ζ=0.5, v=0.2$v_{c,s}$, and F=0.9$F_{cr,s}$.

system are important and must be considered when the axial compression force is close to the critical value.

4.4 Stability of vibrations of a complex double-supported beam system

The model in Fig. 4.10(A) illustrates the uniform motion of a mass subjected to an unchanging vertical force. The mass is moving on top of an infinitely long beam system comprising two deformable beams. The top beam in the system is compressed axially, and it is connected to the lower beam via a viscoelastic layer, whose parameters are k_1 and β_1. The bottom beam, which is not axially stressed, is supported by a viscoelastic foundation, whose parameters are k_2 and β_2. The model in Fig. 4.10(B) also illustrates a mass (subjected to a vertical force) that is uniformly moving along an axially compressed beam. However, in this case, there is no stabilizing beam. Metrikine and Dieterman analyzed this case [7]. However, the model in Fig. 4.10(B) lies on a viscoelastic foundation embodying two layers with disparate stiffness moduli k_1 and k_2, and disparate viscosities β_1 and β_2. This is generally represented as one layer with equivalent parameters: k_e and β_e.

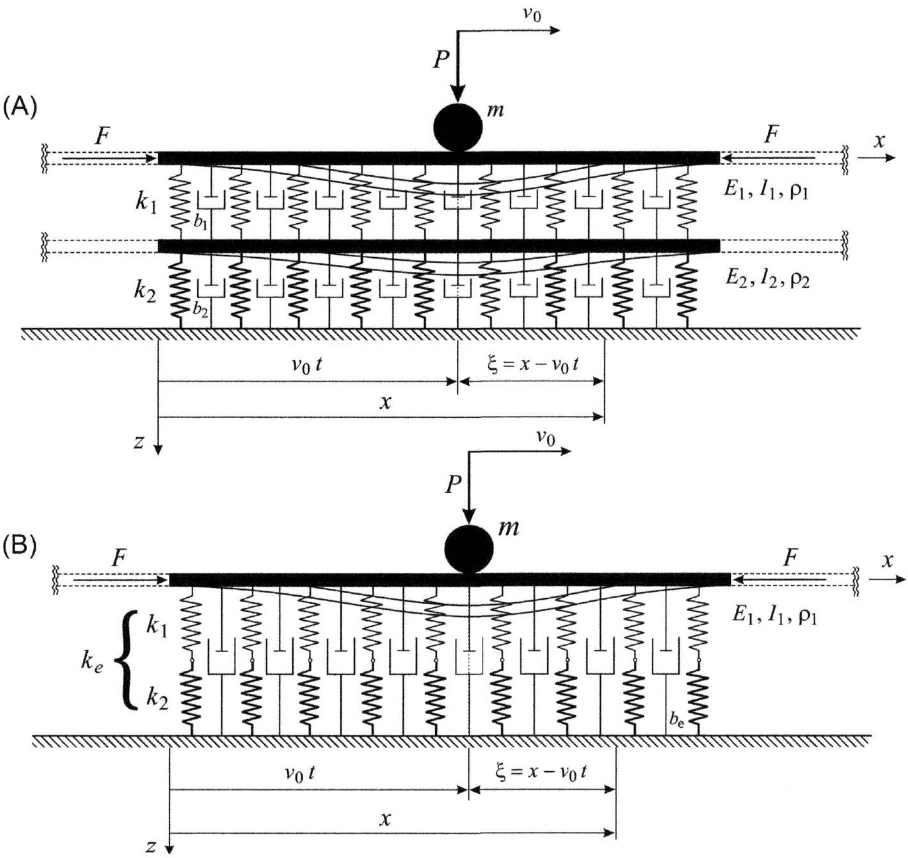

Figure 4.10 Infinite beam systems with uniformly moving mass that is subjected to a constant vertical force. The foundation is viscoelastic. (A) double-supported system and (B) classical beam.

The further examination of the instability is based on the differential equations of motion in the following form:

$$E_1 I_1 \frac{\partial^4 w_1}{\partial x^4} + F \frac{\partial^2 w_1}{\partial x^2} + \rho_1 \frac{\partial^2 w_1}{\partial t^2} + \beta_1 \left(\frac{\partial w_1}{\partial t} - \frac{\partial w_2}{\partial t} \right) + k_1 (w_1 - w_2) = -\left(m \frac{\partial^2 w_1^0}{\partial t^2} + P \right) \delta (x - v_0 t), \tag{4.19}$$

$$E_2 I_2 \frac{\partial^4 w_2}{\partial x^4} + \rho_2 \frac{\partial^2 w_2}{\partial t^2} + \beta_1 \left(\frac{\partial w_2}{\partial t} - \frac{\partial w_1}{\partial t} \right) + \beta_2 \frac{\partial w_2}{\partial t} + k_2 w_2 - k_1 \left(w_1 - w_2 \right) = 0, \tag{4.20}$$

$$w_1^0 (t) = w_1 (v_0 t,\, t), \tag{4.21}$$

in which $w_1^0(t)$ is the transverse displacement of the mass. Adding a moving frame of reference $\xi = x - v_0 t$, Eqs. (4.19–4.21) can be rewritten in the form

$$E_1 I_1 \frac{\partial^4 w_1(\xi,t)}{\partial \xi^4} + F\frac{\partial^2 w_1(\xi,t)}{\partial \xi^2} + \rho_1\left(v_0^2\frac{\partial^2 w_1(\xi,t)}{\partial \xi^2} - 2v_0\frac{\partial^2 w_1(\xi,t)}{\partial \xi \partial t} + \frac{\partial^2 w_1(\xi,t)}{\partial t^2}\right)$$
$$+ \beta_1\left\{\left(\frac{\partial w_1(\xi,t)}{\partial t} - v_0\frac{\partial w_1(\xi,t)}{\partial \xi}\right) - \left(\frac{\partial w_2(\xi,t)}{\partial t} - v_0\frac{\partial w_2(\xi,t)}{\partial \xi}\right)\right\}$$
$$+ k_1(w_1(\xi,t) - w_2(\xi,t)) = -\left(m\frac{\partial^2 w_1(0,t)}{\partial t^2} + P\right)\delta(\xi), \tag{4.22}$$

$$E_2 I_2 \frac{\partial^4 w_2(\xi,t)}{\partial \xi^4} + \rho_2\left(v_0^2\frac{\partial^2 w_2(\xi,t)}{\partial \xi^2} - 2v_0\frac{\partial^2 w_2(\xi,t)}{\partial \xi \partial t} + \frac{\partial^2 w_2(\xi,t)}{\partial t^2}\right)$$
$$+ \beta_2\left(\frac{\partial w_2(\xi,t)}{\partial t} - v_0\frac{\partial w_2(\xi,t)}{\partial \xi}\right)$$
$$- \beta_1\left\{\left(\frac{\partial w_1(\xi,t)}{\partial t} - v_0\frac{\partial w_1(\xi,t)}{\partial \xi}\right) - \left(\frac{\partial w_2(\xi,t)}{\partial t} - v_0\frac{\partial w_2(\xi,t)}{\partial \xi}\right)\right\}$$
$$+ k_2 w_2(\xi,t) - k_1(w_1(\xi,t) - w_2(\xi,t)) = 0. \tag{4.23}$$

Eqs. (4.22) and (4.23) show that the transition to the moving frame of reference eliminates time from the argument of the delta function. The result of this is $w_1^0(t) = w_1(\xi,t)|_{\xi=0}$. Implementing the Laplace integral transform, defined as:

$$\widetilde{w}_{1s}(\xi,s) = \int_0^\infty w_1(\xi,t)e^{-st}dt, \quad \widetilde{w}_{2s}(\xi,s) = \int_0^\infty w_2(\xi,t)e^{-st}dt, \tag{4.24}$$

and assuming the trivial initial conditions, the following system of ordinary differential equations (ODEs) is obtained:

$$E_1 I_1 \frac{\partial^4 \widetilde{w}_{1s}}{\partial \xi^4} + \rho_1 v_0^2 \frac{\partial^2 \widetilde{w}_{1s}}{\partial \xi^2} + F\frac{\partial^2 \widetilde{w}_{1s}}{\partial \xi^2} - 2\rho_1 s v_0 \frac{\partial \widetilde{w}_{1s}}{\partial \xi} + \beta_1 v_0\left(\frac{\partial \widetilde{w}_{2s}}{\partial \xi} - \frac{\partial \widetilde{w}_{1s}}{\partial \xi}\right)+$$
$$+ \beta_1 s(\widetilde{w}_{1s} - \widetilde{w}_{2s}) + \rho_1 s^2 \widetilde{w}_{1s} + k_1(\widetilde{w}_{1s} - \widetilde{w}_{2s}) = -\left(ms^2\widetilde{w}_{1s} + \frac{P}{s}\right)\delta(\xi), \tag{4.25}$$

$$E_2 I_2 \frac{\partial^4 \widetilde{w}_{2s}}{\partial \xi^4} + \rho_2 v_0^2 \frac{\partial^2 \widetilde{w}_{2s}}{\partial \xi^2} - 2\rho_2 s v_0 \frac{\partial \widetilde{w}_{2s}}{\partial \xi} - \left(\beta_1 + \beta_2\right) v_0 \frac{\partial \widetilde{w}_{2s}}{\partial \xi} + \beta_1 v_0 \frac{\partial \widetilde{w}_{1s}}{\partial \xi}+$$
$$+ (\beta_1 + \beta_2)s\widetilde{w}_{2s} - \beta_1 s \widetilde{w}_{1s} + k_2\widetilde{w}_{2s} + k_1(\widetilde{w}_{2s} - \widetilde{w}_{1s}) + \rho_2 s^2 \widetilde{w}_{2s} = 0. \tag{4.26}$$

To eliminate the delta function, the Fourier integral transform is applied in the form

$$\widetilde{\widetilde{w}}_{1k,s}(\bar{k},s) = \int_{-\infty}^{\infty} \widetilde{w}_{1s}(\xi,s)e^{-i\bar{k}\xi}d\xi, \quad \widetilde{\widetilde{w}}_{2k,s}(\bar{k},s) = \int_{-\infty}^{\infty} \widetilde{w}_{2s}(\xi,s)e^{-i\bar{k}\xi}d\xi. \tag{4.27}$$

The following equations are obtained

$$\tilde{\tilde{w}}_{1k,s}[-F\bar{k}^2 + \bar{k}^4E_1I_1 + (s-i\bar{k}v_0)(\beta_1 + \rho_1(s-i\bar{k}v_0)) + k_1]+ \\ + \tilde{\tilde{w}}_{2k,s}(i\bar{k}\beta_1 v_0 - s\beta_1 - k_1) = -\left(ms^2\tilde{w}_{1s}(0, s) + \frac{P}{s}\right), \tag{4.28}$$

$$\tilde{\tilde{w}}_{2k,s}[\bar{k}^4E_2I_2 + (s-i\bar{k}v_0)(\rho_2(s-i\bar{k}v_0) + \beta_1 + \beta_2) + k_1 + k_2]+ \\ + \tilde{\tilde{w}}_{1k,s}(i\bar{k}\beta_1 v_0 - s\beta_1 - k_1) = 0. \tag{4.29}$$

From Eqs. (4.28) and (4.29) with the introduction of the notation in the form

$$D(\bar{k}, s) = E_1I_1\bar{k}^4 - F\bar{k}^2 + (s-i\bar{k}v_0)[\beta_1 + \rho_1(s-i\bar{k}v_0)] + k_1- \\ - \frac{[k_1 + \beta_1(s-i\bar{k}v_0)]^2}{E_2I_2\bar{k}^4 + (s-i\bar{k}v_0)[\rho_2(s-i\bar{k}v_0) + \beta_1 + \beta_2] + k_1 + k_2}, \tag{4.30}$$

and after applying the inverse Fourier transform, the following is obtained

$$\tilde{w}_{1s}(\xi, s) = -\left(ms^2\tilde{w}_{1s}(0, s) + \frac{P}{s}\right)\frac{1}{2\pi}\int_{-\infty}^{\infty}\frac{e^{i\bar{k}\xi}}{D(\bar{k}, s)}\mathrm{d}\bar{k}. \tag{4.31}$$

To determine $\tilde{w}_{1s}(\xi, s)$, the $\tilde{w}_{1s}(0, s)$ should be known. Substituting $\xi = 0$ into Eq. (4.31), the following is obtained

$$\tilde{w}_{1s}(0, s) = -\frac{P}{s\left[ms^2 + \left(\frac{1}{2\pi}\int_{-\infty}^{\infty}\frac{\mathrm{d}\bar{k}}{D(\bar{k}, s)}\right)^{-1}\right]}. \tag{4.32}$$

Substituting Eq. (4.32) into Eq. (4.31) yields

$$\tilde{w}_{1s}(\xi, s) = -\frac{P}{2\pi s}\int_{-\infty}^{\infty}\frac{e^{i\bar{k}\xi}}{D(\bar{k}, s)}\mathrm{d}\bar{k} + \frac{ms^2}{ms^2 + \left(\frac{1}{2\pi}\int_{-\infty}^{\infty}\frac{\mathrm{d}\bar{k}}{D(\bar{k}, s)}\right)^{-1}}\frac{P}{2\pi s}\int_{-\infty}^{\infty}\frac{e^{i\bar{k}\xi}}{D(\bar{k}, s)}\mathrm{d}\bar{k}. \tag{4.33}$$

In Eq. (4.33), the first member describes the beam's deflection under the force ($m = 0$). The second member is related to the beam's deflection under the mass. It can be proved that the second member of Eq. (4.33) can have a pole $s^* = a + ib$, with $a > 0$. Physically, this case implies that the beam's deflection grows exponentially in time; hence, the vibrations of the beam become unstable, and that is thc main point of interest in this analysis. By observing Eq. (4.33), it can be concluded that the beam deflections will become unstable if

$$ms^2 + \left(\frac{1}{2\pi}\int_{-\infty}^{\infty}\frac{\mathrm{d}\bar{k}}{D(\bar{k}, s)}\right)^{-1} = 0. \tag{4.34}$$

Due to the integral in Eq. (4.34), finding the roots s^* directly would be an arduous task. A convenient way to escape from this predicament is to use the D-decomposition method. The basic notion of this method is mapping the imaginary axis of the complex (s)–plane onto the plane of one of the system parameters, which should be regarded as complex. The rule for mapping originates from the characteristic equation, which should be reformulated in a way that explicitly expresses the chosen parameter. When the mapping is completed, a mapped line is obtained. This mapped line is called the D-decomposition curve, and it partitions the parameter plane into domains with a different number of roots with a positive real part. Every D-decomposition curve is usually shaded from the side that is correlated with the right-hand side of the imaginary axis of the complex (s)–plane. Crossing a D-decomposition curve in the direction of the shading will give one extra root with a positive real part.

The D-decomposition of the complex (m)–plane is done in the following way. The mass (m) is correlated with the positive part of the real axis of the (m)–plane. The parameterization of the imaginary axis of the complex (s)–plane is done with the parameter $s = i\Omega$, in which $i = \sqrt{-1}$ and Ω is a real value that is varied $(-\infty < \Omega < +\infty)$. Using the characteristic equation to express m explicitly, and inserting $s = i\Omega$ into it, will give the mapping rule that defines the mapping of the imaginary axis of the complex (s)–plane onto the complex (m)–plane. The mapping rule is

$$m\Omega^2 = \left(\frac{1}{2\pi} \int_{-\infty}^{\infty} \frac{\mathrm{d}\bar{k}}{D(\bar{k},\ \Omega)} \right)^{-1}, \tag{4.35}$$

where

$$D(\bar{k},\ \Omega) = E_1 I_1 \bar{k}^4 - F\bar{k}^2 + (\Omega - \bar{k}v_0)[\rho_1(\bar{k}v_0 - \Omega) + i\beta_l] + k_1 - \frac{(\bar{k}\beta_1 v_0 - \beta_1\Omega + \mathrm{i}k_1)^2}{\rho_2(\Omega - \bar{k}v_0)^2 - i[(\beta_1 + \beta_2)(\Omega - \bar{k}v_0) - i(\bar{k}^4 E_2 I_2 + k_1 + k_2)]}. \tag{4.36}$$

The integral (4.36) can be developed after rationalizing the polynomial and applying the contour integration method; this will produce the expression

$$m\Omega^2 = \left(\frac{1}{2\pi} \int_{-\infty}^{\infty} \frac{\mathrm{Q}(\bar{k},\ \Omega)\mathrm{d}\bar{k}}{P_0(\bar{k} - \bar{k}_1)\ldots(\bar{k} - \bar{k}_8)} \right)^{-1} = \left(i \sum_{j=1}^{8} \mathrm{RES} \frac{Q(\bar{k},\Omega)}{P_0(\bar{k} - \bar{k}_1)\ldots(\bar{k} - \bar{k}_8)} \bigg|_{\bar{k}=\bar{k}_j} \right)^{-1}, \tag{4.37}$$

where $\bar{k}_j$ are the roots of the equation $D(\bar{k},\ \Omega) = 0$, which has a positive imaginary part. An imaginary part is present in all roots, and all of them are simple because of the viscosity of the foundation and the in-between layer. Therefore, the final mapping expression is

$$m = \frac{1}{i\Omega^2} \left(\sum_{j=1}^{8} \frac{\mathrm{Q}(\bar{k}_j,\ \Omega)}{P_0(\bar{k}_j - \bar{k}_1)\ldots(\bar{k}_j - \bar{k}_8)} \right)^{-1}. \tag{4.38}$$

The curves $\mathrm{Re}(m)$ versus $\mathrm{Im}(m)$ in Figs. 4.11 and 4.12 are for the models presented in Fig. 4.10(A and B), and for the given system parameters in [7]. The case when the velocity is small is shown in Fig. 4.11 ($v = 0.5v_{cr}$, where v_{cr} is a critical velocity for a beam without

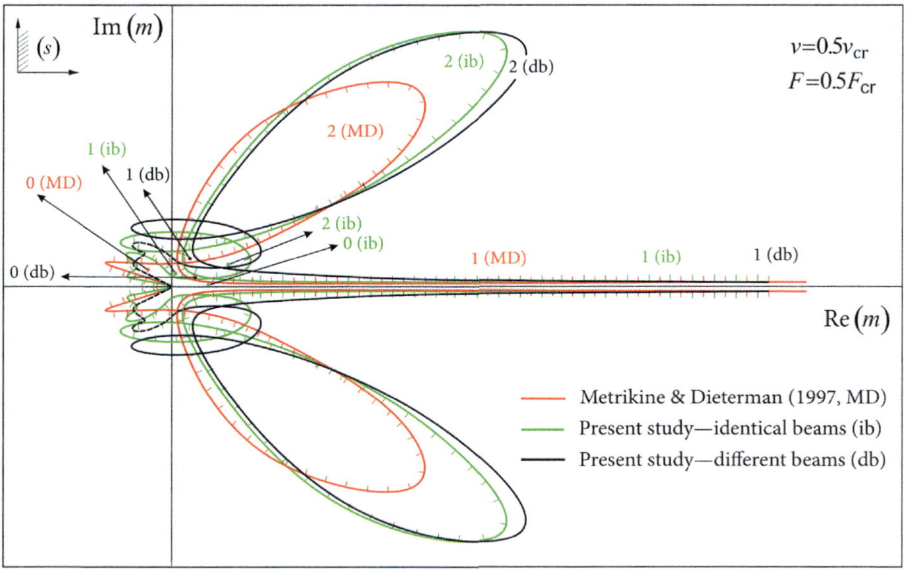

Figure 4.11 D-decomposition. The complex, *(m)*–plane parameters are $v=0.5v_{cr}$ and $F=0.5F_{cr}$.

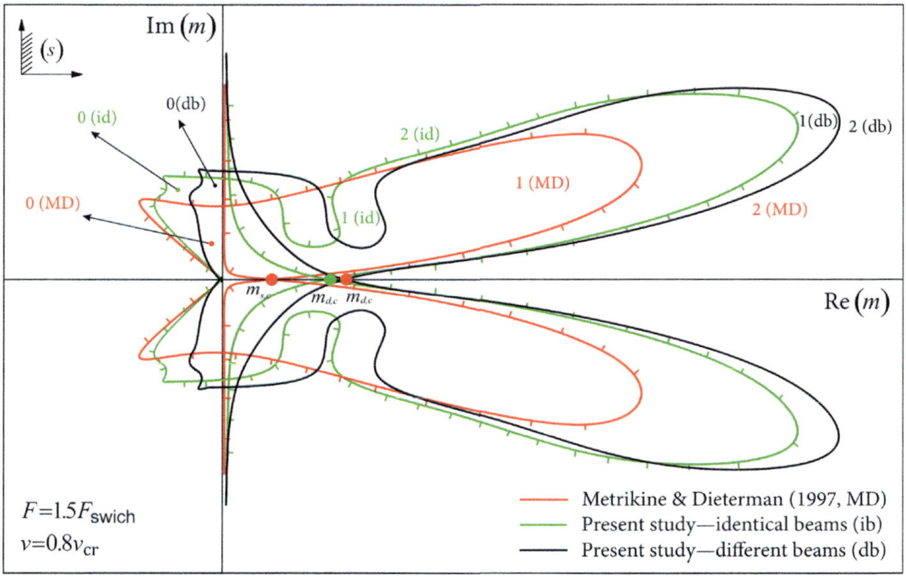

Figure 4.12 D-decomposition. The complex, *(m)*–plane parameters are $v=0.8v_{cr}$ and $F=1.5F_{switch}$.

a stabilizer and with an equivalent stiffness k_e). When the velocity is small, the D-decomposition line does not cross the real axes of the (m)–plane as $\mathrm{Re}(m) > 0$. For all values of the mass, the number of unstable roots between the two models (with and without the stabilizer) is the same. According to the result given in Fig. 4.11, both models are stable for lower velocities $v = 0.5v_{cr}$ and under the influence of the axial force $F = 0.5F_{cr}$.

In Fig. 4.12, the case when the velocity is large is shown. In this case, the velocity is $v = 0.8v_{cr}$, and the intensity of the axial force is $F = 1.5F_{\text{switch}} < F_{cr}$ [4]. It can be seen that the decreased stability of the double-supported system starts with a larger value of mass ($m_{d,c} > m_{s,c}$). Because of this finding, presenting the number of unstable roots for a particular value of m in comparison with the classical model without a stabilizer is significant. The dependence of the critical mass m_{cross} (in the function of velocity) for various values of the compressive axial force is shown in Figs. 4.13 and 4.14. From Fig. 4.13(A), it can be concluded that the stability of the model without the stabilizer and subjection to axial force is lower (with respect to the mass) than the stability of the model with the stabilizer. Additionally, the layer stiffness ratio (ζ) and the stability region decrease proportionally in both models. When subjected to the axial force of intensity $F = F_{\text{switch}}$, the double-supported system has greater stability than the classical system without the stabilizing beam and with the equivalent stiffness k_e; this is shown in Fig. 4.13(B). When the stiffness k_2 of the bottom layer is further reduced, the double-supported system stays stable for an even greater value of the mass; this is expected since the bottom beam compensates for the reduced stiffness. The major finding comes from the result obtained for the intensity of the axial force $F_{cr} > F \geq F_{\text{switch}}$. Within these values, the double-supported model shows much greater stability. This finding begs the question as to what happens when the axial force is $F > F_{\text{switch}}$ for lower layer stiffness $k_2\,(\zeta = 0.5)$. In that situation, the double-supported system also displays better stability properties and permits an even greater mass of a moving object. The dependence of the critical mass for various intensities of the compressive axial force $F = \{0.5F_{cr}, \ \ldots, 0.9F_{cr}\}$ is shown in Fig. 4.14. Since the force is $0.5F_{cr} > F_{\text{switch}}$, Fig. 4.14 shows that the stability regions of the double-supported system are larger for all values when the compressive axial force is present.

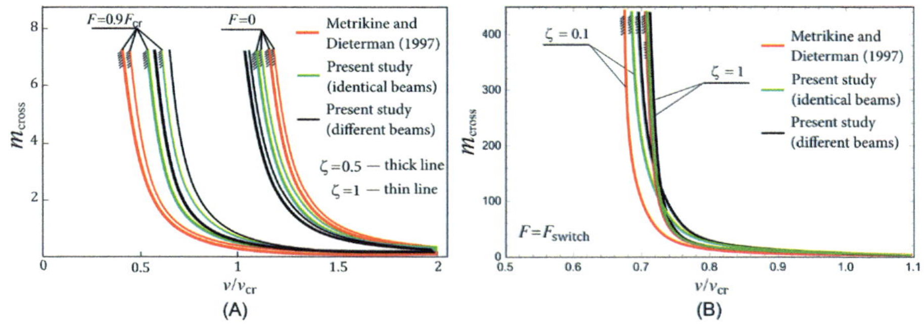

Figure 4.13 Dependence $m_{\text{cross}}(v/v_{\text{cr}})$. The zone above the curve presents the unstable vibrations. (A) $F=0$ and $F=0.9F_{\text{cr}}$ and (B) $F=F_{\text{switch}}$.

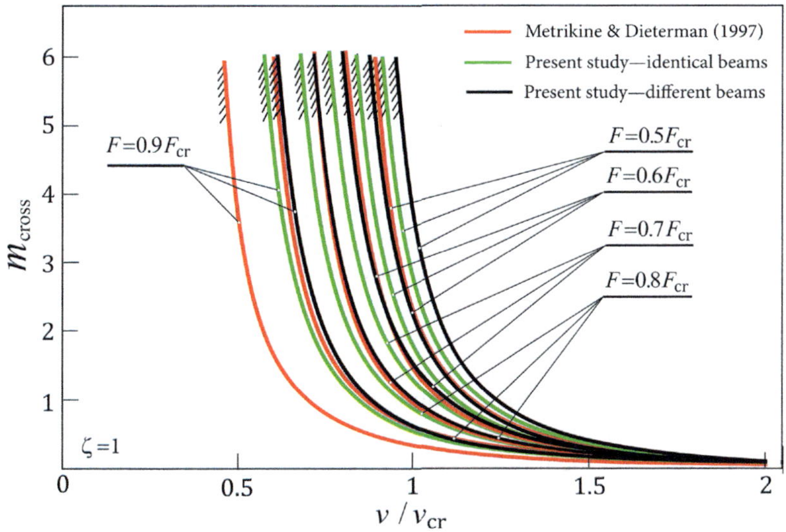

Figure 4.14 Dependence $m_{\text{cross}}(v/v_{\text{cr}})$. The zone above the curve presents the unstable vibrations; $F=\{0.5F_{\text{cr}},\ldots,0.9F_{\text{cr}}\}$.

4.5 Influence of a compressive axial force on harmonic vibrations of the moving mass subjected to a constant vertical force

Deriving the dispersion equation for the frame of reference that moves ($v = v_0$) together with the mass is of no difficulty. This can be done from Eqs. (4.22–4.23) if the compressive axial force is considered for the infinitely small damping.

$$\begin{aligned} &E_1E_2I_1I_2\bar{k}^8 - F\bar{k}^6E_2I_2 - \bar{k}^6E_2I_2\rho_1v_0^2 \\ &\quad + \rho_2(\Omega - \bar{k}v_0)^2[F\bar{k}^2 - E_1I_1\bar{k}^4 + \rho_1(\Omega - \bar{k}v_0)^2 - k_1] \\ &+ 2\bar{k}^5E_2I_2\rho_1v_0\Omega + \bar{k}^4E_1I_1k_1 + \bar{k}^4E_1I_1k_2 + \bar{k}^4E_2I_2k_1 - \bar{k}^4E_2I_2\rho_1\Omega^2 - \bar{k}^2k_1\rho_1v_0^2 \\ &- F\bar{k}^2k_1 - F\bar{k}^2k_2 - \bar{k}^2k_2\rho_1v_0^2 + 2\bar{k}k_1\rho_1v_0\Omega + 2\bar{k}k_2\rho_1v_0\Omega \\ &+ k_1k_2 - k_1\rho_1\Omega^2 - k_2\rho_1\Omega^2 = 0. \end{aligned} \tag{4.39}$$

For the single-beam model, which has an elastic layer with the equivalent stiffness k_e, Fig. 4.10(B), the dispersion equation with the considered influence of the compressive axial force, unlike the equation presented by Verichev and Metrikine [13], has the following form:

$$E_1I_1\bar{k}^4 - F\bar{k}^2 + k_e - \rho_1(\Omega - \bar{k}v_0)^2 = 0. \tag{4.40}$$

The roots of the dispersion equation are real when the velocity of the moving mass is lower than the critical. In addition, when the velocity is lower than the critical, the vibrations of the mass, which is moving on the beam and does not lose contact with the beam, are harmonic. If the vibrations of the mass do not disrupt the waves in the beam, the mass can continue to vibrate harmonically. The critical velocity that depends on the frequency and in which the influence of the axial force is considered can be determined from the discriminants of dispersion Eqs. (4.39) and (4.40).

As the frequency of harmonic vibrations is always smaller than the critical value, which depends on the mass velocity, bifurcation curves are obtained and given in Figs. 4.15–4.20. The area below the curve represents the region of stability. Figs. 4.15–4.17 show the velocity–frequency dependence when the stiffness moduli are of equal values ($k_1 = k_2$). The velocity–frequency function is shown in the dependence of the force F_{switch}, which represents the point of change in stability between the compared models. In Fig. 4.15 ($F = F_{\text{switch}}$), when the frequency of the system is equal to zero, the compared systems (classical and double-supported with identical beams) have the same velocity $v = 526.115$ m/s. This velocity is the intersection point between the curves $v(F_{\text{switch}})$ and simultaneously represents the maximal velocity value at the force $F = F_{\text{switch}}$. When the frequency increases and harmonic vibrations occur (Fig. 4.15), the classical system will show greater stability. It can be seen in Fig. 4.16 that the classical model has better stability properties when the force intensity is within the interval $F \in (0,\ F_{\text{switch}})$. But, when subjected to a force whose intensity is within the interval $F \in (F_{\text{switch}},\ F_{cr})$, the double-supported model has a greater stability area and consequently can handle a greater velocity and vibrational

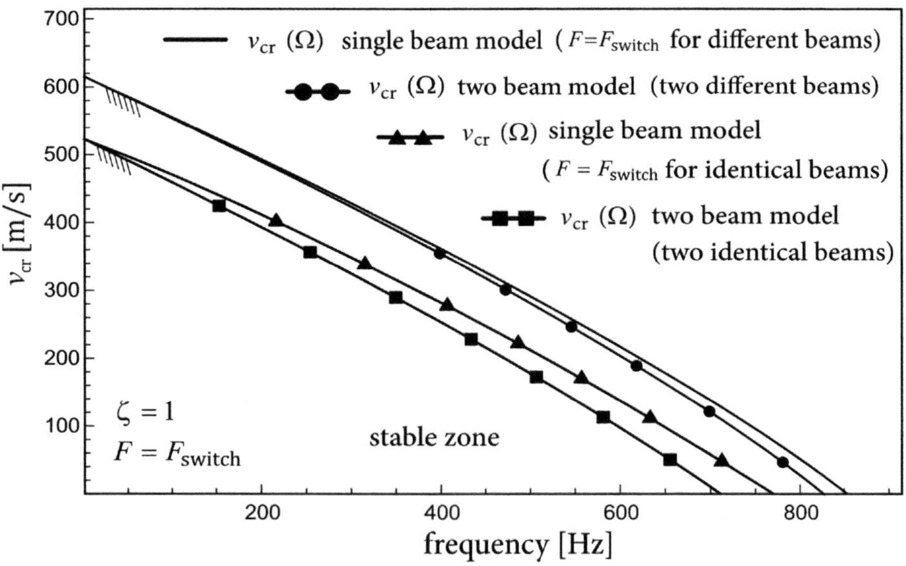

Figure 4.15 Bifurcation curves. Frequency versus velocity; ζ=1 and F=F_{switch}.

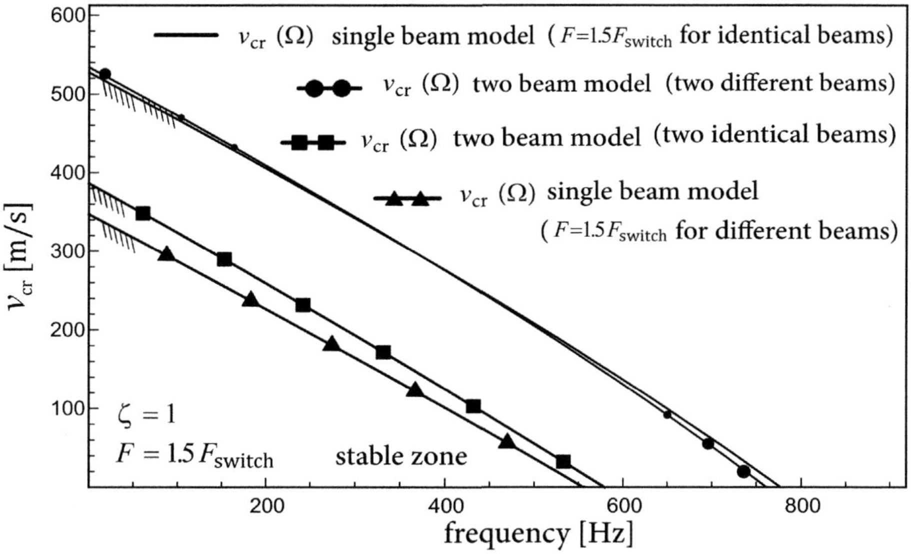

Figure 4.16 Bifurcation curves. Frequency versus velocity; ζ=1 and F=1.5F_{switch}.

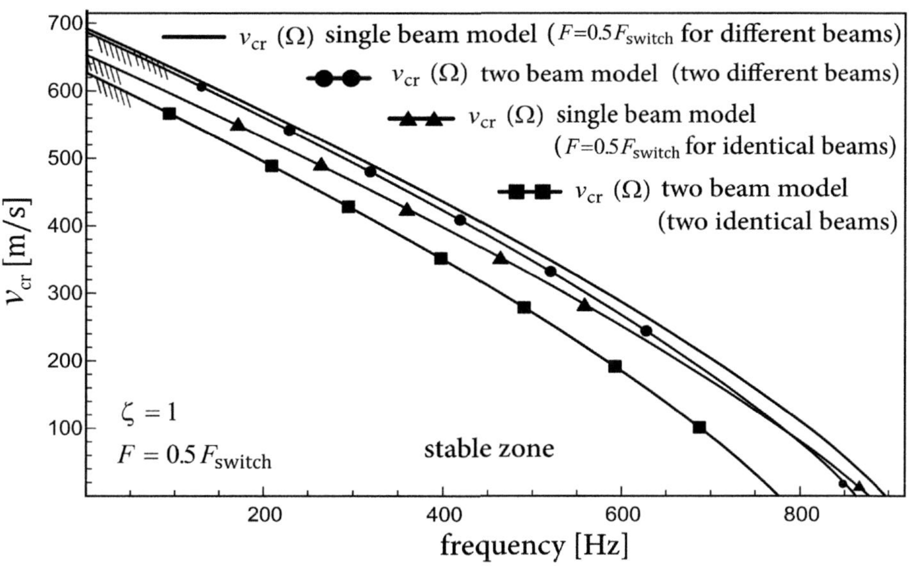

Figure 4.17 Bifurcation curves. Frequency versus velocity; ζ=1 and F=0.5F_{switch}.

frequency. Figs. 4.18–4.20 show the velocity–frequency dependence when the correlation between the stiffness moduli is $k_2 < k_1$. Here, when the force is larger than F_{switch}, the double-supported model also shows better stability properties.

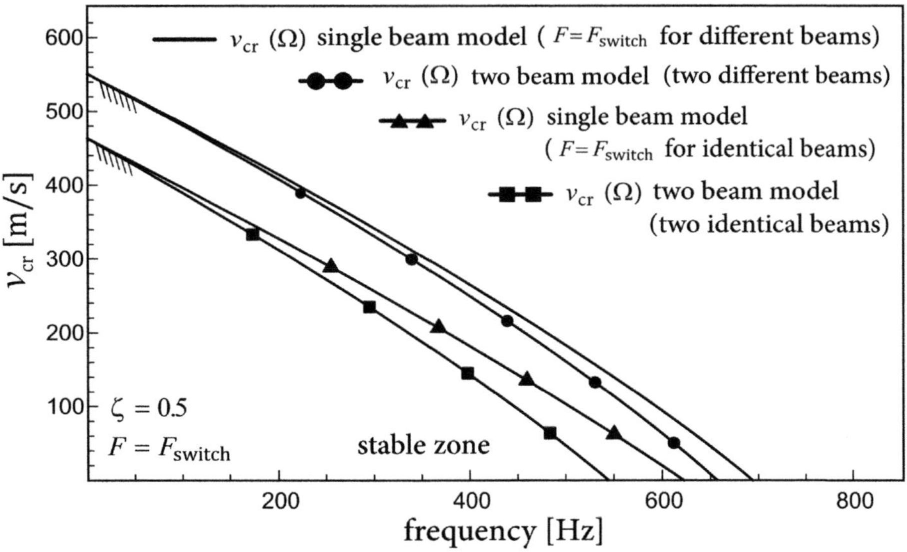

Figure 4.18 Bifurcation curves. Frequency versus velocity; ζ=0.5 and F=F_{switch}.

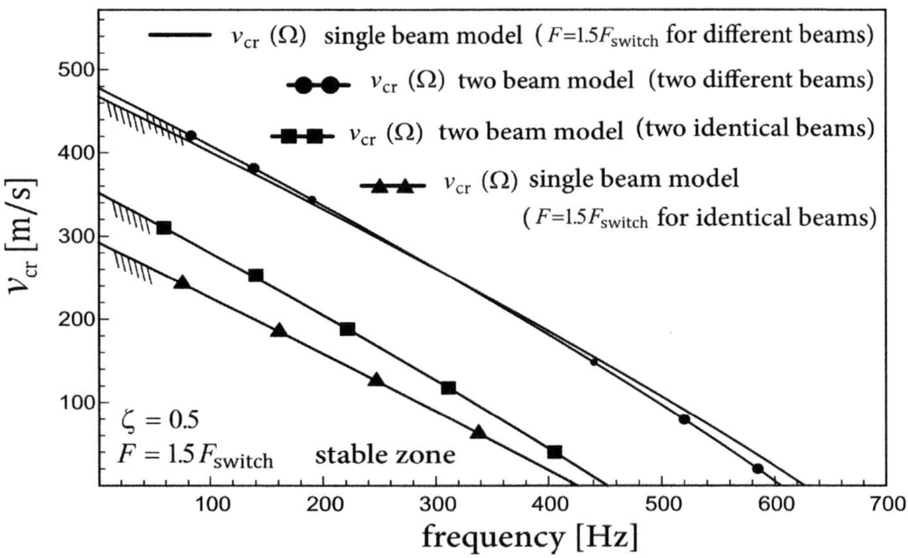

Figure 4.19 Bifurcation curves. Frequency versus velocity; ζ=0.5 and F=1.5F_{switch}.

This was expected since the second support compensates for the general stiffness of the system when the parameter ζ is decreased, making it more stable even in the frequency domain.

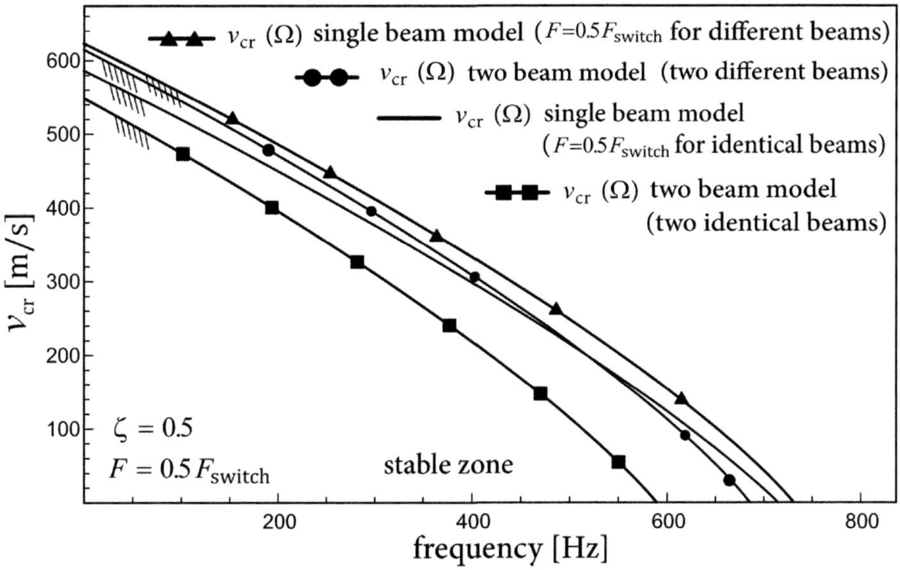

Figure 4.20 Bifurcation curves. Frequency versus velocity; ζ=0.5 and F=0.5F_{switch}.

4.6 Conclusion

The first problem analyzed in this chapter is the response and stability of vibrations caused by a constant vertical force moving uniformly on an axially compressed beam of a complex beam system of infinite length. The second problem was like the first, except in that the vertical force did not act directly on the beam. In the second problem, a mass (which is subjected to a constant vertical force) is moving uniformly on an axially compressed beam of a complex beam system. The basis of these analyses is a complex continuous system; for it, a continuous dynamic model was proposed. The focus of this study was the model that comprises two elastically connected beams with dissimilar geometric and material properties. The presented study includes the newly determined critical velocity and the critical force of the system. Diverse physical parameters were used for the analysis of the regions of stability, which were determined using the D-decomposition method. The attention in the study of the stability domain is given to the effect of the compressive axial forces, the magnitude of varying mass, velocity of the mass (which is subjected to a constant vertical force), stiffness moduli of the elastic support, and different geometric and material properties of the beams. The stability domain was compared with simpler models. It was discovered that the double-beam model could increase the stability domain by changing the characteristics of the lower beam and the lower elastic layer, while such modifications on the single-beam model are not possible.

References

[1] G. Fernandes, E.M. Palmeira, R.C. Gomes, Performance of geosynthetic-reinforced alternative sub-ballast material in a railway track, Geosynthetics International 15 (5) (2008) 311–321.

[2] S.-H. Ju, H.-T. Lin, Analysis of train-induced vibrations and vibration reduction schemes above and below critical Rayleigh speeds by finite element method, Soil Dynamics and Earthquake Engineering 24 (12) (2004) 93–1002.

[3] S.P. Timoshenko, Proc. 2nd International Congress for Applied Mechanics, Zürich, 1927, pp. 1–12.

[4] A.D. Kerr, The continuously supported rail subjected to an axial force and a moving load, International Journal of Mechanical Sciences 14 (1972) 71–78.

[5] J.J. Labra, An axially stressed railroad track on an elastic continuum subjected to a moving load, Acta Mechanica 22 (1975) 113–129.

[6] G.G. Denisov, E.K. Kugusheva, V.V. Novikov, On the problem of the stability of one-dimensional unbounded elastic systems, Journal of Applied Mathematics and Mechanics 49 (4) (1985) 533–537.

[7] A.V. Metrikine, H.A. Dieterman, Instability of vibrations of a mass moving uniformly along an axially compressed beam on a visco-elastic foundation, Journal of Sound and Vibration 201 (5) (1997) 567–576.

[8] T.P. Chang, G.L. Lin, E. Chang, Vibration analysis of a beam with an internal hinge subjected to a random moving oscillator, International Journal of Solids and Structures 43 (2006) 6398–6412.

[9] G. Muscolino, A. Palmeri, Response of beams resting on viscoelastically damped foundation to moving oscillators, International Journal of Solids and Structures 44 (5) (2007) 1317–1336.

[10] P. Kozić, G. Janevski, R. Pavlović, Moment Lyapunov exponents and stochastic stability of a double-beam system under compressive axial loading, International Journal of Solids and Structures 47 (10) (2010) 1435–1442.

[11] V. Stojanović, P. Kozić, G. Janevski, Exact closed-form solutions for the natural frequencies and stability of elastically connected multiple beam systems using Timoshenko and high-order shear deformation theory, Journal of Sound and Vibration 332 (3) (2013) 563–576.

[12] V. Stojanović, P. Kozić, Vibrations and stability of complex beam systems, Springer Tracts in Mechanical Engineering, Springer, 2015.

[13] S.N. Verichev, A.V. Metrikine, Instability of vibrations of a mass that moves uniformly along a beam on a periodically inhomogeneous foundation, Journal of Sound and Vibration 260 (2003) 901–925.

[14] J. Dörr, Der unendliche, federndgebettete Balkenunterdem Einflusseinergleichförmig bewegten Last, Ingenieur-Archiv 14 (3) (1943) 167–192.

[15] J.T. Kenney, Steady-state vibrations of beam on elastic foundation for moving load, Journal of Applied Mechanics 21 (1954) 359–364.

[16] L. Frýba, Vibration of Solids and Structures under Moving Loads, Springer, Netherlands, 1972.

[17] A.R.M. Wolfert, H.A. Dieterman, A.V. Metrikine, Stability of vibrations of two oscillators moving uniformly along a beam on a viscoelastic foundation, Journal of Sound and Vibration 211 (5) (1998) 829–884.

[18] V. Stojanović, P. Kozić, M.D. Petković, Dynamic instability and critical velocity of a mass moving uniformly along a stabilized infinity beam, International Journal of Solids and Structures 108 (2017) 164–174.

[19] L. Ling, J. Han, X. Xiao, X. Jin, Dynamic behavior of an embedded rail track coupled with a tram vehicle, Journal of Vibration and Control (2015) 1–18 .https://doi.org/10.1177/107754631561652.

Stability of vibrations of a system of discrete oscillators that are moving at an overcritical velocity

5

5.1 Introduction

The movement of the oscillators at the critical velocity does not have to be a direct cause for the instability to emerge. The increased traffic and its high speed can generate dynamic effects that influence the stability of mechanical systems. Therefore, a more thorough observation of the varying parameters linked with these dynamic effects has to be performed. What causes instability of vibrations is the phenomenon in which the transfer of the energy into the vibrational energy occurs through the anomalous Doppler waves [1,2]. This often happens when a high-speed train travels faster than the elastic waves propagating in the railway structure. A common thought is that the minimum phase velocity (often called the critical velocity) of elastic waves in a rail is significantly higher than the operational speed of high-speed trains. This may lead to the conclusion that the instability (of a vehicle) caused by the propagation of elastic waves (generated by the vehicle) cannot occur in practice. However, this is not necessarily the case. The influence of at least two effects can cause the instability of a moving train to occur at velocities that are about 250 km/h. One of these is the interaction between the rails and the soft subsoil. The other is the influence of the thermal stresses in the axial direction of the continuously welded rails. As shown in Refs. [3–6], both effects—the interaction with the subsoil and the thermal stresses—may contribute to a significant reduction of the minimum phase velocity of waves in the rails. Instability may occur at relatively low speeds [7]; for the shown example, instability of the rail vehicle occurs at speeds of approximately 250 km/h if the rails are over soft (peat) soil. The research concerning the instability of vibrations of simple mechanical models has instigated further research in this area involving more complex models. Many researchers have studied the dynamic stability and instability of mechanical systems subjected to moving loads; Timoshenko [8] analyzed a problem that involves a constant vertical force moving along the beam and has found an analytical solution that describes the vibrational behavior of this system. An extension of this problem was studied in a few analytical works [3,9,10]; in these works, the beam, besides being acted upon by a moving constant vertical force, is also subjected to compressive axial forces. On the strength of these early works, further extension in the research was done; for example, Denisov [11] studied the regions of dynamic stability by applying the D-decomposition method. Later, Verichev and Metrikine [12] used the D-decomposition method on the problem that involves a beam (supported by the periodically inhomogeneous foundation) on which a mass is moving.

Modeling of Complex Dynamic Systems. DOI: https://doi.org/10.1016/B978-0-443-23942-7.00005-X

Comprehension of dynamic behavior is not possible without knowing the interactions between discrete and continuous systems, as shown in the study by Dimitrovová and Varandas [13]. They studied the transient dynamic response of a beam supported by a foundation that has an abrupt stiffness change and is subjected to a vertical force moving along the beam at a constant velocity. To solve the problem, they used finite integral transformations and linked both parts of the beam by continuity conditions. The above-mentioned theoretical research has led to a comprehensive dynamic description of engineering systems. Dynamical problems from engineering practice regarding high-speed trains can be found in the literature. Therefore, Ju and Lin [14] investigated the issue of fast trains, as mentioned in the previous Chapter 4. The response of a beam subjected to a moving oscillator also has important applications in engineering practices concerning some other similar cases: the vehicle-track interactions [15] and the vehicle–bridge interactions [16]. As far as mechanics is concerned, many researchers have provided a significant contribution and initiative in stability studies. When a strong and deep analysis is required, the number of dimensions that are taken into consideration may need to be increased, and that will increase the complexity of the initially modeled system. On the one side, mention should be made that the reaction of the surface waves, which are excited in the track subsoil, causes the instability of the vehicle/track system. On the other hand, mention should be made that the instability of the vehicle/track system could be solved using two different but also complementary methods. One is qualitative; its objective is to find the conditions when the system is stable or not; this method uses a linear model and the linear critical velocity. The other one is quantitative, and it uses a nonlinear model to describe the instability of vibrations of a moving complex oscillator; this method uses the time-domain analysis to find the solution. Considering that the system can reach instability due to amplitude rise, this will further lead the wheels to lose contact with the rail; the vibration will then have a limited cycle due to the opposing effects between the Doppler waves and the static load. The study presented in this chapter is about a complex, multicoupled process on which the analysis concerning the dynamic stability of vibrations is performed. For this reason, it is important to look back at studies that are about the dynamic stability of coupled systems, in which specific procedures are applied so that regions of stability could be determined [17–20]. In the aforementioned studies, where the contact between the discrete and the continuous structures is in one or two points, the central issue concerning the stability of vibrations and the dynamic response was deciding where the continuous structure is going to be finite. In contrast to that, Zarfam et al. [21] considered the response spectrum of a beam subjected to moving masses; for that multimass effect, they analyzed the variation of a beam structure frequency.

The study presented in this chapter is done on a model that has a complex system of oscillators and multicontact points between the discrete and the continuous system. The infinite beam is described by the Reddy–Bickford beam assumption. The regions of stability are determined according to the complex model that has more than two oscillators. In this sense, the study presents a broadened version of the research by Metrikine et al. [17]. A closed-form analytical solution for the system is discovered and formulated using the special case of a Toeplitz matrix. The system of equations is then solved by the trigonometric method. Verification of the results in closed-form was done with the recurrence formula of the first kind and the n-th order Chebyshev polynomial. In the end, by using the D-decomposition method and the argument principle, the stable regions are determined via numerical calculation for each chosen case and discussed according to the varying number of oscillators and masses. The reported results are useful in engineering practice and could be used for comparison with experimental results specifically intended for the examination of railway transport.

5.2 Dispersion of complex model and characteristic equation

This section presents a mathematical description of the model through a system of equations of motion. The observed model has multiple mechanical oscillators moving along an infinite deformable beam that is continually supported by a viscoelastic foundation. The beam material is assumed to be elastic, isotropic, and homogenous. A high-order shear deformation theory is used for the mathematical description of the behavior of the deformable beam. The critical velocity of the system depends only on the properties of the beam-foundation model and the applied beam theory, so a difference in the results between the critical velocities obtained via different beam theories is to be expected. Based on the Euler–Bernoulli beam assumption, Dimitrovová [22] derived the new formula for the critical velocity of a uniformly moving load on a beam supported by a foundation of finite depth. When the effects of shear deformation and rotary inertia are included in the beam deformation theory, the critical velocity is better approximated, and thus the regions of stability are determined more precisely for different parameters; this may justify the usage of a high-order beam assumption. Also, high-order beam assumptions are better for high-frequency excitations when the wavelength is approaching the thickness of the beam. In the case of using the Reddy–Bickford beam assumption, the displacement field of the beam model can be written as:

$$u(x, z, t) = u^0(x, t) + z\varphi^0(x, t) - \tilde{\alpha} z^3\left(\varphi^0(x, t) + \frac{\partial w^0(x, t)}{\partial x}\right),$$

$$w(x, z, t) = w^0(x, t), \tag{5.1}$$

in which substitutions $\tilde{\alpha}$ and $\tilde{\beta}$ [23] are used in the form $\tilde{\alpha} = 4/h^3$, $\tilde{\beta} = 3\tilde{\alpha}$; x and z are the spatial coordinates in the horizontal and vertical directions, respectively (h represents the thickness of the beam in the z direction), and t represents a time. In Eq. (5.1), the longitudinal displacement is designated with $u^0(x, t)$, the transversal displacement with $w^0(x, t)$, and the independent rotation of the cross-section about the axis parallel to y is designated with $\varphi^0(x, t)$. The longitudinal and shear strains can be written as:

$$\varepsilon_x(x, t) = \frac{\partial u^0}{\partial x} + z\frac{\partial \varphi^0}{\partial x} - \widetilde{\alpha} z^3\left(\frac{\partial \varphi^0}{\partial x} + \frac{\partial^2 w^0}{\partial x^2}\right), \tag{5.2}$$

$$\gamma_{xz}(x, t) = \frac{\partial w^0}{\partial x} + \varphi^0 - \widetilde{\beta} z^2\left(\varphi^0 + \frac{\partial w^0}{\partial x}\right), \tag{5.3}$$

$$\begin{Bmatrix} \sigma_x \\ \tau_{xz} \end{Bmatrix} = \begin{bmatrix} E & 0 \\ 0 & G \end{bmatrix}\begin{Bmatrix} \varepsilon_x \\ \gamma_{xz} \end{Bmatrix}, \tag{5.4}$$

in which E is the Young modulus, and G stands for the shear modulus of elasticity. The model representing the foundation is physically defined via the stiffness modulus k_f, and the damping coefficient ν_f. The reaction of the foundation is

$$r(x, t) = k_f w(x, t) + \nu_f \frac{\partial w(x, t)}{\partial t}. \tag{5.5}$$

The equations of motion are derived from the principle of virtual work, according to which the virtual work of the inertial forces is

$$\delta W_{\text{in}} = -\rho \int_V \left[\frac{\partial^2 w(x, z, t)}{\partial t^2}\delta w(x, z, t) + \frac{\partial^2 u(x, z, t)}{\partial t^2}\delta u(x, z, t)\right]\mathrm{d}V, \tag{5.6}$$

the virtual work of the internal forces is

$$\delta W_V = -\int_V [\sigma_x(x, t)\delta\varepsilon_x(z, t) + \tau_{xz}(x, t)\delta\gamma_{xz}(x, t)]\mathrm{d}V, \tag{5.7}$$

and the virtual work of the external forces is defined by

$$\delta W_{\text{ex}} = \int_L [p_z(x, t) - r(x, t)]\delta w(x, t)\mathrm{d}L. \tag{5.8}$$

The next step would be to introduce the general equation of the virtual work principle, $\delta W_{\text{in}} + \delta W_V + \delta W_{\text{ex}} = 0$, and substitute Eqs. (5.6–5.8) into that equation. Then the Green's theorem is applied successively, integrating the equation by parts and collecting the coefficients of δu, δw, and $\delta\varphi$. The result will turn out to be a set of PDEs. This set of equations mathematically describes the vibrations of the system composed of the Reddy–Bickford beam resting on a viscoelastic foundation, and their form is:

$$\Psi[w, \varphi] + k_f w + \nu_f \frac{\partial w}{\partial t} = p_z(x, t), \tag{5.9a}$$

$$\Phi[w, \varphi] = 0, \tag{5.9b}$$

where

$$\Psi[w, \varphi] = C_1\frac{\partial^2 w}{\partial t^2} - C_2\left(\frac{\partial \varphi}{\partial x} + \frac{\partial^2 w}{\partial x^2}\right) - C_3\frac{\partial^4 w}{\partial x^2 \partial t^2} + C_4\frac{\partial^4 w}{\partial x^4}$$

$$+ C_5 \frac{\partial^3 \varphi}{\partial x \partial t^2} + C_6 \frac{\partial^3 \varphi}{\partial x^3}, \tag{5.9c}$$

$$\Phi[w, \varphi] = D_1 \frac{\partial^2 \varphi}{\partial t^2} - D_2 \frac{\partial^2 \varphi}{\partial x^2} + D_3 \left(\varphi + \frac{\partial w}{\partial x} \right) + D_4 \frac{\partial^3 w}{\partial x \partial t^2} + D_5 \frac{\partial^3 w}{\partial x^3}, \tag{5.9d}$$

in which C_i and D_j are determined and can be found in Ref. [24]. Eqs. (5.9a and 5.9b) present the linearized system of beam equations [23]. As it is shown for the simplest case of a moving mass, substituting $w(x, t) = A_1 \exp\{i[(\Omega - kv)t - kx]\}$, $\varphi(x, t) = A_2 \exp\{i[(\Omega - kv)t - kx]\}$, $i = \sqrt{-1}$, into Eqs. (5.9) produces the same dispersion relation [25] in the following form:

$$k^6 q_3 + k^4 q_2 + k^2 q_1 + q_0 = 0, \tag{5.10}$$

Here, v represents the velocity of the moving mass, k represents the wave numbers that are found as the roots of the dispersion Eq. (5.10), and Ω is the natural frequency. Constants q_0, q_1, q_2, and q_3 are given in Ref. [24]. This dispersion relation takes into account rotational bending effects and shear deformation, for they are important factors in describing the behavior of beams subjected to high-frequency excitation when the wavelength gets close to the thickness of the beam. Using the discriminant of the dispersion Eq. (5.10), after some algebra, the critical velocity of the system can be determined. That is when the roots of the dispersion Eq. (5.10) are not real (these roots are also the wave numbers k). In that instance, the critical velocity can be obtained from the expression:

$$\mathrm{Disc}[k^6 q_3 + k^4 q_2 + k^2 q_1 + q_0] = 0 \Rightarrow$$

$$v_{cr,RB} = v_{ph}^{\min} = \left[\sqrt{\frac{\tilde{p}_1^2}{4} - \tilde{q}_1} - \frac{\tilde{p}_1}{2} \right]^{1/2}. \tag{5.11}$$

Constants $\tilde{p}_1$ and $\tilde{q}_1$ are determined and presented in Ref. [24]. The minimum phase velocity of the waves in the beam, which is equal to the critical velocity in the case when the foundation has no viscous damping, depends on the parameters of the foundation. In the work by Verichev and Metrikine [26], there are detailed results on how the foundation parameters affect the critical velocity. When the stiffness of the foundation is increased, the critical velocity will be higher; increasing the damping coefficient of the foundation leads to a higher magnitude. The change in the foundation parameters will only change the critical velocity, regardless of the type of moving object. The paper by Stojanović et al. [25] contains a comparative example that analyzes the critical velocities related to a simpler beam model. The governing equations describing multiple mechanical oscillators of the same masses m, moving at the same constant velocity v, with the same distance between them d, on an infinite beam supported by the foundation with the stiffness modulus k_0 (Fig. 5.1) can be written as:

$$\Psi[w, \varphi] + k_f w + v_f \frac{\partial w}{\partial t}$$

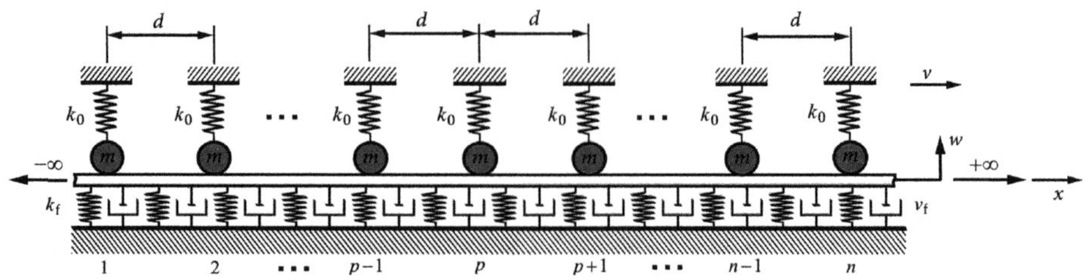

Figure 5.1 Multiple mechanical oscillators moving on a beam. The oscillators are moving at a constant velocity over an infinite beam laid on a foundation.

$$
\begin{aligned}
= -\delta(x - vt)\left(m\frac{d^2 w^{00}}{dt^2} + k_0 w^{01}\right) - \delta(x - vt - d)\left(m\frac{d^2 w^{01}}{dt^2} + k_0 w^{01}\right) \\
- \delta(x - vt - 2d)\left(m\frac{d^2 w^{02}}{dt^2} + k_0 w^{01}\right) - \ldots \\
- \delta[x - vt - (p-1)d]\left(m\frac{d^2 w^{0(p-1)}}{dt^2} + k_0 w^{0(p-1)}\right) \\
- \delta(x - vt - pd)\left(m\frac{d^2 w^{0p}}{dt^2} + k_0 w^{0p}\right) \\
- \delta[x - vt - (p+1)d]\left(m\frac{d^2 w^{0(p+1)}}{dt^2} + k_0 w^{0(p+1)}\right) \\
- \ldots - \delta(x - vt - nd)\left(m\frac{d^2 w^{0n}}{dt^2} + k_0 w^{0n}\right),
\end{aligned}
\tag{5.12}
$$

$$
\Phi[w, \varphi] = 0, \tag{5.13}
$$

where

$$
\begin{gathered}
w^{0p} = w(vt + pd, t), \quad p = 0, 1, 2, \ldots, n, \\
\lim_{|x - vt| \to \infty} w(x, t) = 0, \quad \lim_{|x - vt| \to \infty} \varphi(x, t) = 0.
\end{gathered}
\tag{5.14}
$$

The functions $\Psi[w, \varphi]$ and $\Phi[w, \varphi]$ are given in Eqs. (5.9c and 5.9d), respectively. The following are the displacements: $w(x, t)$ is the vertical deflection of the beam; $w_1^{00}(t)$, $w_1^{01}(t)$, …, $w_2^{0n}(t)$ are the vertical displacements of the masses; $\varphi(x, t)$ is the angle of rotation of the cross-section of the beam. The following are the physical properties: k_f and v_f are the stiffness and the viscosity of the foundations per unit length, respectively; k_0 is the stiffness of the support; n is the number of oscillators. The Dirac delta function is designated with $\delta()$. The centrifugal and Coriolis accelerations are not considered; however, it is good to mention that their effects become noticeable at high speeds. By introducing a moving reference system $\xi = x - vt$, where it is assumed a nonplanar reference from which the deflections are measured, Eqs. (5.12–5.14) can be rewritten in the form:

$$\widetilde{\Psi}[w, \varphi] + k_f w + \nu_f\left(\frac{\partial w(\xi, t)}{\partial t} - v\frac{\partial w(\xi, t)}{\partial \xi}\right) =$$

$$-\delta(\xi)\left(m\frac{\mathrm{d}^2 w^{00}}{\mathrm{d}t^2} + k_0 w^{01}\right) - \delta(\xi - d)\left(m\frac{\mathrm{d}^2 w^{01}}{\mathrm{d}t^2} + k_0 w^{01}\right)$$

$$-\delta(\xi - 2d)\left(m\frac{\mathrm{d}^2 w^{02}}{\mathrm{d}t^2} + k_0 w^{01}\right) - \ldots$$

$$-\delta[\xi - (p-1)d]\left(m\frac{\mathrm{d}^2 w^{0(p-1)}}{\mathrm{d}t^2} + k_0 w^{0(p-1)}\right)$$

$$-\delta(\xi - pd)\left(m\frac{\mathrm{d}^2 w^{0p}}{\mathrm{d}t^2} + k_0 w^{0p}\right) - \delta[\xi - (p+1)d]\left(m\frac{\mathrm{d}^2 w^{0(p+1)}}{\mathrm{d}t^2} + k_0 w^{0(p+1)}\right)$$

$$-\ldots-\delta(\xi - nd)\left(m\frac{\mathrm{d}^2 w^{0n}}{\mathrm{d}t^2} + k_0 w^{0n}\right), \tag{5.15}$$

$$\widetilde{\Phi}[w, \varphi] = 0, \tag{5.16}$$

where

$$w^{0p}(t) = w(pd, t), \quad p = 0,\ 1,\ 2,\ \ldots, n,$$

$$\lim_{|\xi|\to\infty} w(x, t) = 0, \quad \lim_{|\xi|\to\infty} \varphi(x, t) = 0. \tag{5.17}$$

$$\widetilde{\Psi}[w, \varphi] = C_1\left(v^2\frac{\partial^2 w(\xi, t)}{\partial \xi^2} - 2v\frac{\partial^2 w(\xi, t)}{\partial \xi \partial t} + \frac{\partial^2 w(\xi, t)}{\partial t^2}\right)$$

$$-C_2\left(\frac{\partial \varphi(\xi, t)}{\partial \xi} + \frac{\partial^2 w(\xi, t)}{\partial \xi^2}\right)$$

$$-C_3\left(v^2\frac{\partial^4 w(\xi, t)}{\partial \xi^4} - 2v\frac{\partial^4 w(\xi, t)}{\partial \xi^3 \partial t} + \frac{\partial^4 w(\xi, t)}{\partial \xi^2 \partial t^2}\right) + C_4\frac{\partial^4 w(\xi, t)}{\partial \xi^4}$$

$$+C_5\left(v^2\frac{\partial^3 \varphi(\xi, t)}{\partial \xi^3} - 2v\frac{\partial^3 \varphi(\xi, t)}{\partial \xi^2 \partial t} + \frac{\partial^3 \varphi(\xi, t)}{\partial \xi \partial t^2}\right) + C_6\frac{\partial^3 \varphi(\xi, t)}{\partial \xi^3}, \tag{5.18}$$

$$\widetilde{\Phi}[w, \varphi] = D_1\left(v^2\frac{\partial^2 \varphi(\xi, t)}{\partial \xi^2} - 2v\frac{\partial^2 \varphi(\xi, t)}{\partial \xi \partial t} + \frac{\partial^2 \varphi(\xi, t)}{\partial t^2}\right)$$

$$-D_2\frac{\partial^2 \varphi(\xi, t)}{\partial \xi^2} + D_3\left(\varphi(\xi, t) + \frac{\partial w(\xi, t)}{\partial \xi}\right) +$$

$$D_4\left(v^2\frac{\partial^3 w(\xi,t)}{\partial \xi^3} - 2v\frac{\partial^3 w(\xi,t)}{\partial \xi^2 \partial t} + \frac{\partial^3 w(\xi,t)}{\partial \xi \partial t^2}\right) + D_5\frac{\partial^3 w(\xi,t)}{\partial \xi^3} = 0. \tag{5.19}$$

The characteristic equation for the moving system of mechanical oscillators can be analyzed by applying the integral transforms to the system of Eqs. (5.15–5.16).

$$W_s(\xi,s) = \int_0^\infty w(\xi,t)e^{-st}\mathrm{d}t, \quad \Phi_{si}(\xi,s) = \int_0^\infty \varphi(\xi,t)e^{-st}\mathrm{d}t,$$

$$\hat{W}_{k,s}(k,s) = \int_{-\infty}^\infty W_s(\xi,s)e^{-ik\xi}\mathrm{d}\xi, \quad \hat{\Phi}_s(k,s) = \int_{-\infty}^\infty \Phi_s(\xi,s)e^{-ik\xi}\mathrm{d}\xi,$$

$$W^{0p}(s) = \int_0^\infty w^{0p}(t)e^{-st}\mathrm{d}t, \quad i=\sqrt{-1}, \quad p=0,\ 1,\ 2,\ \ldots,n, \tag{5.20}$$

in which s is the Laplace variable. For Eqs. (5.15–5.16), the trivial initial conditions can be assumed—since the governing equations are linear, the initial conditions are insignificant for the stability of the system. Besides, considering that $W^{0p}(s) = W(pd,\ s)$, $(p = 0, 1,\ \ldots, n)$, the application of the transforms results in the system of algebraic equations after the elimination of $\hat{\Phi}_{si}(k,\ s)$ in the present form:

$$F(k,s)\hat{W}_{k,s}(k,s) = -W(0,s)(k_0+ms^2)$$

$$-\ \mathrm{e}^{-id_1k}W(d_1,s)(k_0+ms^2) - \ \ldots\ - e^{-d_nk}W(d_n,s)(k_0+ms^2),$$

$$d_p = pd, \quad p=0,\ 1,\ \ldots,n, \tag{5.21}$$

The function $F(k,\ s)$ can be found in Ref. [24]. By applying the inverse Fourier transform over k to Eq. (5.21), and setting the values $\xi = 0$, $\xi = d_1$, $\xi = d_2$, $\xi = d_3$, $\ldots$, $\xi = d_n$ subsequently, we will produce the following system of algebraic equations with respect to the Laplace-displacements of the contact points $\hat{W}_s(0,\ s)$, $\hat{W}_s(d_1,\ s)$, $\hat{W}_s(d_2,\ s)$, $\ldots$, $\hat{W}_s(d_n,\ s)$.

$$\begin{bmatrix} 1+\mathrm{Int}_0\cdot H & \mathrm{Int}_{+d}\cdot H & \mathrm{Int}_{+d}\cdot H & \mathrm{Int}_{+3d}\cdot H & \cdots & \mathrm{Int}_{+(n-1)d}\cdot H \\ \mathrm{Int}_{-d}\cdot H & 1+\mathrm{Int}_0\cdot H & \mathrm{Int}_{+d}\cdot H & \mathrm{Int}_{+2d}\cdot H & \cdots & \mathrm{Int}_{+(n-2)d}\cdot H \\ \mathrm{Int}_{-2d}\cdot H & \mathrm{Int}_{-d}\cdot H & 1+\mathrm{Int}_0\cdot H & \mathrm{Int}_{+d}\cdot H & \cdots & \mathrm{Int}_{+(n-3)d}\cdot H \\ \mathrm{Int}_{-3d}\cdot H & \mathrm{Int}_{-2d}\cdot H & \mathrm{Int}_{-d}\cdot H & 1+\mathrm{Int}_0\cdot H & \cdots & \mathrm{Int}_{+(n-4)d}\cdot H \\ \cdots & \cdots & \cdots & \cdots & \cdots & \cdots \\ \mathrm{Int}_{-(n-1)d}\cdot H & \mathrm{Int}_{-(n-2)d}\cdot H & \mathrm{Int}_{-(n-3)d}\cdot H & \mathrm{Int}_{-(n-4)d}\cdot H & \cdots & 1+\mathrm{Int}_0\cdot H \end{bmatrix}$$

$$\times \begin{bmatrix} \hat{W}_s(0,s) \\ \hat{W}_s(2d,s) \\ \hat{W}_s(3d,s) \\ \cdots \\ \cdots \\ \cdots \\ \hat{W}_s(nd,s) \end{bmatrix} = \mathbf{0}. \tag{5.22}$$

where

$$H = k_0 + ms^2, \quad \mathrm{Int}_0 = \frac{1}{2\pi}\int_{-\infty}^{\infty}\frac{\mathrm{d}k}{F(k,\ s)}, \quad \mathrm{Int}_{-pd} = \frac{1}{2\pi}\int_{-\infty}^{\infty}\frac{e^{-ikpd}\mathrm{d}k}{F(k,\ s)},$$

$$\mathrm{Int}_{+pd} = \frac{1}{2\pi}\int_{-\infty}^{\infty}\frac{e^{ikpd}\mathrm{d}k}{F(k,\ s)}, \quad (p = 0,\ 1,\ \ldots, n).$$

The result of the application of the inverse Fourier transform is the system of integrals, shown in Eq. (5.22). The system of Eq. (5.22) has a nontrivial solution provided that the determinant of the system is equal to zero.

$$\det\begin{pmatrix} 1+\mathrm{Int}_0\cdot H & \mathrm{Int}_{+d}\cdot H & \mathrm{Int}_{+2d}\cdot H & \mathrm{Int}_{+3d}\cdot H & \cdots & \mathrm{Int}_{+(n-1)d}\cdot H \\ \mathrm{Int}_{-d}\cdot H & 1+\mathrm{Int}_0\cdot H & \mathrm{Int}_{+d}\cdot H & \mathrm{Int}_{+2d}\cdot H & \cdots & \mathrm{Int}_{+(n-2)d}\cdot H \\ \mathrm{Int}_{-2d}\cdot H & \mathrm{Int}_{-d}\cdot H & 1+\mathrm{Int}_0\cdot H & \mathrm{Int}_{+d}\cdot H & \cdots & \mathrm{Int}_{+(n-3)d}\cdot H \\ \mathrm{Int}_{-3d}\cdot H & \mathrm{Int}_{-2d}\cdot H & \mathrm{Int}_{-d}\cdot H & 1+\mathrm{Int}_0\cdot H & \cdots & \mathrm{Int}_{+(n-4)d}\cdot H \\ \cdots & \cdots & \cdots & \cdots & \cdots & \cdots \\ \mathrm{Int}_{-(n-1)d}\cdot H & \mathrm{Int}_{-(n-2)d}\cdot H & \mathrm{Int}_{-(n-3)d}\cdot H & \mathrm{Int}_{-(n-4)d}\cdot H & \cdots & 1+\mathrm{Int}_0\cdot H \end{pmatrix}$$

$$= \det(T_n) = 0. \tag{5.23}$$

The matrix T_n is a member of a group of diagonal-constant matrices named Toeplitz matrices [27]. The eigenvalues of the characteristic Eq. (5.23), with respect to the k_0 stiffness, are of interest here, that is, the characteristic equation of the matrix T_n defines the eigenvalues necessary for analyzing the stability of the model comprising multiple mechanical oscillators. Determining the appropriate eigenmodes is not of major importance from a stability point of view. The instability of vibrations of the complex moving oscillator can appear if at least one root of the characteristic Eq. (5.23) has a positive real part. This equation is an integral equation regarding the Laplace variables; finding its complex roots is of the essence for the model that is here analyzed. An easier way to find a solution to the mathematical problem presented is possible with the use of the D-decomposition method. Denisov et al. [11] defined the idea of this method. Temporarily leaving the physical meaning of the problem, k_0 will be taken as a complex parameter, the complex k_0-plane will be formed, and the D-decomposition method is going to be performed on this plane. Onto this complex plane, the straight line should be mapped $s = i\Omega$. The positive half of the k_0-plane is correlated to the foundation's stiffness. The aim is to find all complex values of k_0, that is, to delineate the lines (in the k_0-plane) that are correlated to the pure imaginary solution $s = i\Omega$ of Eq. (5.23). These lines will partition the k_0-plane into regions with a different number of roots of Eq. (5.23) with respect to the variable s, satisfying the condition $\mathrm{Re}(s) > 0$. For achieving this goal, Eq. (5.23) should be looked at in the following form:

$$f(\Omega,\ k_0) = 0, \tag{5.24}$$

where $s = i\Omega$ and Ω vary from minus to plus infinity. All roots of Eq. (5.24) could be determined with $F(k,\ \Omega)$ defined in Ref. [24]. The contour integration method (the

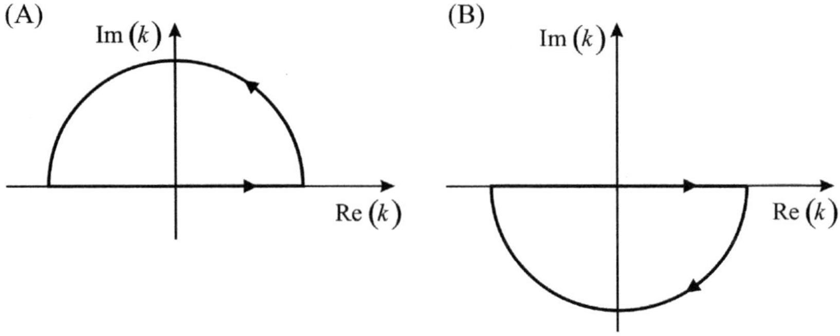

Figure 5.2 Contours of integration. (A) Contour of integration for Int_0 and Int_+ and (B) contour of integration for Int_0.

used contours of integration are shown in Fig. 5.2) is used to develop the integrals that can also be found in Ref. [24] in the following form:

$$\mathrm{Int}_\Delta = \begin{cases} i\sum_{\bar n} \dfrac{(k-k_{\bar n})e^{ik\Delta}}{(k-k_1)\cdots(k-k_6)}\Big|_{k=k_{\bar n}}, & \Delta > 0 \\ i\sum_{\bar n} \dfrac{k-k_{\bar n}}{(k-k_1)\ldots(k-k_6)}\Big|_{k=k_{\bar n}}, & \Delta = 0 \\ -i\sum_{\bar m} \dfrac{(k-k_{\bar m})e^{ik\Delta}}{(k-k_1)\ldots(k-k_6)}\Big|_{k=k_{\bar m}}, & \Delta < 0 \end{cases}, \tag{5.25}$$

in which k_n are the roots with a positive imaginary part, and k_m are the roots with a negative imaginary part of equation $F(k, \Omega) = 0$. After the contour integration method is applied, the short form of the residues may be written as the integration result shown in Eq. (5.25). The full form of all integrals can be found in Ref. [24]. Here presented is the intensive trend for a fast decrease in the values of the complex modules of the integrals $Int_{\pm(n-1)d}$ according to an increase in the number of mechanical oscillators n. Given that all $k_n = k_{n,r} + ik_{n,i}$ have positive imaginary parts $k_{n,i} > 0$, it is obtained that $\exp(ik_n\Delta) = \exp(i(k_{n,r} + ik_{n,i})\Delta) = \exp(ik_{n,r}\Delta)\exp(-k_{n,i}\Delta) \to 0$ when $\Delta \to +\infty$. This further implies that $\mathrm{Int}_\Delta \to 0$ when $\Delta \to +\infty$. In the same manner, it is proven that $\mathrm{Int}_\Delta \to 0$ when $\Delta \to -\infty$, using the fact that k_m have all negative imaginary parts. Based on the previous discussion and Fig. 5.3, many high-order terms $\mathrm{Int}_{k\cdot d}$ can be neglected. These high-order terms are correlated to the off-diagonal entries of the matrix T_n. The values of (Fig. 5.3, shown as a function Δ for the fixed value of Ω) show that the system can be reduced). In that case, the reduced Toeplitz determinant has the following form:

$$\det(T_n^r) =$$

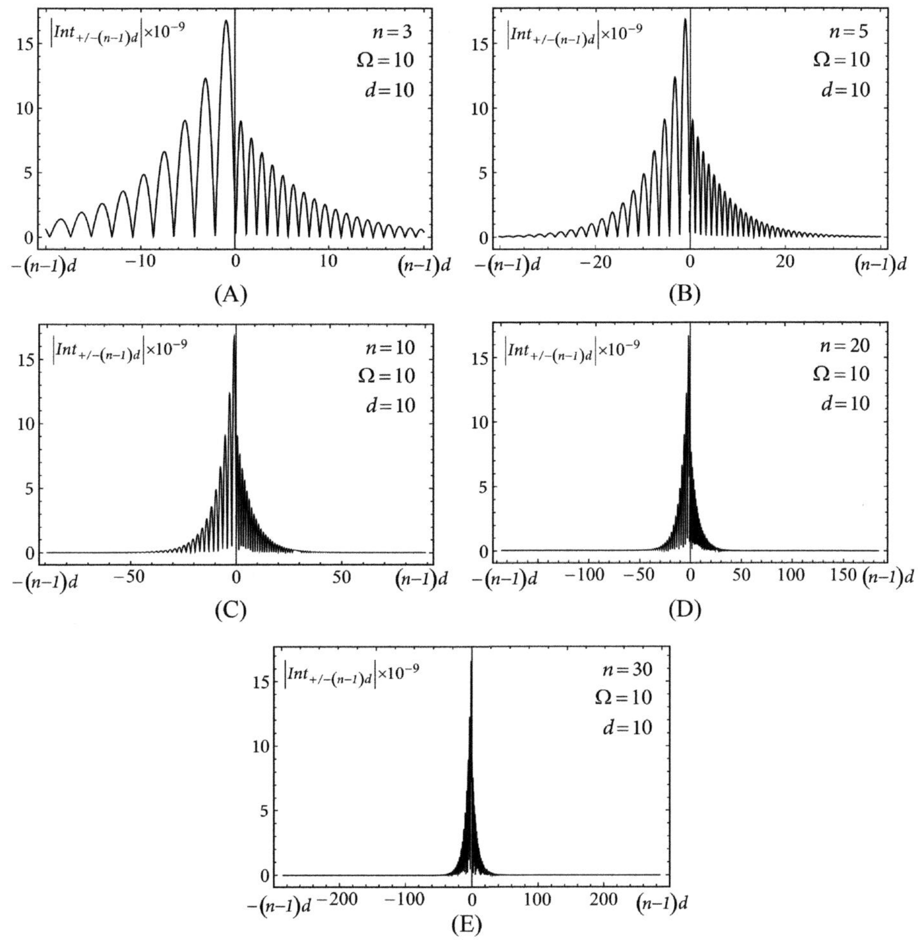

Figure 5.3 Complex modules of integrals. The parameter Ω is fixed, while the number of oscillators is varying: (A) n=3; (B) n=5; (C) n=10; (D) n=20; and (E) n=30.

$$\det\begin{pmatrix} 1+\mathrm{Int}_0\cdot H & \mathrm{Int}_{+d}\cdot H & 0 & 0 & \cdots & 0 \\ \mathrm{Int}_{-d}\cdot H & 1+\mathrm{Int}_0\cdot H & \mathrm{Int}_{+d}\cdot H & 0 & \cdots & 0 \\ 0 & \mathrm{Int}_{-d}\cdot H & 1+\mathrm{Int}_0\cdot H & \mathrm{Int}_{+d}\cdot H & \cdots & 0 \\ 0 & 0 & \mathrm{Int}_{-d}\cdot H & 1+\mathrm{Int}_0\cdot H & \cdots & 0 \\ \cdots & \cdots & \cdots & \cdots & \cdots & \cdots \\ 0 & 0 & 0 & 0 & \cdots & 1+\mathrm{Int}_0\cdot H \end{pmatrix} = 0. \tag{5.26}$$

Obtaining the eigenvalues of the characteristic Eq. (5.26) with respect to k_0 is required for the stability study. Eq. (5.26) can be transformed into a more condensed form using the substitutions α, β, and γ.

$$
\det(T_n^r) = \det \begin{pmatrix}
\alpha & \beta & 0 & \cdots & 0 & 0 & 0 & \cdots & 0 & 0 & 0 \\
\gamma & \alpha & \beta & \cdots & 0 & 0 & 0 & \cdots & 0 & 0 & 0 \\
\cdots & \cdots & \cdots & \cdots & \cdots & \cdots & \cdots & \cdots & \cdots & \cdots & \cdots \\
0 & 0 & 0 & \cdots & \alpha & \beta & 0 & \cdots & 0 & 0 & 0 \\
0 & 0 & 0 & \cdots & \gamma & \alpha & \beta & \cdots & 0 & 0 & 0 \\
0 & 0 & 0 & \cdots & 0 & \gamma & \alpha & \cdots & 0 & 0 & 0 \\
\cdots & \cdots & \cdots & \cdots & \cdots & \cdots & \cdots & \cdots & \cdots & \cdots & \cdots \\
0 & 0 & 0 & \cdots & 0 & 0 & 0 & \cdots & \gamma & \alpha & \beta \\
0 & 0 & 0 & \cdots & 0 & 0 & 0 & \cdots & 0 & \gamma & \alpha
\end{pmatrix} = 0, \tag{5.27}
$$

where

$$
\alpha = 1 + \mathrm{Int}_0(k_0 - m\Omega^2), \quad \beta = \mathrm{Int}_{+d}(k_0 - m\Omega^2), \quad \gamma = \mathrm{Int}_{-d}(k_0 - m\Omega^2). \tag{5.28}
$$

It is easy to see [27] that if two diagonal matrices $\widetilde{C}_n = [\widetilde{c}_{ii}]$ and $\widetilde{D}_n = [\widetilde{d}_{ii}]$ are defined by

$$
\widetilde{c}_{ii} = \frac{\beta^{(j-2)/2}}{\gamma^{(j-1)/2}}, \quad \widetilde{d}_{ii} = \frac{\gamma^{(j-2)/2}}{\beta^{(j-1)/2}}, \quad \mu = \frac{\alpha - \lambda}{\beta^{1/2}\gamma^{1/2}}, \quad j = 1, 2, 3, \ldots, n, \tag{5.29}
$$

then $|\widetilde{C}_n(T_n^r - \lambda I)\widetilde{D}_n| = 0$ becomes

$$
p_{\mathrm{n}}(\mu) = \det \begin{pmatrix}
\mu & 1 & 0 & \cdots & 0 & 0 & 0 & \cdots & 0 & 0 & 0 \\
1 & \mu & 1 & \cdots & 0 & 0 & 0 & \cdots & 0 & 0 & 0 \\
\cdots & \cdots & \cdots & \cdots & \cdots & \cdots & \cdots & \cdots & \cdots & \cdots & \cdots \\
0 & 0 & 0 & \cdots & \mu & 1 & 0 & \cdots & 0 & 0 & 0 \\
0 & 0 & 0 & \cdots & 1 & \mu & 1 & \cdots & 0 & 0 & 0 \\
0 & 0 & 0 & \cdots & 0 & 1 & \mu & \cdots & 0 & 0 & 0 \\
\cdots & \cdots & \cdots & \cdots & \cdots & \cdots & \cdots & \cdots & \cdots & \cdots & \cdots \\
0 & 0 & 0 & \cdots & 0 & 0 & 0 & \cdots & 1 & \mu & 1 \\
0 & 0 & 0 & \cdots & 0 & 0 & 0 & \cdots & 0 & 1 & \mu
\end{pmatrix} = 0. \tag{5.30}
$$

Polynomial equation $p_{\mathrm{n}}(\mu)$ can be solved by the trigonometric method [28]. The matrix Eq. (5.22), according to the allowed transformation of reduced T_n^r for large n, becomes

$$
\begin{bmatrix}
\mu & 1 & 0 & \cdots & 0 & 0 & 0 & \cdots & 0 & 0 & 0 \\
1 & \mu & 1 & \cdots & 0 & 0 & 0 & \cdots & 0 & 0 & 0 \\
\cdots & \cdots & \cdots & \cdots & \cdots & \cdots & \cdots & \cdots & \cdots & \cdots & \cdots \\
0 & 0 & 0 & \cdots & \mu & 1 & 0 & \cdots & 0 & 0 & 0 \\
0 & 0 & 0 & \cdots & 1 & \mu & 1 & \cdots & 0 & 0 & 0 \\
0 & 0 & 0 & \cdots & 0 & 1 & \mu & \cdots & 0 & 0 & 0 \\
\cdots & \cdots & \cdots & \cdots & \cdots & \cdots & \cdots & \cdots & \cdots & \cdots & \cdots \\
0 & 0 & 0 & \cdots & 0 & 0 & 0 & \cdots & 1 & \mu & 1 \\
0 & 0 & 0 & \cdots & 0 & 0 & 0 & \cdots & 0 & 1 & \mu
\end{bmatrix}
\begin{bmatrix}
\hat{W}_s(0, s) \\
\hat{W}_s(d, s) \\
\cdots \\
\hat{W}_s(2d, s) \\
\hat{W}_s(3d, s) \\
\hat{W}_s(4d, s) \\
\cdots \\
\hat{W}_s((n-1)d, s) \\
\hat{W}_s(nd, s)
\end{bmatrix} = 0. \tag{5.31}
$$

By observing the equations (2, 3, ..., $n-1$) of the system (Eq. 5.31), it can be realized that these equations have the same structure. By assuming $\hat{W}_s(pd,\ s) = A_p = N\sin(p\varphi),\ \ p = 0, 1, 2,\ \ldots, n$, and substituting the assumed solutions $N\sin(p\varphi),\ \ p = 2, 3,\ \ldots, n-1$ into the p-th equation of system (Eq. 5.31) yields

$$N\sin(p-1)\varphi + \mu N\sin(p\varphi) + N\sin(p+1)\varphi = 0,\quad p = 2,\ 3,\ \ldots, n-1 \tag{5.32}$$

After some algebra, the following is obtained:

$$(\mu + 2\cos\varphi)N\sin(p\varphi) = 0,\quad p = 2,\ 3,\ \ldots, n-1. \tag{5.33}$$

Eq. (5.33) has to have $N \neq 0$, and $\sin(p\varphi) \neq 0$, because then $A_p \neq 0$, for $p = 2, 3,\ \ldots, n-1$. From there

$$\mu + 2\cos\varphi = 0 \Rightarrow \mu = -2\cos\varphi. \tag{5.34}$$

Eq. (5.34) is a frequency equation. The unknown φ can be determined based on the condition that the assumed solutions satisfy the first and the last equation of the system (Eq. 5.31). When $A_1 = N\sin\varphi$ and $A_2 = N\sin 2\varphi$ are substituted into the first equation of system (Eq. 5.31), the following is obtained:

$$\mu N\sin\varphi - u_n N\sin(2\varphi) = 0 \Rightarrow (\mu + 2\cos\varphi)N\sin\varphi \equiv 0. \tag{5.35}$$

Since it was determined that $N \neq 0$ and $\sin\varphi \neq 0$, Eq. (5.35) will be identical to $\mu + 2\cos\varphi = 0$. When the assumed solutions for A_{n-1} and A_n are substituted into the n-th equation of system (Eq. 5.31), the following is obtained:

$$N\sin(n-1)\varphi + \mu N\sin(n\varphi) = 0. \tag{5.36}$$

After some algebra, the following is obtained:

$$N\sin[(n+1)\varphi] = 0, \tag{5.37}$$

since $N \neq 0$, further periodic solutions are

$$\varphi_s = \frac{s\pi}{n+1},\quad s = 1,\ 2,\ 3,\ \ldots, n. \tag{5.38}$$

The obtained solutions can be verified using the Chebyshev recurrence formula [27] that follows from the expansion of Eq. (5.30). In that case $p_n(\mu)$ satisfies the three-point recurrence relationship

$$p_n(\mu) = \mu p_{n-1}(\mu) - p_{n-2}(\mu),\quad p_1(\mu) = \mu,\quad p_2(\mu) = \mu^2 - 1. \tag{5.39}$$

Considering that $p_1(\mu)$ and $p_2(\mu)$ are the first two Chebyshev polynomials of the first kind, polynomial $p_n(\mu)$ is the Chebyshev polynomial of the first kind and the n-th order. It is known [27] that zeros of $p_n(\mu)$ are

$$\mu_s = 2\cos\frac{s\pi}{n+1} = 2\cos\varphi_s,\quad s = 1,\ 2,\ 3,\ \ldots, n. \tag{5.40}$$

Substituting Eq. (5.28) into Eq. (5.40), and bearing in mind that $\mu = \frac{\alpha - \lambda}{\beta^{1/2}\gamma^{1/2}}$, will produce the $k_{0(s)}$ roots in the following form:

$$k_{0(s)}(\Omega) = \frac{1 - m\Omega^2\left(\text{Int}_0 - 2\sqrt{\text{Int}_{+d}\cdot\text{Int}_{-d}}\cos\left(\frac{s\pi}{n+1}\right)\right)}{1 - \text{Int}_0 + 2\sqrt{\text{Int}_{+d}\cdot\text{Int}_{-d}}\cos\left(\frac{s\pi}{n+1}\right)}, \quad s = 1,\ 2,\ 3,\ \ldots, n. \tag{5.41}$$

Now the D-decomposition curves can be analyzed for both cases—the full and the reduced form of the Toeplitz matrix. Apparently, n curves should be considered, each of them corresponding to one of the n roots of Eq. (5.24) or one of the n roots of Eq. (5.41); the first equation is related to the full form of the Toeplitz matrix, while the latter is related to the reduced form of the Toeplitz matrix. The parametric analysis shows that the D-decomposition curves are useful for the dynamic analysis in the supercritical speed region—when the velocity of the system is above a certain critical velocity, v_{cr}. In the subcritical speed region ($v < v_{cr}$), there seems not to be any crossing point of the D-decomposition curves and the positive part of the real axes. This may indicate that within the subcritical speed region, the number of unstable roots of the characteristic equation (roots with a positive real part) does not vary with the stiffness k_0. However, within the supercritical speed region, the number of unstable roots will start varying when k_0 surpasses a certain critical value of stiffness. The use of the D-decomposition method allows the ascertainment of a relative variation of the number of unstable roots depending on the stiffness k_0, but not the number itself. Solving this is done by determining the number of unstable roots for a single value of stiffness k_0. This can be achieved by using the argument principle. To determine the number of unstable roots, the chosen contour should encircle the right half-plane of the complex variable s, as illustrated in Fig. 5.4. The method consists of setting a fixed value for k_0 and then computating the total change of the argument of the complex expression $f(-is,\ k_0)$ along the entire contour, where f is the function from Eq. (5.24). Per the argument principle, the total change is equal to $2\pi N_{unst}$, where N_{unst} is the number of roots of $f(-is,\ k_0)$ (i.e., Eq. (5.24)) located in the

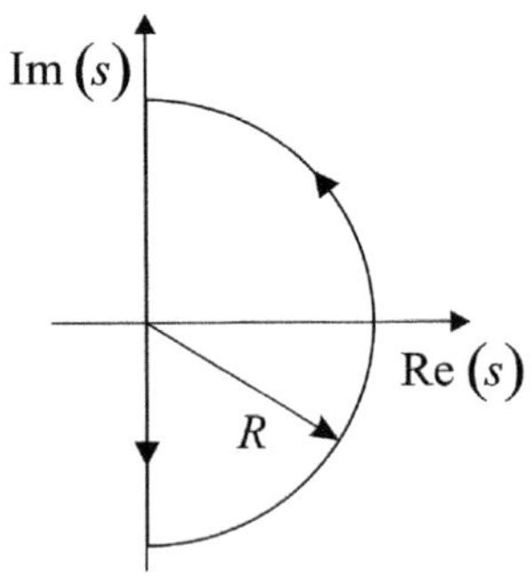

Figure 5.4 Contour of the argument principle. Contour in the (s)-plane according to the argument principle.

positive half-plane,with respect to s. For both parts of the contour (the semicircle and y-axis), some sufficiently small iteration step Δs is chosen, with which the relative change of the argument $\operatorname{Arg}(f(-i(s+\Delta s),\ k_0)/f(-is,\ k_0))$ is computed, and all of these relative changes are summed up. The initial value of the step Δs is selected in such a way as to make the relative change of the modulus and the argument of $f(-is,\ k_0)$ less than 0.001 and 0.1, respectively. Then the step is updated dynamically along the contour, and the initially fixed value is halved until the criterion is not satisfied. If the change in argument between two consecutive values of Ω is greater than 0.1, the size of the step is halved and the calculation is repeated for the new size of the step. Similarly, if the change in an argument is below 0.001, the size of the step is doubled, and the calculation is repeated. The procured expression for $f(-is,\ k_0)$, with the integrals that belong to the type $\mathrm{Int}_{\Delta\pm}$, is exceedingly cumbersome. Because of this, the complete expression for $f(-is,\ k_0)$ is not computed symbolically; this expedites the process of the argument change computation. For each s and k_0, firstly the values of all integrals $\mathrm{Int}_{\Delta\pm}$ are calculated, and then $f(-is,\ k_0)$ is calculated by evaluating the regular (Eq. 5.23) or the reduced (Eq. 5.26) form of the determinant. This procedure speeds up the particular computation of $f(-is,\ k_0)$ by severalfold and reduces memory usage.

5.3 Determination of stable regions

Using the set of model parameters [29] and the above improvements in the D-decomposition method and the argument principle, after the intensive numerical experiments, the D-decomposition curves are obtained. A set of three cases is chosen, in which the effect of varying mass is studied according to the varying number of mechanical oscillators. Up to 30 oscillators were used with a regular Toeplitz matrix and up to 100 with a reduced Toeplitz matrix. It is proven that a further increase in the number of oscillators does not have as a significant influence on the region of stability as does the change in the stiffness k_0. The reduced form of the determinant (Eq. 5.26) causes an error; to decrease this error, a correction parameter is introduced for the upper boundary of the stable region in the form of the function $\chi_{ub}(n,\ m,\ d)$. This function is ascertained by comparing the results between the regular and the reduced form of the Toeplitz matrix for up to 30 oscillators. The regular stable region can be determined directly for up to 30 oscillators. For each varying case of the number of oscillators (for less than 30 oscillators), it is found that the quotient between mass and distance m/d has a linear dependency in the correction parameter function $\chi_{ub}(n,\ m,\ d)$. An effect of the number of oscillators is found to be more prominent in the correction parameter function. Successive comparison of the results from numerous tests has led to the ascertainment of the correction parameter function; this function has the following form:

$$\chi_{ub}(n,\ m,\ d) = 2.56 + \frac{102}{n^3} - 10^{-4}\frac{m}{d} - 2.\ 03\cdot 10^{-4}\frac{n^2}{d^2}. \tag{5.42}$$

The function (Eq. 5.42) produces the results with an average error of 1.5% ($n < 30$). By use of the reduced form of the Toeplitz matrix (for the cases when $n > 30$) and the function (Eq. 5.42), it is possible to obtain the approximate solutions with the predicted error of 1.5%. Since the closed-form solution is unknown, obtaining sufficiently accurate approximate results of the stability regions is important. Some curves that belong to the correction parameter function $\chi_{ub}(n, m, d)$ are depicted in Figs. 5.5 and 5.6 for varying values of concrete parameters; these curves are used in numerical calculations. The correction parameter for the lower boundary of the stable region χ_{lb} is found to fit well for a constant value of 1.1. All stable regions and the occurrences of unstable roots are determined for the selected supercritical velocity $v = 1.12v_{cr}$ as in the case of a model comprising a complex foundation [30]. Variable stable boundaries are documented in Table 5.1 with varying numbers of mechanical oscillators and the variation in their mass. Results documented in Table 5.2 are from a comparative study in which the simpler cases with one and two oscillators are analyzed. Overall, for the more elementary cases (n=1, 2, and 3), the upper stable boundary is always $+\infty$, and that is not the situation in the case for $n > 3$. Based on that and the described argument principle, obtained results show a noteworthy phenomenon; a minute decrease in the stable regions occurs when the number of oscillators increases if the number of oscillators has surpassed thirty. The increase in the mass leads to a decrease in the stable regions. When there are between 10 and 20 oscillators, the stable region's lower stiffness boundary makes a change between 1.79% and 2.5%, according to the increase in mass. This change is smaller when there are between

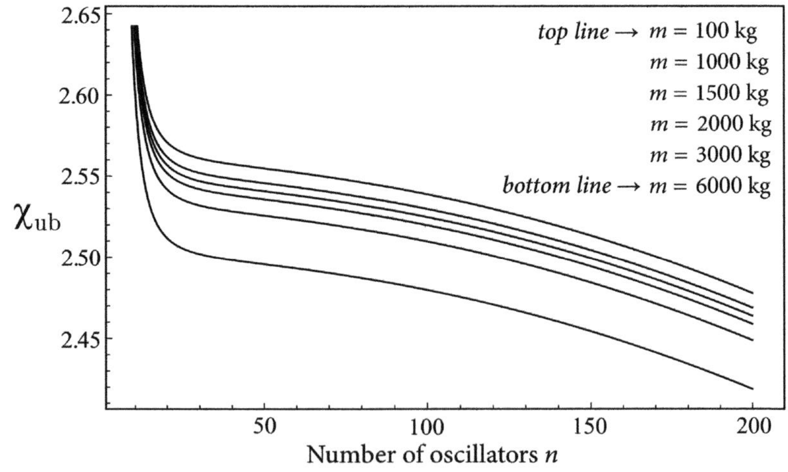

Figure 5.5 Upper boundary correction parameter function $\chi_{ub}(n, m)$. Varying parameters are the number of oscillators (n) and their mass (m), while d=10 m.

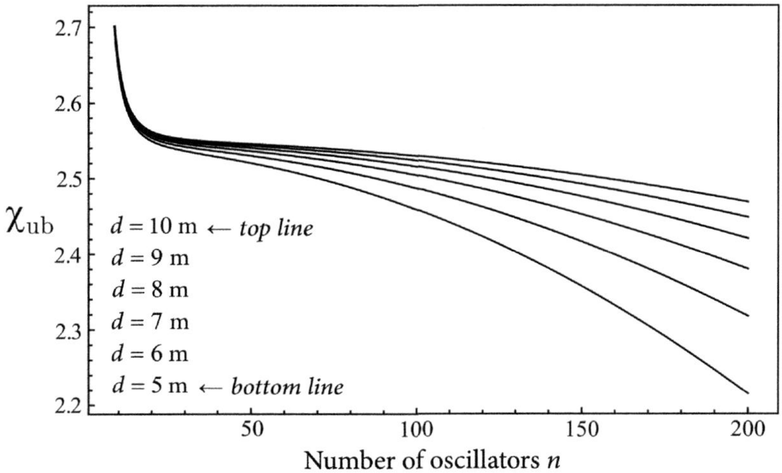

Figure 5.6 Upper boundary correction parameter function $\chi_{ub}(d, n)$. Varying parameters are the number of oscillators (n) and the distance (d), while m=10 kg.

20 and 30 oscillators; the change is between 0.49% and 0.58%. This means that a further increase in the number of oscillators will not affect the realistic detection of a stiffness domain. It is worth noting that an increase in the number of oscillators, regardless of the change in mass, will cause a decrease in the stable region's higher boundary when $n > 30$. However, this stable region's decrease is inconsequential when there are over 20 oscillators. In the general case, the influence of the mass is substantial, regardless of the change in the number of oscillators; in all cases, a mass increase will cause a decrease in the stable regions. In concrete terms, any increase in mass per thousand kilograms, regardless of the number of oscillators, will cause a decrease in the stable region between $6 \cdot 10^7$N/m and $8 \cdot 10^7$N/m, on average. The D-decomposition curves that define the boundaries of stability are documented in Table 5.1.

The diagram of stability depicted in Fig. 5.7 is constructed based on the reported results of stable boundaries given in Table 5.1 (case 1: m=1000 kg). This diagram shows the boundaries of the stable regions as a function of the number of oscillators, where only the regions without unstable roots are selected and depicted. The rise–fall curve effect (regardless of the increase in the number of oscillators) emerges because the anomalous Doppler waves are generated in the beam. A similar rise–fall curve effect (regardless of the increase in velocity) is shown in Fig. 5.7 [26]. There are considerable difficulties in calculations when the number of oscillators is more than 30. Moreover, with the standard Toeplitz matrix, the obtained D-decomposition curves are very complex (even when the number of oscillators is not more than 30), and it is impossible to count the number of unstable roots.

Table 5.1 Stable regions of the regular and reduced Toeplitz matrix cases according to the number of oscillators and varying mass; $k_0 \cdot 10^7$ N/m, $\chi_{ub}(n, m, d)$; χ_{lb}=1.1.

Varying parameter	Type of Toeplitz matrix	$n = 3$	$n = 10$	$n = 20$	$n = 30$	$n = 50$	$n = 100$
m = 1000kg	Regular	{0, +∞}	{0, 251.7}	{0, 215.4}	{0, 211.2}	Difficulties	Difficulties
	Reduced	{0, +∞}	{0, 240.6}	{0, 213.3}	{0, 208.4}	{0, 205.9}	{0, 204.8}
m = 2000kg	Regular	{6.18, +∞}	{7.93, 251.9}	{8.14, 215.7}	{8.18, 210.3}	Difficulties	Difficulties
	Reduced	{6.61, +∞}	{7.96, 241.8}	{8.14, 214.5}	{8.14, 209.6}	{8.15, 207.1}	{8.16, 203.6}
m = 3000kg	Regular	{13.8, +∞}	{16.4, 252.2}	{16.7, 215.9}	{16.8, 210.5}	Difficulties	Difficulties
	Reduced	{14.8, +∞}	{16.5, 243.1}	{16.7, 215.6}	{16.8, 210.8}	{16.9, 208.3}	{16.8, 207.2}

Table 5.2 Stable regions of the single and double oscillatory system; $k_0 \cdot 10^7$ N/m.

Varying parameter	$n = 1$	$n = 2$
$m = 1000$ kg	$\{0, +\infty\}$	$\{0, +\infty\}$
$m = 2000$ kg	$\{1.35, +\infty\}$	$\{4.83, +\infty\}$
$m = 3000$ kg	$\{7.43, +\infty\}$	$\{11.9, +\infty\}$

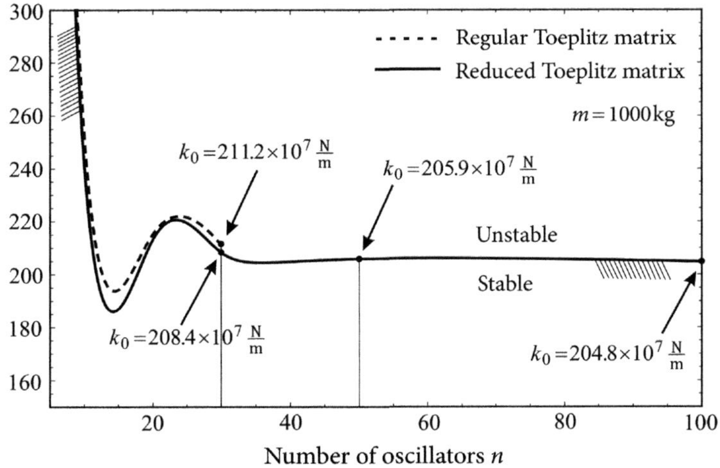

Figure 5.7 Stability regions. Stability regions of the regular and reduced case of Toeplitz matrix according to the number of mechanical oscillators n.

The D-decomposition curves obtained by the regular Toeplitz matrix for $n < 30$ are presented in Figs. 5.8–5.10. When the number of oscillators is substantial, the reduced Toeplitz matrix has the advantage. The procedure where the reduced Toeplitz matrix is utilized can be used for any type and number of oscillators, even if the oscillators belong to the class of complex oscillators. This premise is important, for it enables further expansion and investigation of differently modeled oscillators and oscillators that are coupled in a more complex fashion; that is of technical interest in railway traffic and dynamic stability analysis.

The varying spectrums of the D-decomposition curves for the chosen cases in combination with varying numbers of oscillators and varying masses are presented in Figs. 5.8–5.10. A spectrum of the D-decomposition curves for the case of three oscillators that have masses of 1000 kg each is presented in Fig. 5.8. It can be observed that there are no crossing points with the positive real axis, which implies that the system is stable for all real positive values of k_0. For the case of five

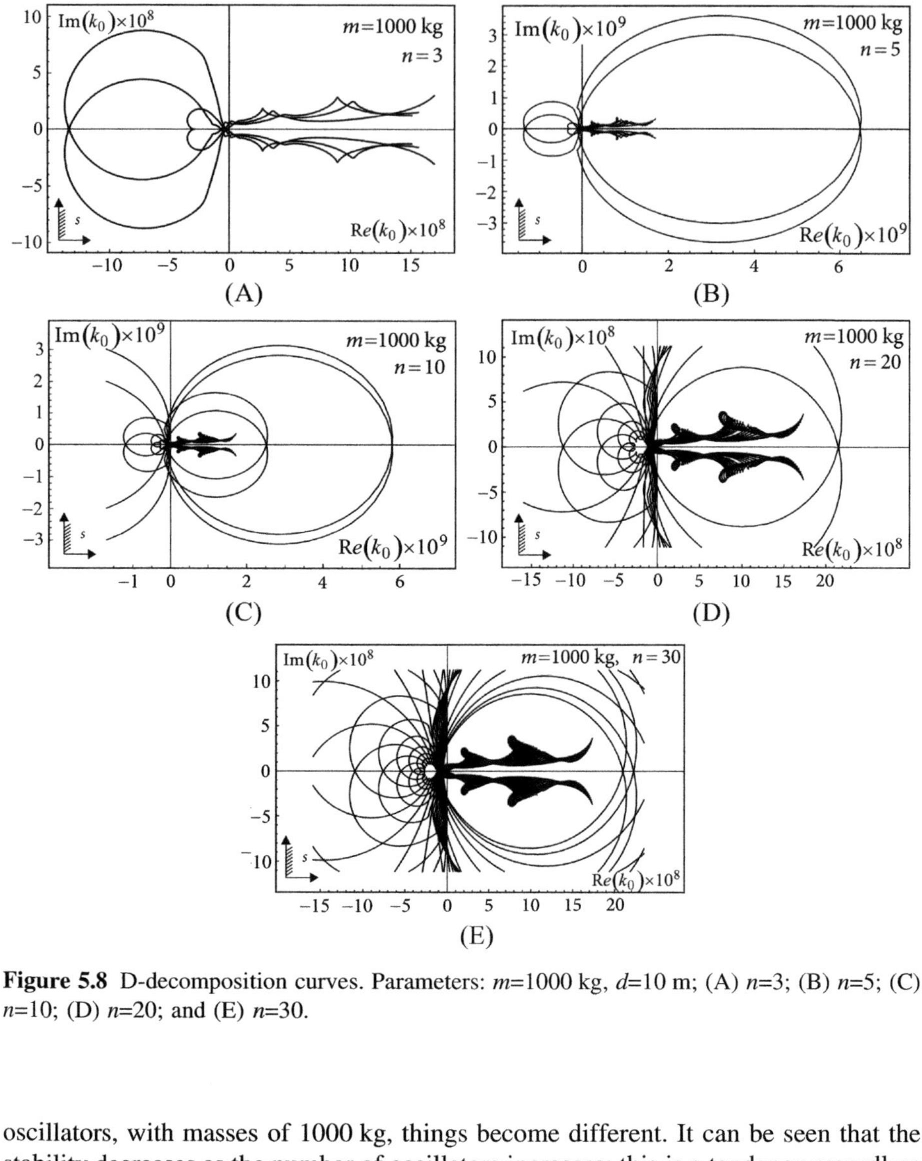

Figure 5.8 D-decomposition curves. Parameters: m=1000 kg, d=10 m; (A) n=3; (B) n=5; (C) n=10; (D) n=20; and (E) n=30.

oscillators, with masses of 1000 kg, things become different. It can be seen that the stability decreases as the number of oscillators increases; this is a tendency regardless of the change in mass.

When the three-oscillator model (Fig. 5.8) is compared with the other two models of this kind, but with increased masses (Figs. 5.9 and 5.10), it can be seen that new crossing points with the real axis appear. This means that the argument principle is needed to count the number of unstable roots (if they exist) belonging to the real axis, which defines the stable regions. Figs. 5.8–5.10 show the sets of the D-decomposition

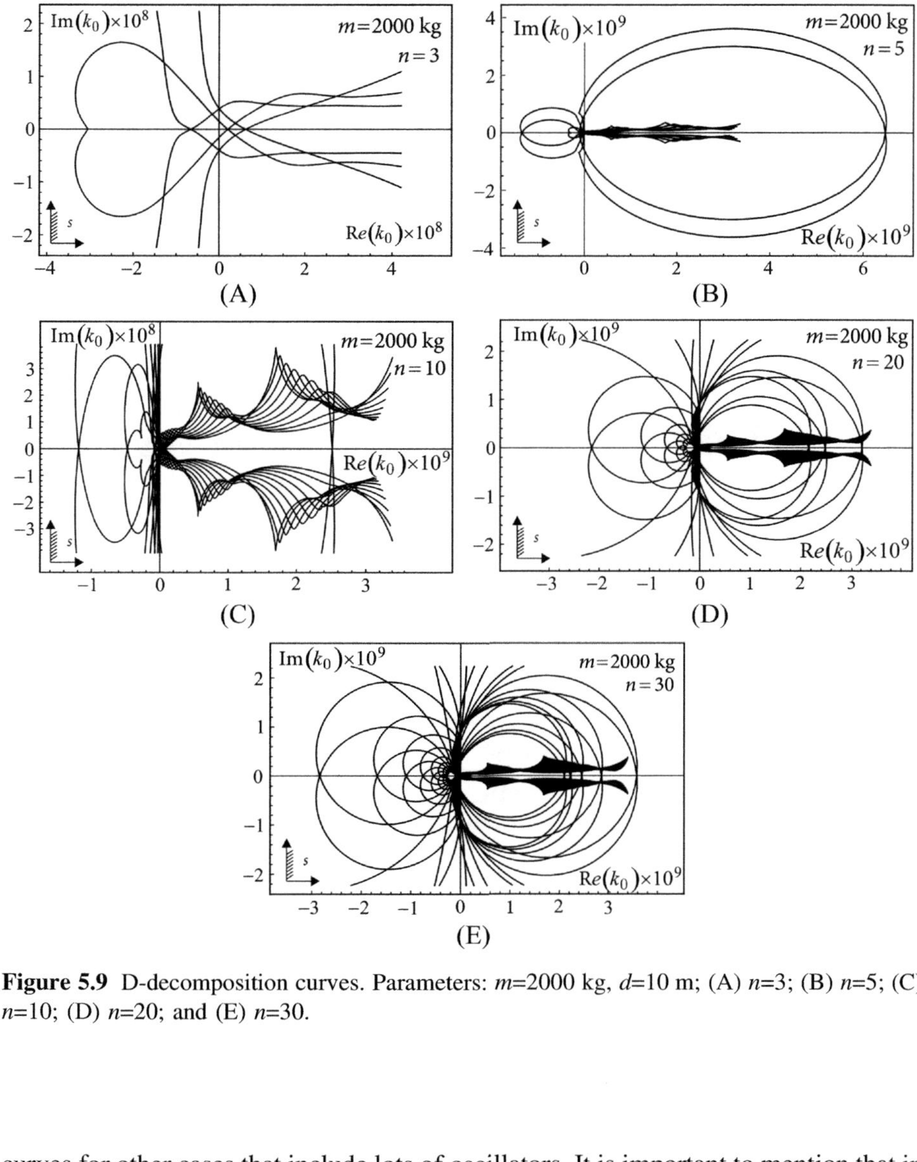

Figure 5.9 D-decomposition curves. Parameters: m=2000 kg, d=10 m; (A) n=3; (B) n=5; (C) n=10; (D) n=20; and (E) n=30.

curves for other cases that include lots of oscillators. It is important to mention that in cases with lots of oscillators, the D-decomposition curves keep a similar path but have different crossing points with the real k_0 axis. A valuable expansion of the analysis concerning the dynamic stability of moving masses is presented in this chapter. The results provided came from intensive analytical and numerical calculations, which cannot be shown here because the expressions are cumbersome and the programming codes are lengthy and intricate.

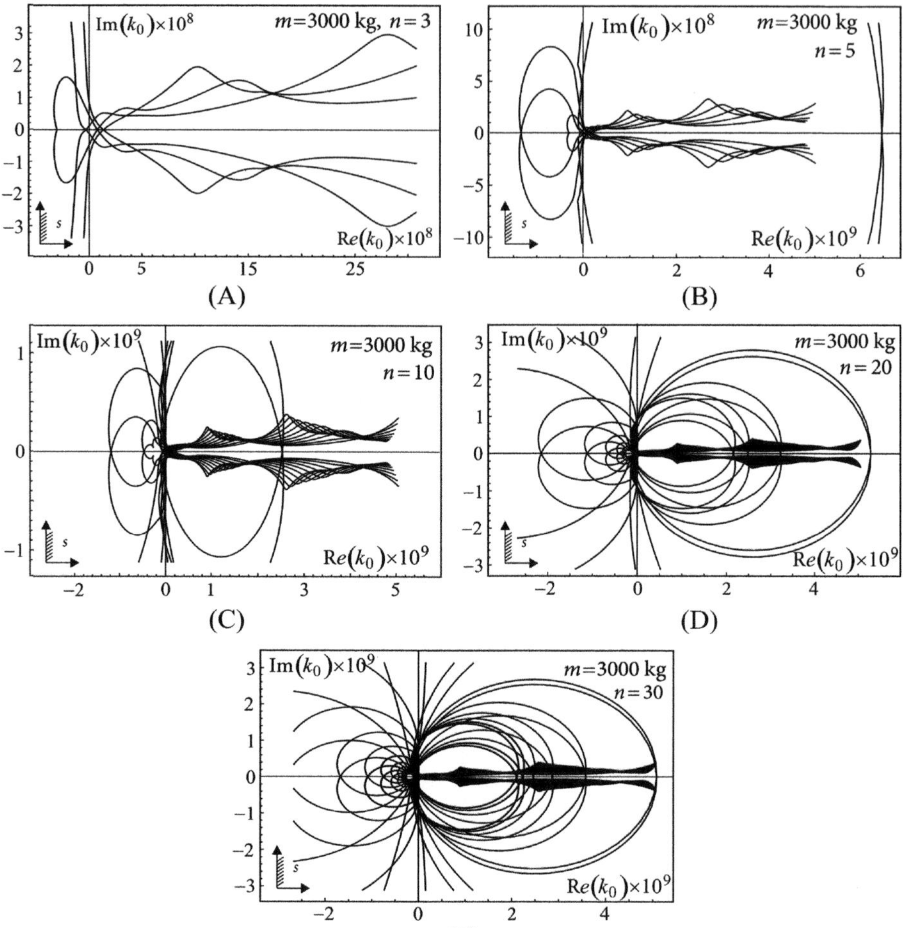

Figure 5.10 D-decomposition curves. Parameters: m=3000 kg, d=10 m; (A) n=3; (B) n=5; (C) n=10; (D) n=20; and (E) n=30.

5.4 Conclusion

The focus of the study presented in this chapter is on a uniformly moving multiple oscillator system. This system is moving along an infinite shear-deformable beam that rests on a viscoelastic foundation. The instability occurs when the necessary condition for that is met; the velocity of the system must exceed a certain critical velocity. When the foundation is elastic (which means it is not viscoelastic), the critical velocity is equal to the minimum phase velocity of waves in the beam. A train that is moving at about 250 km/h will experience instability because the minimum phase velocity has been lowered. This occurs for at least two reasons (explained in the Introduction of this study). One reason is the interaction between the rails and soft subsoil; the other is the thermal stress in the axial direction of continuously welded rails. In that case, the more realistic models could be unstable. The roots of the characteristic equation determine the stability of the model. This characteristic equation is for the interaction between the complex moving oscillator (which is a discrete system) and the beam (which is a continuous system). These roots are studied by subsequently using the D-decomposition method and the argument principle. The D-decomposition method is used to partition the space of the model parameters into domains, which have different numbers of unstable roots. Then the number of unstable roots (in one of these domains and for a specific set of the model parameters) is determined using the argument principle.

It is important to highlight one novel finding in this study: the system has an approximate solution in closed-form, which is formulated using the special case of the Toeplitz matrix. Then the system of equations is solved by the trigonometric method.

After reducing the regular matrix, the roots of the characteristic equation are determined in the closed-form. Verification of the roots was done with the recurrence formula of the first kind and the n-th order Chebyshev polynomial. It is shown that the implemented mathematical procedure leads to getting important results that were previously unknown. This is primarily the case when the structure of the system comprises more than thirty oscillators. The full stable regions are determined and discussed according to the varying mass, distance, and number of oscillators. The collected results are original, and the procedure that is used to determine them can serve as a good example of how a different way can be chosen to solve complex problems.

References

[1] A.V. Metrikine, Unstable lateral oscillations of an object moving uniformly along elastic guide as a result of anomalous Doppler effect, Acoustical Physics 40 (1) (1994) 85–89.

[2] V. Stojanović, P. Kozić, M.D. Petković, Dynamic instability and critical velocity of a mass moving uniformly along a stabilized infinity beam, International Journal of Solids and Structures 108 (2017) 164–174.

[3] A.D. Kerr, The continuously supported rail subjected to an axial force and a moving load, International Journal of Mechanical Sciences 14 (1972) 71–78.

[4] J.J. Labra, An axially stressed railroad track on an elastic continuum subjected to a moving load, Acta Mechanica 22 (1975) 113–129.

[5] A.V. Metrikine, H.A. Dieterman, Instability of vibrations of a mass moving uniformly along an axially compressed beam on a viscoelastic foundation, Journal of Sound and Vibration 201 (5) (1997) 567–576.

[6] X. Sheng, C.J.C. Jones, M. Petyt, Ground vibration generated by a load moving along a railway track, Journal of Sound and Vibration 228 (1) (1999) 129–156.

[7] A.V. Metrikine, K. Popp, Instability of vibrations of an oscillator moving along a beam on an elastic half-space, European Journal of Mechanics—A/Solids 18 (4) (1999) 679–701.

[8] S.P. Timoshenko, Proc. 2nd International Congress for Applied Mechanics, Zürich, 1927, pp. 1–12.

[9] J.T. Kenney, Steady-state vibrations of beam on elastic foundation for moving load, Journal of Applied Mechanics 21 (1954) 359–364.

[10] L. Frýba, Vibration of Solids and Structures under Moving Loads, Springer, Netherlands, 1972.

[11] G.G. Denisov, E.K. Kugusheva, V.V. Novikov, On the problem of the stability of one-dimensional unbounded elastic systems, Journal of Applied Mathematics and Mechanics 49 (4) (1985) 533–537.

[12] S.N. Verichev, A.V. Metrikine, Instability of vibrations of a mass that moves uniformly along a beam on a periodically inhomogeneous foundation, Journal of Sound and Vibration 260 (2003) 901–925.

[13] Z. Dimitrovová, J.N.: Varandas, Critical velocity of a load moving on a beam with a sudden change of foundation stiffness: applications to high-speed trains, Computers & Structures 87 (2009) 1224–1232.

[14] S.-H. Ju, H.-T. Lin, Analysis of train-induced vibrations and vibration reduction schemes above and below critical Rayleigh speeds by finite element method, Soil Dynamics and Earthquake Engineering 24 (12) (2004) 993–1002.

[15] G. Muscolino, A. Palmeri, Response of beams resting on viscoelastically damped foundation to moving oscillators, International Journal of Solids and Structures 44 (5) (2007) 1317–1336.

[16] N. Zhang, H. Xia, Dynamic analysis of coupled railway vehicle–bridge system based on inter-system iteration method, Computers & Structures 114–115 (2013) 26–34.

[17] A.V. Metrikine, A.V. Vostrukhov, A.C.W.M. Vrouwenvelder, Drag experienced by a high-speed train due to excitation of ground vibrations, International Journal of Solids and Structures 48–49 (2001) 8851–8868.

[18] P. Kozić, G. Janevski, R. Pavlović, Moment Lyapunov exponents and stochastic stability of a double-beam system under compressive axial loading, International Journal of Solids and Structures 47 (10) (2010) 1435–1442.

[19] S. Lenci, G. Rega, An asymptotic model for the free vibrations of a two-layer beam, European Journal of Mechanics—A/Solids 42 (2013) 441–453.

[20] M.A. Rezvani, A. Mazraeh, Dynamics and stability analysis of a freight wagon subjective to the railway track and wheelset operational conditions, European Journal of Mechanics—A/Solids 61 (2017) 22–34.

[21] R. Zarfam, A.R. Khaloo, A. Nikkhoo, On the response spectrum of Euler–Bernoulli beams with a moving mass and horizontal support excitation, Mechanics Research Communications 47 (2013) 77–83.

[22] Z. Dimitrovová, Critical velocity of a uniformly moving load on a beam supported by a finite depth foundation, Journal of Sound and Vibration 366 (2016) 325–342.

[23] C.M. Wang, J.N. Reddy, K.H.: Lee, Shear deformable beams and plates, Relationships with Classical Solutions, Elsevier, 2000.

[24] V. Stojanović, M.D. Petković, J. Deng, Stability and vibrations of an overcritical speed moving multiple discrete oscillators along an infinite continuous structure, European Journal of Mechanics—A/Solids 75 (2019) 367–380.

[25] V. Stojanović, M.D. Petković, J. Deng, Instability of vehicle systems moving along an infinite beam on a viscoelastic foundation, European Journal of Mechanics—A/Solids 69 (2018) 238–254.

[26] S.N. Verichev, A.V. Metrikine, Instability of a bogie moving on a flexibly supported Timoshenko beam, Journal of Sound and Vibration 253 (2002) 653–668.

[27] R.T. Gregory, D.L. Karney, A Collection of Matrices for Testing Computational Algorithms, Wiley, New York, 1969.
[28] V. Stojanović, P. Kozić, G. Janevski, Exact closed-form solutions for the natural frequencies and stability of elastically connected multiple beam system using Timoshenko and high-order shear deformation theory, Journal of Sound and Vibration 332 (2013) 563–576.
[29] V. Stojanović, M.D. Petković, Dynamic stability of vibrations and critical velocity of a complex bogie system moving on a flexibly supported infinity track, Journal of Sound and Vibration 434 (2018) 475–501.
[30] V. Stojanović, M.D. Petković, J. Deng, Stability of vibrations of a moving railway vehicle along an infinite complex three-part viscoelastic beam/foundation system, International Journal of Mechanical Sciences 136 (2018) 155–168.

Vibrational benefits of the new stabilizer in moving coupled vehicles

6

6.1 Introduction

Advancements in road and railway traffic have led to high-speed transportation. This implies that in the future, high-speed trains may travel faster than the elastic waves propagating in the railway structure. These elastic waves or Doppler waves can destabilize the system's vibrations [1–3]. The research about the instability of vibrations of simple mechanical models, like in the previously mentioned works, has allowed further investigation into this area. Many researchers have studied the dynamic stability and instability of mechanical systems subjected to moving loads. Timoshenko [4] showed that a moving constant vertical force has a critical velocity at which deflections become sizable. Kerr [5] studied the response of an axially compressed continuously supported infinity beam subjected to a constant vertical force that is in motion. The mechanical model of a beam that is vibrating and is supported by an elastic foundation, subjected to a constant vertical force that is moving, is the basis for further research involving the model of a beam subjected to a moving mass. Thus Denisov et al. [6] identified the region of stability for a mass moving uniformly on a continuously supported beam, by applying the D-decomposition method. Elucidation of this and further analysis of dynamic instability, using the D-decomposition method, can be found in the work by Verichev and Metrikine [7].

In the literature, genuine dynamical problems of engineering practice have been studied with respect to high-speed trains. Therefore Ju and Lin [8] examined the matter of fast trains, as referred to in Chapter 4. Achieving high-speed traffic has been made possible because of theoretical research, which encompasses an array of scientific fields—from civil engineering, through mechanical engineering, to automation, control, and electrical engineering. On the subject of mechanics, many researchers have provided substantial initiative and contributions to this field.

In the aforementioned studies, less complex vehicle models in Refs. [1–3] and [7] were used for the analysis of vibration stability. This chapter is about a study of a complex rail vehicle system that comprises two elastically connected rail cars on five identical supports (made of wheels with suspension elements), which makes this model an extension of the model analyzed by Verichev and Metrikine [7]. Following the same outlined procedure, zones of instability are determined using the mentioned complex rail vehicle model that is moving on a deformable infinity beam, which is supported by a viscoelastic foundation. The beam is mathematically described by a high-order shear deformation theory (the Reddy–Bickford beam assumption). The elastic coupling between the cars is introduced as a technical solution that modifies the region of stability of the rail vehicle system. This is explained in the present analysis, which is based on the study by Stojanović and Petković [9].

Modeling of Complex Dynamic Systems. DOI: https://doi.org/10.1016/B978-0-443-23942-7.00006-1

6.2 Mathematical model and critical velocity

This section is about the mathematical analysis of a complex rail vehicle system that is uniformly traveling on a deformable beam of infinite length, as illustrated in Fig. 6.1. The rail (i.e., the beam) is supported by a viscoelastic foundation and its deformational behavior is mathematically described by a high-order shear deformation theory, specifically the Reddy–Bickford beam assumption [10]. The rail vehicle system (in this study comprised of two rail cars) is modeled as an assembly of two elastically connected rigid bars on identical supports. The scheme of the decomposed vehicle system and the forces acting on it is presented in Fig. 6.2. The connection between the bars (i.e., the rail cars) is realized via an elastic coupling that simulates a rigid or flexible (transverse or/and rotational) connection. The supports are, in fact, wheels with suspension elements. A suspension element comprises a spring and a dashpot in parallel. Suspension elements are on one side attached to the bars (i.e., the bodies of the vehicles), and on the other side, they interact with the beam through the concentrated masses, which are in this case—the wheels. It is assumed that the wheels and the beam are in contact at all times. For the given model, according to dynamic equilibrium (Fig. 6.3), the governing equations may be written as:

$$C_1\frac{\partial^2 w}{\partial t^2} - C_2\left(\frac{\partial\varphi}{\partial x} + \frac{\partial^2 w}{\partial x^2}\right) - C_3\frac{\partial^4 w}{\partial x^2 \partial t^2} + C_4\frac{\partial^4 w}{\partial x^4} + C_5\frac{\partial^3\varphi}{\partial x \ \partial t^2}$$
$$+ C_6\frac{\partial^3\varphi}{\partial x^3} + k_f w + \nu_f\frac{\partial w}{\partial t}$$

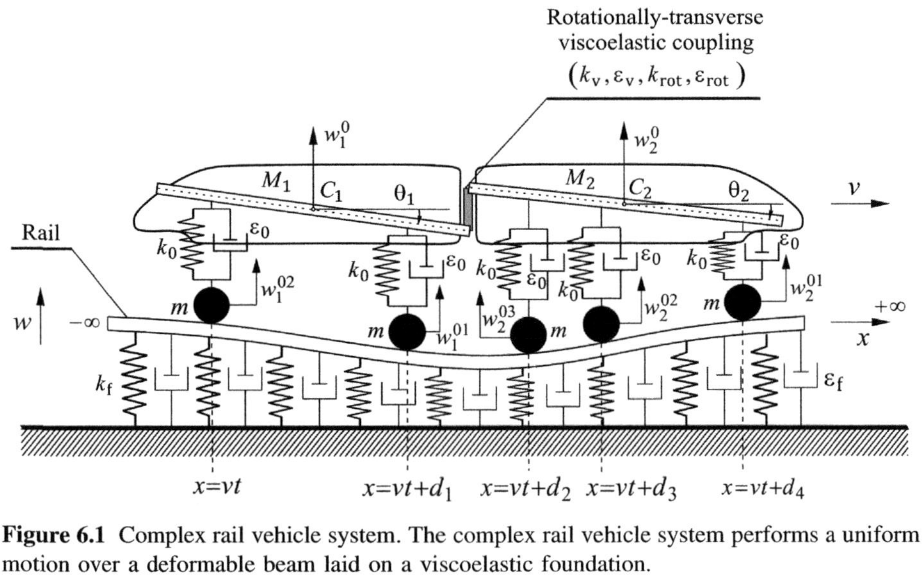

Figure 6.1 Complex rail vehicle system. The complex rail vehicle system performs a uniform motion over a deformable beam laid on a viscoelastic foundation.

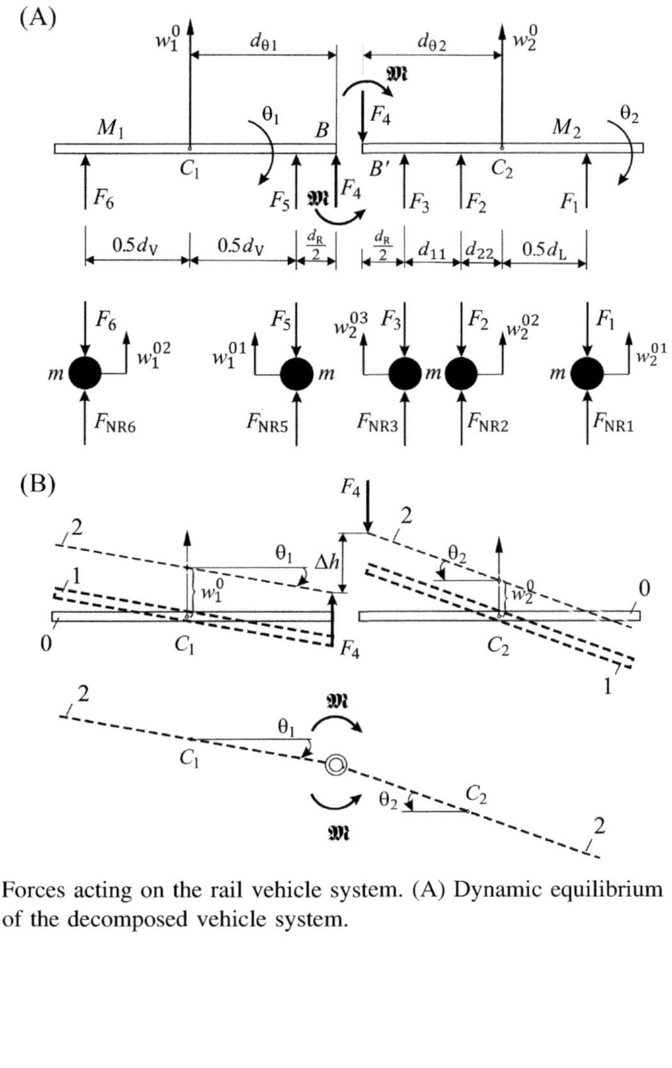

Figure 6.2 Forces acting on the rail vehicle system. (A) Dynamic equilibrium forces, (B) scheme of the decomposed vehicle system.

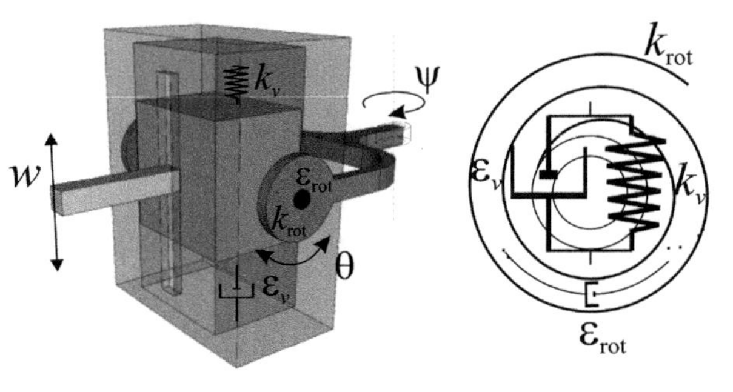

Figure 6.3 Elastic coupling which is used for connecting rail vehicle cars.

$$= -\delta(x - vt - d_4)\left[m\frac{d^2w_2^{01}}{dt^2} + k_0\left(w_2^{01} - w_2^0 + \theta_2\frac{d_L}{2}\right)\right.$$
$$\left.+ \epsilon_0\left(\frac{dw_2^{01}}{dt} - \frac{dw_2^0}{dt} + \frac{d\theta_2}{dt}\frac{d_L}{2}\right)\right]$$
$$- \delta(x - vt - d_3)\left[m\frac{d^2w_2^{02}}{dt^2} + k_0(w_2^{02} - w_2^0 - \theta_2 d_{22})\right.$$
$$\left.+ \epsilon_0\left(\frac{dw_2^{02}}{dt} - \frac{dw_2^0}{dt} - \frac{d\theta_2}{dt}d_{22}\right)\right]$$
$$- \delta(x - vt - d_2)\left[m\frac{d^2w_2^{03}}{dt^2} + k_0\left(w_2^{03} - w_2^0 - \theta_2\frac{d_L}{2}\right)\right.$$
$$\left.+ \epsilon_0\left(\frac{dw_2^{03}}{dt} - \frac{dw_2^0}{dt} - \frac{d\theta_2}{dt}\frac{d_L}{2}\right)\right]$$
$$- \delta(x - vt - d_1)\left[m\frac{d^2w_1^{01}}{dt^2} + k_0\left(w_1^{01} - w_1^0 + \theta_1\frac{d_V}{2}\right)\right.$$
$$\left.+ \epsilon_0\left(\frac{dw_1^{01}}{dt} - \frac{dw_1^0}{dt} - \frac{d\theta_1}{dt}\frac{d_V}{2}\right)\right]$$
$$- \delta(x - vt)\left[m\frac{d^2w_1^{02}}{dt^2} + k_0\left(w_1^{02} - w_1^0 - \theta_1\frac{d_V}{2}\right)\right.$$
$$\left.+ \epsilon_0\left(\frac{dw_1^{02}}{dt} - \frac{dw_1^0}{dt} - \frac{d\theta_1}{dt}\frac{d_V}{2}\right)\right], \tag{6.1}$$

$$D_1\frac{\partial^2\varphi}{\partial t^2} - D_2\frac{\partial^2\varphi}{\partial x^2} + D_3\left(\varphi + \frac{\partial w}{\partial x}\right) + D_4\frac{\partial^3 w}{\partial x\ \ \partial t^2} + D_5\frac{\partial^3 w}{\partial x^3} = 0, \tag{6.2}$$

$$M_2\frac{d^2w_2^0}{dt^2} + k_0\left(w_2^0 - w_2^{01} - \theta_2\frac{d_L}{2}\right) + k_0(w_2^0 - w_2^{02} + \theta_2 d_{22})$$
$$+ k_0\left(w_2^0 - w_2^{03} + \theta_2\frac{d_L}{2}\right) - k_v(w_2^0 - w_1^0 + \theta_2 d_{\theta 2} + \theta_1 d_{\theta 1})$$
$$+ \epsilon_0\left(\frac{dw_2^0}{dt} - \frac{dw_2^{01}}{dt} - \frac{d\theta_2}{dt}\frac{d_L}{2}\right) + \epsilon_0\left(\frac{dw_2^0}{dt} - \frac{dw_2^{02}}{dt} + \frac{d\theta_2}{dt}d_{22}\right)$$

$$+\epsilon_0\left(\frac{dw_2^0}{dt}-\frac{dw_2^{02}}{dt}+\frac{d\theta_2}{dt}\frac{d_L}{2}\right)-\epsilon_v\left(\frac{dw_2^0}{dt}-\frac{dw_1^0}{dt}+\frac{d\theta_2}{dt}d_{\theta 2}+\frac{d\theta_1}{dt}d_{\theta 1}\right)=0, \tag{6.3}$$

$$J_2\frac{d^2\theta_2}{dt^2}-k_0\frac{d_L}{2}\left(w_2^0-w_2^{01}-\theta_2\frac{d_L}{2}\right)+k_0d_{22}\left(w_2^0-w_2^{02}+\theta_2 d_{22}\right)$$
$$+k_0\frac{d_L}{2}\left(w_2^0-w_2^{03}+\theta_2\frac{d_L}{2}\right)-k_v d_{\theta 2}\left(w_2^0-w_1^0+\theta_2 d_{\theta 2}+\theta_1 d_{\theta 1}\right)$$
$$+k_{rot}(\theta_2-\theta_1)-\epsilon_0\frac{d_L}{2}\left(\frac{dw_2^0}{dt}-\frac{dw_2^{01}}{dt}-\frac{d\theta_2}{dt}\frac{d_L}{2}\right)$$
$$+\epsilon_0 d_{22}\left(\frac{dw_2^0}{dt}-\frac{dw_2^{02}}{dt}+\frac{d\theta_2}{dt}d_{22}\right)+\epsilon_0\frac{d_L}{2}\left(\frac{dw_2^0}{dt}-\frac{dw_2^{03}}{dt}+\frac{d\theta_2}{dt}\frac{d_L}{2}\right)$$
$$-\epsilon_v d_{\theta 2}\left(\frac{dw_2^0}{dt}-\frac{dw_1^0}{dt}+\frac{d\theta_2}{dt}d_{\theta 2}+\frac{d\theta_1}{dt}d_{\theta 1}\right)+\epsilon_{rot}\left(\frac{d\theta_2}{dt}-\frac{d\theta_1}{dt}\right)=0, \tag{6.4}$$

$$M_1\frac{d^2w_1^0}{dt^2}+k_v\left(w_2^0-w_1^0+\theta_2 d_{\theta 2}+\theta_1 d_{\theta 1}\right)+k_0\left(w_1^0-w_1^{01}-\theta_1\frac{d_V}{2}\right)$$
$$+k_0\left(w_1^0-w_1^{02}+\theta_1\frac{d_V}{2}\right)+\epsilon_v\left(\frac{dw_2^0}{dt}-\frac{dw_1^0}{dt}+\frac{d\theta_2}{dt}d_{\theta 2}+\frac{d\theta_1}{dt}d_{\theta 1}\right)$$
$$+\epsilon_0\left(\frac{dw_1^0}{dt}-\frac{dw_1^{01}}{dt}-\frac{d\theta_1}{dt}\frac{d_V}{2}\right)+\epsilon_0\left(\frac{dw_1^0}{dt}-\frac{dw_1^{02}}{dt}+\frac{d\theta_1}{dt}\frac{d_V}{2}\right)=0, \tag{6.5}$$

$$J_1\frac{d^2\theta_1}{dt^2}-k_0\frac{d_V}{2}\left(w_1^0-w_1^{01}-\theta_1\frac{d_V}{2}\right)+k_0\frac{d_V}{2}\left(w_1^0-w_1^{02}+\theta_1\frac{d_V}{2}\right)$$
$$-k_v d_{\theta 1}\left(w_2^0-w_1^0+\theta_2 d_{\theta 2}+\theta_1 d_{\theta 1}\right)-k_{rot}(\theta_2-\theta_1)$$
$$-\epsilon_0\frac{d_V}{2}\left(\frac{dw_1^0}{dt}-\frac{dw_1^{01}}{dt}-\frac{d\theta_1}{dt}\frac{d_V}{2}\right)+\epsilon_0\frac{d_V}{2}\left(\frac{dw_1^0}{dt}-\frac{dw_1^{02}}{dt}+\frac{d\theta_1}{dt}\frac{d_V}{2}\right)$$
$$-\epsilon_v d_{\theta 1}\left(\frac{dw_2^0}{dt}-\frac{dw_1^0}{dt}+\frac{d\theta_2}{dt}d_{\theta 2}+\frac{d\theta_1}{dt}d_{\theta 1}\right)-\epsilon_{rot}\left(\frac{d\theta_2}{dt}-\frac{d\theta_1}{dt}\right)=0, \tag{6.6}$$

$$w_2^{01}(t)=w(vt+d_4,t),\ w_2^{02}(t)=w(vt+d_3,t),\ w_2^{03}(t)=w(vt+d_2,t),$$
$$w_1^{01}(t)=w(vt+d_1,t),\ w_1^{02}(t)=w(vt,t),$$
$$\lim_{|x-vt|\to\infty}w(x,t)=0,\quad \lim_{|x-vt|\to\infty}\varphi(x,t)=0, \tag{6.7}$$

in which, the following are the translatory displacements: $w(x, t)$ is the vertical deflection of the beam; $w_1^{01}(t)$, $w_1^{02}(t)$, $w_2^{01}(t)$, $w_2^{02}(t)$, and $w_2^{03}(t)$ are the vertical displacements of the masses; $w_1^{0}(t)$and $w_2^{0}(t)$ are the vertical displacements of the centers of mass of the bars, as shown in Fig. 6.1. The following are the angular displacements: $\varphi(x, t)$ is the angle of rotation of the cross-section of the beam; $\theta_1$and θ_2 are the angles of rotation of the bars around their centers of mass. The following are the physical properties: M_1, M_0, and J_1, J_0 are the masses and moments of inertia of the bars, respectively; k_f and ν_f are the stiffness and the viscosity of the foundation per unit length; k_0 and ϵ_0 are the stiffness and the damping factor of the suspension; k_v and ϵ_v are the stiffness and the damping factor of the elastic coupling in the transversal direction; k_{rot} and ϵ_{rot} are the rotational stiffness and the torsional damping of the elastic coupling. The Dirac delta function is designated with δ. Constants C_i, D_j, $i = 1, . ., 6; j = 1, . ., 5$; are obtained [10] and given in Appendix A 6.1. This high-order shear deformation model is chosen because it has better accuracy than a model based on the Euler– Bernoulli theory and does not require a shear correction factor, which would be the case with a model based on the Timoshenko theory. It can be applied to beams ranging from very thin to moderately thick. By introducing a moving reference system $\xi = x - vt$, Eqs. (6.1–6.6) can be rewritten in the form:

$$
\begin{aligned}
& C_1\left(v^2\frac{\partial^2 w(\xi,t)}{\partial\xi^2} - 2v\frac{\partial^2 w(\xi,t)}{\partial\xi\partial t} + \frac{\partial^2 w(\xi,t)}{\partial t^2}\right) \\
& - C_2\left(\frac{\partial\varphi(\xi,t)}{\partial\xi} + \frac{\partial^2 w(\xi,t)}{\partial\xi^2}\right) - C_3\left(v^2\frac{\partial^4 w(\xi,t)}{\partial\xi^4} - 2v\frac{\partial^4 w(\xi,t)}{\partial\xi^3\partial t} + \frac{\partial^4 w(\xi,t)}{\partial\xi^2\partial t^2}\right) \\
& + C_4\frac{\partial^4 w(\xi,t)}{\partial\xi^4} + C_5\left(v^2\frac{\partial^3\varphi(\xi,t)}{\partial\xi^3} - 2v\frac{\partial^3\varphi(\xi,t)}{\partial\xi^2\partial t} + \frac{\partial^3\varphi(\xi,t)}{\partial\xi\partial t^2}\right) \\
& + C_6\frac{\partial^3\varphi(\xi,t)}{\partial\xi^3} + k_f w + \nu_f\left(\frac{\partial w(\xi,t)}{\partial t} - v\frac{\partial w(\xi,t)}{\partial\xi}\right) \\
& = -\delta(\xi - d_4)\left[m\frac{d^2 w_2^{01}}{dt^2} + k_0\left(w_2^{01} - w_2^{0} + \theta_2\frac{d_L}{2}\right) + \epsilon_0\left(\frac{dw_2^{01}}{dt} - \frac{dw_2^{0}}{dt} + \frac{d\theta_2}{dt}\frac{d_L}{2}\right)\right] \\
& - \delta(\xi - d_3)\left[m\frac{d^2 w_2^{02}}{dt^2} + k_0\left(w_2^{02} - w_2^{0} - \theta_2 d_{22}\right) + \epsilon_0\left(\frac{dw_2^{02}}{dt} - \frac{dw_2^{0}}{dt} - \frac{d\theta_2}{dt}d_{22}\right)\right] \\
& - \delta(\xi - d_2)\left[m\frac{d^2 w_2^{03}}{dt^2} + k_0\left(w_2^{03} - w_2^{0} - \theta_2\frac{d_L}{2}\right) + \epsilon_0\left(\frac{dw_2^{03}}{dt} - \frac{dw_2^{0}}{dt} - \frac{d\theta_2}{dt}\frac{d_L}{2}\right)\right] \\
& - \delta(\xi - d_1)\left[m\frac{d^2 w_1^{01}}{dt^2} + k_0\left(w_1^{01} - w_1^{0} + \theta_1\frac{d_V}{2}\right) + \epsilon_0\left(\frac{dw_1^{01}}{dt} - \frac{dw_1^{0}}{dt} - \frac{d\theta_1}{dt}\frac{d_V}{2}\right)\right]
\end{aligned}
$$

$$- \delta(\xi)\left[m\frac{\mathrm{d}^2 w_1^{02}}{\mathrm{d}t^2} + k_0\left(w_1^{02} - w_1^0 - \theta_1\frac{d_V}{2}\right) + \epsilon_0\left(\frac{\mathrm{d}w_1^{02}}{\mathrm{dt}} - \frac{\mathrm{d}w_1^0}{\mathrm{dt}} - \frac{\mathrm{d}\theta_1}{\mathrm{dt}}\frac{d_V}{2}\right)\right], \tag{6.8}$$

$$D_1\left(v^2\frac{\partial^2\varphi(\xi,t)}{\partial\xi^2} - 2v\frac{\partial^2\varphi(\xi,t)}{\partial\xi\partial t} + \frac{\partial^2\varphi(\xi,t)}{\partial t^2}\right) - D_2\frac{\partial^2\varphi(\xi,t)}{\partial\xi^2}$$
$$+ D_3\left(\varphi(\xi,t) + \frac{\partial w(\xi,t)}{\partial\xi}\right)$$
$$+ D_4\left(v^2\frac{\partial^3 w(\xi,t)}{\partial\xi^3} - 2v\frac{\partial^3 w(\xi,t)}{\partial\xi^2\partial t} + \frac{\partial^3 w(\xi,t)}{\partial\xi\partial t^2}\right) + D_5\frac{\partial^3 w(\xi,t)}{\partial\xi^3} = 0, \tag{6.9}$$

$$M_2\frac{\mathrm{d}^2 w_2^0}{\mathrm{d}t^2} + k_0\left(w_2^0 - w_2^{01} - \theta_2\frac{d_L}{2}\right) + k_0\left(w_2^0 - w_2^{02} + \theta_2 d_{22}\right)$$
$$+ k_0\left(w_2^0 - w_2^{03} + \theta_2\frac{d_L}{2}\right)$$
$$- k_v\left(w_2^0 - w_1^0 + \theta_2 d_{\theta2} + \theta_1 d_{\theta1}\right) + \epsilon_0\left(\frac{\mathrm{d}w_2^0}{\mathrm{dt}} - \frac{\mathrm{d}w_2^{01}}{\mathrm{dt}} - \frac{\mathrm{d}\theta_2}{\mathrm{dt}}\frac{d_L}{2}\right)$$
$$+ \epsilon_0\left(\frac{\mathrm{d}w_2^0}{\mathrm{dt}} - \frac{\mathrm{d}w_2^{02}}{\mathrm{dt}} + \frac{\mathrm{d}\theta_2}{\mathrm{dt}}d_{22}\right) + \epsilon_0\left(\frac{\mathrm{d}w_2^0}{\mathrm{dt}} - \frac{\mathrm{d}w_2^{02}}{\mathrm{dt}} + \frac{\mathrm{d}\theta_2}{\mathrm{dt}}\frac{d_L}{2}\right) = 0, \tag{6.10}$$

$$J_2\frac{\mathrm{d}^2\theta_2}{\mathrm{d}t^2} - k_0\frac{d_L}{2}\left(w_2^0 - w_2^{01} - \theta_2\frac{d_L}{2}\right) + k_0 d_{22}\left(w_2^0 - w_2^{02} + \theta_2 d_{22}\right)$$
$$+ k_0\frac{d_L}{2}\left(w_2^0 - w_2^{03} + \theta_2\frac{d_L}{2}\right) - k_v d_{\theta2}\left(w_2^0 - w_1^0 + \theta_2 d_{\theta2} + \theta_1 d_{\theta1}\right)$$
$$+ k_{rot}\left(\theta_2 - \theta_1\right) - \epsilon_0\frac{d_L}{2}\left(\frac{\mathrm{d}w_2^0}{\mathrm{d}t} - \frac{\mathrm{d}w_2^{01}}{\mathrm{d}t} - \frac{\mathrm{d}\theta_2}{\mathrm{d}t}\frac{d_L}{2}\right)$$
$$+ \epsilon_0 d_{22}\left(\frac{\mathrm{d}w_2^0}{\mathrm{d}t} - \frac{\mathrm{d}w_2^{02}}{\mathrm{d}t} + \frac{\mathrm{d}\theta_2}{\mathrm{d}t}d_{22}\right) + \epsilon_0\frac{d_L}{2}\left(\frac{\mathrm{d}w_2^0}{\mathrm{d}t} - \frac{\mathrm{d}w_2^{03}}{\mathrm{d}t} + \frac{\mathrm{d}\theta_2}{\mathrm{d}t}\frac{d_L}{2}\right)$$
$$- \epsilon_v d_{\theta2}\left(\frac{\mathrm{d}w_2^0}{\mathrm{d}t} - \frac{\mathrm{d}w_1^0}{\mathrm{d}t} + \frac{\mathrm{d}\theta_2}{\mathrm{d}t}d_{\theta2} + \frac{\mathrm{d}\theta_1}{\mathrm{d}t}d_{\theta1}\right) + \epsilon_{rot}\left(\frac{\mathrm{d}\theta_2}{\mathrm{d}t} - \frac{\mathrm{d}\theta_1}{\mathrm{d}t}\right) = 0, \tag{6.11}$$

$$M_1\frac{\mathrm{d}^2 w_1^0}{\mathrm{d}t^2} + k_v\left(w_2^0 - w_1^0 + \theta_2 d_{\theta2} + \theta_1 d_{\theta1}\right) + k_0\left(w_1^0 - w_1^{01} - \theta_1\frac{d_V}{2}\right)$$
$$+ k_0\left(w_1^0 - w_1^{02} + \theta_1\frac{d_V}{2}\right) + \epsilon_v\left(\frac{\mathrm{d}w_2^0}{\mathrm{d}t} - \frac{\mathrm{d}w_1^0}{\mathrm{d}t} + \frac{\mathrm{d}\theta_2}{\mathrm{d}t}d_{\theta2} + \frac{\mathrm{d}\theta_1}{\mathrm{d}t}d_{\theta1}\right)$$

$$+\epsilon_0\left(\frac{dw_1^0}{dt}-\frac{dw_1^{01}}{dt}-\frac{d\theta_1}{dt}\frac{d_V}{2}\right)+\epsilon_0\left(\frac{dw_1^0}{dt}-\frac{dw_1^{02}}{dt}+\frac{d\theta_1}{dt}\frac{d_V}{2}\right)=0, \tag{6.12}$$

$$J_1\frac{d^2\theta_1}{dt^2}-k_0\frac{d_V}{2}\left(w_1^0-w_1^{01}-\theta_1\frac{d_V}{2}\right)+k_0\frac{d_V}{2}\left(w_1^0-w_1^{02}+\theta_1\frac{d_V}{2}\right)$$

$$-k_v d_{\theta 1}(w_2^0-w_1^0+\theta_2 d_{\theta 2}+\theta_1 d_{\theta 1})-k_{rot}(\theta_2-\theta_1)$$

$$-\epsilon_0\frac{d_V}{2}\left(\frac{dw_1^0}{dt}-\frac{dw_1^{01}}{dt}-\frac{d\theta_1}{dt}\frac{d_V}{2}\right)+\epsilon_0\frac{d_V}{2}\left(\frac{dw_1^0}{dt}-\frac{dw_1^{02}}{dt}+\frac{d\theta_1}{dt}\frac{d_V}{2}\right)$$

$$-\epsilon_v d_{\theta 1}\left(\frac{dw_2^0}{dt}-\frac{dw_1^0}{dt}+\frac{d\theta_2}{dt}d_{\theta 2}+\frac{d\theta_1}{dt}d_{\theta 1}\right)-\epsilon_{rot}\left(\frac{d\theta_2}{dt}-\frac{d\theta_1}{dt}\right)=0, \tag{6.13}$$

$$w_2^{01}(t)=w(d_4,t),\ w_2^{02}(t)=w(d_3,t),\ w_2^{03}(t)=w(d_2,t),$$

$$w_1^{01}(t)=w(d_1,t),\ w_1^{02}(t)=w(0,t),$$

$$\lim_{|\xi|\to\infty}w(x,t)=0,\ \lim_{|\xi|\to\infty}\varphi(x,t)=0, \tag{6.14}$$

To obtain the system's critical velocity [3], the dispersion equation has to be presented in the reference system that moves together with the rail vehicle. For the undamped case, the dispersion equation can be obtained in the following form:

$$k^6q_3+k^4q_2+k^2q_1+q_0=0, \tag{6.15}$$

where

$$q_0=D_3k_f,\ q_1=D_2k_f-(C_1D_3+D_1k_f)v^2,$$

$$q_2=C_1D_1v^4+C_2(D_2+D_5)+D_3(C_4-C_6)$$
$$-v^2[C_1D_2+C_2(D_1-D_4)+D_3(C_3+C_5)],$$

$$q_3=v^2(C_5D_5+C_6D_4-C_3D_2-C_4D_1)+v^4(C_3D_1+C_5D_4)+C_4D_2+C_6D_5.$$

The roots of the dispersion equation are real when the velocity of the moving mass is lower than the critical velocity. Besides, when the velocity is lower than the critical, the vibrations of the mass—that is moving on the beam and does not lose contact with the beam—are harmonic. If the vibrations of the mass do not disrupt the waves in the beam, then the mass can continue to vibrate harmonically. In mathematical terms, this all means that the wave numbers k that are found as the roots of the dispersion Eq. (6.15) may not be real. Using the discriminant of the dispersion Eq. (6.15), after some algebra, the critical velocity of the system can be determined:

$$v_{cr}=\sqrt{-\frac{\tilde{p}_1}{2}+\sqrt{\frac{\tilde{p}_1^2}{4}-\tilde{q}_1}}, \tag{6.16}$$

where v_{cr} represents the minimum phase velocity of waves in the beam (described by the Reddy–Bickford beam assumption) on the elastic foundation, and where

$$\widetilde{p}_1 = \frac{1}{2}[c_1 + \sqrt{c_1^2 - 4c_2 + 4\chi_1}],\ \widetilde{q}_1 = \frac{1}{2}\left[\chi_1 + \frac{c_1\chi_1 - 2c_3}{\sqrt{c_1^2 - 4c_2 + 4\chi_1}}\right]$$

$$\chi_1 = \left(\frac{\Delta_1 - \sqrt{\Delta_1^2 - 4\Delta_0^3}}{2}\right)^{1/3},\ \Delta_0 = b^2 - 3ac,\ \Delta_1 = 2b^3 - 9abc + 27a^2d.$$

Expressions for constants a, b, c, d, c_1, c_2, and c_3 are determined, but are somewhat hefty, and because of their size, they cannot be displayed here in this chapter. The authors can be contacted for these expressions. After the analytical determination of the system's critical velocity, the characteristic equation—for vibrations of the moving vehicle on the beam—can be analyzed by applying the integral transforms to the system of Eqs. (6.8–6.13):

$$W_s(\xi, s) = \int_0^\infty w(\xi, t)e^{-st}dt,\ \Phi_s(\xi, s) = \int_0^\infty \varphi(\xi, t)e^{-st}dt,$$

$$\hat{W}_{k,s}(k, s) = \int_{-\infty}^\infty W_s(\xi, s)e^{-ik\xi}d\xi,\ \hat{\Phi}_s(k, s) = \int_{-\infty}^\infty \Phi_s(\xi, s)e^{-ik\xi}d\xi,$$

$$W_1^{01}(s) = \int_0^\infty w_1^{01}(t)e^{-st}dt,\ W_1^{02}(s) = \int_0^\infty w_1^{02}(t)e^{-st}dt,$$

$$W_2^{01}(s) = \int_0^\infty w_2^{01}(t)e^{-st}dt,\ W_2^{02}(s) = \int_0^\infty w_2^{02}(t)e^{-st}dt,$$

$$W_2^{03}(s) = \int_0^\infty w_2^{03}(t)e^{-st}dt,\ W_1^{0}(s) = \int_0^\infty w_1^{0}(t)e^{-st}dt,$$

$$W_2^{0}(s) = \int_0^\infty w_2^{0}(t)e^{-st}dt,$$

$$\Theta_1 = \Theta_1(s) = \int_0^\infty \theta_1(t)e^{-st}dt,\ \Theta_2 = \Theta_2(s) = \int_0^\infty \theta_2(t)e^{-st}dt,\ i = \sqrt{-1}.$$

For the problem Eqs. (6.8–6.13), the trivial initial conditions can be assumed—since the governing equations are linear, the initial conditions do not influence the system's stability. Also, considering that $W_1^{02}(s) = W(0, s)$, $W_1^{01}(s) = W(d_1, s)$, $W_2^{03}(s) = W(d_2, s)$, $W_2^{02}(s) = W(d_3, s)$, and $W_2^{01}(s) = W(d_4, s)$, the application of the transforms results in the system of algebraic equations after the elimination $\hat{\Phi}_{si}(k, s)$ in the present form:

$$F(k, s)\hat{W}_{k,s}(k, s) =$$

$$e^{-id_2k}\left[-k_0W(d_2, s) - ms^2W(d_2, s) - s\varepsilon_0W(d_2, s)\right.$$

$$\left. + k_0W_2^0(s) + s\varepsilon_0W_2^0(s) + \frac{1}{2}\Theta_2k_0d_L + \frac{1}{2}\Theta_2s\varepsilon_0d_L\right]$$

$$
\begin{aligned}
&+ e^{-id_3k}\Big[-k_0W(d_3, s) - ms^2W(d_3, s) - s\epsilon_0W(d_3, s) \\
&+ k_0W_2^0(s) + s\epsilon_0W_2^0(s) - d_{11}\Theta_2k_0 + \frac{1}{2}\Theta_2k_0d_L + \frac{1}{2}\Theta_2s\epsilon_0d_L - d_{11}\Theta_2s\epsilon_0\Big] \\
&+ e^{-id_4k}\Big[-k_0W(d_4, s) - ms^2W(d_4, s) - s\epsilon_0W(d_4, s) + k_0W_2^0(s) \\
&+ s\epsilon_0W_2^0(s) - \frac{1}{2}\Theta_2k_0d_L - \frac{1}{2}\Theta_2s\epsilon_0d_L\Big] \\
&+ e^{-id_1k}\Big[-k_0W(d_1, s) - ms^2W(d_1, s) - s\epsilon_0W(d_1, s) \\
&+ k_0W_1^0(s) + s\epsilon_0W_1^0(s) - \frac{1}{2}d_1\Theta_1k_0 - \frac{1}{2}d_1\Theta_1s\epsilon_0\Big] \\
&- k_0W(0, s) - ms^2W(0, s) - s\epsilon_0W(0, s) + k_0W_1^0(s) \\
&+ s\epsilon_0W_1^0(s) + \frac{1}{2}d_1\Theta_1k_0 + \frac{1}{2}d_1\Theta_1s\epsilon_0,
\end{aligned} \tag{6.17}
$$

$$
\begin{aligned}
&\frac{1}{2}\Theta_2s\epsilon_0d_L - k_0W(d_3, s) - k_0W(d_4, s) - s\epsilon_0W(d_2, s) - s\epsilon_0W(d_3, s) - s\epsilon_0W(d_4, s) \\
&+ k_vW_1^0(s) - k_vW_2^0(s) + 3k_0W_2^0(s) + M_2s^2W_2^0(s) + s\epsilon_vW_1^0(s) - s\epsilon_vW_2^0(s) \\
&+ 3s\epsilon_0W_2^0(s) - d_{11}\Theta_2k_0 + \frac{1}{2}\Theta_2k_0d_L - \Theta_1d_{\theta1}k_v - \Theta_2d_{\theta2}k_v - k_0W(d_2, s) \\
&- \Theta_1sd_{\theta1}\epsilon_v - \Theta_2sd_{\theta2}\epsilon_v - d_{11}\Theta_2s\epsilon_0 = 0,
\end{aligned} \tag{6.18}
$$

$$
\begin{aligned}
&d_{11}k_0W(d_3, s) - \frac{1}{2}k_0d_LW(d_2, s) - \frac{1}{2}k_0d_LW(d_3, s) + \frac{1}{2}k_0d_LW(d_4, s) \\
&- \frac{1}{2}s\epsilon_0d_LW(d_2, s) - \frac{1}{2}s\epsilon_0d_LW(d_3, s) + \frac{1}{2}s\epsilon_0d_LW(d_4, s) + d_{11}s\epsilon_0W(d_3, s) \\
&+ \frac{1}{2}k_0d_LW_2^0(s) + d_{\theta2}k_vW_1^0(s) - d_{\theta2}k_vW_2^0(s) - d_{11}k_0W_2^0(s) \\
&+ \frac{1}{2}s\epsilon_0d_LW_2^0(s) + sd_{\theta2}\epsilon_vW_1^0(s) - sd_{\theta2}\epsilon_vW_2^0(s) - d_{11}s\epsilon_0W_2^0(s) \\
&+ d_{11}^2\Theta_2k_0 + \frac{3}{4}\Theta_2k_0d_L^2 - d_{11}\Theta_2k_0d_L - \Theta_1d_{\theta1}d_{\theta2}k_v - \Theta_2d_{\theta2}^2k_v \\
&+ \frac{3}{4}\Theta_2s\epsilon_0d_L^2 - d_{11}\Theta_2s\epsilon_0d_L - \Theta_1sd_{\theta1}d_{\theta2}\epsilon_v - \Theta_2sd_{\theta2}^2\epsilon_v \\
&+ d_{11}^2\Theta_2s\epsilon_0 + \Theta_2J_2s^2 - \Theta_1k_{\mathrm{rot}} + \Theta_2k_{\mathrm{rot}} - \Theta_1s\epsilon_{\mathrm{rot}} + \Theta_2s\epsilon_{\mathrm{rot}} = 0,
\end{aligned} \tag{6.19}
$$

$$- k_0 W(d_1, s) - s\epsilon_0 W(d_1, s) - k_0 W(0, s) - s\epsilon_0 W(0, s) - k_v W_1^0(s) + k_v W_2^0(s) + 2k_0 W_1^0(s) + M_1 s^2 W_1^0(s) - s\epsilon_v W_1^0(s) + s\epsilon_v W_2^0(s) + 2s\epsilon_0 W_1^0(s) + \Theta_1 d_{\theta 1} k_v + \Theta_2 d_{\theta 2} k_v + \Theta_1 s d_{\theta 1} \epsilon_v + \Theta_2 s d_{\theta 2} \epsilon_v = 0, \tag{6.20}$$

$$\frac{1}{2} d_1 k_0 W(d_1, s) - \frac{1}{2} d_1 k_0 W(0, s) - \frac{1}{2} d_1 s\epsilon_0 W(0, s) + \frac{1}{2} d_1 s\epsilon_0 W(d_1, s) + d_{\theta 1} k_v W_1^0(s) - d_{\theta 1} k_v W_2^0(s) + s d_{\theta 1} \epsilon_v W_1^0(s) - s d_{\theta 1} \epsilon_v W_2^0(s) + \frac{1}{2} d_1^2 \Theta_1 k_0 - \Theta_2 d_{\theta 1} d_{\theta 2} k_v - \Theta_1 d_{\theta 1}^2 k_v - \Theta_2 s d_{\theta 1} d_{\theta 2} \epsilon_v - \Theta_1 s d_{\theta 1}^2 \epsilon_v + \frac{1}{2} d_1^2 \Theta_1 s\epsilon_0 + \Theta_1 J_1 s^2 + \Theta_1 k_{\mathrm{rot}} - \Theta_2 k_{\mathrm{rot}} + \Theta_1 s\epsilon_{\mathrm{rot}} - \Theta_2 s\epsilon_{\mathrm{rot}} = 0 \tag{6.21}$$

where

$$F(k, s) = \Big[ik^3 D_1 v^3 v_f - ik^3 D_2 v v_f - 3k^2 s D_1 v^2 v_f + k^2 s D_2 v_f - 3iks^2 D_1 v v_f - ikD_3 v v_f + s^3 D_1 v_f + sD_3 v_f + k^6 C_3 D_1 v^4 - k^6 C_3 D_2 v^2 - k^6 C_4 D_1 v^2 + k^6 C_4 D_2 + k^6 C_5 D_4 v^4 + k^6 C_5 D_5 v^2 + k^6 C_6 D_4 v^2 + k^6 C_6 D_5 + 4ik^5 s C_3 D_1 v^3 - 2ik^5 s C_3 D_2 v - 2ik^5 s C_4 D_1 v + 4ik^5 s C_5 D_4 v^3 + 2ik^5 s C_5 D_5 v + 2ik^5 s C_6 D_4 v - 6k^4 s^2 C_3 D_1 v^2 + k^4 s^2 C_3 D_2 + k^4 s^2 C_4 D_1 - 6k^4 s^2 C_5 D_4 v^2 - k^4 s^2 C_5 D_5 - k^4 s^2 C_6 D_4 + k^4 C_1 D_1 v^4 - k^4 C_1 D_2 v^2 - k^4 C_2 D_1 v^2 + k^4 C_2 D_2 + k^4 C_2 D_4 v^2 + k^4 C_2 D_5 - k^4 C_3 D_3 v^2 + k^4 C_4 D_3 - k^4 C_5 D_3 v^2 - k^4 C_6 D_3 - 4ik^3 s^3 C_3 D_1 v - 4ik^3 s^3 C_5 D_4 v + 4ik^3 s C_1 D_1 v^3 - 2ik^3 s C_1 D_2 v - 2ik^3 s C_2 D_1 v + 2ik^3 s C_2 D_4 v - 2ik^3 s C_3 D_3 v - 2ik^3 s C_5 D_3 v + k^2 s^4 C_3 D_1 + k^2 s^4 C_5 D_4 - 6k^2 s^2 C_1 D_1 v^2 + k^2 s^2 C_1 D_2 + k^2 s^2 C_2 D_1 - k^2 s^2 C_2 D_4 + k^2 s^2 C_3 D_3 + k^2 s^2 C_5 D_3 - k^2 C_1 D_3 v^2 - k^2 D_1 k_f v^2 + k^2 D_2 k_f - 4iks^3 C_1 D_1 v - 2iks C_1 D_3 v - 2iks D_1 k_f v + s^4 C_1 D_1 + s^2 C_1 D_3 + s^2 D_1 k_f + D_3 k_f \Big] / [k^2 D_1(-v^2) + k^2 D_2 - 2iks D_1 v + s^2 D_1 + D_3]. \tag{6.22}$$

Starting from the system of Eqs. (6.17–6.21) and applying to it the inverse Fourier transform over k, then eliminating W_1^0, W_2^0, Θ_1, andΘ_2, and setting the values $\xi = 0$, $\xi = d_1$, $\xi = d_2$, $\xi = d_3$, $\xi = d_4$ subsequently, will produce a system of five algebraic equations—with respect to the Laplace-displacements of the contact points $\hat{W}_s(0, s)$, $\hat{W}_s(d_1, s)$, $\hat{W}_s(d_2, s)$, $\hat{W}_s(d_3, s)$, and $\hat{W}_s(d_4, s)$

$$[\mathbf{u}] \times [\hat{\boldsymbol{W}}]^{\mathrm{T}} = 0 \Leftrightarrow$$

$$\begin{bmatrix} u_{11} & \cdots & u_{15} \\ \vdots & \ddots & \vdots \\ u_{51} & \cdots & u_{55} \end{bmatrix} \times \left\lfloor \hat{W}_s(0,s) \;\; \hat{W}_s(d_1,s) \;\; \hat{W}_s(d_2,s) \;\; \hat{W}_s(d_3,s) \;\; \hat{W}_s(d_4,s) \right\rfloor^{T} = 0. \tag{6.23}$$

The obtained system of Eq. (6.23) is a system that contains 21 integrals, which is a consequence of the inverse Fourier transform. Here are the forms of the integrals:

$$\mathrm{Int}_0 = \frac{1}{2\pi}\int_{-\infty}^{\infty}\frac{\mathrm{d}k}{F(k,s)},\ \mathrm{Int}_{-d_1} = \frac{1}{2\pi}\int_{-\infty}^{\infty}\frac{e^{-ikd_1}\mathrm{d}k}{F(k,s)},\ \mathrm{Int}_{-d_2} = \frac{1}{2\pi}\int_{-\infty}^{\infty}\frac{e^{-ikd_2}\mathrm{d}k}{F(k,s)},$$

$$\mathrm{Int}_{-d_3} = \frac{1}{2\pi}\int_{-\infty}^{\infty}\frac{e^{-ikd_3}\mathrm{d}k}{F(k,s)},\ \mathrm{Int}_{-d_4} = \frac{1}{2\pi}\int_{-\infty}^{\infty}\frac{e^{-ikd_4}\mathrm{d}k}{F(k,s)},$$

$$\mathrm{Int}_{-d_1d_4} = \frac{1}{2\pi}\int_{-\infty}^{\infty}\frac{e^{ik(d_1-d_4)}\mathrm{d}k}{F(k,s)},\ \mathrm{Int}_{-d_1d_2} = \frac{1}{2\pi}\int_{-\infty}^{\infty}\frac{e^{ik(d_1-d_2)}\mathrm{d}k}{F(k,s)},$$

$$\mathrm{Int}_{+d_1} = \frac{1}{2\pi}\int_{-\infty}^{\infty}\frac{e^{ikd_1}\mathrm{d}k}{F(k,s)},\ \mathrm{Int}_{-d_1d_3} = \frac{1}{2\pi}\int_{-\infty}^{\infty}\frac{e^{ik(d_1-d_3)}\mathrm{d}k}{F(k,s)},$$

$$\mathrm{Int}_{-d_2d_4} = \frac{1}{2\pi}\int_{-\infty}^{\infty}\frac{e^{ik(d_2-d_4)}\mathrm{d}k}{F(k,s)},\ \mathrm{Int}_{-d_2d_3} = \frac{1}{2\pi}\int_{-\infty}^{\infty}\frac{e^{ik(d_2-d_3)}\mathrm{d}k}{F(k,s)},$$

$$\mathrm{Int}_{+d_2d_1} = \frac{1}{2\pi}\int_{-\infty}^{\infty}\frac{e^{ik(d_2-d_1)}\mathrm{d}k}{F(k,s)},\ \mathrm{Int}_{+d_2} = \frac{1}{2\pi}\int_{-\infty}^{\infty}\frac{e^{ikd_2}\mathrm{d}k}{F(k,s)},$$

$$\mathrm{Int}_{-d_3d_4} = \frac{1}{2\pi}\int_{-\infty}^{\infty}\frac{e^{ik(d_3-d_4)}\mathrm{d}k}{F(k,s)},\ \mathrm{Int}_{+d_3d_2} = \frac{1}{2\pi}\int_{-\infty}^{\infty}\frac{e^{ik(d_3-d_2)}\mathrm{d}k}{F(k,s)},$$

$$\mathrm{Int}_{+d_3d_1} = \frac{1}{2\pi}\int_{-\infty}^{\infty}\frac{e^{ik(d_3-d_1)}\mathrm{d}k}{F(k,s)},\ \mathrm{Int}_{+d_3} = \frac{1}{2\pi}\int_{-\infty}^{\infty}\frac{e^{ikd_3}\mathrm{d}k}{F(k,s)},$$

$$\mathrm{Int}_{+d_4d_1} = \frac{1}{2\pi}\int_{-\infty}^{\infty}\frac{e^{ik(d_4-d_1)}\mathrm{d}k}{F(k,s)},\ \mathrm{Int}_{+d_4d_2} = \frac{1}{2\pi}\int_{-\infty}^{\infty}\frac{e^{ik(d_4-d_2)}\mathrm{d}k}{F(k,s)},$$

$$\mathrm{Int}_{+d_4} = \frac{1}{2\pi}\int_{-\infty}^{\infty}\frac{e^{ikd_4}\mathrm{d}k}{F(k,s)},\ \mathrm{Int}_{+d_4d_3} = \frac{1}{2\pi}\int_{-\infty}^{\infty}\frac{e^{ik(d_4-d_3)}\mathrm{d}k}{F(k,s)}. \tag{6.24}$$

The system of Eqs. (6.17–6.21) does have a nontrivial solution if and only if the condition:

$$\det[\mathbf{u}] = 0 \tag{6.25}$$

is satisfied. By examining the eigenvalues of the characteristic Eq. (6.25), the stability of the model can be studied, and that is done and presented in the following sections.

6.3 Dynamic stability analysis

If there is at least one root of the characteristic Eq. (6.25) with a positive real part, then the vibrations of the vehicle are unstable—hence, that is the criterion for determining the instability of vibrations. This equation is an integral equation with respect to the Laplace variable s. To solve the presented mathematical problem more easily, the D-decomposition method is applied. Denisov et al. [6] are the first to have used and described the concept of this method, and since then other researchers have used it. In this study, the k_0-plane is where the D-decomposition is performed, and the positive half of the real axis in the k_0-plane is correlated with the suspension's stiffness. Taking into account Eq. (6.25), the mapping rule onto this plane satisfies the following equation:

$$k_0 = f(\Omega), \tag{6.26}$$

where the parameter Ω is varied from minus to plus infinity and for $s = i\Omega$. Function f can be found from the characteristic Eq. (6.25) where $F(k, \ \Omega)$ and the integrals now have the following forms:

$$\begin{gathered}
\mathrm{Int}_0 = \frac{1}{2\pi}\int_{-\infty}^{\infty}\frac{\mathrm{d}k}{F(k,\Omega)}, \ \mathrm{Int}_{-\mathrm{d}_1} = \frac{1}{2\pi}\int_{-\infty}^{\infty}\frac{e^{-ikd_1}\mathrm{d}k}{F(k,\Omega)}, \ \mathrm{Int}_{-\mathrm{d}_2} = \frac{1}{2\pi}\int_{-\infty}^{\infty}\frac{e^{-ikd_2}\mathrm{d}k}{F(k,\Omega)}, \\
\mathrm{Int}_{-\mathrm{d}_3} = \frac{1}{2\pi}\int_{-\infty}^{\infty}\frac{e^{-ikd_3}\mathrm{d}k}{F(k,\Omega)}, \ \mathrm{Int}_{-\mathrm{d}_4} = \frac{1}{2\pi}\int_{-\infty}^{\infty}\frac{e^{-ikd_4}\mathrm{d}k}{F(k,\Omega)}, \\
\mathrm{Int}_{-\mathrm{d}_1\mathrm{d}_4} = \frac{1}{2\pi}\int_{-\infty}^{\infty}\frac{e^{ik(d_1-d_4)}\mathrm{d}k}{F(k,\Omega)}, \ \mathrm{Int}_{-\mathrm{d}_1\mathrm{d}_2} = \frac{1}{2\pi}\int_{-\infty}^{\infty}\frac{e^{ik(d_1-d_2)}\mathrm{d}k}{F(k,\Omega)}, \\
\mathrm{Int}_{+\mathrm{d}_1} = \frac{1}{2\pi}\int_{-\infty}^{\infty}\frac{e^{ikd_1}\mathrm{d}k}{F(k,\Omega)}, \ \mathrm{Int}_{-\mathrm{d}_1\mathrm{d}_3} = \frac{1}{2\pi}\int_{-\infty}^{\infty}\frac{e^{ik(d_1-d_3)}\mathrm{d}k}{F(k,\Omega)}, \\
\mathrm{Int}_{-\mathrm{d}_2\mathrm{d}_4} = \frac{1}{2\pi}\int_{-\infty}^{\infty}\frac{e^{ik(d_2-d_4)}\mathrm{d}k}{F(k,\Omega)}, \ \mathrm{Int}_{-\mathrm{d}_2\mathrm{d}_3} = \frac{1}{2\pi}\int_{-\infty}^{\infty}\frac{e^{ik(d_2-d_3)}\mathrm{d}k}{F(k,\Omega)}, \\
\mathrm{Int}_{+\mathrm{d}_2\mathrm{d}_1} = \frac{1}{2\pi}\int_{-\infty}^{\infty}\frac{e^{ik(d_2-d_1)}\mathrm{d}k}{F(k,\Omega)}, \ \mathrm{Int}_{+\mathrm{d}_2} = \frac{1}{2\pi}\int_{-\infty}^{\infty}\frac{e^{ikd_2}\mathrm{d}k}{F(k,\Omega)}, \\
\mathrm{Int}_{-\mathrm{d}_3\mathrm{d}_4} = \frac{1}{2\pi}\int_{-\infty}^{\infty}\frac{e^{ik(d_3-d_4)}\mathrm{d}k}{F(k,\Omega)}, \ \mathrm{Int}_{+\mathrm{d}_3\mathrm{d}_2} = \frac{1}{2\pi}\int_{-\infty}^{\infty}\frac{e^{ik(d_3-d_2)}\mathrm{d}k}{F(k,\Omega)}, \\
\mathrm{Int}_{+\mathrm{d}_3\mathrm{d}_1} = \frac{1}{2\pi}\int_{-\infty}^{\infty}\frac{e^{ik(d_3-d_1)}\mathrm{d}k}{F(k,\Omega)}, \ \mathrm{Int}_{+\mathrm{d}_3} = \frac{1}{2\pi}\int_{-\infty}^{\infty}\frac{e^{ikd_3}\mathrm{d}k}{F(k,\Omega)}, \\
\mathrm{Int}_{+\mathrm{d}_4\mathrm{d}_1} = \frac{1}{2\pi}\int_{-\infty}^{\infty}\frac{e^{ik(d_4-d_1)}\mathrm{d}k}{F(k,\Omega)}, \ \mathrm{Int}_{+\mathrm{d}_4\mathrm{d}_2} = \frac{1}{2\pi}\int_{-\infty}^{\infty}\frac{e^{ik(d_4-d_2)}\mathrm{d}k}{F(k,\Omega)}, \\
\mathrm{Int}_{+\mathrm{d}_4} = \frac{1}{2\pi}\int_{-\infty}^{\infty}\frac{e^{ikd_4}\mathrm{d}k}{F(k,\Omega)}, \ \mathrm{Int}_{+\mathrm{d}_4\mathrm{d}_3} = \frac{1}{2\pi}\int_{-\infty}^{\infty}\frac{e^{ik(d_4-d_3)}\mathrm{d}k}{F(k,\Omega)},
\end{gathered} \tag{6.27}$$

$$
\begin{aligned}
F(k,\Omega) = \Big[& ik^3D_1v^3v_f - ik^3D_2vv_f - 3ik^2D_1v^2\Omega v_f + ik^2D_2\Omega v_f + 3ikD_1v\Omega^2v_f \\
& - ikD_3vv_f - iD_1\Omega^3v_f + iD_3\Omega v_f + k^6C_3D_1v^4 - k^6C_3D_2v^2 - k^6C_4D_1v^2 + k^6C_4D_2 \\
& + k^6C_5D_4v^4 + k^6C_5D_5v^2 + k^6C_6D_4v^2 + k^6C_6D_5 - 4k^5C_3D_1v^3\Omega + 2k^5C_3D_2v\Omega \\
& + 2k^5C_4D_1v\Omega - 4k^5C_5D_4v^3\Omega - 2k^5C_5D_5v\Omega - 2k^5C_6D_4v\Omega + k^4C_1D_1v^4 - k^4C_1D_2v^2 \\
& - k^4C_2D_1v^2 + k^4C_2D_2 + k^4C_2D_4v^2 + k^4C_2D_5 + 6k^4C_3D_1v^2\Omega^2 - k^4C_3D_2\Omega^2 \\
& - k^4C_3D_3v^2 - k^4C_4D_1\Omega^2 + k^4C_4D_3 - k^4C_5D_3v^2 + 6k^4C_5D_4v^2\Omega^2 + k^4C_5D_5\Omega^2 \\
& - k^4C_6D_3 + k^4C_6D_4\Omega^2 - 4k^3C_1D_1v^3\Omega + 2k^3C_1D_2v\Omega + 2k^3C_2D_1v\Omega - 2k^3C_2D_4v\Omega \\
& - 4k^3C_3D_1v\Omega^3 + 2k^3C_3D_3v\Omega + 2k^3C_5D_3v\Omega - 4k^3C_5D_4v\Omega^3 + 6k^2C_1D_1v^2\Omega^2 \\
& - k^2C_1D_2\Omega^2 - k^2C_1D_3v^2 - k^2C_2D_1\Omega^2 + k^2C_2D_4\Omega^2 + k^2C_3D_1\Omega^4 \\
& - k^2C_3D_3\Omega^2 - k^2C_5D_3\Omega^2 + k^2C_5D_4\Omega^4 - k^2D_1k_fv^2 + k^2D_2k_f \\
& - 4kC_1D_1v\Omega^3 + 2kC_1D_3v\Omega + 2kD_1k_fv\Omega + C_1D_1\Omega^4 \\
& - C_1D_3\Omega^2 - D_1k_f\Omega^2 + D_3k_f \Big] / [k^2D_1(-v^2) + k^2D_2 + 2kD_1v\Omega - D_1\Omega^2 + D_3].
\end{aligned}
\tag{6.28}
$$

It is feasible to find the function $f(\Omega)$ after the integrals are evaluated by the contour integration method presented in Section 1.6 and Ref. [7]. According to the applied contours of integration, which are shown in Fig. 6.2, the outcome of integration is the following set:

$$
\mathrm{Int}_0 = i\sum_{\mathrm{n}} \frac{k-k_n}{(k-k_1)\ldots(k-k_6)}\bigg|_{k=k_n}, \quad \mathrm{Int}_{-\mathrm{d}_1} = -i\sum_{\mathrm{n}} \frac{(k-k_m)e^{-ikd_1}}{(k-k_1)\ldots(k-k_6)}\bigg|_{k=k_m},
$$

$$
\mathrm{Int}_{-\mathrm{d}_2} = -i\sum_{\mathrm{n}} \frac{(k-k_m)e^{-ikd_2}}{(k-k_1)\ldots(k-k_6)}\bigg|_{k=k_m}, \quad \mathrm{Int}_{-\mathrm{d}_3} = -i\sum_{\mathrm{n}} \frac{(k-k_m)e^{-ikd_3}}{(k-k_1)\ldots(k-k_6)}\bigg|_{k=k_m},
$$

$$
\mathrm{Int}_{-\mathrm{d}_4} = -i\sum_{\mathrm{n}} \frac{(k-k_m)e^{-ikd_4}}{(k-k_1)\ldots(k-k_6)}\bigg|_{k=k_m},
$$

$$
\mathrm{Int}_{-\mathrm{d}_1\mathrm{d}_4} = -i\sum_{\mathrm{n}} \frac{(k-k_m)e^{ik(d_1-d_4)}}{(k-k_1)\ldots(k-k_6)}\bigg|_{k=k_m},
$$

$$
\mathrm{Int}_{-\mathrm{d}_1\mathrm{d}_2} = -i\sum_{\mathrm{n}} \frac{(k-k_m)e^{ik(d_1-d_2)}}{(k-k_1)\ldots(k-k_6)}\bigg|_{k=k_m},
$$

$$\mathrm{Int}_{+\mathrm{d}_1} = i\sum_{\mathrm{n}} \frac{(k-k_n)e^{ikd_1}}{(k-k_1)\ldots(k-k_6)}\Bigg|_{k=k_n},$$

$$\mathrm{Int}_{-\mathrm{d}_1\mathrm{d}_3} = -i\sum_{\mathrm{n}} \frac{(k-k_m)e^{ik(d_1-d_3)}}{(k-k_1)\ldots(k-k_6)}\Bigg|_{k=k_m},$$

$$\mathrm{Int}_{-\mathrm{d}_2\mathrm{d}_4} = -i\sum_{\mathrm{n}} \frac{(k-k_m)e^{ik(d_2-d_4)}}{(k-k_1)\ldots(k-k_6)}\Bigg|_{k=k_m},$$

$$\mathrm{Int}_{-\mathrm{d}_2\mathrm{d}_3} = -i\sum_{\mathrm{n}} \frac{(k-k_m)e^{ik(d_2-d_3)}}{(k-k_1)\ldots(k-k_6)}\Bigg|_{k=k_m},$$

$$\mathrm{Int}_{+\mathrm{d}_2\mathrm{d}_1} = i\sum_{\mathrm{n}} \frac{(k-k_n)e^{ik(d_2-d_1)}}{(k-k_1)\ldots(k-k_6)}\Bigg|_{k=k_n},$$

$$\mathrm{Int}_{+\mathrm{d}_2} = i\sum_{\mathrm{n}} \frac{(k-k_n)e^{ikd_2}}{(k-k_1)\ldots(k-k_6)}\Bigg|_{k=k_n},$$

$$\mathrm{Int}_{-\mathrm{d}_3\mathrm{d}_4} = -i\sum_{\mathrm{n}} \frac{(k-k_m)e^{ik(d_3-d_4)}}{(k-k_1)\ldots(k-k_6)}\Bigg|_{k=k_m},$$

$$\mathrm{Int}_{+\mathrm{d}_3\mathrm{d}_2} = i\sum_{\mathrm{n}} \frac{(k-k_n)e^{ik(d_3-d_2)}}{(k-k_1)\ldots(k-k_6)}\Bigg|_{k=k_n},$$

$$\mathrm{Int}_{+\mathrm{d}_3\mathrm{d}_1} = i\sum_{\mathrm{n}} \frac{(k-k_n)e^{ik(d_3-d_1)}}{(k-k_1)\ldots(k-k_6)}\Bigg|_{k=k_n},$$

$$\mathrm{Int}_{+\mathrm{d}_3} = i\sum_{\mathrm{n}} \frac{(k-k_n)e^{ikd_3}}{(k-k_1)\ldots(k-k_6)}\Bigg|_{k=k_n}, \quad \mathrm{Int}_{+\mathrm{d}_4\mathrm{d}_1} = i\sum_{\mathrm{n}} \frac{(k-k_n)e^{ik(d_4-d_1)}}{(k-k_1)\ldots(k-k_6)}\Bigg|_{k=k_n},$$

$$\mathrm{Int}_{+\mathrm{d}_4\mathrm{d}_2} = i\sum_{\mathrm{n}} \frac{(k-k_n)\mathrm{e}^{ik(d_4-d_2)}}{(k-k_1)\ldots(k-k_6)}\Bigg|_{k=k_n}, \quad \mathrm{Int}_{+\mathrm{d}_4} = i\sum_{\mathrm{n}} \frac{(k-k_n)\mathrm{e}^{ikd_4}}{(k-k_1)\ldots(k-k_6)}\Bigg|_{k=k_n},$$

$$\mathrm{Int}_{+\mathrm{d}_4\mathrm{d}_3} = i\sum_{\mathrm{n}} \frac{(k-k_n)e^{ik(d_4-d_3)}}{(k-k_1)\ldots(k-k_6)}\Bigg|_{k=k_n}, \tag{6.29}$$

in which k_n are the roots with a positive imaginary part, and k_m are the roots with a negative imaginary part of equation $F(k,\ \Omega) = 0$. The analysis of the D-decomposition curves can now be done. Five curves should be examined; each one of them is correlated with one of the five roots of Eq. (6.26). The parametric analysis leads to the finding that the D-decomposition curves are of interest for dynamic analysis for the case in which the velocity is in the supercritical speed region, that is, when the complex vehicle system is moving at a velocity that is greater than a certain critical velocity v_{cr}. When the speed is in the subcritical region ($v < v_{cr}$), the D-decomposition curves do not cross the positive part of the real axes at any point. This means that when $v < v_{cr}$, the number of unstable roots of the characteristic equation (roots with a positive real part) does not vary with the stiffness of the suspension (k_0). However, when the $v > v_{cr}$, the number of unstable roots will change when k_0 surpasses the certain critical value of the stiffness. The use of the D-decomposition method allows the ascertainment of a relative variation of the number of unstable roots depending on the stiffness k_0, but not the number itself. Thus finding the number of unstable roots for any value of the stiffness k_0, is what remains to be done. This can be achieved by using the argument principle [7]. To determine the number of unstable roots, the chosen contour should encircle the right half-plane of the complex variable s, as illustrated in Fig. 6.4.

In the upcoming analysis, the complex rail vehicle system is assumed to be composed of a locomotive and wagon that are connected via elastic coupling. Car bodies are represented by rigid bars, beam represents rail, and viscoelastic foundation represents ground. The following set of model parameters was used to obtain the D-decomposition curves in this study:

$$\rho = 7849\,\text{kg},\ A = 7.0687 \times 10^{-3}\text{m}^2,$$

$$I = 3.055 \times 10^{-5}\,\text{m}^4,\ E = 2 \times 10^{11}\text{Pa},$$

$$G = 7.813 \times 10^{10}\text{Pa},\ k_f = 10^8 \frac{\text{N}}{\text{m}^2},\ \nu_f = 10^8 \frac{\text{N·s}}{\text{m}^2},$$

$$J_1 = 598 \times 10^3 \frac{\text{kg·m}}{\text{s}^2},\ J_2 = \eta J_1,\ M_1 = 20 \times 10^3\,\text{kg},\ M_2 = \zeta M_1,$$

$$\eta = \zeta = (0.5; 1; 2),\ m = 10^3\,\text{kg},\ \varepsilon_0 = 86 \times 10^3 \frac{\text{N·s}}{\text{m}},$$

$$d_V = d_L = 15.7\,\text{m},\ d_R = 1\,\text{m},\ d_{\theta 1} = d_{\theta 2} = 8.37\,\text{m}. \tag{6.30}$$

6.4 Influence of the elastic coupling on dynamic stability

This section is about the analysis of the influence of elastic coupling on dynamic stability. The elastic coupling, which links the bars (i.e., the rail cars), is defined by the parameters k_v, k_{rot}, ε_v, andε_{rot}. The D-decomposition curves and the number of unstable roots are determined for varying values of these parameters. The D-decomposition curves for one pair (k_{v1}, k_{rot1}; k_{v2}, k_{rot2}) of the chosen parameters that are given in Table 6.1 and for the supercritical velocity $v = 1.12\, v_{cr}$, can be seen in Figs. 6.4 and 6.5. In these figures, the D-decomposition curves are displayed separately to show the critical stiffness values without numbering the unstable roots due to the return D-decomposition lines of the other roots. Besides, an integrated

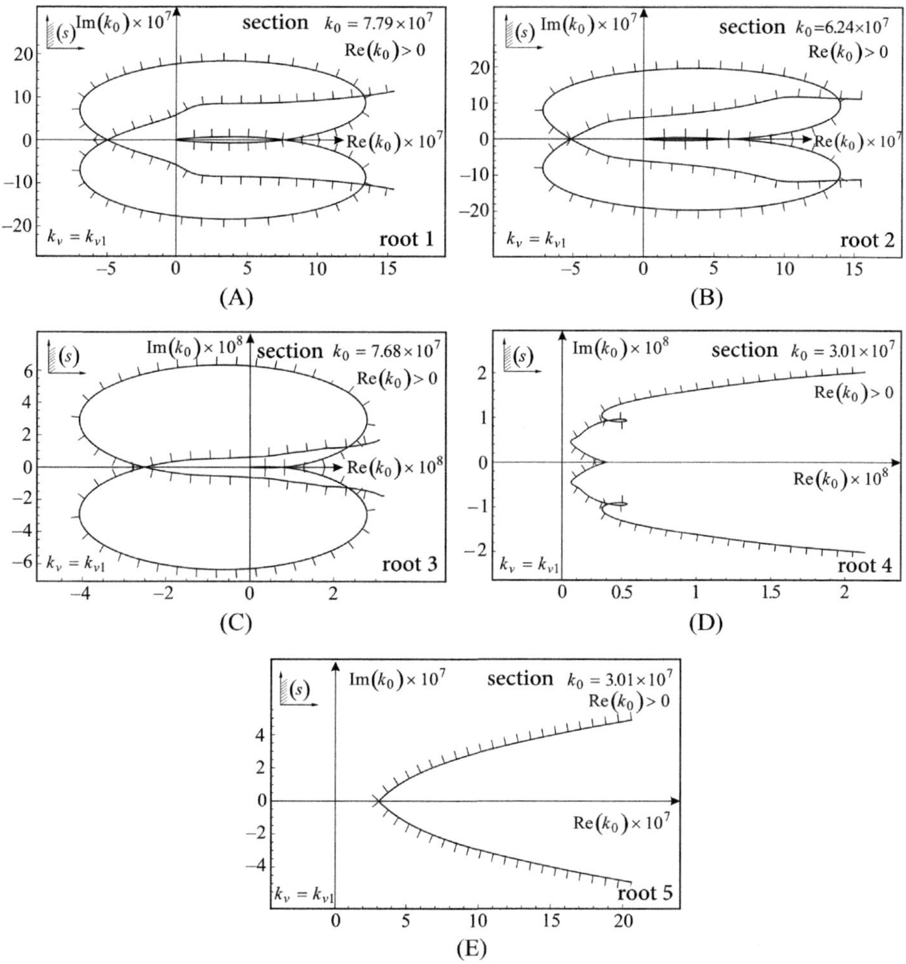

Figure 6.4 D-decomposition curves for k_v=k_{v1} (N/m). (A) root 1; (B) root 2; (C) root 3; (D) root 4; (E) root 5. The parameters and stability domains are presented in Table 6.1.

diagram with all the curves put in one place is difficult to analyze. Integrated diagrams for the chosen parameters that are given in Tables 6.1–6.3 can be seen in Fig. 6.6. For all the examples, the number of unstable roots is given in Tables 6.1–6.8. As can be seen (Tables 6.1–6.3), the elastic coupling with smaller stiffness in the transverse direction k_v (and damping $\varepsilon_v \in (0,\ 8.6)$ kN·s/m) increases the stability region for the stiffness region $k_0 \in (2.74,\ 3.01) \times 10^7$N/m. However, in engineering practice, these couplings are usually rigid. So, for the case of higher stiffness (simulation of rigid connection) of the elastic coupling in the transverse direction k_v and damping $\varepsilon_v = 10$ N·s/m; $\varepsilon_v = 8600$ N·s/m, stability region appears for $k_0 \in (4.17,\ 6.24) \times 10^7$N/m; $k_0 \in (4.08,\ 6.24) \times 10^7$N/m, respectively. This finding opens a possibility that allows the region of stability to be changed by

Table 6.1 The number of unstable roots. Influence of stiffness k_v on stability regions; ε_v=10 (N·s)/m, v=1.12v_{cr}, k_{v1}=2×10^{10} N/m, k_{v2}=10^5 N/m, k_{rot1}=k_{rot2}=10 (N·m)/rad, ε_{rot}=10 (N·m·s)/rad, unit for k_0 is 10^7 N/m.

<table>
<tr><th>k_0</th><th>0</th><th>0.016</th><th>2.74</th><th>3.01</th><th>4.17</th><th>5.11</th><th>6.24</th><th>7.68</th><th>7.79</th><th>9.01</th><th>9.55</th><th>9.99</th><th>21.6</th><th>+∞</th></tr>
<tr><td>k_{v1}, k_{rot1}</td><td>2</td><td>2|2</td><td>2|2</td><td>2|0</td><td>0|0</td><td>0|0</td><td>0|2</td><td>2|4</td><td>4|6</td><td>6|6</td><td>6|6</td><td>6|6</td><td>6|6</td><td>6</td></tr>
<tr><td>k_{v2}, k_{rot2}</td><td>3</td><td>3|2</td><td>2|0</td><td>0|0</td><td>0|2</td><td>2|4</td><td>4|4</td><td>4|4</td><td>4|4</td><td>4|2</td><td>2|4</td><td>4|6</td><td>6|7</td><td>7</td></tr>
</table>

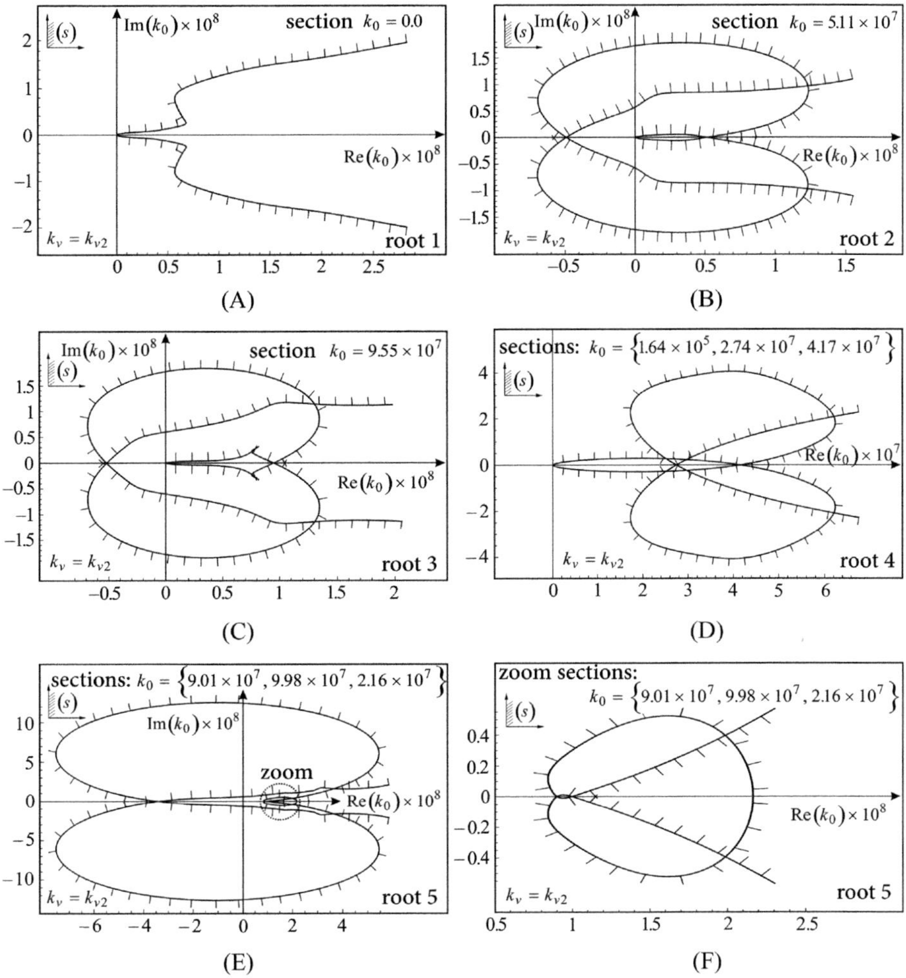

Figure 6.5 D-decomposition curves for k_v=k_{v2} (N/m). (A) root 1; (B) root 2; (C) root 3; (D) root 4; (E) root 5; (F) magnified segment of root 5. The parameters and stability domains are presented in Table 6.1.

changing the parameters of the elastic coupling. For practical applications, this is important because the stiffness of the suspension k_0 cannot be changed, and the system can become unstable when a certain desired velocity is reached. If this occurs, by modifying the parameters of the elastic coupling, the system can become stable again for the new velocity.

The influence of the rotational stiffness of the coupling k_{rot} is shown in Table 6.4; the D-decomposition curves for all the roots are shown in Fig. 6.7. As the numbers of roots for each case are determined, it can be seen that the stable zones appear for different regions of the suspension's stiffness k_0. In conjunction with the coupling's

Table 6.2 The number of unstable roots. Influence of stiffness k_v on stability regions; ε_v=8.6 (kN·s)/m, $v=1.12v_{cr}$, $k_{v1}=2\times10^{10}$ N/m, $k_{v2}=10^5$ N/m, $k_{rot1}=k_{rot2}$=10 (N·m)/rad, ε_{rot}=10 (N·m·s)/rad, unit for k_0 is 10^7 N/m.

<table>
<tr><th>k_0</th><th>0</th><th>0.016</th><th>2.74</th><th>3.01</th><th>4.08</th><th>4.83</th><th>6.24</th><th>7.68</th><th>7.79</th><th>9.06</th><th>9.47</th><th>10.2</th><th>21.6</th><th>+∞</th></tr>
<tr><td>k_{v1}, k_{rot1}</td><td>2</td><td>2|2</td><td>2|2</td><td>2|0</td><td>0|0</td><td>0|0</td><td>0|2</td><td>2|4</td><td>4|6</td><td>6|6</td><td>6|6</td><td>6|6</td><td>6|6</td><td>6</td></tr>
<tr><td>k_{v2}, k_{rot2}</td><td>3</td><td>3|2</td><td>2|0</td><td>0|0</td><td>0|2</td><td>2|4</td><td>4|4</td><td>4|4</td><td>4|4</td><td>4|2</td><td>2|4</td><td>4|6</td><td>6|7</td><td>7</td></tr>
</table>

Table 6.3 The number of unstable roots. Influence of stiffness k_v on stability regions; ε_v=86 (kN·s)/m, v=1.12v_{cr}, k_{v1}=2×10^{10} N/m, k_{v2}=10^5 N/m, k_{rot1}=k_{rot2}=10 (N·m)/rad, ε_{rot}=10 (N·m·s)/rad, unit for k_0 is 10^7 N/m.

<table>
<tr><th>k_0</th><th>0</th><th>0.016</th><th>2.72</th><th>3.01</th><th>3.50</th><th>6.24</th><th>7.68</th><th>7.79</th><th>8.82</th><th>10.6</th><th>11.1</th><th>21.6</th><th>+∞</th></tr>
<tr><td>k_{v1}, k_{rot1}</td><td>2</td><td>2|2</td><td>2|2</td><td>2|0</td><td>0|0</td><td>0|2</td><td>2|4</td><td>4|6</td><td>6|6</td><td>6|6</td><td>6|6</td><td>6|6</td><td>6</td></tr>
<tr><td>k_{v2}, k_{rot2}</td><td>3</td><td>3|4</td><td>4|2</td><td>2|2</td><td>2|4</td><td>4|4</td><td>4|4</td><td>4|4</td><td>4|6</td><td>6|4</td><td>4|6</td><td>6|7</td><td>7</td></tr>
</table>

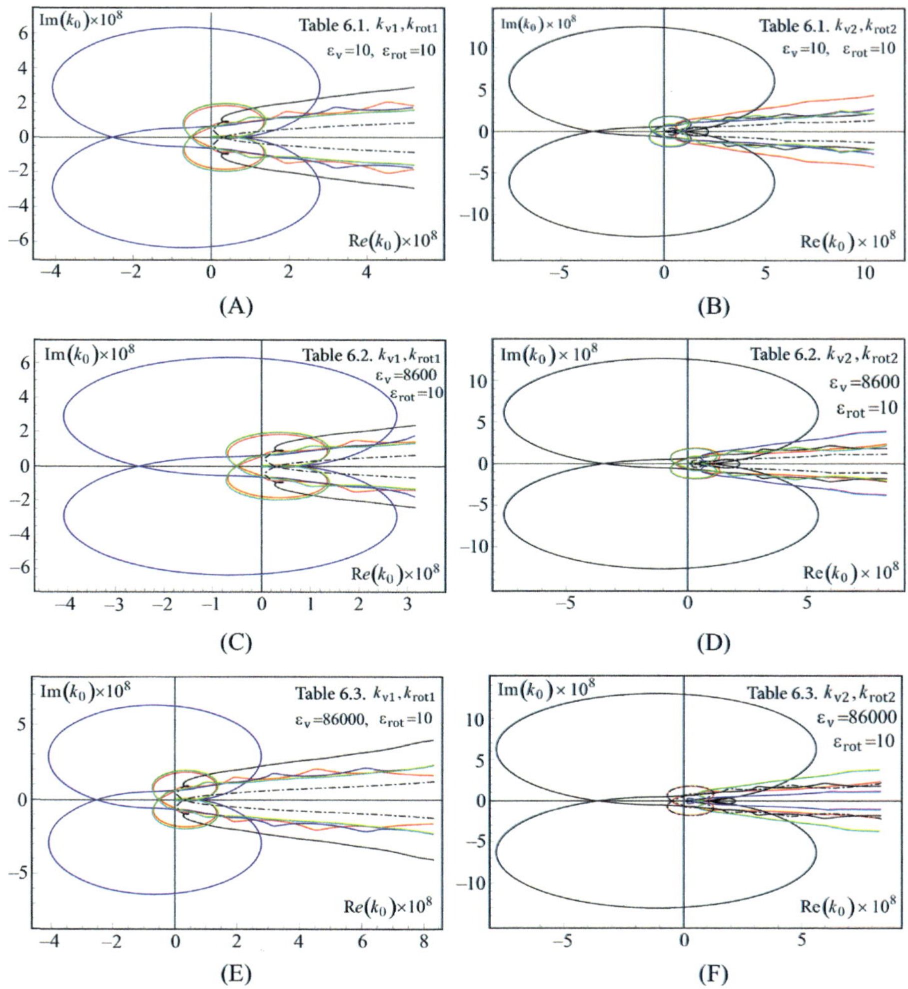

Figure 6.6 D-decomposition curves (N/m). The parameters and stability domains are presented in Tables 6.1–6.3. (A) Table 6.1, first row; (B) Table 6.1, second row; (C) Table 6.2 first row; (D) Table 6.2, second row; (E) Table 6.3, first row; (F) Table 6.3, second row.

transverse stiffness k_v, it is possible to change the locations of the regions of stability in relation to the unchangeable value of k_0.

For the chosen parameters of the elastic coupling given in Tables 6.5–6.7 and for the supercritical velocity $v = 1.39\ v_{cr}$, the D-decomposition curves for all the roots are presented in Fig. 6.8. An increase in velocity leads to a decrease in stability, that is, the stable regions. Owning to the high velocity, this outcome was expected.

The D-decomposition curves that are delineated in Fig. 6.9 are for the supercritical velocity $v = 1.39\ v_{cr}$ and the varying rotational stiffness k_{rot} of the elastic coupling

Table 6.4 The number of unstable roots. Influence of stiffness k_{rot} on stability regions; $v=1.12v_{cr}$, $k_{v1}=2\times10^{10}$ N/m, $k_{v2}=10^{5}$ N/m, $k_{rot1}=2\times10^{10}$ (N·m)/rad, $k_{rot2}=10^{5}$ (N·m)/rad, $\varepsilon_v=10$ (N·s)/m, $\varepsilon_{rot}=10$ (N·m·s)/rad, unit for k_0 is 10^{7} N/m.

k_0	0	0.01	1.4	2.2	2.9	3.0	3.4	4.1	6.1	6.2	6.8	6.85	7.6	7.7	8.7	9.6	20.0	+∞
k_{v1}, k_{rot2}	2	2|2	2|2	2|0	0|0	0|2	2|2	2|2	2|2	2|4	4|4	4|4	4|6	6|6	6|6	6|6	6|6	6
k_{v2}, k_{rot1}	3	3|2	2|2	2|2	2|0	0|0	0|0	0|2	2|2	2|2	2|4	4|4	4|4	4|4	4|2	2|4	4|5	5
k_{v1}, k_{rot1}	2	2|2	2|2	2|2	2|2	2|2	2|0	0|0	0|2	2|2	2|2	2|4	4|4	4|4	4|4	4|4	4|4	4

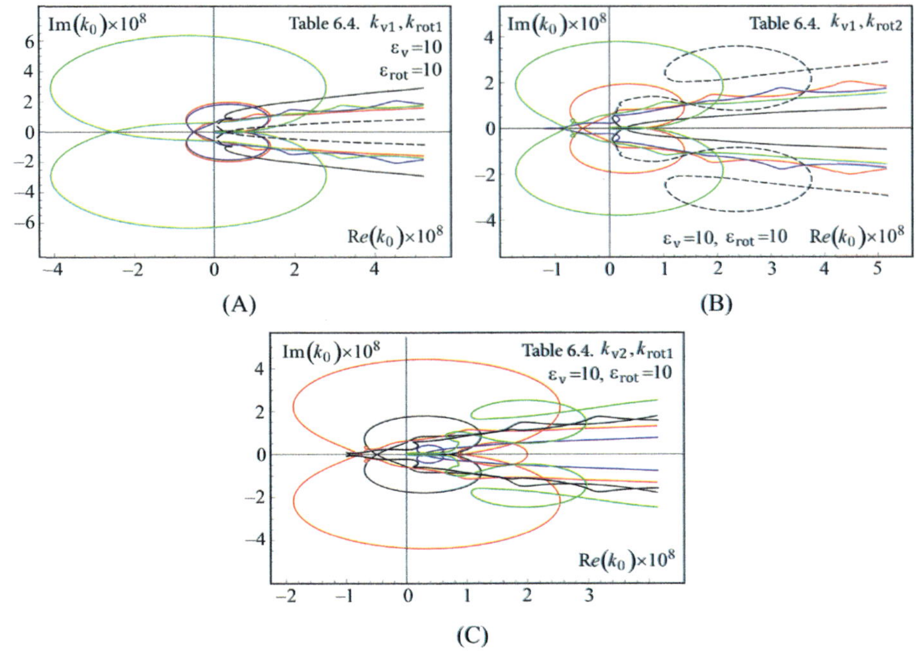

Figure 6.7 D-decomposition curves (N/m). The parameters and stability domains are presented in Table 6.4. (A) Table 6.4, first row; (B) Table 6.4, second row; (C) Table 6.4, third row.

given in Table 6.8. The same effect arises as in the aforementioned case (Tables 6.5–6.7) but with some new instability zones.

6.5 Influence of the mass ratio between the connected rail cars on dynamic stability

In this section, the influence of two masses M_1 and M_2 (the mass of the wagon and the locomotive, respectively) is analyzed. When $M_2 > M_1$, it is noticeable (Tables 6.9 and 6.10) that the system is more unstable than in the case when $M_2 = M_1$. In the opposite scenario ($M_1 > M_2$), the overall conclusion is the same, but for some other stiffness regions. For the compared cases of $M_2 < M_1$ and $M_2 = M_1$, the stable zone occurs for the interval $k_0 \in (2.74,\ 3.82) \times 10^7$N/m. For the compared cases of $M_2 > M_1$ and $M_2 = M_1$, the stable zone occurs for the interval $k_0 \in (2.82,\ 3.85) \times 10^7$N/m. The D-decomposition curves in Fig. 6.10 are for the supercritical velocity $v = 1.12\ v_{cr}$, two different mass ratios and parameters of the elastic coupling are given in Tables 6.9 and 6.10.

Table 6.5 The number of unstable roots. Influence of stiffness k_v on stability regions; ε_v=10 (N·s)/m, v=1.39v_{cr}, k_{v1}=2×10^{10} N/m, k_{v2}=10^5 N/m, k_{rot1}=k_{rot2}=10^{10} (N·m)/rad, ε_{rot}=10 (N·m·s)/rad, unit for k_0 is 10^7 N/m.

<table>
<tr><th>k_0</th><th>0</th><th>0.016</th><th>6.705</th><th>10.12</th><th>10.34</th><th>10.44</th><th>11.86</th><th>12.24</th><th>14.08</th><th>14.10</th><th>15.34</th><th>17.46</th><th>18.96</th><th>24.30</th><th>26.06</th><th>36.25</th><th>42.55</th><th>42.57</th><th>42.93</th><th>44.27</th><th>45.27</th><th>46.43</th><th>49.77</th><th>72.94</th><th>+∞</th></tr>
<tr><td>k_{v1}, k_{rot1}</td><td>8</td><td>8|8</td><td>8|8</td><td>8|10</td><td>10|10</td><td>10|10</td><td>10|10</td><td>10|10</td><td>10|11</td><td>11|9</td><td>9|11</td><td>11|11</td><td>11|12</td><td>12|12</td><td>12|10</td><td>10|8</td><td>8|8</td><td>8|6</td><td>6|6</td><td>6|6</td><td>6|8</td><td>8|6</td><td>6|8</td><td>8|6</td><td>6</td></tr>
<tr><td>k_{v2}, k_{rot2}</td><td>9</td><td>9|8</td><td>8|10</td><td>10|10</td><td>10|12</td><td>12|14</td><td>14|12</td><td>12|10</td><td>10|10</td><td>10|10</td><td>10|10</td><td>10|11</td><td>11|11</td><td>11|13</td><td>13|13</td><td>13|13</td><td>13|11</td><td>11|11</td><td>11|9</td><td>9|7</td><td>7|7</td><td>7|7</td><td>7|7</td><td>7|7</td><td>7</td></tr>
</table>

Table 6.6 The number of unstable roots. Influence of stiffness k_v on stability regions; ε_v=8.6 (kN·s)/m, v=1.39v_{cr}, k_{v1}=2×10^{10} N/m, k_{v2}=10^5 N/m, k_{rot1}=k_{rot2}=10^{10} (N·m)/rad, ε_{rot}=10 (N·m·s)/rad, unit for k_0 is 10^7 N/m.

k_0	0	0.016	6.879	10.10	10.12	10.26	11.86	12.23	14.08	14.10	15.34	17.46	18.96	24.27	26.06	36.25	42.56	42.57	42.94	44.35	45.27	46.43	49.77	72.94	+∞
k_{v1}, k_{rot1}	8	8|8	8|8	8|10	10|10	10|10	10|10	10|10	10|11	11|9	9|11	11|11	11|12	12|12	12|10	10|8	8|8	8|6	6|6	6|6	6|8	8|6	6|8	8|6	6
k_{v2}, k_{rot2}	9	9|8	8|10	10|10	10|12	12|14	14|12	12|10	10|10	10|10	10|10	10|11	11|11	11|13	13|13	13|13	13|11	11|11	11|9	9|7	7|7	7|7	7|7	7|7	7

Table 6.7 The number of unstable roots. Influence of stiffness k_v on stability regions; ε_v=86 (kN·s)/m, $v=1.39v_{cr}$, $k_{v1}=2\times10^{10}$ N/m, $k_{v2}=10^5$ N/m, $k_{rot1}=k_{rot2}=10^{10}$ (N·m)/rad, ε_{rot}=10 (N·m·s)/rad, unit for k_0 is 10^7 N/m.

k_0	0	0.016	0.689	1.819	9.051	9.438	10.12	11.58	11.86	14.08	14.10	15.34	17.46	18.96	24.11	26.06	36.26	42.55	42.57	42.94	44.90	45.27	46.43	49.77	72.94	+∞
k_{v1}, k_{rot1}	8	8∣8	8∣8	8∣8	8∣8	8∣8	8∣10	10∣10	10∣10	10∣10	10∣10	10∣11	11∣11	11∣9	9∣9	9∣11	11∣12	12∣12	12∣10	10∣10	10∣10	10∣8	8∣6	6∣8	8∣6	6
k_{v2}, k_{rot2}	9	9∣10	10∣8	8∣10	10∣12	12∣14	14∣14	14∣12	12∣10	10∣10	10∣10	10∣10	10∣11	11∣11	11∣13	13∣13	13∣13	13∣11	11∣11	11∣9	9∣7	7∣7	7∣7	7∣7	7∣7	7

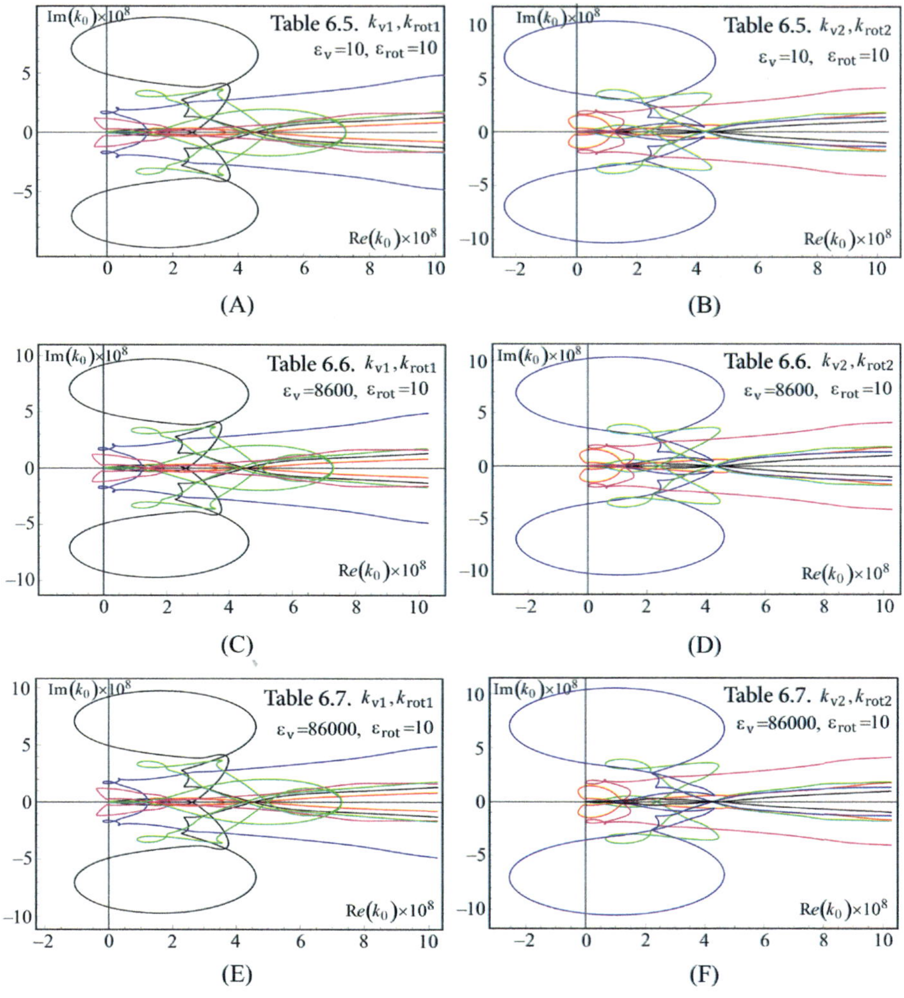

Figure 6.8 D-decomposition curves (N/m). The parameters and stability domains are presented in Tables 6.5–6.7. (A) Table 6.5, first row; (B) Table 6.5, second row; (C) Table 6.6, first row; (D) Table 6.6, second row; (E) Table 6.7, first row; (F) Table 6.7, second row.

6.6 Influence of the wheelbases on dynamic stability

The stability domains for different wheelbase sizes are given in Tables 6.11 and 6.12. For the selected supercritical velocity, the smallest wheelbase ($d_V = d_L = 10.7$ m) has an interesting effect on the stability domains. In that case, the stable domain is nonexistent. The stable domain increases if the wheelbase is increased ($d_V = d_L = 15.7$ m) and it appears within the interval $k_0 \in (2.74,\ 4.17) \times 10^7$N/m. For $d_V = d_L = 20.7$ m it is within the interval $k_0 \in (1.73,\ 3.90) \times 10^7$N/m. The

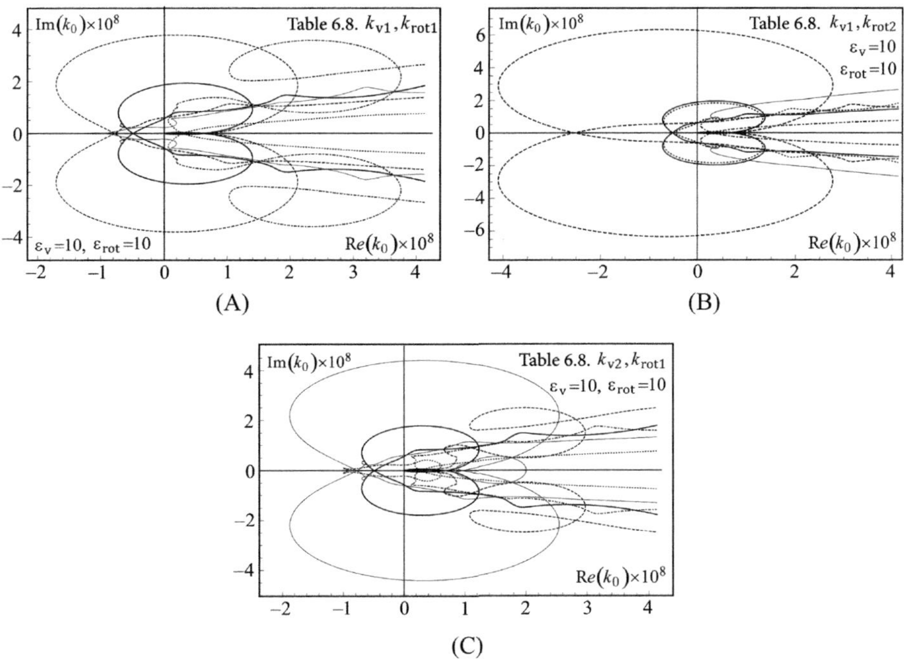

(A) (B) (C)

Figure 6.9 D-decomposition curves (N/m). The parameters and stability domains are presented in Table 6.8. (A) Table 6.8, first row; (B) Table 6.8, second row; (C) Table 6.8, third row.

D-decomposition curves in Fig. 6.11 are for the supercritical velocity $v = 1.12\, v_{cr}$, while parameters of the elastic coupling are given in Tables 6.11 and 6.12.

Finding the relation between coupling stiffness, vehicle suspension stiffness, and vehicle critical velocity can be achieved by implementing the D-decomposition method, Fig. 6.12 illustrates this relation. The figure also shows the boundaries of the stability domain for four different cases concerning the stiffness of the elastic coupling. As it can be noticed, with a rigid connection between the bars the system is more stable when the supercritical velocity is smaller than $v = 1.08\, v_{cr}$. With a rigid rotational but elastic transverse connection, the system behaves more stably if $v > 1.08\, v_{cr}$. This leads to the general conclusion that the elastic coupling between the bars can change the zone of stability, which can have a useful purpose in engineering practice.

6.7 Conclusion

The analysis in this chapter was about the stability of vibrations of a complex rail vehicle system (comprising two rail cars) that is uniformly moving along a continuous structure (a deformable beam of infinite length, supported by a viscoelastic foundation). The vehicle system was approximated with a model

Table 6.8 The number of unstable roots. Influence of stiffness k_{rot} on stability regions; v=1.39v_{cr}, k_{v1}=2×10^{10} N/m, k_{v2}=10^5 N/m, k_{rot1}=2 × 10^{10} (N·m)/rad, k_{rot2}=10^5 (N·m)/rad, ε_{v}=10 (N·s)/m, ε_{rot}=10 (N·m·s)/rad, unit for k_0 is 10^7 N/m.

<table>
<tr><th>k_0</th><th>0</th><th>0.0158</th><th>6.28</th><th>8.76</th><th>9.54</th><th>9.95</th><th>10.12</th><th>10.65</th><th>12.09</th><th>12.10</th><th>13.61</th><th>14.09</th><th>14.10</th><th>14.23</th><th>14.68</th><th>15.34</th><th>18.09</th><th>18.19</th><th>18.47</th><th>18.96</th><th>20.08</th><th>26.06</th><th>35.85</th><th>36.25</th><th>41.80</th><th>41.96</th><th>42.58</th><th>42.74</th><th>43.62</th><th>45.281</th><th>45.285</th><th>46.28</th><th>46.44</th><th>49.77</th><th>55.44</th><th>61.48</th><th>72.94</th><th>226.7</th><th>+∞</th></tr>
<tr><td>k_{v1}, k_{rot1}</td><td>8</td><td>8|8</td><td>8|8</td><td>8|8</td><td>8|8</td><td>8|8</td><td>8|10</td><td>10|10</td><td>10|10</td><td>10|10</td><td>10|10</td><td>10|10</td><td>10|10</td><td>10|10</td><td>10|10</td><td>10|11</td><td>11|11</td><td>11|11</td><td>11|11</td><td>11|9</td><td>9|9</td><td>9|11</td><td>11|11</td><td>11|12</td><td>12|12</td><td>12|12</td><td>12|10</td><td>10|10</td><td>10|10</td><td>10|8</td><td>8|8</td><td>8|8</td><td>8|6</td><td>6|8</td><td>8|8</td><td>8|8</td><td>8|6</td><td>6|6</td><td>6</td></tr>
<tr><td>k_{v2}, k_{rot1}</td><td>9</td><td>9|8</td><td>8|10</td><td>10|12</td><td>12|12</td><td>12|12</td><td>12|12</td><td>12|12</td><td>12|12</td><td>12|12</td><td>12|12</td><td>12|12</td><td>12|12</td><td>12|12</td><td>12|12</td><td>12|12</td><td>12|14</td><td>14|15</td><td>15|13</td><td>13|13</td><td>13|13</td><td>13|13</td><td>13|13</td><td>13|13</td><td>13|13</td><td>13|11</td><td>11|9</td><td>9|7</td><td>7|7</td><td>7|7</td><td>7|7</td><td>7|7</td><td>7|7</td><td>7|7</td><td>7|7</td><td>7|7</td><td>7|7</td><td>7|7</td><td>7</td></tr>
<tr><td>k_{v1}, k_{rot2}</td><td>8</td><td>8|8</td><td>8|8</td><td>8|8</td><td>8|10</td><td>10|12</td><td>12|12</td><td>12|10</td><td>10|10</td><td>10|10</td><td>10|12</td><td>12|12</td><td>12|12</td><td>12|10</td><td>10|11</td><td>11|11</td><td>11|11</td><td>11|11</td><td>11|11</td><td>11|11</td><td>11|13</td><td>13|13</td><td>13|11</td><td>11|11</td><td>11|9</td><td>9|9</td><td>9|9</td><td>9|9</td><td>9|9</td><td>9|9</td><td>9|7</td><td>7|5</td><td>5|5</td><td>5|5</td><td>5|7</td><td>7|5</td><td>5|5</td><td>5|6</td><td>6</td></tr>
</table>

Table 6.9 The number of unstable roots. Influence of mass ratio on stability regions; $v=1.12v_{cr}$, $k_v=10^5$ N/m, $k_{rot}=10$ (N·m)/rad, $\varepsilon_v=10$ (N·s)/m, $\varepsilon_{rot}=10$ (N·m·s)/rad, unit for k_0 is 10^7 N/m.

<table>
<tr><th>k_0</th><th>0</th><th>0.016</th><th>2.74</th><th>2.79</th><th>3.82</th><th>4.17</th><th>5.11</th><th>7.56</th><th>8.41</th><th>9.01</th><th>9.55</th><th>9.9</th><th>10.8</th><th>21.6</th><th>22.3</th><th>+∞</th></tr>
<tr><td>$M_1=2M_2$</td><td>3</td><td>3|2</td><td>2|2</td><td>2|0</td><td>0|2</td><td>2|2</td><td>2|2</td><td>2|4</td><td>4|6</td><td>6|6</td><td>6|6</td><td>6|6</td><td>6|4</td><td>4|5</td><td>5|7</td><td>7</td></tr>
<tr><td>$M_1=M_2$</td><td>3</td><td>3|2</td><td>2|0</td><td>0|0</td><td>0|0</td><td>0|2</td><td>2|4</td><td>4|4</td><td>4|4</td><td>4|2</td><td>2|4</td><td>4|6</td><td>6|6</td><td>6|7</td><td>7|7</td><td>7</td></tr>
</table>

Table 6.10 The number of unstable roots. Influence of mass ratio on stability regions; v=1.12v_{cr}, k_v=10^5 N/m, k_{rot}=10 (N·m)/rad, ε_v=10 (N·s)/m, ε_{rot}=10 (N·m·s)/rad, unit for k_0 is 10^7 N/m.

<table>
<tr><th>k_0</th><th>0</th><th>0.016</th><th>2.74</th><th>2.82</th><th>3.85</th><th>4.17</th><th>5.11</th><th>6.35</th><th>8.93</th><th>9.01</th><th>9.55</th><th>9.99</th><th>21.6</th><th>+∞</th></tr>
<tr><td>M_2=2M_1</td><td>3</td><td>3|2</td><td>2|2</td><td>2|0</td><td>0|2</td><td>2|2</td><td>2|2</td><td>2|4</td><td>4|6</td><td>6|6</td><td>6|6</td><td>6|6</td><td>6|7</td><td>7</td></tr>
<tr><td>M_2=M_1</td><td>3</td><td>3|2</td><td>2|0</td><td>0|0</td><td>0|0</td><td>0|2</td><td>2|4</td><td>4|4</td><td>4|4</td><td>4|2</td><td>2|4</td><td>4|6</td><td>6|7</td><td>7</td></tr>
</table>

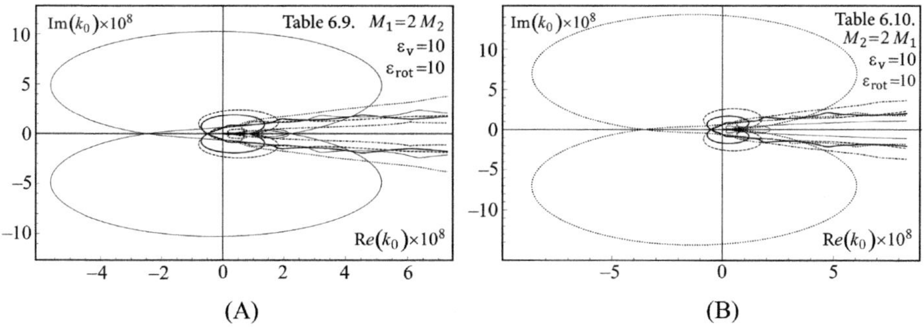

Figure 6.10 D-decomposition curves (N/m). The parameters and stability domains are presented in Tables 6.9 and 6.10. (A) Table 6.9; (B) Table 6.10.

consisting of two elastically connected rigid bars. The improved dynamic model is proposed for this complex system. The presented research has a newly determined critical velocity of the system. With the help of the D-decomposition method, new stability regions were determined and reviewed for varying physical parameters. For the study of the stability domain, the emphasis was on the effect of the elastic coupling, the velocity of the vehicle, the ratio between the masses of the rail cars, and different wheelbase sizes. The primary goal of this study—besides determining the critical velocity of the system—is proving that the proposed elastic coupling (transverse or/and rotational) between the rail cars can act as a stabilizer. An elastic coupling like this can increase the stability region for determined zones—these zones are defined by the stiffness of the suspension. Otherwise speaking, by adjusting the physical parameters of the elastic coupling that connects the train's cars, the stability zone can be increased (in comparison to the conventional train models). The study has determined the stability domain boundaries for varying cases of elastic coupling parameters and varying supercritical velocities of the complex rail vehicle system.

Appendix A 6.1

$$C_1 = \rho bh;\ C_2 = \frac{1}{80}b\beta^2Gh^5 - \frac{1}{6}b\beta Gh^3 + bGh;\ C_3 = \frac{1}{448}\alpha^2bh^7\rho;\ C_4 = \frac{1}{448}\alpha^2bEh^7;$$

$$C_5 = \frac{1}{80}\alpha bh^5\rho - \frac{1}{448}\alpha^2bh^7\rho;\ C_6 = \frac{1}{448}\alpha^2bEh^7 - \frac{1}{80}\alpha bEh^5;$$

$$D_1 = \frac{1}{448}\alpha^2bh^7\rho - \frac{1}{40}\alpha bh^5\rho + \frac{1}{12}bh^3\rho;\ D_2 = \frac{1}{448}\alpha^2bEh^7 - \frac{1}{40}\alpha bEh^5 + \frac{1}{12}bEh^3;$$

$$D_3 = \frac{1}{80}b\beta^2Gh^5 - \frac{1}{6}b\beta Gh^3 + bGh;\ D_4 = \frac{1}{448}\alpha^2bh^7\rho - \frac{1}{80}\alpha bh^5\rho;$$

Table 6.11 The number of unstable roots. Influence of the wheelbase size d on stability regions; v=1.12v_{cr}, k_v=10^5 N/m, k_{rot}=10 (N·m)/rad, ε_v=10 (N·s)/m, ε_{rot}=10 (N·m·s)/rad, unit for k_0 is 10^7 N/m.

<table>
<tr><th>k_0</th><th>0</th><th>0.0162</th><th>0.0163</th><th>2.47</th><th>2.74</th><th>2.96</th><th>4.17</th><th>5.11</th><th>5.76</th><th>5.96</th><th>9.01</th><th>9.55</th><th>9.99</th><th>21.6</th><th>67.7</th><th>88.9</th><th>+∞</th></tr>
<tr><td>d_V=d_L=10.7 m</td><td>3</td><td>3|2</td><td>2|2</td><td>2|4</td><td>4|4</td><td>4|2</td><td>2|2</td><td>2|2</td><td>2|4</td><td>4|6</td><td>6|6</td><td>6|6</td><td>6|6</td><td>6|6</td><td>6|7</td><td>7|5</td><td>5</td></tr>
<tr><td>d_V=d_L=15.7 m</td><td>3</td><td>3|3</td><td>3|2</td><td>2|2</td><td>2|0</td><td>0|0</td><td>0|2</td><td>2|4</td><td>4|4</td><td>4|4</td><td>4|2</td><td>2|4</td><td>4|6</td><td>6|7</td><td>7|7</td><td>7|7</td><td>7</td></tr>
</table>

Table 6.12 The number of unstable roots. Influence of the wheelbase size d on stability regions; v=1.12v_{cr}, k_v=10^5 N/m, k_{rot}=10 (N·m)/rad, ε_v=10 (N·s)/m, ε_{rot}=10 (N·m·s)/rad, unit for k_0 is 10^7 N/m.

k_0	0	0.0163	0.0164	1.73	2.74	3.90	3.91	4.17	5.11	9.01	9.21	9.55	9.99	12.9	14.9	21.6	53.4	+∞
d_V=d_L=20.7 m	3	3|3	3|2	2|0	0|0	0|2	2|4	4|4	4|4	4|4	4|6	6|6	6|6	6|8	8|9	9|9	9|7	7
d_V=d_L=15.7 m	3	3|2	2|2	2|2	2|0	0|0	0|0	0|2	2|4	4|2	2|2	2|4	4|6	6|6	6|6	6|7	7|7	7

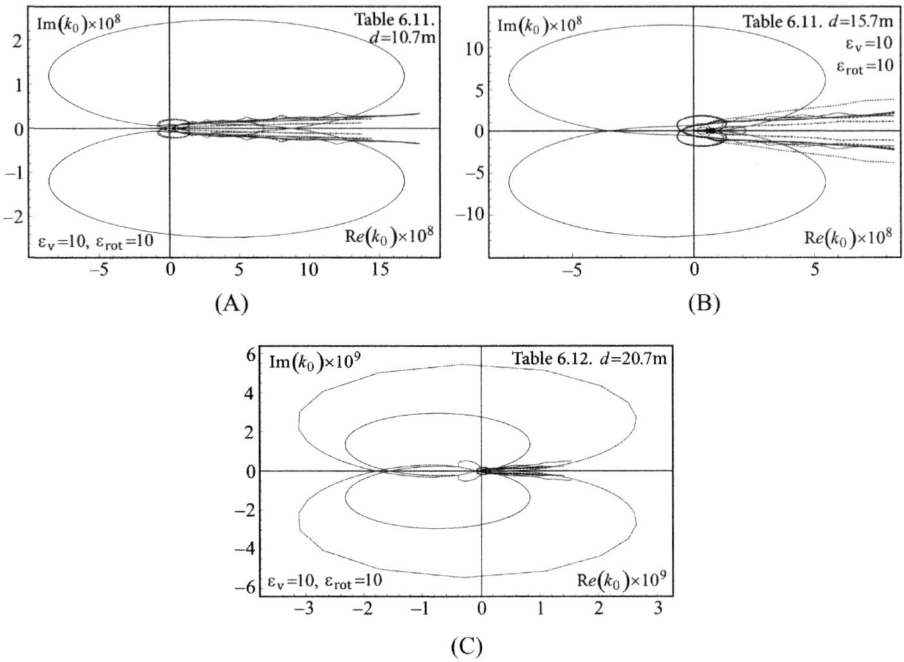

Figure 6.11 D-decomposition curves (N/m). The parameters and stability domains are presented in Tables 6.11 and 6.12. (A) Tables 6.11, first row; (B) Table 6.11, second row; (C) Table 6.12.

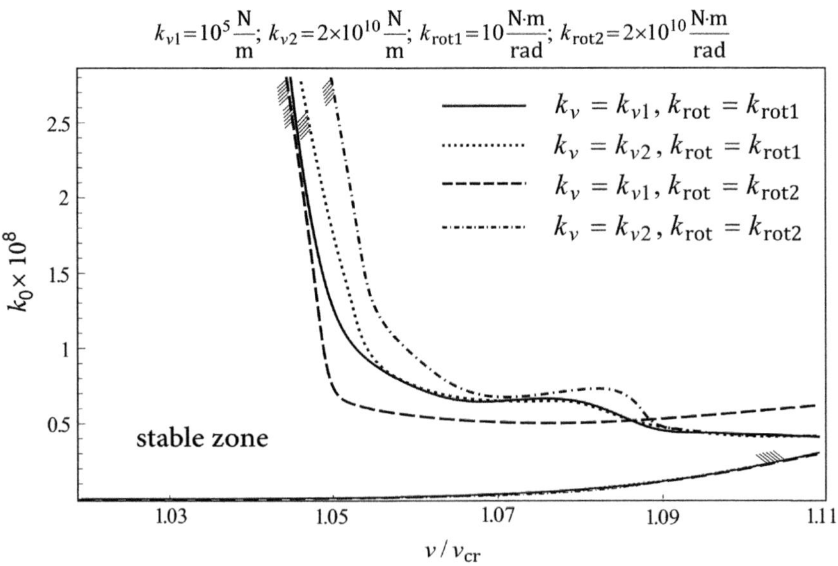

Figure 6.12 Boundaries of the stability domain. The boundaries are for four different cases of elastic coupling parameters; ε_v=8.6 (kN·s)/m, ε_{rot}=10 (N·m·s)/rad.

$$D_5 = \frac{1}{80}\alpha bEh^5 - \frac{1}{448}\alpha^2 bEh^7, \ \alpha = \frac{4}{h^3}, \ \beta = 3\alpha.$$

References

[1] A.V. Metrikine, Unstable lateral oscillations of an object moving uniformly along elastic guide as a result of anomalous Doppler effect, Acoustical Physics 40 (1) (1994) 85–89.

[2] A.V. Metrikine, H.A. Dieterman, Instability of vibrations of a mass moving uniformly along an axially compressed beam on a visco-elastic foundation, Journal of Sound and Vibration 201 (5) (1997) 567–576.

[3] S.N. Verichev, A.V. Metrikine, Instability of vibrations of a mass that moves uniformly along a beam on a periodically inhomogeneous foundation, Journal of Sound and Vibration 260 (2003) 901–925.

[4] S.P. Timoshenko, Proceedings of the 2nd International Congress for Applied Mechanics, Zürich (1927) 1–12.

[5] A.D. Kerr, The continuously supported rail subjected to an axial force and a moving load, International Journal of Mechanical Sciences 14 (1972) 71–78.

[6] G.G. Denisov, E.K. Kugusheva, V.V. Novikov, On the problem of the stability of one-dimensional unbounded elastic systems, Journal of Applied Mathematics and Mechanics 49 (4) (1985) 533–537.

[7] S.N. Verichev, A.V. Metrikine, Instability of a bogie moving on a flexibly supported Timoshenko beam, Journal of Sound and Vibration 253 (2002) 653–668.

[8] S. HJu, H.T. Lin, Analysis of train-induced vibrations and vibration reduction schemes above and below critical Rayleigh speeds by finite element method, Soil Dynamics and Earthquake Engineering 24 (2004) 993–1002.

[9] V. Stojanović, M.D. Petković, J. Deng, Dynamic stability of vibrations and critical velocity of a complex bogie system moving on a flexibly supported infinity track, Journal of Sound and Vibration 434 (2018) 475–501.

[10] V. Stojanović, M.D. Petković, Nonlinear dynamic analysis of damaged Reddy–Bickford beams supported on an elastic Pasternak foundation, Journal of Sound and Vibration 385 (2016) 239–266.

Dynamics and stability of a complex rail vehicle system

7

7.1 Introduction

Modern railway traffic is reaching ever-higher speeds. This encourages scientists to study the complex dynamic models that are now under the influence of several physical phenomena, which interact with one another. Physical phenomena such as vibrations of beams, the Doppler effect, or the moving mass problem are widely known today, but their mutual interactions are still not fully understood. Untroubled transportation is the main reason for considering such problems. To solve them, an analysis of complex dynamic models is required. For conducting such an analysis, first, an insight into the earlier studies is a requisite.

Many researchers have studied the dynamic stability and instability of mechanical systems subjected to moving loads. Timoshenko [1] showed that a moving constant vertical force has a critical velocity at which deflections become considerable. Kerr [2] studied the response of an axially compressed continuously supported infinite beam subjected to a constant vertical force that is in motion. Early analytical works [3–5] in the field of vibrations of solids and structures have motivated researchers to contemplate the complex dynamic behavior of mechanical systems, such as the ones presented here. The mechanical model of a beam that is vibrating and is supported by an elastic foundation, subjected to a constant vertical force that is moving, is the basis for further research involving the model of a beam subjected to a moving mass. Thus Denisov et al. [6] have identified the region of stability for a mass moving uniformly on a continuously supported beam, by applying the D-decomposition method. Some new studies point out that understanding the interactions between basic continuous systems is essential for understanding the dynamic behavior of complex structures, as shown in the analysis by Serpilli and Lenci [7]. They analyzed the mechanical behavior of a complex structure that comprises laminated strips made of three layers and filled with a material that is elastic, isotropic, and homogenous. Their considered problem is a mathematical description of the frequently used engineering models for the multilayer panels. It is an example of a system in which complex dynamic behavior is a product of interactions among the more basic continuous models. Later, Lenci and Rega [8] simplified the mathematical problem of complex structures using the asymptotic development method. This is done by using the possibility to diminish the minor effects; which led to a considerable decrease in the number of parameters and the number of equations (from 14 down to 2). Then the computation of the limited natural frequencies and the normal modes was executed without difficulty. The interaction between a discrete and a continuous system has an important role in the stability analysis of complex structures. Thus Stojanović et al. [9] presented an analysis of the dynamic stability of this kind of structure; as a means to solve the problem, they used the D-decomposition method and the argument principle. They

Modeling of Complex Dynamic Systems. DOI: https://doi.org/10.1016/B978-0-443-23942-7.00007-3

also determined the conditions that make the structure behave in a more stable fashion since knowing the structure's stability regions for the given parameters is important in engineering practice. The same method was applied here to find the stability regions, although the focus was on the problems closer to practical engineering—the moving vehicles. In the literature, veritable dynamical problems from engineering practice have been studied, with respect to high-speed trains. Therefore Ju and Lin [10] investigated the phenomena occurring with fast trains, as mentioned in the Chapter 4. Achieving high-speed traffic has been made possible because of theoretical research, which encompasses an array of scientific fields—from civil engineering, through mechanical engineering, to automation, control, and electrical engineering. On the subject of mechanics, many researchers have contributed to this field. With that in mind, Rezvani and Mazraeh [11] analyzed the dynamics and stability in an interesting problem in which the behavior of a railway freight wagon was modified by varying the axle box clearances. They examined the influence of axle box clearances on vehicle curving and theorized about the possibility of chaotic solutions.

This chapter is about a study of two types of complex rail vehicle systems. It is based on a study realized by Stojanović et al. [12]. The first type is modeled as a system of three elastically connected rail cars, where two types of elastic connections between the cars were considered: the traditional one and the one proposed here that allows some transversal displacement. The vehicle of the second type is a single-car vehicle with additional secondary suspension. For both types of vehicles, and both types of elastic connections, zones of instability for various supercritical velocities are determined.

7.2 Dispersion relation and critical velocity

Karttunen and von Hertzen [13] considered one of the more complex models that include a discrete load moving over a continuous structure. In it, a viscoelastic circular cylinder cover (which is attached to an elastic foundation) is subjected to a circumferentially moving load. The complexity of their model comes from the continuous system, which comprises a cylinder and a foundation. Contrary to that study, the complexity of the model can also stem from the moving discrete system, like in the mechanical systems that are analyzed here. The first step is to derive equations that correlate with the beam theory that was used to describe the continuous system. For further analysis of critical velocity, two different beam theories are used: the Euler–Bernoulli beam assumption and the high-order shear deformation beam assumption. Since the system's critical velocity depends only on the properties of the beam (together with the employed beam theory) and the properties of the foundation, it is realistic to expect some difference between the determined critical velocities. The high-order beam theories take into account rotary inertia and shear deformation; therefore they are more accurate but computationally more demanding. Using a high-order beam theory in this study is justified because it leads to a more accurate determination of stability regions for varying parameters. During the motion of a rail vehicle, a part of the kinetic energy can be transferred to the rails and foundation in the form of vibration energy through anomalous Doppler waves [14]. These waves are radiated to the rails only when the vehicle is traveling above the supercritical speed. When this speed is surpassed, the continuous system of rails gets high-frequency excitation from the discrete complex system of coupled oscillators that are moving along the rails. The parametric study of the stability of the model presented here is based on the amplification of the vehicle's own vibrations, from the vibrations coming back from the rails. The high-order beam model takes into account rotational bending effects and shear deformation [15–17], as they are important factors in describing the behavior of beams subjected to high-frequency excitation when the wavelength is close to the thickness of the beam. The material from which the beam is made is assumed to be elastic, isotropic, and homogeneous. The foundation is represented as a viscoelastic layer. Using a high-order shear deformation theory, particularly the Reddy–Bickford beam assumption, the model's displacement field can be written as

$$u(x, z, t) = u^0(x, t) + z\varphi^0(x, t) - \alpha z^3\left(\varphi^0(x, t) + \frac{\partial w^0(x, t)}{\partial x}\right),$$
$$w(x, z, t) = w^0(x, t), \tag{7.1}$$

in which $\alpha = 4/h^3$, $\beta = 3\alpha$, h is the thickness of the beam. In Eq. (7.1), the longitudinal displacement is designated with $u^0(x, t)$, the transversal displacement with $w^0(x, t)$, and the independent rotation of the cross-section about the axis parallel to y is designated with $\varphi^0(x, t)$ (Fig. 7.1).

The longitudinal and shear strains have the following mathematical form

$$\varepsilon_x(x, t) = \frac{\partial u^0}{\partial x} + z\frac{\partial \varphi^0}{\partial x} - \alpha z^3\left(\frac{\partial \varphi^0}{\partial x} + \frac{\partial^2 w^0}{\partial x^2}\right), \tag{7.2}$$

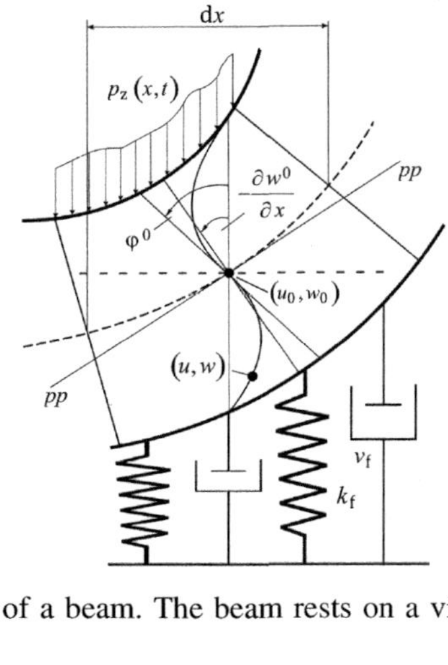

Figure 7.1 Elementary part of a beam. The beam rests on a viscoelastic foundation.

$$\gamma_{xz}(x,t) = \frac{\partial w^0}{\partial x} + \varphi^0 - \beta z^2\left(\varphi^0 + \frac{\partial w^0}{\partial x}\right), \tag{7.3}$$

$$\begin{Bmatrix} \sigma_x \\ \tau_{xz} \end{Bmatrix} = \begin{bmatrix} E & 0 \\ 0 & G \end{bmatrix} \begin{Bmatrix} \varepsilon_x \\ \gamma_{xz} \end{Bmatrix}, \tag{7.4}$$

in which E is the Young modulus, and G stands for the shear modulus of elasticity. The model of the foundation is physically defined via the stiffness modulus k_f, and the damping coefficient ν_f. The reaction of the foundation is

$$r(x,t) = k_f w(x,t) + \nu_f \frac{\partial w(x,t)}{\partial t}. \tag{7.5}$$

The equations of motion are derived from the principle of virtual work, according to which the virtual work of the inertial forces is

$$\delta W_{in} = -\rho \int_V \left[\frac{\partial^2 w(x,z,t)}{\partial t^2}\delta w(x,z,t) + \frac{\partial^2 u(x,z,t)}{\partial t^2}\delta u(x,z,t)\right]\mathrm{d}V, \tag{7.6}$$

the virtual work of the internal forces is

$$\delta W_V = -\int_V [\sigma_x(x,t)\delta\varepsilon_x(z,t) + \tau_{xz}(x,t)\delta\gamma_{xz}(x,t)]\mathrm{d}V, \tag{7.7}$$

and the virtual work of the external forces is defined by

$$\delta W_{ex} = \int_L [p_z(x,t) - r(x,t)]\delta w(x,t)\mathrm{d}L. \tag{7.8}$$

The next step would be to introduce the general equation of the virtual work principle; $\delta W_{in} + \delta W_V + \delta W_{ex} = 0$, and substitute the Eqs. (7.6–7.8) into that equation. Then the Green theorem is applied successively; integrating the equation by parts and collecting the coefficients of δu, δw, and $\delta\varphi$. The result will turn out to be a set of PDEs. This set of equations mathematically describes the vibrations of the system composed of the Reddy–Bickford beam resting on a viscoelastic foundation, and their form is

$$C_1\frac{\partial^2 w}{\partial t^2} - C_2\left(\frac{\partial\varphi}{\partial x} + \frac{\partial^2 w}{\partial x^2}\right) - C_3\frac{\partial^4 w}{\partial x^2\partial t^2} + C_4\frac{\partial^4 w}{\partial x^4} + C_5\frac{\partial^3\varphi}{\partial x\partial t^2} + C_6\frac{\partial^3\varphi}{\partial x^3}$$
$$+ k_f w + \nu_f\frac{\partial w}{\partial t} = p_z(x,t), \tag{7.9a}$$

$$D_1\frac{\partial^2\varphi}{\partial t^2} - D_2\frac{\partial^2\varphi}{\partial x^2} + D_3\left(\varphi + \frac{\partial w}{\partial x}\right) + D_4\frac{\partial^3 w}{\partial x\partial t^2} + D_5\frac{\partial^3 w}{\partial x^3} = 0, \tag{7.9b}$$

in which C_i, D_j; $i = 1, ..,6$; $j = 1, ..,5$; are determined and can be found in [12]. System of Eqs. (7.9a) and (7.9b) can be acquired by reducing the nonlinear terms in the set of nonlinear PDEs [18]. When instead of the high-order shear deformation beam assumption, the simpler Euler–Bernoulli beam assumption is used, the mathematical description of the vibrations will be as follows

$$EI\frac{\partial^4 w}{\partial x^4} + \rho A\frac{\partial^2 w}{\partial t^2} + k_f w + \nu_f\frac{\partial w}{\partial t} = p_z(x, t). \tag{7.10}$$

A moving rail vehicle system has complex dynamical behavior, which is described later. To understand this dynamic, it is customary to consider the kinematics of waves emitted by a mass (m) that is moving on a beam at a constant velocity (v) [9,19], as shown in Fig. 7.2. In that case, Eq. (7.10) becomes

$$EI\frac{\partial^4 w}{\partial x^4} + \rho A_{cs}\frac{\partial^2 w}{\partial t^2} + k_f w + \nu_f\frac{\partial w}{\partial t} = -\delta(x - vt)m\frac{d^2 w_0}{dt^2}, \tag{7.11}$$

in which $\delta()$ represents the Dirac delta function. As clarified in Ref. [19], the wave-number k and frequency ω of the waves, emitted by the mass, satisfy the dispersion relation, which is obtained by substituting $w(x, t) = A\exp\{i[(\Omega - kv)t - kx]\}$, $i = \sqrt{-1}$ into Eq. (7.11). For the undamped case, the following is obtained

$$EIk^4 + k_f w - \rho A_{cs}(\Omega - kv)^2 = 0. \tag{7.12}$$

The roots of the dispersion equation are real when the velocity of the moving mass is lower than the critical. Besides, when the velocity is lower than the critical, the vibrations of the mass (that is, moving on the beam without losing contact with the beam) are harmonic. If the vibrations of the mass do not disrupt the waves in the beam, the mass can continue to vibrate harmonically. Using the discriminant of dispersion Eq. (7.12), the critical velocity (the minimum phase velocity of waves in the beam) can be determined from the condition.

$$\mathrm{Disc}_k[EIk^4 + k_f w - \rho A_{cs}(\Omega - kv)^2] = 0 \Rightarrow v_{cr,E} = v_{ph}^{\min} = \left[\frac{4EIk_f}{(\rho A_{cs})^2}\right]^{0.25}. \tag{7.13}$$

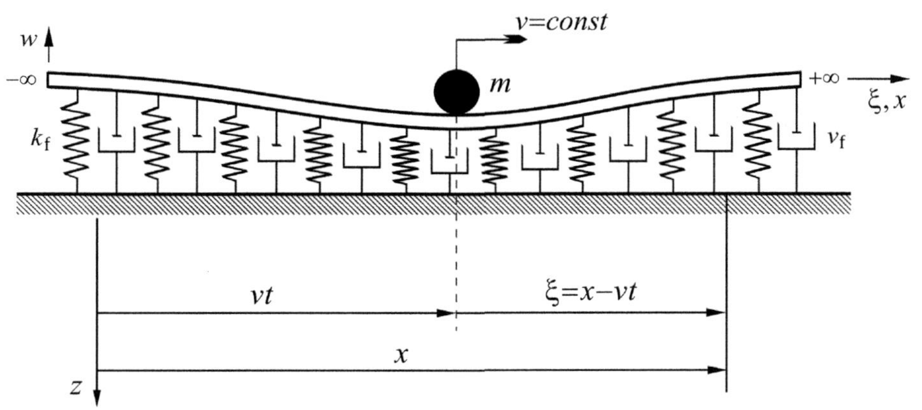

Figure 7.2 Moveable mass on a beam. The beam is resting on a viscoelastic foundation, while the mass is moving at a constant velocity.

Kerr [2] obtained the same minimum phase velocity of waves in the beam presented in Eq. (7.13). Tracing the same method and substituting $w(x, t) = A_1 \exp\{i[(\Omega - kv)t - kx]\}$, $\varphi(x, t) = A_2 \exp\{i[(\Omega - kv)t - kx]\}$ into Eqs. (7.9a) and (7.9b), dispersion relation can be obtained, and its mathematical form is

$$k^6 q_3 + k^4 q_2 + k^2 q_1 + q_0 = 0, \tag{7.14}$$

in which constants q_0, q_1, q_2, and q_3 are given in Ref. [12]. This dispersion relation takes into account rotational bending effects and shear deformation, for they are important factors in describing the behavior of beams subjected to high-frequency excitation when the wavelength gets close to the thickness of the beam. Using the discriminant of the dispersion Eq. (7.14), after some algebra, the critical velocity of the system can be determined. That is, when the roots of the dispersion Eq. (7.14) are not real (these roots are also the wavenumbers k). In that instance, the critical velocity can be obtained from the form

$$\mathrm{Disc}_k[k^6 q_3 + k^4 q_2 + k^2 q_1 + q_0] = 0 \Rightarrow$$
$$v_{cr,RB} = v_{ph}^{\min} = \left[\sqrt{\frac{\widetilde{p}_1^2}{4} - \widetilde{q}_1} - \frac{\widetilde{p}_1}{2}\right]^{0.5}. \tag{7.15}$$

Constants $\widetilde{p}_1$ and $\widetilde{q}_1$ are determined and presented in Ref. [12]. For the beam models considered, the ratio between the critical velocities $\zeta = v_{cr,E}/v_{cr,RB}$ in the function of the foundation stiffness k_f is presented in Fig. 7.3. There, the critical velocity determined using the Euler–Bernoulli beam assumption is designated with $v_{cr,E}$, while the critical velocity determined using the Reddy–Bickford beam assumption is designated with $v_{cr,RB}$.

The ratio between the critical velocities (Fig. 7.3) was calculated using the system parameters given in Refs. [9,19]. When the stiffness of the foundation increases, the defined critical velocity ratio ζ also increases. The high-order shear deformation beam assumption gives lower values for the critical velocity than the Euler–Bernoulli beam assumption.

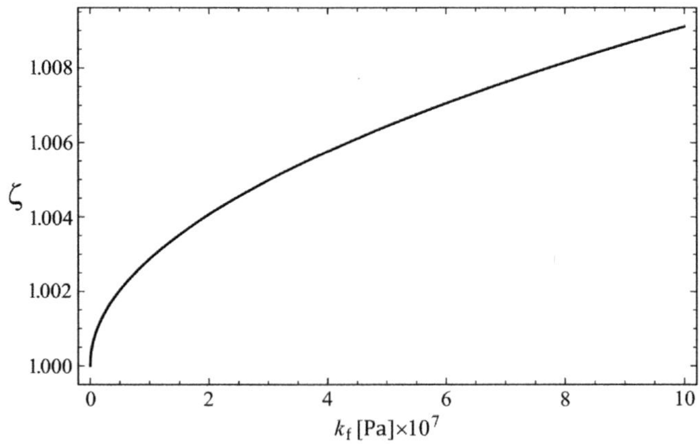

Figure 7.3 Relation between ratio ζ and parameter k_f. Ratio ζ is the ratio between two different beam theories, parameter k_f is the foundation stiffness parameter.

7.3 Modeling of the complex rail vehicle systems and the characteristic equations

Two complex rail vehicle systems are included in this analysis (Fig. 7.4). The first type is modeled as a system of three elastically connected rail cars. The rail car model is basic; it consists of a rigid bar supported by two supports that are composed of suspension elements and wheels. Two types of elastic connections between the cars are considered: the traditional, and the one proposed here that allows some transversal displacement (Fig. 7.5). The second type of vehicle comprises only one rail car, but with two rows of suspension. The car's body (i.e., the rigid bar) is supported via the secondary suspension onto two bogies. The bogies' supports are composed of wheels with primary suspension elements. A suspension element comprises a spring and a dashpot in parallel. The primary suspension elements are on one side attached to the rigid bar, while on the other side, they interact with the rail through the wheels (i.e., the masses m). The rail car in the second system has four contact points with the infinite beam—Fig. 7.4B. It is assumed that the wheels and the beam (i.e., the rail) are in contact at all times. The scheme of a decomposed vehicle system, and the dynamic equilibrium forces that are acting on it, can be seen in Fig. 7.6.

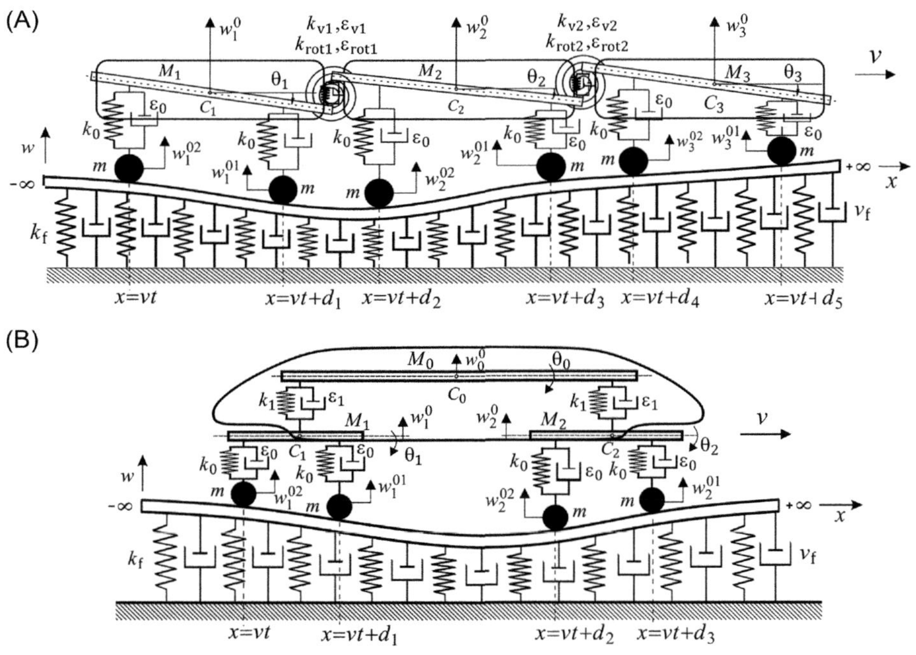

Figure 7.4 Uniform motion of a complex vehicle system. The vehicle travels on an infinite beam that is placed on a viscoelastic foundation; (A) vehicle system comprising three rail cars linked together via elastic couplings; (B) vehicle system with primary and secondary suspension.

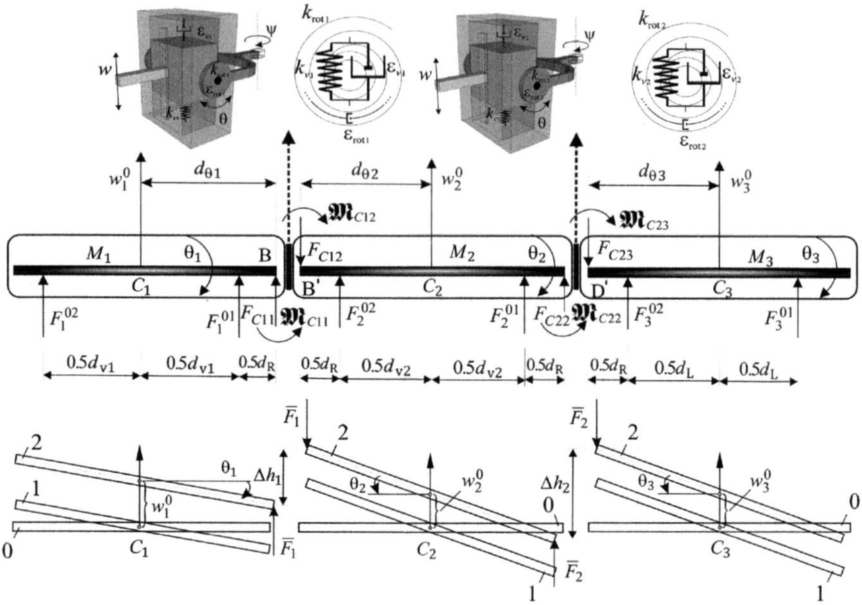

$F_1^{02}=-k_0[w_1^{02}-(w_1^0+0.5\,\theta_1 d_{V1})]$, $F_2^{02}=-k_0[w_2^{02}-(w_2^0+0.5\,\theta_2 d_{V2})]$, $F_3^{02}=-k_0[w_3^{02}-(w_3^0+0.5\,\theta_3 d_{L})]$,

$F_1^{01}=-k_0[w_1^{01}-(w_1^0-0.5\,\theta_1 d_{V1})]$, $F_2^{01}=-k_0[w_2^{01}-(w_2^0-0.5\,\theta_2 d_{V2})]$, $F_3^{01}=-k_0[w_3^{01}-(w_3^0-0.5\,\theta_3 d_{L})]$,

$F_{C11}=-F_{C12}=-k_{V1}[w_2^0-w_1^0+\theta_2 d_{\theta 2}+\theta_1 d_{\theta 1}]$, $F_{C2}^2=-F_{C2}^3=-k_{V2}[w_3^0-w_2^0+\theta_3 d_{\theta 3}+\theta_2 d_{\theta 2}]$,

$\mathfrak{M}_{C11}=-\mathfrak{M}_{C12}=-k_{\mathrm{rot}}(\theta_2-\theta_1)$, $\mathfrak{M}_{C22}=-\mathfrak{M}_{C23}=-k_{\mathrm{rot}}(\theta_3-\theta_2)$,

Figure 7.5 Scheme of a triplex vehicle system. The scheme shows a decomposed vehicle system with its viscoelastic couplings.

The governing equations, for the model in Fig. 7.4A, according to the dynamic equilibrium (Fig. 7.5) may be written in the reduced form as

$$\Phi_1[w,\varphi] = -\delta(x-vt-d_5)\left(m\frac{\mathrm{d}^2 w_3^{01}}{\mathrm{d}t^2}+\mathscr{L}_{31}^1\right)-\delta(x-vt-d_4)\left(m\frac{\mathrm{d}^2 w_3^{02}}{\mathrm{d}t^2}+\mathscr{L}_{31}^2\right)$$

$$-\delta(x-vt-d_3)\left(m\frac{\mathrm{d}^2 w_2^{01}}{\mathrm{d}t^2}+\mathscr{L}_{22}^1\right)-\delta(x-vt-d_2)\left(m\frac{\mathrm{d}^2 w_2^{02}}{\mathrm{d}t^2}+\mathscr{L}_{22}^2\right)$$

$$-\delta(x-vt-d_1)\left(m\frac{\mathrm{d}^2 w_1^{01}}{\mathrm{d}t^2}+\mathscr{L}_{13}^1\right)-\delta(x-vt)\left(m\frac{\mathrm{d}^2 w_1^{02}}{\mathrm{d}t^2}+\mathscr{L}_{13}^2\right), \tag{7.16}$$

$$\Phi_2[w,\varphi]=0, \tag{7.17}$$

$$M_3\frac{\mathrm{d}^2 w_3^0}{\mathrm{d}t^2}-\mathscr{L}_{31}^1-\mathscr{L}_{31}^2-\mathscr{L}_{23}=0, \tag{7.18}$$

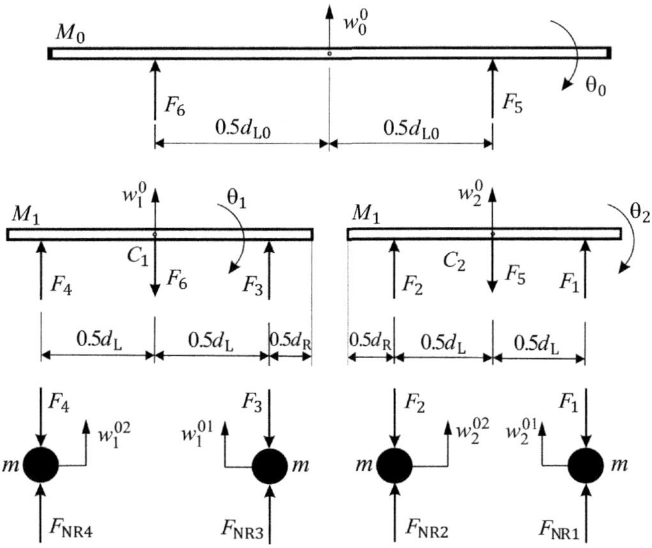

Figure 7.6 Scheme of a decomposed vehicle system. The elements are decomposed from the primary and secondary suspension, the dynamic equilibrium forces are indicated.

$$J_3\frac{d^2\theta_3}{dt^2}+\frac{d_L}{2}(\mathscr{L}_{31}^1-\mathscr{L}_{31}^2)-\mathscr{L}_{23}d_{\theta 3}+\mathfrak{M}_{23}=0, \tag{7.19}$$

$$M_2\frac{d^2w_2^0}{dt^2}-\mathscr{L}_{22}^1-\mathscr{L}_{22}^2+\mathscr{L}_{23}-\mathscr{L}_{12}=0, \tag{7.20}$$

$$J_2\frac{d^2\theta_2}{dt^2}+(\mathscr{L}_{22}^1-\mathscr{L}_{22}^2)\frac{d_{V2}}{2}-(\mathscr{L}_{23}+\mathscr{L}_{12})d_{\theta 2}-\mathfrak{M}_{23}+\mathfrak{M}_{12}=0, \tag{7.21}$$

$$M_1\frac{d^2w_1^0}{dt^2}-\mathscr{L}_{13}^1-\mathscr{L}_{13}^2+\mathscr{L}_{12}=0, \tag{7.22}$$

$$J_1\frac{d^2\theta_1}{dt^2}+(\mathscr{L}_{13}^1-\mathscr{L}_{13}^2)\frac{d_{V1}}{2}-\mathscr{L}_{12}d_{\theta 1}-\mathfrak{M}_{12}=0, \tag{7.23}$$

where

$$\mathscr{L}_{ik}^j=k_0\left[w_i^{0j}-\left(w_i^0-\frac{\theta_i}{2}\tilde{d}_k(-1)^{j+1}\right)\right]+\epsilon_0\frac{d}{dt}\left[w_i^{0j}-\left(w_i^0-\frac{\theta_i}{2}\tilde{d}_k(-1)^{j+1}\right)\right],$$

$$\mathscr{L}_{ij}=k_{vi}[w_j^0-w_i^0+\theta_j d_{\theta j}+\theta_i d_{\theta i}]+\epsilon_{vi}\frac{d}{dt}[w_j^0-w_i^0+\theta_j d_{\theta j}+\theta_i d_{\theta i}],$$

$$\mathfrak{M}_{ij}=k_{roti}[\theta_j-\theta_i]+\epsilon_{roti}[\theta_j-\theta_i],$$

$$w_3^{01}(t)=w(vt+d_5,t),\ \ w_3^{02}(t)=w(vt+d_4,t),\ \ w_2^{01}(t)=w(vt+d_3,t),$$

$$w_2^{02}(t) = w(vt + d_2, t),\ \ w_1^{01}(t) = w(vt + d_1, t),\ \ w_1^{02}(t) = w(vt, t),$$

$$\lim_{|x-vt|\to\infty} w(x, t) = 0, \quad \lim_{|x-vt|\to\infty} \varphi(x, t) = 0,$$

$$\widetilde{d}_1 = d_L,\ \widetilde{d}_2 = d_{V2},\ \widetilde{d}_3 = d_{V1}, \tag{7.24}$$

and where $\Phi_1[w, \varphi]$ and $\Phi_2[w, \varphi]$ present left-hand sides of Eqs. (7.9a) and (7.9b), respectively. The following are the translatory displacements: $w(x, t)$ is the vertical deflection of the beam; $w_1^{01}(t)$, $w_1^{02}(t)$, $w_2^{01}(t)$, $w_2^{02}(t)$, $w_3^{01}(t)$, and $w_3^{02}(t)$ are the vertical displacements of the masses; $w_1^{0}(t)$, $w_2^{0}(t)$, and $w_3^{0}(t)$ are the vertical displacements of the centers of mass of the bars (the car's main body is represented by a bar). The following are the angular displacements: $\varphi(x, t)$ is the angle of rotation of the cross-section of the beams; θ_1, θ_2, and θ_3 are the angles of rotation of the bars around their centers of mass. The following are the physical properties: M_1, M_2, M_3 and J_1, J_2, J_3 are the masses and moments of inertia of the bars, respectively; k_f and ν_f are the stiffness and the viscosity of the foundation per unit length, respectively; k_0 and ϵ_0 are the stiffness and the damping factor of the suspension; k_{v1}, k_{v2}, and ϵ_{v1}, ϵ_{v2} are the stiffnesses and the damping factors of the elastic couplings in the transversal direction; k_{rot1}, k_{rot2} and ϵ_{rot1}, ϵ_{rot2} are the rotational stiffness and the torsional damping of the elastic couplings. The Dirac delta-function is designated with $\delta()$. By introducing a moving reference system $\xi = x - vt$, Eqs. (7.16) and (7.17) can be rewritten in the form

$$C_1\left(v^2\frac{\partial^2 w(\xi, t)}{\partial \xi^2} - 2v\frac{\partial^2 w(\xi, t)}{\partial \xi \partial t} + \frac{\partial^2 w(\xi, t)}{\partial t^2}\right)$$

$$- C_2\left(\frac{\partial \varphi(\xi, t)}{\partial \xi} + \frac{\partial^2 w(\xi, t)}{\partial \xi^2}\right) - C_3\left(v^2\frac{\partial^4 w(\xi, t)}{\partial \xi^4} - 2v\frac{\partial^4 w(\xi, t)}{\partial \xi^3 \partial t} + \frac{\partial^4 w(\xi, t)}{\partial \xi^2 \partial t^2}\right)$$

$$+ C_4\frac{\partial^4 w(\xi, t)}{\partial \xi^4} + C_5\left(v^2\frac{\partial^3 \varphi(\xi, t)}{\partial \xi^3} - 2v\frac{\partial^3 \varphi(\xi, t)}{\partial \xi^2 \partial t} + \frac{\partial^3 \varphi(\xi, t)}{\partial \xi \partial t^2}\right)$$

$$+ C_6\frac{\partial^3 \varphi(\xi, t)}{\partial \xi^3} + k_f w + \nu_f\left(\frac{\partial w(\xi, t)}{\partial t} - v\frac{\partial w(\xi, t)}{\partial \xi}\right)$$

$$= -\delta(\xi - d_5)\left(m\frac{d^2 w_3^{01}}{dt^2} + \mathscr{L}_{31}^1\right) - \delta(\xi - d_4)\left(m\frac{d^2 w_3^{02}}{dt^2} + \mathscr{L}_{31}^2\right)$$

$$- \delta(\xi - d_3)\left(m\frac{d^2 w_2^{01}}{dt^2} + \mathscr{L}_{22}^1\right) - \delta(\xi - d_2)\left(m\frac{d^2 w_2^{02}}{dt^2} + \mathscr{L}_{22}^2\right)$$

$$- \delta(\xi - d_1)\left(m\frac{d^2 w_1^{01}}{dt^2} + \mathscr{L}_{13}^1\right) - \delta(\xi)\left(m\frac{d^2 w_1^{02}}{dt^2} + \mathscr{L}_{13}^2\right), \tag{7.25}$$

$$D_1\left(v^2\frac{\partial^2\varphi(\xi,t)}{\partial\xi^2} - 2v\frac{\partial^2\varphi(\xi,t)}{\partial\xi\partial t} + \frac{\partial^2\varphi(\xi,t)}{\partial t^2}\right) - D_2\frac{\partial^2\varphi(\xi,t)}{\partial\xi^2}$$

$$+ D_3\left(\varphi(\xi,t) + \frac{\partial w(\xi,t)}{\partial\xi}\right)$$

$$+ D_4\left(v^2\frac{\partial^3 w(\xi,t)}{\partial\xi^3} - 2v\frac{\partial^3 w(\xi,t)}{\partial\xi^2\partial t} + \frac{\partial^3 w(\xi,t)}{\partial\xi\partial t^2}\right) + D_5\frac{\partial^3 w(\xi,t)}{\partial\xi^3} = 0, \tag{7.26}$$

$$w_3^{01}(t) = w(d_5,t),\ w_3^{02}(t) = w(d_4,t),\ w_2^{01}(t) = w(d_3,t),$$

$$w_2^{02}(t) = w(d_2,t),\ w_1^{01}(t) = w(d_1,t),\ w_1^{02}(t) = w(0,t),$$

$$\lim_{|\xi|\to\infty} w(\xi,t) = 0,\ \lim_{|\xi|\to\infty}\varphi(\xi,t) = 0. \tag{7.27}$$

The characteristic equation for vibrations, which is here applied on the rail vehicle that is moving along the beam, can be analyzed by applying the integral transforms to the system of Eqs. (7.18–7.23) and Eqs. (7.25–7.26):

$$W_s(\xi,s) = \int_0^\infty w(\xi,t)e^{-st}\mathrm{d}t,\ \Phi_s(\xi,s) = \int_0^\infty \varphi(\xi,t)e^{-st}\mathrm{d}t,$$

$$\hat{W}_{k,s}(k,s) = \int_{-\infty}^\infty W_s(\xi,s)e^{-ik\xi}\mathrm{d}\xi,\ \hat{\Phi}_s(k,s) = \int_{-\infty}^\infty \Phi_s(\xi,s)e^{-ik\xi}\mathrm{d}\xi,$$

$$W_{\tilde{i}}^{0\tilde{j}}(s) = \int_0^\infty w_{\tilde{i}}^{0\tilde{j}}(t)e^{-st}\mathrm{d}t,\ W_{\tilde{i}}^{0}(s) = \int_0^\infty w_{\tilde{i}}^{0}(t)e^{-st}\mathrm{d}t,$$

$$\Theta_{\tilde{i}}(s) = \int_0^\infty \theta_{\tilde{i}}(t)e^{-st}\mathrm{d}t,\ \tilde{i} = 1,..,3;\ \tilde{j} = 1,\ 2;\ i = \sqrt{-1}.$$

The trivial initial conditions can be adopted for the problem—since the governing equations are linear, the initial conditions do not influence the system's stability. Also, considering that $W_1^{02}(s) = W(0,s)$, $W_1^{01}(s) = W(d_1,s)$, $W_2^{02}(s) = W(d_2,s)$, $W_2^{01}(s) = W(d_3,s)$, $W_3^{02}(s) = W(d_4,s)$, and $W_3^{01}(s) = W(d_5,s)$, the application of the results of the transforms after some algebra in the system of algebraic equations after the elimination $\hat{\Phi}_s(k,s)$ in the present form

$$\left(\sum_{\tilde{\tilde{j}}=0}^{6}\tilde{a}_{\tilde{\tilde{j}}}k^{\tilde{\tilde{j}}}\right)\cdot\left(\sum_{\tilde{\tilde{i}}=0}^{2}\tilde{b}_{\tilde{\tilde{i}}}k^{\tilde{\tilde{i}}}\right)^{-1}\cdot\hat{W}_{k,s}(k,s) = e^{-id_4k}\cdot\left[k_0W_3^0(s) - k_0W(d_4,s) - ms^2W(d_4,s)\right.$$

$$\left. - s\epsilon_0W(d_4,s) + s\epsilon_0W_3^0(s) + \frac{1}{2}\Theta_3k_0d_L + \frac{1}{2}\Theta_3s\epsilon_0d_L\right]$$

$$+ e^{-id_5k}\left(-k_0W(d_5,s) - ms^2W(d_5,s) - s\epsilon_0W(d_5,s)\right.$$

$$\left. + k_0W_3^0(s) + s\epsilon_0W_3^0(s) - \frac{1}{2}\Theta_3k_0d_L - \frac{1}{2}\Theta_3s\epsilon_0d_L\right)$$

$$+ e^{-id_1k}\Bigg(-k_0W(d_1, s) - ms^2W(d_1, s) - s\epsilon_0W(d_1, s)$$

$$+ k_0W_1^0(s) + s\epsilon_0W_1^0(s) - \frac{1}{2}\Theta_1k_0d_{V1} - \frac{1}{2}\Theta_1s\epsilon_0d_{V1}\Bigg)$$

$$+ e^{-id_2k}\Bigg(-k_0W(d_2, s) - ms^2W(d_2, s) - s\epsilon_0W(d_2, s)$$

$$+ k_0W_2^0(s) + s\epsilon_0W_2^0(s) + \frac{1}{2}\Theta_2k_0d_{V2} + \frac{1}{2}\Theta_2s\epsilon_0d_{V2}\Bigg)$$

$$+ e^{-id_3k}\Bigg(-k_0W(d_3, s) - ms^2W(d_3, s) - s\epsilon_0W(d_3, s)$$

$$+ k_0W_2^0(s) + s\epsilon_0W_2^0(s) - \frac{1}{2}\Theta_2k_0d_{V2} - \frac{1}{2}\Theta_2s\epsilon_0d_{V2}\Bigg) - k_0W(0, s) - ms^2W(0, s)$$

$$- s\epsilon_0W(0, s) + k_0W_1^0(s) + s\epsilon_0W_1^0(s) + \frac{1}{2}\Theta_1k_0d_{V1} + \frac{1}{2}\Theta_1s\epsilon_0d_{V1}, \tag{7.28}$$

$$k_{v2}W_2^0(s) - k_0W(d_4, s) - k_0W(d_5, s) - s\epsilon_0W(d_4, s) - s\epsilon_0W(d_5, s)$$

$$- k_{v2}W_3^0(s) + 2k_0W_3^0(s) + M_3s^2W_3^0(s) + s\epsilon_{v2}W_2^0(s)$$

$$- s\epsilon_{v2}W_3^0(s) + 2s\epsilon_0W_3^0(s) - \Theta_2d_{\theta2}k_{v2} - \Theta_3d_{\theta3}k_{v2}$$

$$- \Theta_2sd_{\theta2}\epsilon_{v2} - \Theta_3sd_{\theta3}\epsilon_{v2} = 0, \tag{7.29}$$

$$\frac{1}{2}k_0d_LW(d_5, s) - \frac{1}{2}k_0d_LW(d_4, s) - \frac{1}{2}s\epsilon_0d_LW(d_4, s) + \frac{1}{2}s\epsilon_0d_LW(d_5, s)$$

$$+ d_{\theta3}k_{v2}W_2^0(s) - d_{\theta3}k_{v2}W_3^0(s) + sd_{\theta3}\epsilon_{v2}W_2^0(s) - sd_{\theta3}\epsilon_{v2}W_3^0(s) + \frac{1}{2}\Theta_3k_0d_L^2$$

$$- \Theta_2d_{\theta2}d_{\theta3}k_{v2} - \Theta_3d_{\theta3}^2k_{v2} + \frac{1}{2}\Theta_3s\epsilon_0d_L^2 - \Theta_2sd_{\theta2}d_{\theta3}\epsilon_{v2} - \Theta_3sd_{\theta3}^2\epsilon_{v2}$$

$$+ \Theta_3J_3s^2 - \Theta_2k_{\text{rot2}} + \Theta_3k_{\text{rot2}} - \Theta_2s\epsilon_{\text{rot2}} + \Theta_3s\epsilon_{\text{rot2}} = 0, \tag{7.30}$$

$$k_{v1}W_1^0(s) - k_0W(d_2, s) - k_0W(d_3, s) - s\epsilon_0W(d_2, s) - s\epsilon_0W(d_3, s) - k_{v1}W_2^0(s)$$

$$- k_{v2}W_2^0(s) + k_{v2}W_3^0(s) + 2k_0W_2^0(s) + M_2s^2W_2^0(s) + s\epsilon_{v1}W_1^0(s) - s\epsilon_{v1}W_2^0(s)$$

$$- s\epsilon_{v2}W_2^0(s) + s\epsilon_{v2}W_3^0(s) + 2s\epsilon_0W_2^0(s) - \Theta_1d_{\theta1}k_{v1} - \Theta_2d_{\theta2}k_{v1} + \Theta_2d_{\theta2}k_{v2}$$

$$+ \Theta_3d_{\theta3}k_{v2} - \Theta_1sd_{\theta1}\epsilon_{v1} - \Theta_2sd_{\theta2}\epsilon_{v1} + \Theta_2sd_{\theta2}\epsilon_{v2} + \Theta_3sd_{\theta3}\epsilon_{v2} = 0, \tag{7.31}$$

$$\frac{1}{2}k_0 d_{V2} W(d_3, s) - \frac{1}{2}k_0 d_{V2} W(d_2, s) - \frac{1}{2}s\epsilon_0 d_{V2} W(d_2, s)$$

$$+ \frac{1}{2}s\epsilon_0 d_{V2} W(d_3, s) + d_{\theta 2}k_{v1}W_1^0(s) - d_{\theta 2}k_{v1}W_2^0(s) + d_{\theta 2}k_{v2}W_2^0(s)$$

$$- d_{\theta 2}k_{v2}W_3^0(s) + sd_{\theta 2}\epsilon_{v1}W_1^0(s) - sd_{\theta 2}\epsilon_{v1}W_2^0(s)$$

$$+ sd_{\theta 2}\epsilon_{v2}W_2^0(s) - sd_{\theta 2}\epsilon_{v2}W_3^0(s) - \Theta_1 d_{\theta 1}d_{\theta 2}k_{v1} - \Theta_2 d_{\theta 2}^2 k_{v1}$$

$$- \Theta_3 d_{\theta 2}d_{\theta 3}k_{v2} - \Theta_2 d_{\theta 2}^2 k_{v2} + \frac{1}{2}\Theta_2 k_0 d_{V2}^2 - \Theta_1 s d_{\theta 1}d_{\theta 2}\epsilon_{v1} - \Theta_2 s d_{\theta 2}^2\epsilon_{v1}$$

$$- \Theta_3 s d_{\theta 2}d_{\theta 3}\epsilon_{v2} - \Theta_2 s d_{\theta 2}^2\epsilon_{v2} + \frac{1}{2}\Theta_2 s\epsilon_0 d_{V2}^2 + \Theta_2 J_2 s^2 - \Theta_1 k_{rot1} + \Theta_2 k_{rot1}$$

$$+ \Theta_2 k_{rot2} - \Theta_3 k_{rot2} - \Theta_1 s\epsilon_{rot1} + \Theta_2 s\epsilon_{rot1} + \Theta_2 s\epsilon_{rot2} - \Theta_3 s\epsilon_{rot2} = 0, \quad (7.32)$$

where $\tilde{a}_{\tilde{\tilde{j}}}$ ($\tilde{\tilde{j}} = 0, .,6$) and $\tilde{b}_{\tilde{\tilde{i}}}$ ($\tilde{\tilde{i}} = 0,2$) are given in Ref. [12]. Examining the system of Eqs. (7.28–7.32) it can be observed that the mathematical complexity of the system does not originate from the beam theory that is used because the continuous beam reactions reside only in the expression: $\left(\Sigma_{\tilde{\tilde{j}}=0}^{6} \tilde{a}_{\tilde{\tilde{j}}} k^{\tilde{\tilde{j}}}\right)\cdot\left(\Sigma_{\tilde{\tilde{i}}=0}^{2} \tilde{b}_{\tilde{\tilde{i}}} k^{\tilde{\tilde{i}}}\right)^{-1}$, on the left-hand side of Eq. (7.28). Starting from the system of Eqs. (7.28–7.32) and applying to it the inverse Fourier transform over k, then eliminating W_1^0, W_2^0, W_3^0, Θ_1, Θ_2, and Θ_3, and setting the values $\xi = 0$, $\xi = d_1$, $\xi = d_2$, $\xi = d_3$, $\xi = d_4$, and $\xi = d_5$ subsequently, will bring about the following system of algebraic equations—with respect to the Laplace-displacements of the contact points $\hat{W}_s(0, s)$, $\hat{W}_s(d_1, s)$, $\hat{W}_s(d_2, s)$, $\hat{W}_s(d_3, s)$, $\hat{W}_s(d_4, s)$, and $\hat{W}_s(d_5, s)$

$$[\mathbf{u}]\times[\hat{W}]^{\mathrm{T}} = 0 \Leftrightarrow$$

$$\begin{bmatrix} u_{11} & \cdots & u_{16} \\ \vdots & \ddots & \vdots \\ u_{61} & \cdots & u_{66} \end{bmatrix} \times \left\lfloor \hat{W}_s(0, s) \quad \hat{W}_s(d_1, s) \quad \hat{W}_s(d_2, s) \quad \hat{W}_s(d_3, s) \quad \hat{W}_s(d_4, s) \quad \hat{W}_s(d_5, s) \right\rfloor^T$$

$$= 0. \quad (7.33)$$

The obtained system Eq. (7.33) is a system of integrals, which is a consequence of the inverse Fourier transform. The integrals are provided in Ref. [12]. The system of Eqs. (7.28–7.32) has a nontrivial solution if and only if the condition

$$\det[\mathbf{u}] = 0. \quad (7.34)$$

is satisfied. By examining the eigenvalues of the characteristic Eq. (7.34), the stability of the model can be studied, and that is done and presented in the following sections by applying the D-decomposition method and the argument principle.

The model of a vehicle shown in Fig. 7.4B contains two rows of suspension—the primary and the secondary suspension. Characteristics of this system and its dynamic stability are also analyzed in the upcoming sections. The governing equations, for the model in Fig. 7.4B, according to the dynamic equilibrium (Fig. 7.6) may be written in the reduced form as

$$\Phi_1[w,\varphi]=$$

$$-\delta(x-vt-d_3)\left(m\frac{d^2w_2^{01}}{dt^2}+\tilde{\mathscr{L}}_{21}\right)-\delta(x-vt-d_2)\left(m\frac{d^2w_2^{02}}{dt^2}+\tilde{\mathscr{L}}_{22}\right)$$

$$-\delta(x-vt-d_1)\left(m\frac{d^2w_1^{01}}{dt^2}+\tilde{\mathscr{L}}_{11}\right)-\delta(x-vt)\left(m\frac{d^2w_1^{02}}{dt^2}+\tilde{\mathscr{L}}_{12}\right), \tag{7.35}$$

$$\Phi_2[w,\varphi]=0, \tag{7.36}$$

$$M_1\frac{d^2w_2^0}{dt^2}-\tilde{\mathscr{L}}_{21}-\tilde{\mathscr{L}}_{22}+\tilde{\mathscr{L}}_2=0, \tag{7.37}$$

$$J_1\frac{d^2\theta_2}{dt^2}+\left(\tilde{\mathscr{L}}_{21}-\tilde{\mathscr{L}}_{22}\right)\frac{d_L}{2}=0, \tag{7.38}$$

$$M_1\frac{d^2w_1^0}{dt^2}-\tilde{\mathscr{L}}_{11}-\tilde{\mathscr{L}}_{12}+\tilde{\mathscr{L}}_1=0, \tag{7.39}$$

$$J_1\frac{d^2\theta_1}{dt^2}+\left(\tilde{\mathscr{L}}_{11}-\tilde{\mathscr{L}}_{12}\right)\frac{d_L}{2}=0, \tag{7.40}$$

$$M_0\frac{d^2w_0^0}{dt^2}-\tilde{\mathscr{L}}_1-\tilde{\mathscr{L}}_2=0, \tag{7.41}$$

$$J_0\frac{d^2\theta_0}{dt^2}+\left(\tilde{\mathscr{L}}_2-\tilde{\mathscr{L}}_1\right)\frac{d_{L0}}{2}=0, \tag{7.42}$$

$$w_2^{01}(t)=w(vt+d_3,t),\ \ w_2^{02}(t)=w(vt+d_2,t),$$

$$w_1^{01}(t)=w(vt+d_1,t),\ \ w_1^{02}(t)=w(vt,t),$$

$$\lim_{|x-vt|\to\infty} w(x,t)=0,\quad \lim_{|x-vt|\to\infty}\varphi(x,t)=0, \tag{7.43}$$

where

$$\tilde{\mathscr{L}}_{ij}=k_0\left(w_i^{0j}-w_i^0+\theta_i\frac{d_L}{2}(-1)^{j+1}\right)+\epsilon_0\frac{d}{dt}\left(w_i^{0j}-w_i^0+\theta_i\frac{d_L}{2}(-1)^{j+1}\right),$$

$$\tilde{\mathscr{L}}_i=k_1\left(w_i^0-w_0^0+\theta_0\frac{d_{L0}}{2}\right)+\epsilon_1\frac{d}{dt}\left(w_i^0-w_0^0+\theta_0\frac{d_{L0}}{2}\right),$$

in which the following are the translatory displacements: $w(x,t)$ is the vertical deflection of the beam; $w_1^{01}(t)$, $w_1^{02}(t)$, $w_2^{01}(t)$, and $w_2^{02}(t)$, are the vertical displacements of the masses (i.e., the wheels); $w_1^0(t)$, $w_2^0(t)$, and $w_0^0(t)$ are the vertical displacements of the mass centers of the bogies' bodies and the car's main body, respectively. The following are the angular displacements: θ_1, θ_2, and θ_0 are the angles of rotation of the bogies' bodies and the car's main body, respectively, around their centers of mass. The following are the physical properties: M_1, M_0 and J_1, J_0 are the masses and the moments of inertia of the bogie's body and the car's main body; k_1 and ϵ_1 are the stiffness and damping factors of the secondary suspension, respectively. Following the same procedure, as described in the previous model, the system of algebraic equations is obtained

$$\left(\sum_{\tilde{\tilde{j}}=0}^{6} \tilde{a}_{\tilde{\tilde{j}}} k^{\tilde{\tilde{j}}}\right)\cdot\left(\sum_{\tilde{\tilde{i}}=0}^{2} \tilde{b}_{\tilde{\tilde{i}}} k^{\tilde{\tilde{i}}}\right)^{-1}\cdot \hat{W}_{k,s}(k,s) = e^{-id_1 k}\Big[k_0 W_1^0(s) - k_0 W(d_1,s)$$
$$- ms^2 W(d_1,s) - s\epsilon_0 W(d_1,s) + s\epsilon_0 W_1^0(s) - \frac{1}{2}\Theta_1 k_0 d_L - \frac{1}{2}\Theta_1 s\epsilon_0 d_L\Big]$$
$$+ e^{-id_2 k}\Big(-k_0 W(d_2,s) - ms^2 W(d_2,s) - s\epsilon_0 W(d_2,s) + k_0 W_2^0(s)$$
$$+ s\epsilon_0 W_2^0(s) + \frac{1}{2}\Theta_2 k_0 d_L + \frac{1}{2}\Theta_2 s\epsilon_0 d_L\Big) + e^{-id_3 k}\Big(-k_0 W(d_3,s) - ms^2 W(d_3,s)$$
$$- s\epsilon_0 W(d_3,s) + k_0 W_2^0(s) + s\epsilon_0 W_2^0(s) - \frac{1}{2}\Theta_2 k_0 d_L - \frac{1}{2}\Theta_2 s\epsilon_0 d_L\Big) - k_0 W(0,s)$$
$$- ms^2 W(0,s) - s\epsilon_0 W(0,s) + k_0 W_1^0(s) + s\epsilon_0 W_1^0(s) + \frac{1}{2}\Theta_1 k_0 d_L + \frac{1}{2}\Theta_1 s\epsilon_0 d_L, \tag{7.44}$$

$$M_1 s^2 W_2^0(s) - k_0 W(d_2,s) - k_0 W(d_3,s) - s\epsilon_0 W(d_2,s) - s\epsilon_0 W(d_3,s)$$
$$- k_1 W_0^0(s) + 2k_0 W_2^0(s) + k_1 W_2^0(s) - s\epsilon_1 W_0^0(s)$$
$$+ 2s\epsilon_0 W_2^0(s) + s\epsilon_1 W_2^0(s) + \frac{1}{2}\Theta_0 k_1 d_{L0} + \frac{1}{2}\Theta_0 s\epsilon_1 d_{L0} = 0, \tag{7.45}$$

$$\Theta_2 J_1 s^2 - \frac{1}{2}k_0 d_L W(d_2,s) + \frac{1}{2}k_0 d_L W(d_3,s) - \frac{1}{2}s\epsilon_0 d_L W(d_2,s)$$
$$+ \frac{1}{2}s\epsilon_0 d_L W(d_3,s) + \frac{1}{2}\Theta_2 k_0 d_L^2 + \frac{1}{2}\Theta_2 s\epsilon_0 d_L^2 = 0, \tag{7.46}$$

$$M_1 s^2 W_1^0(s) - k_0 W(d_1,s) - s\epsilon_0 W(d_1,s) - k_0 W(0,s) - s\epsilon_0 W(0,s)$$
$$- k_1 W_0^0(s) + 2k_0 W_1^0(s) + k_1 W_1^0(s) - s\epsilon_1 W_0^0(s)$$
$$+ 2s\epsilon_0 W_1^0(s) + s\epsilon_1 W_1^0(s) - \frac{1}{2}\Theta_0 k_1 d_{L0} - \frac{1}{2}\Theta_0 s\epsilon_1 d_{L0} = 0, \tag{7.47}$$

$$\Theta_1 J_1 s^2 - \frac{1}{2} k_0 d_L W(0, s) + \frac{1}{2} k_0 d_L W(d_1, s) - \frac{1}{2} s\epsilon_0 d_L W(0, s)$$
$$+ \frac{1}{2} s\epsilon_0 d_L W(d_1, s) + \frac{1}{2}\Theta_1 k_0 d_L^2 + \frac{1}{2}\Theta_1 s\epsilon_0 d_L^2 = 0, \qquad (7.48)$$

$$M_0 s^2 W_0^0(s) + 2k_1 W_0^0(s) - k_1 W_1^0(s) - k_1 W_2^0(s)$$
$$+ 2s\epsilon_1 W_0^0(s) - s\epsilon_1 W_1^0(s) - s\epsilon_1 W_2^0(s) = 0, \qquad (7.49)$$

$$\Theta_0 J_0 s^2 - \frac{1}{2} k_1 d_{L0} W_1^0(s) + \frac{1}{2} k_1 d_{L0} W_2^0(s) - \frac{1}{2} s\epsilon_1 d_{L0} W_1^0(s) + \frac{1}{2} s\epsilon_1 d_{L0} W_2^0(s)$$
$$+ \frac{1}{2}\Theta_0 k_1 d_{L0}^2 + \frac{1}{2}\Theta_0 s\epsilon_1 d_{L0}^2 = 0. \qquad (7.50)$$

Beginning from the system of Eqs. (7.44–7.50) and applying to it the inverse Fourier transform over k, then eliminating W_0^0, W_1^0, W_2^0, Θ_0, Θ_1, and Θ_2, and setting the values $\xi = 0$, $\xi = d_1$, $\xi = d_2$, $\xi = d_3$ subsequently, will bring about the resulting system of algebraic equations—with respect to the Laplace-displacements of the contact points $\hat{W}_s(0, s)$, $\hat{W}_s(d_1, s)$, $\hat{W}_s(d_2, s)$, and $\hat{W}_s(d_3, s)$

$$[\tilde{\mathbf{u}}]\times\lfloor \widetilde{\hat{W}} \rfloor^{\mathrm{T}} = 0 \Leftrightarrow$$

$$\begin{bmatrix} \tilde{u}_{11} & \cdots & \tilde{u}_{14} \\ \vdots & \ddots & \vdots \\ \tilde{u}_{41} & \cdots & \tilde{u}_{44} \end{bmatrix} \times \left\lfloor \widetilde{\hat{W}_s}(0, s) \;\; \widetilde{\hat{W}_s}(d_1, s) \;\; \widetilde{\hat{W}_s}(d_2, s) \;\; \widetilde{\hat{W}_s}(d_3, s) \right\rfloor^{\mathrm{T}} = 0. \qquad (7.51)$$

The obtained system of Eq. (7.51) is a system of integrals, which is a consequence of the inverse Fourier transform. The integrals are provided in Ref. [12]. The system of Eqs. (7.44–7.50) does have a nontrivial solution if and only if the determinant of the matrix $[\mathbf{u}]$ satisfies $\det[\tilde{\mathbf{u}}] = 0$. The stability of this and the previous model is studied by investigating the eigenvalues of the characteristic equations: $\det[\mathbf{u}] = 0$; $\det[\tilde{\mathbf{u}}] = 0$, in the upcoming sections.

7.4 Dynamic stability analysis

If there is at least one root of the characteristic Eq. (7.34) with a positive real part, then the vibrations of the vehicle are unstable—this is the criterion for determining the instability of vibrations. The mentioned equation is an integral equation with respect to the Laplace variables. To solve the presented mathematical problem more easily, the D-decomposition method is applied. Denisov et al. [6] are the first to have used and described the concept of this method. Departing, for now, from the physical meaning of the problem, k_0 will be viewed as a complex parameter. This means that the k_0-plane is where the D-decomposition is performed, and the positive half of the real axis in the k_0-plane is correlated with the stiffness of the vehicle's suspension. The aim is to find all complex values k_0 (which means to delineate the lines in the

k_0-plane) that are correlated to the pure imaginary solution $s = i\Omega$ of Eq. (7.34). These lines will partition the k_0-plane into regions with a different number of roots of Eq. (7.34) with respect to the variable s, satisfying the condition $\mathrm{Re}(s) > 0$. To achieve this goal, Eq. (7.34) should be looked at in the following form

$$f(\Omega,\ k_0) = 0, \tag{7.52}$$

where $s = i\Omega$. The function f is a 6th-degree polynomial with respect to k_0. The next step is to find all the roots of Eq. (7.52) with respect to the variable k_0, when the parameter Ω is varied from minus to plus infinity. Function $F(k,\ \Omega)$, integrals, and constants are given in Ref. [12]. Consonant to the used contours of integration, which are shown in Fig. 5.2, the integration result may be written as

$$\begin{aligned}
\mathrm{Int}_0 &= i\sum\nolimits_n \frac{k-k_n}{(k-k_1)\cdots(k-k_6)}\bigg|_{k=k_n},\\
\mathrm{Int}_{\Delta-} &= -i\sum\nolimits_m \frac{(k-k_m)e^{ik\Delta}}{(k-k_1)\cdots(k-k_6)}\bigg|_{k=k_m},\ \Delta<0,\\
\mathrm{Int}_{\Delta+} &= i\sum\nolimits_n \frac{(k-k_n)e^{ik\Delta}}{(k-k_1)\cdots(k-k_6)}\bigg|_{k=k_n},\ \Delta>0,
\end{aligned} \tag{7.53}$$

in which k_n are the roots with a positive imaginary part, and k_m are the roots with a negative imaginary part of equation $F(k,\ \Omega) = 0$. After this, the D-decomposition curves can be analyzed. Six curves should be examined; each one of them is correlated to one of the six roots of Eq. (7.52).

The parametric analysis leads to the finding that the D-decomposition curves are of interest for dynamic analysis for the case in which the velocity is in the supercritical speed region, that is, when the vehicle system is moving at a velocity that is greater than a certain critical velocity v_{cr}. When the speed is in the subcritical region ($v < v_{cr}$), the D-decomposition curves do not cross the positive part of the real axes at any point. This means that when $v < v_{cr}$, the number of unstable roots of the characteristic equation (roots with a positive real part) does not vary with the stiffness of the suspension (k_0). However, when the $v > v_{cr}$, the number of unstable roots will change when k_0 surpasses a certain critical value of stiffness. The use of the D-decomposition method allows a relative variation of the number of unstable roots (depending on the stiffness k_0) to be determined, but not finding the number itself. Thus finding the number of unstable roots for any value of the stiffness k_0, is what remains to be done. This can be achieved by using the argument principle.

To determine the number of unstable roots, the chosen contour should encircle the right half-plane of the complex variable s, as is illustrated in Fig. 5.4. The method consists of setting a fixed value for k_0, and then computation of the total change of the argument of the complex expression $f(-is,\ k_0)$ along the entire contour, where f is the function from Eq. (7.52). Under the argument principle, the total change is equal to $2\pi N_{unst}$, where N_{unst} is the number of roots of $f(-is,\ k_0)$ (i.e., Eq. (7.46)) located in the positive half-plane, with respect to s. For both parts of the contour (the semicircle and y-axis), some sufficiently small iteration step Δs is chosen, with which the relative change of the argument—$\mathrm{Arg}(f(-i(s+\Delta s),\ k_0)/f(-is,\ k_0))$—is computed, and all of these relative changes are summed up. The size of the iteration step Δs is dynamically adopted along the contour.

The set of the model parameters used by Verichev and Metrikine [19] and Stojanović et al. [9] was used here for obtaining the D-decomposition curves for the model from Fig. 7.4A

$$\rho = 7849\,\text{kg},\ A = 7.0687 \times 10^{-3}\,\text{m}^2,\ I = 3.055 \times 10^{-5}\,\text{m}^4,$$

$$E = 2 \times 10^{11}\text{Pa},\ G = 7.813 \times 10^{10}\text{Pa},\ k_f = 10^8\,\frac{\text{N}}{\text{m}^2},\ \nu_f = 10^4\,\frac{\text{N·s}}{\text{m}^2}$$

$$J_i = 598 \times 10^3\,\frac{\text{kg·m}}{\text{s}^2},\ M_i = 20 \times 10^3\,\text{kg},$$

$$m = 10^3\text{kg},\ \varepsilon_0 = 86 \times 10^3\,\frac{\text{N·s}}{\text{m}},$$

$$d_V = d_L = 15.7\,\text{m},\ d_R = 1\,\text{m},\ d_{\theta\text{i}} = 8.3\,\text{m}.\ i = (1,\ 2,\ 3). \tag{7.54}$$

and for the model from Fig. 7.4B, the following set used by Yang et al. [20]

$$k_1 = 8 \times 10^5\,\frac{\text{N}}{\text{m}},\ \varepsilon_1 = 10^5\,\frac{\text{N·s}}{\text{m}},\ \varepsilon_0 = 86 \times 10^3\,\frac{\text{N·s}}{\text{m}}$$

$$J_0 = 2.1 \times 10^6\,\frac{\text{kg·m}}{\text{s}^2},\ J_1 = 3200\,\frac{\text{kg·m}}{\text{s}^2},$$

$$M_0 = 32 \times 10^3\text{kg},\ M_1 = 3.3 \times 10^3\,\text{kg},$$

$$m = 10^3\text{kg},\ d_L = 1\,\text{m},\ d_{L0} = 14.7\,\text{m}. \tag{7.55}$$

7.5 Influence of the viscoelastic coupling on dynamic stability of a triple vehicle system

This section is about the analysis of the influence of elastic coupling on dynamic stability. The elastic coupling, which links the bars (i.e., the rail cars), is defined by the parameters $k_{vj},\ k_{rotj},\ \varepsilon_{vj},\ \varepsilon_{rotj},\ j = 1,\ 2$. For varying values of these parameters, the D-decomposition curves and the number of unstable roots are computed. Fig. 7.7 shows the obtained D-decomposition curves for each root separately. In these figures, the D-decomposition curves are displayed separately to illustrate the critical stiffnesses without numbering the unstable roots due to the returning D-decomposition lines of the other roots. An integrated diagram with all the curves is difficult to analyze. The rules for counting the unstable roots were described by Stojanović et. al. [9] where the moving mass problem was studied. Here, presented are three cases with different stiffness values of the elastic coupling. The first case relates to the conventional connection between cars; the connection already present in technical practice that does not allow transverse deflections in the contact points ($k_{v1} = 2 \times 10^{10}$ N/m; $k_{v2} = 2 \times 10^{10}$ N/m). In the second case, one specific elastic

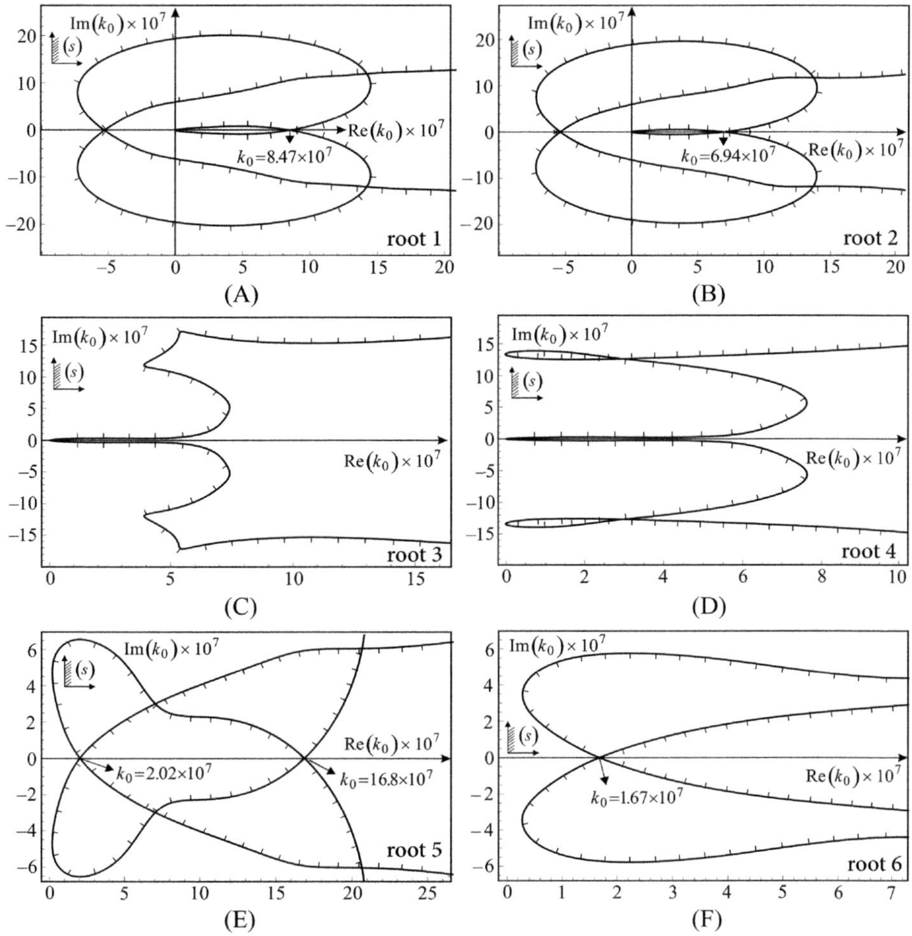

Figure 7.7 D-decomposition curves. (A–F) Roots 1–6 for Case 1 (N/m).

coupling with decreased stiffness in the transversal direction is chosen ($k_{v1} = 2 \times 10^{10}$N/m; $k_{v2} = 10^5$N/m). In the third case, two specific elastic couplings with decreased stiffness in the transversal direction are chosen ($k_{v1} = 10^5$N/m; $k_{v2} = 10^5$N/m). The results obtained for the first case are reported in Fig. 7.7. The sections of the D-decompositions curves and real axis present the real values of the stiffness of the support where the number of unstable roots changes.

To properly determine the number of unstable roots in the system, the D-decomposition curves must be integrated into one diagram. For the supercritical velocity $v = 1.12\ v_{cr}$, the number of unstable roots for each interval of k_0, according to Fig. 7.7 are given in Table 7.1. The integrated diagrams of the D-decomposition curves for all three cases are given in Fig. 7.8.

As mentioned, only the sections with the real axis are of interest, considering that the next step in the stability analysis is determining the exact number of unstable

Table 7.1 The number of unstable roots. Influence of stiffness k_v on stability regions; v=1.12v_{cr}, k_{v1}=2×10^{10} N/m, k_{v2}=10^5 N/m, k_{rot1}=k_{rot2}=k_{rot3}=10 (N·m)/rad, ε_{v1}=ε_{v2}=ε_{v3}=10 (N·s)/m, ε_{rot1}=ε_{rot2}=ε_{rot3}=10 (N·m·s)/rad, unit for k_0 is 10^7 N/m.

k_0	**0**	**0.2064**	**0.2127**	**0.2195**	**1.6795**	**1.6951**	**1.6978**	**1.9527**	**1.9785**	**2.0246**	**2.2028**	**2.7047**	**3.5283**	**4.4776**	**4.5235**	**6.5060**	**6.9499**	**8.4703**	**10.151**	**16.8158**	**25.1663**	**+∞**
Case 1	4	4|4	4|4	4|4	4|4	4|2	2|2	2|2	2|2	2|0	0|0	0|0	0|0	0|0	0|0	0|0	0|2	2|4	4|4	4|6	6|6	6
Case 2	5	5|5	5|4	4|4	4|4	4|4	4|2	2|2	2|0	0|0	0|0	0|0	0|2	2|4	4|4	4|4	4|4	4|4	4|6	6|6	6|7	7
Case 3	6	6|5	5|5	5|4	4|2	2|2	2|2	2|0	0|0	0|0	0|2	2|4	4|4	4|4	4|6	6|8	8|8	8|8	8|8	8|8	8|8	8

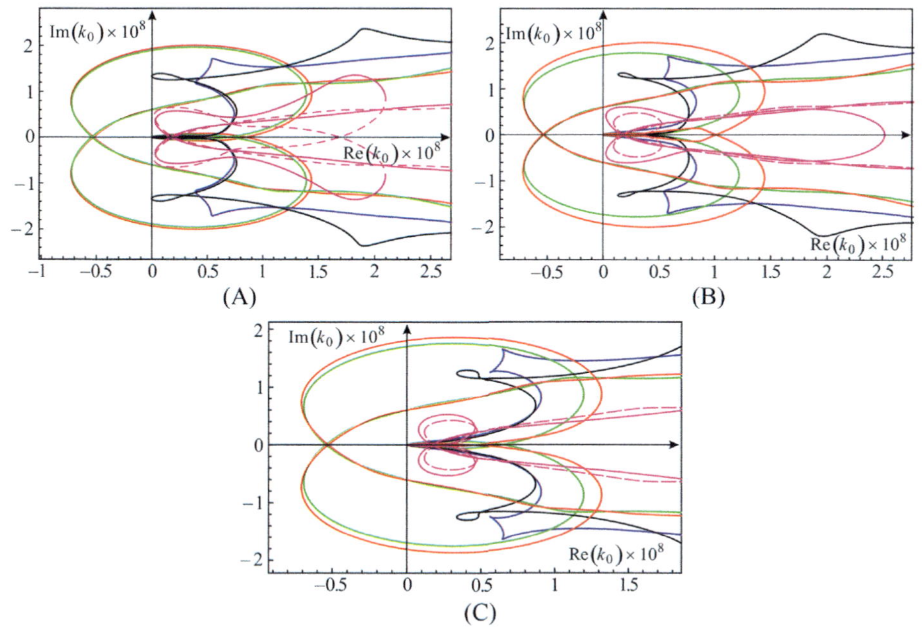

Figure 7.8 D-decomposition curves (N/m). The parameters and stability domains are presented in Table 7.1. (A) Case 1; (B) Case 2; (C) Case 3.

roots for each segment of k_0. This can be achieved with the argument principle, which is described earlier. As can be seen in Table 7.1, for the selected set of parameters, a stable region without unstable roots still exists, even though the speed is in the supercritical region. The system with two proposed specific elastic couplings (the third case) has a stable region within the interval $k_0 \in (1.95,\ 1.97) \times 10^7$ N/m. The system with one specific elastic coupling (the second case) is stable within the interval $k_0 \in (1.97,\ 2.02) \times 10^7$ N/m. Within the mentioned intervals, the system with traditional couplings is unstable. This leads to the conclusion that the region of stability can be changed by changing the parameters of the elastic couplings. Put differently, the system with fixed parameters can be made stable for varying velocities, even for the speed in the supercritical region. The dependence of the critical stiffness of the suspension on the vehicle speed can also be determined with the D-decomposition method. Examples of this dependence are shown in Fig. 7.9. The boundaries of the stability domain for the three mentioned cases are also presented.

From Fig. 7.9A, it can be seen that the system belonging to the first case has a larger stability region in comparison with the system associated with the second case, but the second case system can be stable for lower values of the suspension stiffness and higher velocities. Fig. 7.9B shows the comparison between the stable regions associated with the first and third cases. A similar conclusion as in the previous explanation can be drawn.

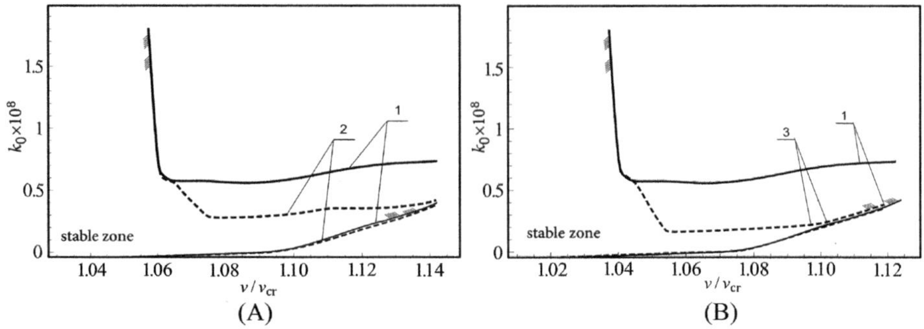

Figure 7.9 Boundaries of the stability domain. (A) For two different cases of elastic couplings (Case 1 and Case 2). (B) For two different cases of elastic couplings (Case 1 and Case 3).

Table 7.2 Varying damping coefficient of the secondary suspension.

Varying parameter	$\varepsilon_1 \left[\frac{\text{N}\cdot\text{s}}{\text{m}}\right]$
Case 4	10^5
Case 5	2×10^5
Case 6	10^6

7.6 Influence of the secondary suspension on dynamic stability of a vehicle that has four contact points with the rail

7.6.1 Influence of the damping factor (ε_1), produced by the secondary suspension

In this section, the influence of the damping, generated by the secondary suspension of the vehicle, on stability is analyzed. The varying damping parameter is given in Table 7.2. The integrated D-decomposition curves are given in Fig. 7.10. It is indisputable that an increase in the damping coefficient of the secondary suspension brings about a decrease in dynamic stability. The completely stable region in Case 4 is within the interval $k_0 \in (1.74,\ 3.07) \times 10^7$ N/m. The stable region in Case 5 is within the interval $k_0 \in (1.791,\ 2.96) \times 10^7$ N/m. Case 6 encompasses only the interval $k_0 \in (1.94,\ 2.89) \times 10^7$ N/m. The other intervals with various levels of instability are given in Table 7.3. The level of instability is determined based on the number of unstable roots. The system in Case 6 is notably unstable outside the zones of complete stability, in comparison with other cases.

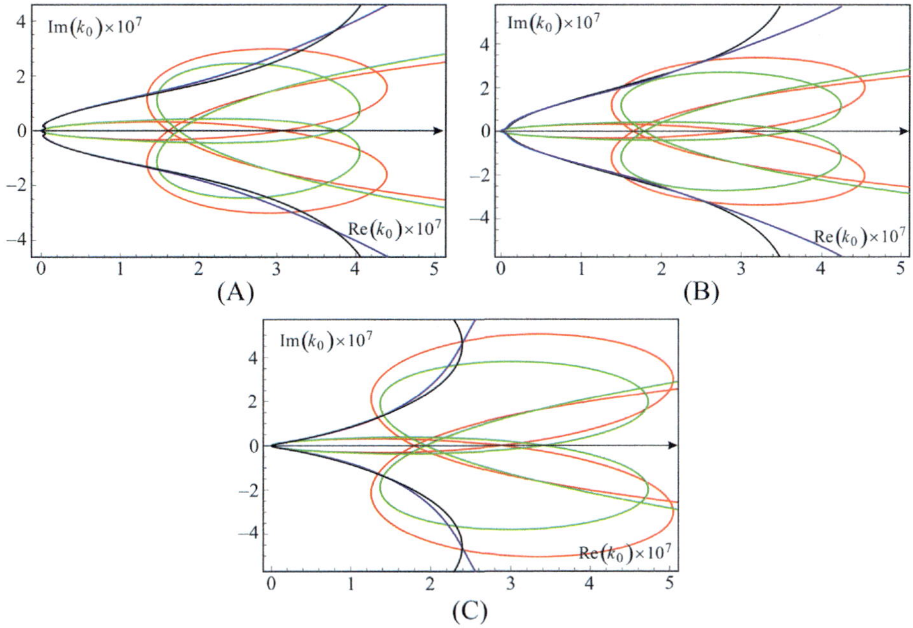

Figure 7.10 D-decomposition curves (N/m). The parameters and stability domains are presented in Table 7.2. (A) Case 4; (B) Case 5; (C) Case 6.

7.6.2 Influence of the secondary suspension's stiffness (k_1)

In Table 7.4, the stability domains for different values of the stiffness of the secondary suspension are presented. If the secondary suspension is a rigid connection, then the change in stability domains is interesting. The example with the rigid connection is chosen because of the minuscule change in stability regions by varying the stiffness of the secondary suspension. In the rigid connection scenario (Case 8), the stable domain belongs to the interval $k_0 \in (1.89,\ 3.83) \times 10^7$ N/m. Compared to a scenario with lower stiffness (Case 7), it is noticeable that the rigid connection produces a completely stable region in the interval $k_0 \in (3.05,\ 3.83) \times 10^7$ N/m, while case 7 in this region has two and four unstable roots. However, an elastic connection (Case 7) possesses better stability characteristics in the lower intervals $k_0 \in (1.60,\ 3.05) \times 10^7$ N/m. A set of the complex D-decomposition curves, based on which the stability regions are determined, are presented in Fig. 7.11.

7.7 Influence of the supercritical velocity and varying parameters of the secondary suspension on boundaries of the stability domain

With the D-decomposition method, it is possible to find the dependence of the critical stiffness (of the primary suspension) on the velocity (of the vehicle) for varying

Table 7.3 The number of unstable roots. The level of the instability of k_0 intervals according to the number of unstable roots (Cases 4–6, v=1.12v_{cr}), unit for k_0 is 10^7 N/m.

k_0	**0**	**1.60**	**1.65**	**1.74**	**1.791**	**1.798**	**1.94**	**2.89**	**2.96**	**3.07**	**3.42**	**3.65**	**3.75**	**+∞**
Case 4	4	4|2	2|2	2|0	0|0	0|0	0|0	0|0	0|0	0|2	2|2	2|2	2|4	4
Case 5	4	4|4	4|2	2|2	2|0	0|0	0|0	0|0	0|2	2|2	2|2	2|4	4|4	4
Case 6	4	4|4	4|4	4|4	4|4	4|2	2|0	0|2	2|2	2|2	2|4	4|4	4|4	4

Table 7.4 The number of unstable roots. The level of the instability of k_0 intervals according to the number of unstable roots (Cases 7 and 8, v=1.12v_{cr}), unit for k_0 is 10^7 N/m.

k_0	0	1.60	1.74	1.76	1.89	3.05	3.72	3.83	3.91	4.86	9.71	+∞
Case 7	4	4|2	2|0	0|0	0|0	0|2	2|4	4|4	4|4	4|4	4|4	4
Case 8	4	4|4	4|4	4|2	2|0	0|0	0|0	0|2	2|4	4|6	6|4	4

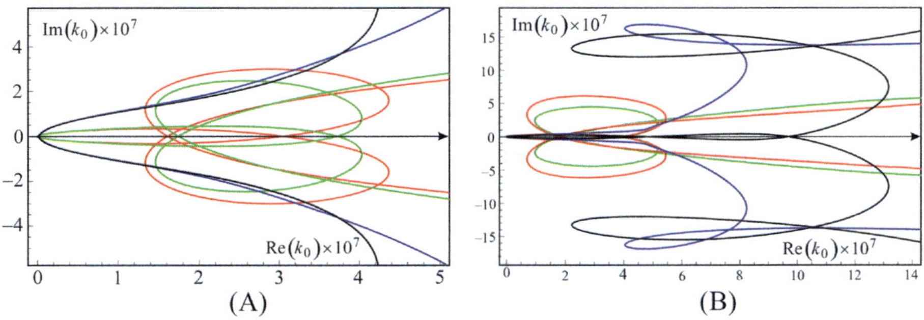

Figure 7.11 D-decomposition curves (N/m). The parameters and stability domains are presented in Table 7.4. (A) Case 7; (B) Case 8 (rigid connection).

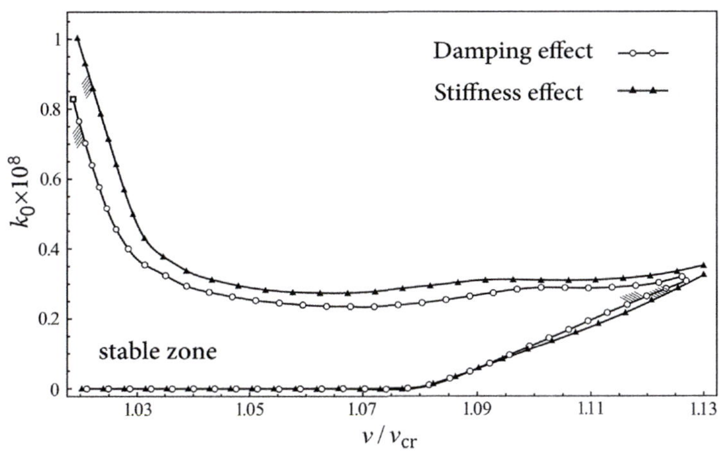

Figure 7.12 Boundaries of the stability domain. Case 6—damping factor (ε_1=10^6 N·s/m); Case 8—stiffness factor (k_1=∞).

parameters (the damping and stiffness parameters of the secondary suspension), as with the results previously reported (Fig. 7.9). This type of dependence is shown in Fig. 7.12; with indicated boundaries of the stability domain for two different cases. An interesting observation is that the influence of velocity on stability depends on the varying parameters of the secondary suspension. This means that the zones of stability can be changed by modifying the parameters of the secondary suspension. The increase in velocity in both cases (Fig. 7.12) causes a decrease in the stable region's size. Due to the results presented and discussed, analyzing the influence of any parameter for each velocity is important.

7.8 Conclusion

The analysis presented in this chapter deals with the vibration stability of two types of complex vehicle systems, both traveling on an infinitely long rail that is continuously supported by a viscoelastic foundation. For solving the mathematical problems that arise during the analysis, the D-decomposition method and the argument principle were applied. These methods were used to partition the space of the model parameters into domains and to determine the number of unstable roots in one of these domains. The principal contribution of these analyses, concerning the first vehicle system, is the finding that specific viscoelastic couplings (used to link the vehicle elements) can have a beneficial influence on the vibrational stability of the vehicle. For the second vehicle type (the vehicle with two rows of suspension), the secondary suspension can also have positive modifying effects on the regions of stability. In both cases, varying values of the supercritical velocity were considered.

When the proposed specific viscoelastic couplings are used, a new stable region (interval without unstable roots) can appear at the location where the system with the traditional couplings is unstable.

An increase in dynamic stability appears when the damping of the secondary suspension is decreased.

When the secondary suspension is a rigid connection (i.e., $k_1 \to \infty$), a completely stable region appears at the location where the suspension with lower k_1 stiffness has two and four unstable roots. Contrary to that, an elastic suspension (lower k_1 stiffness) has better stability characteristics on the lower intervals of the primary suspension stiffness (k_0).

The interesting phenomena that occur during the travel of the complex vehicle system on the rail of an infinite length, which is considered in the present study, are essential for the analysis on a parameter basis. Because even the minute changes in parameters are accountable for the extensive changes in dynamic stability.

References

[1] S.P. Timoshenko, Method of analysis of statical and dynamical stresses in rail. In: Proceedings of the 2nd International Congress for Applied Mechanics, Zürich, Switzerland, 1927, pp. 1–12.

[2] A.D. Kerr, The continuously supported rail subjected to an axial force and a moving load, International Journal of Mechanical Sciences 14 (1972) 71–78.

[3] J. Dörr, Der unendliche, federnd gebettete Balken unter dem Einflu einer gleichförmig bewegten Last, Ingenieur-Archiv 14 (3) (1943) 167–192.

[4] J.T. Kenney, Steady-state vibrations of beam on elastic foundation for moving load, Journal of Applied Mechanics 21 (1954) 359–364.
[5] L. Frýba, Vibration of Solids and Structures under Moving Loads, Springer, Netherlands, 1972.
[6] G.G. Denisov, E.K. Kugusheva, V.V. Novikov, On the problem of the stability of one-dimensional unbounded elastic systems, Journal of Applied Mathematics and Mechanics 49 (4) (1985) 533–537.
[7] M. Serpilli, S. Lenci, Limit models in the analysis of three different layered elastic strips, European Journal of Mechanics—A/Solids 27 (2) (2008) 247–268.
[8] S. Lenci, G. Rega, An asymptotic model for the free vibrations of a two-layer beam, European Journal of Mechanics—A/Solids 42 (2013) 441–453.
[9] V. Stojanović, P. Kozić, M.D. Petković, Dynamic instability and critical velocity of a mass moving uniformly along a stabilized infinity beam, International Journal of Solids and Structures 108 (2017) 164–174.
[10] S.-H. Ju, H.-T. Lin, Analysis of train-induced vibrations and vibration reduction schemes above and below critical Rayleigh speeds by finite element method, Soil Dynamics and Earthquake Engineering 24 (12) (2004) 993–1002.
[11] M.A. Rezvani, A. Mazraeh, Dynamics and stability analysis of a freight wagon subjective to the railway track and wheelset operational conditions, European Journal of Mechanics—A/Solids 61 (2017) 22–34.
[12] V. Stojanović, M.D. Petković, J. Deng, Instability of vehicle systems moving along an infinite beam on a viscoelastic foundation, European Journal of Mechanics/A Solids 69 (2018) 238–254.
[13] A.T. Karttunen, R. von Hertzen, Steady-state vibration of a viscoelastic cylinder cover subjected to moving loads, European Journal of Mechanics—A/Solids 58 (2016) 202–210.
[14] A.V. Metrikine, Unstable lateral oscillations of an object moving uniformly along elastic guide as a result of anomalous Doppler effect, Acoustical Physics 40 (1) (1994) 85–89.
[15] K. Chandrashekhara, K. Krisnamurty, S. Roy, Free vibration of composite beams including rotary inertia and shear deformation, Composite Structures 14 (1990) 269–279.
[16] S.J. Song, A.M. Waas, Effects of shear deformation on buckling and free vibration of laminated composite beams, Composite Structures 37 (1997) 33–43.
[17] J.N. Reddy, Mechanics of Laminated Composite Plates and Shells: Theory and Analysis, second ed., CRC Press, 2003.
[18] V. Stojanović, M.D. Petković, Nonlinear dynamic analysis of damaged Reddy–Bickford beams supported on an elastic Pasternak foundation, Journal of Sound and Vibration 385 (2016) 239–266.
[19] S.N. Verichev, A.V. Metrikine, Instability of vibrations of a mass that moves uniformly along a beam on a periodically inhomogeneous foundation, Journal of Sound and Vibration 260 (2003) 901–925.
[20] X. Yang, S. Gu, S. Zhou, J. Yang, Y. Zhou, S. Lian, Effect of track irregularity on the dynamic response of a slab track under a high-speed train based on the composite track element method, Applied Acoustics 99 (2015) 72–84.

Modeling of a three-part viscoelastic foundation and its effect on dynamic stability

8

8.1 Introduction

High-speed transportation comes from progress in road and railway traffic. For achieving further progress, a more extensive analysis of the dynamic effects of the varying physical parameters on the stability of such mechanical systems is needed. The instability of vibrations may occur due to the phenomenon in which energy is transferred into vibrational energy through Doppler waves [1,2]. This often happens when a high-speed train travels faster than the elastic waves that propagate in the railway structure. Further study in this area was made possible on account of research concerning the instability of vibrations of simple mechanical models. Many researchers have studied the dynamic stability and instability of mechanical systems subjected to moving loads. Timoshenko [3] showed that a moving constant vertical force has a critical velocity at which deflections become substantial. Kerr [4] studied the response of an axially compressed continuously supported beam of infinite length subjected to a constant vertical force that is in motion. The mechanical model of a beam that is vibrating and is supported by an elastic foundation, subjected to a constant vertical force that is moving, is the basis for further research involving the model of a beam subjected to a moving mass. Thus Denisov et al. [5] identified the stability region for a mass that moves uniformly on a continuously supported beam by applying the D-decomposition method. Clarification of this and further analysis of dynamic instability, using the D-decomposition method, can be found in the work of Verichev and Metrikine [6]. Certain studies suggest that continuous mechanical systems can have a very complex dynamic response. Comprehension of dynamic behavior is not possible without knowing the interactions between discrete and continuous systems, as shown in the study by Dimitrovová and Varandas [7]. They studied the transient dynamic response of a beam supported on a foundation that has abrupt stiffness change and is subjected to a vertical force moving at a constant velocity. To solve the problem, they used finite integral transformations and linked both parts of the beam by continuity conditions.

High-speed trains are interesting for their genuine dynamic problems, which are documented in the literature. Like in the study concerning fast trains done by Ju and Lin [8], as referred to in Chapter 4. Achieving high-speed traffic has been made possible because of theoretical research, which encompasses an array of scientific fields, for example, civil engineering, mechanical engineering, automation, control, and electrical engineering. Zhang and Xia [9] presented noteworthy results from the engineering practice concerning the dynamic analysis of a coupled vehicle-bridge system. Using the intersystem iteration method, they got the dynamic response of the

Modeling of Complex Dynamic Systems. DOI: https://doi.org/10.1016/B978-0-443-23942-7.00008-5

bridge subsystem and the vehicle subsystem separately. The results they got were sufficiently close to the results obtained from the commonly used time-step iteration method, just with fewer programming predicaments. Contrary to that study, the complexity of the model can also come from the continuous system, so improvements in methods for these systems are needed. Therefore for the cases where vibrations are induced by a moving mass, Dimitrovová [10] suggested an iterative procedure for the calculation of the vibrational frequency. This kind of mechanical system—the system with moving properties—is studied here. Many researchers have provided contributions and initiatives for further research relating to the mechanics and stability of such systems. Additionally, it should be mentioned that the instability of the vehicle/track system comes from the reaction of the surface waves excited in the track subsoil. This phenomenon could occur in the case of lines crossing zones with soft soils and such instability has been reported in the past [11]. However, it should be mentioned that the instability of the vehicle/track system could be solved with two diverse—yet complementary—methods. One is qualitative and aims to find the conditions when the system is stable or unstable; this is done by determining the linear critical velocity and using a linear model. The other one is quantitative, which uses a nonlinear model, with it the stability is determined by the time-domain analysis. When the system becomes unstable, the amplitude increases to the point at which the wheels lose contact with the rail, and the vibration has a limited cycle due to the contrary effects of abnormal Doppler waves and static load. Applying bifurcation theory to the nonlinear stability analysis reveals that the system becomes unstable at the nonlinear critical velocity, which is lower than the linear critical velocity [12–14]. In the aforementioned studies, the stability of vibrations and dynamic response analyses were done for less complex train models [6] or finite continuous beams [9]. This chapter is about a study of a complex rail vehicle model that has two rows of suspension and four contact points with the rail. It is based on a study done by Stojanović et al. [15]. A Reddy–Bickford beam assumption is used for the mathematical description of the deformable rail. Parametric analysis is performed and stability regions are examined according to varying physical parameters.

8.2 Dispersion relation and critical velocity

The first step is to derive equations that are associated with the beam theory that was used to describe the continuous system (the beam/foundation system). For further analysis of critical velocity, two different beam theories are used: the Euler–Bernoulli beam assumption and the high-order shear deformation beam assumption. Since the system's critical velocity depends only on the properties of the beam (together with the applied beam assumption) and the properties of the foundation, it is realistic to expect some difference between the determined critical velocities. The high-order beam theories take into account rotary inertia and shear deformation; therefore they are more accurate but computationally more demanding. Using a high-order beam theory in this study is justified because it leads to a more accurate determination of stability regions for varying parameters. The high-order beam model takes into account rotational bending effects and shear deformation, for they are important factors in describing the behavior of beams subjected to high-frequency excitation when the wavelength gets close to the thickness of the beam. The material from which the beam is made is assumed to be elastic, isotropic, and homogenous. The foundation is defined as a viscoelastic layer. Using a high-order shear deformation theory, or the Reddy–Bickford beam assumption, the model displacement field can be written as:

$$u(x, z, t) = u^0(x, t) + z\varphi^0(x, t) - \alpha z^3\left(\varphi^0(x, t) + \frac{\partial w^0(x, t)}{\partial x}\right),$$

$$w(x, z, t) = w^0(x, t). \tag{8.1}$$

In Eq. (8.1), the longitudinal displacement is designated with $u^0(x, t)$, the transversal displacement with $w^0(x, t)$, the independent rotation of the cross-section about the axis parallel to y is designated with $\varphi^0(x, t)$, and the factor is $\alpha = 4/h^3$ (Fig. 7.1).

The longitudinal and shear strains have the following mathematical form:

$$\varepsilon_x(x, t) = \frac{\partial u^0}{\partial x} + z\frac{\partial \varphi^0}{\partial x} - \alpha z^3\left(\frac{\partial \varphi^0}{\partial x} + \frac{\partial^2 w^0}{\partial x^2}\right), \tag{8.2}$$

$$\gamma_{xz}(x, t) = \frac{\partial w^0}{\partial x} + \varphi^0 - \beta z^2\left(\varphi^0 + \frac{\partial w^0}{\partial x}\right), \tag{8.3}$$

$$\begin{Bmatrix} \sigma_x \\ \tau_{xz} \end{Bmatrix} = \begin{bmatrix} E & 0 \\ 0 & G \end{bmatrix}\begin{Bmatrix} \varepsilon_x \\ \gamma_{xz} \end{Bmatrix}, \tag{8.4}$$

in which E is the Young modulus, G stands for the shear modulus of elasticity, and the value of the factor is $\beta = 3\alpha = 12/h^3$. The model describing the foundation is physically defined via the stiffness modulus k_f, and the damping coefficient ν_f (for the case of viscoelastic foundation). The reaction of the foundation is:

$$r(x, t) = k_f w(x, t) + \nu_f \frac{\partial w(x, t)}{\partial t}. \tag{8.5}$$

The equations of motion are derived from the principle of virtual work, according to which the virtual work of the inertial forces is:

$$\delta W_{in} = -\rho \int_V \left[\frac{\partial^2 w(x, z, t)}{\partial t^2}\delta w(x, z, t) + \frac{\partial^2 u(x, z, t)}{\partial t^2}\delta u(x, z, t)\right]\mathrm{d}V, \tag{8.6}$$

the virtual work of the internal forces is:

$$\delta W_V = -\int_V [\sigma_x(x, t)\delta\varepsilon_x(z, t) + \tau_{xz}(x, t)\delta\gamma_{xz}(x, t)]\mathrm{d}V, \tag{8.7}$$

and the virtual work of the external forces is defined by

$$\delta W_{ex} = \int_L [p_z(x, t) - r(x, t)]\delta w(x, t)\mathrm{d}L. \tag{8.8}$$

The next step would be to introduce the general equation of the virtual work principle; $\delta W_{in} + \delta W_V + \delta W_{ex} = 0$, and substitute the Eqs. (8.6–8.8) into that equation. Then the Green theorem is applied successively—integrating the equation by parts and collecting the coefficients of δu, δw, and $\delta\varphi$. The result will turn out to be a set of partial differential equations (PDEs). This set of equations mathematically describes the vibrations of the system comprising the Reddy–Bickford beam that is resting on a viscoelastic foundation, and their form is:

$$\Psi[w, \varphi] + k_f w + \nu_f \frac{\partial w}{\partial t} = p_z(x, t), \tag{8.9a}$$

$$\Phi[w, \varphi] = 0, \tag{8.9b}$$

where

$$\Psi[w, \varphi] = C_1 \frac{\partial^2 w}{\partial t^2} - C_2\left(\frac{\partial \varphi}{\partial x} + \frac{\partial^2 w}{\partial x^2}\right) - C_3 \frac{\partial^4 w}{\partial x^2 \partial t^2} + C_4 \frac{\partial^4 w}{\partial x^4} + C_5 \frac{\partial^3 \varphi}{\partial x \partial t^2} + C_6 \frac{\partial^3 \varphi}{\partial x^3}, \tag{8.9c}$$

$$\Phi[w, \varphi] = D_1 \frac{\partial^2 \varphi}{\partial t^2} - D_2 \frac{\partial^2 \varphi}{\partial x^2} + D_3\left(\varphi + \frac{\partial w}{\partial x}\right) + D_4 \frac{\partial^3 w}{\partial x \partial t^2} + D_5 \frac{\partial^3 w}{\partial x^3}, \tag{8.9d}$$

in which C_i, D_j are determined and given in Ref. [15]. System of Eqs. (8.9a–d) can be acquired by reducing the nonlinear terms in the set of nonlinear PDEs [16]. When instead of the high-order shear deformation beam assumption, the simpler Euler–Bernoulli beam assumption is used, the mathematical description of the vibrations will be as follows:

$$EI\frac{\partial^4 w}{\partial x^4} + \rho A \frac{\partial^2 w}{\partial t^2} + k_f w + \nu_f \frac{\partial w}{\partial t} = p_z(x, t). \tag{8.10}$$

A moving rail vehicle system has complex dynamical behavior, which is described later. To understand this dynamic, it is customary to consider the kinematics of waves emitted by a mass m that is moving on a beam at a constant velocity v, as shown in Fig. 7.2. In that case, Eq. (8.10) becomes:

$$EI\frac{\partial^4 w}{\partial x^4} + \rho A_{cs} \frac{\partial^2 w}{\partial t^2} + k_f w + \nu_f \frac{\partial w}{\partial t} = -\delta(x - vt) m \frac{d^2 w_0}{dt^2}, \tag{8.11}$$

in which $\delta(\,)$ represents the Dirac delta function. As it was clarified in Ref. [6], the wave number k and frequency ω of the waves, emitted by the mass, satisfy the dispersion relation, which is obtained by substituting $w(x, t) = A\exp\{i[(\Omega - kv)t - kx]\}$, $i = \sqrt{-1}$ into Eq. (8.11). For the undamped case, the following is obtained:

$$EIk^4 + k_f w - \rho A_{cs}(\Omega - kv)^2 = 0. \tag{8.12}$$

The roots of the dispersion equation are real when the velocity of the moving mass is lower than the critical. In addition, when the velocity is lower than the critical, the vibrations of the mass (that is traveling on the beam and does not lose contact with the beam) are harmonic. If the vibrations of the mass do not disrupt the waves in the beam, the mass can continue to vibrate harmonically. Using the discriminant of dispersion Eq. (8.12), the critical velocity (the minimum phase velocity of waves in the beam) can be determined from the condition:

$$\mathrm{Disc}_k\left[EIk^4 + k_f w - \rho A_{cs}(\Omega - kv)^2\right] = 0 \Rightarrow v_{cr,E} = v_{ph}^{\min} = \left[\frac{4EIk_f}{(\rho A_{cs})^2}\right]^{0.25}. \tag{8.13}$$

Kerr [4] obtained the same minimum phase velocity of waves in the beam presented in Eq. (8.13). Tracing the same method and substituting $w(x,t) = A_1 \exp\{i[(\Omega - kv)t - kx]\}$, $\varphi(x,t) = A_2 \exp\{i[(\Omega - kv)t - kx]\}$ into Eqs. (8.9a–d), dispersion relation can be obtained and its mathematical form is:

$$k^6 q_3 + k^4 q_2 + k^2 q_1 + q_0 = 0, \tag{8.14}$$

in which constants q_0, q_1, q_2, and q_3 are given in Ref. [15]. This dispersion relation takes into account rotational bending effects and shear deformation, for they are important factors in describing the behavior of beams subjected to high-frequency excitation when the wavelength gets close to the thickness of the beam. Using the discriminant of the dispersion Eq. (8.14), after some algebra, the critical velocity of the system can be determined. That is, when the roots of the dispersion Eq. (8.14) are not real (these roots are also the wave numbers k). In that instance, the critical velocity can be obtained from the form:

$$\mathrm{Disc}_k\left[k^6 q_3 + k^4 q_2 + k^2 q_1 + q_0\right] = 0 \Rightarrow v_{cr,RB} = v_{ph}^{\min} = \left[\sqrt{\frac{\tilde{p}_1^2}{4} - \tilde{q}_1} - \frac{\tilde{p}_1}{2}\right]^{0.5}. \tag{8.15}$$

Constants $\tilde{p}_1$ and $\tilde{q}_1$ are determined and presented in Ref. [15]. For the considered beam models, the ratio between the critical velocities $\zeta = v_{cr,E}/v_{cr,RB}$ in the function of the foundation stiffness k_f is presented in Fig. 7.3. There, the critical velocity determined using the Euler–Bernoulli beam assumption is designated with $v_{cr,E}$, while the critical velocity determined using the Reddy–Bickford beam assumption is designated with $v_{cr,RB}$.

The ratio between the critical velocities (Fig. 7.3) was calculated using the system parameters given in Refs. [2,6]. When the stiffness of the foundation increases, the defined critical velocity ratio ζ also increases. The high-order shear deformation beam assumption gives lower values for the critical velocity than the Euler–Bernoulli beam assumption. Following the same procedure, it is possible to obtain the critical velocity in the case of the complex three-part viscoelastic beam/foundation model depicted in Fig. 8.1. In this model, the rail is represented as a supported infinite beam, both the slab and the base are considered as supported free beams, and the connections of the track structure are modeled as viscoelastic spring-damping elements. The slab track system consists of the rail, pad, slab, CA mortar, and base as shown in Fig. 8.1A. In this configuration, the velocity ratio is $\zeta = v_{cr,E}/v_{cr,RB} = 1.0086$; for the parameters available in Ref. [17].

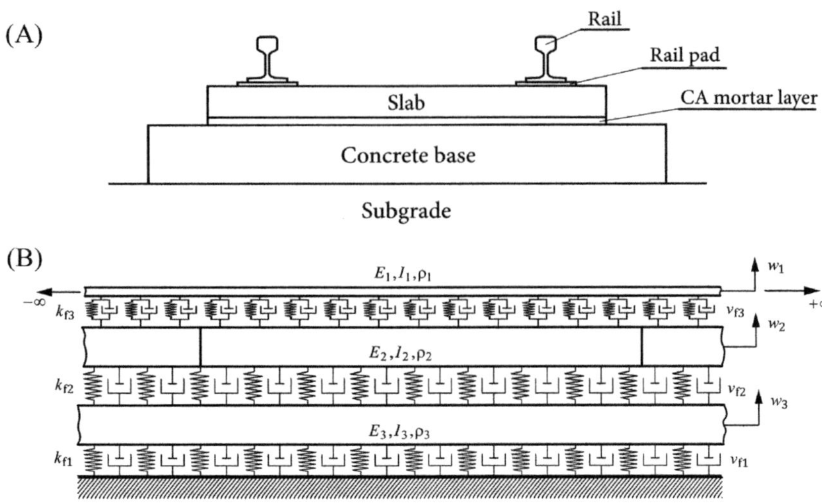

Figure 8.1 Slab track system. (A) Real-world structure comprising a rail, pad, slab, CA mortar, and base; (B) Mechanical interpretation of the track structure modeled as an infinite beam that is resting on viscoelastic spring-damping elements.

8.3 Modeling of railway vehicle system and characteristic equation

The moving system, that is, the railway vehicle is shown in Fig. 8.2. The following is assumed about the model: the vehicle and the railway track are symmetric about the vertical axis; the longitudinal motion of the track does not affect the tangent track; only half of the coupling system is analyzed in circulation; the vehicle system (moving with constant velocity v) and the track system are in continuous contact; the rail is represented as a deformable beam with shear accounted using a high-order beam assumption. The vehicle is an assembly of a rigid bar (representing the car body) that is supported by the secondary suspension on two bogies; the bogies are in contact with the track via four wheelsets (two for each bogie). A wheelset comprises a wheel and a primary suspension element. A suspension element comprises a spring and a dashpot, in parallel. The governing equations in the reduced form according to the dynamic equilibrium (Fig. 8.3) may be written as:

$$\Psi_1[w_1, \varphi_1] + k_{f3}(w_1 - w_2) + \nu_{f3}\left(\frac{\partial w_1}{\partial t} - \frac{\partial w_2}{\partial t}\right) =$$

$$- \delta(x - vt - d_3)\left(m\frac{\mathrm{d}^2 w_2^{01}}{\mathrm{d}t^2} + \tilde{\mathscr{L}}_{21}\right) - \delta(x - vt - d_2)\left(m\frac{\mathrm{d}^2 w_2^{02}}{\mathrm{d}t^2} + \tilde{\mathscr{L}}_{22}\right)$$

$$- \delta(x - vt - d_1)\left(m\frac{\mathrm{d}^2 w_1^{01}}{\mathrm{d}t^2} + \tilde{\mathscr{L}}_{11}\right) - \delta(x - vt)\left(m\frac{\mathrm{d}^2 w_1^{02}}{\mathrm{d}t^2} + \tilde{\mathscr{L}}_{12}\right), \tag{8.16}$$

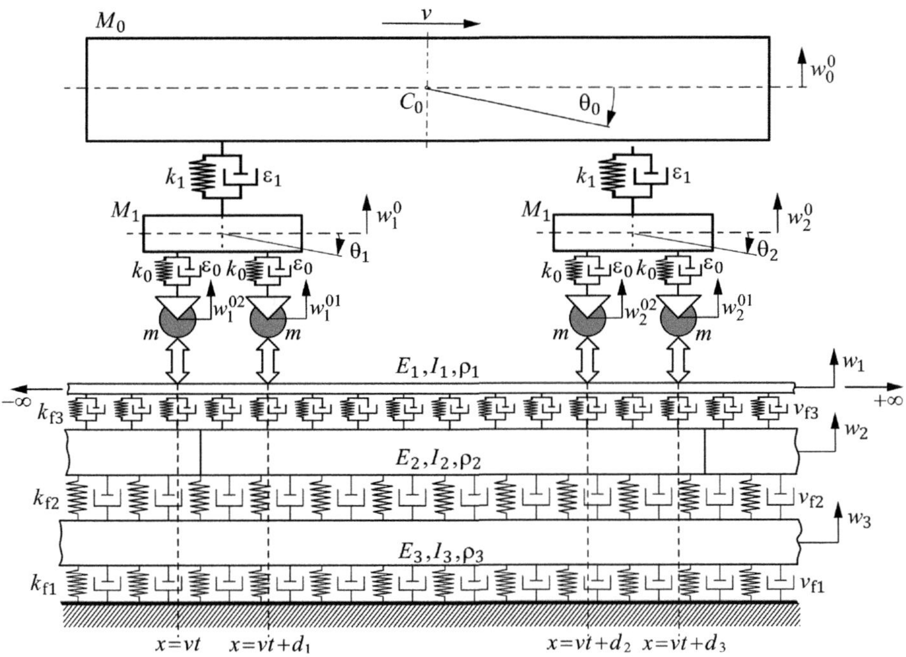

Figure 8.2 Uniform motion of a railway vehicle. The vehicle has primary and secondary suspension while the beam is laid on a complex foundation system.

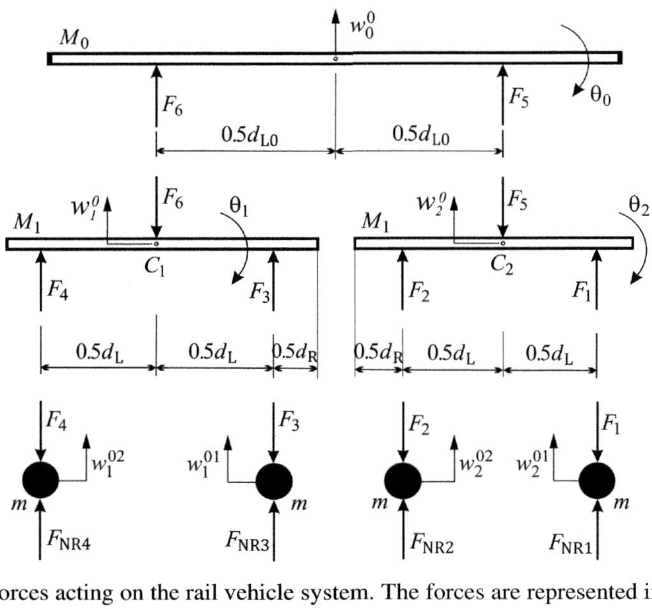

Figure 8.3 Forces acting on the rail vehicle system. The forces are represented in the dynamic equilibrium and the vehicle system is decomposed based on its primary and secondary suspension.

$$\Phi_1[w_1, \varphi_1] = 0, \tag{8.17}$$

$$\Psi_2[w_2, \varphi_2] + k_{f2}(w_2 - w_3) - k_{f3}(w_1 - w_2) + \nu_{f2}\left(\frac{\partial w_2}{\partial t} - \frac{\partial w_3}{\partial t}\right) - \nu_{f3}\left(\frac{\partial w_1}{\partial t} - \frac{\partial w_2}{\partial t}\right) = 0, \tag{8.18}$$

$$\Phi_2[w_2, \varphi_2] = 0, \tag{8.19}$$

$$\Psi_3[w_3, \varphi_3] + k_{f1}w_3 - k_{f2}(w_2 - w_3) + \nu_{f1}\frac{\partial w_3}{\partial t} - \nu_{f2}\left(\frac{\partial w_2}{\partial t} - \frac{\partial w_3}{\partial t}\right) = 0, \tag{8.20}$$

$$\Phi_3[w_3, \varphi_3] = 0, \tag{8.21}$$

$$M_1\frac{d^2w_2^0}{dt^2} - \tilde{\mathscr{L}}_{21} - \tilde{\mathscr{L}}_{22} + \tilde{\mathscr{L}}_2 = 0, \tag{8.22}$$

$$J_1\frac{d^2\theta_2}{dt^2} + (\tilde{\mathscr{L}}_{21} - \tilde{\mathscr{L}}_{22})\frac{d_L}{2} = 0, \tag{8.23}$$

$$M_1\frac{d^2w_1^0}{dt^2} - \tilde{\mathscr{L}}_{11} - \tilde{\mathscr{L}}_{12} + \tilde{\mathscr{L}}_1 = 0, \tag{8.24}$$

$$J_1\frac{d^2\theta_1}{dt^2} + (\tilde{\mathscr{L}}_{11} - \tilde{\mathscr{L}}_{12})\frac{d_L}{2} = 0, \tag{8.25}$$

$$M_0\frac{d^2w_0^0}{dt^2} - \tilde{\mathscr{L}}_1 - \tilde{\mathscr{L}}_2 = 0, \tag{8.26}$$

$$J_0\frac{d^2\theta_0}{dt^2} + (\tilde{\mathscr{L}}_2 - \tilde{\mathscr{L}}_1)\frac{d_{L0}}{2} = 0, \tag{8.27}$$

$$w_2^{01}(t) = w(vt + d_3, t), \; w_2^{02}(t) = w(vt + d_2, t),$$
$$w_1^{01}(t) = w(vt + d_1, t), \; w_1^{02}(t) = w(vt, t),$$
$$\lim_{|x-vt|\to\infty} w_i(x, t) = 0, \quad \lim_{|x-vt|\to\infty} \varphi_i(x, t) = 0, \tag{8.28}$$

where

$$\tilde{\mathscr{L}}_{ij} = k_0\left(w_i^{0j} - w_i^0 + \theta_i\frac{d_L}{2}(-1)^{j+1}\right) + \epsilon_0\frac{d}{dt}\left(w_i^{0j} - w_i^0 + \theta_i\frac{d_L}{2}(-1)^{j+1}\right),$$

$$\tilde{\mathscr{L}}_i = k_1\left(w_i^0 - w_0^0 + \theta_0\frac{d_{L0}}{2}\right) + \epsilon_1\frac{d}{dt}\left(w_i^0 - w_0^0 + \theta_0\frac{d_{L0}}{2}\right),$$

and where $\Psi_i[w_i, \varphi_i]$ and $\Phi_i[w_i, \varphi_i]$, $(i = 1, 2, 3)$, are presented with Eqs. (8.9c, 8.9d), respectively. The following are the translatory displacements: $w_i(x, t)$ is the vertical deflection of the beam; $w_1^{01}(t)$, $w_1^{02}(t)$, $w_2^{01}(t)$, and $w_2^{02}(t)$, are the vertical displacements of the masses (wheels); $w_1^0(t)$, and $w_2^0(t)$ are the vertical displacements of the centers of mass of the bars that represent the bogies' bodies, and $w_0^0(t)$ is the vertical displacement of the bar that represents the car's body. The following are the angular displacements: $\varphi_i(x, t)$ is the rotation angle of the beams' cross-section; Θ_0, Θ_1, and Θ_2 are the angles of rotation of the bars around their centers of mass. The following are the physical properties: M_0, M_1 and J_0, J_1 are the masses and moments of inertia of the bars, respectively; k_{fi} and ν_{fi} are the stiffness moduli and the viscosity shear moduli of the foundation per unit length, respectively; k_0 and ϵ_0 are the stiffness modulus and the damping factor of the primary suspension; k_1 and ϵ_1 are the stiffness modulus and the damping factor of the secondary suspension. The Dirac delta function is designated with $\delta(\,)$. Introducing a moving reference system $\xi = x - vt$, Eqs. (8.16–8.21) can be rewritten in the form:

$$\begin{aligned}
&\tilde{\Psi}_1[w_1, \varphi_1] + k_{f3}(w_1 - w_2)\\
&+ \nu_{f3}\left[\left(\frac{\partial w_1(\xi, t)}{\partial t} - v\frac{\partial w_1(\xi, t)}{\partial \xi}\right) - \left(\frac{\partial w_2(\xi, t)}{\partial t} - v\frac{\partial w_2(\xi, t)}{\partial \xi}\right)\right] =\\
&- \delta(\xi - d_3)\left(m\frac{\mathrm{d}^2 w_2^{01}}{\mathrm{d}t^2} + \tilde{\mathscr{L}}_{21}\right) - \delta(\xi - d_2)\left(m\frac{\mathrm{d}^2 w_2^{02}}{\mathrm{d}t^2} + \tilde{\mathscr{L}}_{22}\right)\\
&- \delta(\xi - d_1)\left(m\frac{\mathrm{d}^2 w_1^{01}}{\mathrm{d}t^2} + \tilde{\mathscr{L}}_{11}\right) - \delta(\xi)\left(m\frac{\mathrm{d}^2 w_1^{02}}{\mathrm{d}t^2} + \tilde{\mathscr{L}}_{12}\right),
\end{aligned} \tag{8.29}$$

$$\tilde{\Phi}_1[w_1, \varphi_1] = 0, \tag{8.30}$$

$$\begin{aligned}
&\tilde{\Psi}_2[w_2, \varphi_2] + k_{f2}(w_2 - w_3) - k_{f3}(w_1 - w_2)\\
&+ \nu_{f2}\left[\left(\frac{\partial w_2(\xi, t)}{\partial t} - v\frac{\partial w_2(\xi, t)}{\partial \xi}\right) - \left(\frac{\partial w_3(\xi, t)}{\partial t} - v\frac{\partial w_3(\xi, t)}{\partial \xi}\right)\right]\\
&- \nu_{f3}\left[\left(\frac{\partial w_1(\xi, t)}{\partial t} - v\frac{\partial w_1(\xi, t)}{\partial \xi}\right) - \left(\frac{\partial w_2(\xi, t)}{\partial t} - v\frac{\partial w_2(\xi, t)}{\partial \xi}\right)\right] = 0,
\end{aligned} \tag{8.31}$$

$$\tilde{\Phi}_2[w_2, \varphi_2] = 0, \tag{8.32}$$

$$\begin{aligned}
&\tilde{\Psi}_3[w_3, \varphi_3] + k_{f1}w_3 - k_{f2}(w_2 - w_3) + \nu_{f1}\left(\frac{\partial w_3(\xi, t)}{\partial t} - v\frac{\partial w_3(\xi, t)}{\partial \xi}\right)\\
&- \nu_{f2}\left[\left(\frac{\partial w_2(\xi, t)}{\partial t} - v\frac{\partial w_2(\xi, t)}{\partial \xi}\right) - \left(\frac{\partial w_3(\xi, t)}{\partial t} - v\frac{\partial w_3(\xi, t)}{\partial \xi}\right)\right] = 0,
\end{aligned} \tag{8.33}$$

$$\tilde{\Phi}_3[w_3, \varphi_3] = 0, \tag{8.34}$$

where

$$\widetilde{\Psi}_i[w_i, \varphi_i]=$$

$$C_{i1}\left(v^2\frac{\partial^2 w_i(\xi,t)}{\partial\xi^2}-2v\frac{\partial^2 w_i(\xi,t)}{\partial\xi\partial t}+\frac{\partial^2 w_i(\xi,t)}{\partial t^2}\right)-C_{i2}\left(\frac{\partial\varphi_i(\xi,t)}{\partial\xi}+\frac{\partial^2 w_i(\xi,t)}{\partial\xi^2}\right)$$

$$-C_{i3}\left(v^2\frac{\partial^4 w_i(\xi,t)}{\partial\xi^4}-2v\frac{\partial^4 w_i(\xi,t)}{\partial\xi^3\partial t}+\frac{\partial^4 w_i(\xi,t)}{\partial\xi^2\partial t^2}\right)+C_{i4}\frac{\partial^4 w_i(\xi,t)}{\partial\xi^4}$$

$$+C_{i5}\left(v^2\frac{\partial^3\varphi_i(\xi,t)}{\partial\xi^3}-2v\frac{\partial^3\varphi_i(\xi,t)}{\partial\xi^2\partial t}+\frac{\partial^3\varphi_i(\xi,t)}{\partial\xi\partial t^2}\right)+C_{i6}\frac{\partial^3\varphi_i(\xi,t)}{\partial\xi^3}, \tag{8.35}$$

$$\widetilde{\Phi}_i[w_i,\varphi_i]=D_{i1}\left(v^2\frac{\partial^2\varphi_i(\xi,t)}{\partial\xi^2}-2v\frac{\partial^2\varphi_i(\xi,t)}{\partial\xi\partial t}+\frac{\partial^2\varphi_i(\xi,t)}{\partial t^2}\right)$$

$$-D_{i2}\frac{\partial^2\varphi_i(\xi,t)}{\partial\xi^2}+D_{i3}\left(\varphi_i(\xi,t)+\frac{\partial w_i(\xi,t)}{\partial\xi}\right)$$

$$+D_{i4}\left(v^2\frac{\partial^3 w_i(\xi,t)}{\partial\xi^3}-2v\frac{\partial^3 w_i(\xi,t)}{\partial\xi^2\partial t}+\frac{\partial^3 w_i(\xi,t)}{\partial\xi\partial t^2}\right)+D_{i5}\frac{\partial^3 w_i(\xi,t)}{\partial\xi^3}=0, \tag{8.36}$$

$$w_2^{01}(t)=w_1(d_3,t),\ w_2^{02}(t)=w_1(d_2,t),$$

$$w_1^{01}(t)=w_1(d_1,t),\ w_1^{02}(t)=w_1(0,t),$$

$$\lim_{|\xi|\to\infty}w_i(x,t)=0,\ \lim_{|\xi|\to\infty}\varphi_i(x,t)=0. \tag{8.37}$$

It is possible to analyze the characteristic equation for vibrations of the moving railway vehicle along the beam by applying the integral transforms to the system of Eqs. (8.22–8.27) and Eqs. (8.29–8.34):

$$W_{si}(\xi,s)=\int_0^\infty w_i(\xi,t)e^{-st}\mathrm{d}t,\ \Phi_{si}(\xi,s)=\int_0^\infty\varphi_i(\xi,t)e^{-st}\mathrm{d}t,$$

$$\hat{W}_{k,si}(k,s)=\int_{-\infty}^\infty W_{si}(\xi,s)e^{-ik\xi}\mathrm{d}\xi,\ \hat{\Phi}_{si}(k,s)=\int_{-\infty}^\infty\Phi_{si}(\xi,s)e^{-ik\xi}\mathrm{d}\xi,$$

$$W_1^{01}(s)=\int_0^\infty w_1^{01}(t)e^{-st}\mathrm{d}t,\ W_1^{02}(s)=\int_0^\infty w_1^{02}(t)e^{-st}\mathrm{d}t,$$

$$W_2^{01}(s)=\int_0^\infty w_2^{01}(t)e^{-st}\mathrm{d}t,\ W_2^{02}(s)=\int_0^\infty w_2^{02}(t)e^{-st}\mathrm{d}t,$$

$$W_0^0(s)=\int_0^\infty w_0^0(t)e^{-st}\mathrm{d}t,\ W_1^0(s)=\int_0^\infty w_1^0(t)e^{-st}\mathrm{d}t,\ W_2^0(s)=\int_0^\infty w_2^0(t)e^{-st}\mathrm{d}t,$$

$$\Theta_0=\Theta_0(s)=\int_0^\infty\theta_0(t)e^{-st}\mathrm{d}t,\ \Theta_1=\Theta_1(s)=\int_0^\infty\theta_1(t)e^{-st}\mathrm{d}t,$$

$\Theta_2 = \Theta_2(s) = \int_0^\infty \theta_2(t)e^{-st}dt$, (in index: $i = 1, 2, 3$), (in other cases: $i = \sqrt{-1}$). By assuming that the initial conditions to the problem Eq. (8.22–8.27) and Eq. (8.29–8.34) are trivial (the initial conditions do not affect the system stability since the governing equations are linear) and bearing in mind that $W_1^{02}(s) = W(0, s)$, $W_1^{01}(s) = W(d_1, s)$, $W_2^{02}(s) = W(d_2, s)$, and $W_2^{01}(s) = W(d_3, s)$, the application of the transforms results in the system of algebraic equations after the elimination $\hat{\Phi}_{si}(k, s)$ in the present form:

$$F(k, s)\hat{W}_{k,s}(k, s)=$$

$$\begin{aligned}
&\frac{1}{2}\Big\{e^{-id_1k}\big\{2W_1^0(s)(k_0 + s\epsilon_0) - 2W(d_1, s)[k_0 + s(ms + \epsilon_0)] - \Theta_1 d_L(k_0 + s\epsilon_0)\big\} \\
&+ e^{-id_2k}\big\{2W_2^0(s)(k_0 + s\epsilon_0) - 2W(d_2, s)[k_0 + s(ms + \epsilon_0)] + \Theta_2 d_L(k_0 + s\epsilon_0)\big\} \\
&+ e^{-id_3k}\{2W_2^0(s)(k_0 + s\epsilon_0) - 2W(d_3, s)[k_0 + s(ms + \epsilon_0)] \\
&- \Theta_2 d_L(k_0 + s\epsilon_0)\} - 2k_0W(0, s) - 2ms^2W(0, s) \\
&- 2s\epsilon_0W(0, s) + 2k_0W_1^0(s) + 2s\epsilon_0W_1^0(s) + \Theta_1k_0d_L + \Theta_1s\epsilon_0d_L\Big\},
\end{aligned} \tag{8.38}$$

$$\frac{1}{2}[\Theta_0 d_{L0}(k_1 + s\epsilon_1) - 2(k_0 + s\epsilon_0)(W(d_2, s) + W(d_3, s)) - 2W_0^0(s)(k_1 + s\epsilon_1)] + W_2^0(s)[2k_0 + k_1 + s(M_1s + 2\epsilon_0 + \epsilon_1)] = 0, \tag{8.39}$$

$$d_L(k_0 + s\epsilon_0)[W(d_3, s) - W(d_2, s) + \Theta_2 d_L] + 2\Theta_2J_1s^2 = 0, \tag{8.40}$$

$$\frac{1}{2}[-2(k_0 + s\epsilon_0)(W(d_1, s) + W(0, s)) - 2W_0^0(s)(k_1 + s\epsilon_1) - \Theta_0 d_{L0}(k_1 + s\epsilon_1)] + W_1^0(s)[2k_0 + k_1 + s(M_1s + 2\epsilon_0 + \epsilon_1)] = 0, \tag{8.41}$$

$$d_L(k_0 + s\epsilon_0)(W(d_1, s) - W(0, s) + \Theta_1 d_L) + 2\Theta_1J_1s^2 = 0, \tag{8.42}$$

$$k_1(2W_0^0(s) - W_1^0(s) - W_2^0(s)) + s[M_0sW_0^0(s) + \epsilon_1(2W_0^0(s) - W_1^0(s) - W_2^0(s))] = 0, \tag{8.43}$$

$$d_{L0}(k_1 + s\epsilon_1)(W_2^0(s) - W_1^0(s) + \Theta_0 d_{L0}) + 2\Theta_0J_0s^2 = 0, \tag{8.44}$$

with $F(k, s)$ presented in Ref. [15]. The three-part beam-foundation reaction is mathematically presented with the function $F(k, s)$. Starting from the system of Eqs. (8.38–8.44) and applying to it the inverse Fourier transform over k, then eliminating W_0^0, W_1^0, W_2^0, and Θ_0, Θ_1, Θ_2, and setting the values $\xi = 0$, $\xi = d_1$, $\xi = d_2$, $\xi = d_3$, subsequently, will produce a system of algebraic equations—with respect to the Laplace-displacements of the contact points $\hat{W}_s(0, s)$, $\hat{W}_s(d_1, s)$, $\hat{W}_s(d_2, s)$, and $\hat{W}_s(d_3, s)$

$$[\mathbf{u}] \times \lfloor \hat{W} \rfloor^{\mathrm{T}} = 0 \Leftrightarrow$$

$$\begin{bmatrix} u_{11} & \cdots & u_{14} \\ \vdots & \ddots & \vdots \\ u_{41} & \cdots & u_{44} \end{bmatrix} \times \lfloor \hat{W}_s(0, s) \hat{W}_s(d_1, s) \hat{W}_s(d_2, s) \hat{W}_s(d_3, s) \rfloor^{\mathrm{T}} = 0. \tag{8.45}$$

The system obtained from Eq. (8.45) is a system of integrals, which is a consequence of the inverse Fourier transform; these integrals can be found in Ref. [15]. The system of equations Eqs. (8.38–8.44) possesses a nontrivial solution provided that the determinant of the matrix **G** satisfies the condition:

$$\det \mathbf{G} = 0 \tag{8.46}$$

In the following sections, the stability of the model is analyzed by investigating the eigenvalues of the characteristic Eq. (8.46).

8.4 Dynamic stability analysis

The vibrations of the vehicle are unstable if at least one root of the characteristic Eq. (8.46) with a positive real part exists—this criterion is used for determining the instability of vibrations. Eq. (8.32) is an integral equation with respect to the Laplace variable s, finding its complex roots is not a trivial problem, especially for the model that is here studied. To solve the presented mathematical problem more easily, the D-decomposition method is applied. Denisov et al. [5] have described the concept of this method. Departing, for now, from the physical meaning of the problem, k_0 will be viewed as a complex parameter and a straight line $s = i\Omega$ will be mapped onto the complex plane—the k_0-plane. The parameter Ω is varied from minus to plus infinity. This means that the k_0-plane is where the D-decomposition is performed, that is, it is assumed temporarily that k_0 is the complex variable, and the positive half of the k_0-plane is correlated to the stiffness of the primary suspension. The aim is to find all complex values of k_0 (i.e., to delineate the lines in the k_0-plane) that are correlated to the pure imaginary solution $s = i\Omega$ of Eq. (8.46). These lines will partition the k_0-plane into regions with a different number of roots of Eq. (8.46) with respect to the variable s, satisfying the condition $\mathrm{Re}(s) > 0$. To achieve this goal, Eq. (8.46) should be looked at in the following form

$$f(\Omega, k_0) = 0, \tag{8.47}$$

where $s = i\Omega$. The next step is to find all the roots of Eq. (8.47) with respect to the variable k_0. Function $F(k, \Omega)$ and integrals are presented in Ref. [15]. With the help of the contour integration method (whose contours of integration are shown in Fig. 5.2), the integrals presented in Ref. [15] can be developed. This produces the expressions:

$$\mathrm{Int}_0 = i\sum\nolimits_{\mathrm{n}} \frac{k-k_n}{(k-k_1)\cdots(k-k_6)}\Big|_{k=k_n},$$
$$\mathrm{Int}_{\Delta-} = -i\sum\nolimits_{\mathrm{m}} \frac{(k-k_m)e^{ik\Delta}}{(k-k_1)\cdots(k-k_6)}\Big|_{k=k_m},\ \Delta < 0,$$
$$\mathrm{Int}_{\Delta+} = i\sum\nolimits_{\mathrm{n}} \frac{(k-k_n)e^{ik\Delta}}{(k-k_1)\cdots(k-k_6)}\Big|_{k=k_n},\ \Delta > 0, \tag{8.48}$$

in which k_n are the roots with a positive imaginary part, and k_m are the roots with a negative imaginary part of equation $F(k,\ \Omega) = 0$. To solve the integrals given in Ref. [15] more easily, the contour integration method is applied. The short-written form of the residues after the contour integration method is applied is given in Eq. (8.48). The entire form of all expressions Eq. (8.48) is given in Ref. [15]. After this, the D-decomposition curves can be analyzed. Five curves should be examined, each correlated to one of the five roots of Eq. (8.47). The parametric analysis leads to a finding that the D-decomposition curves are of interest for dynamic analysis for the case in which the velocity is in the supercritical speed region, that is, when the complex vehicle system is moving at a velocity that is greater than a certain critical velocity v_{cr}. When the speed is in the subcritical region ($v < v_{cr}$), the D-decomposition curves do not cross the positive part of the real axes at any point. This means that when the $v < v_{cr}$, the number of unstable roots of the characteristic equation (roots with a positive real part) does not vary with the stiffness of the rail vehicle supports (k_0). However, when the $v > v_{cr}$, the number of unstable roots will change when k_0 surpasses a certain critical value of stiffness. The use of the D-decomposition method allows the determination of a relative variation of the number of unstable roots depending on the stiffness k_0, but not the number itself. Therefore the task that remains is to find the number of unstable roots for any value of stiffness k_0. This can be achieved by using the argument principle. To determine the number of unstable roots, the chosen contour should encircle the right half-plane of the complex variable s, as is illustrated in Fig. 5.4. The method consists of setting a fixed value for k_0, and then computation of the total change of the argument of the complex expression $f(-is,\ k_0)$, along the entire contour, where f is the function from Eq. (8.47). By the argument principle, the total change is equal to $2\pi N_{unst}$, where N_{unst} is the number of roots of $f(-is,\ k_0)$ (i.e., Eq. (8.46)) located in the positive half-plane, with respect to s. For both parts of the contour (the semicircle and y-axis), some sufficiently small iteration step Δs is chosen, with which the relative change of the argument $\mathrm{Arg}(f(-i(s+\Delta s),\ k_0)/f(-is,\ k_0))$ is computed, and all of these relative changes are summed up. The size of the iteration step Δs is dynamically adopted along the contour. The procured expression for $f(-is,\ k_0)$, with the integrals that belong to the type $Int_{\Delta\pm}$, is exceedingly cumbersome. To expedite the computing of the argument change, the complete expression for $f(-is,\ k_0)$ is not computed symbolically. Instead of that, for each s and k_0, firstly the values of all integrals $\mathrm{Int}_{\Delta\pm}$ are computed, and then $f(-is,\ k_0)$ is computed by evaluating the determinant Eq. (8.46).

This procedure speeds up the particular computation of $f(-is, k_0)$ by severalfold, and reduces the needed amount of working memory.

8.5 Parametric study

8.5.1 A review of considered cases and the referent example

In this section, several comparative examples are analyzed. The D-decomposition curves are obtained for the same set of model parameters used in Ref. [17]; these parameters are used to examine the reference example (Case 1), results are in Table 8.1. Other parameters for comparative examples used in other cases (Cases 2–6) are given in Table 8.2. In the upcoming sections, various factors of influence have been examined and discussed based on the dynamic stability of the system. The different factors of influence that are analyzed are the damping factor created by the secondary suspension (Case 2), the stiffness of the secondary suspension (Case 3), the mass of the bogies (Case 4), the wheelbase size (Case 5), and the influence that the position of the supports has (Case 6). The obtained D-decomposition curves for the referent Case 1 (diagrams for each root separately and the integrated diagram)

Table 8.1 The number of unstable roots. Level of the instability of k_0 intervals according to the number of unstable roots (Case 1, v=1.12v_{cr}), unit for k_0 is 10^7 N/m.

k_0	0	2.573	3.067	5.074	5.076	+∞
Case 1	4	4|2	2|0	0|4	4|4	4

Table 8.2 Varying parameters representing damping of the secondary suspension (ε_1); stiffness of the secondary suspension (k_1); the mass of the bogies (M_1); the size of the wheelbase (d_{L0}); and the position of the supports (d_L).

	Varying parameter	Values
Case 2.1	$\varepsilon_1\left[\frac{N\cdot s}{m}\right]$	10^5
Case 2.2		2×10^5
Case 2.3		10^6
Case 3.1	$k_1\left[\frac{N}{m}\right]$	2×10^4
Case 3.2		∞
Case 4.1	M_1 [kg]	3.3×10^3
Case 4.2		4.3×10^3
Case 4.3		8.3×10^3
Case 5.1	d_{L0} [m]	10.7
Case 5.2		20.7
Case 5.3		30.7
Case 6.1	d_L [m]	2
Case 6.2		6

can be seen in Fig. 8.4. Here, the D-decomposition curves are displayed separately for the purpose of showing the critical stiffness without numbering the unstable roots because of the return lines of the other roots. Besides, an integrated diagram with all the curves is difficult to analyze. For all the cases, the number of unstable roots is determined according to the argument principle for complex structures. The D-decomposition curves that are delineated in Fig. 8.4A–D are for supercritical velocity $v = 1.12\, v_{cr}$. The separately shown D-decomposition curves can show the sections with the real axis of the roots. What they cannot show are the regions with a number of unstable roots. For this reason, the integrated diagram is given in Fig. 8.4E. It is apparent that the stable zone (the zone that lacks unstable roots) belongs to the stiffness region $k_0 \in (3.067,\ 5.074) \times 10^7$ N/m. It should be pointed out that for the

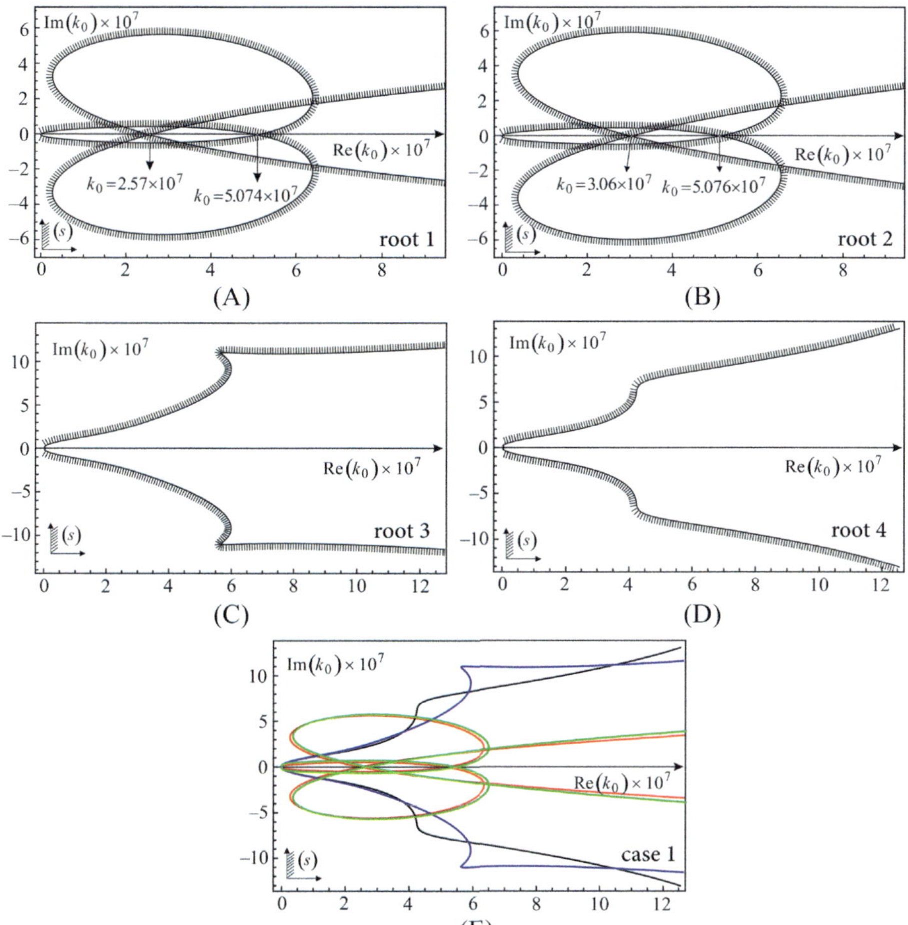

Figure 8.4 D-decomposition curves for Case 1 (N/m). (A) root 1, (B) root 2, (C) root 3, (D) root 4, (E) integrated diagram for D-decomposition curves. The stability domains for this reference example are presented in Table 8.1.

chosen supercritical velocity, other k_0 intervals do have unstable roots in varying numbers. The level of instability is determined on the basis of these numbers.

8.5.2 Influence of the damping—produced by the secondary suspension—on the stability domain

In this section, the influence of the damping, produced by the secondary suspension of the vehicle, on stability is analyzed. The varying parameter is given in Table 8.2. The integrated D-decomposition curves are given in Fig. 8.5. As can be seen, an increase in the value of the damping coefficient of the secondary suspension causes a decrease in dynamic stability. The completely stable region in Case 2.1 is within the interval $k_0 \in (3.07,\ 5.074) \times 10^7$ N/m. The stable region in Case 2.2 exists within the interval $k_0 \in (3.11,\ 4.78) \times 10^7$ N/m, while the stable region in Case 2.3 encompasses only the segment where $k_0 \in (3.61,\ 4.34) \times 10^7$ N/m. The other intervals with various levels of instability are shown in Table 8.3. The level of instability of an interval is determined on the basis of the number of unstable roots within the interval. The system in Case 2.3 is not in the stable region in comparison with the other cases outside the zones of complete stability.

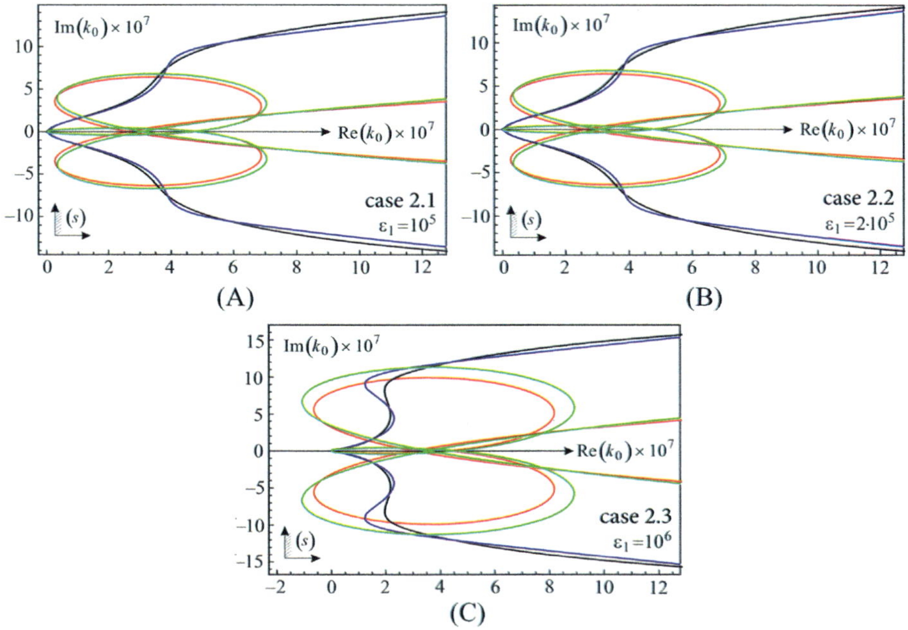

Figure 8.5 D-decomposition curves (N/m). The varying parameter analyzed is the damping of the secondary suspension (ε_1). All roots are reported and sections with real axis present relevant changes of stability according to the stiffness of the supports. The stability domains are presented in Table 8.3. (A) Case 2.1; (B) Case 2.2; (C) Case 2.3.

Table 8.3 The number of unstable roots. The level of the instability of k_0 intervals according to the number of unstable roots (Case 2, v=1.12v_{cr}), unit for k_0 is 10^7 N/m.

<table>
<tr><th>k_0</th><th>0</th><th>2.58</th><th>2.62</th><th>3.03</th><th>3.07</th><th>3.11</th><th>3.61</th><th>4.34</th><th>4.55</th><th>4.78</th><th>4.79</th><th>5.074</th><th>5.076</th><th>+∞</th></tr>
<tr><td>Case 2.1</td><td>4</td><td>4|2</td><td>2|2</td><td>2|2</td><td>2|0</td><td>0|0</td><td>0|0</td><td>0|0</td><td>0|0</td><td>0|0</td><td>0|0</td><td>0|4</td><td>4|4</td><td>4</td></tr>
<tr><td>Case 2.2</td><td>4</td><td>4|4</td><td>4|2</td><td>2|2</td><td>2|2</td><td>2|0</td><td>0|0</td><td>0|0</td><td>0|0</td><td>0|4</td><td>4|4</td><td>4|4</td><td>4|4</td><td>4</td></tr>
<tr><td>Case 2.3</td><td>4</td><td>4|4</td><td>4|4</td><td>4|2</td><td>2|2</td><td>2|2</td><td>2|0</td><td>0|2</td><td>2|4</td><td>4|4</td><td>4|4</td><td>4|4</td><td>4|4</td><td>4</td></tr>
</table>

8.6 Influence of the secondary suspension stiffness on the stability domain

The stability domains for varying values of the stiffness of the secondary suspension are shown in Table 8.4. If the secondary suspension is a rigid connection, then the change in stability domains is interesting. The example with the rigid connection is chosen because of the minuscule change in stability regions by varying the stiffness of the secondary suspension. In the rigid connection scenario, the stable domain belongs to the interval $k_0 \in (4.29,\ 4.54) \times 10^7$ N/m. Compared to a scenario with lower stiffness (Case 3.1), it can be seen that the rigid connection leads to instability in the intervals $k_0 \in (3.07,\ 4.29;\ 4.54,\ 5.02) \times 10^7$ N/m, while Case 3.1 here has a stable region. Fig. 8.6 shows a set of complex D-decomposition curves by which the stability regions are determined.

8.7 Influence of the bogie's mass on dynamic stability

In this section, the influence of the mass M_1 of the bogies is analyzed. The stability domains for different values of this mass are given in Table 8.5. If the mass is

Table 8.4 The number of unstable roots. The level of the instability of k_0 intervals according to the number of unstable roots (Case 3, v=1.12v_{cr}), unit for k_0 is 10^7 N/m.

k_0	0	2.58	3.07	3.59	4.29	4.54	5.02	5.03	5.86	+∞
Case 3.1	4	4|2	2|0	0|0	0|0	0|0	0|4	4|4	4|4	4
Case 3.2	4	4|4	4|4	4|2	2|0	0|2	2|2	2|2	2|4	4

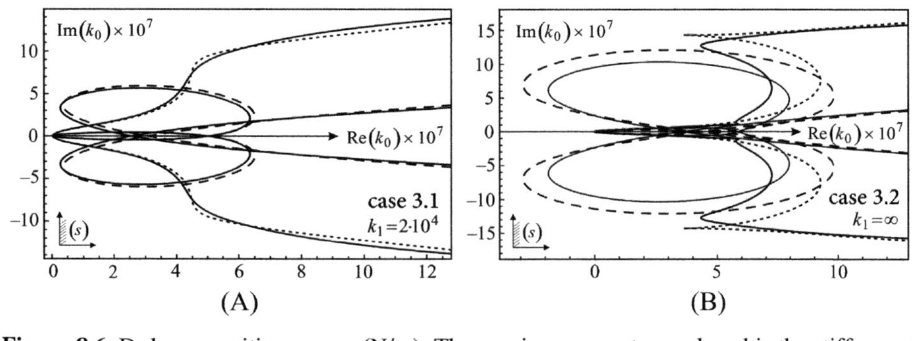

Figure 8.6 D-decomposition curves (N/m). The varying parameter analyzed is the stiffness of the secondary suspension (k_1). All roots are reported and sections with real axis present relevant changes of stability according to the stiffness of the supports. The stability domains are presented in Table 8.4. (A) Case 3.1; (B) Case 3.2.

Table 8.5 The number of unstable roots. The level of the instability of k_0 intervals according to the number of unstable roots (Case 4, v=1.12v_{cr}), unit for k_0 is 10^7 N/m.

<table>
<tr><th>k_0</th><th>0</th><th>2.58</th><th>2.78</th><th>3.04</th><th>3.06</th><th>3.07</th><th>3.21</th><th>3.32</th><th>3.84</th><th>4.52</th><th>4.53</th><th>5.074</th><th>5.076</th><th>+∞</th></tr>
<tr><td>Case 4.1</td><td>4</td><td>4|2</td><td>2|2</td><td>2|2</td><td>2|2</td><td>2|0</td><td>0|0</td><td>0|0</td><td>0|0</td><td>0|0</td><td>0|0</td><td>0|4</td><td>4|4</td><td>4</td></tr>
<tr><td>Case 4.2</td><td>4</td><td>4|4</td><td>4|2</td><td>2|2</td><td>2|2</td><td>2|2</td><td>2|2</td><td>2|0</td><td>0|0</td><td>0|4</td><td>4|4</td><td>4|4</td><td>4|4</td><td>4</td></tr>
<tr><td>Case 4.3</td><td>4</td><td>4|4</td><td>4|4</td><td>4|8</td><td>8|8</td><td>8|8</td><td>8|6</td><td>6|6</td><td>6|4</td><td>4|4</td><td>4|4</td><td>4|4</td><td>4|4</td><td>4</td></tr>
</table>

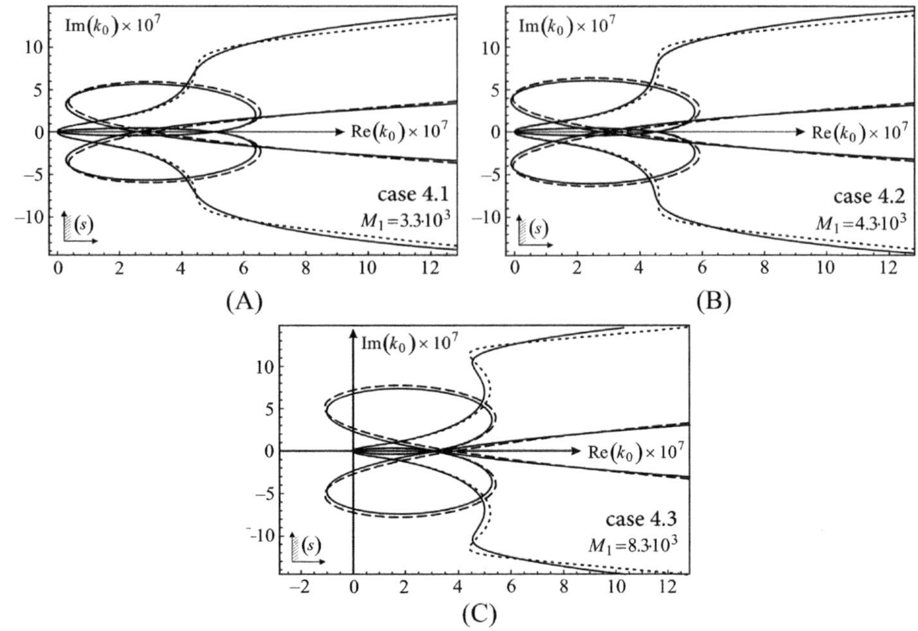

Figure 8.7 D-decomposition curves (N/m). The varying parameter analyzed is the mass of the bogie (M_1). All roots are reported and sections with real axis present relevant changes of stability according to the stiffness of the supports. The stability domains are presented in Table 8.5. (A) Case 4.1; (B) Case 4.2; (C) Case 4.3.

increasing, then knowing the change in stability domains is important. An increase in this mass leads to a decrease in stable regions when the comparison is made with Case 4.1, which has the lowest mass. In that case, the stable domain belongs to the interval $k_0 \in (3.07,\ 5.074) \times 10^7$ N/m. The set of D-decomposition curves is depicted in Fig. 8.7. The change in damping produced by the secondary suspension has a similar influence on the stability as this change in mass.

8.8 Influence of the wheelbase size on dynamic stability

The stability domains for different wheelbase sizes are given in Table 8.6. For the selected supercritical velocity and the smallest wheelbase (Case 5.1), the stable domain is minute and encompasses only the segment $k_0 \in (3.2,\ 4.81) \times 10^7$ N/m. The stable domain changes if the wheelbase size is increased and it appears for $k_0 \in (2.87,\ 5.02) \times 10^7$ N/m (Case 5.2) and $k_0 \in (2.91,\ 5.02) \times 10^7$ N/m (Case 5.3). Fig. 8.8 shows the D-decomposition curves from which the results in Table 8.6 are obtained.

Table 8.6 The number of unstable roots. The level of the instability of k_0 intervals according to the number of unstable roots (Case 5, v=1.12v_{cr}), unit for k_0 is 10^7 N/m.

<table>
<tr><th>k_0</th><th>0</th><th>2.31</th><th>2.75</th><th>2.79</th><th>2.87</th><th>2.91</th><th>3.20</th><th>4.81</th><th>5.02</th><th>5.04</th><th>5.10</th><th>5.39</th><th>+∞</th></tr>
<tr><td>Case 5.1</td><td>4</td><td>4|2</td><td>2|2</td><td>2|2</td><td>2|2</td><td>2|2</td><td>2|0</td><td>0|2</td><td>2|2</td><td>2|2</td><td>2|2</td><td>2|4</td><td>4</td></tr>
<tr><td>Case 5.2</td><td>4</td><td>4|4</td><td>4|4</td><td>4|2</td><td>2|0</td><td>0|0</td><td>0|0</td><td>0|0</td><td>0|2</td><td>2|2</td><td>2|4</td><td>4|4</td><td>4</td></tr>
<tr><td>Case 5.3</td><td>4</td><td>4|</td><td>4|2</td><td>2|2</td><td>2|2</td><td>2|0</td><td>0|0</td><td>0|0</td><td>0|2</td><td>2|4</td><td>4|4</td><td>4|4</td><td>4</td></tr>
</table>

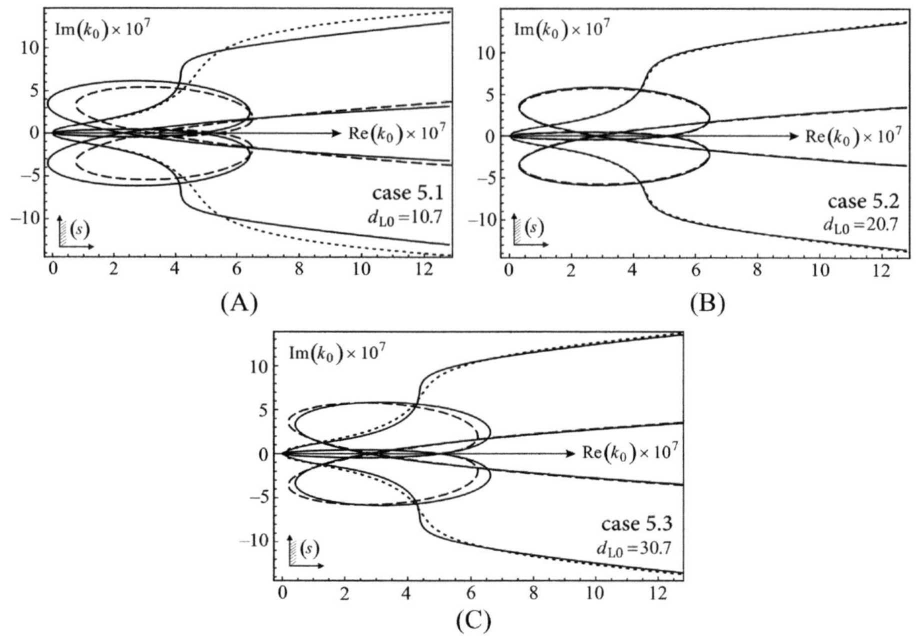

Figure 8.8 D-decomposition curves (N/m). The varying parameter analyzed is the wheelbase size (d_{L0}). All roots are reported and sections with real axis present relevant changes of stability according to the stiffness of the supports. The stability domains are presented extendedin Table 8.6. (A) Case 5.1; (B) Case 5.2; (C) Case 5.3.

Table 8.7 The number of unstable roots. The level of the instability of k_0 intervals according to the number of unstable roots (Case 6, v=1.12v_{cr}), unit for k_0 is 10^7 N/m.

k_0	0	3.47	3.67	6.83	7.17	7.22	+∞
Case 6.1	4	4|2	2|0	0|0	0|2	2|4	4
Case 6.2	0	0|0	0|0	0|2	2|2	2|2	2

8.9 Influence of the position of the supports on dynamic stability

The position of the supports can have a significant influence on the dynamic stability of complex vehicles. A complex mechanical system (compared to a simple system) produces varying stability regions. As mentioned, the model of the vehicle that is here studied has four contact points with the rail on which it is traveling. When comparing Cases 6.1 and 6.2, a noticeable shift in the stable region can be seen, as well as the overlap of stable regions in the interval $k_0 \in (3.67,\ 6.83) \times 10^7$ N/m. Table 8.7 presents all the levels of instability determined for the varying positions of the

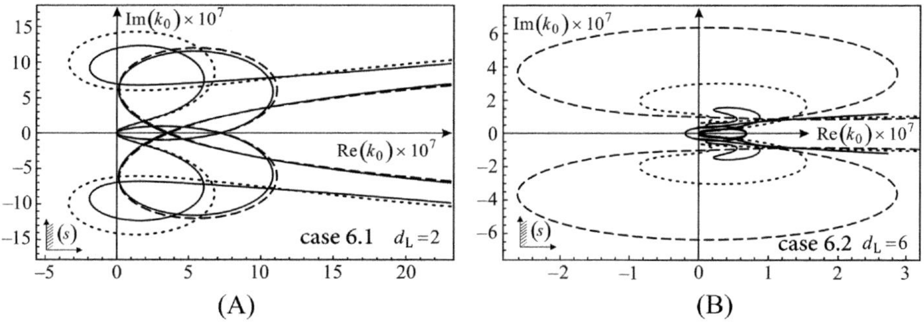

Figure 8.9 D-decomposition curves (N/m). The varying parameter analyzed is the position of the supports (d_L). All roots are reported and sections with real axis present relevant changes of stability according to the stiffness of the supports. The stability domains are presented in Table 8.7. (A) Case 6.1; (B) Case 6.2.

supports. As shown in Fig. 8.9, the considered model contains cumbersome D-decomposition curves.

8.10 Conclusion

This chapter was about the stability of vibrations of a complex and discrete rail vehicle system that is moving along a continuous structure. This continuous structure is a deformable three-part beam/foundation system of an infinite length. The instability eventuates when the velocity of the vehicle exceeds a specific critical velocity. In the absence of viscosity in the beam foundation, this velocity is equal to the minimum phase velocity of waves in the beam on the elastic foundation. The stability of the model is determined by the roots of the characteristic equation for the vehicle that interacts with the beam. Analyses of these roots are done by subsequent use of the D-decomposition method and the argument principle. The D-decomposition method has been implemented to partition the model space into domains with a diverse number of unstable roots. Following this, the argument principle has been used on a set of model parameters to identify the number of unstable roots in one of these domains. The emphasis (of the parametric study of the stability domain) was placed on the following determinants that affect stability: the influence of the damping and stiffness of the secondary suspension of the railway vehicle, the influence of the mass of the bogies, the influence of the size of the wheelbase, and the influence of the positions of the supports.

The stability of the model has been shown to depend on the damping in the secondary suspension and the mass of the bogies. An increase in the damping of the secondary suspension and the mass of the bogies leads to a decrease in dynamic stability. A decrease in the stiffness of the secondary suspensions leads to an increase in dynamic stability. The wheelbase size is interesting because the optimal chosen brings about an increase in the dynamic stability (there is no general rule about dynamic stability by increase or decrease of the wheelbase) as in the case of the positions of the supports. Stability regions are determined for each case and documented by the depicted D-decomposition diagrams.

References

[1] A.V. Metrikine, Unstable lateral oscillations of an object moving uniformly along elastic guide as a result of anomalous Doppler effect, Acoustical Physics 40 (1) (1994) 85–89.

[2] V. Stojanović, P. Kozić, M.D. Petković, Dynamic instability and critical velocity of a mass moving uniformly along a stabilized infinity beam, International Journal of Solids and Structures 108 (2017) 164–174.

[3] S. Timoshenko, Method of analysis of statical and dynamical stresses in rail, in: Proceedings of the 2nd International Congress for Applied Mechanics, Zürich, Switzerland, 1927, pp. 1–12.

[4] A.D. Kerr, The continuously supported rail subjected to an axial force and a moving load, International Journal of Mechanical Sciences 14 (1) (1972) 71–78.

[5] G.G. Denisov, E.K. Kugusheva, V.V. Novikov, On the problem of the stability of one-dimensional unbounded elastic systems, Journal of Applied Mathematics and Mechanics 49 (4) (1985) 533–537.

[6] S.N. Verichev, A.V. Metrikine, Instability of a railway vehicle moving on a flexibly supported Timoshenko beam, Journal of Sound and Vibration 253 (3) (2002) 653–668.

[7] Z. Dimitrovová, J.N. Varandas, Critical velocity of a load moving on a beam with a sudden change of foundation stiffness: applications to high-speed trains, Computers & Structures 87 (19–20) (2009) 1224–1232.

[8] S.H. Ju, H.T. Lin, Analysis of train-induced vibrations and vibration reduction schemes above and below critical Rayleigh speeds by finite element method, Soil Dynamics and Earthquake Engineering 24 (12) (2004) 993–1002.

[9] N. Zhang, H. Xia, Dynamic analysis of coupled railway vehicle–bridge system based on inter-system iteration method, Computers & Structures 114–115 (2013) 26–34.

[10] Z. Dimitrovová, New semi-analytical solution for a uniformly moving mass on a beam on a two-parameter visco-elastic foundation, International Journal of Mechanical Sciences 127 (2017) 142–162.

[11] C. Madshus, A.M. Kaynia, High-speed railway lines on soft ground: dynamic behaviour at critical train speed, Journal of Sound and Vibration 231 (2000) 689–701.

[12] T. Mazilu, M. Dumitriu, C. Tudorache, On the dynamics of interaction between a moving mass and an infinite one-dimensional elastic structure at the stability limit, Journal of Sound and Vibration 330 (2011) 3729–3743.

[13] T. Mazilu, M. Dumitriu, C. Tudorache, Instability of an oscillator moving along a Timoshenko beam on viscoelastic foundation, Nonlinear Dynamics 67 (2012) 1273–1293.

[14] T. Mazilu, Instability of a train of oscillators moving along a beam on a viscoelastic foundation, Journal of Sound and Vibration 332 (2013) 4597–4619.

[15] V. Stojanović, M.D. Petković, J. Deng, Stability of vibrations of a moving railway vehicle along an infinite complex three-part viscoelastic beam/foundation system, International Journal of Mechanical Sciences 136 (2018) 155–168.

[16] V. Stojanović, M.D. Petković, Nonlinear dynamic analysis of damaged Reddy–Bickford beams supported on an elastic Pasternak foundation, Journal of Sound and Vibration 385 (2016) 239–266.

[17] X. Yang, S. Gu, S. Zhou, J. Yang, Y. Zhou, S. Lian, Effect of track irregularity on the dynamic response of a slab track under a high-speed train based on the composite track element method, Applied Acoustics 99 (2015) 72–84.

9 Vibrational instability in a complex moving object: innovative approaches to elastically damped connections between car body components and supports

9.1 Introduction

The examination of vibrations in beam structures induced by moving loads has played a pivotal role in shaping the design of modern railway lines. The advancements in road and railway traffic have facilitated high-speed transportation, necessitating a more thorough analysis of the dynamic phenomena related to train-track-soil interaction. Consequently, ongoing investigations into moving loads remain a crucial focus in contemporary research. In the case where the velocity of a high-speed train exceeds the velocity of elastic waves that are spread out through the railway structure, vibration instability may arise due to the Doppler waves [1,2]. Previous research on the instability of vibrations in simple mechanical models paved the way for further studies. Timoshenko [3] demonstrated that a moving constant vertical force has a critical velocity, causing significant deflections when reached. Kerr [4] examined the response of an axially compressed continuously supported beam to a moving vertical force. Nelson and Conver [5] examined the dynamic stability of the lateral response in a simply supported Bernoulli-Euler beam loaded with a continuous series of equally spaced mass particles. They employed the Galerkin method to derive a set of approximate governing equations of motion with periodic coefficients. Floquet theory was then applied to analyze the parametric regions of stability, visually represented in graphical form. Subsequent studies, such as those by Takemiya [6] and Kim [7], explored the dynamic stability and instability of mechanical systems subjected to moving loads, incorporating varying effects. Attar et al. [8] presented a discrete model to study the dynamic behavior of beam-like structures under moving loads in the presence of surface defects. Denisov et al. [9] identified stability regions for a mass moving uniformly on a continuously supported beam using the D-decomposition method. This approach charts regions in the complex plane where the characteristic equation possesses either stable or unstable roots for a specified value of the moving object's velocity. More dynamic instability analysis, using the D-decomposition method, can be found in the work by Verichev and Metrikine [10,11], Mazilu [12], Mazilu et al. [13], and Stojanović et al. [14]. Dimitrovová and Varandas [15] explored the transient dynamic response of a beam supported on a foundation with a sudden stiffness change and subjected to a force moving with a constant velocity, utilizing finite integral transformations. Then Dimitrovová [16] proposed a repetitive procedure

Modeling of Complex Dynamic Systems. DOI: https://doi.org/10.1016/B978-0-443-23942-7.00009-7

for determining oscillation frequencies induced by a moving mass, and Erbas et al. [17] conducted a three-dimensional analysis of near-resonant regimes of a point load moving uniformly along the surface of a coated elastic half-space. Svedholm et al. [18] presented valuable results related to loads moving on beams of finite length. On the other hand, Lee and Renshaw [19] developed a new solution technique for solving the moving mass problem in nonconservative, linear, distributed parameter systems using complex eigenfunction expansions. The mentioned researched problems have ample applications, so further expansion of the dynamic analysis to the problems related to plates can be of significance. Thereby, Sorentino and Catania [20] analyzed plates subjected to distributed moving loads, and Song et al. [21] presented a parametric study of the dynamic response of sandwich plates under moving loads.

Many studies investigated concrete problems from the engineering practice, like the high-speed trains, as seen in Ju and Lin [22], as already mentioned in Chapter 4. Nevertheless, theoretical research, spanning civil engineering, mechanical engineering, automation, control, and electrical engineering, has played a crucial role in enabling high-speed traffic. The swift development of high-speed railways has led to a considerable focus on the dynamic response of bridges under the influence of moving vehicle loads. Consequently, numerous studies addressing the dynamic behavior of bridges subjected to the moving loads of running vehicles [23–27] have been conducted. Zhang and Xia [28] presented dynamic analysis results for a coupled moving oscillator/bridge system, utilizing the intersystem iteration method. While examining the stability of high-speed trains, railway track models are essential tools for studying their dynamic behavior, influencing the safety and comfort of railway transport. In the literature, one-, two-, and three-layer models can be found, and a comparison between layered models of a railway track is presented in Dimitrovová [29]. The widely used two-layer model of the railway track is employed in Dimitrovová's research [30,31], and in Dimitrovová and Mazilu [32], the interaction of a mass moving uniformly on an infinite beam on a three-layer viscoelastic foundation is analyzed with the objective of determining the critical velocity. In Koziol [33], the wavelet transform is utilized to examine the impact of moving forces on a double-beam model. The Adomian decomposition is employed to address the nonlinearity present in the viscoelastic layer that connects the two beams. The model is subsequently refined into a two-layer model by Czyczula et al. [34] and expanded to incorporate nonlinearity in the foundation in Koziol and Pilecki [35]. Ouzizi et al. [36] elaborated on the nonlinear dynamic analysis of beams on the fractional order viscoelastic nonlinear foundation subjected to a moving load with variable speed. In our research, we observe a complex oscillatory system moving along a complexly modeled finite beam-foundation system, which is the subject of the study in Stojanović and Petković [37].

As seen from previous works, numerous papers have used simple configurations to represent a railway vehicle, like a single moving mass [38–40], two moving masses [41], a two-mass oscillator [42], or a train of two-mass oscillators [43,44]. However, some researchers have delved into the development of a complex model for a moving oscillator that more closely resembles the actual train model. Yau et al. [45] examined the vibration of simple and three-span continuous beams under the passage of high-speed trains. To simulate certain commercially available high-speed trains, they incorporated two vehicle models in their numerical study. In the research by Dai et al. [46], the focus was on the dynamic analysis of vehicle-track coupling based on a double-beam track model. The vehicle model included the car body, bogie frame, and wheelset. Lei and Noda [47] provide analyzes of the dynamic response of a vehicle and track coupling system with random irregularities in the vertical profile of the track, and Kouroussis et al. [48] present a discrete modeling approach for the vertical track-soil coupling in vehicle-track dynamics. Then Stojanović et al. [49,50] examined the instability of a bogie traversing a complex beam-layer structure and the dynamics of moving coupled objects. They utilized the D-decomposition technique and the argument principle to identify the region of instability within the parameter space of the system.

This chapter provides an in-depth stability analysis of vibrations in a complex mechanical oscillator moving on an infinite continuous beam-foundation system. The study investigates the influence of various additional stabilizers on the stability of the standard complex moving system. It is essential to note that these modified models, showcasing specific advantages in terms of stability, have the potential to introduce new technical solutions or enhancements to existing practical models in the technical domain. One of the newly examined models involves a stabilizer with an adjustable vertical position connected to the car's body, while the second model features a stabilizer connected to the bogies. In summary, this paper thoroughly explores the vibrational stability of a coupled moving oscillator uniformly traveling on a complexly modeled infinite high-order shear deformable coupled beam system on a viscoelastic base. The focus is specifically on the influence of various types of additional stabilizers. After determining stability limits for both cases of an additional stabilizer, it was observed that connecting the stabilizer to the car body results in larger stability regions. In this scenario, significant changes to the stiffness and damping parameters can be made with minimal alteration to the stability region. In contrast, when the stabilizer is connected to the bogies, changes in the stiffness and damping parameters can have a more substantial impact and manipulation of the stability region. This effect is particularly pronounced in heavier car bodies, although stability regions are consistently smaller in all cases considered.

9.2 Modeling of the complex coupled models and characteristic equation

The model that we observe presents a complexly interconnected moving oscillator, upon which the analyzes in this research are grounded, and capable of simulating a standard railway vehicle system. This system moves along a complexly modeled, flexibly supported infinite high-order deformable beam-layer coupled structure resting on a viscoelastic foundation. The study is concentrated on discerning the variance between two proposed types of stabilizers under identical system parameters. The first model incorporates a stabilizer with an adjustable vertical position connected to the car's body, as depicted in Fig. 9.1, while the second model involves a stabilizer

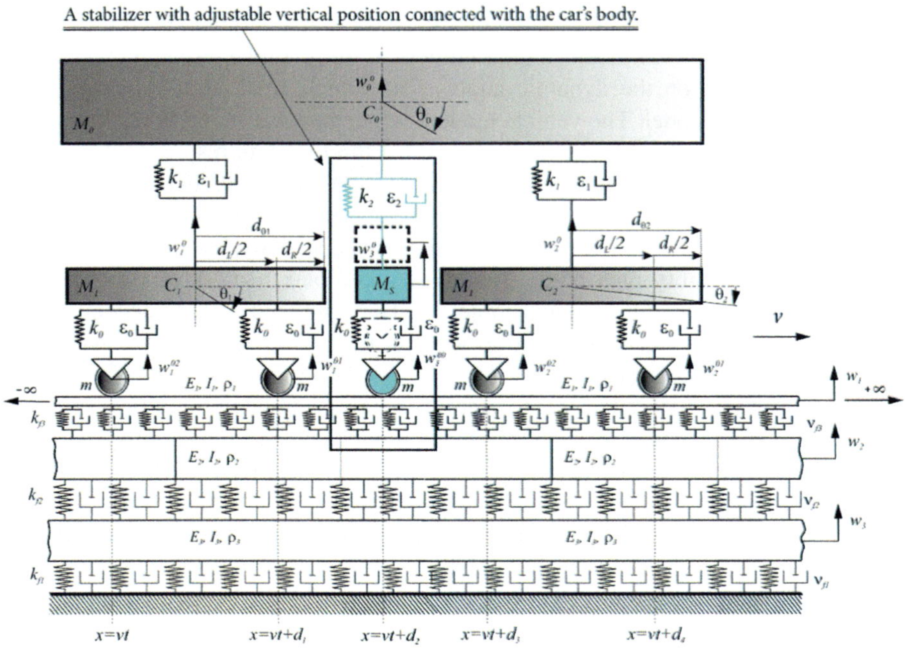

Figure 9.1 Complex coupled moving oscillator. The stabilizer is connected to the body of the car.

connected to the bogies, as illustrated in Fig. 9.2. A complex three-part viscoelastic beam/foundation model that is being used can be seen in Fig. 8.1. The equations for a single high-order shear deformable beam interacting with the bogie were derived and can be found in Stojanović and Petković [51].

The interaction of the infinite coupled high-order shear deformable beam-layer mechanical system, as depicted in Figs. 9.1 and 9.2, is defined by the set of Eqs. (9.1–9.3). These equations mathematically define the vibrations of the Reddy-Bickford beam-layer system. The constants C_{ik} and D_{ij} which have been derived and presented in Stojanović and Petković [51] specifically for one beam (requiring additional indices for each distinct beam), serve as a foundational framework applicable to all other beams based on the parameters provided later in Results. The parameters E and G represent Young's modulus and the shear modulus of the beam material, while ρ signifies the mass density of the beam material.

$$\Psi_1 + k_{f3}(w_1 - w_2) + \nu_{f3}\left(\frac{\partial w_1}{\partial t} - \frac{\partial w_2}{\partial t}\right) = p_{1/2}(x,t), \quad \Phi_1 = 0, \tag{9.1}$$

$$\Psi_2 + k_{f2}(w_2 - w_3) - k_{f3}(w_1 - w_2) + \nu_{f2}\left(\frac{\partial w_2}{\partial t} - \frac{\partial w_3}{\partial t}\right) - \nu_{f3}\left(\frac{\partial w_1}{\partial t} - \frac{\partial w_2}{\partial t}\right) = 0,$$

$$\Phi_2 = 0, \tag{9.2}$$

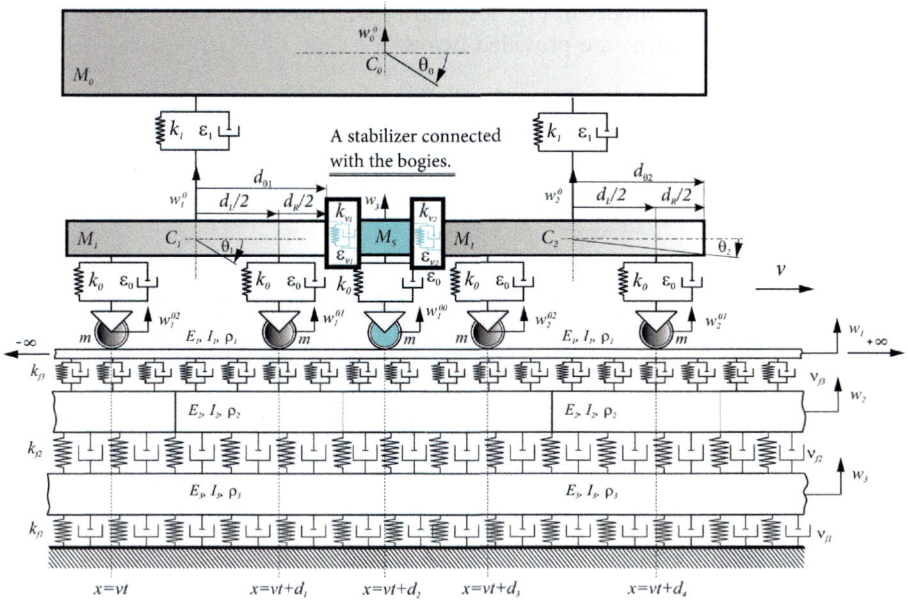

Figure 9.2 Complex coupled moving oscillator. The stabilizer connected to the bogies.

$$\Psi_3 + k_{f1}w_3 - k_{f2}(w_2 - w_3) + \nu_{f1}\frac{\partial w_3}{\partial t} - \nu_{f2}\left(\frac{\partial w_2}{\partial t} - \frac{\partial w_3}{\partial t}\right) = 0,$$

$$\Phi_3 = 0, \tag{9.3}$$

where:

$$\Psi_i = C_{i1}\frac{\partial^2 w_i}{\partial t^2} - C_{i2}\left(\frac{\partial \varphi_i}{\partial x} + \frac{\partial^2 w_i}{\partial x^2}\right) - C_{i3}\frac{\partial^4 w_i}{\partial x^2 \partial t^2}$$

$$+ C_{i4}\frac{\partial^4 w_i}{\partial x^4} + C_{i5}\frac{\partial^3 \varphi_i}{\partial x \partial t^2} + C_{i6}\frac{\partial^3 \varphi_i}{\partial x^3}, \tag{9.4}$$

$$\Phi_i = D_{i1}\frac{\partial^2 \varphi_i}{\partial t^2} - D_{i2}\frac{\partial^2 \varphi_i}{\partial x^2} + D_{i3}\left(\varphi_i + \frac{\partial w_i}{\partial x}\right)$$

$$+ D_{i4}\frac{\partial^3 w_i}{\partial x \partial t^2} + D_{i5}\frac{\partial^3 w_i}{\partial x^3}, \quad i = 1,\ 2,\ 3. \tag{9.5}$$

where $w_i(x, t)$ are vertical deflections of the beams and $\varphi_i(x, t)$ are the angles of rotation of the cross-section of the beams. The parameters k_{fi} and ν_{fi} denote the stiffness moduli and viscosity shear moduli of the foundation per unit length. In Eq. (9.1), the term $p_{1/2}(x, t)$ represents the response of the bogie system, specifically

$p_1(x, t)$ is related to the model in Fig. 9.1, and $p_2(x, t)$ is linked to the model in Fig. 9.2. Their respective forms are provided below:

$$
\begin{aligned}
p_1(x,t) = &-\delta(x - vt - d_4)\left(m\frac{\mathrm{d}^2 w_2^{01}}{\mathrm{d}t^2} + \left[k_0\left(w_2^{01} - w_2^0 + \theta_2\frac{d_L}{2}\right)\right.\right.\\
&\left.\left.+ \epsilon_0\frac{\mathrm{d}}{\mathrm{dt}}\left(w_2^{01} - w_2^0 + \theta_2\frac{d_L}{2}\right)\right]\right)\\
&-\delta(x - vt - d_3)\left(m\frac{\mathrm{d}^2 w_2^{02}}{\mathrm{d}t^2} + \left[k_0\left(w_2^{02} - w_2^0 - \theta_2\frac{d_L}{2}\right)\right.\right.\\
&\left.\left.+ \epsilon_0\frac{\mathrm{d}}{\mathrm{dt}}\left(w_2^{02} - w_2^0 - \theta_2\frac{d_L}{2}\right)\right]\right)\\
&-\delta(x - vt - d_2)\left(m\frac{\mathrm{d}^2 w_1^{00}}{\mathrm{d}t^2} + \left[k_0\left(w_1^{00} - w_3^0\right) + \epsilon_0\frac{\mathrm{d}}{\mathrm{dt}}\left(w_1^{00} - w_3^0\right)\right]\right)\\
&-\delta(x - vt - d_1)\left(m\frac{\mathrm{d}^2 w_1^{01}}{\mathrm{d}t^2} + \left[k_0\left(w_1^{01} - w_1^0 + \theta_1\frac{d_L}{2}\right)\right.\right.\\
&\left.\left.+ \epsilon_0\frac{\mathrm{d}}{\mathrm{dt}}\left(w_1^{01} - w_1^0 + \theta_1\frac{d_L}{2}\right)\right]\right)\\
&-\delta(x - vt)\left(m\frac{\mathrm{d}^2 w_1^{02}}{\mathrm{d}t^2} + \left[\begin{array}{c} k_0\left(w_1^{02} - w_1^0 - \theta_1\frac{d_L}{2}\right)\\ + \epsilon_0\frac{\mathrm{d}}{\mathrm{dt}}\left(w_1^{02} - w_1^0 - \theta_1\frac{d_L}{2}\right)\end{array}\right]\right),
\end{aligned}
\tag{9.6}
$$

$$
\begin{aligned}
p_2(x,t) = &-\delta(x - vt - d_4)\left(m\frac{\mathrm{d}^2 w_2^{01}}{\mathrm{d}t^2} + \left[k_0\left(w_2^{01} - w_2^0 + \theta_2\frac{d_L}{2}\right)\right.\right.\\
&\left.\left.+ \epsilon_0\frac{\mathrm{d}}{\mathrm{dt}}\left(w_2^{01} - w_2^0 + \theta_2\frac{d_L}{2}\right)\right]\right)\\
&-\delta(x - vt - d_3)\left(m\frac{\mathrm{d}^2 w_2^{02}}{\mathrm{d}t^2} + \left[k_0\left(w_2^{02} - w_2^0 - \theta_2\frac{d_L}{2}\right)\right.\right.\\
&\left.\left.+ \epsilon_0\frac{\mathrm{d}}{\mathrm{dt}}\left(w_2^{02} - w_2^0 - \theta_2\frac{d_L}{2}\right)\right]\right)\\
&-\delta(x - vt - d_2)\left(m\frac{\mathrm{d}^2 w_1^{00}}{\mathrm{d}t^2} + \left[k_0\left(w_1^{00} - w_3\right) + \epsilon_0\frac{\mathrm{d}}{\mathrm{dt}}\left(w_1^{00} - w_3\right)\right]\right)
\end{aligned}
$$

$$- \delta(x - vt - d_1)\left(m\frac{d^2 w_1^{01}}{dt^2} + \left[k_0\left(w_1^{01} - w_1^0 + \theta_1\frac{d_L}{2} \right) + \epsilon_0\frac{d}{dt}\left(w_1^{01} - w_1^0 + \theta_1\frac{d_L}{2} \right) \right] \right)$$

$$- \delta(x - vt)\left(m\frac{d^2 w_1^{02}}{dt^2} + \left[k_0\left(w_1^{02} - w_1^0 - \theta_1\frac{d_L}{2} \right) + \epsilon_0\frac{d}{dt}\left(w_1^{02} - w_1^0 - \theta_1\frac{d_L}{2} \right) \right] \right) \quad (9.7)$$

The Dirac delta-function is presented with $\delta(\,)$. In the case where we have a stabilizer with an adjustable vertical position connected to the car's body, and when using the term (9.6) for the response of the bogie system, the full form of the dynamic equilibrium of the coupled moving oscillator can be described by adding equations:

$$M_1\frac{d^2 w_1^0}{dt^2} - \left[k_0\left(w_1^{01} - w_1^0 + \theta_1\frac{d_L}{2} \right) + \epsilon_0\frac{d}{dt}\left(w_1^{01} - w_1^0 + \theta_1\frac{d_L}{2} \right) \right]$$
$$- \left[k_0\left(w_1^{02} - w_1^0 - \theta_1\frac{d_L}{2} \right) + \epsilon_0\frac{d}{dt}\left(w_1^{02} - w_1^0 - \theta_1\frac{d_L}{2} \right) \right]$$
$$+ \left[k_1\left(w_1^0 - w_0^0 - \theta_0\frac{d_{L0}}{2} \right) + \epsilon_1\frac{d}{dt}\left(w_1^0 - w_0^0 - \theta_0\frac{d_{L0}}{2} \right) \right] = 0, \quad (9.8)$$

$$J_1\frac{d^2\theta_1}{dt^2} + \frac{d_L}{2}\left[k_0\left(w_1^{01} - w_1^0 + \theta_1\frac{d_L}{2} \right) + \epsilon_0\frac{d}{dt}\left(w_1^{01} - w_1^0 + \theta_1\frac{d_L}{2} \right) \right]$$
$$- \frac{d_L}{2}\left[k_0\left(w_1^{02} - w_1^0 - \theta_1\frac{d_L}{2} \right) + \epsilon_0\frac{d}{dt}\left(w_1^{02} - w_1^0 - \theta_1\frac{d_L}{2} \right) \right] = 0, \quad (9.9)$$

$$M_1\frac{d^2 w_2^0}{dt^2} - \left[k_0\left(w_2^{01} - w_2^0 + \theta_2\frac{d_L}{2} \right) + \epsilon_0\frac{d}{dt}\left(w_2^{01} - w_2^0 + \theta_2\frac{d_L}{2} \right) \right]$$
$$- \left[k_0\left(w_2^{02} - w_2^0 - \theta_2\frac{d_L}{2} \right) + \epsilon_0\frac{d}{dt}\left(w_2^{02} - w_2^0 - \theta_2\frac{d_L}{2} \right) \right]$$
$$+ \left[k_1\left(w_2^0 - w_0^0 + \theta_0\frac{d_{L0}}{2} \right) + \epsilon_1\frac{d}{dt}\left(w_2^0 - w_0^0 + \theta_0\frac{d_{L0}}{2} \right) \right] = 0, \quad (9.10)$$

$$J_1\frac{d^2\theta_2}{dt^2} + \frac{d_L}{2}\left[k_0\left(w_2^{01} - w_2^0 + \theta_2\frac{d_L}{2} \right) + \epsilon_0\frac{d}{dt}\left(w_2^{01} - w_2^0 + \theta_2\frac{d_L}{2} \right) \right]$$
$$- \frac{d_L}{2}\left[k_0\left(w_2^{02} - w_2^0 - \theta_2\frac{d_L}{2} \right) + \epsilon_0\frac{d}{dt}\left(w_2^{02} - w_2^0 - \theta_2\frac{d_L}{2} \right) \right] = 0, \quad (9.11)$$

$$M_0 \frac{\partial^2 w_0^0}{\partial t^2} - \left[k_1\left(w_2^0 - w_0^0 + \theta_0 \frac{d_{L0}}{2}\right) + \epsilon_1 \frac{\mathrm{d}}{\mathrm{dt}}\left(w_2^0 - w_0^0 + \theta_0 \frac{d_{L0}}{2}\right)\right]$$

$$- \left[k_1\left(w_1^0 - w_0^0 - \theta_0 \frac{d_{L0}}{2}\right) + \epsilon_1 \frac{\mathrm{d}}{\mathrm{dt}}\left(w_1^0 - w_0^0 - \theta_0 \frac{d_{L0}}{2}\right)\right]$$

$$- \left[k_2 (w_3^0 - w_0^0) + \epsilon_2 \frac{\mathrm{d}}{\mathrm{dt}} (w_3^0 - w_0^0)\right] = 0, \tag{9.12}$$

$$J_0 \frac{\mathrm{d}^2 \theta_0}{\mathrm{d}t^2} + \frac{d_{L0}}{2}\left[k_1\left(w_2^0 - w_0^0 + \theta_0 \frac{d_{L0}}{2}\right) + \epsilon_1 \frac{\mathrm{d}}{\mathrm{dt}}\left(w_2^0 - w_0^0 + \theta_0 \frac{d_{L0}}{2}\right)\right]$$

$$- \frac{d_{L0}}{2}\left[k_1\left(w_1^0 - w_0^0 - \theta_0 \frac{d_{L0}}{2}\right) + \epsilon_1 \frac{\mathrm{d}}{\mathrm{dt}}\left(w_1^0 - w_0^0 - \theta_0 \frac{d_{L0}}{2}\right)\right] = 0, \tag{9.13}$$

$$M_s \frac{\mathrm{d}^2 w_3^0}{\mathrm{d}t^2} - \left[k_0 (w_1^{00} - w_3^0) + \epsilon_0 \frac{\mathrm{d}}{\mathrm{dt}} (w_1^{00} - w_3^0)\right]$$

$$+ \left[k_2 (w_3^0 - w_0^0) + \epsilon_2 \frac{\mathrm{d}}{\mathrm{dt}} (w_3^0 - w_0^0)\right] = 0, \tag{9.14}$$

$$w_1^{01}(t) = w(vt + d_1, t), \quad w_3^{00}(t) = w(vt + d_2, t), \quad w_1^{02}(t) = w(vt, t),$$

$$w_2^{02}(t) = w(vt + d_3, t), \quad w_2^{01}(t) = w(vt + d_4, t),$$

$$\lim_{|x - vt| \to \infty} w_i(x, t) = 0, \quad \lim_{|x - vt| \to \infty} \varphi_i(x, t) = 0, \tag{9.15}$$

and in cases where a stabilizer is connected to the bogies, and when the response of the bogie system is described with (9.7) the governing equations are:

$$M_1 \frac{\mathrm{d}^2 w_1^0}{\mathrm{d}t^2} - \left[k_0\left(w_1^{01} - w_1^0 + \theta_1 \frac{d_L}{2}\right) + \epsilon_0 \frac{\mathrm{d}}{\mathrm{dt}}\left(w_1^{01} - w_1^0 + \theta_1 \frac{d_L}{2}\right)\right]$$

$$- \left[k_0\left(w_1^{02} - w_1^0 - \theta_1 \frac{d_L}{2}\right) + \epsilon_0 \frac{\mathrm{d}}{\mathrm{dt}}\left(w_1^{02} - w_1^0 - \theta_1 \frac{d_L}{2}\right)\right]$$

$$+ \left[k_{v1} (w_3^0 - w_1^0 + \theta_1 d_{\theta 1}) + \epsilon_{v1} \frac{\mathrm{d}}{\mathrm{dt}} (w_3^0 - w_1^0 + \theta_1 d_{\theta 1})\right]$$

$$+ \left[k_1\left(w_1^0 - w_0^0 - \theta_0 \frac{d_{L0}}{2}\right) + \epsilon_1 \frac{\mathrm{d}}{\mathrm{dt}}\left(w_1^0 - w_0^0 - \theta_0 \frac{d_{L0}}{2}\right)\right] = 0, \tag{9.16}$$

$$J_1 \frac{\mathrm{d}^2 \theta_1}{\mathrm{d}t^2} + \frac{d_L}{2}\left[k_0\left(w_1^{01} - w_1^0 + \theta_1 \frac{d_L}{2}\right) + \epsilon_0 \frac{\mathrm{d}}{\mathrm{dt}}\left(w_1^{01} - w_1^0 + \theta_1 \frac{d_L}{2}\right)\right]$$

$$- \frac{d_L}{2}\left[k_0\left(w_1^{02} - w_1^0 - \theta_1 \frac{d_L}{2}\right) + \epsilon_0 \frac{\mathrm{d}}{\mathrm{dt}}\left(w_1^{02} - w_1^0 - \theta_1 \frac{d_L}{2}\right)\right]$$

$$- d_{\theta 1}\left[k_{v1}(w_3^0 - w_1^0 + \theta_1 d_{\theta 1}) + \epsilon_{v1}\frac{\mathrm{d}}{\mathrm{dt}}(w_3^0 - w_1^0 + \theta_1 d_{\theta 1})\right] = 0, \tag{9.17}$$

$$M_1\frac{\mathrm{d}^2 w_2^0}{\mathrm{d}t^2} - \left[k_0\left(w_2^{01} - w_2^0 + \theta_2\frac{d_L}{2}\right) + \epsilon_0\frac{\mathrm{d}}{\mathrm{dt}}\left(w_2^{01} - w_2^0 + \theta_2\frac{d_L}{2}\right)\right]$$

$$- \left[k_0\left(w_2^{02} - w_2^0 - \theta_2\frac{d_L}{2}\right) + \epsilon_0\frac{\mathrm{d}}{\mathrm{dt}}\left(w_2^{02} - w_2^0 - \theta_2\frac{d_L}{2}\right)\right]$$

$$- \left[k_{v2}(w_2^0 - w_3^0 + \theta_2 d_{\theta 2}) + \epsilon_{v2}\frac{\mathrm{d}}{\mathrm{dt}}(w_2^0 - w_3^0 + \theta_2 d_{\theta 2})\right]$$

$$+ \left[k_1\left(w_2^0 - w_0^0 + \theta_0\frac{d_{L0}}{2}\right) + \epsilon_1\frac{\mathrm{d}}{\mathrm{dt}}\left(w_2^0 - w_0^0 + \theta_0\frac{d_{L0}}{2}\right)\right] = 0, \tag{9.18}$$

$$J_1\frac{\mathrm{d}^2\theta_2}{\mathrm{d}t^2} + \frac{d_L}{2}\left[k_0\left(w_2^{01} - w_2^0 + \theta_2\frac{d_L}{2}\right) + \epsilon_0\frac{\mathrm{d}}{\mathrm{dt}}\left(w_2^{01} - w_2^0 + \theta_2\frac{d_L}{2}\right)\right]$$

$$- \frac{d_L}{2}\left[k_0\left(w_2^{02} - w_2^0 - \theta_2\frac{d_L}{2}\right) + \epsilon_0\frac{\mathrm{d}}{\mathrm{dt}}\left(w_2^{02} - w_2^0 - \theta_2\frac{d_L}{2}\right)\right]$$

$$- d_{\theta 2}\left[k_{v2}(w_2^0 - w_3^0 + \theta_2 d_{\theta 2}) + \epsilon_{v2}\frac{\mathrm{d}}{\mathrm{dt}}(w_2^0 - w_3^0 + \theta_2 d_{\theta 2})\right] = 0, \tag{9.19}$$

$$M_0\frac{\partial^2 w_0^0}{\partial t^2} - \left[k_1\left(w_2^0 - w_0^0 + \theta_0\frac{d_{L0}}{2}\right) + \epsilon_1\frac{\mathrm{d}}{\mathrm{dt}}\left(w_2^0 - w_0^0 + \theta_0\frac{d_{L0}}{2}\right)\right]$$

$$- \left[k_1\left(w_1^0 - w_0^0 - \theta_0\frac{d_{L0}}{2}\right) + \epsilon_1\frac{\mathrm{d}}{\mathrm{dt}}\left(w_1^0 - w_0^0 - \theta_0\frac{d_{L0}}{2}\right)\right], \tag{9.20}$$

$$J_0\frac{\mathrm{d}^2\theta_0}{\mathrm{d}t^2} + \frac{d_{L0}}{2}\left[k_1\left(w_2^0 - w_0^0 + \theta_0\frac{d_{L0}}{2}\right) + \epsilon_1\frac{\mathrm{d}}{\mathrm{dt}}\left(w_2^0 - w_0^0 + \theta_0\frac{d_{L0}}{2}\right)\right]$$

$$- \frac{d_{L0}}{2}\left[k_1\left(w_1^0 - w_0^0 - \theta_0\frac{d_{L0}}{2}\right) + \epsilon_1\frac{\mathrm{d}}{\mathrm{dt}}\left(w_1^0 - w_0^0 - \theta_0\frac{d_{L0}}{2}\right)\right] = 0, \tag{9.21}$$

$$M_s\frac{\mathrm{d}^2 w_3^0}{\mathrm{d}t^2} - \left[k_0(w_1^{00} - w_3^0) + \epsilon_0\frac{\mathrm{d}}{\mathrm{dt}}(w_1^{00} - w_3^0)\right]$$

$$- \left[k_{v1}(w_3^0 - w_1^0 + \theta_1 d_{\theta 1}) + \epsilon_{v1}\frac{\mathrm{d}}{\mathrm{dt}}(w_3^0 - w_1^0 + \theta_1 d_{\theta 1})\right]$$

$$+ \left[k_{v2}(w_2^0 - w_3^0 + \theta_2 d_{\theta 2}) + \epsilon_{v2}\frac{\mathrm{d}}{\mathrm{dt}}(w_2^0 - w_3^0 + \theta_2 d_{\theta 2})\right], \tag{9.22}$$

$$w_1^{01}(t) = w(vt + d_1, t), \quad w_3^{00}(t) = w(vt + d_2, t), \quad w_1^{02}(t) = w(vt, t),$$

$$w_2^{02}(t) = w(vt + d_3, t), \quad w_2^{01}(t) = w(vt + d_4, t),$$

$$\lim_{|x-vt|\to\infty} w_i(x, t) = 0, \quad \lim_{|x-vt|\to\infty} \varphi_i(x, t) = 0, \tag{9.23}$$

where $w_1^{01}(t)$, $w_1^{02}(t)$, $w_2^{01}(t)$, $w_2^{02}(t)$, and $w_1^{00}(t)$ are representing the vertical displacements of the masses (wheels); $w_1^0(t)$ and $w_2^0(t)$ denote the vertical displacements of the centers of mass of the bars that are simulating the bogies, and $w_0^0(t)$ stands for the vertical displacement of the bar representing the car's body. θ_0, θ_1, and θ_2 are the angles of rotation of the bars around their centers of mass. M_0, M_1 and J_0, J_1 correspond to the masses and moments of inertia of the car body and bogies. M_s represents the mass of the stabilizer, which functions as an additional mechanical oscillator. The parameters k_0, ϵ_0, k_1, and ϵ_1 are the stiffness modulus and damping factors of the supports, while $k_{v1,2}$, $\epsilon_{v1,2}$ and k_2, ϵ_2 correspond to the stiffness modulus and damping factor of the elastic coupling presented in Figs. 9.1 and 9.2 involving the stabilizer. By introducing a moving reference system $\xi = x - vt$, Eqs. (9.1–9.3) can be rewritten in the form:

$$\Psi_1 + k_{f3}(w_1 - w_2) + \nu_{f3}\begin{bmatrix} \left(\frac{\partial w_1(\xi,t)}{\partial t} - v\frac{\partial w_1(\xi,t)}{\partial \xi}\right) \\ -\left(\frac{\partial w_2(\xi,t)}{\partial t} - v\frac{\partial w_2(\xi,t)}{\partial \xi}\right)\end{bmatrix} = \overline{p}_{1/2}(\xi, t), \quad \overline{\Phi}_1 = 0, \tag{9.24}$$

$$\Psi_2 + k_{f2}(w_2 - w_3) - k_{f3}(w_1 - w_2)$$

$$+ \nu_{f2}\left[\left(\frac{\partial w_2(\xi, t)}{\partial t} - v\frac{\partial w_2(\xi, t)}{\partial \xi}\right) - \left(\frac{\partial w_3(\xi, t)}{\partial t} - v\frac{\partial w_3(\xi, t)}{\partial \xi}\right)\right]$$

$$- \nu_{f3}\left[\left(\frac{\partial w_1(\xi, t)}{\partial t} - v\frac{\partial w_1(\xi, t)}{\partial \xi}\right) - \left(\frac{\partial w_2(\xi, t)}{\partial t} - v\frac{\partial w_2(\xi, t)}{\partial \xi}\right)\right] = 0,$$

$$\overline{\Phi}_2 = 0, \tag{9.25}$$

$$\Psi_3 + k_{f1}w_3 - k_{f2}(w_2 - w_3) + \nu_{f1}\left(\frac{\partial w_3(\xi, t)}{\partial t} - v\frac{\partial w_3(\xi, t)}{\partial \xi}\right)$$

$$- \nu_{f2}\left[\left(\frac{\partial w_2(\xi, t)}{\partial t} - v\frac{\partial w_2(\xi, t)}{\partial \xi}\right) - \left(\frac{\partial w_3(\xi, t)}{\partial t} - v\frac{\partial w_3(\xi, t)}{\partial \xi}\right)\right] = 0,$$

$$\overline{\Phi}_3 = 0, \tag{9.26}$$

where,

$$\overline{\Psi}_i = C_{i1}\left(v^2\frac{\partial^2 w_i(\xi,t)}{\partial\xi^2} - 2v\frac{\partial^2 w_i(\xi,t)}{\partial\xi\partial t} + \frac{\partial^2 w_i(\xi,t)}{\partial t^2}\right)$$
$$- C_{i2}\left(\frac{\partial\varphi_i(\xi,t)}{\partial\xi} + \frac{\partial^2 w_i(\xi,t)}{\partial\xi^2}\right)$$
$$- C_{i3}\left(v^2\frac{\partial^4 w_i(\xi,t)}{\partial\xi^4} - 2v\frac{\partial^4 w_i(\xi,t)}{\partial\xi^3\partial t} + \frac{\partial^4 w_i(\xi,t)}{\partial\xi^2\partial t^2}\right) + C_{i4}\frac{\partial^4 w_i(\xi,t)}{\partial\xi^4}$$
$$+ C_{i5}\left(v^2\frac{\partial^3\varphi_i(\xi,t)}{\partial\xi^3} - 2v\frac{\partial^3\varphi_i(\xi,t)}{\partial\xi^2\partial t} + \frac{\partial^3\varphi_i(\xi,t)}{\partial\xi\partial t^2}\right) + C_{i6}\frac{\partial^3\varphi_i(\xi,t)}{\partial\xi^3}, \tag{9.27}$$

$$\overline{\Phi}_i = D_{i1}\left(v^2\frac{\partial^2\varphi_i(\xi,t)}{\partial\xi^2} - 2v\frac{\partial^2\varphi_i(\xi,t)}{\partial\xi\partial t} + \frac{\partial^2\varphi_i(\xi,t)}{\partial t^2}\right) - D_{i2}\frac{\partial^2\varphi_i(\xi,t)}{\partial\xi^2}$$
$$+ D_{i3}\left(\varphi_i(\xi,t) + \frac{\partial w_i(\xi,t)}{\partial\xi}\right) + D_{i4}\left(v^2\frac{\partial^3 w_i(\xi,t)}{\partial\xi^3} - 2v\frac{\partial^3 w_i(\xi,t)}{\partial\xi^2\partial t} + \frac{\partial^3 w_i(\xi,t)}{\partial\xi\partial t^2}\right)$$
$$+ D_{i5}\frac{\partial^3 w_i(\xi,t)}{\partial\xi^3} = 0, \tag{9.28}$$

where, in the case of a moving oscillator with a stabilizer connected to the car's body:

$$\overline{p}_1(x,t) = -\delta(\xi - d_4)\left(m\frac{\mathrm{d}^2 w_2^{01}}{\mathrm{d}t^2} + \left[\begin{array}{c} k_0\left(w_2^{01} - w_2^0 + \theta_2\frac{d_L}{2}\right) \\ +\,\epsilon_0\frac{\mathrm{d}}{\mathrm{d}t}\left(w_2^{01} - w_2^0 + \theta_2\frac{d_L}{2}\right)\end{array}\right]\right)$$
$$- \delta(\xi - d_3)\left(m\frac{\mathrm{d}^2 w_2^{02}}{\mathrm{d}t^2} + \left[k_0\left(w_2^{02} - w_2^0 - \theta_2\frac{d_L}{2}\right) + \epsilon_0\frac{\mathrm{d}}{\mathrm{d}t}\left(w_2^{02} - w_2^0 - \theta_2\frac{d_L}{2}\right)\right]\right)$$
$$- \delta(\xi - d_2)\left(m\frac{\mathrm{d}^2 w_1^{00}}{\mathrm{d}t^2} + \left[k_0(w_1^{00} - w_3^0) + \epsilon_0\frac{\mathrm{d}}{\mathrm{d}t}(w_1^{00} - w_3^0)\right]\right)$$
$$- \delta(\xi - d_1)\left(m\frac{\mathrm{d}^2 w_1^{01}}{\mathrm{d}t^2} + \left[k_0\left(w_1^{01} - w_1^0 + \theta_1\frac{d_L}{2}\right) + \epsilon_0\frac{\mathrm{d}}{\mathrm{d}t}\left(w_1^{01} - w_1^0 + \theta_1\frac{d_L}{2}\right)\right]\right)$$
$$- \delta(\xi)\left(m\frac{\mathrm{d}^2 w_1^{02}}{\mathrm{d}t^2} + \left[k_0\left(w_1^{02} - w_1^0 - \theta_1\frac{d_L}{2}\right) + \epsilon_0\frac{\mathrm{d}}{\mathrm{d}t}\left(w_1^{02} - w_1^0 - \theta_1\frac{d_L}{2}\right)\right]\right), \tag{9.29}$$

$$w_2^{01}(t) = w(d_4,t),\quad w_2^{02}(t) = w(d_3,t),\quad w_1^{00}(t) = w(d_2,t),$$
$$w_1^{01}(t) = w(d_1,t),\quad w_1^{02}(t) = w(0,t), \tag{9.30}$$

and in the case of a moving oscillator with a stabilizer connected to the bogies.

$$\overline{p}_2(x,t) = -\delta(\xi - d_4)\left(m\frac{d^2w_2^{01}}{dt^2} + \left[\begin{array}{c} k_0\left(w_2^{01} - w_2^0 + \theta_2\frac{d_L}{2}\right) \\ + \epsilon_0\frac{d}{dt}\left(w_2^{01} - w_2^0 + \theta_2\frac{d_L}{2}\right)\end{array}\right]\right)$$

$$- \delta(\xi - d_3)\left(m\frac{d^2w_2^{02}}{dt^2} + \left[k_0\left(w_2^{02} - w_2^0 - \theta_2\frac{d_L}{2}\right) + \epsilon_0\frac{d}{dt}\left(w_2^{02} - w_2^0 - \theta_2\frac{d_L}{2}\right)\right]\right)$$

$$- \delta(\xi - d_2)\left(m\frac{d^2w_1^{00}}{dt^2} + \left[k_0(w_1^{00} - w_3) + \epsilon_0\frac{d}{dt}(w_1^{00} - w_3)\right]\right)$$

$$- \delta(\xi - d_1)\left(m\frac{d^2w_1^{01}}{dt^2} + \left[k_0\left(w_1^{01} - w_1^0 + \theta_1\frac{d_L}{2}\right) + \epsilon_0\frac{d}{dt}\left(w_1^{01} - w_1^0 + \theta_1\frac{d_L}{2}\right)\right]\right)$$

$$- \delta(\xi)\left(m\frac{d^2w_1^{02}}{dt^2} + \left[k_0\left(w_1^{02} - w_1^0 - \theta_1\frac{d_L}{2}\right) + \epsilon_0\frac{d}{dt}\left(w_1^{02} - w_1^0 - \theta_1\frac{d_L}{2}\right)\right]\right), \tag{9.31}$$

$$w_2^{01}(t) = w(d_4, t), \quad w_2^{02}(t) = w(d_3, t), \quad w_1^{00}(t) = w(d_2, t),$$
$$w_1^{01}(t) = w(d_1, t), \quad w_1^{02}(t) = w(0, t), \tag{9.32}$$

$$\lim_{|\xi|\to\infty} w_i(x, t) = 0, \quad \lim_{|\xi|\to\infty} \varphi_i(x, t) = 0. \tag{9.33}$$

The analysis of vibration characteristic equations can be conducted by applying integral transforms to the system of Eqs. (9.8–9.15), (9.16–9.23), and (9.24–9.26):

$$W_{si}(\xi, s) = \int_0^\infty w_i(\xi, t)e^{-st}dt,$$

$$\Phi_{si}(\xi, s) = \int_0^\infty \varphi_i(\xi, t)e^{-st}dt, \quad \widehat{W}_{p,si}(p, s) = \int_{-\infty}^\infty W_{si}(\xi, s)e^{-ip\xi}d\xi$$

$$\widehat{\Phi}_{si}(p, s) = \int_{-\infty}^\infty \Phi_{si}(\xi, s)e^{-ip\xi}d\xi, \quad W_1^{01}(s) = \int_0^\infty w_1^{01}(t)e^{-st}dt,$$

$$W_1^{02}(s) = \int_0^\infty w_1^{02}(t)e^{-st}dt, \quad W_1^{00}(s) = \int_0^\infty w_1^{00}(t)e^{-st}dt$$

$$W_2^{01}(s) = \int_0^\infty w_2^{01}(t)e^{-st}dt, \quad W_2^{02}(s) = \int_0^\infty w_2^{02}(t)e^{-st}dt,$$

$$W_1^0(s) = \int_0^\infty w_1^0(t)e^{-st}dt, \quad \Theta_1(s) = \int_0^\infty \theta_1(t)e^{-st}dt,$$

$$W_2^0(s) = \int_0^\infty w_2^0(t)e^{-st}dt, \quad \Theta_2(s) = \int_0^\infty \theta_2(t)e^{-st}dt,$$

$$W_0^0(s) = \int_0^\infty w_0^0(t)e^{-st}\mathrm{d}t, \quad \Theta_0(s) = \int_0^\infty \theta_0(t)e^{-st}\mathrm{d}t,$$

$$W_3^0(s) = \int_0^\infty w_3^0(t)e^{-st}\mathrm{d}t, \tag{9.34}$$

Assuming that the initial conditions specified in Eqs. (9.15) and (9.23) are negligible (as they do not affect system stability due to the linearity of the governing equations) and given that $W_1^{02}(s) = W(0, s)$, $W_1^{01}(s) = W(d_1, s)$, $W_1^{00}(s) = W(d_2, s)$, $W_2^{02}(s) = W(d_3, s)$, $W_2^{01}(s) = W(d_4, s)$ applying the transforms leads to the system of algebraic equations in its current form, after eliminating $\widehat{\Phi}_{si}$, for the system presented in Fig. 9.1.

$$F(p, s)\widehat{W}_{p,s}(p, s) =$$

$$\mathrm{e}^{-\mathrm{i}d_1k}\begin{bmatrix} k_0W_1^0(s) + s\epsilon_0W_1^0(s) - \frac{1}{2}\Theta_1k_0d_L - \frac{1}{2}\Theta_1s\epsilon_0d_L \\ -k_0\mathrm{W}(d_1, s) - ms^2\mathrm{W}(d_1, s) - s\epsilon_0\mathrm{W}(d_1, s) \end{bmatrix}$$

$$+ \mathrm{e}^{-\mathrm{i}d_2k}\begin{bmatrix} k_0W_0^0(s) + s\epsilon_0W_0^0(s) - k_0\mathrm{W}(d_2, s) \\ -ms^2\mathrm{W}(d_2, s) - s\epsilon_0\mathrm{W}(d_2, s) \end{bmatrix}$$

$$+ \mathrm{e}^{-\mathrm{i}d_3k}\begin{bmatrix} k_0W_2^0(s) + s\epsilon_0W_2^0(s) + \frac{1}{2}\Theta_2k_0d_L + \frac{1}{2}\Theta_2s\epsilon_0d_L \\ -k_0\mathrm{W}(d_3, s) - ms^2\mathrm{W}(d_3, s) - s\epsilon_0\mathrm{W}(d_3, s) \end{bmatrix}$$

$$+ \mathrm{e}^{-\mathrm{i}d_4k}\begin{bmatrix} k_0W_2^0(s) + s\epsilon_0W_2^0(s) - \frac{1}{2}\Theta_2k_0d_L - \frac{1}{2}\Theta_2s\epsilon_0d_L \\ -k_0\mathrm{W}(d_4, s) - ms^2\mathrm{W}(d_4, s) - s\epsilon_0\mathrm{W}(d_4, s) \end{bmatrix}$$

$$- k_0W(0, s) - ms^2W(0, s) - s\epsilon_0W(0, s) + k_0W_1^0(s)$$

$$+ s\epsilon_0W_1^0(s) + \frac{1}{2}\Theta_1k_0d_L + \frac{1}{2}\Theta_1s\epsilon_0d_L, \tag{9.35}$$

$$M_1s^2W_2^0(s) - k_0W(d_3, s) - k_0W(d_4, s) - s\epsilon_0W(d_3, s) - s\epsilon_0\mathrm{W}(d_4, s) + 2k_0W_2^0(s)$$

$$+ 2s\epsilon_0W_2^0(s) + k_1W_2^0(s) + s\epsilon_1W_2^0(s) - k_1W_0^0(s)$$

$$- s\epsilon_1W_0^0(s) + \frac{1}{2}\theta_0k_1d_{L0} + \frac{1}{2}\theta_0s\epsilon_1d_{L0} = 0, \tag{9.36}$$

$$\Theta_2J_1s^2 - \frac{1}{2}k_0d_LW(d_3, s) + \frac{1}{2}k_0d_LW(d_4, s) - \frac{1}{2}s\epsilon_0d_LW(d_3, s)$$

$$+ \frac{1}{2}s\epsilon_0d_LW(d_4, s) + \frac{1}{2}\Theta_2k_0d_L^2 + \frac{1}{2}\Theta_2s\epsilon_0d_L^2 = 0, \tag{9.37}$$

$$M_1s^2W_1^0(s) - k_0W(d_1, s) - s\epsilon_0W(d_1, s) - k_0W(0, s) - s\epsilon_0W(0, s)$$

$$+ 2k_0W_1^0(s) + 2s\epsilon_0W_1^0(s) + k_1W_1^0(s) + s\epsilon_1W_1^0(s)$$

$$- k_1 W_0^0(s) - s\epsilon_1 W_0^0(s) - \frac{1}{2}\theta_0 k_1 d_{L0} - \frac{1}{2}\theta_0 s\epsilon_1 d_{L0} = 0, \tag{9.38}$$

$$\Theta_1 J_1 s^2 - \frac{1}{2}k_0 d_L W(0, s) + \frac{1}{2}k_0 d_L W(d_1, s) - \frac{1}{2}s\epsilon_0 d_L W(0, s)$$
$$+ \frac{1}{2}s\epsilon_0 d_L W(d_1, s) + \frac{1}{2}\Theta_1 k_0 d_L^2 + \frac{1}{2}\Theta_1 s\epsilon_0 d_L^2 = 0, \tag{9.39}$$

$$M_0 s^2 W_0^0(s) + 2k_1 W_0^0(s) - k_1 W_1^0(s) - k_1 W_2^0(s) + k_2 W_0^0(s) - k_2 W_3^0(s)$$
$$+ 2s\epsilon_1 W_0^0(s) - s\epsilon_1 W_1^0(s) - s\epsilon_1 W_2^0(s) + s\epsilon_2 W_0^0(s) - s\epsilon_2 W_3^0(s) = 0, \tag{9.40}$$

$$\theta_0 J_0 s^2 - \frac{1}{2}k_1 d_{L0} W_1^0(s) + \frac{1}{2}k_1 d_{L0} W_2^0(s) - \frac{1}{2}s\epsilon_1 d_{L0} W_1^0(s)$$
$$+ \frac{1}{2}s\epsilon_1 d_{L0} W_2^0(s) + \frac{1}{2}\theta_0 k_1 d_{L0}^2 + \frac{1}{2}\theta_0 s\epsilon_1 d_{L0}^2 = 0, \tag{9.41}$$

$$M_s s^2 W_3^0(s) - k_0 W(d_2, s) - s\epsilon_0 W(d_2, s) + k_0 W_3^0(s)$$
$$+ k_2 W_3^0(s) + s\epsilon_0 W_3^0(s) + s\epsilon_2 W_3^0(s) - k_2 W_0^0(s) - s\epsilon_2 W_0^0(s) = 0, \tag{9.42}$$

and for the model presented in Fig. 9.2, the system of algebraic equations is:

$$F(p, s)\widehat{W}_{p,s}(p, s) =$$

$$e^{-id_1 k}\begin{bmatrix} k_0 W_1^0(s) + s\epsilon_0 W_1^0(s) - \frac{1}{2}\Theta_1 k_0 d_L - \frac{1}{2}\Theta_1 s\epsilon_0 d_L \\ - k_0 \mathrm{W}(d_1, s) - ms^2 \mathrm{W}(d_1, s) - s\epsilon_0 \mathrm{W}(d_1, s) \end{bmatrix}$$
$$+ e^{-id_2 k}[k_0 W_0^0(s) + s\epsilon_0 W_0^0(s) - k_0 \mathrm{W}(d_2, s) - ms^2 \mathrm{W}(d_2, s) - s\epsilon_0 \mathrm{W}(d_2, s)]$$
$$+ e^{-id_3 k}\begin{bmatrix} k_0 W_2^0(s) + s\epsilon_0 W_2^0(s) + \frac{1}{2}\Theta_2 k_0 d_L + \frac{1}{2}\Theta_2 s\epsilon_0 d_L \\ - k_0 \mathrm{W}(d_3, s) - ms^2 \mathrm{W}(d_3, s) - s\epsilon_0 \mathrm{W}(d_3, s) \end{bmatrix}$$
$$+ e^{-id_4 k}\begin{bmatrix} k_0 W_2^0(s) + s\epsilon_0 W_2^0(s) - \frac{1}{2}\Theta_2 k_0 d_L - \frac{1}{2}\Theta_2 s\epsilon_0 d_L \\ - k_0 \mathrm{W}(d_4, s) - ms^2 \mathrm{W}(d_4, s) - s\epsilon_0 \mathrm{W}(d_4, s) \end{bmatrix}$$
$$- k_0 W(0, s) - ms^2 W(0, s) - s\epsilon_0 W(0, s) + k_0 W_1^0(s) + s\epsilon_0 W_1^0(s)$$
$$+ \frac{1}{2}\Theta_1 k_0 d_L + \frac{1}{2}\Theta_1 s\epsilon_0 d_L, \tag{9.43}$$

$$M_1 s^2 W_2^0(s) - k_0 W(d_3, s) - k_0 W(d_4, s) - s\epsilon_0 W(d_3, s) - s\epsilon_0 \mathrm{W}(d_4, s) + 2k_0 W_2^0(s)$$
$$+ 2s\epsilon_0 W_2^0(s) + k_1 W_2^0(s) + s\epsilon_1 W_2^0(s) + k_{v2} W_3^0(s) - k_{v2} W_2^0(s) + s\epsilon_{v2} W_3^0(s)$$
$$- s\epsilon_{v2} W_2^0(s) - \Theta_2 d_{\theta 2} k_{v2} - \Theta_2 s d_{\theta 2}\epsilon_{v2} - k_1 W_0^0(s) - s\epsilon_1 W_0^0(s)$$
$$+ \frac{1}{2}\theta_0 k_1 d_{L0} + \frac{1}{2}\theta_0 s\epsilon_1 d_{L0} = 0, \tag{9.44}$$

$$\Theta_2 J_1 s^2 - \frac{1}{2}k_0 d_L W(d_3, s) + \frac{1}{2}k_0 d_L W(d_4, s) - \frac{1}{2}s\epsilon_0 d_L W(d_3, s)$$

$$+ \frac{1}{2}s\epsilon_0 d_L W(d_4, s) + d_{\theta 2}k_{v2}W_3^0(s) - d_{\theta 2}k_{v2}W_2^0(s) + d_{\theta 2}s\epsilon_{v2}W_3^0(s)$$

$$- d_{\theta 2}s\epsilon_{v2}W_2^0(s) - \Theta_2 d_{\theta 2}^2 k_{v2} - \Theta_2 d_{\theta 2}^2 s\epsilon_{v2} + \frac{1}{2}\Theta_2 k_0 d_L^2 + \frac{1}{2}\Theta_2 s\epsilon_0 d_L^2 = 0, \quad (9.45)$$

$$M_1 s^2 W_1^0(s) - k_0 W(d_1, s) - s\epsilon_0 W(d_1, s) - k_0 W(0, s) - s\epsilon_0 W(0, s)$$

$$+ 2k_0 W_1^0(s) + 2s\epsilon_0 W_1^0(s) + k_1 W_1^0(s) + s\epsilon_1 W_1^0(s)$$

$$- k_1 W_0^0(s) - s\epsilon_1 W_0^0(s) + k_{v1}W_3^0(s) - k_{v1}W_1^0(s) + s\epsilon_{v1}W_3^0(s)$$

$$- s\epsilon_{v1}W_1^0(s) + \Theta_1 d_{\theta 1}k_{v1} + \Theta_1 s d_{\theta 1}\epsilon_{v1} - \frac{1}{2}\theta_0 k_1 d_{L0} - \frac{1}{2}\theta_0 s\epsilon_1 d_{L0} = 0, \quad (9.46)$$

$$\Theta_1 J_1 s^2 - \frac{1}{2}k_0 d_L W(0, s) + \frac{1}{2}k_0 d_L W(d_1, s) - \frac{1}{2}s\epsilon_0 d_L W(0, s) + \frac{1}{2}s\epsilon_0 d_L W(d_1, s)$$

$$- d_{\theta 1}k_{v1}W_3^0(s) + d_{\theta 1}k_{v1}W_1^0(s) - d_{\theta 1}s\epsilon_{v1}W_3^0(s) + d_{\theta 1}s\epsilon_{v1}W_1^0(s)$$

$$- \Theta_1 d_{\theta 1}^2 k_{v1} - \Theta_1 d_{\theta 1}^2 s\epsilon_{v1} + \frac{1}{2}\Theta_1 k_0 d_L^2 + \frac{1}{2}\Theta_1 s\epsilon_0 d_L^2 = 0, \quad (9.47)$$

$$M_0 s^2 W_0^0(s) + 2k_1 W_0^0(s) - k_1 W_1^0(s) - k_1 W_2^0(s)$$

$$+ 2s\epsilon_1 W_0^0(s) - s\epsilon_1 W_1^0(s) - s\epsilon_1 W_2^0(s) = 0, \quad (9.48)$$

$$\theta_0 J_0 s^2 - \frac{1}{2}k_1 d_{L0} W_1^0(s) + \frac{1}{2}k_1 d_{L0} W_2^0(s) - \frac{1}{2}s\epsilon_1 d_{L0} W_1^0(s)$$

$$+ \frac{1}{2}s\epsilon_1 d_{L0} W_2^0(s) + \frac{1}{2}\theta_0 k_1 d_{L0}^2 + \frac{1}{2}\theta_0 s\epsilon_1 d_{L0}^2 = 0, \quad (9.49)$$

$$M_s s^2 W_3^0(s) - k_0 W(d_2, s) - s\epsilon_0 W(d_2, s) + k_0 W_3^0(s) + s\epsilon_0 W_3^0(s) - k_{v1}W_3^0(s)$$

$$+ k_{v1}W_1^0(s) - k_{v2}W_3^0(s) + k_{v2}W_2^0(s) - s\epsilon_{v1}W_3^0(s) - s\epsilon_{v2}W_3^0(s) + s\epsilon_{v1}W_1^0(s)$$

$$+ s\epsilon_{v2}W_2^0(s) - \Theta_1 d_{\theta 1}k_{v1} + \Theta_2 d_{\theta 2}k_{v2} - \Theta_1 s d_{\theta 1}\epsilon_{v1} + \Theta_2 s d_{\theta 2}\epsilon_{v2} = 0, \quad (9.50)$$

where $F(p, s)$ is the complex beam-layer reaction. Starting with the system of Eqs. (9.35–9.42) and (9.43–9.50) and applying the inverse Fourier transform over k, then eliminating W_0^0, W_1^0, W_2^0, Θ_0, Θ_1, Θ_2 and setting the values $\xi = 0$, $\xi = d_1$, $\xi = d_2$,

$\xi = d_3$, $\xi = d_4$ subsequently, will yield a system of algebraic equations concerning the Laplace displacements of the contact points $w_s(0, s)$, $w_s(d_1, s)$, ..., $w_s(d_4, s)$

$$\boldsymbol{G_1} \cdot \boldsymbol{w_s}[(0,\ d_1,\ d_2,\ d_3,\ d_4),\ s] = 0 \ \Leftrightarrow \ [G_1]_{5x5}\begin{Bmatrix} w_s(0,\ s) \\ \vdots \\ w_s(d_4,\ s) \end{Bmatrix}_{5x1} = \begin{Bmatrix} 0 \\ \vdots \\ 0 \end{Bmatrix}_{5x1}, \tag{9.51}$$

$$\boldsymbol{G_2} \cdot \boldsymbol{w_s}[(0,\ d_1,\ d_2,\ d_3,\ d_4),\ s] = 0 \ \Leftrightarrow \ [G_2]_{5x5}\begin{Bmatrix} w_s(0,\ s) \\ \vdots \\ w_s(d_4,\ s) \end{Bmatrix}_{5x1} = \begin{Bmatrix} 0 \\ \vdots \\ 0 \end{Bmatrix}_{5x1}. \tag{9.52}$$

The existence of a nontrivial solution for the model 1 Eqs. (9.35–9.42), involving a stabilizer connected to the car's body, and the model 2 Eqs. (9.43–9.50), featuring a stabilizer connected to the bogies is contingent upon the determinants of the matrices $[G_1]_{5x5}$ and $[G_2]_{5x5}$ satisfy

$$\det \mathbf{G}_{1/2} = 0. \tag{9.53}$$

In the following sections, we assess the stability of the model by analyzing the eigenvalues associated with the characteristic Eq. (9.53).

9.3 Solution method and stability criteria

The examination of stability involves determining the characteristics of the roots of the characteristic equation (Eq. 9.53). If at least one root possesses a positive real part, the vibrations of the complex moving oscillator are deemed unstable. This serves as the criterion for identifying instability in the vibrations. The specified equation is integral with respect to Laplace variables. To simplify the resolution of the mathematical problem, the D-decomposition method is employed, as introduced and described by Denisov et al. [9]. Temporarily setting aside the physical interpretation of the problem, the parameter k_0 is considered a complex parameter, and a mapping of the straight line onto the complex plane k_0 is conducted with $s = i\Omega$ (where Ω ranges from minus to plus infinity). This implies that the D-decomposition is carried out in the k_0-plane, where the positive half of the real axis in the k_0-plane corresponds to the stiffness of the vehicle's suspension. The objective is to identify all complex values of k_0 —effectively delineating lines in the k_0-plane—that correlates with the pure imaginary solution $s = i\Omega$ of Eq. (9.53). These lines will divide the k_0-plane into regions with varying numbers of roots of Eq. (9.53) concerning the variables, satisfying the condition $\text{Re}(s) > 0$. To achieve this objective, Eq. (9.53) is reformulated in the following manner:

$$f(\Omega,\ k_0) = 0, \tag{9.54}$$

where $s = i\Omega$. The next step is to find all the roots of Eq. (9.53) with respect to the variable k_0. The integration of these integrals can be accomplished through the contour integration technique [9,10] resulting in the subsequent outcome. The analysis delineates the bogie response, characterized by a specific coupled structure, through the function $F(\Omega,p)$, which encompasses roots.

$$\begin{aligned}
\mathrm{Int}_0 &= i\sum\nolimits_n \frac{k-k_n}{(k-k_1)\cdots(k-k_6)}\bigg|_{k=k_n}, \\
\mathrm{Int}_{\Delta-} &= -i\sum\nolimits_m \frac{(k-k_m)e^{ik\mathscr{R}}}{(k-k_1)\cdots(k-k_6)}\bigg|_{k=k_m}, \quad \mathscr{R}<0, \\
\mathrm{Int}_{\Delta+} &= i\sum\nolimits_n \frac{(k-k_n)e^{ik\mathscr{R}}}{(k-k_1)\cdots(k-k_6)}\bigg|_{k=k_n}, \quad \mathscr{R}>0,
\end{aligned} \tag{9.55}$$

The roots with positive imaginary parts are denoted as k_n, and roots with negative imaginary parts, denoted as k_m, of equation $F(k, \ \Omega) = 0$ are examined. The compact representation of the residues, acquired through contour integration (9.55), is available in the same structure and format in Appendices A6 and A7 of Stojanović et al. [14]. This reference provides the comprehensive form of all expressions (9.55) after the application of the contour integration method the D-decomposition curves are analyzed. Five curves should be examined due to the 5×5 determinants of matrices $\mathbf{G}_1$ and $\mathbf{G}_2$ given in Eqs. (9.51) and (9.52), each of these roots is correlated to one of the five roots of Eq. (9.54). The parametric analysis reveals that D-decomposition curves are significant for dynamic analysis when the velocity is in the supercritical speed region, that is, when the complex moving oscillator is moving at a velocity surpassing the critical velocity. The critical velocity is identified as demonstrated in Stojanović et al. [14] and the methodology is not reiterated here. In the subcritical region ($v < v_{cr}$), the D-decomposition curves do not cross the positive part of the real axis at any point. This implies that when $v < v_{cr}$, the number of unstable roots of the characteristic equation (roots with a positive real part) remains constant regardless of the stiffness k_0. However, when $v > v_{cr}$, the number of unstable roots starts varying the moment k_0 surpasses the critical value of stiffness. The use of the D-decomposition method allows determination of the relative variation of the number of unstable roots depending on the stiffness k_0, but not the number itself. To find the number of unstable roots for any given value of stiffness k_0, the argument principle is employed. This involves fixing a value for k_0 and computing the total change of the argument of the complex expression $f(-is, k_0)$ along the entire contour, where f is the function from Eq. (9.54). According to the argument principle, the total change is equal to $2\pi N_{un}$, where $2\pi N_{un}$ is the number of roots of $f(-is, k_0)$ located in the positive half-plane with respect to s. For both parts of the contour (the semicircle and y-axis), a sufficiently small iteration step Δs is chosen, with which the relative change of the argument $\mathrm{Arg}(f(-i(s+\Delta s), k_0)/f(-is, k_0))$ is computed, and all of these relative changes are summed up. The size of the iteration step Δs is dynamically adjusted along the contour. The expression for $f(-is, k_0)$ with integrals of the type $Int_{\Delta\pm}$ is complex. To expedite the computation of the argument change, the complete expression for $f(-is, k_0)$ is not computed symbolically. Instead, for each s and k_0, the values of all integrals $Int_{\Delta\pm}$ are computed first, and then $f(-is, k_0)$ is computed by evaluating the determinant (9.53). This approach significantly accelerates the computation of $f(-is, k_0)$ and requires less utilization of working memory.

9.4 Results and discussion

In this segment, we examined a complex mechanical system, and we established D-decomposition curves based on a specific set of geometric and material parameters. Subsequently, the D-decomposition curves and the count of unstable roots are computed for a supercritical velocity with a value of $v = v_{cr}$. The discussion on the dynamic stability of the system revolves around the extent of instability, arising due to the velocity being within the supercritical range.

$$E_1 = 2 \times 10^{11}\,\mathrm{N/m^2}, \quad G_1 = 7.813 \times 10^{10}\,\mathrm{N/m^2},$$

$$\rho_1 = 7849\,\mathrm{kg/m^3}, \quad I_1 = 3.055 \times 10^{-5}\,\mathrm{m^4},$$

$$E_2 = 3.5 \times 10^{10}\,\mathrm{N/m^2}, \quad G_2 = 1.4583 \times 10^{10}\,\mathrm{N/m^2},$$

$$\rho_2 = 2500\,\mathrm{kg/m^3}, \quad I_2 = 6.6875 \times 10^{-4}\,\mathrm{m^4},$$

$$E_3 = 3.3 \times 10^{10}\,\mathrm{N/m^2}, \quad G_3 = 1.375 \times 10^{10}\,\mathrm{N/m^2},$$

$$\rho_3 = 2500\,\mathrm{kg/m^3}, \quad I_3 = 1.7 \times 10^{-3}\,\mathrm{m^4},$$

$$k_{f1} = 6 \times 10^{7}\,\mathrm{N/m^2}, \quad k_{f2} = 1.25 \times 10^{9}\,\mathrm{N/m^2}, \quad k_{f3} = 6 \times 10^{7}\,\mathrm{N/m^2},$$

$$\nu_{f1} = 9 \times 10^{4}\,\mathrm{N{\cdot}s/m^2}, \quad \nu_{f2} = 8.3 \times 10^{4}\,\mathrm{N{\cdot}s/m^2}, \quad \nu_{f3} = 3.625 \times 10^{4}\,\mathrm{N{\cdot}s/m^2},$$

$$M_1 = 3333\,\mathrm{kg}, \quad m = 10^{3}\,\mathrm{kg}, \quad \epsilon_0 = 86 \times 10^{3}\,\mathrm{N{\cdot}s/m^2},$$

$$d_L = 15.7\,\mathrm{m}, \quad d_R = 1\,\mathrm{m}, \quad J = 598 \times 10^{3}\,\mathrm{kg{\cdot}m/s^2}. \tag{9.56}$$

First, we consider the model depicted in Fig. 9.1. This model incorporates a stabilizer with an adjustable vertical position connected to the car's body using a spring of stiffness k_2 and damping element ϵ_2. Ten distinct cases have been examined, and the varied parameters are listed in Table 9.1. The corresponding D-decomposition curve diagrams are illustrated in Figs. 9.3–9.13. The highlighted stable regions are presented on each of the D-decomposition curves, where no unstable roots are present. Conversely, the remaining segments delineate primary stiffness regions k_0 characterized by the presence of multiple unstable roots. Figs. 9.3–9.7 depict D-decomposition curves for the case where the car body mass is equal to 32,000 kg. First, we consider the change in stability regions for the first two cases listed in Table 9.1, depicted in Figs. 9.3 and 9.4. These are the cases where the spring stiffness values, k_1 and k_2, are equal, while the damping coefficient is ϵ_2 varied. It can be observed that the stability region decreases very slightly when the damping coefficient ϵ_2 is higher compared to ϵ_1 (Case 2). Additionally, in this case, the stability region shifts and occurs for slightly higher values of the stiffness k_0, since the stability region for Case 1 commences when the stiffness k_0 is equal to

Table 9.1 System parameters for the stabilizer connected to the car's body.

	M_S[kg]	M_0[kg]	k_1[N/m]	k_2[N/m]	ϵ_1]	ϵ_2[Ns/m^2]
Case 1	2000	32,000	$8 \cdot 10^5$	$8 \cdot 10^5$	$2 \cdot 10^5$	$1 \cdot 10^5$
Case 2	2000	32,000	$8 \cdot 10^5$	$8 \cdot 10^5$	$2 \cdot 10^5$	$4 \cdot 10^5$
Case 3	2000	32,000	$8 \cdot 10^5$	$4 \cdot 10^5$	$2 \cdot 10^5$	$2 \cdot 10^5$
Case 4	2000	32,000	$8 \cdot 10^5$	$16 \cdot 10^5$	$2 \cdot 10^5$	$2 \cdot 10^5$
Case 5	750	32,000	$8 \cdot 10^5$	$8 \cdot 10^5$	$2 \cdot 10^5$	$2 \cdot 10^5$
Case 6	2000	6400	$8 \cdot 10^5$	$8 \cdot 10^5$	$2 \cdot 10^5$	$1 \cdot 10^5$
Case 7	2000	6400	$8 \cdot 10^5$	$8 \cdot 10^5$	$2 \cdot 10^5$	$4 \cdot 10^5$
Case 8	2000	6400	$8 \cdot 10^5$	$4 \cdot 10^5$	$2 \cdot 10^5$	$2 \cdot 10^5$
Case 9	2000	6400	$8 \cdot 10^5$	$16 \cdot 10^5$	$2 \cdot 10^5$	$2 \cdot 10^5$
Case 10	720	6400	$8 \cdot 10^5$	$8 \cdot 10^5$	$2 \cdot 10^5$	$2 \cdot 10^5$

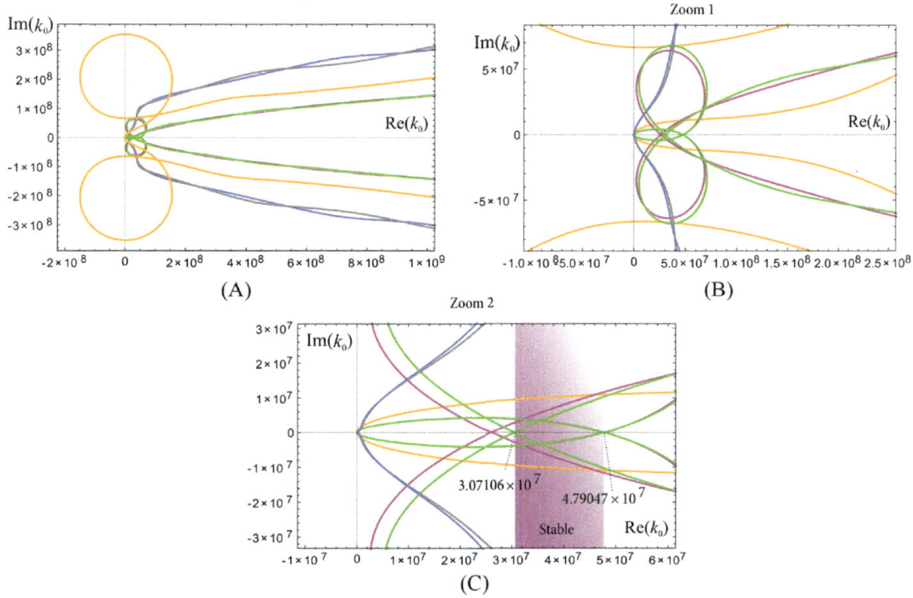

Figure 9.3 D-decomposition curves for Case 1, for the system parameters from Table 9.1. (A) The entire diagram, (B) magnified area, (C) magnified area with a marked stable interval for the determined k_0 boundary.

3.07106×10^7 N/m, while in Case 2, it occurs when k_0 reaches the value of 3.07501×10^7 N/m.

Figs. 9.5 and 9.6 illustrate the D-decomposition curves for Cases 3 and 4 (Table 9.1), representing scenarios where the damping coefficients, ϵ_1 and ϵ_2, are now equal and constant, while the stiffness of the support, k_2, varies. It can be observed that there

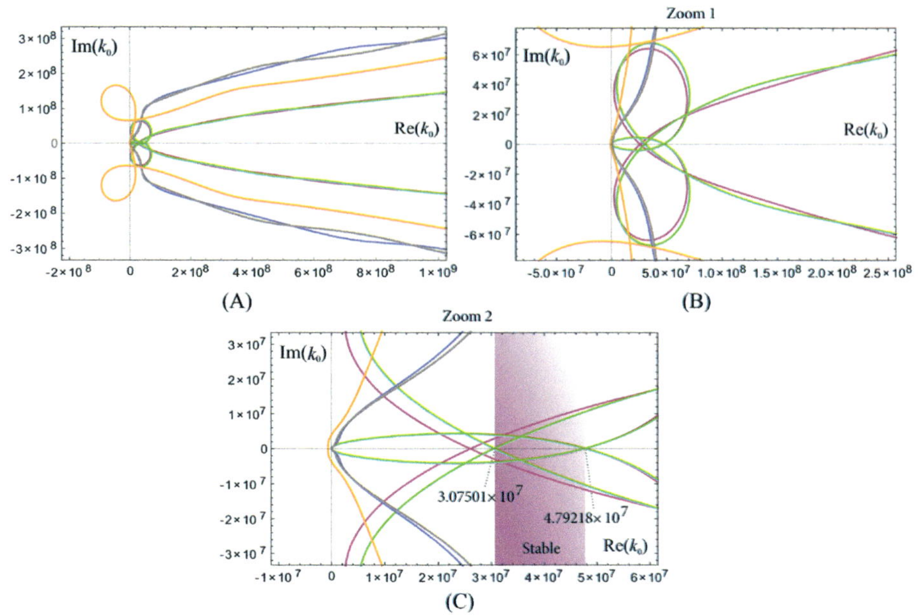

Figure 9.4 D-decomposition curves for Case 2, for the system parameters from Table 9.1. (A) The entire diagram, (B) magnified area, (C) magnified area with a marked stable interval for the determined k_0 boundary.

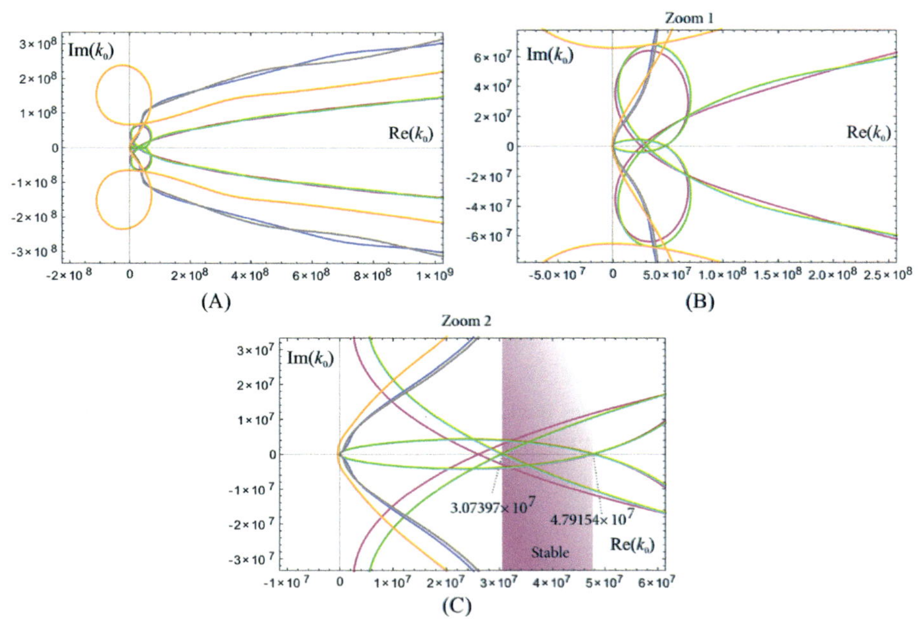

Figure 9.5 D-decomposition curves for Case 3, for the system parameters from Table 9.1. (A) The entire diagram, (B) magnified area, (C) magnified area with a marked stable interval for the determined k_0 boundary.

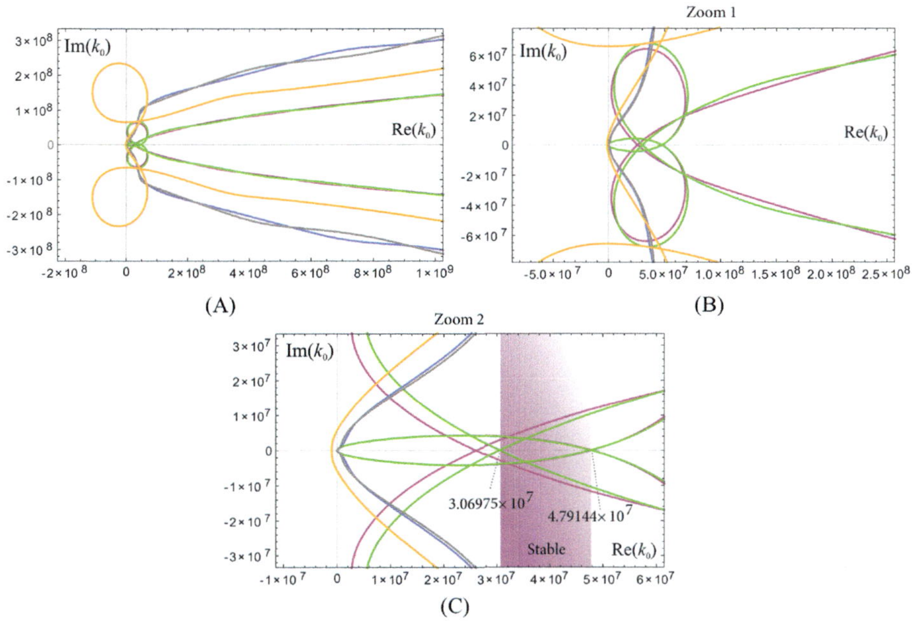

Figure 9.6 D-decomposition curves for Case 4, for the system parameters from Table 9.1. (A) The entire diagram, (B) magnified area, (C) magnified area with a marked stable interval for the determined k_0 boundary.

is a slight increase in the stability region when k_2 is twice the value of k_1 compared to the case where k_2 is half the value of k_1. Although this increase is not significantly large, it is greater compared to the first two cases involving variations in damping coefficients. Furthermore, there is a slight shift of the stability region when the stiffness coefficient k_2 is increased, where stability now starts from slightly lower values of k_0 as can be seen in Figs. 9.5 and 9.6.

In Fig. 9.7, we examine a scenario where both stiffness coefficients, k_1 and k_2, along with damping coefficients, ϵ_1 and ϵ_2, remain equal and constant. However, in this specific instance, the mass of the stabilizer decreases to 750 kg, in contrast to the prior cases where it was 2000 kg. Despite this alteration, no significant change in the stability region, that is, the region devoid of unstable roots, compared to all preceding cases. Nonetheless, in this particular scenario, the stable region exhibits a slight expansion, and the stable zone can be achieved with even lower values of primary suspension stiffness k_0 compared to the previous cases, starting at k_0= 3.05286 N/m.

In all previously presented cases, the mass of the car body is assumed to be 32,000 kg. It can be observed that, in the case of such a stabilizer connected to a car body weighing 32,000 kg, as illustrated in Fig. 9.1, variations in parameters may result in a minor change in the size of the stability region, as well as a slight shift in its position. In this scenario, altering the stiffness and damping parameters does not significantly manipulate the stability region. However, it shows us that it is possible

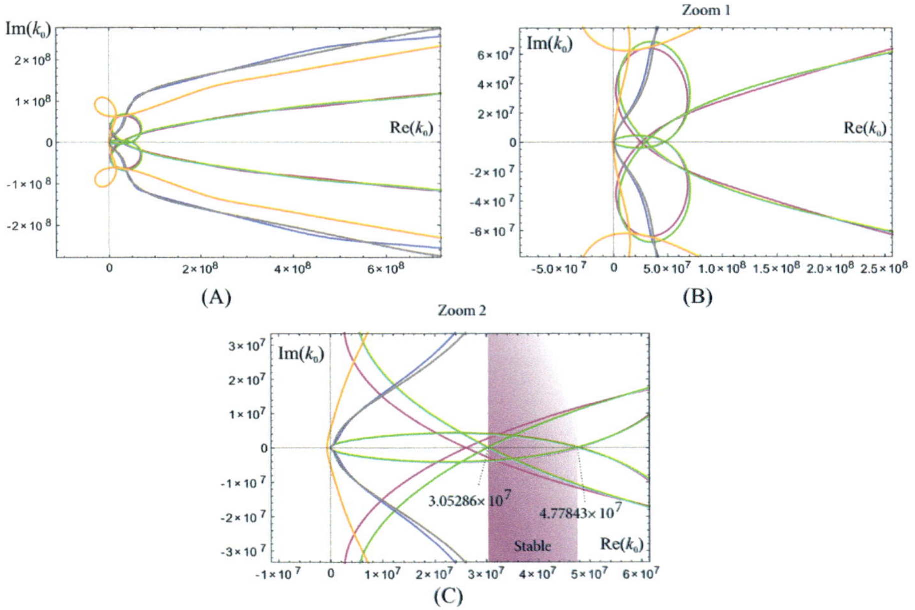

Figure 9.7 D-decomposition curves for Case 5, for the system parameters from Table 9.1. (A) The entire diagram, (B) magnified area, (C) magnified area with a marked stable interval for the determined k_0 boundary.

to vary the stiffness and damping of the support while maintaining the position and size of the stability region.

Now we are considering cases where it is assumed that the mass of the car body is twice as large as in the previously examined examples, and the results obtained for these cases are shown in Figs. 9.8–9.12. Firstly, cases are depicted where the stiffness values k_1 and k_2 remain the same while varying ϵ_2. Case 6, illustrated in Fig. 9.8, shows the D-decomposition curves when ϵ_2 is twice smaller than ϵ_1, while Case 7, where ϵ_2 is twice larger than ϵ_1, is shown in Fig. 9.9. If we compare these two cases with the results shown in Figs. 9.3 and 9.4 (Cases 1 and 2, Table 9.1), we see that for higher car body mass, the stability region slightly increases, and the overall stability region shifts to lower values for the primary suspension stiffness k_0. By comparing Cases 6 and 7, it can also be noticed that for cases with higher car body mass, the trend of a slight reduction in the stability region continues due to the increase in the damping coefficient ϵ_2, as well as a shift of the overall stability region to slightly higher values of the primary stiffness moduli k_0. In this scenario, with the car body mass being twice as large, these changes are somewhat more noticeable.

Next, we are examining how the variation of stiffness k_2 affects the stability of the system when the car's body mass is 6400 kg. The results are shown in Figs. 9.10 and 9.11, for the parameters given in Table 9.1 for Cases 8 and 9. Case 10 is depicted in Fig. 9.12. Comparing the results, it can be observed that in the case of a heavier car body, changes in the stiffness coefficient k_2 have a negligible impact on stability.

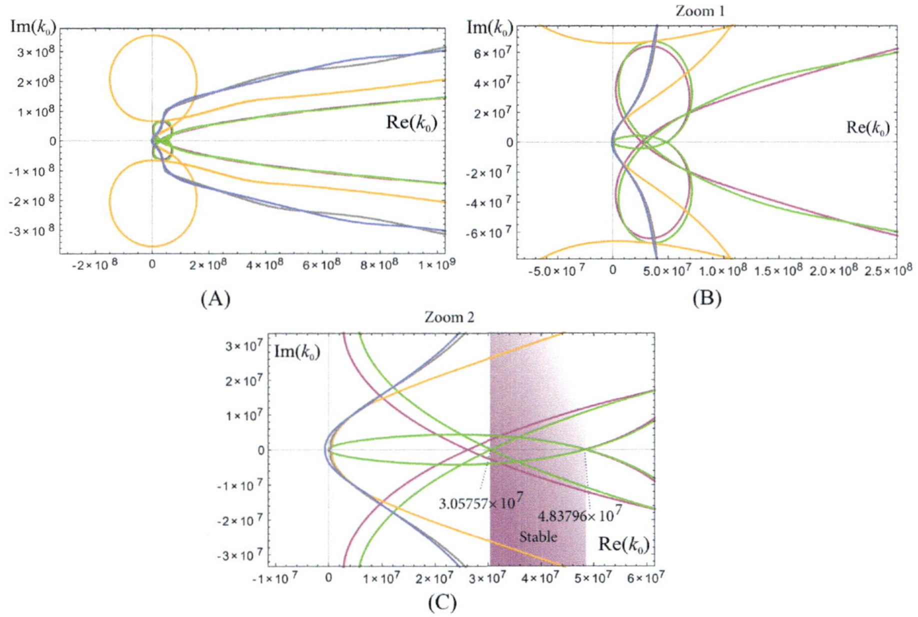

Figure 9.8 D-decomposition curves for Case 6, for the system parameters from Table 9.1. (A) The entire diagram, (B) magnified area, (C) magnified area with a marked stable interval for the determined k_0 boundary.

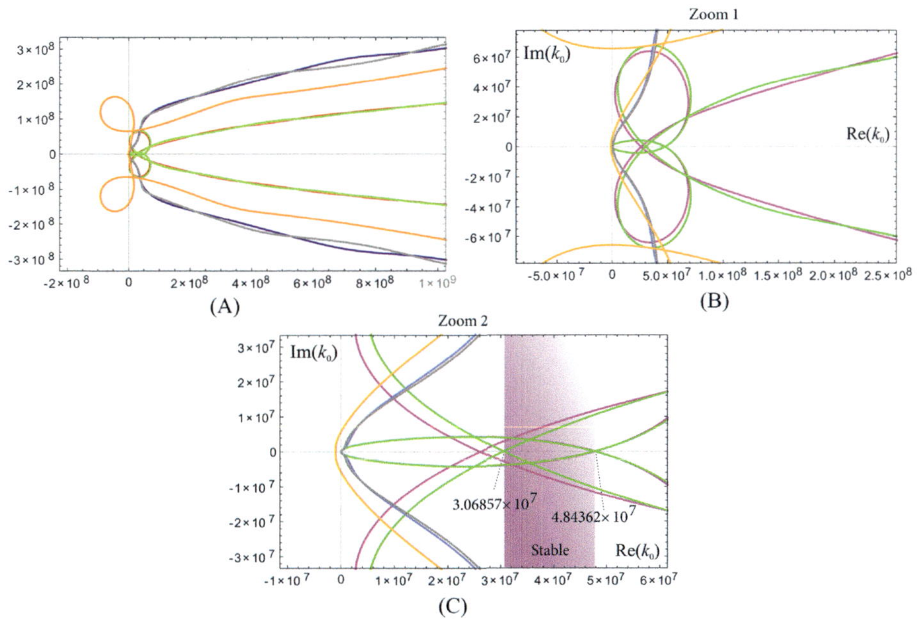

Figure 9.9 D-decomposition curves for Case 7, for the system parameters from Table 9.1. (A) The entire diagram, (B) magnified area, (C) magnified area with a marked stable interval for the determined k_0 boundary.

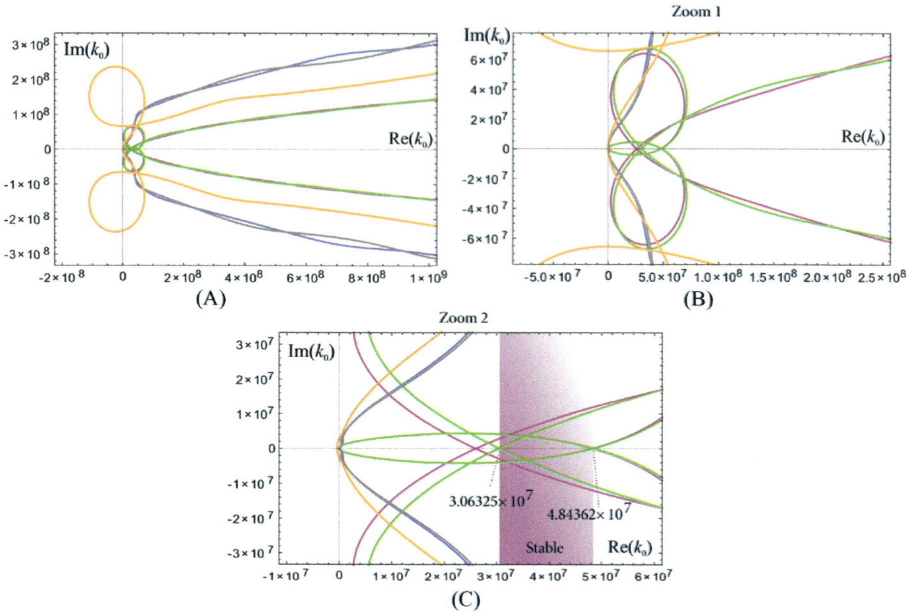

Figure 9.10 D-decomposition curves for Case 8, for the system parameters from Table 9.1. (A) The entire diagram, (B) magnified area, (C) magnified area with a marked stable interval for the determined k_0 boundary.

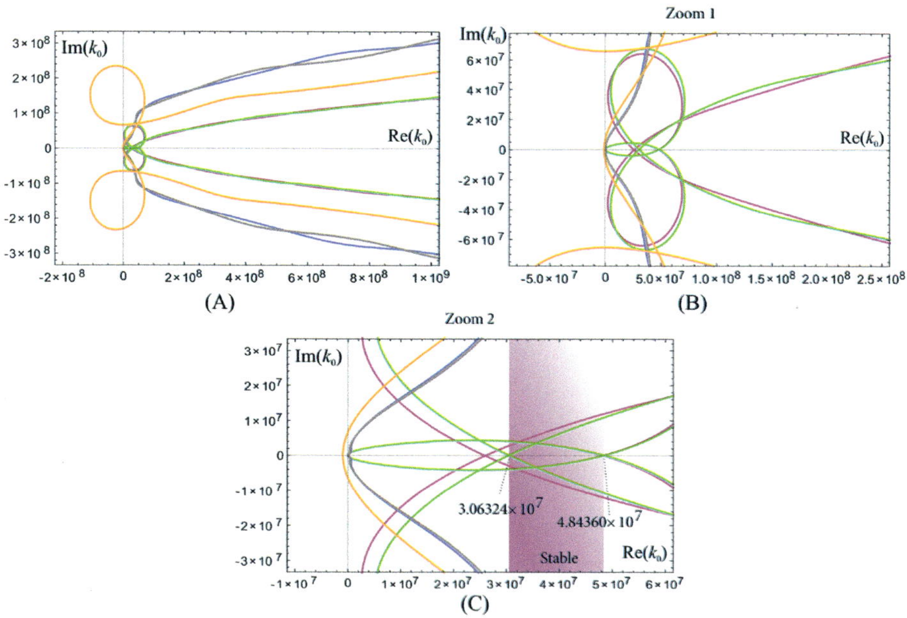

Figure 9.11 D-decomposition curves for Case 9, for the system parameters from Table 9.1. (A) The entire diagram, (B) magnified area, (C) magnified area with a marked stable interval for the determined k_0 boundary.

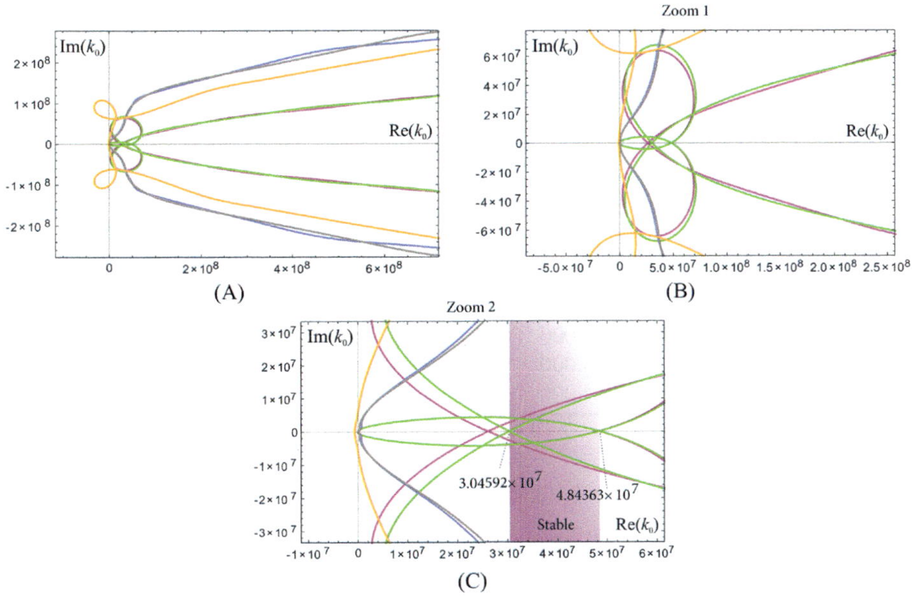

Figure 9.12 D-decomposition curves for Case 10, for the system parameters from Table 9.1. (A) The entire diagram, (B) magnified area, (C) magnified area with a marked stable interval for the determined k_0 boundary.

However, the results obtained for higher car body mass are again more favorable in terms of stability compared to the results obtained for the case of lower car body masses (Cases 3 and 4, Table 9.1), Considering that the stability region is larger for Cases 8 and 9 than for Cases 3 and 4. The last case we are examining for the stabilizer connected to the car body is when the stiffness and damping coefficients, k_1 and k_2, as well as ϵ_1 and ϵ_2, are mutually equal, but the mass of the stabilizer is reduced to 750 kg. In this case, the stability region is the largest compared to previously considered cases for the same car body mass, and it also starts with the smallest values of k_0 (Case 10, Table 9.1).

Comparing the results obtained for a lower car body mass and a higher car body mass, we can conclude that changes in the stiffness and damping parameters, k_2 and ϵ_2, again do not significantly affect the stability of the system. However, the stability region is larger for Cases 6–10 compared to the results obtained for Cases 1–5. This indicates that in the case of a larger car body mass, such a connection with the stabilizer provides an improved stability region, which is also stable and does not vary significantly with changes in the stiffness and damping parameters.

If we assess the "level" of instability by considering the number of unstable roots in each specific part of the region k_0 [N/m]: {;}, {number of unstable roots}, Model 1, presented in Fig. 9.1, exhibits a specific number of unstable roots based on the cases outlined in Table 9.1:

Case 1: {0; 2.6205 × 10^7; 3.07106 × 10^7; 4.79047 × 10^7; 4.79955 × 10^7; + ∞}, **{4,2,0,4,4};**

Table 9.2 System parameters for the stabilizer connected to the bogies.

	M_S[kg]	M_0[kg]	k_{v1}[N/m]	k_{v1} [N/m]	ϵ_v[Ns/m^2]
Case 1	2000	32,000	$0.5 \cdot 10^5$	$0.5 \cdot 10^5$	0
Case 2	2000	32,000	$2.5 \cdot 10^5$	$2.5 \cdot 10^5$	0
Case 3	2000	32,000	$2 \cdot 10^5$	$3 \cdot 10^5$	0
Case 4	2000	32,000	$3 \cdot 10^5$	$2 \cdot 10^5$	0
Case 5	2000	32,000	$2.5 \cdot 10^5$	$2.5 \cdot 10^5$	8600
Case 6	2000	64,000	$2.5 \cdot 10^5$	$2.5 \cdot 10^5$	4300
Case 7	2000	64,000	$2.5 \cdot 10^5$	$2.5 \cdot 10^5$	17200
Case 8	2000	64,000	$2.75 \cdot 10^5$	$2.75 \cdot 10^5$	8600
Case 9	2000	64,000	$2 \cdot 10^5$	$2 \cdot 10^5$	8600
Case 10	720	64,000	$3 \cdot 10^5$	$3 \cdot 10^5$	8600

Case 2: $\{0;\ 2.6215\times 10^7;\ 3.07501\times 10^7;\ 4.79218\times 10^7;\ 4.80912\times 10^7;\ +\infty\}$, **{4,2,0,4,4}**;

Case 3: $\{0;\ 2.62089\times 10^7;\ 3.07397\times 10^7;\ 4.79154\times 10^7;\ 4.7999\times 10^7;\ +\infty\}$, **{4,2,0,4,4}**;

Case 4: $\{0;\ 2.62089\times 10^7;\ 3.06975\times 10^7;\ 4.79144\times 10^7;\ 4.7983\times 10^7;\ +\infty\}$, **{4,2,0,4,4}**;

Case 5: $\{0;\ 2.61758\times 10^7;\ 3.05286\times 10^7;\ 4.77843\times 10^7;\ 4.80906\times 10^7;\ +\infty\}$, **{4,2,0,4,4}**;

Case 6: $\{0;\ 2.60853\times 10^7;\ 3.05757\times 10^7;\ 4.83796\times 10^7;\ 4.85056\times 10^7;\ +\infty\}$, **{4,2,0,2,4}**;

Case 7: $\{0;\ 2.61192\times 10^7;\ 3.06857\times 10^7;\ 4.84362\times 10^7;\ 4.86013\times 10^7;\ +\infty\}$, **{4,2,0,2,4}**;

Case 8: $\{0;\ 2.61157\times 10^7;\ 3.06325\times 10^7;\ 4.84362\times 10^7;\ 4.86107\times 10^7;\ +\infty\}$, **{4,2,0,2,4}**;

Case 9: $\{0;\ 2.61156\times 10^7;\ 3.06324\times 10^7;\ 4.8436\times 10^7;\ 4.86092\times 10^7;\ +\infty\}$, **{4,2,0,2,4}**;

Case 10: $\{0;\ 2.6121\times 10^7;\ 3.0459\times 10^7;\ 4.84363\times 10^7;\ 4.85405\times 10^7;\ +\infty\}$, **{4,2,0,4,4}**;

From the displayed numbers of unstable roots within the intervals of k_0, we can conclude that for stabilizer connected to the car's body, even the number of unstable roots varies very little, as well as the intervals in which they are located.

Now we present results obtained for the system parameters listed in Table 9.2. All these scenarios relate to the stabilizer depicted in Fig. 9.2, where the stabilizer is connected to the bogies, and the connection is realized through springs with stiffness k_{v1} and k_{v1} and dampers with damping coefficients $\epsilon_{v1} = \epsilon_{v2} = \epsilon_v$.

The first four scenarios explore situations where the damping coefficient ϵ_v is set to zero, while simultaneously varying the stiffness coefficients k_{v1} and k_{v2} to analyze their impact on the system's stability. From Figs. 9.13–9.16, it can be observed that changing these stiffness coefficients does not significantly affect the size or position of the stability region. We initially examined two scenarios where k_{v1} and k_{v2} are

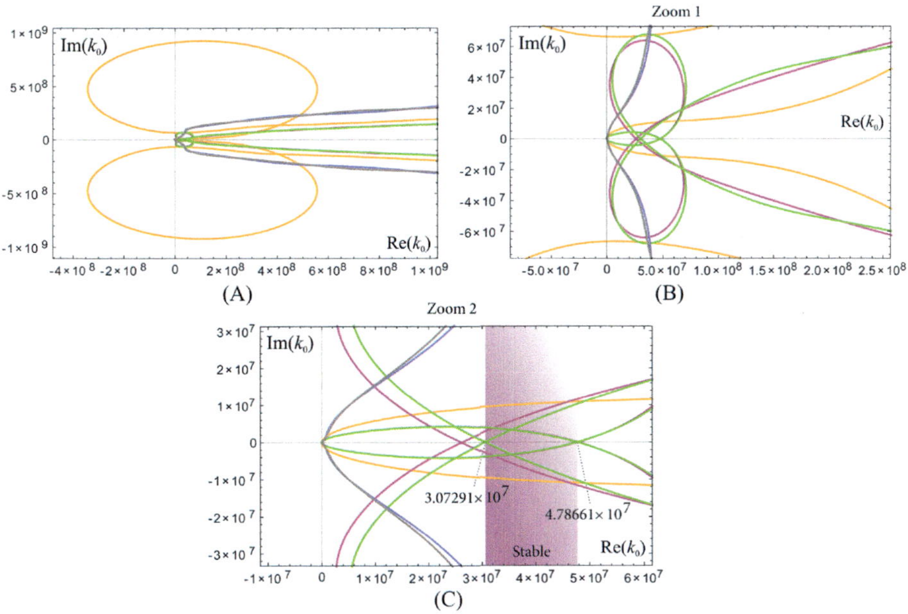

Figure 9.13 D-decomposition curves for Case 1, for the system parameters from Table 9.2. (A) The entire diagram, (B) magnified area, (C) magnified area with a marked stable interval for the determined k_0 boundary.

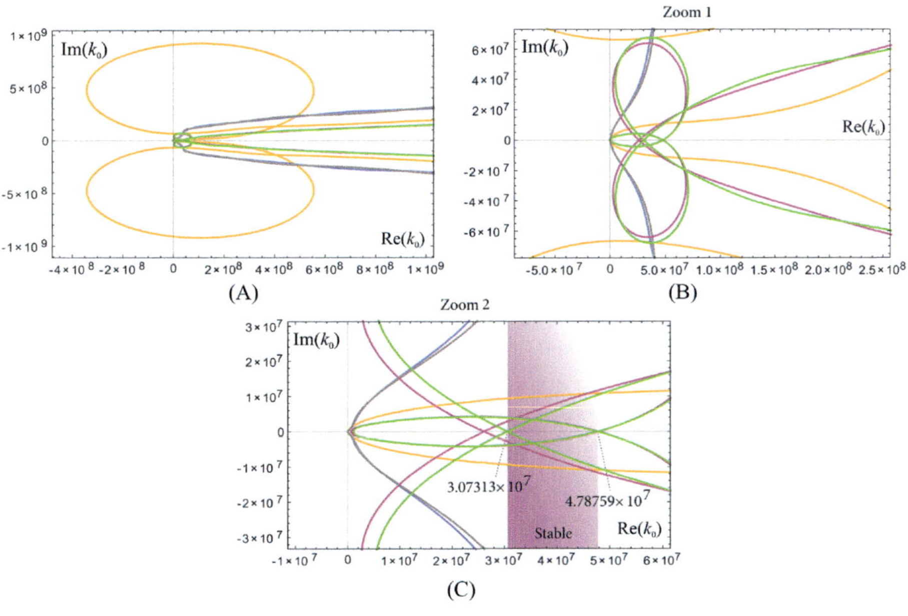

Figure 9.14 D-decomposition curves for Case 2, for the system parameters from Table 9.2. (A) The entire diagram, (B) magnified area, (C) magnified area with a marked stable interval for the determined k_0 boundary.

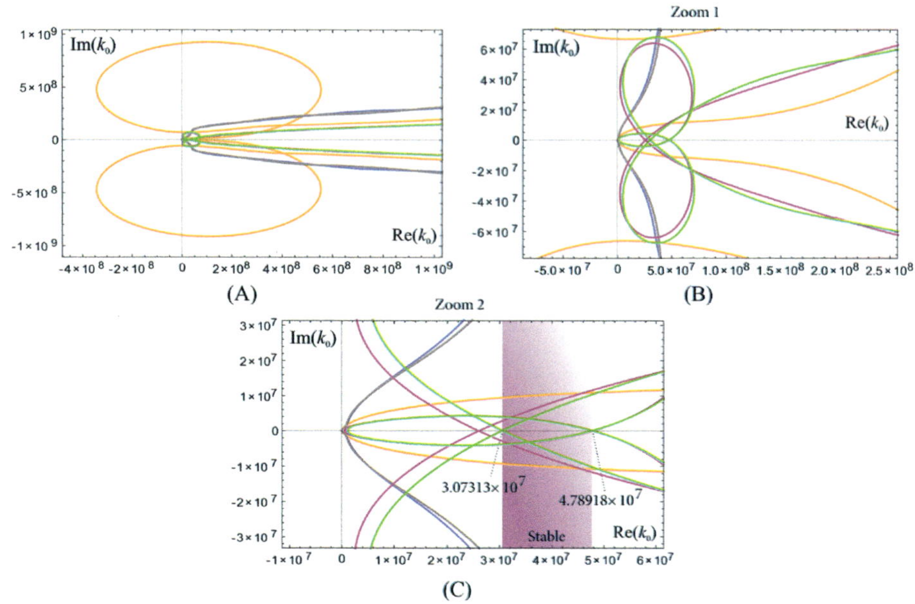

Figure 9.15 D-decomposition curves for Case 3, for the system parameters from Table 9.2. (A) The entire diagram, (B) magnified area, (C) magnified area with a marked stable interval for the determined k_0 boundary.

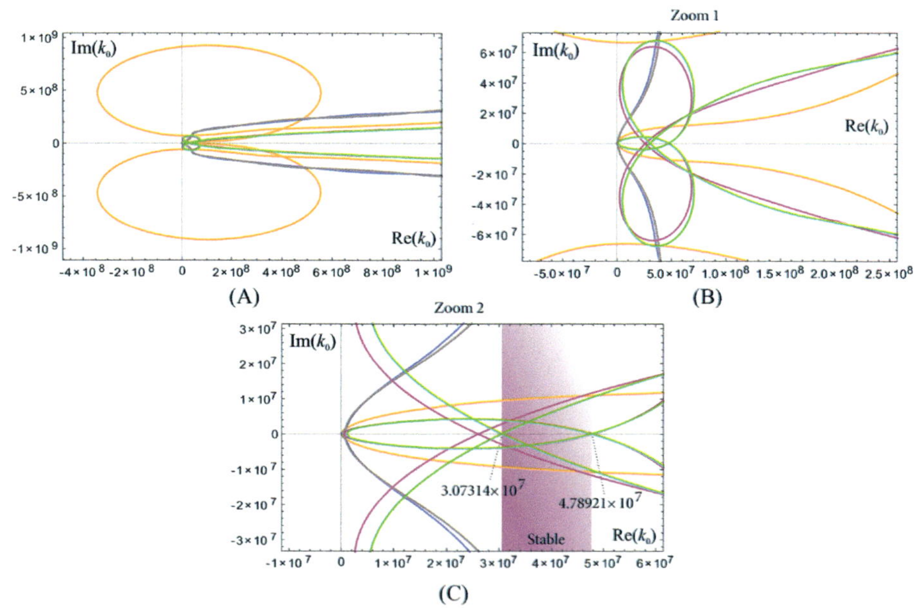

Figure 9.16 D-decomposition curves for Case 4, for the system parameters from Table 9.2. (A) The entire diagram, (B) magnified area, (C) magnified area with a marked stable interval for the determined k_0 boundary.

equal (Figs. 9.13 and 9.14), with initial values of 0.5×10^5 N/m and then 2.5×10^5 N/m. We see that by increasing the stiffness coefficients $k_{\nu 1}$ and $k_{\nu 2}$, the stability results are almost identical, as there is just a slightly larger stability region obtained for higher values of $k_{\nu 1}$ and $k_{\nu 2}$. Subsequently, we investigated whether setting different values for these stiffness coefficients $k_{\nu 1}$ and $k_{\nu 2}$ would result in different stability outcomes (Figs. 9.15 and 9.16), and it is evident that there are very small changes in the stability region compared to Cases 1 and 2 (Table 9.2), although a somewhat larger stability region is obtained for Cases 3 and 4 when the stiffness values are mutually different. In all presented results, the stable region starts for almost the same values of k_0. It is also noticeable that stability is not significantly affected by weather $k_{\nu 1}$ is greater or smaller than $k_{\nu 2}$. Stability was also examined for some other parameters, but as they did not yield significant changes, the authors did not present all the results. Based on the presented results, which correspond to the case of a car body mass of 32,000 kg, it can be concluded that this type of stabilizer achieves a very slight reduction in stability region compared to the previously proposed stabilizer for the same car body weight (Case 1–5, Table 9.1.)

Additional scenarios, whose parameter combinations are given in Table 9.2 (Cases 5–10), pertain to situations where the car body weight is doubled, now amounting to 64,000 kg. Initially, three cases were examined—Cases 5–7—where the stiffness coefficients $k_{\nu 1}$ and $k_{\nu 2}$ were set to be equal and constant, while varying the damping coefficient ϵ_ν. The obtained results are shown in Figs. 9.17–9.19, revealing a clear trend of significant reduction in the stability region with increasing ϵ_ν. This change is

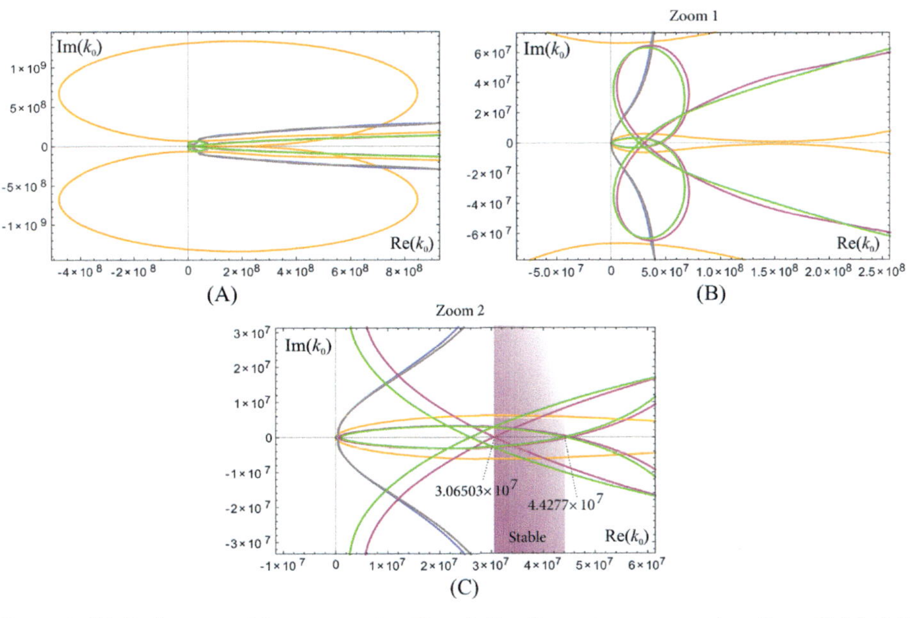

Figure 9.17 D-decomposition curves for Case 5, for the system parameters from Table 9.2. (A) The entire diagram, (B) magnified area, (C) magnified area with a marked stable interval for the determined k_0 boundary.

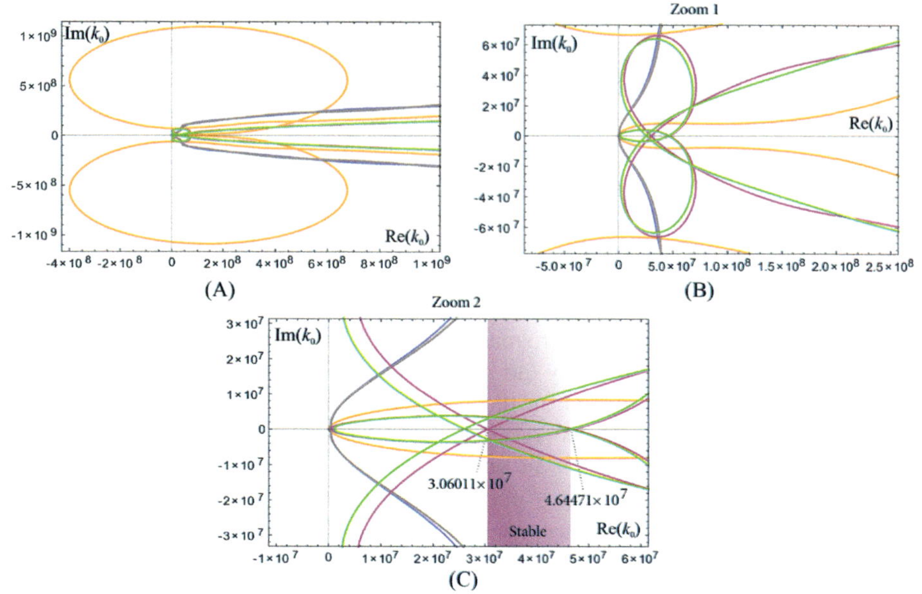

Figure 9.18 D-decomposition curves for Case 6, for the system parameters from Table 9.2. (A) The entire diagram, (B) magnified area, (C) magnified area with a marked stable interval for the determined k_0 boundary.

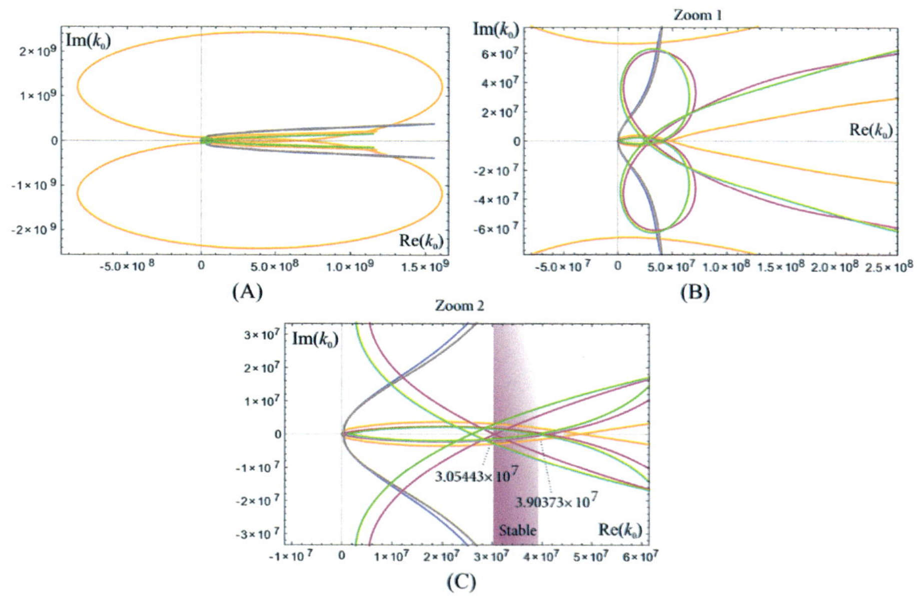

Figure 9.19 D-decomposition curves for Case 7, for the system parameters from Table 9.2. (A) The entire diagram, (B) magnified area, (C) magnified area with a marked stable interval for the determined k_0 boundary.

more pronounced between Case 5 (Fig. 9.17), chosen as a reference, and Case 7 (Fig. 9.19), where ϵ_v is twice as large, compared to the change between Cases 5 and 6 (Fig. 9.18), where ϵ_v is twice as small as the reference value. The worst results in terms of stability were obtained in Case 7, depicted in Fig. 9.19, where a very clear reduction in the stability region can be observed.

Finally, we consider Cases 8–10, as shown in Table 9.2. In these scenarios, the damping coefficient ϵ_v is set as constant, while the stiffness coefficients k_{v1} and k_{v2} are varied. In Case 8 (Fig. 9.20), these two values are set equal and increased compared to the previously observed cases. In this instance, a slight reduction in the stability region is observed compared to the referent Case 5, but also, the stability region starts from slightly lower values of k_0. However, in Cases 9 and 10 (Figs. 9.21 and 9.22), where k_{v1} and k_{v2} are set to be different, a mild increase in the stability region is noted compared to the referent case. Notably, practically the same results in terms of stability results are obtained when k_{v1} is greater than k_{v2}, as well as vice versa (Cases 9 and 10, Table 9.2). In Cases 8–10, it was also observed that the stability region shifted for slightly smaller values of the coefficient k_0.

Referring to the "level" of instability, determined by the number of unstable roots for each specific section of the region k_0 [N/m]: {;},{number of unstable roots}, Model 2, presented in Fig. 9.2, exhibits varying degrees of instability based on the cases outlined in Table 9.2.

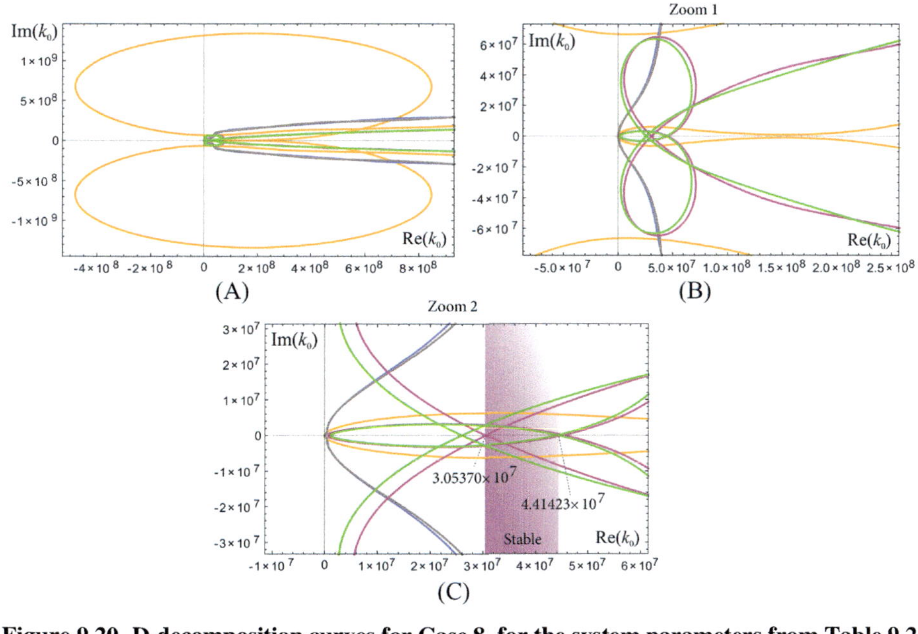

Figure 9.20 D-decomposition curves for Case 8, for the system parameters from Table 9.2. (A) The entire diagram, (B) magnified area, (C) magnified area with a marked stable interval for the determined k_0 boundary.

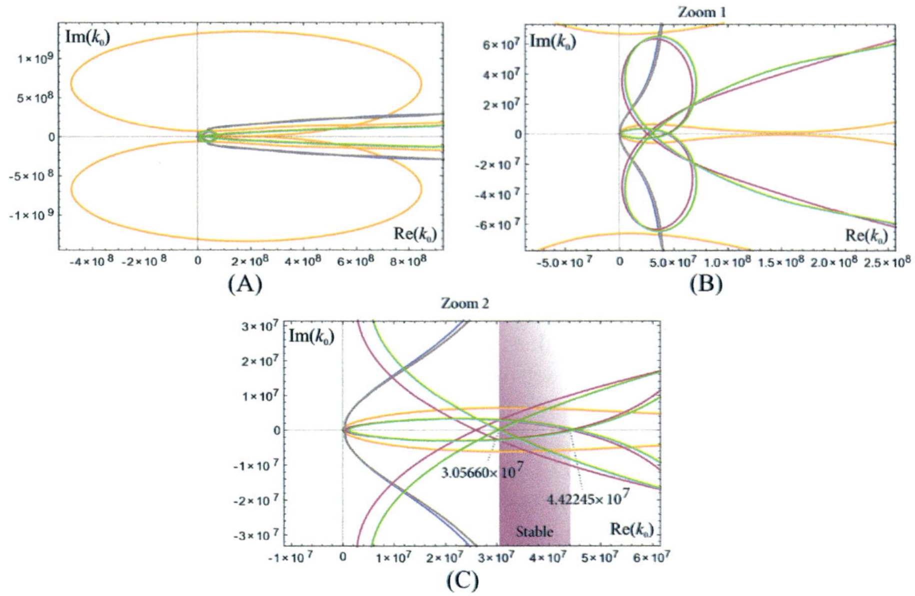

Figure 9.21 D-decomposition curves for Case 9, for the system parameters from Table 9.2. (A) The entire diagram, (B) magnified area, (C) magnified area with a marked stable interval for the determined k_0 boundary.

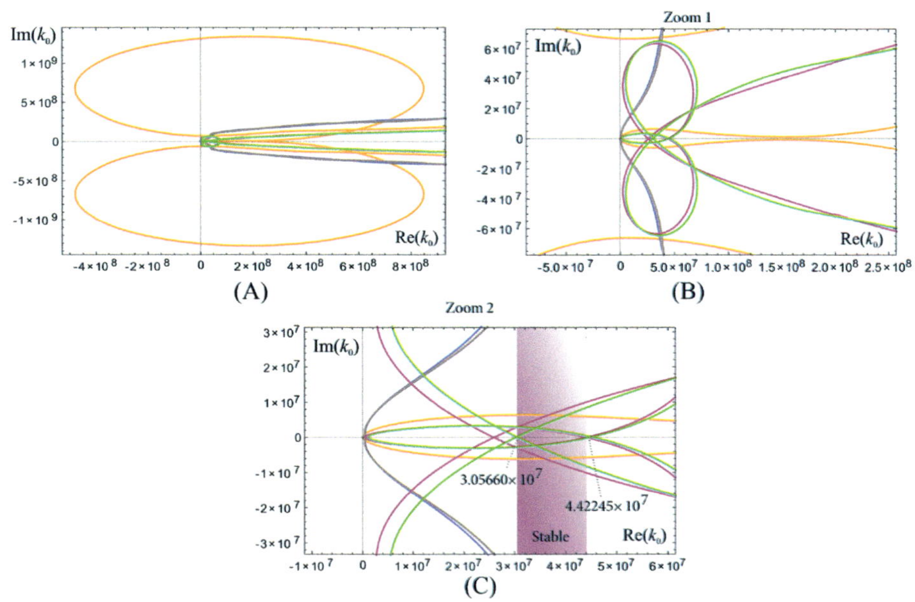

Figure 9.22 D-decomposition curves for Case 10, for the system parameters from Table 9.2. (A) The entire diagram, (B) magnified area, (C) magnified area with a marked stable interval for the determined k_0 boundary.

Case 1: {0; 125030; 225091; 2.62028×10^7; 3.07291×10^7; 4.78661×10^7; 4.79785×10^7; $+\infty$}, **{6,5,4,2,0,4,4}**;

Case 2: {0; 625741; 1.12728×10^6; 2.62017×10^7; 3.07313×10^7; 4.78759×10^7; 4.80418×10^7; $+\infty$}, **{6,5,4,2,0,4,4}**;

Case 3: {0; 574582; 1.17854×10^6; 2.62017×10^7; 3.07313×10^7; 4.78918×10^7; 4.80272×10^7; $+\infty$}, **{6,5,4,2,0,4,4}**;

Case 4: {0; 574582; 1.17854×10^6; 2.62017×10^7; 3.07313×10^7; 4.78918×10^7; 4.80272×10^7; $+\infty$}, **{6,5,4,2,0,4,4}**;

Case 5: {0; 625741; 1.12728×10^6; 2.61194×10^7; 3.06503×10^7; 4.4277×10^7; 4.54386×10^7; $+\infty$}, **{6,5,4,2,0,2,4}**;

Case 6: {0; 625741; 1.12728×10^6; 2.6085×10^7; 3.06011×10^7; 4.64471×10^7; 4.70399×10^7; $+\infty$}, **{6,5,4,2,0,2,4}**

Case 7: {0; 625741; 1.12728×10^6; 2.59622×10^7; 3.05443×10^7; 3.90373×10^7; 4.10839×10^7; 4.7845×10^7; 6.65063×10^8; $+\infty$}, **{6,5,4,2,0,2,4,6,4}**

Case 8: {0; 688397; 1.24026×10^6; 2.60138×10^7; 3.0537×10^7; 4.41423×10^7; 4.52977×10^7; $+\infty$}, **{6,5,4,2,0,2,4}**

Case 9: {0; 574582; 1.17854×10^6; 2.60383×10^7; 3.0566×10^7; 4.4245×10^7; 4.53051×10^7; $+\infty$}, **{6,5,4,2,0,2,4}**

Case 10: {0; 574582; 1.17854×10^6; 2.60383×10^7; 3.0566×10^7; 4.4245×10^7; 4.53051×10^7; $+\infty$}, **{6,5,4,2,0,2,4}**

In conclusion, we can infer that in the scenario where the car body mass is doubled, with a stabilizer connected to the bogies, manipulation of the stability region is more achievable through parameter variation, especially the damping coefficient ε_v. However, when comparing the stability regions for this type of stabilizer with results for the case where the stabilizer is connected to the car body, which was the initially considered scenario, it is evident that connection to the car body yields the best results in terms of stability. Specifically, in all cases where the car body mass is doubled (Cases 6–10, Table 9.1), the stability regions encompass those obtained for scenarios with the alternative proposed stabilizer (Cases 5–10, Table 9.2). Based on the obtained results, it can be concluded from the comparative analysis of these two proposed stabilizers that the first proposed configuration, shown in Fig. 9.1, provides better stability outcomes for both larger and smaller car body weights, compared to the second proposed configuration.

9.5 Conclusion

This study focused on analyzing the stability of vibrations in a complex moving oscillator traversing a continuous and infinite three-part viscoelastic beam-foundation system. The investigation emphasized distinguishing stability variations between two distinct models, both featuring additional stabilizers—one connected to the car's body with an adjustable vertical position, and another connected to the bogies. This comparison sheds light on the differences in vibration stability within these systems. A critical condition for instability is the surpassing of the critical velocity, equivalent to the minimum phase velocity of waves in the beam. The characteristic equation for the complex oscillator that is moving on a continuous infinite beam laid on an elastic foundation is determined. Subsequently, the roots of the characteristic equation were calculated, and the model's stability was assessed based on these roots. The examination of roots involved the sequential application of the D-decomposition method and the argument principle. The D-decomposition method was employed to partition the model's space into domains with varying numbers of unstable roots. Subsequently, the argument principle was applied to a set of model parameters to determine the number of unstable roots within one of these domains. After establishing stability limits for both cases of an additional stabilizer, it was observed that connecting the stabilizer to the car body resulted in larger stability regions. In this scenario, significant adjustments to stiffness and damping parameters could be made with minimal impact on the stability region. In contrast, when the stabilizer was connected to the bogies, changes in stiffness and damping parameters had a more pronounced impact and manipulation of the stability region, particularly noticeable in heavier car bodies, although stability regions were consistently smaller in all cases considered.

References

[1] L. Fryba, Vibration of Solids and Structures under Moving Loads, Telford, London, 1999.

[2] V. Stojanović, P. Kozić, M.D. Petković, Dynamic instability and critical velocity of a mass moving uniformly along a stabilized infinity beam, International Journal of Solids and Structures 108 (2017) 164–174.

[3] S. Timoshenko, Method of analysis of statical and dynamical stresses in rail, Proceedings of the 2nd International Congress for Applied Mechanics, Zürich, Switzerland. (1927) 1–12.

[4] A.D. Kerr, The continuously supported rail subjected to an axial force and a moving load, International Journal of Mechanical Sciences 14 (1972) 71–78.

[5] H.D. Nelson, R.A. Conover, R. A., Dynamic stability of a beam carrying moving masses, Journal of Applied Mechanics 38 (1971) 1003–1006.

[6] H. Takemiya, Time-domain thin layer method for computing transient response due to sudden/moving loads, Journal of Mechanics of Materials and Structures 3 (2008).

[7] S.M. Kim, Stability and dynamic response of Rayleigh beam-columns on an elastic foundation under moving loads of constant amplitude and harmonic variation, Engineering Structures 27 (2005) 869–880.

[8] M. Attar, A. Karrech, K. Regenauer-Lieb, A lattice spring model for dynamic analysis of damaged beam-type structures under moving loads, European Journal of Mechanics - A/Solids 60 (2016) 196–207.

[9] G.G. Denisov, E.K. Kugusheva, V.V. Novikov, On the problem of the stability of one-dimensional unbounded elastic systems, Journal of Applied Mathematics and Mechanics 49 (1985) 533–537.

[10] S. Verichev, A. Metrikine, Stability of a moving vehicle on an infinite beam, Proceedings of the Xxviii Summer School on Actual Problems in Mechanics, St. Petersburg, Russia, 2001, pp. 54–65.

[11] S. Verichev, A. Metrikine, Instability of a bogie moving on a flexibly supported Timoshenko beam, Journal of Sound and Vibration 253 (2002) 653–668.

[12] T. Mazilu, Instability of a train of oscillators moving along a beam on a viscoelastic foundation, Journal of Sound and Vibration 332 (2013) 4597–4619.
[13] T. Mazilu, M. Dumitriu, C. Tudorache, Instability of an oscillator moving along a Timoshenko beam on viscoelastic foundation, Nonlinear Dynamics 67 (2012) 1273–1293.
[14] V. Stojanović, M.D. Petković, J. Deng, Stability and vibrations of an overcritical speed moving multiple discrete oscillators along an infinite continuous structure, European Journal of Mechanics, A/Solids 75 (2019) 367–380.
[15] Z. Dimitrovová, J.N. Varandas, Critical velocity of a load moving on a beam with a sudden change of foundation stiffness: applications to high-speed trains, Computers & Structures 87 (2009) 1224–1232.
[16] Z. Dimitrovová, New semi-analytical solution for a uniformly moving mass on a beam on a two-parameter visco-elastic foundation, International Journal of Mechanical Sciences 127 (2017) 142–162.
[17] B. Erbas, J. Kaplunov, D.A. Prikazchikov, O. Sahin, The near-resonant regimes of a moving load in a three-dimensional problem for a coated elastic half-space, Mathematics and Mechanics of Solids 22 (2016) 89–100.
[18] C. Svedholm, A. Zangeneh, C. Pacoste, S. François, R. Karoumi, Vibration of damped uniform beams with general end conditions under moving loads, Engineering Structures 126 (2016) 40–52.
[19] K.Y. Lee, A.A. Renshaw, Solution of the moving mass problem using complex eigenfunction expansions, Journal of Applied Mechanics. 67 (2000) 823–827.
[20] S. Sorrentino, G. Catania, Dynamic analysis of rectangular plates crossed by distributed moving loads, Mathematics and Mechanics of Solids 23 (2017) 1291–1302.
[21] Q. Song, Z. Liu, J. Shi, Y. Wan, Parametric study of the dynamic response of sandwich plate under moving loads, Thin-Walled Structures 123 (2018) 82–99.
[22] S.H. Ju, H.T. Lin, Analysis of train-induced vibrations and vibration reduction schemes above and below critical Rayleigh speeds by finite element method, Soil Dynamics and Earthquake Engineering 24 (2004) 993–1002.
[23] H. Xia, N. Zhang, Dynamic analysis of railway bridge under high-speed trains, Computers & Structures 83 (2005) 1891–1901.
[24] L. Gong, M.S. Cheung, Computer simulation of dynamic interactions between vehicle and long-span box girder bridges, Tsinghua Science and Technology 13 (2008) 71–77.
[25] H. Yoon, W.J. Chin, J.-R. Cho, J.Y. Kang, J. Kwark, Comparison of measured and dynamic analysis vertical accelerations of high-speed railway bridges crossed by KTX train, Engineering 5 (2013) 756–763.
[26] V.H. Tran, The impact effect of highway bridge due to heavy vehicle, IABSE Symposium: Engineering for Progress, Nature and People, (2014), pp. 3039–3046.
[27] J. Zhan, Z. Wang, X. Kong, H. Xia, C. Wang, H. Xiang, A drive-by frequency identification method for simply supported railway bridges using dynamic responses of passing two-axle vehicles, Journal of Bridge Engineering 26 (2021) 1–10.
[28] N. Zhang, H. Xia, Dynamic analysis of coupled railway vehicle–bridge system based on inter-system iteration method, Computers & Structures 114–115 (2013) 26–34.
[29] Z. Dimitrovová, On the critical velocity of moving force and instability of moving mass in layered railway track models by semianalytical approaches, Vibration 6 (2023) 113–146.
[30] Z. Dimitrovová, Two-layer model of the railway track: analysis of the critical velocity and instability of two moving proximate masses, International Journal of Mechanical Sciences 217 (2022) 107042.

[31] Z. Dimitrovová, Instability of vibrations of mass(es) moving uniformly on a two-layer track model: parameters leading to irregular cases and associated implications for railway design, Applied Sciences 13 (2023) 12356.
[32] Z. Dimitrovová, T. Mazilu, Semi-analytical approach and Green's function method: a comparison in the analysis of the interaction of a moving mass on an infinite beam on a three-layer viscoelastic foundation at the stability limit—the effect of damping of foundation materials, Materials 17 (2024) 279.
[33] P. Koziol, Wavelet approximation of the Adomian's decomposition applied to a nonlinear problem of a double-beam response subject to a series of moving loads, Journal of Theoretical and Applied Mechanics 52 (2014) 687–697.
[34] W. Czyczula, P. Koziol, D. Kudla, S. Lisowski, Analytical evaluation of track response in the vertical direction due to a moving load, Journal of Vibration and Control 23 (2016) 2989–3006.
[35] P. Koziol, R. Pilecki, Nonlinear double-beam system dynamics, Archives of Civil Engineering 67 (2021) 337–353.
[36] A. Ouzizi, F. Abdoun, L. Azrar, Nonlinear dynamics of beams on nonlinear fractional viscoelastic foundation subjected to moving load with variable speed, Journal of Sound and Vibration 523 (2022) 116730.
[37] V. Stojanović, M.D. Petković, Moment Lyapunov exponents and stochastic stability of a three-dimensional system on elastic foundation using a perturbation approach, Journal of Applied Mechanics 80 (2013) 051009.
[38] D.G. Duffy, The response of an infinite railroad track to a moving, vibrating mass, Journal of Applied Mechanics. 57 (1990) 66–73.
[39] J.-S. Chen, S.-Y. Chen, W.-Z. Hsu, Effects of geometric nonlinearity on the response of a long beam on viscoelastic foundation to a moving mass, Journal of Sound and Vibration 497 (2021) 115961.
[40] J.-S. Chen, W.-Z. Hsu, Stability of an infinite beam on visco-elastic foundation under a uniformly moving distributed mass, Journal of Sound and Vibration 521 (2022) 116684.
[41] Z. Dimitrovová, Dynamic interaction and instability of two moving proximate masses on a beam on a Pasternak viscoelastic foundation, Applied Mathematical Modelling 100 (2021) 192–217.
[42] A.V. Metrikine, S.N. Verichev, J. Blaauwendraad, Stability of a two-mass oscillator moving on a beam supported by a visco-elastic half-space, International Journal of Solids and Structures 42 (2005) 1187–1207.
[43] B. Yang, H. Gao, S. Liu, Vibrations of a multi-span beam structure carrying many moving oscillators, International Journal of Structural Stability and Dynamics 18 (2018) 1850125.
[44] S. Roy, G. Chakraborty, A. DasGupta, Coupled dynamics of a viscoelastically supported infinite string and a number of discrete mechanical systems moving with uniform speed, Journal of Sound and Vibration 415 (2018) 184–209.
[45] J.-D. Yau, Y.-B. Yang, S.-R. Kuo, Impact response of high speed rail bridges and riding comfort of rail cars, Engineering Structures 21 (1999) 836–844.
[46] Y. Qi, H. Dai, J. Yang, K. Xu, Dynamic analysis of vehicle track coupling based on double beam track model, Shock and Vibration (2018) 7469894.
[47] X. Lei, N.-A. Noda, Analyses of dynamic response of vehicle and track coupling system with random irregularity of track vertical profile, Journal of Sound and Vibration 258 (2002) 147–165.

[48] G. Kouroussis, G. Gazetas, I. Anastasopoulos, C. Conti, O. Verlinden, Discrete modelling of vertical track–soil coupling for vehicle–track dynamics, Soil Dynamics and Earthquake Engineering 31 (2011) 1711–1723.

[49] V. Stojanović, J. Deng, M.D. Petković, D. Milić, Non-stability of a bogie moving along a specific infinite complex flexibly beam-layer structure, Engineering Structures 295 (2023) 116788.

[50] V. Stojanović, J. Deng, D. Milić, M.D. Petković, Dynamics of moving coupled objects with stabilizers and unconventional couplings, Journal of Sound and Vibration 570 (2024) 118020.

[51] V. Stojanović, M.D. Petković, Nonlinear dynamic analysis of damaged Reddy-Bickford beams supported on an elastic Pasternak foundation, Journal of Sound and Vibration 385 (2016) 239–266.

Part III

Linear and nonlinear vibrations: Stabilizing phenomena and applications

Nonlinear amplitude analysis of shear deformable beams supported by an elastic foundation with variable discontinuity

10

10.1 Introduction

Somewhere from the mid-2000s onwards, there has been a notable interest in vibrations of beams laid on elastic foundations. Such an increase in research is the product of developments in engineering. In the scenarios when a foundation is made of distinct types of layers, the beam response can become intricate. That depends on the mathematical model used to describe vibratory motion. Furthermore, the exact equations used for the linear regime do not provide satisfactory predictions of the beam behavior under real circumstances [1–3]. Mamandiet et al. [4] analyzed nonlinear effects in the frequency domain. The research concerning the nonlinear regime was broadened with the problem of a moving load [3,5]. Kim and Cho [6] studied the vibration of a shear beam-column laid on an elastic foundation. Their work gave a sound understanding of the linear dynamic behavior of those kinds of beams. Vibrations in their system occurred while the system was exposed to moving loads of two types—either a moving load of constant amplitude or a moving load of harmonic amplitude—in both cases, the velocity of advance is constant. A more sophisticated linear model [7], comprising a beam on a viscoelastic foundation subjected to moving SDoF oscillators, is interesting from the aspect of comprehending the behavior of such systems; in it, several internal variables were introduced to describe the frequency-dependent performance of the viscoelastic foundation.

Equations derived from the analysis of linear systems are not suitable for describing deflections of large beams, because geometric nonlinearities cause a motion that is disparate from the motion present in linear models. Chang and Liu [8] performed the nonlinear deterministic and nonlinear random vibration analysis of a model comprising a beam on an elastic foundation exposed to a moving load. To solve the nonlinear system of differential equations, they used the implicit direct integration method. Shear deformations and rotary inertia were ignored, while the effects of longitudinal deflections and inertia were included according to the Bernoulli–Euler hypothesis. Sapountzakis and Kampitsis [3] developed a boundary element method (BEM) for geometrically nonlinear behavior. That BEM is intended for shear-deformable beams, weighted by moving loads, laid on a tensionless nonlinear viscoelastic foundation defined by three parameters. That study demonstrated the advantage of the finite element method (FEM) when nonlinear manifestations in such problems are studied. In addition, the nonlinear effects of the foundation were studied in different types of structures. Malekzadeh and Vosoughi [9] analyzed the behavior of

Modeling of Complex Dynamic Systems. DOI: https://doi.org/10.1016/B978-0-443-23942-7.00010-3

composite thin beams; for large amplitude free vibration analysis, they used the differential quadrature method. Demeio and Lenci [10] for their study of nonlinear resonances of a semiinfinite cable laid on a nonlinear elastic foundation, used the multiple time scale method.

The p-version finite element method (p-FEM) can be adequately used for interpreting the vibrational effects of the various models in space [11–21]. Stojanović et al. [22] studied the vibrational effects of a shear-deformable beam including geometric nonlinearity of vibrations, nonlinearity in the foundation, and discontinuity in the foundation. In that study, they used the improved version of p-FEM, which is more efficient for solving problems in which discontinuity in the foundation exists. Verified results of that study can be found in the mentioned reference, in which it can be observed that the discontinuity tends to change the familiar dynamic behavior—particularly around the discontinuity. The physical model presented in this chapter has a discontinuity in the foundation. To improve the efficiency of the p-FEM, which is used in a problem that has a steep change in the domain (discontinuity in the foundation), two shape functions are applied to the problem, and their use depends on the location and width of the discontinuity. The validated p-FEM for damaged beams is used here for its easy generalization to any size and position of the discontinuity in the foundation.

The analysis in this chapter is based on the study conducted by Stojanović [23]. Its prime objects of analysis are shear deformable beams with various sizes and locations of the discontinuity and various stiffness moduli for the two foundation types (Winkler and Pasternak). Matters of interest are the influence of discontinuity on natural frequencies, cross-section rotation mode shapes, new asymmetrical general transverse mode shapes, and nonlinear deflections—deformations with maximal deflections. All of this is done for the steady-state regime under concentrated harmonic excitations. Equations of motion are obtained with the use of the principle of virtual work and numerically solved by the Newmark-beta method [24]. Validation of the model was performed by comparing the calculated values of natural frequencies with the results acquired from the ANSYS® FEM software [25].

10.2 Mathematical model and equations of motion

The beam length is designated with L, the width with b, and h is for thickness—in general, the beam has a rectangular cross-section. The discontinuity is demarcated by l_1 and l_2 (Fig. 10.1). By setting adequate values for l_1 and l_2, this model can be used for any size and location of the discontinuity. The beam is made of a material that is assumed to be elastic, isotropic, and homogeneous. The foundation is an elastic layer that can be of Winkler or Pasternak type—by ignoring the shear, the Pasternak type will be reduced to the Winkler type foundation. The global coordinate system can be seen in Fig. 10.1.

Furthermore, Fig. 10.1 shows a nondimensional axis ξ that acts as a local coordinate system for the FEM. In p-FEM (like in FEM) the number of elements is predominately defined by the geometry of the structure that is analyzed; for example,

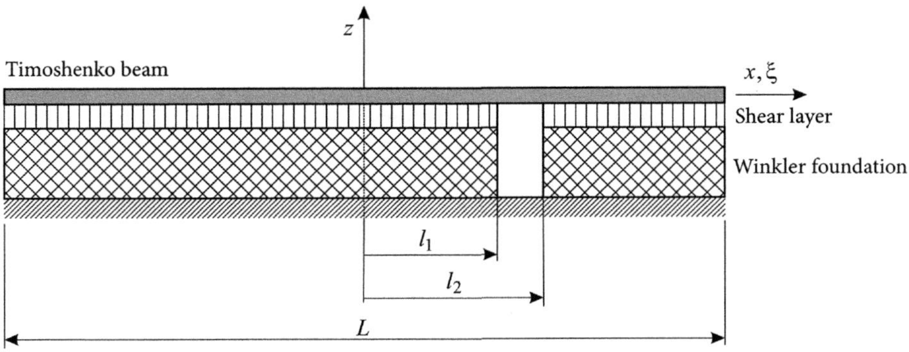

Figure 10.1 Beam supported by elastic foundation. The foundation has a variable discontinuity.

in a plane portal frame comprising three straight beams, three beam elements are used [14]. However, steep localized changes in the domain (discontinuities) require more elements. The displacement components along x and z directions are designated as u and w, respectively. Using the Timoshenko beam assumption, the model's displacement field can be written as Eq. (10.1)

$$u(x, z, t) = u^0(x, t) + z\theta^0(x, t), \quad w(x, z, t) = w^0(x, t). \tag{10.1}$$

The independent rotations of the cross-sections about the axis parallel to y are designated by $\theta^0(x, t)$. The Timoshenko beam assumption, which is the first-order shear deformation theory, is adopted here for the deformation model, for it is more accurate than the Bernoulli–Euler beam assumption, and is still not too computationally demanding. It applies to beams spanning from very thin to moderately thick since the p-FEM does not incur shear locking. A von Kármán type of approach is chosen, which means the longitudinal and shear strains have the following form:

$$\varepsilon_x(x, t) = u^0_{,x}(x, t) + \frac{1}{2}\left(w^0_{,x}(x, t)\right)^2 + z\theta^0_{,x}(x, t), \tag{10.2}$$

$$\gamma_{xz}(x, t) = w^0_{,x}(x, t) + \theta^0(x, t). \tag{10.3}$$

The comma in the subscript followed by a variable means that the function is the partial derivative of the variable. In each element, the vector $\mathbf{d}_0(\xi, t)$, which is formed from the displacement components, may be formulated as a combination of the hierarchical shape functions

$$\mathbf{d}_0(\xi, t) = \mathbf{N}(\xi)\mathbf{q}(t) \Leftrightarrow$$

$$\begin{Bmatrix} u^0(\xi, t) \\ w^0(\xi, t) \\ \theta^0(\xi, t) \end{Bmatrix} = \begin{bmatrix} N^u(\xi)^{\mathrm{T}} & \mathbf{0} & \mathbf{0} \\ \mathbf{0} & N^w(\xi)^{\mathrm{T}} & \mathbf{0} \\ \mathbf{0} & \mathbf{0} & N^\theta(\xi)^{\mathrm{T}} \end{bmatrix} \begin{Bmatrix} \mathbf{q}_u(t) \\ \mathbf{q}_w(t) \\ \mathbf{q}_\theta(t) \end{Bmatrix}, \tag{10.4}$$

where $\mathbf{q_u}(t)$ is the vector of generalized longitudinal displacements, $\mathbf{q_w}(t)$ is the vector of generalized transverse displacements, and $\mathbf{q_\theta}(t)$ is the vector of generalized rotations about the y axis. $\mathbf{N}(\xi)$ is the matrix of shape functions, comprised of row vectors of longitudinal, transverse, and rotation shape functions. The shape functions that represent the discontinuity can be identified by the superscript "d". The vectors of the longitudinal, transverse, and rotation shape functions have the following form, respectively:

$$N^{w}(\xi)^{\mathrm{T}} = \left| f_1^{\mathrm{d}}(L, l_c, \xi) f_2^{\mathrm{d}}(L, l_c, \xi) \,\middle|\, f_1(\xi) f_2(\xi) f_3(\xi) \ldots f_{\mathrm{pw}}(\xi) \right|,$$

$$N^{w}(\xi)^{\mathrm{T}} = \left| f_1^{\mathrm{d}}(L, l_c, \xi) f_2^{\mathrm{d}}(L, l_c, \xi) \,\middle|\, f_1(\xi) f_2(\xi) f_3(\xi) \ldots f_{\mathrm{pw}}(\xi) \right|,$$

$$N^{\theta}(\xi)^{\mathrm{T}} = \left| f_1^{\mathrm{d}}(L, l_c, \xi) f_2^{\mathrm{d}}(L, l_c, \xi) \,\middle|\, f_1(\xi) f_2(\xi) f_3(\xi) \ldots f_{\mathrm{p}\theta}(\xi) \right|, \tag{10.5}$$

in which the counts of longitudinal, transverse, and rotation-shape functions are; $pu + 2$, $pw + 2$, and $p\theta + 2$. The two additional ones (+2) are added due to the steep variations in the structure, that is, the discontinuity. The $f_{\mathrm{r}}(\xi)$ shape functions (where $r > 4$) have the value of zero, also the derivative at the element limits has the value of zero, hence they are internal to each element. These shape functions have been substantially used in p-FEM models of beams, plates, and shells [10–18]. The third-order polynomials $f_1(\xi) - f_4(\xi)$, commonly known by the term Hermite cubics [21], are incorporated when the boundary conditions let the motion occur or connect the elements. It should be verified that the maximum value of the shape function $f_1^{\mathrm{d}}(L, l_{\mathrm{c}}, \xi)$ is in the middle of the discontinuity. At the same time, the shape function $f_2^{\mathrm{d}}(L, l_{\mathrm{c}}, \xi)$ has the maximum derivative at the same point. The middle of the discontinuity is; $l_c = (l_1 + l_2)/2$. The two newly introduced shape functions represent a modification of the basic shape functions to allow the analysis of a geometric element with discontinuity to be considered in four parts, thereby increasing the accuracy and sensitivity of the p-FEM method. The first geometric segment is analyzed by both newly introduced functions from the beginning of the discontinuity. The second geometric segment follows from the start of the discontinuity to the midpoint, where one function reaches its maximum value and the other is zero. The third geometric segment extends from the midpoint to the end of the discontinuity. The fourth segment stretches from the end of the discontinuity to the end of the beam or structure. These four geometric segments are essential as the newly introduced functions consist of two Hermite cubics joined at the midpoint of the discontinuity.

As the material is linearly elastic and isotropic, the following relation holds

$$\boldsymbol{\sigma} = \mathbf{D}\boldsymbol{\varepsilon} \qquad \begin{Bmatrix} \sigma_x \\ \tau_{xz} \end{Bmatrix} = \begin{bmatrix} E & 0 \\ 0 & \lambda G \end{bmatrix} \begin{Bmatrix} \varepsilon_x \\ \gamma_{xz} \end{Bmatrix}, \tag{10.6}$$

where $\boldsymbol{D}$ is the matrix of moduli of elasticity; $\boldsymbol{\sigma}$ is the vector of the nonzero stresses; $\boldsymbol{\varepsilon}$ is the vector of the related strains; E is the Young modulus and G is the shear modulus. The Timoshenko beam assumption comes with the shear correction factor, which is included in the shear stiffness. Different values have been suggested for this factor. Since the beams in this study have a rectangular cross-section, the adopted value for the shear correction factor will be; $\lambda = (5 + 5\nu)/(6 + 5\nu)$, where ν represents the Poisson ratio. That expression has been reported to give suitable values for the shear correction factor in Reference [26]. Also, it gives results that are at least as good as other options when theoretical and experimental results are compared [27]. The longitudinal strain is:

$$\varepsilon_x = \lfloor 1 \quad z \rfloor \left(\begin{Bmatrix} \varepsilon_0^{\mathrm{p}} \\ \varepsilon_0^{\mathrm{b}} \end{Bmatrix} + \begin{Bmatrix} \varepsilon_{\mathrm{L}}^{\mathrm{p}} \\ 0 \end{Bmatrix} \right). \tag{10.7}$$

In Eq. (10.7), the strain was fragmented into two linear strains; longitudinal $\varepsilon_0^{\mathrm{p}}$ and bending $z\varepsilon_0^{\mathrm{b}}$; and one geometric nonlinear longitudinal strain $\varepsilon_{\mathrm{L}}^{\mathrm{p}}$. In the present model, they can be written as:

$$\varepsilon_0^{\mathrm{p}} = \mathbf{N}_{,\xi}^{\mathrm{u}\,\mathrm{T}} \mathbf{q}_{\mathbf{u}} \frac{2}{L}, \quad \varepsilon_0^{b} = \mathbf{N}_{,\xi}^{\theta\,\mathrm{T}} \mathbf{q}_{\theta} \frac{2}{L}, \quad \varepsilon_L^{p} = \frac{2}{L^2} \mathbf{q}_{\mathbf{w}}^{\mathrm{T}} \mathbf{N}_{,\xi}^{\mathrm{w}} \mathbf{N}_{,\xi}^{\mathrm{w}\,\mathrm{T}} \mathbf{q}_{\mathbf{w}},$$

$$\gamma_{xz} = \left\lfloor \frac{2}{L} \mathbf{N}_{,\xi}^{\mathrm{w}\,\mathrm{T}} \quad \mathbf{N}^{\theta\mathrm{T}} \right\rfloor \begin{Bmatrix} \mathbf{q}_{\mathbf{w}} \\ \mathbf{q}_{\theta} \end{Bmatrix}, \tag{10.8}$$

where the prime (') symbol denotes the first derivative of the function with respect to the coordinate ξ. Integration of the normal stress, the shear stress, and the moment generated by the normal stress about the axis parallel to y, will produce two forces and a moment per unit width. Substituting Eq. (10.8) into equations that define forces and moment results in:

$$\begin{Bmatrix} T_x \\ M_x \end{Bmatrix} = \begin{bmatrix} A & B \\ B & D \end{bmatrix} \left(\begin{Bmatrix} \varepsilon_0^{\mathrm{p}} \\ \varepsilon_0^{\mathrm{b}} \end{Bmatrix} + \begin{Bmatrix} \varepsilon_{\mathrm{L}}^{\mathrm{p}} \\ 0 \end{Bmatrix} \right), \quad Q_x = \frac{Eh\lambda}{2(1+\nu)} \gamma_{xz},$$

$$T_x = \int_{-h/2}^{h/2} \sigma_x \mathrm{d}z, \quad Q_x = \int_{-h/2}^{h/2} \tau_{xz} \mathrm{d}z, \quad M_x = \int_{-h/2}^{h/2} \sigma_x z \mathrm{d}z, \tag{10.9}$$

where A, B, and D are the extension, coupling, and bending coefficients, respectively:

$$A = \int_z E\mathrm{d}z = Eh, \quad B = \int_z zE\mathrm{d}z = 0, \quad D = \int_z z^2 E\mathrm{d}z = \frac{Eh^3}{12}. \tag{10.10}$$

The model that represents the foundation [3] is mathematically defined by the linear Winkler modulus k_w, the nonlinear Winkler modulus k_{wNL}, the Pasternak (shear) modulus k_p, and the dumping coefficient c in the following form:

$$p_x(x,t) = k_w w(x,t) + k_{wNL} w^3(x,t) - k_p \frac{\partial^2 w(x,t)}{\partial x^2} + c\frac{\partial w(x,t)}{\partial t}, \quad \text{for } p_x(x,t) > 0. \tag{10.11}$$

In this case; $p_x(x,t) \le 0 \Longrightarrow p_x(x,t) = 0$.

The equations of motion are derived from the principle of virtual work, according to which

$$\delta W_{\mathrm{in}} + \delta W_{\mathrm{V}} + \delta W_{\mathrm{ex}} = 0, \tag{10.12}$$

where items δW_{in}, δW_{V}, and δW_{ex} are virtual works done by inertia, internal, and external forces, respectively, as a result of a virtual displacement with components $\delta\mathbf{d}$ in the form:

$$\delta\mathbf{d} = \begin{Bmatrix} \delta\text{u} \\ \delta\text{w} \\ \delta\theta \end{Bmatrix} = \mathbf{N}\,\delta\mathbf{q}. \tag{10.13}$$

The virtual work of internal forces is:

$$\delta W_{\text{V}} = -\int_V \delta\varepsilon^{\text{T}}\sigma \text{d}V =$$

$$= -b\int_L \left(\begin{Bmatrix} \delta\varepsilon_0^p \\ \delta\varepsilon_0^b \end{Bmatrix}^{\text{T}} + \begin{Bmatrix} \delta\varepsilon_L^p \\ 0 \end{Bmatrix}^{\text{T}} \right) \begin{bmatrix} A & B \\ B & D \end{bmatrix} \left(\begin{Bmatrix} \varepsilon_0^p \\ \varepsilon_0^b \end{Bmatrix} + \begin{Bmatrix} \varepsilon_L^p \\ 0 \end{Bmatrix} \right) dx = \delta\mathbf{q}^{\text{T}}\mathbf{K}(\mathbf{q})\mathbf{q}, \tag{10.14}$$

where $\delta\varepsilon$ is the virtual strains vector. The virtual work of inertia forces is:

$$\delta W_{\text{in}} = -\int_V \rho\delta\mathbf{d}^{\text{T}}\ddot{\mathbf{d}}\text{d}V = -b\int_{-\frac{h}{2}}^{\frac{h}{2}}\int_{-L/2}^{L/2} \rho(\delta u\ddot{u} + \delta w\ddot{w})\text{d}x\text{d}z = \delta\mathbf{q}^{\text{T}}\mathbf{M}\ddot{\mathbf{q}}. \tag{10.15}$$

The virtual work of external forces is:

$$\delta W_{\text{ex}} = \int_L [P_w^j(t)\delta(x - x_j) + P_w^d(x,t) - k_w w(x,t)]\delta w(x,t)$$

$$+ \left[k_p \frac{\partial^2 w(x,t)}{\partial x^2}\right]\delta w(x,t) + [P_u^j(t)\delta(x - x_j) + P_u^d(x,t)]\delta u(x,t)$$

$$+ [M^j(t)\delta(x - x_j) + M^d(x,t)]\delta\theta(x,t) = \left\{\delta q_u \;\; \delta q_w \;\; \delta q_\theta\right\}^{\text{T}} \begin{Bmatrix} \mathbf{F}_{\mathbf{u}}^{\text{E}} \\ \mathbf{F}_{\mathbf{u}}^{\text{E}} \\ \mathbf{M}^{\text{E}} \end{Bmatrix}, \tag{10.16}$$

where $\ddot{\mathbf{d}}$ are the components of acceleration of the beam's point. Implementing the virtual work principle, and including the nonlinear effect of the foundation, the following equations of motion are obtained:

$$\begin{bmatrix} \mathbf{M}^{\text{l}} & 0 & 0 \\ 0 & \mathbf{M}^{\text{b}} & 0 \\ 0 & 0 & \mathbf{M}^{\text{r}} \end{bmatrix} \begin{Bmatrix} \ddot{\mathbf{q}}_{\mathbf{u}} \\ \ddot{\mathbf{q}}_{\mathbf{w}} \\ \ddot{\mathbf{q}}_{\theta} \end{Bmatrix} + \alpha \begin{bmatrix} \mathbf{K}_\ell^{\text{l}} & 0 & 0 \\ 0 & \mathbf{K}_\ell^{\gamma_{11}} + \mathbf{K}_\ell^{\text{W}} - \mathbf{K}_\ell^{\text{P}} & \mathbf{K}_\ell^{\gamma_{12}} \\ 0 & \mathbf{K}_\ell^{\gamma_{21}} & \mathbf{K}_\ell^{\text{b}} + \mathbf{K}_\ell^{\gamma_{22}} \end{bmatrix} \begin{Bmatrix} \dot{\mathbf{q}}_{\mathbf{u}} \\ \dot{\mathbf{q}}_{\mathbf{w}} \\ \dot{\mathbf{q}}_{\theta} \end{Bmatrix}$$

$$+ \beta \begin{bmatrix} \mathbf{M}^{\text{l}} & 0 & 0 \\ 0 & \mathbf{M}^{\text{b}} & 0 \\ 0 & 0 & \mathbf{M}^{\text{r}} \end{bmatrix} \begin{Bmatrix} \dot{\mathbf{q}}_{\mathbf{u}} \\ \dot{\mathbf{q}}_{\mathbf{w}} \\ \dot{\mathbf{q}}_{\theta} \end{Bmatrix} + \begin{bmatrix} \mathbf{K}_\ell^{\text{l}} & 0 & 0 \\ 0 & \mathbf{K}_\ell^{\gamma_{11}} + \mathbf{K}_\ell^{\text{W}} - \mathbf{K}_\ell^{\text{P}} & \mathbf{K}_\ell^{\gamma_{12}} \\ 0 & \mathbf{K}_\ell^{\gamma_{21}} & \mathbf{K}_\ell^{\text{b}} + \mathbf{K}_\ell^{\gamma_{22}} \end{bmatrix} \begin{Bmatrix} \mathbf{q}_{\mathbf{u}} \\ \mathbf{q}_{\mathbf{w}} \\ \mathbf{q}_{\theta} \end{Bmatrix}$$

$$+ \begin{bmatrix} 0 & \mathbf{K}_{\text{n}\ell}^{2}(\mathbf{q}_{\mathbf{w}}) & 0 \\ \mathbf{K}_{\text{n}\ell}^{3}(\mathbf{q}_{\mathbf{w}}) & \mathbf{K}_{\text{n}\ell}^{4}(\mathbf{q}_{\mathbf{w}}) + \mathbf{K}_{\text{n}\ell}^{\text{W}}(\mathbf{q}_{\mathbf{w}}) & 0 \\ 0 & 0 & 0 \end{bmatrix} \begin{Bmatrix} \mathbf{q}_{\mathbf{u}} \\ \mathbf{q}_{\mathbf{w}} \\ \mathbf{q}_{\theta} \end{Bmatrix} = \begin{Bmatrix} \mathbf{F}_{\mathbf{u}}^{\text{E}} \\ \mathbf{F}_{\mathbf{w}}^{\text{E}} \\ \mathbf{M}^{\text{E}} \end{Bmatrix}. \tag{10.17}$$

$\left\{\mathbf{F}_{\mathbf{u}}^{\mathbf{E}}(t)\ \ \mathbf{F}_{\mathbf{w}}^{\mathbf{E}}(t)\ \ \mathbf{M}^{\mathrm{E}}(t)\right\}^{\mathrm{T}}$ represents the vector of generalized external forces. Matrices of type $\mathbf{M}$ and $\mathbf{K}_{\ell}$ (mass and stiffness) are constant matrices, which make the linear terms in the equations of motion. Matrices $\mathbf{K}_{\ell}^{\mathrm{W}}$ and $\mathbf{K}_{\ell}^{\mathrm{P}}$ describe the linear effects of the foundation with discontinuity. Matrices $\mathbf{K}_{\mathrm{n}\ell}^{\mathbf{2}}$ and $\mathbf{K}_{\mathrm{n}\ell}^{\mathbf{3}}$ have a linear dependency on the solution, which makes quadric nonlinear terms. Matrix $\mathbf{K}_{\mathrm{n}\ell}^{\mathbf{4}}$ has a quadric dependency with the solution, which makes cubic nonlinearity. Finally, matrix $\mathbf{K}_{\mathrm{n}\ell}^{\mathrm{W}}$ has a cubic dependency with the solution, which defines the nonlinearity of the fourth-order. The superscripts; l, b, r, and γ represent—respectively—longitudinal, bending, cross-section rotation, and shear effects. Utilizing the Rayleigh proportional type of damping, where α and β are damping coefficients, the equations of motion can be written in a more condensed form:

$$\mathbf{M}\ddot{\mathbf{q}}(t) + \alpha(\mathbf{K}_{\ell})\dot{\mathbf{q}}(t) + \beta(\mathbf{M})\dot{\mathbf{q}}(t) + (\mathbf{K}_{\ell} + \mathbf{K}_{\mathrm{n}\ell}(\mathbf{q}(t)))\mathbf{q}(t) = \mathbf{F}(t). \quad (10.18)$$

The mass, the stiffness, and the matrices that represent the foundation with the discontinuity are given in the following expressions:

$$\mathbf{M}^{\mathrm{l}} = \frac{\rho hbl}{2}\int_{-1}^{\frac{2l_c}{L}} \mathbf{N}^{\mathrm{u1}}\mathbf{N}^{\mathrm{u1T}}\mathrm{d}\xi + \frac{\rho hbl}{2}\int_{\frac{2l_c}{L}}^{1} \mathbf{N}^{\mathrm{u2}}\mathbf{N}^{\mathrm{u2T}}\mathrm{d}\xi, \quad (10.19)$$

$$\mathbf{M}^{\mathrm{b}} = \frac{\rho hbl}{2}\int_{-1}^{\frac{2l_c}{L}} \mathbf{N}^{\mathrm{w1}}\mathbf{N}^{\mathrm{w1T}}\mathrm{d}\xi + \frac{\rho hbl}{2}\int_{\frac{2l_c}{L}}^{1} \mathbf{N}^{\mathrm{w2}}\mathbf{N}^{\mathrm{w2T}}\mathrm{d}\xi, \quad (10.20)$$

$$\mathbf{M}^{\mathrm{r}} = \frac{\rho hbl}{2}\int_{-1}^{\frac{2l_c}{L}} \mathbf{N}^{\theta 1}\mathbf{N}^{\theta 1\mathrm{T}}\mathrm{d}\xi + \frac{\rho hbl}{2}\int_{\frac{2l_c}{L}}^{1} \mathbf{N}^{\theta 2}\mathbf{N}^{\theta 2\mathrm{T}}\mathrm{d}\xi, \quad (10.21)$$

$$\mathbf{K}^{\mathrm{l}} = \frac{2Ehb}{L}\int_{-1}^{\frac{2l_c}{L}} \frac{\mathrm{d}\mathbf{N}^{\mathrm{u1}}}{\mathrm{d}\xi}\frac{\mathrm{d}\mathbf{N}^{\mathrm{u1T}}}{\mathrm{d}\xi}\mathrm{d}\xi + \frac{2Ehb}{L}\int_{\frac{2l_c}{L}}^{1} \frac{\mathrm{d}\mathbf{N}^{\mathrm{u2}}}{\mathrm{d}\xi}\frac{\mathrm{d}\mathbf{N}^{\mathrm{u2T}}}{\mathrm{d}\xi}\mathrm{d}\xi, \quad (10.22)$$

$$\mathbf{K}^{\gamma_{11}} = \frac{2\lambda Ghb}{L}\int_{-1}^{\frac{2l_{1c}}{L}} \frac{\mathrm{d}\mathbf{N}^{\mathrm{w1}}}{\mathrm{d}\xi}\frac{\mathrm{d}\mathbf{N}^{\mathrm{w1T}}}{\mathrm{d}\xi}\mathrm{d}\xi + \frac{2\lambda Ghb}{L}\int_{\frac{2l_c}{L}}^{1} \frac{\mathrm{d}\mathbf{N}^{\mathrm{w2}}}{\mathrm{d}\xi}\frac{\mathrm{d}\mathbf{N}^{\mathrm{w2T}}}{\mathrm{d}\xi}\mathrm{d}\xi, \quad (10.23)$$

$$\mathbf{K}^{\gamma_{12}} = \lambda Ghb\int_{-1}^{\frac{2l_c}{L}} \frac{\mathrm{d}\mathbf{N}^{\mathrm{w1}}}{\mathrm{d}\xi}\mathbf{N}^{\theta 1\mathrm{T}}\mathrm{d}\xi + \lambda Ghb\int_{\frac{2l_c}{L}}^{1} \frac{\mathrm{d}\mathbf{N}^{\mathrm{w2}}}{\mathrm{d}\xi}\mathbf{N}^{\theta 2\mathrm{T}}\mathrm{d}\xi, \quad (10.24)$$

$$\mathbf{K}^{\gamma_{21}} = \lambda Ghb\int_{-1}^{\frac{2l_c}{L}} \mathbf{N}^{\theta 1}\frac{\mathrm{d}\mathbf{N}^{\mathrm{w1T}}}{\mathrm{d}\xi}\mathrm{d}\xi + \lambda Ghb\int_{\frac{2l_c}{L}}^{1} \mathbf{N}^{\theta 2}\frac{\mathrm{d}\mathbf{N}^{\mathrm{w2T}}}{\mathrm{d}\xi}\mathrm{d}\xi, \quad (10.25)$$

$$\mathbf{K}^{\gamma_{22}} = \frac{\lambda GhbL}{2}\int_{-1}^{\frac{2l_c}{L}} \mathbf{N}^{\theta 1\mathrm{T}}\mathbf{N}^{\theta 1}\mathrm{d}\xi + \frac{\lambda GhbL}{2}\int_{\frac{2l_c}{L}}^{1} \mathbf{N}^{\theta 2\mathrm{T}}\mathbf{N}^{\theta 2}\mathrm{d}\xi, \quad (10.26)$$

$$\mathbf{K}^{\mathrm{W}} = \frac{L}{2}k_w\int_{-1}^{\frac{2l_1}{L}} \mathbf{N}^{\mathrm{w1}}\mathbf{N}^{\mathrm{w1T}}\mathrm{d}\xi + \frac{L}{2}k_{\mathrm{w}}\int_{\frac{2l_2}{L}}^{1} \mathbf{N}^{\mathrm{w2}}\mathbf{N}^{\mathrm{w2T}}\mathrm{d}\xi, \quad (10.27)$$

$$\mathbf{K}^{\mathrm{P}} = \frac{2}{L}k_{\mathrm{p}}\int_{-1}^{\frac{2l_1}{L}} \frac{\mathrm{d}\mathbf{N}^{\mathrm{w1}}}{\mathrm{d}\xi}\frac{\mathrm{d}\mathbf{N}^{\mathrm{w1T}}}{\mathrm{d}\xi}\mathrm{d}\xi + \frac{2}{L}k_{\mathrm{p}}\int_{\frac{2l_2}{L}}^{1} \frac{\mathrm{d}\mathbf{N}^{\mathrm{w2}}}{\mathrm{d}\xi}\frac{\mathrm{d}\mathbf{N}^{\mathrm{w2T}}}{\mathrm{d}\xi}\mathrm{d}\xi, \tag{10.28}$$

$$\mathbf{K}_{\mathrm{n}}^{2}(\mathbf{q}_{\mathbf{w}}) = \frac{2Ebh}{L^2}\int_{-1}^{\frac{2l_c}{L}} \frac{\mathrm{d}w_0}{\mathrm{d}\xi}\frac{\mathrm{d}\mathbf{N}^{\mathrm{u1}}}{\mathrm{d}\xi}\frac{\mathrm{d}\mathbf{N}^{\mathrm{w1T}}}{\mathrm{d}\xi}\mathrm{d}\xi + \frac{2Ebh}{L^2}\int_{\frac{2l_c}{L}}^{1} \frac{\mathrm{d}w_0}{\mathrm{d}\xi}\frac{\mathrm{d}\mathbf{N}^{\mathrm{u2}}}{\mathrm{d}\xi}\frac{\mathrm{d}\mathbf{N}^{\mathrm{w2T}}}{\mathrm{d}\xi}\mathrm{d}\xi, \tag{10.29}$$

$$\mathbf{K}_{\mathrm{n}}^{3}(\mathbf{q}_{\mathbf{w}}) = 2\mathbf{K}_{\mathrm{n}}^{2\mathrm{T}}(\mathbf{q}_{\mathbf{w}}), \tag{10.30}$$

$$\mathbf{K}_{\mathrm{n}}^{4}(\mathbf{q}_{\mathbf{w}}) = \frac{4Ebh}{L^3}\int_{-1}^{\frac{2l_c}{L}} \left(\frac{\mathrm{d}w_0}{\mathrm{d}\xi}\right)^2\frac{\mathrm{d}\mathbf{N}^{\mathrm{w1}}}{\mathrm{d}\xi}\frac{\mathrm{d}\mathbf{N}^{\mathrm{w1T}}}{\mathrm{d}\xi}\mathrm{d}\xi + \frac{4Ebh}{L^3}\int_{\frac{2l_c}{L}}^{1} \left(\frac{\mathrm{d}w_0}{\mathrm{d}\xi}\right)^2\frac{\mathrm{d}\mathbf{N}^{\mathrm{w2}}}{\mathrm{d}\xi}\frac{\mathrm{d}\mathbf{N}^{\mathrm{w2T}}}{\mathrm{d}\xi}\mathrm{d}\xi, \tag{10.31}$$

$$\mathbf{K}_{\mathrm{n}}^{2}(\mathbf{q}_{\mathbf{w}}) = \frac{2Ebh}{L^2}\int_{-1}^{\frac{2l_c}{L}} \frac{\mathrm{d}w_0}{\mathrm{d}\xi}\frac{\mathrm{d}\mathbf{N}^{\mathrm{u1}}}{\mathrm{d}\xi}\frac{\mathrm{d}\mathbf{N}^{\mathrm{w1T}}}{\mathrm{d}\xi}\mathrm{d}\xi + \frac{2Ebh}{L^2}\int_{\frac{2l_c}{L}}^{1} \frac{\mathrm{d}w_0}{\mathrm{d}\xi}\frac{\mathrm{d}\mathbf{N}^{\mathrm{u2}}}{\mathrm{d}\xi}\frac{\mathrm{d}\mathbf{N}^{\mathrm{w2T}}}{\mathrm{d}\xi}\mathrm{d}\xi, \tag{10.32}$$

$$\mathbf{K}_{\mathrm{n}\ell}^{W}(\mathbf{q}_{\mathbf{w}}) = \frac{L}{2}k_{w}\int_{-1}^{\frac{2l_1}{L}} (w_0)^3\mathbf{N}^{\mathrm{w1T}}\mathbf{N}^{\mathrm{w1}}\mathrm{d}\xi + \frac{L}{2}k_{\mathrm{w}}\int_{\frac{2l_2}{L}}^{1} (w_0)^3\mathbf{N}^{\mathrm{w2T}}\mathbf{N}^{\mathrm{w2}}\mathrm{d}\xi, \tag{10.33}$$

where $\mathbf{N}^{\mathrm{u1}}$, $\mathbf{N}^{\mathrm{w1}}$, and $\mathbf{N}^{\theta1}$ belong to the domain $-1 \le \xi \le 2l_c/L$, and $\mathbf{N}^{\mathrm{u2}}$, $\mathbf{N}^{\mathrm{w2}}$, and $\mathbf{N}^{\theta2}$ belong to the domain $2l_c/L \le \xi \le 1$.

10.3 Natural frequencies and general modes of vibration

In the upcoming analysis, the p-FEM will be used on various clamped-clamped beams, whose geometric and material properties are based on the models from Reference [22], and the data used in References [15,28]

$$L = 406\ \mathrm{mm},\quad b = 20\ \mathrm{mm},\quad h = 2\ \mathrm{mm},\quad E = 7.172\ \mathrm{GPa},\quad \rho = 2800\ \mathrm{kgm}^{-3}.$$

The transverse natural frequencies for different parameters of discontinuity are given in Table 10.1. Four cases of a deformable beam on an elastic foundation without and with discontinuity are shown in Figs. 10.2–10.5.

The natural frequencies for the presented cases (Figs. 10.2–10.5) with clamped-clamped beams, and various values of the foundation stiffness moduli are given in Tables 10.1–10.8. For the case of a simply supported beam without a discontinuity in the foundation, natural frequencies can be obtained easily from the frequency Eq. (10.27) for the Timoshenko beam assumption by putting $m = 1$ as given in reference [29].

For verification and comparison of the calculated numerical results, the commercial software package ANSYS® [25] is used. In the ANSYS® FEM analysis, the h-version element BEAM189 is used, and with it, shear effects and rotary inertia are considered.

Table 10.1 Natural frequencies [Hz] for Case 1: k_w=50×10^5 Pa, k_p={$E/[2(1+\nu)]$}×10^{-8} GPa.

Mode	Winkler		Pasternak	
	ANSYS BEAM189 100 elements	***p*-FEM (10sf)**	**ANSYS BEAM189 100 elements**	***p*-FEM (10sf)**
1	342.12	342.143	348.41	348.580
2	378.56	378.581	400.08	400.252
3	478.80	478.813	515.38	515.515
4	658.80	658.825	704.25	705.176

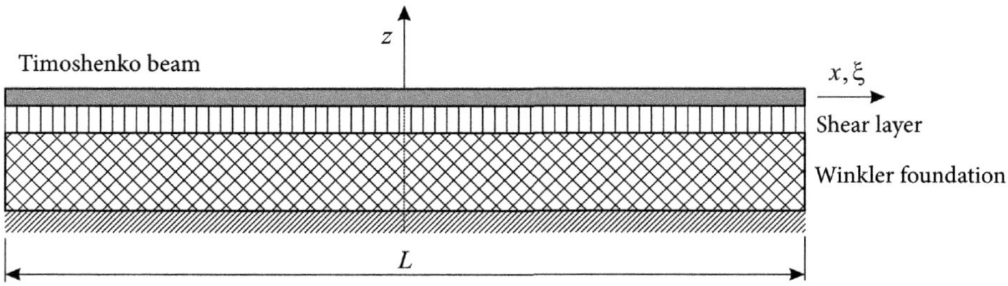

Figure 10.2 Beam supported by elastic foundation. The foundation is without discontinuity; Case 1.

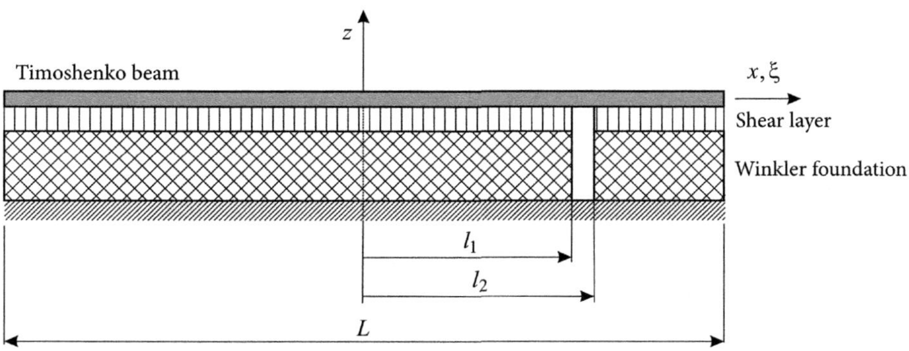

Figure 10.3 Beam supported by elastic foundation. The foundation has a discontinuity; Case 2: l_1=68 mm, l_2=75 mm.

The selected element has three nodes with 6 degrees of freedom (DOF) at each node; although, for a single in-plane vibration analysis only 3 DOF per node are required. BEAM189 is an element established on the Timoshenko beam assumption, and it is appropriate for the analysis of thin to moderately thick beam structures. The presented results obtained from ANSYS® are calculated with 100 elements, which is more than what is needed to obtain the convergence of natural frequencies. The minimum

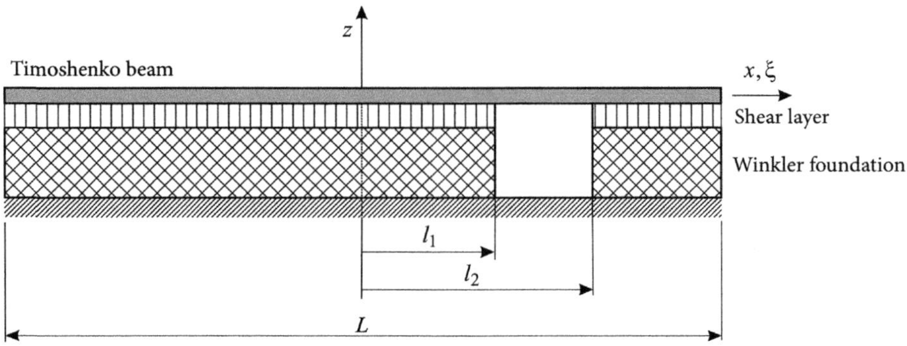

Figure 10.4 Beam supported by elastic foundation. The foundation has a discontinuity; Case 3: l_1=68 mm, l_2=105 mm.

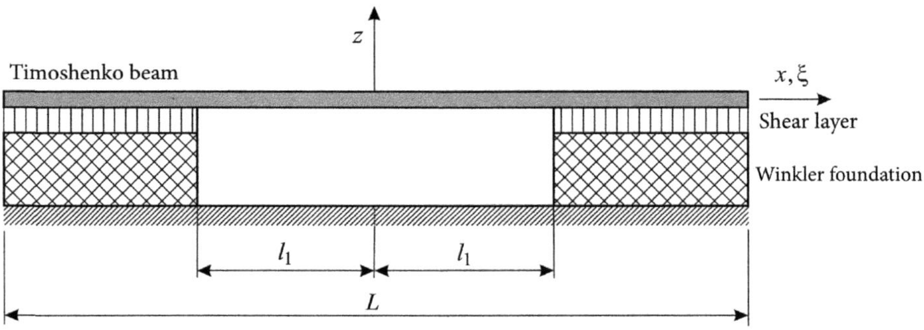

Figure 10.5 Beam supported by elastic foundation. The foundation has a discontinuity; Case 4: l_1=l_2=L/6.

Table 10.2 Natural frequencies [Hz] for Case 2: k_w=50×10^5 Pa, k_p={E/[2(1+v)]}×10^{-8} GPa.

Mode	Winkler		Pasternak	
	ANSYS BEAM189 100 elements	***p*-FEM (10sf)**	**ANSYS BEAM189 100 elements**	***p*-FEM (10sf)**
1	335.25	336.856	342.24	343.172
2	371.12	373.230	392.14	394.122
3	475.32	478.194	511.15	514.841
4	651.41	657.607	697.85	703.212

number of needed elements is about 30. This fine mesh is generated for the purpose of obtaining a reference model, which is used for the validation of the p-FEM model. The comparisons shown in Tables 10.1–10.8 lead to the conclusion that the formulation based on p-FEM, which is explained here, provides accurate results based on good

Table 10.3 Natural frequencies [Hz] for Case 3: k_w=50×10^5 Pa, k_p={E/[2(1+ν)]}×10^{-8} GPa.

Mode	Winkler		Pasternak	
	ANSYS BEAM189 100 elements	***p*-FEM (10sf)**	**ANSYS BEAM189 100 elements**	***p*-FEM (10sf)**
1	302.11	303.557	307.85	308.048
2	358.28	360.483	343.92	376.128
3	466.61	469.768	501.12	504.365
4	650.44	655.575	695.78	700.760

Table 10.4 Natural frequencies [Hz] for Case 4: k_w=50×10^5 Pa, k_p={E/[2(1+ν)]}×10^{-8} GPa.

Mode	Winkler		Pasternak	
	ANSYS BEAM189 100 elements	***p*-FEM (10sf)**	**ANSYS BEAM189 100 elements**	***p*-FEM (10sf)**
1	154.42	156.687	149.74	151.748
2	321.22	324.779	334.36	337.723
3	451.65	455.105	477.21	481.135
4	624.98	629.271	653.89	658.659

Table 10.5 Natural frequencies [Hz] for Case 1: k_w=100×10^5 Pa, k_p={E/[2(1+ν)]}×10^{-8} GPa.

Mode	Winkler		Pasternak	
	ANSYS BEAM189 100 elements	***p*-FEM (10sf)**	**ANSYS BEAM189 100 elements**	***p*-FEM (10sf)**
1	478.65	479.729	484.13	484.342
2	505.02	506.354	521.44	522.755
3	583.36	585.082	613.37	615.480
4	736.87	739.658	779.69	781.227

agreement with the results obtained from the analysis of ANSYS® h-FEM. Changes in the natural frequencies provide a depiction of all models and structures. However, the local effects—the effects produced by the discontinuity with a specific location in the foundation—are neither included nor visible. For this reason, it is prudent to present local changes in natural mode shapes; general natural mode shapes for all cases of discontinuity in the foundation can be seen in Figs. 10.6–10.11.

It is indisputable that discontinuity causes a change in the natural transverse and rotation mode shape. It should be stressed that the asymmetrical mode shapes occur

Table 10.6 Natural frequencies [Hz] for Case 2: k_w=100×10^5 Pa, k_p={E/[2(1+v)]}×10^{-8} GPa.

Mode	Winkler		Pasternak	
	ANSYS BEAM189 100 elements	***p*-FEM (10sf)**	**ANSYS BEAM189 100 elements**	***p*-FEM (10sf)**
1	470.17	470.830	475.86	475.723
2	498.23	499.339	513.21	514.772
3	582.77	584.124	612.27	614.543
4	734.84	737.585	775.74	778.432

Table 10.7 Natural frequencies [Hz] for Case 3: k_w=100×10^5 Pa, k_p={E/[2(1+v)]}×10^{-8} GPa.

Mode	Winkler		Pasternak	
	ANSYS BEAM189 100 elements	***p*-FEM (10sf)**	**ANSYS BEAM189 100 elements**	***p*-FEM (10sf)**
1	404.12	404.331	407.28	407.470
2	489.38	490.169	499.71	500.495
3	571.46	573.282	600.69	602.120
4	731.15	734.282	772.47	775.305

Table 10.8 Natural frequencies [Hz] for Case 4: k_w=100×10^5 Pa, k_p={E/[2(1+v)]}×10^{-8} GPa.

Mode	Winkler		Pasternak	
	ANSYS BEAM189 100 elements	***p*-FEM (10sf)**	**ANSYS BEAM189 100 elements**	***p*-FEM (10sf)**
1	187.09	187.662	182.87	183.094
2	412.77	413.341	421.05	422.009
3	551.98	553.035	572.58	574.135
4	689.18	692.563	716.88	719.877

when the discontinuity is present only on one side of the beam. For Cases 2 and 3 (nonsymmetrical discontinuity), it was troublesome to envision the mode shapes in higher modes based on the type of discontinuity. For that reason, the obtained results are important in understanding the vibrations of the adopted models.

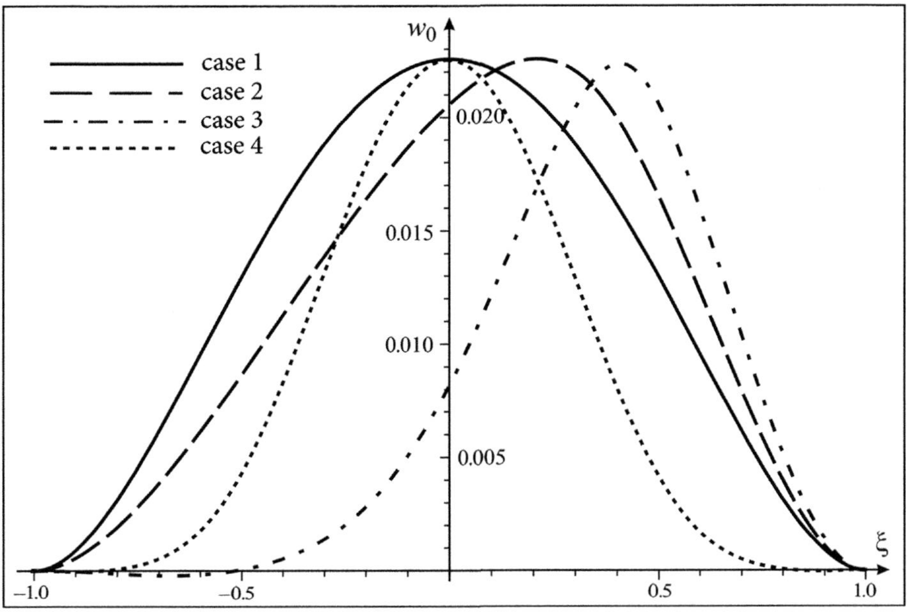

Figure 10.6 Components of mode shapes. Transverse components for mode 1.

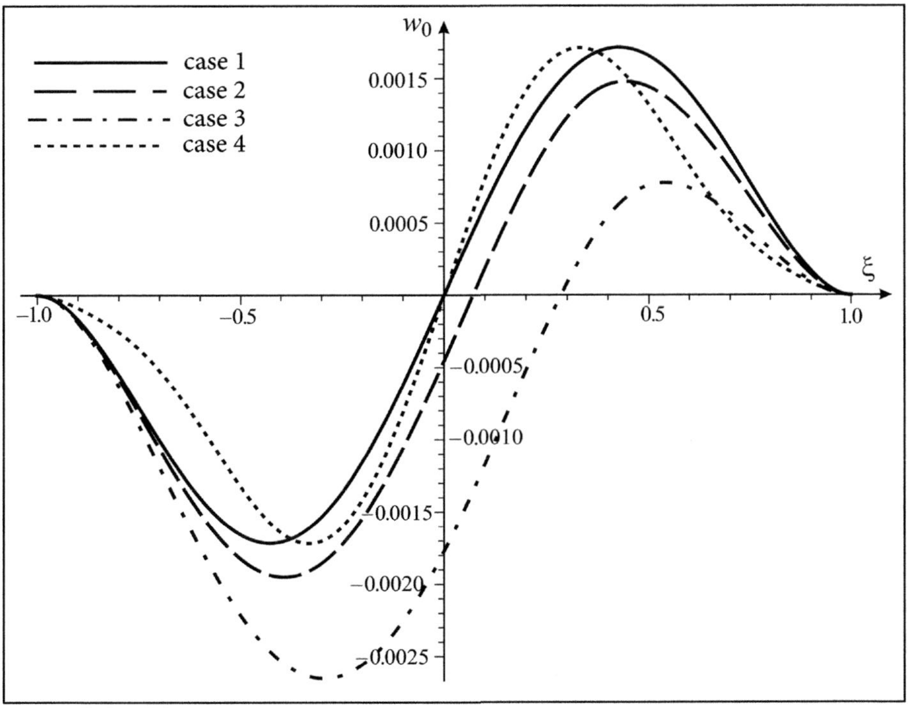

Figure 10.7 Components of mode shapes. Transverse components for mode 2.

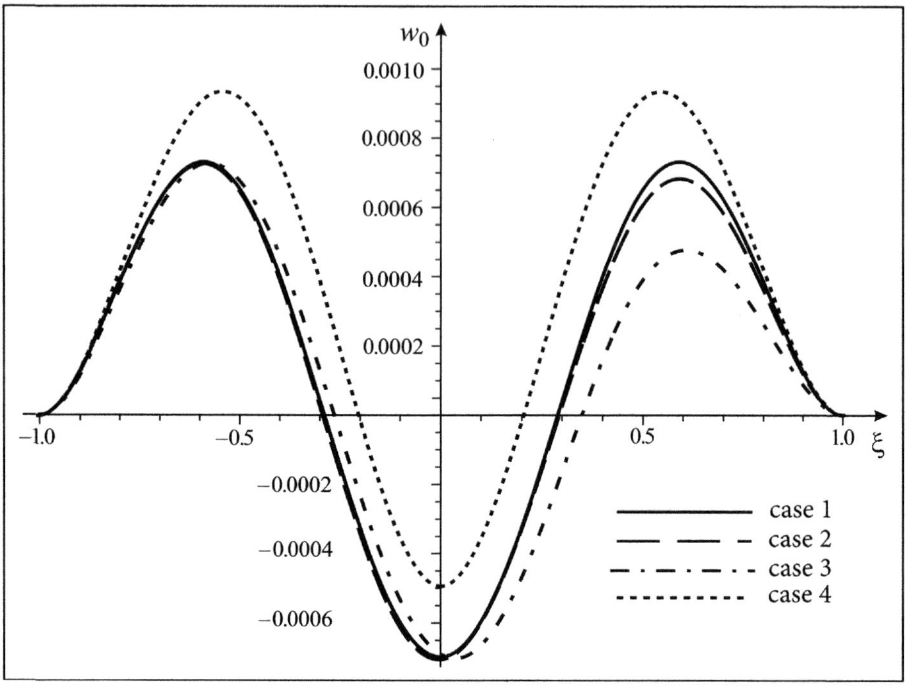

Figure 10.8 Components of mode shapes. Transverse components for mode 3.

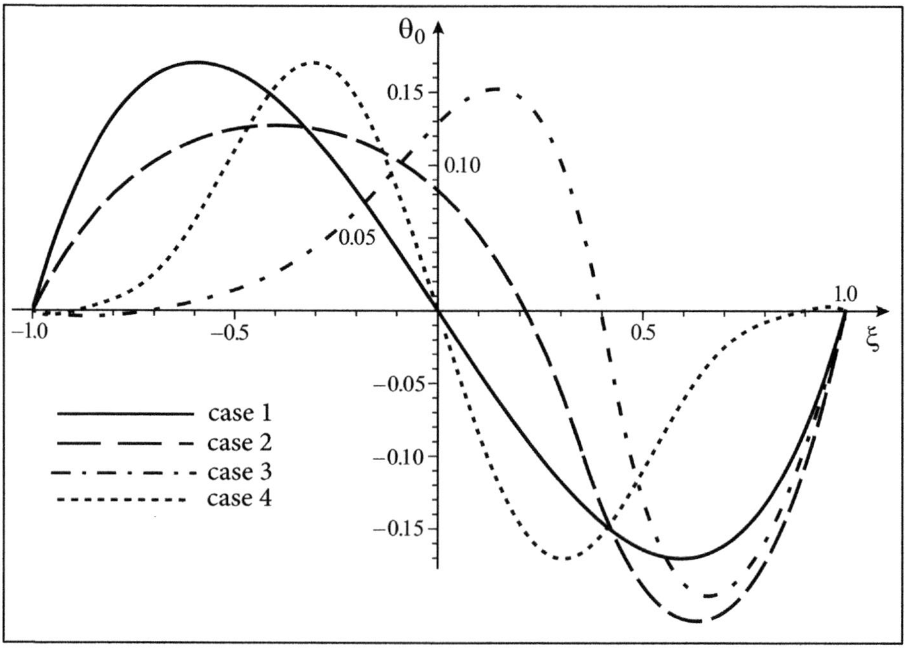

Figure 10.9 Components of mode shapes. Rotation components for mode 1.

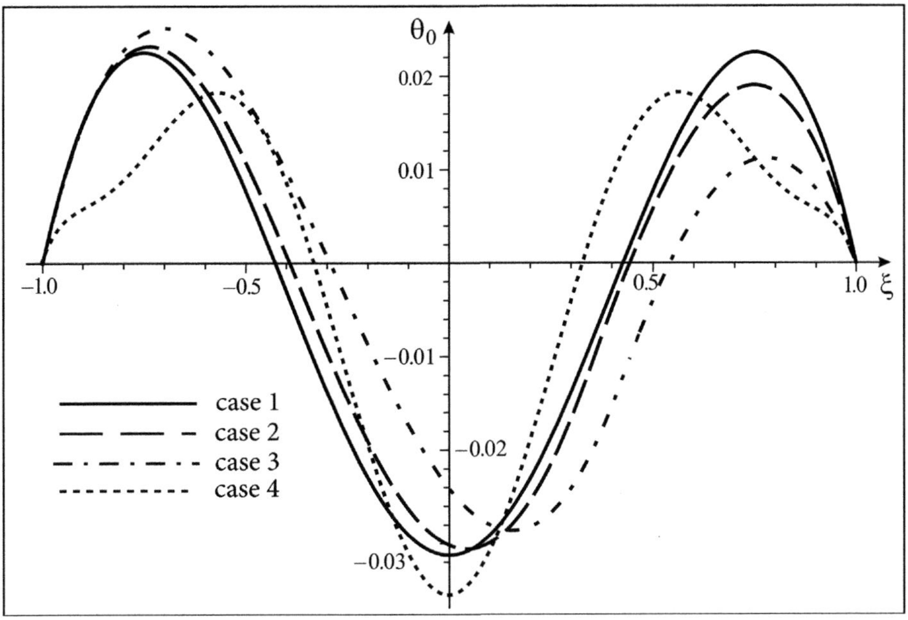

Figure 10.10 Components of mode shapes. Rotation components for mode 2.

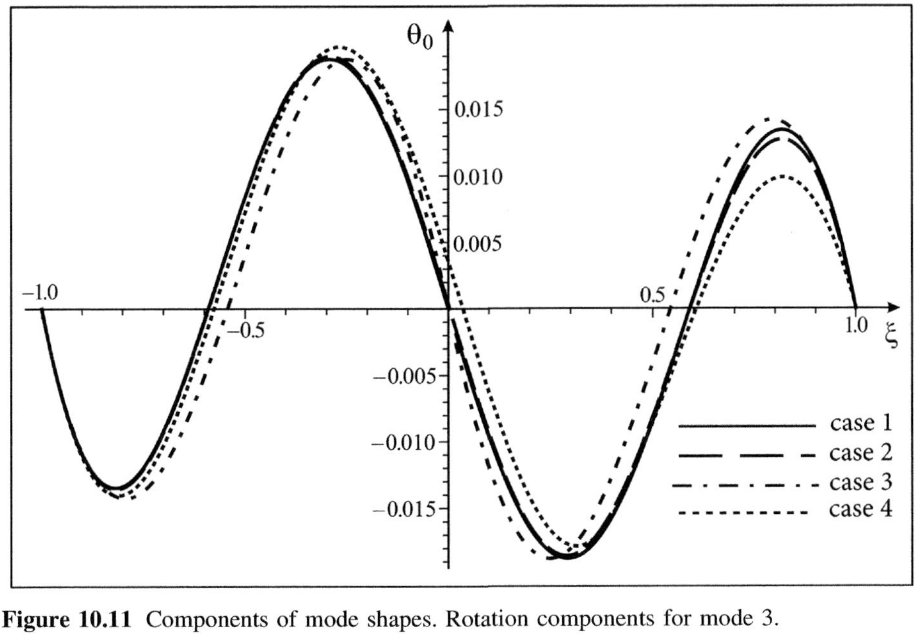

Figure 10.11 Components of mode shapes. Rotation components for mode 3.

10.4 Geometric nonlinear vibrations

This section addresses the analysis of forced nonlinear vibrations, which are analyzed via numerical simulations in the time domain. Solving the equations of motion is done by the Newmark-beta method. The results are presented for various stiffness moduli and different sizes and positions of discontinuity. The applied force varies harmonically in time, according to the function; $f(t) = F\cos(\omega_e t)$, where F is the excitation amplitude and ω_e the excitation frequency. Amplitudes of the beam's response to harmonic excitations were measured in laboratory experiments in Reference [28] and numerically analyzed in Reference [15], with the properties listed in the previous section and without an elastic foundation. For a beam without a foundation, the transverse nondimensional displacement of the beam's middle point, calculated with the present model just without a foundation, is presented in Fig. 10.12. A harmonic force with an amplitude of 0.134 N was applied at the middle point of the beam. Ten transverse shape functions were used in the p-FEM. In Fig. 10.12, the ordinate represents the values of the beam's middle point transverse displacement (divided by the thickness h), during a period of vibration $w(0, t)$. The abscissa represents the values of a nondimensional frequency. The response of the beam that lacks the foundation (Fig. 10.12) is in agreement with the one computed by means of the shooting method and p-FEM [15].

Beam deformations, that is, transverse deflections, with marked maximal transverse deflection points, for the four cases and varied stiffness moduli of the foundation are presented in Figs. 10.13 and 10.14. For Cases 2 and 3, the deflections are asymmetrical, which results from the nonsymmetrical geometrical position of the discontinuity in the foundation, which can be seen in Figs. 10.13B and C and 10.14B and C. Responses for lower stiffness modulus of the foundation are presented in Fig. 10.13, while responses in Fig. 10.14 are for higher stiffness. The lower stiffness modulus of the foundation will cause lower deflections of the beam, compared to a situation when the stiffness modulus is higher. For the chosen shear modulus, the deflections are larger when the elastic foundation is of Pasternak type, compared to a

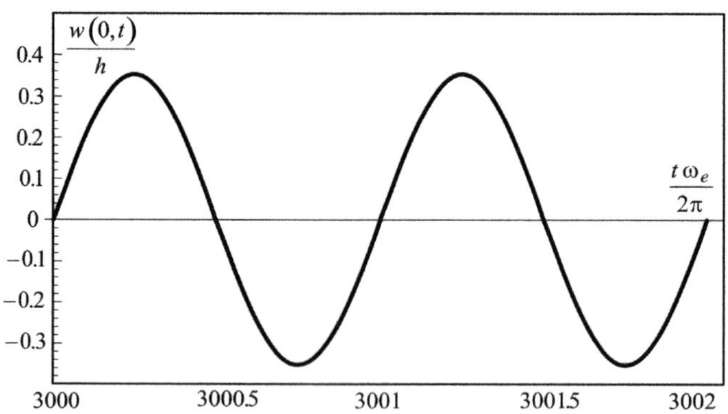

Figure 10.12 Transverse nondimensional displacement of the middle point for a beam without foundation. Parameters: x=0, F=0.134 N, ω_e=1.02 ω_{11}.

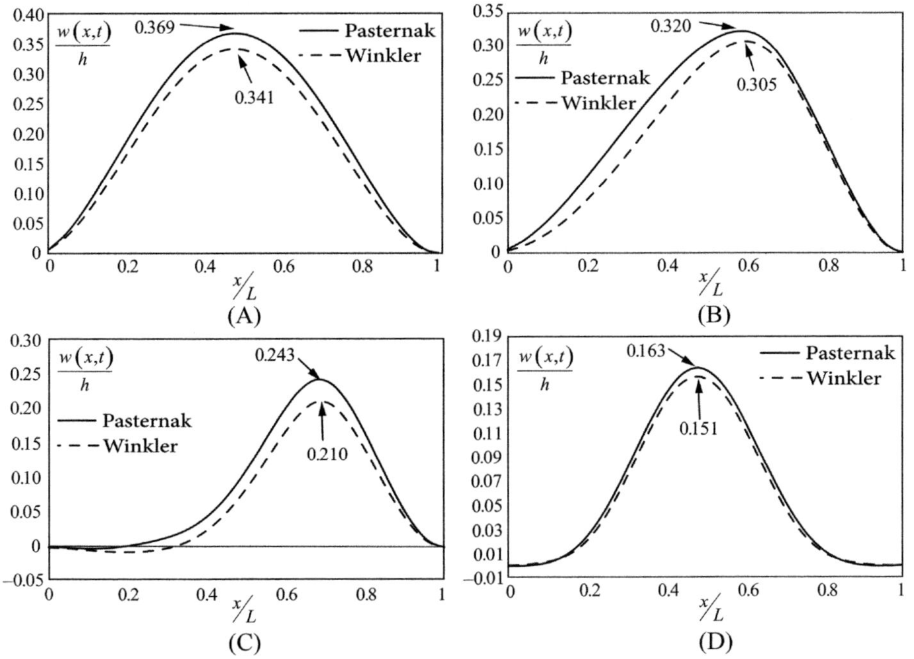

Figure 10.13 Maximal deflection of the beam. The steady-state regime; F=0.134 N, ω_e=1.02 ω_{11}, k_w=100×10^5 Pa. (A) Case 1, (B) Case 2, (C) Case 3, and (D) Case 4.

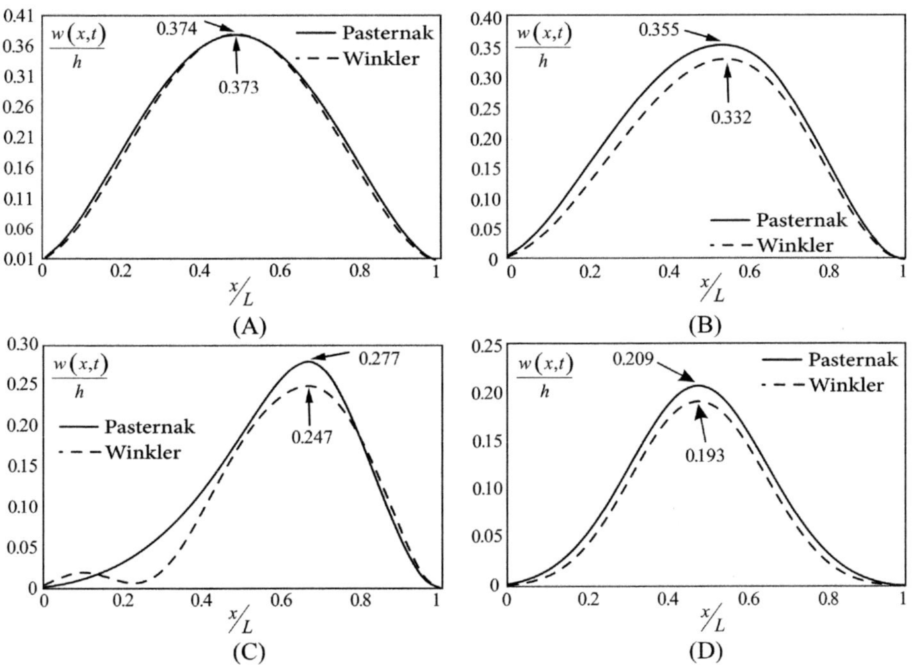

Figure 10.14 Maximal deflection of the beam. The steady-state regime; F=0.134 N, ω_e=1.02 ω_{11}, k_w=50×10^5 Pa. (A) Case 1, (B) Case 2, (C) Case 3, and (D) Case 4.

case where it is of Winkler type. The adopted p-FEM can also be used on moderately thick beams, with discontinuities in the foundation.

10.5 Conclusions

The focus of the study in this chapter is on geometric nonlinear vibrations of shear deformable beams that are laid on a nonlinear elastic foundation, which has a variable discontinuity. The p-FEM method established on the Timoshenko beam assumption is used and employed in the adopted model. The major novelties of this study are the obtained frequencies, the general vibration mode shapes in the linear regime, and the effect of the beam geometric nonlinearity of vibrations coupled with the effects of the nonlinear foundation with a discontinuity on the maximal deflections caused by the steady-state forced vibrations. The developed matrices can be generalized for any discontinuity in the foundation just by changing the boundaries of the discontinuity.

The developed p-FEM method displays good computational quality, and the reason for this is the shape functions for the beams with discontinuity [20], which are implemented into the presented model that comprises a foundation with discontinuity. The geometric nonlinearity of vibrations is considered, and a study concerning its effect on the dynamic behavior caused by external harmonic forces is performed. As seen in the diagrams of maximal deflections, geometric nonlinearity, together with a nonlinear foundation that has varying positions of discontinuity—particularly, when the position of discontinuity is nonsymmetrical in relation to the middle of the beam—led to asymmetric vibrations regarding the vertical mid-axis. It is important to emphasize the significance of geometric nonlinearity in understanding the vibrational character of the adopted model.

References

[1] T.M. Wang, J.E. Stephens, Natural frequencies of Timoshenko beams on Pasternak foundations, Journal of Sound and Vibration 51 (1977) 149–155.

[2] L. Sun, A closed-form solution of a Bernoulli-Euler beam on a viscoelastic foundation under harmonic line loads, Journal of Sound and Vibration 242 (2001) 619–627.

[3] E.J. Sapountzakis, A.E. Kampitsis, Nonlinear response of shear deformable beams on tensionless nonlinear viscoelastic foundation under moving loads, Journal of Sound and Vibration 330 (2011) 5410–5426.

[4] A. Mamandi, M.H. Kargarnovin, S. Farsi, Dynamic analysis of a simply supported beam resting on a nonlinear elastic foundation under compressive axial load using nonlinear normal modes techniques under three-to-one internal resonance condition, Nonlinear Dynamics 70 (2012) 1147–1172.

[5] J.-S. Chen, Y.-K. Chen, Steady state and stability of a beam on a damped tensionless foundation under a moving load, International Journal of Non-Linear Mechanics 46 (2011) 180–185.

[6] S.-M. Kim, Y.-H. Cho, Vibration and dynamic buckling of shear beam-columns on elastic foundation under moving harmonic loads, International Journal of Solids and Structures 43 (2006) 393–412.

[7] G. Muscolino, A. Palmeri, Response of beams resting on viscoelastically damped foundation to moving oscillators, International Journal of Solids and Structures 44 (2007) 1317–1336.

[8] T.-P. Chang, Y.-N. Liu, Dynamic finite element analysis of a nonlinear beam subjected to a moving load, International Journal of Solids and Structures 33 (1996) 1673–1688.

[9] P. Malekzadeh, A.R. Vosoughi, DQM large amplitude vibration of composite beams on nonlinear elastic foundations with restrained edges, Communications in Nonlinear Science and Numerical Simulation 14 (2009) 906–915.

[10] L. Demeio, S. Lenci, Nonlinear resonances of a semi-infinite cable on a nonlinear elastic foundation, Communications in Nonlinear Science and Numerical Simulation 18 (2013) 785–798.

[11] B. Szabó, I. Babuska, Finite Element Analysis, John Wiley & Sons, 1991.

[12] P. Ribeiro, A p-version, first-order shear deformation, finite element for geometrically non-linear vibration of curved beams, International Journal for Numerical Methods in Engineering 61 (2004) 2696–2715.

[13] P. Ribeiro, Free periodic vibrations of beams with large displacements and initial plastic strains, International Journal of Mechanical Sciences 52 (2010) 1407–1418.

[14] P. Ribeiro, Hierarchical finite element analyses of geometrically non-linear vibration of beams and plane frames, Journal of Sound and Vibration 246 (2001) 225–244.

[15] P. Ribeiro, Non-linear forced vibrations of thin/thick beams and plates by the finite element and shooting methods, Computers & Structures 82 (2004) 1413–1423.

[16] W. Han, M. Petyt, Linear vibration analysis of laminated rectangular plates using the hierarchical finite element method, part 1: free vibration analysis, Computers & Structures 61 (1996) 705–712.

[17] N.S. Bardell, The application of symbolic computing to the hierarchical finite element method, International Journal for Numerical Methods in Engineering 28 (1989) 1181–1204.

[18] N.S. Bardell, R.S. Langley, J.M. Dunsdon, G.S. Aglietti, An h-p finite element vibration analysis of open conical sandwich panels and conical sandwich frusta, Journal of Sound and vibration 226 (1999) 345–377.

[19] P. Ribeiro, A hierarchical finite element for geometrically non-linear vibration of doubly curved, moderately thick isotropic shallow shells, International Journal for Numerical Methods in Engineering 56 (2003) 715–738.

[20] S. Stoykov, P. Ribeiro, Nonlinear forced vibrations and static deformations of 3D beams with rectangular cross-section: The influence of warping, shear deformation, and longitudinal displacements, International Journal of Mechanical Sciences 52 (2010) 1505–1521.

[21] A. Houmat, Free vibration analysis of membranes using the h-p version of the finite element method, Journal of Sound and Vibration 282 (2005) 401–410.

[22] V. Stojanović, P. Ribeiro, S. Stoykov, Non-linear vibration of Timoshenko damaged beams by a new p-version finite element method, Computers & Structures 120 (2013) 107–119.

[23] V. Stojanović, Geometrically nonlinear vibrations of beams supported by a nonlinear elastic foundation with variable discontinuity, Communications in Nonlinear Science and Numerical Simulation 28 (2015) 66–80.

[24] M. Petyt, Introduction to Finite Element Vibration Analysis, Cambridge University Press, Cambridge, 1990.

[25] ANSYS® Workbench User's Guide (2009).

[26] T. Kaneko, On Timoshenko's correction for shear in vibrating beams, Journal of Physics D 8 (1975) 1927–1936.

[27] J.R. Hutchinson, Shear coefficients for Timoshenko beam theory, Journal of Applied Mechanics 68 (2001) 87–92.
[28] H. Wolfe, An experimental investigation of nonlinear behaviour of beams and plates excited to high levels of dynamic response (Ph.D. thesis), University of Southampton, 1995.
[29] V. Stojanović, P. Kozić, G. Janevski, Exact closed-form solutions for the natural frequencies and stability of elastically connected multiple beam systems using Timoshenko and high order shear deformation theory, Journal of Sound and Vibration 332 (2013) 563–576.

Nonlinear vibrational characteristics of damaged beams resting on a Pasternak foundation

11

11.1 Introduction

The study presented in this chapter concerns the geometrically nonlinear free and forced vibrations of damaged nonrigid beams laid on a nonlinear Pasternak foundation. For a mathematical description of deformations, a high-order shear deformation theory is used. The derived equations of motion can be applied to beams subjected to the joint influence of concentrated or arbitrarily distributed transverse loading and axial loading. This study aims to equalize the frequency and displacement response between a damaged and an intact beam, or, otherwise stated, make the damaged beam behave like the intact one. That is done by analyzing the effects of the foundation stiffness on the aforementioned response quantities. According to the analysis, the functions that describe the stiffness of the foundation depend on the location and size of the damage. For the analysis of geometrically nonlinear vibrations of damaged beams described via the Reddy-Bickford beam assumption, an advanced *p*-version finite element method is developed. This study gives a clear picture of the nonlinear dynamical behavior of four types of beams: an intact beam, a damaged beam, an intact beam on an elastic foundation, and a damaged beam on an elastic foundation. Furthermore, that chapter gives the derivation of a novel set of two nonlinear partial differential equations, and in them, only the axial and transverse displacements figure. The equations of forced nonlinear vibrations are solved in the time domain with the help of the Newmark-beta integration method. Analyses of free vibrations are performed (by means of harmonic balance and the use of continuation methods), and backbone curves are constructed. Both applied methods are thoroughly detailed in Chapter 2.

Damaged structures can appear in engineering; therefore, the knowledge of their dynamic behavior is of significance because these damages induce altered dynamical responses. To predict the behavior of damaged structures, an accurate mathematical model is needed, often with a reduced number of degrees of freedom. Better models lead to better results and a better understanding of the vibrational phenomena of damaged beams. A notable problem that may occur is the natural phenomenon of resonance, which leads to greater damage. The materialization of physical damage can result from a design defect or a flaw in the production process. Taking into account this, additional stabilization of the dynamic motion can have a beneficial effect on the system, which is weighted by dynamic stresses.

The work carried out here builds on other studies related to damaged structures. Concretely, it describes the dynamic motion of a damaged beam laid on an immovable nonlinear elastic foundation. There is a vast research literature concerning the

Modeling of Complex Dynamic Systems. DOI: https://doi.org/10.1016/B978-0-443-23942-7.00011-5

vibrations of beams with cracks. As a preface to the present work, the following text gives a summary of the existing knowledge.

Dimarogonas [1] studied the influence of a crack on the surface of a shaft. In that scenario, the crack is subjected to the alternating opening-closing regime, in which the time function depends on the speed of rotation and amplitude of vibration. The crack introduces local flexibility, which affects the vibration response of the shaft. If the general motion is considered, the cracked section of the shaft described in the form of the local stiffness matrix leads to a coupled system, whereas for a shaft without a crack, the system is decoupled. Cawley and Adams [2] tested the integrity of structures using a nondestructive method: measurements of natural frequencies. They demonstrated how damage in a structure could be located, measured, and quantified by making measurements at a single point in the structure. In works [3–11], various methods for estimation of the location of changes in a structure are presented; in them, as a starting point, the natural frequencies determined from a damaged system are used. Friswell et al. [3] used model updating to rate the stiffness at all potentially damaged locations simultaneously. In the method they have used, the ratio between the natural frequencies from measured and analytical data is of interest. Narkis [4], in a study on the identification of crack location on a beam, simulated the effect of a crack by an equivalent spring. From that study, of an approximate model, algebraic equations emerged that give values of natural frequencies that concur with the natural frequencies of a cracked beam. Sinha et al. [5] introduced a simplified model for locating cracks in beams that undergo transverse vibrations. That technique uses Euler–Bernoulli beam elements and adds slight modifications to the local flexibility near the cracks. Crack locations and sizes are then estimated by using the model and minimizing the difference between the predicted and measured natural frequencies iteratively by model updating. The introduced method has shown excellent agreement with the experimental data.

Often, minute changes in the configuration of vibratory systems can cause problems in the implementation of analytical methods. Therefore, solving those kinds of systems is often performed by a commercial finite element analysis (FEA) software or by applying simplified approaches to the vibratory analysis of such systems. Simplified analysis of natural frequencies, mode shapes, and damping factors can be used for estimating the effects of cracks on vibrations [6–12]. For instance, Fernandez–Saez et al. [13] made a simplified procedure for calculating the natural frequencies of damaged beams with the Euler–Bernoulli beam assumption. In that method, the crack in a beam is defined through a hinge and an elastic spring. The transverse deflection of the beam with a crack is defined by adding polynomial functions to the functions of the intact beam. The results obtained by that procedure are close to those obtained by the finite element method (FEM). Lee and Chung [14], using the natural frequency data obtained by FEM simulations, have developed a procedure to estimate the location and size of the crack in beam structures. A stiffness reduction described crack damage at the element level, and there were no changes to the mass matrix. Christides and Barr [15] designed a model (based on the Euler–Bernoulli beam assumption and improved by experiments) for the bending of beams with symmetric cracks. They integrated the changes in stress brought about by the crack through a local function that was hypothesized to have an exponential decay with the distance from the crack. Stojanović et al. [16] developed a new p-FEM (based on the Timoshenko beam assumption) mainly meant to solve geometrically nonlinear vibrations of damaged beams. In that study, they derived the new displacement shape functions—like the functions of the damaged location—that have led to more efficient models, that is, improved accuracy at a lower computational cost. That work also shows the coupling between longitudinal vibrations and cross-sectional rotation, the sudden change of direction of longitudinal displacements, and a particular change in the cross-section rotation at the damaged location. The work presented in this chapter extends the results related to the mentioned model of damaged beams by adding a nonlinear elastic foundation. The model of a beam on a nonlinear elastic foundation symbolizes many mechanical systems, for example, vibrating machines on elastic foundations, transversally supported pipelines, etc. The problem with a beam laid on a nonlinear elastic foundation can be found in many studies. Coşkun [17] studied the response of a finite beam on a two-parameter tensionless Pasternak foundation, subjected to a harmonic load. Ma et al. [18,19] investigated the static response of an infinite beam laid on a tensionless two-parameter Pasternak foundation, weighted down by complex transverse loads, including its weight. The latter problem they have analyzed is about the buckling of an infinite thin plate, laid on a tensionless Winkler foundation and burdened with shearing loads. Nayfeh and Mook [20] studied beams laid on nonlinear spring beads and analyzed direct methods for constructing normal nonlinear modes. Shaw and Pierre [21] used a method established on the concept of invariant manifolds of dynamical systems to obtain normal nonlinear modes of beams laid on nonlinear elastic foundations. Nagaya and Kato [22] studied the transient response of magnetically levitated, multispan beams. Kanaka and Venkateswara [23] studied the effects of nonlinear elastic foundations on vibration and buckling problems. Studies [20–23] show that deflections of larger beams cause geometric nonlinearities in vibrations, leading to motions different from the ones investigated in linear systems. Chang and Liu [24] have done a nonlinear vibration analysis of a beam on an elastic foundation under a moving load. They neglected the effects of shear deformations and rotary inertia but included the effects of longitudinal deformations and inertia via the Euler–Bernoulli hypothesis. Sapountzakis and Kampitsis [25] studied the nonlinear response of intact shear-deformable beams subjected to moving loads laid on a nonlinear viscoelastic foundation without tension. For those kinds of beams, they developed a boundary element method (BEM) and showed the advantage of using FEM for analyzing nonlinear occurrences in such problems. The effects of the nonlinear foundation were the focus of some studies. Malekzadeh and Vosoughi [26] analyzed the free large-amplitude vibrations of thin composite beams with the help of an accurate and efficient differential quadrature (DQ) method. Demeio and Lenci [27] did a study of the nonlinear resonance of a semiinfinite cable laid on a nonlinear elastic foundation by employing the multiple time scales (MTS) method.

Structural design nowadays, with consistent improvements and development in mechanics, requires computational modeling. One of the most effective computational methods is p-FEM [28–39]. In scenarios involving specific structures, improvement or change of the method is needed so it can apply to such a specific dynamic model. In such a situation, it is of the essence to validate the obtained results by means of commercial software or by an experiment. Stojanović [40] studied the effects of a nonlinear elastic foundation with variable discontinuities on beam vibrations (based on the Timoshenko beam assumption) by employing an advanced formulation of p-FEM. In that work, two p-FEM shape functions for beam elements are introduced: one for a damaged beam and the other for a damaged beam on an elastic foundation, due to steep changes in the cross-sectional area [16]. Having in mind the importance of knowing the dynamic behavior of damaged structures, the study presented in this chapter defines how the elastic foundation—in the role of a stabilizer—provides stabilization to a damaged beam. Analysis of the problem is done using a high-order shear deformation theory, and the elastic foundation is generalized to the Pasternak type. The results calculated for diverse models (intact, damaged, intact on a foundation, and damaged on a foundation) are compared with the results of the models that are solved using the Timoshenko beam assumption and classical Winkler elastic layering. The selection of the elastic foundation stiffness moduli is done to match the fundamental frequency of the damaged beam (which is connected to the elastic foundation) with the fundamental frequency of the intact beam. However, the effort to equalize the fundamental frequencies leads to a noticeable decrease in maximal deflection in the steady-state nonlinear regime of vibrations of the damaged beams. To equalize the displacements, an additional stiffness correction factor (for the elastic foundation) is added. The solution is given through the stiffness function for the Pasternak elastic foundation. According to it, the stiffness depends on the depth and location of the damage. The deformed shapes of the considered beam models are presented. Geometrically nonlinear free vibrations are analyzed with the harmonic balance and continuation methods. As it happens, the nonlinear elastic foundation, with a specific value of stiffness, can produce stabilizing effects for the free vibrations of damaged beams.

11.2 Mathematical model of intact Reddy–Bickford beam on nonlinear Pasternak foundation

The beam length is designated with L, the width with b, and h is the thickness—in general, the beam has a rectangular cross-section. The beam is made of a material that is assumed to be elastic, isotropic, and homogenous. The foundation is an elastic layer of the Pasternak type—by ignoring the shear, the Pasternak foundation will be reduced to the Winkler foundation. The physical model with a global coordinate system xyz is shown in Fig. 11.1.

Fig. 11.1 shows a nondimensional axis ξ that acts as a local coordinate system for the FEM. The displacement components along the x and z directions are designated with u and w, respectively. Using a high-order shear deformation theory, in this case, the Reddy–Bickford beam assumption, the model displacement field can be written as:

$$u(x, z, t) = u^0(x, t) + z\theta^0(x, t) - \alpha z^3\left(\theta^0(x, t) + \frac{\partial w^0(x, t)}{\partial x}\right),$$

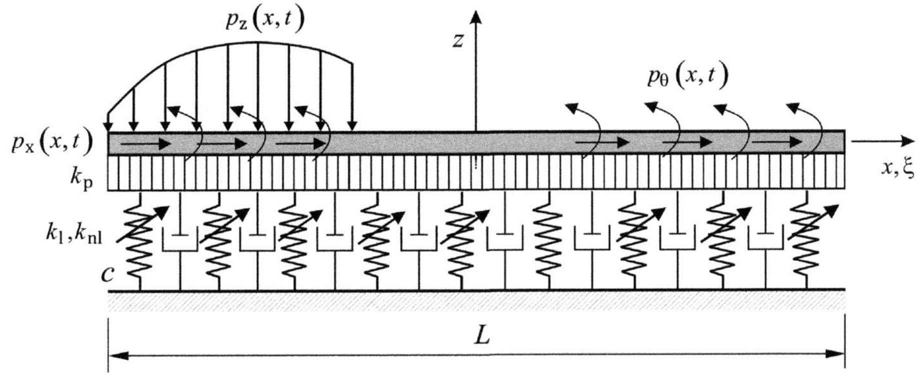

Figure 11.1 Infinite beam. The beam is laid on a nonlinear Pasternak foundation.

$$w(x, z, t) = w^0(x, t). \tag{11.1}$$

where $\alpha = 4/h^3$, $\beta = 3\alpha$. The longitudinal displacement is designated with $u^0(x, t)$, the transversal displacement with $w^0(x, t)$, and the independent rotation of the cross-section about the axis parallel to y is designated with $\theta^0(x, t)$ (Fig. 7.1, in which, the letter φ is used instead of the letter θ). This high-order shear deformation model is chosen because it has better accuracy than a model based on the Euler–Bernoulli beam assumption and does not require a shear correction factor, which would be the case with a model based on the Timoshenko beam assumption. Besides, it is not overly difficult for computation.

The chosen high-order theory can be applied to beams ranging from very thin to moderately thick. A von Kármán type of approach is chosen, which means the longitudinal and shear strains have the following form:

$$\varepsilon_x(x, t) = \frac{\partial u^0}{\partial x} + \frac{1}{2}\left(\frac{\partial w^0}{\partial x}\right)^2 + z\frac{\partial \theta^0}{\partial x} - \alpha z^3\left(\frac{\partial \theta^0}{\partial x} + \frac{\partial^2 w^0}{\partial x^2}\right), \tag{11.2}$$

$$\gamma_{xz}(x, t) = \frac{\partial w^0}{\partial x} + \theta^0 - \beta z^2\left(\theta^0 + \frac{\partial w^0}{\partial x}\right), \tag{11.3}$$

$$\begin{Bmatrix} \sigma_x \\ \tau_{xz} \end{Bmatrix} = \begin{bmatrix} E & 0 \\ 0 & G \end{bmatrix}\begin{Bmatrix} \varepsilon_x \\ \gamma_{xz} \end{Bmatrix}, \tag{11.4}$$

The model of the foundation is mathematically defined via stiffness modulus k_l, the nonlinear modulus k_{nl}, the shear (Pasternak) modulus k_p, and the dumping coefficient c (for the case of a viscoelastic foundation). The reaction has the following form:

$$p_{sz}(x, t) = k_l w(x, t) + k_{nl} w^3(x, t) - k_p\frac{\partial^2 w(x, t)}{\partial x^2} + c\frac{\partial w(x, t)}{\partial t}, \quad p_{sz}(x, t) \geq 0. \tag{11.5}$$

It is worth mentioning that this model (nonlinear viscoelastic Pasternak foundation) has more familiar characteristics for engineers and is also easy to control in applications [41]. By using a nonlinear viscoelastic Pasternak foundation, more accurate results are obtained because shear effects are not neglected, which gives a model that is more faithful to reality.

The equations of motion are derived from the principle of virtual work, according to which the virtual work of the inertial forces is

$$\delta W_{in} = -\rho b \int_0^L \int_{-h/2}^{h/2} \left[\frac{\partial^2 w(x, z, t)}{\partial t^2} \delta w(x, z, t) + \frac{\partial^2 u(x, z, t)}{\partial t^2} \delta u(x, z, t) \right] \mathrm{d}x\mathrm{d}z, \tag{11.6}$$

The virtual work of internal forces is

$$\delta W_V = -b \int_0^L \int_{-h/2}^{h/2} [\sigma_x(x, t)\delta\varepsilon_x(z, t) + \tau_{xz}(x, t)\delta\gamma_{xz}(x, t)] \mathrm{d}x\mathrm{d}z, \tag{11.7}$$

And the virtual work of the external forces is given by

$$\begin{aligned} \delta W_{ex} = \int_0^L & \{ [P_w^j(t)\delta(x - x_j) + P_w^d(x, t) - k_w w(x, t)]\delta w(x, t) \\ & + k_p \frac{\partial^2 w(x, t)}{\partial x^2} \delta w(x, t) + [P_u^j(t)\delta(x - x_j) + P_u^d(x, t) - k_u u(x, t)]\delta u(x, t) \\ & + [M^j(t)\delta(x - x_j) + M^j(x, t) - k_\theta \theta(x, t)]\delta\theta(x, t) \} \mathrm{d}x. \end{aligned} \tag{11.8}$$

The next step would be to introduce the general equation of the virtual work principle, $\delta W_{in} + \delta W_V + \delta W_{ex} = 0$, and substitute the Eqs. (11.6–11.8) into that equation. Then the Green theorem is applied successively—integrating the equation by parts and collecting the coefficients of δu, δw, and $\delta\theta$. Taking into account the expression (Eq. 11.4), the result will turn out to be a set of nonlinear partial differential equations (PDEs). That set of equations represents the vibration of an intact Reddy–Bickford beam resting on a nonlinear elastic Pasternak foundation, and they have the following form:

$$B_1 \frac{\partial^2 u(x, t)}{\partial t^2} - B_2 \frac{\partial w(x, t)}{\partial x} \frac{\partial^2 w(x, t)}{\partial x^2} - B_3 \frac{\partial^2 u(x, t)}{\partial x^2} = p_x(x, t), \tag{11.9a}$$

$$\begin{aligned} & C_1 \frac{\partial^2 w(x, t)}{\partial t^2} - C_2 \frac{\partial^4 w(x, t)}{\partial x^2 \partial t^2} + C_3 \frac{\partial^3 \theta(x, t)}{\partial x \partial t^2} - C_4 \frac{\partial u(x, t)}{\partial x} \frac{\partial^2 w(x, t)}{\partial x^2} \\ & - C_5 \frac{\partial^2 u(x, t)}{\partial x^2} \frac{\partial w(x, t)}{\partial x} - C_6 \frac{\partial^2 w(x, t)}{\partial x^2} \left(\frac{\partial w(x, t)}{\partial x} \right)^2 + C_7 \frac{\partial^2 w(x, t)}{\partial x^2} \\ & + C_8 \frac{\partial^4 w(x, t)}{\partial x^4} - C_9 \frac{\partial \theta(x, t)}{\partial x} + C_{10} \frac{\partial^3 \theta(x, t)}{\partial x^3} + p_{sz}(x, t) = p_z(x, t), \end{aligned} \tag{11.9b}$$

$$D_1\frac{\partial^2\theta(x,t)}{\partial t^2}-D_2\frac{\partial^2\theta(x,t)}{\partial x^2}+D_3\frac{\partial^3 w(x,t)}{\partial x\partial t^2}$$

$$+D_4\frac{\partial w(x,t)}{\partial x}+D_5\frac{\partial^3 w(x,t)}{\partial x^3}+D_6\theta(x,t)=p_\theta(x,t), \tag{11.9c}$$

where

$$B_1=bh\rho,\quad B_2=bEh,\quad B_3=B_2,$$

$$C_1=B_1,\quad C_2=\frac{1}{448}\alpha^2bh^7\rho,\quad C_3=\frac{1}{80}\alpha bh^5\rho-\frac{1}{448}\alpha^2bh^7\rho,$$

$$C_4=B_2,\quad C_5=B_2,\quad C_6=\frac{3}{2}B_2,$$

$$C_7=\frac{1}{80}b\beta^2Gh^5-\frac{1}{6}b\beta Gh^3+bGh,\quad C_8=\frac{1}{448}\alpha^2bEh^7,$$

$$C_9=-C_7,\quad C_{10}=\frac{1}{448}\alpha^2bEh^7-\frac{1}{80}\alpha bEh^5,$$

$$D_1=\frac{1}{448}\alpha^2bh^7\rho-\frac{1}{40}\alpha bh^5\rho+\frac{1}{12}bh^3\rho,$$

$$D_2=-\frac{1}{448}\alpha^2bEh^7+\frac{1}{40}\alpha bEh^5-\frac{1}{12}bEh^3,\quad D_3=\frac{1}{448}\alpha^2bh^7\rho-\frac{1}{80}\alpha bh^5\rho,$$

$$D_4=C_7,\quad D_5=\frac{1}{80}\alpha bEh^5-\frac{1}{448}\alpha^2bEh^7,\quad D_6=C_7.$$

Combining the Eqs. (11.9b) and (11.9c), when $p_\theta(x,t)=0$, the following differential equations with respect to u, w are derived:

$$B_1\frac{\partial^2 u(x,t)}{\partial t^2}-B_2\frac{\partial w(x,t)}{\partial x}\frac{\partial^2 w(x,t)}{\partial x^2}-B_3\frac{\partial^2 u(x,t)}{\partial x^2}=p_x(x,t), \tag{11.10a}$$

$$-\frac{1}{D_6}\Bigg[C_2D_1\frac{\partial^6 w(x,t)}{\partial x^2\partial t^4}+C_3D_3\frac{\partial^6 w(x,t)}{\partial x^2\partial t^4}-C_8D_1\frac{\partial^6 w(x,t)}{\partial x^4\partial t^2}-C_2D_2\frac{\partial^6 w(x,t)}{\partial x^4\partial t^2}$$

$$+C_{10}D_3\frac{\partial^6 w(x,t)}{\partial x^4\partial t^2}+C_3D_5\frac{\partial^6 w(x,t)}{\partial x^4\partial t^2}-C_7D_1\frac{\partial^4 w(x,t)}{\partial x^2\partial t^2}+C_1D_2\frac{\partial^4 w(x,t)}{\partial x^2\partial t^2}$$

$$-C_9D_3\frac{\partial^4 w(x,t)}{\partial x^2\partial t^2}+C_3D_4\frac{\partial^4 w(x,t)}{\partial x^2\partial t^2}+C_2D_6\frac{\partial^4 w(x,t)}{\partial x^2\partial t^2}-C_1D_1\frac{\partial^4 w(x,t)}{\partial t^4}$$

$$-C_1D_6\frac{\partial^2 w(x,t)}{\partial t^2}+(C_8D_2+C_{10}D_5)\frac{\partial^6 w(x,t)}{\partial x^6}+\frac{\partial^4 w(x,t)}{\partial x^4}$$

$$+(C_9D_4+C_7D_6)\frac{\partial^2 w(x,t)}{\partial x^2}+D_2\frac{\partial^2 p_{sz}(x,t)}{\partial x^2}-D_2\frac{\partial^2 p_z(x,t)}{\partial x^2}$$

$$- 2C_6D_2\left(\frac{\partial^2 w\,(x,t)}{\partial x^2}\right)^3 + C_4D_1\frac{\partial^3 w\,(x,t)}{\partial x\partial t^2}\frac{\partial^2 w\,(x,t)}{\partial x^2}$$

$$+ 2C_6D_1\left(\frac{\partial^2 w\,(x,t)}{\partial x\partial t}\right)^2\frac{\partial^2 w\,(x,t)}{\partial x^2} + 2C_6D_1\frac{\partial w\,(x,t)}{\partial x}\frac{\partial^3 w\,(x,t)}{\partial x\partial t^2}\frac{\partial^2 w\,(x,t)}{\partial x^2}$$

$$+ C_4D_6\frac{\partial w(x,t)}{\partial x}\frac{\partial^2 w(x,t)}{\partial x^2} + C_6D_6\left(\frac{\partial w(x,t)}{\partial x}\right)^2\frac{\partial^2 w(x,t)}{\partial x^2}$$

$$- 3C_4D_2\frac{\partial^3 w\,(x,t)}{\partial x^3}\frac{\partial^2 w\,(x,t)}{\partial x^2} - 2C_5D_2\frac{\partial^3 u\,(x,t)}{\partial x^3}\frac{\partial^2 w\,(x,t)}{\partial x^2}$$

$$- 6C_6D_2\frac{\partial w\,(x,t)}{\partial x}\frac{\partial^3 w\,(x,t)}{\partial x^3}\frac{\partial^2 w\,(x,t)}{\partial x^2} + 2C_4D_1\frac{\partial^2 w\,(x,t)}{\partial x\partial t}\frac{\partial^3 w\,(x,t)}{\partial x^2\partial t}$$

$$+ C_4D_1\frac{\partial w\,(x,t)}{\partial x}\frac{\partial^4 w\,(x,t)}{\partial x^2\partial t^2} + C_5D_1\frac{\partial w\,(x,t)}{\partial x}\frac{\partial^4 u\,(x,t)}{\partial x^2\partial t^2}$$

$$+ 2C_5D_1\frac{\partial^3 u\,(x,t)}{\partial x^2\partial t}\frac{\partial^2 w\,(x,t)}{\partial x\partial t} + C_5D_1\frac{\partial^2 u\,(x,t)}{\partial x^2}\frac{\partial^3 w\,(x,t)}{\partial x\partial t^2}$$

$$+ 4C_6D_1\frac{\partial w\,(x,t)}{\partial x}\frac{\partial^2 w\,(x,t)}{\partial x\partial t}\frac{\partial^3 w\,(x,t)}{\partial x^2\partial t} + C_6D_1\left(\frac{\partial w\,(x,t)}{\partial x}\right)^2\frac{\partial^4 w\,(x,t)}{\partial x^2\partial t^2}$$

$$+ C_5D_6\frac{\partial^2 u\,(x,t)}{\partial x^2}\frac{\partial w\,(x,t)}{\partial x} - C_4D_2\frac{\partial w\,(x,t)}{\partial x}\frac{\partial^4 w\,(x,t)}{\partial x^4}$$

$$- C_5D_2\frac{\partial^4 u\,(x,t)}{\partial x^4}\frac{\partial w\,(x,t)}{\partial x}$$

$$\left.- C_5D_2\frac{\partial^2 u(x,t)}{\partial x^2}\frac{\partial^3 w(x,t)}{\partial x^3} - C_6D_2\left(\frac{\partial w(x,t)}{\partial x}\right)^2\frac{\partial^4 w(x,t)}{\partial x^4}\right]$$

$$+ \frac{D_1}{D_6}\left(\frac{\partial^2 p_{sz}\,(x,t)}{\partial t^2} - \frac{\partial^2 p_z\,(x,t)}{\partial t^2}\right) + p_{sz}\,(x,t) = p_z\,(x,t). \tag{11.10b}$$

Eqs. (11.10a) and (11.10b) form the governing differential equations for a Reddy–Bickford beam laid on a nonlinear viscoelastic Pasternak foundation. The usual bending moment and shear force have the following forms:

$$M_{xx} = \int_{-b/2}^{b/2}\int_{-h/2}^{h/2} z\sigma_x \mathrm{d}z\mathrm{d}y,\quad Q_x = \int_{-b/2}^{b/2}\int_{-h/2}^{h/2} \tau_{xz}\mathrm{d}z\mathrm{d}y, \tag{11.11}$$

and P_{xx} and R_x are the higher-order stress resultants.

$$P_{xx} = \int_{-b/2}^{b/2}\int_{-h/2}^{h/2} z^3\sigma_x \mathrm{d}z\mathrm{d}y,\quad R_x = \int_{-b/2}^{b/2}\int_{-h/2}^{h/2} z^2\tau_{xz}\mathrm{d}z\mathrm{d}y. \tag{11.12}$$

The boundary conditions for the Reddy–Bickford theory are of the following form:

$$\begin{Bmatrix} w \\ \frac{\mathrm{d}w}{\mathrm{d}x} \\ \theta \end{Bmatrix} \text{ or } \begin{Bmatrix} \widetilde{V}_x = \alpha\frac{\mathrm{d}P_{xx}}{\mathrm{d}x} + \widetilde{Q}_x \\ \alpha P_{xx} \\ \widetilde{M}_{xx} \end{Bmatrix}, \tag{11.13}$$

in which $\widetilde{V}_x$ represents the effective shear force; $\widetilde{M}_{xx} = M_{xx} - \alpha P_{xx}$, $\widetilde{Q}_x = Q_x - \beta R_x$. For the theory based on the Reddy–Bickford beam assumption [42], the boundary conditions state that the primary (or kinematic) variables (w, $\frac{\mathrm{d}w}{\mathrm{d}x}$, θ) or else the secondary (or force) variables ($\widetilde{V}_x$, αP_{xx}, $\widetilde{M}_{xx}$) must be equal to zero. That is different from the Timoshenko beam assumption, where the only requirement is the specification of the two primary variables—w and θ. The displacement field contains a polynomial variation of the transverse shear strain (and thus shear stress) and the disappearance of transverse shear strain (and again, correspondingly, shear stress) on the top and bottom planes of a beam (Fig. 7.1). An additional condition appears from this. The Euler–Bernoulli and Timoshenko beam assumptions have two boundary conditions at each end of the beam (overall four boundary conditions), which makes them fourth-order beam theories. The Reddy–Bickford beam theory has three boundary conditions at each end of the beam (amounting to six boundary conditions), thus making it a sixth-order beam theory. These boundary conditions also influence Eqs. (11.10a and 11.10b), given as:

$$\alpha_1 u(x, t) + \alpha_2 \frac{1}{2} EA \left[\frac{\partial u(x, t)}{\partial x} + \left(\frac{\partial w(x, t)}{\partial x} \right)^2 \right] = \alpha_3, \tag{11.14}$$

$$\beta_1 w(x, t) + \beta_2 \widetilde{V}_x(x, t) = \beta_3, \quad \gamma_1 \theta(x, t) + \gamma_2 \widetilde{M}_{xx}(x, t) = \gamma_3 \tag{11.15a,b}$$

at the beam ends $x = 0;\ l$ together with the initial conditions.

$$u(x, 0) = \widetilde{u}_0(x), \quad \dot{u}(x, 0) = \dot{\widetilde{u}}_0(x),$$

$$w(x, 0) = \widetilde{w}_0(x), \quad \dot{w}(x, 0) = \dot{\widetilde{w}}_0(x), \tag{11.16}$$

in which $\widetilde{u}_0(x)$, $\dot{\widetilde{u}}_0(x)$, $\widetilde{w}_0(x)$, and $\dot{\widetilde{w}}_0(x)$ are the prescribed functions. Combining Eqs. (11.13, 11.10a and 11.10b), after applying some algebra, equations for the effective shear force $\widetilde{V}_x$, the bending moment $\widetilde{M}_{xx}$, and the higher-order shear resultants P_{xx} can be obtained—due to their extensiveness they are not shown here but are defined by Stojanović and Petković [43]. Eqs. (11.14 and 11.15) define the most general nonlinear boundary conditions correlated with the problem in question and can include constraints or elastic support. Functions specified at the beam's ends are denoted with α_j, β_j, γ_j ($j = 1, 2, 3$). It is perceivable that all types of the common boundary conditions (free, simply supported, guided edge, or clamped) can be derived from these equations by adequately specifying the functions at the beam's ends (e.g., for a clamped edge it is $\alpha_1 = \beta_1 = \gamma_1 = 1;\ \alpha_2 = \beta_2 = \gamma_2 = \alpha_3 = \beta_3 = \gamma_3 = 0$).

11.3 Modeling of damage and the p-FEM formulation

In the p-FEM, the number of elements depends mainly on the geometry of the analyzed structure. Localized steep variations (discontinuity, which means damage) require an increased number of elements for their proper defining. In each element, the vector $\mathbf{d}_0(\xi, t)$, which is formed from the displacement components, may be formulated by combining the hierarchical shape functions.

$$\mathbf{d}_0(\xi, t) = \mathbf{N}(\xi)\mathbf{q}(t) \Leftrightarrow$$

$$\begin{Bmatrix} u^0(\xi, t) \\ w^0(\xi, t) \\ \theta^0(\xi, t) \end{Bmatrix} = \begin{bmatrix} N^u(\xi)^{\mathrm{T}} & \mathbf{0} & \mathbf{0} \\ \mathbf{0} & N^w(\xi)^{\mathrm{T}} & \mathbf{0} \\ \mathbf{0} & \mathbf{0} & N^\theta(\xi)^{\mathrm{T}} \end{bmatrix} \begin{Bmatrix} \mathbf{q}_\mathbf{u}(t) \\ \mathbf{q}_\mathbf{w}(t) \\ \mathbf{q}_\theta(t) \end{Bmatrix}, \tag{11.17}$$

in which $\mathbf{q}_\mathbf{u}(t)$ represents the vector of generalized longitudinal displacements; $\mathbf{q}_\mathbf{w}(t)$ is the vector of generalized transverse displacements; and $\mathbf{q}_\theta(t)$ is the vector of generalized rotations around the y axis. $\mathbf{N}(\xi)$ represents the matrix of shape functions, comprising row vectors of longitudinal, transverse, and rotation shape functions. The shape functions that represent the discontinuity can be identified by the superscript "d". The vectors of the longitudinal, transverse, and rotation shape functions have the following form:

$$\begin{aligned} N^u(\xi)^T &= \lfloor f_1^d(L, l_c, \xi) f_2^d(L, l_c, \xi) \mid f_1(\xi) f_2(\xi) f_3(\xi) \dots f_{pu}(\xi) \rfloor, \\ N^w(\xi)^T &= \lfloor f_1^d(L, l_c, \xi) f_2^d(L, l_c, \xi) \mid f_1(\xi) f_2(\xi) f_3(\xi) \dots f_{pw}(\xi) \rfloor, \\ N^\theta(\xi)^T &= \lfloor f_1^d(L, l_c, \xi) f_2^d(L, l_c, \xi) \mid f_1(\xi) f_2(\xi) f_3(\xi) \dots f_{p\theta}(\xi) \rfloor, \end{aligned} \tag{11.18}$$

in which the counts of longitudinal, transverse, and rotation shape functions are $pu + 2$, $pw + 2$, and $p\theta + 2$. Two additional (+2) are added due to steep variations and damage to the structure. The $f_r(\xi)$ shape functions (where $r > 4$) have the value of zero; also the derivative at the element limits has the value of zero; hence they are internal to each element [16]. These shape functions have been widely used in p-FEM models of beams, plates, and shells. Since the focus here is on clamped-clamped beams, hierarchical sets of shape functions are required solely. The chosen displacement shape functions are defined in nondimensional coordinates, as laid out in the following expression:

$$f_{r-2}(\xi) = \sum_{n=0}^{\mathrm{INT}(r/2)} \frac{(-1)^n (2r - 2n - 7)!!}{2^n n! (r - 2n - 1)!} \xi^{r-2n-1}, \quad r > 4, \quad \xi = \frac{2x}{L}, \tag{11.19}$$

where, for example, $r!! = r(r-2)\dots(2\, or\, 1)$, $0!! = (-1)!! = 1$ and $\mathrm{INT}(r/2)$ indicate the integer part of $r/2$. The Rodrigues form of Legendre polynomials [44] is used for the derivation of a set of shape functions—the clamped boundary conditions are defined by these shape functions. The third-order polynomials $f_1(\xi)$ $-f_4(\xi)$, commonly known by the term Hermite cubics, are incorporated when the boundary conditions permit the motion to take place or to connect the elements.

It should be verified that the maximum value of the shape function $f_1^d\,(L,\ l_c,\ \xi)$ is in the middle of the discontinuity. At the same time, the shape function $f_2^d\,(L,\ l_c,\ \xi)$ has the maximum derivative at the same point. The middle of the discontinuity is $l_c = (l_1 + l_2)/2$. As the material is isotropic and linearly elastic, the following relation, associated with the design shown in Fig. 11.2 holds:

$$\boldsymbol{\sigma} = \mathbf{D}\boldsymbol{\varepsilon} \qquad \begin{Bmatrix} \sigma_x \\ \tau_{xz} \end{Bmatrix} = \begin{bmatrix} E & 0 \\ 0 & G \end{bmatrix} \begin{Bmatrix} \varepsilon_x \\ \gamma_{xz} \end{Bmatrix}, \tag{11.20}$$

in which $\boldsymbol{D}$ represents the matrix of moduli of elasticity; $\boldsymbol{\sigma}$ is the vector of the non-zero stresses; $\boldsymbol{\varepsilon}$ is the vector of the strains. Substituting the expressions from the vector (Eq. 11.17) into Eqs. (11.2) and (11.3), and using the relation between global and local coordinates $\xi = 2x/L$ specified in Eq. (11.19), the longitudinal and shear strains have the following expression:

$$\varepsilon_x = \lfloor 1 \quad z \quad z^2 \quad z^3 \rfloor \left\{ \begin{Bmatrix} \varepsilon_0^l \\ \varepsilon_0^b \\ 0 \\ \varepsilon_0^h \end{Bmatrix} + \begin{Bmatrix} \varepsilon_{nl}^l \\ 0 \\ 0 \\ 0 \end{Bmatrix} \right\}, \tag{11.21a}$$

$$\gamma_{xz} = \lfloor \gamma_w \quad \gamma_\theta \rfloor \begin{Bmatrix} \mathbf{q_w} \\ \mathbf{q_\theta} \end{Bmatrix}. \tag{11.21b}$$

In Eq. (11.21a), the longitudinal strain is broken down into three linear components and one nonlinear strain component. The linear longitudinal strain; $\varepsilon_0^l = \frac{2}{L}\mathbf{N}_{,\xi}^{u\,\mathrm{T}}\mathbf{q_u}$. The bending linear strain; $z\varepsilon_0^b = z\frac{2}{L}\mathbf{N}_{(,\xi)}^{\theta\mathrm{T}}\mathbf{q_\theta}$. The higher-order strain resultant; $z^3\varepsilon_0^h = -z^3\alpha\left(\frac{2}{L}\mathbf{N}_{(,\xi)}^{\theta\mathrm{T}}\mathbf{q_\theta} + \frac{4}{L^2}\mathbf{N}_{(,\xi)}^{w\mathrm{T}}\mathbf{q_w}\right)$. The geometrically nonlinear longitudinal strain is $\varepsilon_{nl}^l = \frac{2}{L^2}\mathbf{q_w}^{\mathrm{T}}\mathbf{N}_{,\xi}^{w}\mathbf{N}_{,\xi}^{w\,\mathrm{T}}\mathbf{q_w}$. In Eq. (11.21b), the shear strain is broken down into two

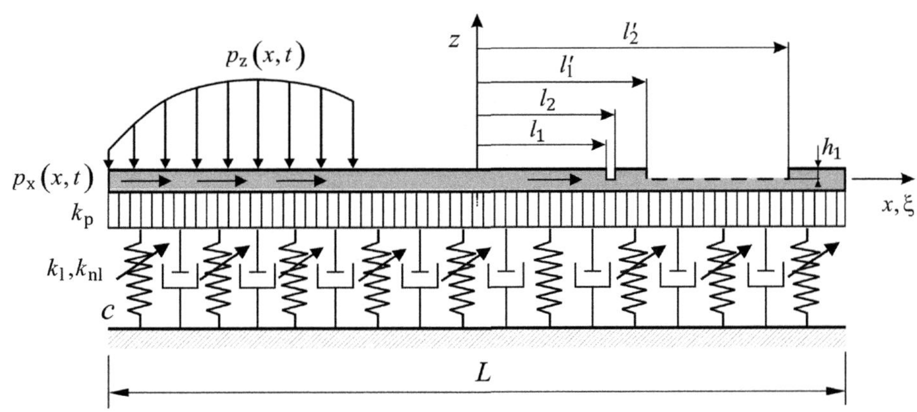

Figure 11.2 Infinite beam with damage. The beam laid on a nonlinear Pasternak foundation.

linear parts: $\gamma_w \mathbf{q_w} = \frac{2}{L}\mathbf{N}^{wT}_{,\xi}\mathbf{q_w}$ and $\gamma_\theta \mathbf{q_\Theta} = \mathbf{N}^{\theta T}\mathbf{q_\Theta}$. The equations of motion of damaged Reddy–Bickford beams laid on a Pasternak foundation are acquired using the principle of virtual work, as a result of a virtual displacement with components $\boldsymbol{\delta d}$ in the form:

$$\delta \mathbf{d} = \{\delta \mathrm{u} \quad \delta \mathrm{w} \quad \delta\theta\}^{\mathrm{T}} = \mathbf{N}\delta\mathbf{q}. \tag{11.22}$$

The virtual work of internal forces is

$$\delta W_V = -\int_V \delta\boldsymbol{\varepsilon}^{\mathrm{T}}\boldsymbol{\sigma}\mathrm{d}V=$$

$$-b\left(\int_{-L/2}^{l_1}\int_{-h/2}^{h/2}\delta\boldsymbol{\varepsilon}^{\mathrm{T}}\boldsymbol{\sigma}\mathrm{d}x\mathrm{d}z + \int_{l_1}^{l_2}\int_{-h/2}^{(h/2)-h_1}\delta\boldsymbol{\varepsilon}^{\mathrm{T}}\boldsymbol{\sigma}\mathrm{d}x\mathrm{d}z + \int_{l_2}^{L/2}\int_{-h/2}^{h/2}\delta\boldsymbol{\varepsilon}^{\mathrm{T}}\boldsymbol{\sigma}\mathrm{d}x\mathrm{d}z\right)=$$

$$= \delta\mathbf{q}^{\mathrm{T}}\mathbf{K}(\mathbf{q})\mathbf{q}, \tag{11.23}$$

where $\delta\boldsymbol{\varepsilon}$ is the vector constructed of virtual strains. The peculiarity of this expression of virtual work comes from the change in geometry, which is the result of damage. The first member in the expression $(\int_{-L/2}^{l_1}\int_{-h/2}^{h/2}\delta\boldsymbol{\varepsilon}^{\mathrm{T}}\boldsymbol{\sigma}\mathrm{d}x\mathrm{d}z)$ represents the part of the beam from the left side to the beginning of the damage, along the x-axis (Fig. 11.2). The second member $(\int_{l_1}^{l_2}\int_{-h/2}^{(h/2)-h_1}\delta\boldsymbol{\varepsilon}^{\mathrm{T}}\boldsymbol{\sigma}\mathrm{d}x\mathrm{d}z)$ represents the part of the beam from the beginning of the damage to the end of the damage. The last member $(\int_{l_2}^{L/2}\int_{-h/2}^{h/2}\delta\boldsymbol{\varepsilon}^{\mathrm{T}}\boldsymbol{\sigma}\mathrm{d}x\mathrm{d}z)$ defines the final remaining part of the beam. For the same reason, the virtual work of the inertia forces also comprises three parts:

$$\delta W_{in} = -\int_V \rho\delta\mathbf{d}^{\mathrm{T}}\ddot{\mathbf{d}}\mathrm{d}V=$$

$$-b\int_{-h/2}^{h/2}\int_{-L/2}^{l_1}\rho(\delta u\ddot{u} + \delta w\ddot{w})\mathrm{d}x\mathrm{d}z - b\int_{-h/2}^{(h/2)-h_1}\int_{l_1}^{l_2}\rho(\delta u\ddot{u} + \delta w\ddot{w})\mathrm{d}x\mathrm{d}z$$

$$-b\int_{-h/2}^{h/2}\int_{l_2}^{L/2}\rho(\delta u\ddot{u} + \delta w\ddot{w})\mathrm{d}x\mathrm{d}z = \delta\mathbf{q}^{\mathrm{T}}\mathbf{M}\ddot{\mathbf{q}}. \tag{11.24}$$

The virtual work of external forces is

$$\delta W_{ex} = \int_L [P_w^j(t)\delta(x - x_j) + P_w^d(x, t) - k_w w(x, t)]\delta w(x, t)$$

$$+ [P_u^j(t)\delta(x - x_j) + P_u^d(x, t)]\delta u(x, t) + [M^j(t)\delta(x - x_j) + M^d(x, t)]\delta\theta(x, t)=$$

$$= \left\{\delta q_u \quad \delta q_w \quad \delta q_\theta\right\}^{\mathrm{T}}\begin{Bmatrix}\mathbf{F}_u^{\mathrm{E}}\\ \mathbf{F}_w^{\mathrm{E}}\\ \mathbf{M}^{\mathrm{E}}\end{Bmatrix}, \tag{11.25}$$

where $\ddot{\boldsymbol{d}}$ are the components of the acceleration of a point belonging to the beam. Implementing the virtual work principle and including the nonlinear effect of the foundation, the following equations of motion of a stabilized beam (a damaged beam on a foundation, the third case in the upcoming analysis) are obtained:

$$
\begin{aligned}
&\left[\begin{bmatrix} \mathbf{M}_{11} & 0 & 0 \\ 0 & \mathbf{M}_{22} & \mathbf{M}_{23} \\ 0 & \mathbf{M}_{23}^{\mathrm{T}} & \mathbf{M}_{33} \end{bmatrix} + \begin{bmatrix} 0 & \mathbf{M}^{lt} & \mathbf{M}^{lr} \\ \mathbf{M}^{lt\mathrm{T}} & 0 & 0 \\ \mathbf{M}^{lr\mathrm{T}} & 0 & 0 \end{bmatrix}\right] \begin{Bmatrix} \ddot{\mathbf{q}}_{\mathbf{u}} \\ \ddot{\mathbf{q}}_{\mathbf{w}} \\ \ddot{\mathbf{q}}_{\theta} \end{Bmatrix} \\
&+ \alpha \left(\left[\begin{bmatrix} \mathbf{K}_{11} & 0 & 0 \\ 0 & \mathbf{K}_{22} + \mathbf{K}_{\ell}^{W} - \mathbf{K}_{\ell}^{P} & \mathbf{K}_{23} \\ 0 & \mathbf{K}_{23}^{\mathrm{T}} & \mathbf{K}_{33} \end{bmatrix} + \begin{bmatrix} 0 & \mathbf{K}^{lt} & \mathbf{K}^{lr} \\ \mathbf{K}^{lt\mathrm{T}} & 0 & 0 \\ \mathbf{K}^{lr\mathrm{T}} & 0 & 0 \end{bmatrix}\right]\right) \begin{Bmatrix} \dot{\mathbf{q}}_{\mathbf{u}} \\ \dot{\mathbf{q}}_{\mathbf{w}} \\ \dot{\mathbf{q}}_{\theta} \end{Bmatrix} \\
&+ \beta \left(\left[\begin{bmatrix} \mathbf{M}_{11} & 0 & 0 \\ 0 & \mathbf{M}_{22} & \mathbf{M}_{23} \\ 0 & \mathbf{M}_{23}^{\mathrm{T}} & \mathbf{M}_{33} \end{bmatrix} + \begin{bmatrix} 0 & \mathbf{M}^{lt} & \mathbf{M}^{lr} \\ \mathbf{M}^{lt\mathrm{T}} & 0 & 0 \\ \mathbf{M}^{lr\mathrm{T}} & 0 & 0 \end{bmatrix}\right]\right) \begin{Bmatrix} \dot{\mathbf{q}}_{\mathbf{u}} \\ \dot{\mathbf{q}}_{\mathbf{w}} \\ \dot{\mathbf{q}}_{\theta} \end{Bmatrix} \\
&+ \left[\begin{bmatrix} \mathbf{K}_{11} & 0 & 0 \\ 0 & \mathbf{K}_{22} + \mathbf{K}_{\ell}^{W} - \mathbf{K}_{\ell}^{P} & \mathbf{K}_{23} \\ 0 & \mathbf{K}_{23}^{\mathrm{T}} & \mathbf{K}_{33} \end{bmatrix} + \begin{bmatrix} 0 & \mathbf{K}^{lt} & \mathbf{K}^{lr} \\ \mathbf{K}^{lt\mathrm{T}} & 0 & 0 \\ \mathbf{K}^{lr\mathrm{T}} & 0 & 0 \end{bmatrix}\right] \begin{Bmatrix} \mathbf{q}_{\mathbf{u}} \\ \mathbf{q}_{\mathbf{w}} \\ \mathbf{q}_{\theta} \end{Bmatrix} \\
&+ \left[\begin{bmatrix} 0 & \mathbf{K}_{n\ell}^{2}(\mathbf{q}_{\mathbf{w}}) & 0 \\ \mathbf{K}_{n\ell}^{3}(\mathbf{q}_{\mathbf{w}}) & \mathbf{K}_{n\ell}^{4}(\mathbf{q}_{\mathbf{w}}) + \mathbf{K}_{n\ell}^{W}(\mathbf{q}_{\mathbf{w}}) & 0 \\ 0 & 0 & 0 \end{bmatrix} + \begin{bmatrix} 0 & 0 & 0 \\ 0 & 0 & \mathbf{K}_{n\ell}^{br}(\mathbf{q}_{\mathbf{w}}) \\ 0 & \mathbf{K}_{n\ell}^{br}(\mathbf{q}_{\mathbf{w}})^{\mathrm{T}} & 0 \end{bmatrix}\right] \begin{Bmatrix} \mathbf{q}_{\mathbf{u}} \\ \mathbf{q}_{\mathbf{w}} \\ \mathbf{q}_{\theta} \end{Bmatrix} \\
&= \begin{Bmatrix} \mathbf{F}_{\mathbf{u}}^{\mathrm{E}} & \mathbf{F}_{\mathbf{w}}^{\mathrm{E}} & \mathbf{M}^{\mathrm{E}} \end{Bmatrix}^{\mathrm{T}}.
\end{aligned}
\tag{11.26}
$$

$\begin{Bmatrix} \mathbf{F}_{\mathbf{u}}^{\mathrm{E}}(t) & \mathbf{F}_{\mathbf{w}}^{\mathrm{E}}(t) & \mathbf{M}^{\mathrm{E}}(t) \end{Bmatrix}^{\mathrm{T}}$ is the vector of generalized external forces. Matrices of type $\mathbf{M}$ and $\mathbf{K}$ (mass and stiffness) are constant matrices, which make the linear terms in the equations of motion. Matrices $\mathbf{K}_{\ell}^{W}$ and $\mathbf{K}_{\ell}^{P}$ describe the linear effects of the foundation. Matrices $\mathbf{K}_{\mathrm{n}\ell}^{\mathbf{2}}$ and $\mathbf{K}_{\mathrm{n}\ell}^{\mathbf{3}}$ have a linear dependency with the solution, which makes quadric nonlinear terms. Matrix $\mathbf{K}_{\mathrm{n}\ell}^{\mathbf{4}}$ has a quadric dependency with the solution—which makes the cubic nonlinearity. Matrix $\mathbf{K}_{\mathrm{n}\ell}^{W}$ has a cubic dependency with the solution. The superscripts; l, b, r, and γ are—respectively—longitudinal, bending, cross-section rotation, and shear effects. The superscript lr is the longitudinal-rotation coupling that takes place because of the damage. Using the Rayleigh proportional type of damping, where α and β are damping coefficients, the equations of motion can be written in a more condensed form:

$$
\mathbf{M}\ddot{\mathbf{q}}(t) + \alpha(\mathbf{K}_{\ell})\dot{\mathbf{q}}(t) + \mathbf{B}(\mathbf{M})\dot{\mathbf{q}}(t) + (\mathbf{K}_{\ell} + \mathbf{K}_{\mathrm{n}\ell}(\mathbf{q}(t)))\mathbf{q}(t) = \mathbf{F}(t). \tag{11.27}
$$

The mass and stiffness matrices are fully developed and given in Appendix 1. One consequence of damage is the change in the structure of the matrices. Here is an explanation of how one of the matrices (matrix $\mathbf{K}_{11}$) is obtained. Substituting Eq. (11.17) in Eq. (11.23) one member of the global linear stiffness matrix is obtained;

during this, the correlation between global and local coordinates and the composition of the two imported shape functions for the damaged beam [16] is considered. Here, that member takes place 11 (longitudinal and longitudinal) in the matrix $\mathbf{K}_\ell$ given in Eq. (11.27).

$$\mathbf{K}_{11} = \frac{2Ehb}{L}\int_{-1}^{2l_1/L} \frac{\mathrm{d}\mathbf{N}^{\mathbf{u}1\mathrm{T}}}{\mathrm{d}\xi}\frac{\mathrm{d}\mathbf{N}^{\mathbf{u}1}}{\mathrm{d}\xi}\mathrm{d}\xi + \frac{2Eb}{L}\int_{2l_1/L}^{(2l_1+x_c)/L}\int_{-h/2}^{h/2-h_1} \frac{\mathrm{d}\mathbf{N}^{\mathbf{u}1\mathrm{T}}}{\mathrm{d}\xi}\frac{\mathrm{d}\mathbf{N}^{\mathbf{u}1}}{\mathrm{d}\xi}\mathrm{z}^2\mathrm{d}z\mathrm{d}\xi$$
$$+ \frac{2Eb}{L}\int_{\frac{(2l_1+x_c)}{L}}^{\frac{2l_2}{L}}\int_{-\frac{h}{2}}^{h/2-h_1} \frac{\mathrm{d}\mathbf{N}^{\mathbf{u}2\mathrm{T}}}{\mathrm{d}\xi}\frac{\mathrm{d}\mathbf{N}^{\mathbf{u}2}}{\mathrm{d}\xi}\mathrm{z}^2\mathrm{d}z\mathrm{d}\xi$$
$$+ \frac{2Ehb}{L}\int_{2l_2/L1}^{1} \frac{\mathrm{d}\mathbf{N}^{\mathbf{u}2\mathrm{T}}}{\mathrm{d}\xi}\frac{\mathrm{d}\mathbf{N}^{\mathbf{u}2}}{\mathrm{d}\xi}\mathrm{d}\xi. \tag{11.28}$$

Using the same procedure, other members of the global stiffness matrix $\mathbf{K}_\ell$ are obtained. Equations of motion for an intact beam—which is the first case—have the following form:

$$\begin{bmatrix} \mathbf{M}_{11} & 0 & 0 \\ 0 & \mathbf{M}_{22} & \mathbf{M}_{23} \\ 0 & \mathbf{M}_{23}^{\mathbf{T}} & \mathbf{M}_{33} \end{bmatrix}\begin{Bmatrix} \ddot{\mathbf{q}}_{\mathbf{u}} \\ \ddot{\mathbf{q}}_{\mathbf{w}} \\ \ddot{\mathbf{q}}_{\theta} \end{Bmatrix} + \alpha\left(\begin{bmatrix} \mathbf{K}_{11} & 0 & 0 \\ 0 & \mathbf{K}_{22} & \mathbf{K}_{23} \\ 0 & \mathbf{K}_{23}^{\mathbf{T}} & \mathbf{K}_{33} \end{bmatrix}\right)\begin{Bmatrix} \dot{\mathbf{q}}_{\mathbf{u}} \\ \dot{\mathbf{q}}_{\mathbf{w}} \\ \dot{\mathbf{q}}_{\theta} \end{Bmatrix}$$
$$+\beta\left(\begin{bmatrix} \mathbf{M}_{11} & 0 & 0 \\ 0 & \mathbf{M}_{22} & \mathbf{M}_{23} \\ 0 & \mathbf{M}_{23}^{\mathbf{T}} & \mathbf{M}_{33} \end{bmatrix}\right)\begin{Bmatrix} \dot{\mathbf{q}}_{\mathbf{u}} \\ \dot{\mathbf{q}}_{\mathbf{w}} \\ \dot{\mathbf{q}}_{\theta} \end{Bmatrix} + \begin{bmatrix} \mathbf{K}_{11} & 0 & 0 \\ 0 & \mathbf{K}_{22} & \mathbf{K}_{23} \\ 0 & \mathbf{K}_{23}^{\mathbf{T}} & \mathbf{K}_{33} \end{bmatrix}\begin{Bmatrix} \mathbf{q}_{\mathbf{u}} \\ \mathbf{q}_{\mathbf{w}} \\ \mathbf{q}_{\theta} \end{Bmatrix}$$
$$+\begin{bmatrix} 0 & \mathbf{K}_{n\ell}^{2}(\mathbf{q}_{\mathbf{w}}) & 0 \\ \mathbf{K}_{n\ell}^{3}(\mathbf{q}_{\mathbf{w}}) & \mathbf{K}_{n\ell}^{4}(\mathbf{q}_{\mathbf{w}}) & 0 \\ 0 & 0 & 0 \end{bmatrix}\begin{Bmatrix} \mathbf{q}_{\mathbf{u}} \\ \mathbf{q}_{\mathbf{w}} \\ \mathbf{q}_{\theta} \end{Bmatrix} = \begin{Bmatrix} \mathbf{F}_{u}^{\mathbf{E}} \\ \mathbf{F}_{w}^{\mathbf{E}} \\ \mathbf{M}^{\mathbf{E}} \end{Bmatrix}. \tag{11.29}$$

Equations of motion for a damaged beam—which is the second case—have the following form:

$$\left[\begin{bmatrix} \mathbf{M}_{11} & 0 & 0 \\ 0 & \mathbf{M}_{22} & \mathbf{M}_{23} \\ 0 & \mathbf{M}_{23}^{\mathbf{T}} & \mathbf{M}_{33} \end{bmatrix} + \begin{bmatrix} 0 & \mathbf{M}^{lt} & \mathbf{M}^{lr} \\ \mathbf{M}^{lt\mathrm{T}} & 0 & 0 \\ \mathbf{M}^{lr\mathrm{T}} & 0 & 0 \end{bmatrix}\right]\begin{Bmatrix} \ddot{\mathbf{q}}_{\mathbf{u}} \\ \ddot{\mathbf{q}}_{\mathbf{w}} \\ \ddot{\mathbf{q}}_{\theta} \end{Bmatrix}$$
$$+\alpha\left(\left[\begin{bmatrix} \mathbf{K}_{11} & 0 & 0 \\ 0 & \mathbf{K}_{22} & \mathbf{K}_{23} \\ 0 & \mathbf{K}_{23}^{\mathbf{T}} & \mathbf{K}_{33} \end{bmatrix} + \begin{bmatrix} 0 & \mathbf{K}^{lt} & \mathbf{K}^{lr} \\ \mathbf{K}^{lt\mathrm{T}} & 0 & 0 \\ \mathbf{K}^{lr\mathrm{T}} & 0 & 0 \end{bmatrix}\right]\right)\begin{Bmatrix} \dot{\mathbf{q}}_{\mathbf{u}} \\ \dot{\mathbf{q}}_{\mathbf{w}} \\ \dot{\mathbf{q}}_{\theta} \end{Bmatrix}$$
$$+\beta\left(\left[\begin{bmatrix} \mathbf{M}_{11} & 0 & 0 \\ 0 & \mathbf{M}_{22} & \mathbf{M}_{23} \\ 0 & \mathbf{M}_{23}^{\mathbf{T}} & \mathbf{M}_{33} \end{bmatrix} + \begin{bmatrix} 0 & \mathbf{M}^{lt} & \mathbf{M}^{lr} \\ \mathbf{M}^{lt\mathrm{T}} & 0 & 0 \\ \mathbf{M}^{lr\mathrm{T}} & 0 & 0 \end{bmatrix}\right]\right)\begin{Bmatrix} \dot{\mathbf{q}}_{\mathbf{u}} \\ \dot{\mathbf{q}}_{\mathbf{w}} \\ \dot{\mathbf{q}}_{\theta} \end{Bmatrix}$$

$$+\left[\begin{bmatrix}\mathbf{K}_{11} & 0 & 0\\ 0 & \mathbf{K}_{22} & \mathbf{K}_{23}\\ 0 & \mathbf{K}_{23}^{\mathrm{T}} & \mathbf{K}_{33}\end{bmatrix}+\begin{bmatrix}0 & \mathbf{K}^{lt} & \mathbf{K}^{lr}\\ \mathbf{K}^{lt\mathrm{T}} & 0 & 0\\ \mathbf{K}^{lr\mathrm{T}} & 0 & 0\end{bmatrix}\right]\begin{Bmatrix}\mathbf{q}_{\mathrm{u}}\\ \mathbf{q}_{\mathrm{w}}\\ \mathbf{q}_{\theta}\end{Bmatrix}$$

$$+\left[\begin{bmatrix}0 & \mathbf{K}_{n\ell}^{2}(\mathbf{q}_{\mathrm{w}}) & 0\\ \mathbf{K}_{n\ell}^{3}(\mathbf{q}_{\mathrm{w}}) & \mathbf{K}_{n\ell}^{4}(\mathbf{q}_{\mathrm{w}}) & 0\\ 0 & 0 & 0\end{bmatrix}+\begin{bmatrix}0 & 0 & 0\\ 0 & 0 & \mathbf{K}_{n\ell}^{br}(\mathbf{q}_{\mathrm{w}})\\ 0 & \mathbf{K}_{n\ell}^{br}(\mathbf{q}_{\mathrm{w}})^{\mathrm{T}} & 0\end{bmatrix}\right]\begin{Bmatrix}\mathbf{q}_{\mathrm{u}}\\ \mathbf{q}_{\mathrm{w}}\\ \mathbf{q}_{\theta}\end{Bmatrix}$$

$$=\begin{Bmatrix}\mathbf{F}_{\mathrm{u}}^{\mathrm{E}} & \mathbf{F}_{\mathrm{w}}^{\mathrm{E}} & \mathbf{M}^{\mathrm{E}}\end{Bmatrix}^{\mathrm{T}}. \tag{11.30}$$

Equations of motion for the fourth case—an intact beam on a foundation—have the following form:

$$\begin{bmatrix}\mathbf{M}_{11} & 0 & 0\\ 0 & \mathbf{M}_{22} & \mathbf{M}_{23}\\ 0 & \mathbf{M}_{23}^{\mathrm{T}} & \mathbf{M}_{33}\end{bmatrix}\begin{Bmatrix}\ddot{\mathbf{q}}_{\mathrm{u}}\\ \ddot{\mathbf{q}}_{\mathrm{w}}\\ \ddot{\mathbf{q}}_{\theta}\end{Bmatrix}+\alpha\left(\begin{bmatrix}\mathbf{K}_{11} & 0 & 0\\ 0 & \mathbf{K}_{22}+\mathbf{K}_{\ell}^{W}-\mathbf{K}_{\ell}^{P} & \mathbf{K}_{23}\\ 0 & \mathbf{K}_{23}^{\mathrm{T}} & \mathbf{K}_{33}\end{bmatrix}\right)\begin{Bmatrix}\dot{\mathbf{q}}_{\mathrm{u}}\\ \dot{\mathbf{q}}_{\mathrm{w}}\\ \dot{\mathbf{q}}_{\theta}\end{Bmatrix}$$

$$+\beta\left(\begin{bmatrix}\mathbf{M}_{11} & 0 & 0\\ 0 & \mathbf{M}_{22} & \mathbf{M}_{23}\\ 0 & \mathbf{M}_{23}^{\mathrm{T}} & \mathbf{M}_{33}\end{bmatrix}\right)\begin{Bmatrix}\dot{\mathbf{q}}_{\mathrm{u}}\\ \dot{\mathbf{q}}_{\mathrm{w}}\\ \dot{\mathbf{q}}_{\theta}\end{Bmatrix}+\begin{bmatrix}\mathbf{K}_{11} & 0 & 0\\ 0 & \mathbf{K}_{22}+\mathbf{K}_{\ell}^{W}-\mathbf{K}_{\ell}^{P} & \mathbf{K}_{23}\\ 0 & \mathbf{K}_{23}^{\mathrm{T}} & \mathbf{K}_{33}\end{bmatrix}\begin{Bmatrix}\mathbf{q}_{\mathrm{u}}\\ \mathbf{q}_{\mathrm{w}}\\ \mathbf{q}_{\theta}\end{Bmatrix}$$

$$+\begin{bmatrix}0 & \mathbf{K}_{n\ell}^{2}(\mathbf{q}_{\mathrm{w}}) & 0\\ \mathbf{K}_{n\ell}^{3}(\mathbf{q}_{\mathrm{w}}) & \mathbf{K}_{n\ell}^{4}(\mathbf{q}_{\mathrm{w}})+\mathbf{K}_{n\ell}^{W}(\mathbf{q}_{\mathrm{w}}) & 0\\ 0 & 0 & 0\end{bmatrix}\begin{Bmatrix}\mathbf{q}_{\mathrm{u}}\\ \mathbf{q}_{\mathrm{w}}\\ \mathbf{q}_{\theta}\end{Bmatrix}=\begin{Bmatrix}\mathbf{F}_{\mathrm{u}}^{\mathrm{E}}\\ \mathbf{F}_{\mathrm{w}}^{\mathrm{E}}\\ \mathbf{M}^{\mathrm{E}}\end{Bmatrix}. \tag{11.31}$$

These four cases—Eqs. (11.26), Eqs. (11.29–11.31) —are analyzed and the results are discussed.

11.4 Natural frequencies and the stiffness function

In the upcoming analysis, the p-FEM will be used on various clamped–clamped (CC) beams, with the following geometric and material properties: $L = 406$ mm, $b = 20$ mm, $h = 20.3$ mm, $E = 7.172$ GPa, $\rho = 2800\ \mathrm{kg{\cdot}m^{-3}}$. These geometric and material properties for the beam were chosen since the results relating to this beam performance were experimentally verified in comparison with the p-FEM [16] for the instance with damage, and for the instance without damage [31]. The beam is experimentally tested in the nonlinear frequency–amplitude dependence. It is to be noted that the results obtained by the improved p-FEM (with two shape functions adjusted for the location of the damage) of the damaged beam [16] are compared with the experimental results for the third case of the damaged beam, submitted in Table 11.1 in Ref. [5]. Agreement with the experimental results, submitted in Table 5 in [5], are shown in Table 11.1 in Ref. [16] using ANSYS® commercial software, and

Table 11.1 Natural frequencies [Hz] for the four cases of the Timoshenko beam models (Winkler foundation); l_1=68 mm, l_2=75.5 mm, k_1=36.628·10^4 Pa; depth of damage: h_1=9 mm.

Mode	Intact beam [16]	Damaged beam [16]	Damaged beam on the elastic foundation	Intact beam on the elastic foundation
1	629.74	623.15	629.74	636.51
2	1700.75	1620.58	1623.12	1703.12
3	3246.35	3227.21	3228.46	3247.59
4	5198.96	5047.80	5048.55	5199.73

in it BEAM189 elements (beam was defined by 100 elements). Some of the comparable results are; ANSYS=74.923 Hz, Classical p-FEM=75.012 Hz, Improved p-FEM=74.924 Hz, Experimental Ref. [5] =74.688 Hz. After enough numerical tests are performed using the adopted p-FEM on damaged beams, the same method with the same number of shape functions is used on the beam that is experimentally tested [45] (using the same geometric and material properties, the results are compared in the nonlinear vibrations). Damage to this experimentally tested beam is proportional to damaged beams that have also been experimentally tested [5]. It should be pointed out that in the present study, a Pasternak foundation is used and it is also analyzed with the same number of shape functions (degrees of freedom). The model with the Pasternak foundation is tested using a large number of degrees of freedom (100 elements). For this, the commercial software ANSYS® is used, and the acquired results agree with the results in [40], which are used for verification.

Considering that the damage reduces the value of natural frequencies, by choosing the adequate stiffness of the elastic foundation, the first natural frequency can be settled somewhere between the natural frequency of the damaged beam on the elastic foundation and the intact beam without the elastic foundation. By doing a large number of numerical experiments, the presented functions are determined. Using the obtained numerical results, interpolation curves are computed, and their functions have the following forms.

11.4.1 Stiffness functions for compensation of the Timoshenko beam model laid on a Winkler foundation

Location of damage: l_1 = 28 mm; l_2 = 35.5 mm

$$f_{stiff,1}(p) = -630. \times 10^4 p^4 - \frac{2.12 \times 10^9}{1.93 \times 10^9 p^2 - 1.82 \times 10^9} - 1.03 \times 10^7 p$$

$$- \frac{1.01 \times 10^9}{1.93 \times 10^9 p - 1.87 \times 10^9} + 1.66 \times 10^7. \qquad (11.32)$$

Location of damage: $l_1 = 38$ mm; $l_2 = 45.5$ mm

$$f_{stiff,2}(p) = 3.54 \times 10^7 p^4 - 1.06 \times 10^8 p^3 + 7.08 \times 10^7 p^2 - 0.9375p. \tag{11.33}$$

Location of damage: $l_1 = 48$ mm; $l_2 = 55.5$ mm

$$f_{stiff,3}(p) = -719. \times 10^4 p^3 - \frac{220.}{p^3} - 2.69 \times 10^7 p^2 + 3.98 \times 10^7 p - 569. \times 10^4. \tag{11.34}$$

Location of damage: $l_1 = 58$ mm; $l_2 = 65.5$ mm

$$f_{stiff,4}(p) = -8.07 \times 10^7 p - \frac{4.51 \times 10^7}{p} + 1.25 \times 10^8. \tag{11.35}$$

Location of damage: $l_1 = 68$ mm; $l_2 = 75.5$ mm

$$f_{stiff,5}(p) = -8.44 \times 10^7 p + \frac{761. \times 10^3}{4.88 \times 10^7 - 4.93 \times 10^7 p} - \frac{4.88 \times 10^7}{p} + 1.33 \times 10^8. \tag{11.36}$$

11.4.2 Stiffness functions for compensation of the Reddy–Bickford beam model laid on a Pasternak foundation have the following forms

Location of damage: $l_1 = 28$ mm; $l_2 = 35.5$ mm

$$f_{stiff,1}(p) = 4.90 \times 10^8 p^4 - 1.81 \times 10^9 p^3 + 2.49 \times 10^9 p^2 - 1.52 \times 10^9 p + 3.61 \times 10^8, \tag{11.37}$$

Location of damage: $l_1 = 38$ mm; $l_2 = 45.5$ mm

$$f_{stiff,2}(p) = 5.49 \times 10^8 p^4 - 1.98 \times 10^9 p^3 + 2.65 \times 10^9 p^2 - 1.57 \times 10^9 p + 3.60 \times 10^8, \tag{11.38}$$

Location of damage: $l_1 = 48$ mm; $l_2 = 55.5$ mm

$$f_{stiff,3}(p) = 5.14 \times 10^8 p^4 - 1.88 \times 10^9 p^3 + 2.55 \times 10^9 p^2 - 1.53 \times 10^9 p + 3.54 \times 10^8, \tag{11.39}$$

Location of damage: $l_1 = 58$ mm; $l_2 = 65.5$ mm

$$f_{stiff,4}(p) = 4.49 \times 10^8 p^4 - 1.57 \times 10^9 p^3 + 2.01 \times 10^9 p^2 - 1.13 \times 10^9 p + 2.42 \times 10^8, \tag{11.40}$$

Location of damage: $l_1 = 68$ mm; $l_2 = 75.5$ mm

$$f_{stiff,5}(p) = 1.71 \times 10^9 p^4 - 6.23 \times 10^9 p^3 + 8.45 \times 10^9 p^2 - 5.07 \times 10^9 p + 1.14 \times 10^9. \tag{11.41}$$

Some selected functions are displayed as a dependency of the required foundation stiffness with respect to the ratio of natural frequencies between the damaged and the intact beam. The depth of damage is included through the ratio of natural frequencies between the damaged and the intact beam. Fig. 11.3 shows these functions for the five selected damage locations. According to the functions shown in Fig. 11.3, it is feasible to select a specific value of the stiffness of the elastic foundation that will make the damaged beam on the elastic foundation have the same fundamental natural frequency as the intact beam. On the other hand, when it comes to nonlinear forced vibrations, where there is a possibility of interaction among the higher modes of oscillations, a correction factor is needed. The purpose of that correction factor is to increase the stiffness of the elastic support so that the maximum amplitudes of vibrations of the damaged beam on the elastic foundation are drawn closer to the maximum amplitudes of the intact beam. Based on the numerical experiments, the value of the correction factor is determined to be $k_{cor} = 106.34$. Based on the further conducted nonlinear analysis of four cases shown in Fig. 11.3, the natural frequencies for the diverse models are presented in Tables 11.1 and 11.2.

As was mentioned, a certain value of stiffness of the elastic foundation can increase the first natural frequency of the damaged beam on the elastic foundation,

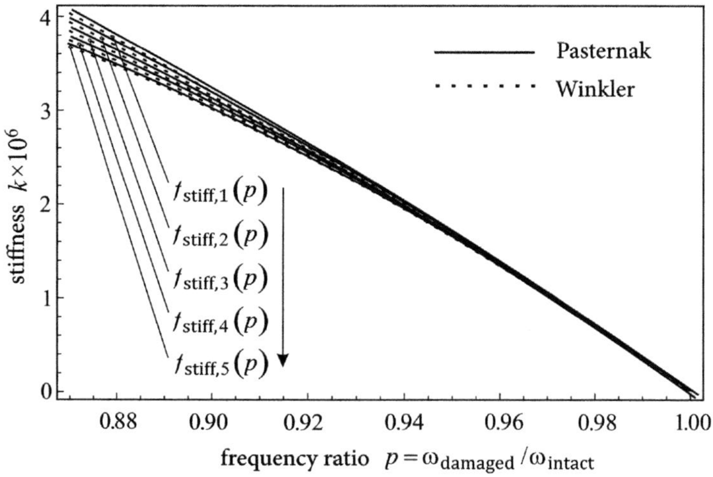

Figure 11.3 Stiffness of the elastic support. The stiffness needs to change with the change of frequency ratio between the damaged and the intact beam.

Table 11.2 Natural frequencies [Hz] for the four cases of the Reddy–Bickford beam models (Pasternak foundation); l_1=68 mm, l_2=75.5 mm, k_1=37.182×10^4 Pa; depth of damage: h_1=9 mm.

Mode	Intact beam [16]	Damaged beam [16]	Damaged beam on the elastic foundation	Intact beam on the elastic foundation
1	629.73	626.14	629.74	636.14
2	1698.87	1642.46	1622.78	1701.24
3	3240.82	3230.97	3231.21	3242.05
4	5187.14	5083.14	5062.14	5187.90

and by that, equate it with the first natural frequency of the intact beam. For vibrations in the nonlinear regime, there is a need for the correction factor of stiffness $k_{nl} = k_{cor}k_l$. This factor draws closer to each other the maximal amplitudes of vibrations between the damaged beam on the elastic foundation and the intact beam without a foundation. Analysis of the nonlinear vibrations of the intact beam on the elastic foundation (case four) is presented to give a comparison of some general aspects of the vibrations of the beams on the elastic foundations.

11.5 Forced geometrically nonlinear vibrations in the time domain and the Newmark method

In this section, forced nonlinear vibrations are examined by means of numerical simulations in the time domain. For all four cases (Tables 11.1 and 11.2 and Figs. 11.1 and 11.2), the equations of motion are solved by the Newmark method. The results obtained are for the models of a damaged Timoshenko beam [16] laid on a Winkler nonlinear foundation and a damaged Reddy–Bickford beam laid on a Pasternak nonlinear foundation. The applied force changes harmonically in time—$f(t) = F\cos(\omega_e t)$ —where ω_e is the excitation frequency and F is the excitation amplitude. For the case of an intact beam without an elastic foundation (with the properties as listed in the previous section), amplitudes of the response to harmonic excitations were measured in laboratory experiments [5,45] and numerically analyzed [16,31]. In Fig. 11.4, the deformed shapes and maximal deflections of the beams for different models are shown. It can be noted that the maximal amplitude of the damaged beam laid on the elastic foundation is nearly equal to the case of the intact beam. This suggests that the damaged beam may vibrate similarly to the intact beam if the right stiffness of the elastic foundation is determined.

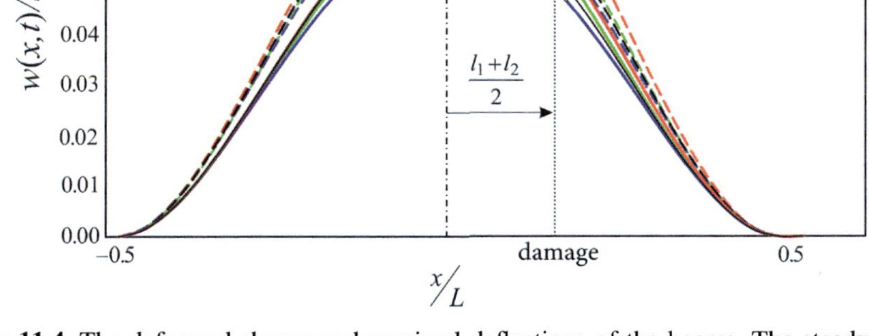

Figure 11.4 The deformed shapes and maximal deflections of the beams. The steady-state regime at x=0 for F=10 N and ω_e=1.02 ω_{11}.

Although the peak value of the amplitude of the damaged beam on the elastic foundation is near the peak amplitude of the intact beam, in the damaged region of the beam, some deviations in deflections still exist. The elastic foundation causes mitigation of the amplitude, so the intact beam on the elastic foundation has a smaller amplitude of vibration compared to the intact beam without the foundation. The time histories of the aforementioned four beam models are presented in Figs. 11.5–11.7. A conclusion emerges from the results: the relevant difference among the amplitudes between the damaged and intact beam, Fig. 11.5, is fully compensated for by the elastic foundation (see Fig. 11.6). In addition, better approximations are obtained via the Reddy–Bickford theory. In Fig. 11.6 (for forced vibrations of the damaged beam on the elastic foundation), it is clear that there is no steep increase in the amplitude at the beginning. Fig. 11.7 demonstrates the reduction of the maximum amplitude—for the cases of the intact beam—when the elastic foundation is added (the foundation has the same elastic stiffness as the foundation that was used for the dynamical balancing of the damaged beam).

The phase plots for the four studied cases are presented in Figs. 11.8 and 11.9, for the location in the middle of the damage $x = (l_1 + l_2)/2$. The plots for the intact and damaged beam without a foundation are in Fig. 11.8, while the plots for the intact and damaged beam laid on an elastic foundation (the Timoshenko beam is laid on the Winkler layer, and the Reddy–Bickford beam is laid on the Pasternak layer) are in Fig. 11.9. Furthermore, Fig. 11.8 indicates the difference in deviation between the top and bottom of the beam. The reason for this asymmetrical response is the location of the damage, which is at the upper part of the beam. This is not an evident effect when the phase plot is observed at $x = 0$. In Fig. 11.9, it can be seen that the stabilization properties of the elastic foundation are lowering the speed of movement of the points

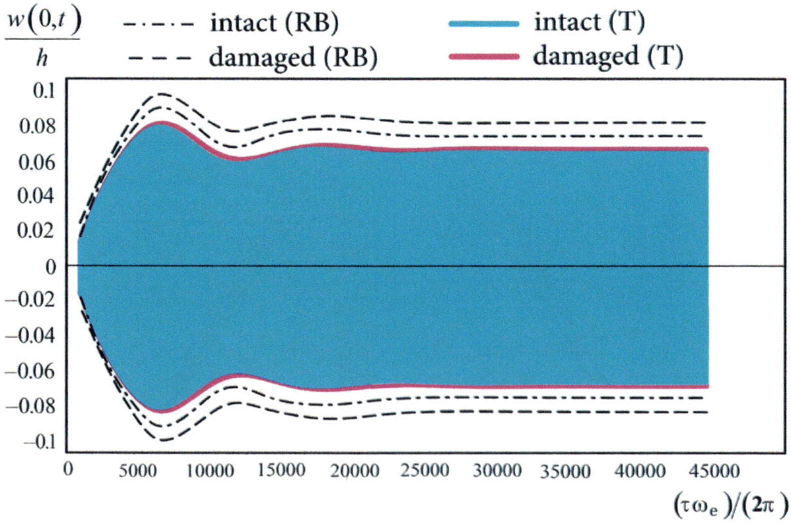

Figure 11.5 Time histories. The intact and damaged Timoshenko beam and the intact and damaged Reddy–Bickford beam at x=0 midpoint excitation for F=10 N and ω_e=1.02 ω_{11}.

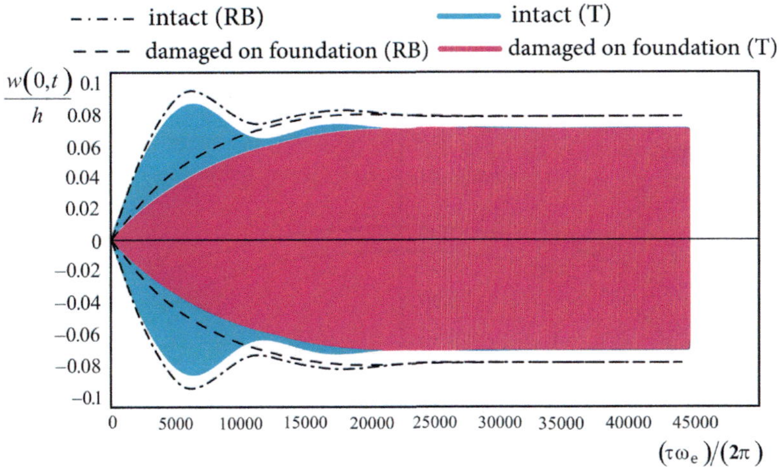

Figure 11.6 Time histories. The intact and damaged Timoshenko beam on the Winkler foundation and the intact and damaged Reddy–Bickford beam on the Pasternak foundation at x=0 midpoint excitation for F=10 N and ω_e=1.02 ω_{11}.

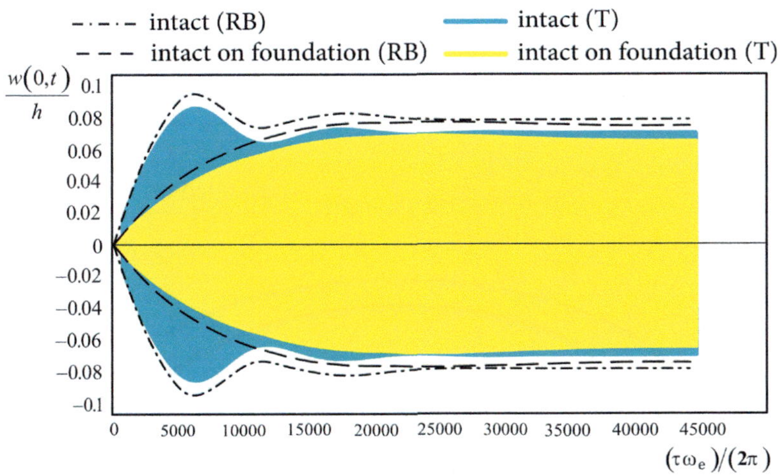

Figure 11.7 Time histories. The intact Timoshenko beam without foundation and on the Winkler foundation, and the intact Reddy–Bickford beam without foundation and on the Pasternak foundation at x=0 midpoint excitation for F=10 N and ω_e=1.02 ω_{11}.

located at the site of the damage, and the speed is very close to the speed of the same point on the intact beam.

The time histories in the steady-state regime of the compared models are shown in Figs. 11.10–11.12, and they are for the location in the middle of the damage $x = L_1 = (l_1 + l_2)/2$. As mentioned, the compared models are the intact and damaged

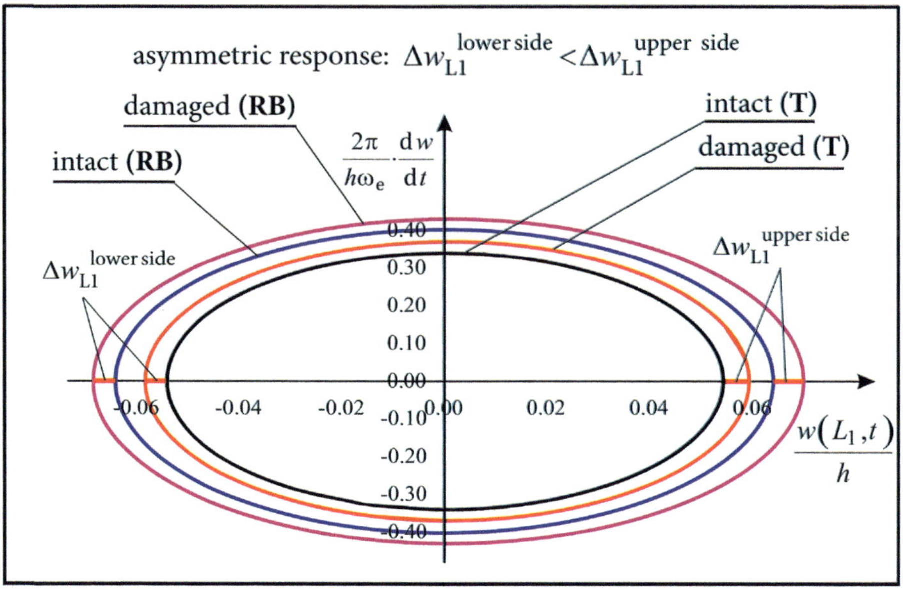

Figure 11.8 Phase plots. The intact and damaged beam in the steady-state regime at $x=(l_1+ l_2)/2$ for $F=10$ N and $\omega_e=1.02\ \omega_{11}$.

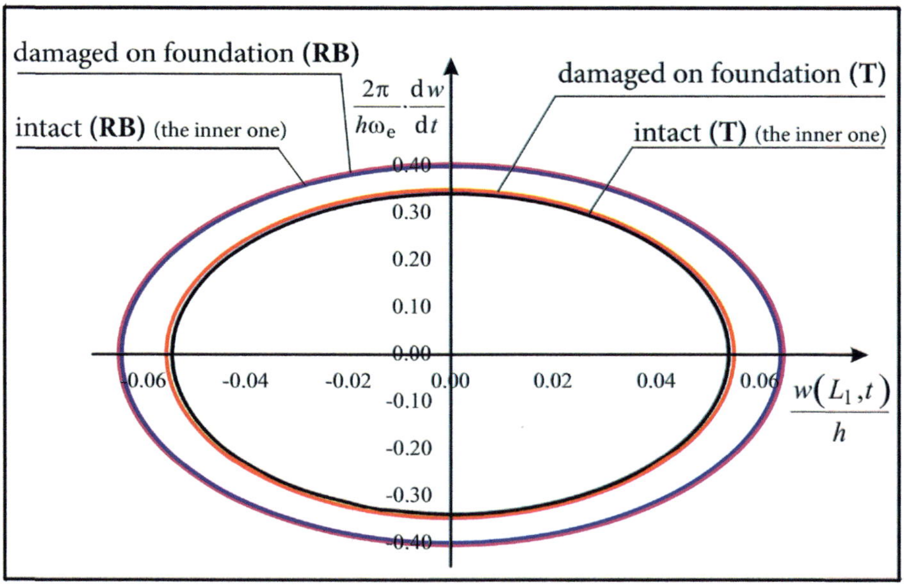

Figure 11.9 Phase plots. The intact and damaged beam on a foundation in the steady-state regime at $x=(l_1+ l_2)/2$ for $F=10$ N and $\omega_e=1.02\ \omega_{11}$.

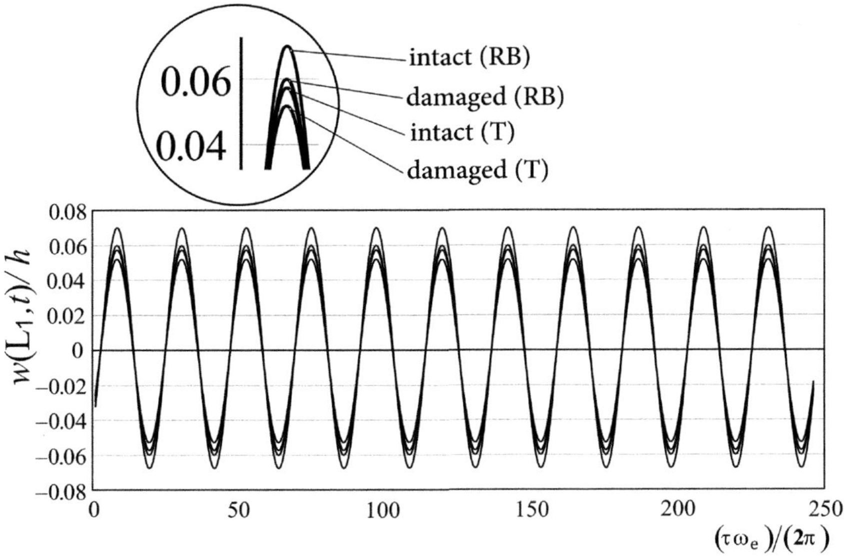

Figure 11.10 Time histories. The intact and damaged beam in the steady-state regime at $x=(l_1+l_2)/2$ mid-point excitation for F=10 N and ω_e=1.02 ω_{11}.

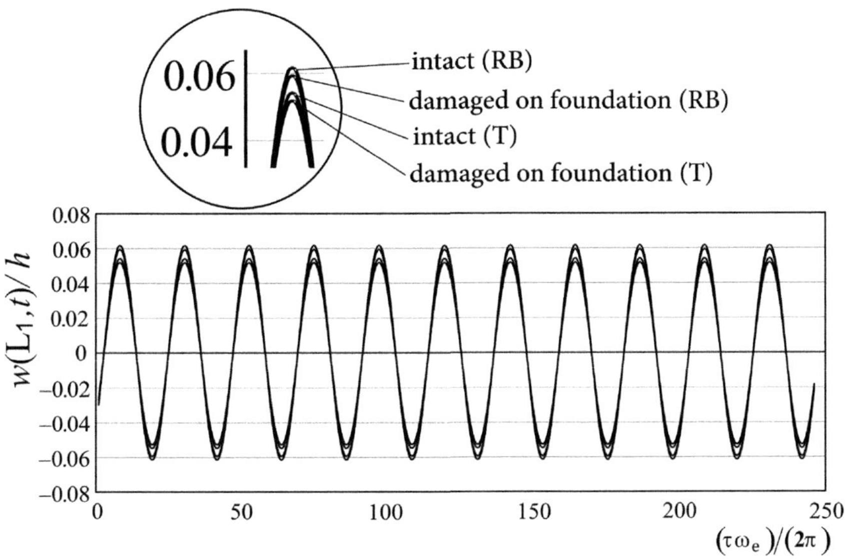

Figure 11.11 Time histories. The intact and damaged beam on the foundation in the steady-state regime at $x=(l_1+l_2)/2$ mid-point excitation for F=10 N and ω_e=1.02 ω_{11}.

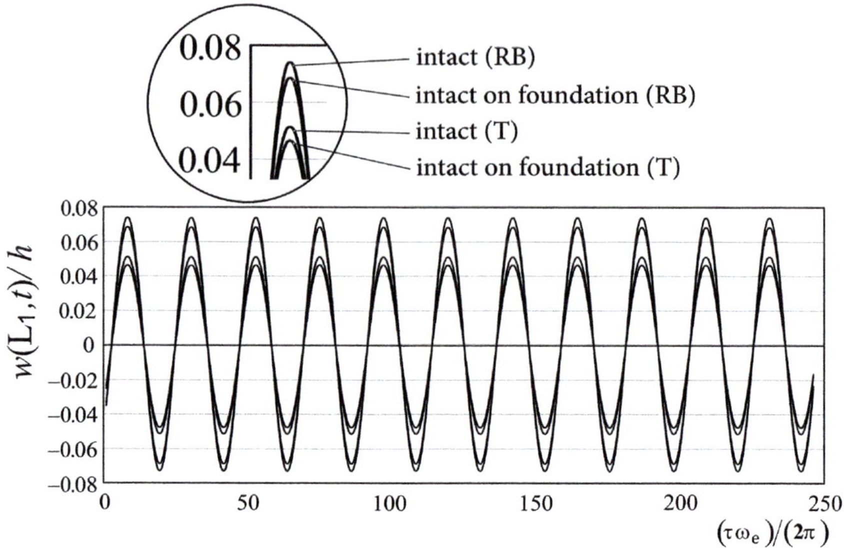

Figure 11.12 Time histories. The intact and damaged beam on the foundation in the steady-state regime at $x=(l_1+l_2)/2$ mid-point excitation for F=10 N and ω_e=1.02 ω_{11}.

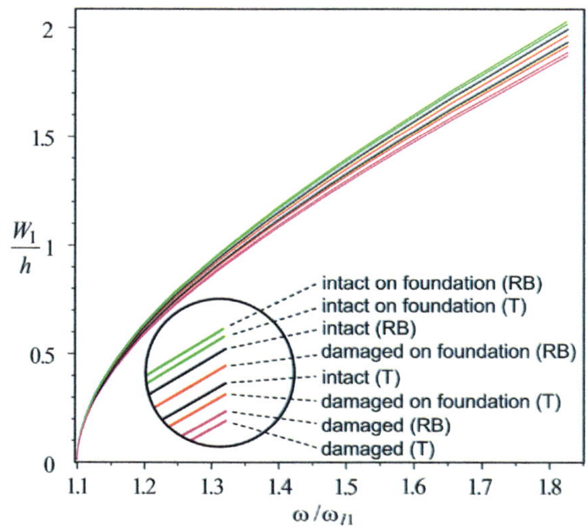

Figure 11.13 Backbone curves. The four models of the CC beam: intact, damaged, damaged on the foundation, and intact on the foundation beam. Amplitude W_1 calculated at x=0.

beams, the intact and damaged beams on a foundation, and the intact beams without and with an elastic foundation, respectively. The separate steady-state mode of vibration shown in this way provides a better understanding of the conclusions drawn from the diagrams (Figs. 11.5–11.7). Better approximations are made using the

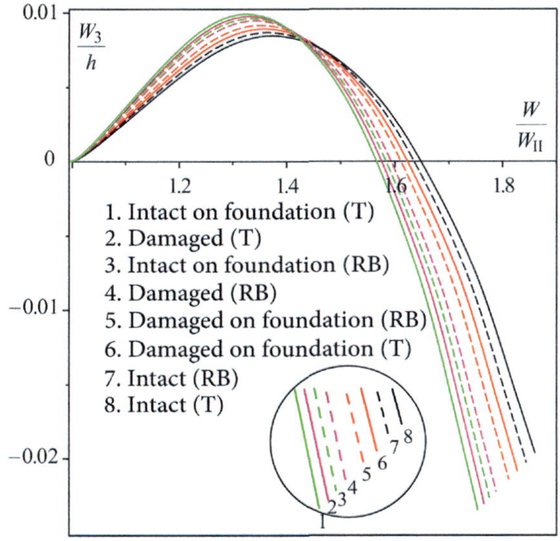

Figure 11.14 Backbone curves. The four models of the CC beam: intact, damaged, damaged on the foundation, and intact on the foundation beam. Amplitude W_3 calculated at x=0.

Reddy–Bickford beam assumption and a nonlinear Pasternak foundation. It should be pointed out that by selecting an appropriate stiffness for the support, it is feasible to obtain almost the same deflections as those of the damaged and intact beams. With a change in external force, there are no qualitative changes in the nonlinear response among intact and damaged beams on a foundation. This is significant because the determined functions of stiffness compensation presented in Fig. 11.3 are in relation to the level of damage and apply to any external concentrated harmonic force acting on the beam.

11.6 Geometrically nonlinear free vibrations in the frequency domain and the Continuation method

The focus of this section is the nonlinear free vibration analysis of the damaged Reddy–Bickford beam laid on a nonlinear Pasternak foundation. When it comes to the interaction of higher modes, stabilization properties of the foundation do exist. This is confirmed by the results obtained from the same examples of beams shown in the earlier section. For free vibration analysis, the harmonic balance method is used. The system of nonlinear algebraic equations with frequency variation is solved with the continuation method. The linearlized updated-mode method (LUM) method [46] is not very fitting for achieving convergence of multiharmonic free vibrations. For that reason, a continuation method is put to use in most of the analyses. Two main loops make the continuation method. A predictor of the solution is defined in the external loop. Prediction of the vector of generalized coordinates is acquired through the last two determined vectors of the curve, $\{\mathbf{w}\}_i$ and $\{\mathbf{w}\}_{i-1}$, as follows:

$$\{\mathbf{w}\}_{i+1} = \{\mathbf{w}\}_i + \Delta\{\mathbf{w}\}_{i+1},\ \Delta\{w\}_{i+1} = (\{\mathbf{w}\}_i - \{\mathbf{w}\}_{i-1})\frac{\mathrm{d}wa}{\Delta w}. \tag{11.42}$$

where $\Delta\{\mathbf{w}\}_{i+1}$ and Δw are the amplitudes of the vectors $(\{\mathbf{w}\}_i - \{\mathbf{w}\}_{i-1})$, and d$wa$ is the amplitude of the first increment vector. A prediction ω_{i+1}^2, which is the square of the frequency of vibration, is calculated using the equation $\omega_{i+1}^2 = \omega_i^2 + \Delta\omega_0^2$, where $\Delta\omega_0^2 = \pm s/\sqrt{\{\delta\mathbf{w}\}_1^{\mathrm{T}}\{\delta\mathbf{w}\}_1}$. The sign is chosen to be the same as the previous increment unless the determinant $|\mathbf{J}|$, of the Jacobian of $\{F\}$, $[\mathbf{J}]$, has changed the sign. In the last case, the sign reversal is applied. In the following equation:

$$[\mathbf{J}]\{\delta\mathbf{w}\}_1 = [\mathbf{M}]\{\mathbf{w}\}_{i+1}, \tag{11.43}$$

$\{\delta\mathbf{w}\}_1$ is defined according to the arc length, s. In an internal loop, the approximated solution gets corrected. If a system consisting of a damaged beam on an elastic foundation (not affected by external forces $\delta W_{ex} = 0$) is taken into consideration, then by applying the virtual work principle, the following is obtained:

$$\left[\begin{bmatrix} \mathbf{M}_{11} & \mathbf{0} & \mathbf{0} \\ \mathbf{0} & \mathbf{M}_{22} & \mathbf{M}_{23} \\ \mathbf{0} & \mathbf{M}_{23}^{\mathrm{T}} & \mathbf{M}_{33} \end{bmatrix} + \begin{bmatrix} \mathbf{0} & \mathbf{M}^{lt} & \mathbf{M}^{lr} \\ \mathbf{M}^{lt\mathrm{T}} & \mathbf{0} & \mathbf{0} \\ \mathbf{M}^{lr\mathrm{T}} & \mathbf{0} & \mathbf{0} \end{bmatrix}\right]\begin{Bmatrix} \ddot{\mathbf{q}}_{\mathbf{u}} \\ \ddot{\mathbf{q}}_{\mathbf{w}} \\ \ddot{\mathbf{q}}_{\theta} \end{Bmatrix}$$
$$+\left[\begin{bmatrix} \mathbf{K}_{11} & \mathbf{0} & \mathbf{0} \\ \mathbf{0} & \mathbf{K}_{22} + \mathbf{K}^{W} - \mathbf{K}^{P} & \mathbf{K}_{23} \\ \mathbf{0} & \mathbf{K}_{23}^{\mathrm{T}} & \mathbf{K}_{33} \end{bmatrix} + \begin{bmatrix} \mathbf{0} & \mathbf{K}^{lt} & \mathbf{K}^{lr} \\ \mathbf{K}^{lt\mathrm{T}} & \mathbf{0} & \mathbf{0} \\ \mathbf{K}^{lr\mathrm{T}} & \mathbf{0} & \mathbf{0} \end{bmatrix}\right]\begin{Bmatrix} \mathbf{q}_{\mathbf{u}} \\ \mathbf{q}_{\mathbf{w}} \\ \mathbf{q}_{\theta} \end{Bmatrix}$$
$$+\left[\begin{bmatrix} \mathbf{0} & \mathbf{K}_{n\ell}^{2}(\mathbf{q}_{\mathbf{w}}) & \mathbf{0} \\ \mathbf{K}_{n\ell}^{3}(\mathbf{q}_{\mathbf{w}}) & \mathbf{K}_{n\ell}^{4}(\mathbf{q}_{\mathbf{w}}) + \mathbf{K}_{n\ell}^{W}(\mathbf{q}_{\mathbf{w}}) & \mathbf{0} \\ \mathbf{0} & \mathbf{0} & \mathbf{0} \end{bmatrix} + \begin{bmatrix} \mathbf{0} & \mathbf{0} & \mathbf{0} \\ \mathbf{0} & \mathbf{0} & \mathbf{K}_{n\ell}^{br}(\mathbf{q}_{\mathbf{w}}) \\ \mathbf{0} & \mathbf{K}_{n\ell}^{br}(\mathbf{q}_{\mathbf{w}})^{\mathrm{T}} & \mathbf{0} \end{bmatrix}\right]\begin{Bmatrix} \mathbf{q}_{\mathbf{u}} \\ \mathbf{q}_{\mathbf{w}} \\ \mathbf{q}_{\theta} \end{Bmatrix}$$
$$= \{\mathbf{0}\ \ \mathbf{0}\ \ \mathbf{0}\}^{\mathrm{T}}. \tag{11.44}$$

The corrections $\{\delta\mathbf{w}\}$ and $\delta\omega^2$ are acquired by applying Newton's method to Eq. (11.44).

$$[\,\mathbf{J}\,]\{\delta\mathbf{w}\}_1 - [\mathbf{M}]\{\mathbf{w}\}_{i+1}\delta\omega^2 = -\{\mathbf{F}\}. \tag{11.45}$$

Variations in frequency are taken into account in Eq. (11.45), and, for that reason, there is a need for one more equation. This is achieved by setting the distance between the two consecutive points of the frequency response function curve (the arc-length s) to a fixed value $s^2 = \Delta\{\mathbf{w}\}_{i+1}$. The products $[\mathbf{M}^{l}]\{\ddot{\mathbf{q}}_{\mathbf{u}}\}$, $[\mathbf{M}^{lr}]\{\ddot{\mathbf{q}}_{\theta}\}$, and

$[\mathbf{M}^{lr\mathrm{T}}]\{\ddot{\mathbf{q}}_{\mathbf{u}}\}$ express minute quantities and do not affect the solution in a nonlinear model [30] and, therefore, can be discarded.

$$\begin{bmatrix} \mathbf{m}_{11} & \mathbf{0} \\ \mathbf{0} & \mathbf{m}_{22} \end{bmatrix} \begin{Bmatrix} \ddot{\mathbf{q}}_{\mathbf{w}} \\ \ddot{\mathbf{q}}_{\theta} \end{Bmatrix} + \begin{bmatrix} \mathbf{k}_{11} & \mathbf{k}_{12} \\ \mathbf{k}_{21} & \mathbf{k}_{22} \end{bmatrix} \begin{Bmatrix} \mathbf{q}_{\mathbf{w}} \\ \mathbf{q}_{\theta} \end{Bmatrix} + \begin{bmatrix} \mathbf{knl}_{11}(\mathbf{q}_{\mathbf{w}}) & \mathbf{knl}_{12}(\mathbf{q}_{\mathbf{w}}) \\ \mathbf{knl}_{21}(\mathbf{q}_{\mathbf{w}}) & \mathbf{0} \end{bmatrix} \begin{Bmatrix} \mathbf{q}_{\mathbf{w}} \\ \mathbf{q}_{\theta} \end{Bmatrix} = \begin{Bmatrix} \mathbf{0} \\ \mathbf{0} \end{Bmatrix}. \tag{11.46}$$

where

$$\mathbf{m}_{11} = \mathbf{M}_{22},\ \mathbf{m}_{12} = \mathbf{M}_{23},\ \mathbf{k}_{11} = \widetilde{\mathbf{K}}_{22} - \mathbf{K}^{lt\mathrm{T}}\mathbf{K}_{11}^{-1}\mathbf{K}^{lt},$$

$$\mathbf{k}_{12} = \mathbf{K}_{23} - \mathbf{K}^{lt\mathrm{T}}\mathbf{K}_{11}^{-1}\mathbf{K}^{lr},$$

$$\mathbf{knl}_{11}(\mathbf{q}_{\mathbf{w}}) = \mathbf{K}_{n\ell}^{4}(\mathbf{q}_{\mathbf{w}}) - \mathbf{K}^{lt\mathrm{T}}\mathbf{K}_{11}^{-1}\mathbf{K}_{n\ell}^{2}(\mathbf{q}_{\mathbf{w}}) - \mathbf{K}_{n\ell}^{3}(\mathbf{q}_{\mathbf{w}})\mathbf{K}_{11}^{-1}\mathbf{K}^{lt} - \mathbf{K}_{n\ell}^{3}(\mathbf{q}_{\mathbf{w}})\mathbf{K}_{11}^{-1}\mathbf{K}_{n\ell}^{2}(\mathbf{q}_{\mathbf{w}}),$$

$$\mathbf{knl}_{12}(\mathbf{q}_{\mathbf{w}}) = \mathbf{K}_{n\ell}^{br} - \mathbf{K}_{n\ell}^{3}(\mathbf{q}_{\mathbf{w}})\mathbf{K}_{11}^{-1}\mathbf{K}^{lr},$$

$$\mathbf{m}_{21} = \mathbf{M}_{23}^{\mathrm{T}},\ \mathbf{m}_{22} = \mathbf{M}_{33},$$

$$\mathbf{k}_{21} = \mathbf{K}_{23}{}^{\mathrm{T}} - \mathbf{K}^{lr\mathrm{T}}\mathbf{K}_{11}^{-1}\mathbf{K}^{lt},\ \ \mathbf{k}_{22} = \mathbf{K}_{33} - \mathbf{K}^{lr\mathrm{T}}\mathbf{K}_{11}^{-1}\mathbf{K}^{lr},$$

$$\mathbf{knl}_{21}(\mathbf{q}_{\mathbf{w}}) = \mathbf{K}_{n\ell}^{lr\mathrm{T}}(\mathbf{q}_{\mathbf{w}}) - \mathbf{K}^{lr\mathrm{T}}\mathbf{K}_{11}^{-1}\mathbf{K}_{n\ell}^{lr\mathrm{T}}(\mathbf{q}_{\mathbf{w}}).$$

The vector of generalized coordinates and their second derivative (acceleration) can be represented as a summation of the first three members of a trigonometric series (based on the results of the Fourier spectrum, the first three members are enough) in the following form:

$$\begin{Bmatrix} \mathbf{q}_{\mathbf{w}}(t) \\ \mathbf{q}_{\theta}(t) \end{Bmatrix} = \frac{1}{2}\begin{Bmatrix} Q_{W_{c0}} \\ Q_{\theta_{c0}} \end{Bmatrix} + \cos(\omega t)\begin{Bmatrix} Q_{W_{c1}} \\ Q_{\theta_{c1}} \end{Bmatrix} + \cos(2\omega t)\begin{Bmatrix} Q_{W_{c2}} \\ Q_{\theta_{c2}} \end{Bmatrix} + \cos(3\omega t)\begin{Bmatrix} Q_{W_{c3}} \\ Q_{\theta_{c3}} \end{Bmatrix}, \tag{11.47}$$

$$\begin{Bmatrix} \ddot{\mathbf{q}}_{\mathbf{w}}(t) \\ \ddot{\mathbf{q}}_{\theta}(t) \end{Bmatrix} = -\omega^2\cos(\omega t)\begin{Bmatrix} Q_{W_{c1}} \\ Q_{\theta_{c1}} \end{Bmatrix} - 4\omega^2\cos(2\omega t)\begin{Bmatrix} Q_{W_{c2}} \\ Q_{\theta_{c2}} \end{Bmatrix} - 9\omega^2\cos(3\omega t)\begin{Bmatrix} Q_{W_{c3}} \\ Q_{\theta_{c3}} \end{Bmatrix}. \tag{11.48}$$

where ω represents the natural frequency of the system. The vector of new unknowns has become

$$\{\mathbf{Q}^{\mathrm{HBM}}\} = \left\{Q_{W_{c0}}\ \ Q_{W_{c1}}\ \ Q_{W_{c2}}\ \ Q_{W_{c3}}\ \ Q_{\theta_{c0}}\ \ Q_{\theta_{c1}}\ \ Q_{\theta_{c2}}\ \ Q_{\theta_{c3}}\right\}^{\mathrm{T}}. \tag{11.49}$$

If the expressions (Eq. 11.47) and (Eq. 11.48) are substituted into the initial set of equations of motion (Eq. 11.46) the following is obtained:

$$(-\omega^2\mathbf{M}^{\mathrm{HBM}} + \mathbf{KL}^{\mathrm{HBM}} + \mathbf{KNL}^{\mathrm{HBM}})\{\mathbf{Q}^{\mathrm{HBM}}\} = \{\mathbf{0}\}, \tag{11.50}$$

where

$$\mathbf{M}^{\mathrm{HBM}} = \begin{bmatrix} 0 & 0 & 0 & 0 & 0 & 0 & 0 & 0 \\ 0 & \mathbf{m}_{11} & 0 & 0 & 0 & 0 & 0 & 0 \\ 0 & 0 & 4\mathbf{m}_{11} & 0 & 0 & 0 & 0 & 0 \\ 0 & 0 & 0 & 9\mathbf{m}_{11} & 0 & 0 & 0 & 0 \\ 0 & 0 & 0 & 0 & 0 & 0 & 0 & 0 \\ 0 & 0 & 0 & 0 & 0 & \mathbf{m}_{22} & 0 & 0 \\ 0 & 0 & 0 & 0 & 0 & 0 & 4\mathbf{m}_{22} & 0 \\ 0 & 0 & 0 & 0 & 0 & 0 & 0 & 9\mathbf{m}_{22} \end{bmatrix},$$

$$\mathbf{KL}^{\mathrm{HBM}} = \begin{bmatrix} (1/2)\mathbf{k}_{11} & 0 & 0 & 0 & (1/2)\mathbf{k}_{12} & 0 & 0 & 0 \\ 0 & \mathbf{k}_{11} & 0 & 0 & 0 & \mathbf{k}_{12} & 0 & 0 \\ 0 & 0 & \mathbf{k}_{11} & 0 & 0 & 0 & \mathbf{k}_{12} & 0 \\ 0 & 0 & 0 & \mathbf{k}_{11} & 0 & 0 & 0 & \mathbf{k}_{12} \\ (1/2)\mathbf{k}_{12}^{T} & 0 & 0 & 0 & (1/2)\mathbf{k}_{22} & 0 & 0 & 0 \\ 0 & \mathbf{k}_{12}^{T} & 0 & 0 & 0 & \mathbf{k}_{22} & 0 & 0 \\ 0 & 0 & \mathbf{k}_{12}^{T} & 0 & 0 & 0 & \mathbf{k}_{22} & 0 \\ 0 & 0 & 0 & \mathbf{k}_{12}^{T} & 0 & 0 & 0 & \mathbf{k}_{22} \end{bmatrix},$$

$$\mathbf{KNL}^{\mathrm{HBM}} =$$

$$\begin{bmatrix} \mathbf{KNL1}_{11} & \mathbf{KNL1}_{12} & \mathbf{KNL1}_{13} & \mathbf{KNL1}_{14} & \mathbf{KNL2}_{11} & \mathbf{KNL2}_{12} & \mathbf{KNL2}_{13} & \mathbf{KNL2}_{14} \\ \mathbf{KNL1}_{21} & \mathbf{KNL1}_{22} & \mathbf{KNL1}_{23} & \mathbf{KNL1}_{24} & \mathbf{KNL2}_{21} & \mathbf{KNL2}_{22} & \mathbf{KNL2}_{23} & \mathbf{KNL2}_{24} \\ \mathbf{KNL1}_{31} & \mathbf{KNL1}_{32} & \mathbf{KNL1}_{33} & \mathbf{KNL1}_{34} & \mathbf{KNL2}_{31} & \mathbf{KNL2}_{32} & \mathbf{KNL2}_{33} & \mathbf{KNL2}_{34} \\ \mathbf{KNL1}_{41} & \mathbf{KNL1}_{42} & \mathbf{KNL1}_{43} & \mathbf{KNL1}_{44} & \mathbf{KNL2}_{41} & \mathbf{KNL2}_{42} & \mathbf{KNL2}_{43} & \mathbf{KNL2}_{44} \\ \mathbf{KNL3}_{11} & \mathbf{KNL3}_{12} & \mathbf{KNL3}_{13} & \mathbf{KNL3}_{14} & 0 & 0 & 0 & 0 \\ \mathbf{KNL3}_{21} & \mathbf{KNL3}_{22} & \mathbf{KNL3}_{23} & \mathbf{KNL3}_{24} & 0 & 0 & 0 & 0 \\ \mathbf{KNL3}_{31} & \mathbf{KNL3}_{32} & \mathbf{KNL3}_{33} & \mathbf{KNL3}_{34} & 0 & 0 & 0 & 0 \\ \mathbf{KNL3}_{41} & \mathbf{KNL3}_{42} & \mathbf{KNL3}_{43} & \mathbf{KNL3}_{44} & 0 & 0 & 0 & 0 \end{bmatrix}$$

$$\mathbf{KNL1} = \mathbf{knl}_{11}(\mathbf{q}_w),\ \mathbf{KNL2} = \mathbf{knl}_{12}(\mathbf{q}_w),\ \mathbf{KNL3} = \mathbf{knl}_{21}(\mathbf{q}_w).$$

Matrix $\mathbf{KNL}^{\mathbf{HBM}}$ is given in detail in Appendix 2.

Equations for free geometrically nonlinear vibrations for the rest of the considered models, shown in Figs. 11.1 and 11.2, can be obtained following the same procedure that involves applying the virtual work principle. The amplitude-frequency characteristics of the model's first and third harmonics are presented in Figs. 11.13 and 11.14. It is unequivocal that a nonlinear foundation produces stabilization. One more proof of that is the amplitude-frequency characteristic, which is very similar to that of an intact beam.

11.7 Conclusions

The focus of this chapter was on the model of a Reddy–Bickford beam laid on a nonlinear Pasternak foundation. From there, a system of nonlinear PDEs is created. For the studied model, only numerical solutions were obtainable. To solve the problem of the geometrically nonlinear response of damaged Reddy–Bickford beams laid on a nonlinear viscoelastic foundation, a p-FEM is developed. This p-FEM applies to general boundary conditions, addresses the influence of shear deformation and rotary inertia, and can be used on beams that sustain moderately large deflections. The general PDEs of motion are implemented in the Reddy–Bickford beam assumption; the beam is assumed to be subjected to the joint action of a concentrated or arbitrarily distributed transverse loading and axial loading. The major contribution of the findings presented in this chapter comes from the analysis of making the damaged beams more stable. Stabilization is done by adding a nonlinear foundation (with adequate stiffness) to the damaged beam.

Accounting for the damage and the use of the high-order theory has led to the discovery of coupling effects that occur between the transversal and longitudinal deformations in the linear mass and stiffness matrices.

It was proved that by using a nonlinear viscoelastic foundation as a stabilizer, the maximal transversal deflections of damaged beams could be very close to the maximal transversal deflections of intact beams.

In this regard, a special function for the stiffness of the elastic or viscoelastic layer and its correction factor for a nonlinear regime can be determined. The function will depend on the change in fundamental natural frequency, which results from damage.

A better approximation is achieved using the Reddy–Bickford beam assumption and a Pasternak layer as a foundation. This is concluded by comparing the numerical results of the adopted mathematical models.

Using the harmonic balance and continuation methods, a set of backbone curves of the first and third harmonics, in the frequency domain, are determined. The results justify the use of the elastic layer for the end of achieving a very similar dynamic behavior between the damaged and the intact beams.

The interaction between the higher modes, which is shown through the amplitude-frequency relation, opens the space for further studies. Probably better approximations can be found for stiffness functions. These new stiffness functions may lead to an almost full overlap of results between the damaged and intact beams in the nonlinear mode of vibration.

Appendix 1

$$\mathbf{M}_{11} = \frac{\rho hbl}{2}\int_{-1}^{2l_1/L} \mathbf{N}^{u1\mathrm{T}}\mathbf{N}^{u1}\mathrm{d}\xi + \frac{\rho bl}{2}\int_{2l_1/L}^{(2l_1+x_c)/L}\int_{-h/2}^{\left(\frac{h}{2}\right)-h_1} \mathbf{N}^{u1\mathrm{T}}\mathbf{N}^{u1}\mathrm{d}z\mathrm{d}\xi$$

$$+ \frac{\rho bl}{2}\int_{(2l_1+x_c)/L}^{2l_2/L}\int_{-h/2}^{\left(\frac{h}{2}\right)-h_1} \mathbf{N}^{u2\mathrm{T}}\mathbf{N}^{u2}\mathrm{d}z\mathrm{d}\xi + \frac{\rho hbl}{2}\int_{2l_2/L1}^{1} \mathbf{N}^{u2\mathrm{T}}\mathbf{N}^{u2}\mathrm{d}\xi,$$

$$\mathbf{M}^{lt} = -\rho b\alpha \int_{-\frac{h}{2}}^{\frac{h}{2}} \int_{-1}^{\frac{2l_1}{L}} z^3 \mathbf{N}^{u1\mathrm{T}} \frac{\mathrm{d}\mathbf{N}^{w1}}{\mathrm{d}\xi} \mathrm{d}\xi \mathrm{d}z - \rho b\alpha \int_{2l_1/L}^{(2l_1+x_c)/L} \int_{-h/2}^{\frac{h}{2}-h_1} z^3 \mathbf{N}^{u1\mathrm{T}} \frac{\mathrm{d}\mathbf{N}^{w1}}{\mathrm{d}\xi} \mathrm{d}z \mathrm{d}\xi$$

$$- \rho b\alpha \int_{(2l_1+x_c)/L}^{2l_2/L} \int_{-h/2}^{\frac{h}{2}-h_1} z^3 \mathbf{N}^{u2\mathrm{T}} \frac{\mathrm{d}\mathbf{N}^{w2}}{\mathrm{d}\xi} \mathrm{d}z \mathrm{d}\xi - \rho b\alpha \int_{2l_2/L}^{1} \int_{-h/2}^{h/2} z^3 \mathbf{N}^{u2\mathrm{T}} \frac{\mathrm{d}\mathbf{N}^{w2}}{\mathrm{d}\xi} \mathrm{d}z \mathrm{d}\xi,$$

$$\mathbf{M}^{lr(1)} = \frac{\rho bL}{2} \int_{-h/2}^{h/2} \int_{-1}^{2l_1/L} z\mathbf{N}^{u1\mathrm{T}} \mathbf{N}^{\theta 1} \mathrm{d}\xi \mathrm{d}z + \frac{\rho bL}{2} \int_{2l_1/L}^{(2l_1+x_c)/L} \int_{-h/2}^{\frac{h}{2}-h_1} z\mathbf{N}^{u1\mathrm{T}} \mathbf{N}^{\theta 1} \mathrm{d}\xi \mathrm{d}z$$

$$+ \frac{\rho bL}{2} \int_{(2l_1+x_c)/L}^{2l_2/L} \int_{-h/2}^{\frac{h}{2}-h_1} z\mathbf{N}^{u2\mathrm{T}} \mathbf{N}^{\theta 2} \mathrm{d}\xi \mathrm{d}z + \frac{\rho bL}{2} \int_{2l_2/L}^{1} \int_{-h/2}^{h/2} z\mathbf{N}^{u2\mathrm{T}} \mathbf{N}^{\theta 2} \mathrm{d}\xi \mathrm{d}z,$$

$$\mathbf{M}^{lr(2)} = \frac{\rho bL\alpha}{2} \int_{-h/2}^{h/2} \int_{-1}^{2l_1/L} z^3\mathbf{N}^{u1\mathrm{T}} \mathbf{N}^{\theta 1} \mathrm{d}\xi \mathrm{d}z + \frac{\rho bL\alpha}{2} \int_{2l_1/L}^{(2l_1+x_c)/L} \int_{-h/2}^{\frac{h}{2}-h_1} z^3\mathbf{N}^{u1\mathrm{T}} \mathbf{N}^{\theta 1} \mathrm{d}\xi \mathrm{d}z$$

$$+ \frac{\rho bL\alpha}{2} \int_{(2l_1+x_c)/L}^{2l_2/L} \int_{-h/2}^{\frac{h}{2}-h_1} z\mathbf{N}^{u2\mathrm{T}} \mathbf{N}^{\theta 2} \mathrm{d}\xi \mathrm{d}z + \frac{\rho bL\alpha}{2} \int_{\frac{2l_2}{L}}^{1} \int_{-\frac{h}{2}}^{\frac{h}{2}} z\mathbf{N}^{u2\mathrm{T}} \mathbf{N}^{\theta 2} \mathrm{d}\xi \mathrm{d}z,$$

$$\mathbf{M}^{lr} = \mathbf{M}^{lr(1)} - \mathbf{M}^{lr(2)},$$

$$\mathbf{M}_{22} = \frac{\rho hbl}{2} \int_{-1}^{2l_1/L} \mathbf{N}^{w1\mathrm{T}} \mathbf{N}^{w1} \mathrm{d}\xi + \frac{\rho bl}{2} \int_{2l_1/L}^{(2l_1+x_c)/L} \int_{-h/2}^{\frac{h}{2}-h_1} \mathbf{N}^{w1\mathrm{T}} \mathbf{N}^{w1} \mathrm{d}z \mathrm{d}\xi$$

$$+ \frac{\rho bl}{2} \int_{(2l_1+x_c)/L}^{2l_2/L} \int_{-h/2}^{\frac{h}{2}-h_1} \mathbf{N}^{w2\mathrm{T}} \mathbf{N}^{w2} \mathrm{d}z \mathrm{d}\xi + \frac{\rho hbl}{2} \int_{2l_2/L1}^{1} \mathbf{N}^{w2\mathrm{T}} \mathbf{N}^{w2} \mathrm{d}\xi,$$

$$\mathbf{M}_{23}^{(1)} = \rho b\alpha^2 \int_{-\frac{h}{2}}^{\frac{h}{2}} \int_{-1}^{\frac{2l_1}{L}} z^6 \frac{\mathrm{d}\mathbf{N}^{w1}}{\mathrm{d}\xi}^{\mathrm{T}} \mathbf{N}^{\theta 1} \mathrm{d}\xi \mathrm{d}z + \rho b\alpha^2 \int_{2l_1/L}^{(2l_1+x_c)/L} \int_{-h/2}^{\frac{h}{2}-h_1} z^6 \frac{\mathrm{d}\mathbf{N}^{w1}}{\mathrm{d}\xi}^{\mathrm{T}} \mathbf{N}^{\theta 1} \mathrm{d}\xi \mathrm{d}z$$

$$+ \rho b\alpha^2 \int_{(2l_1+x_c)/L}^{2l_2/L} \int_{-h/2}^{\frac{h}{2}-h_1} z^6 \frac{\mathrm{d}\mathbf{N}^{w2}}{\mathrm{d}\xi}^{\mathrm{T}} \mathbf{N}^{\theta 2} \mathrm{d}\xi \mathrm{d}z + \rho b\alpha^2 \int_{2l_2/L}^{1} \int_{-h/2}^{h/2} z^6 \frac{\mathrm{d}\mathbf{N}^{w2}}{\mathrm{d}\xi}^{\mathrm{T}} \mathbf{N}^{\theta 2} \mathrm{d}\xi \mathrm{d}z,$$

$$\mathbf{M}_{23}^{(2)} = \rho b\alpha \int_{-\frac{h}{2}}^{\frac{h}{2}} \int_{-1}^{\frac{2l_1}{L}} z^4 \frac{\mathrm{d}\mathbf{N}^{\mathrm{w1}}}{\mathrm{d}\xi}^{\mathrm{T}} \mathbf{N}^{\theta 1} \mathrm{d}\xi \mathrm{d}z + \rho b\alpha \int_{2l_1/L}^{(2l_1+x_c)/L} \int_{-h/2}^{\frac{h}{2}-h_1} z^4 \frac{\mathrm{d}\mathbf{N}^{\mathrm{w1}}}{\mathrm{d}\xi}^{\mathrm{T}} \mathbf{N}^{\theta 1} \mathrm{d}\xi \mathrm{d}z$$

$$+ \rho b\alpha \int_{(2l_1+x_c)/L}^{2l_2/L} \int_{-h/2}^{\frac{h}{2}-h_1} z^4 \frac{\mathrm{d}\mathbf{N}^{\mathrm{w2}}}{\mathrm{d}\xi}^{\mathrm{T}} \mathbf{N}^{\theta 2} \mathrm{d}\xi \mathrm{d}z + \rho b\alpha \int_{2l_2/L}^{1} \int_{-h/2}^{h/2} z^4 \frac{\mathrm{d}\mathbf{N}^{\mathrm{w2}}}{\mathrm{d}\xi}^{\mathrm{T}} \mathbf{N}^{\theta 2} \mathrm{d}\xi \mathrm{d}z,$$

$$\mathbf{M}_{23} = \mathbf{M}_{23}^{(1)} - \mathbf{M}_{23}^{(2)}, \quad \mathbf{M}_{32} = \mathbf{M}_{23}^{\mathrm{T}},$$

$$\mathbf{M}_{33}^{(1)} = \frac{\rho b l}{2} \int_{-h/2}^{h/2} \int_{-1}^{2l_1/L} z^2 \mathbf{N}^{\theta 1 \mathrm{T}} \mathbf{N}^{\theta 1} \mathrm{d}\xi \mathrm{d}z + \frac{\rho b l}{2} \int_{2l_1/L}^{(2l_1+x_c)/L} \int_{-h/2}^{\frac{h}{2}-h_1} z^2 \mathbf{N}^{\theta 1 \mathrm{T}} \mathbf{N}^{\theta 1} \mathrm{d}\xi \mathrm{d}z$$

$$+ \frac{\rho b l}{2} \int_{(2l_1+x_c)/L}^{2l_2/L} \int_{-h/2}^{\frac{h}{2}-h_1} z^2 \mathbf{N}^{\theta 2 \mathrm{T}} \mathbf{N}^{\theta 2} \mathrm{d}\xi \mathrm{d}z + \frac{\rho b l}{2} \int_{2l_2/L}^{1} \int_{-h/2}^{h/2} z^2 \mathbf{N}^{\theta 2 \mathrm{T}} \mathbf{N}^{\theta 2} \mathrm{d}\xi \mathrm{d}z,$$

$$\mathbf{M}_{33}^{(2)} = \rho b\alpha L \int_{-h/2}^{h/2} \int_{-1}^{2l_1/L} z^4 \mathbf{N}^{\theta 1 \mathrm{T}} \mathbf{N}^{\theta 1} \mathrm{d}\xi \mathrm{d}z + \rho b\alpha L \int_{2l_1/L}^{(2l_1+x_c)/L} \int_{-h/2}^{\frac{h}{2}-h_1} z^4 \mathbf{N}^{\theta 1 \mathrm{T}} \mathbf{N}^{\theta 1} \mathrm{d}\xi \mathrm{d}z$$

$$+ \rho b\alpha L \int_{(2l_1+x_c)/L}^{2l_2/L} \int_{-h/2}^{\frac{h}{2}-h_1} z^4 \mathbf{N}^{\theta 2 \mathrm{T}} \mathbf{N}^{\theta 2} \mathrm{d}\xi \mathrm{d}z + \rho b\alpha L \int_{2l_2/L}^{1} \int_{-h/2}^{h/2} z^4 \mathbf{N}^{\theta 2 \mathrm{T}} \mathbf{N}^{\theta 2} \mathrm{d}\xi \mathrm{d}z,$$

$$\mathbf{M}_{33}^{(3)} = \frac{\rho b L \alpha^2}{2} \int_{-\frac{h}{2}}^{\frac{h}{2}} \int_{-1}^{\frac{2l_1}{L}} z^6 \mathbf{N}^{\theta 1 \mathrm{T}} \mathbf{N}^{\theta 1} \mathrm{d}\xi \mathrm{d}z + \frac{\rho b L \alpha^2}{2} \int_{2l_1/L}^{(2l_1+x_c)/L} \int_{-h/2}^{\frac{h}{2}-h_1} z^6 \mathbf{N}^{\theta 1 \mathrm{T}} \mathbf{N}^{\theta 1} \mathrm{d}\xi \mathrm{d}z$$

$$+ \frac{\rho b L \alpha^2}{2} \int_{(2l_1+x_c)/L}^{2l_2/L} \int_{-h/2}^{\frac{h}{2}-h_1} z^6 \mathbf{N}^{\theta 2 \mathrm{T}} \mathbf{N}^{\theta 2} \mathrm{d}\xi \mathrm{d}z + \frac{\rho b L \alpha^2}{2} \int_{2l_2/L}^{1} \int_{-h/2}^{h/2} z^6 \mathbf{N}^{\theta 2 \mathrm{T}} \mathbf{N}^{\theta 2} \mathrm{d}\xi \mathrm{d}z,$$

$$\mathbf{M}_{33} = \mathbf{M}_{33}^{(1)} - \mathbf{M}_{33}^{(2)} + \mathbf{M}_{33}^{(3)},$$

$$\mathbf{K}_{11} = \frac{2Ehb}{L} \int_{-1}^{2l_1/L} \frac{\mathrm{d}\mathbf{N}^{\mathrm{u1T}}}{\mathrm{d}\xi} \frac{\mathrm{d}\mathbf{N}^{\mathrm{u1}}}{\mathrm{d}\xi} \mathrm{d}\xi + \frac{2Eb}{L} \int_{2l_1/L}^{(2l_1+x_c)/L} \int_{-h/2}^{\frac{h}{2}-h_1} \frac{\mathrm{d}\mathbf{N}^{\mathrm{u1T}}}{\mathrm{d}\xi} \frac{\mathrm{d}\mathbf{N}^{\mathrm{u1}}}{\mathrm{d}\xi} z^2 \mathrm{d}z \mathrm{d}\xi$$

$$+\frac{2Eb}{L}\int_{(2l_1+x_c)/L}^{2l_2/L}\int_{-h/2}^{\frac{h}{2}-h_1}\frac{\mathrm{d}\mathbf{N}^{\mathbf{u2}\mathrm{T}}}{\mathrm{d}\xi}\frac{\mathrm{d}\mathbf{N}^{\mathbf{u2}}}{\mathrm{d}\xi}\mathrm{z}^2\mathrm{d}z\mathrm{d}\xi+\frac{2Ehb}{L}\int_{2l_2/L1}^{1}\frac{\mathrm{d}\mathbf{N}^{\mathbf{u2}\mathrm{T}}}{\mathrm{d}\xi}\frac{\mathrm{d}\mathbf{N}^{\mathbf{u2}}}{\mathrm{d}\xi}\mathrm{d}\xi,$$

$$\mathbf{K}^{lt}=-\frac{4Eb\alpha}{L^2}\int_{-\frac{h}{2}}^{\frac{h}{2}}\int_{-1}^{\frac{2l_1}{L}}\mathrm{z}^3\frac{\mathrm{d}\mathbf{N}^{\mathbf{u1}\mathrm{T}}}{\mathrm{d}\xi}\frac{\mathrm{d}^2\mathbf{N}^{\mathbf{w1}}}{\mathrm{d}\xi^2}\mathrm{d}\xi\mathrm{d}z-\frac{4Eb\alpha}{L^2}\int_{2l_1/L}^{(2l_1+x_c)/L}\int_{-h/2}^{\frac{h}{2}-h_1}\mathrm{z}^3\frac{\mathrm{d}\mathbf{N}^{\mathbf{u1}\mathrm{T}}}{\mathrm{d}\xi}\frac{\mathrm{d}^2\mathbf{N}^{\mathbf{w1}}}{\mathrm{d}\xi^2}\mathrm{d}\xi\mathrm{d}z$$

$$-\frac{4Eb\alpha}{L^2}\int_{\frac{(2l_1+x_c)}{L}}^{\frac{2l_2}{L}}\int_{-\frac{h}{2}}^{\frac{h}{2}-h_1}\mathrm{z}^3\frac{\mathrm{d}\mathbf{N}^{\mathbf{u2}\mathrm{T}}}{\mathrm{d}\xi}\frac{\mathrm{d}^2\mathbf{N}^{\mathbf{w2}}}{\mathrm{d}\xi^2}\mathrm{d}\xi\mathrm{d}z-\frac{4Eb\alpha}{L^2}\int_{2l_2/L}^{1}\int_{-h/2}^{h/2}\mathrm{z}^3\frac{\mathrm{d}\mathbf{N}^{\mathbf{u2}\mathrm{T}}}{\mathrm{d}\xi}\frac{\mathrm{d}^2\mathbf{N}^{\mathbf{w2}}}{\mathrm{d}\xi^2}\mathrm{d}\xi\mathrm{d}z,$$

$$\mathbf{K}^{lr(1)}=\frac{2Eb}{L}\int_{-\frac{h}{2}}^{\frac{h}{2}}\int_{-1}^{\frac{2l_1}{L}}\mathrm{z}\frac{\mathrm{d}\mathbf{N}^{\mathbf{u1}\mathrm{T}}}{\mathrm{d}\xi}\frac{\mathrm{d}\mathbf{N}^{\theta 1}}{\mathrm{d}\xi}\mathrm{d}\xi\mathrm{d}z+\frac{2Eb}{L}\int_{2l_1/L}^{(2l_1+x_c)/L}\int_{-h/2}^{\frac{h}{2}-h_1}\mathrm{z}\frac{\mathrm{d}\mathbf{N}^{\mathbf{u1}\mathrm{T}}}{\mathrm{d}\xi}\frac{\mathrm{d}\mathbf{N}^{\theta 1}}{\mathrm{d}\xi}\mathrm{d}\xi\mathrm{d}z$$

$$+\frac{2Eb}{L}\int_{\frac{(2l_1+x_c)}{L}}^{\frac{2l_2}{L}}\int_{-\frac{h}{2}}^{\frac{h}{2}-h_1}\mathrm{z}\frac{\mathrm{d}\mathbf{N}^{\mathbf{u2}\mathrm{T}}}{\mathrm{d}\xi}\frac{\mathrm{d}\mathbf{N}^{\theta 2}}{\mathrm{d}\xi}\mathrm{d}\xi\mathrm{d}z+\frac{2Eb}{L}\int_{2l_2/L}^{1}\int_{-h/2}^{h/2}\mathrm{z}\frac{\mathrm{d}\mathbf{N}^{\mathbf{u2}\mathrm{T}}}{\mathrm{d}\xi}\frac{\mathrm{d}\mathbf{N}^{\theta 2}}{\mathrm{d}\xi}\mathrm{d}\xi\mathrm{d}z,$$

$$\mathbf{K}^{lr(2)}=\frac{2Eb\alpha}{L}\int_{-\frac{h}{2}}^{\frac{h}{2}}\int_{-1}^{\frac{2l_1}{L}}\mathrm{z}^3\frac{\mathrm{d}\mathbf{N}^{\mathbf{u1}\mathrm{T}}}{\mathrm{d}\xi}\frac{\mathrm{d}\mathbf{N}^{\theta 1}}{\mathrm{d}\xi}\mathrm{d}\xi\mathrm{d}z+\frac{2Eb\alpha}{L}\int_{2l_1/L}^{(2l_1+x_c)/L}\int_{-h/2}^{\frac{h}{2}-h_1}\mathrm{z}^3\frac{\mathrm{d}\mathbf{N}^{\mathbf{u1}\mathrm{T}}}{\mathrm{d}\xi}\frac{\mathrm{d}\mathbf{N}^{\theta 1}}{\mathrm{d}\xi}\mathrm{d}\xi\mathrm{d}z$$

$$+\frac{2Eb\alpha}{L}\int_{\frac{(2l_1+x_c)}{L}}^{\frac{2l_2}{L}}\int_{-\frac{h}{2}}^{\frac{h}{2}-h_1}\mathrm{z}^3\frac{\mathrm{d}\mathbf{N}^{\mathbf{u2}\mathrm{T}}}{\mathrm{d}\xi}\frac{\mathrm{d}\mathbf{N}^{\theta 2}}{\mathrm{d}\xi}\mathrm{d}\xi\mathrm{d}z+\frac{2Eb\alpha}{L}\int_{2l_2/L}^{1}\int_{-h/2}^{h/2}\mathrm{z}^3\frac{\mathrm{d}\mathbf{N}^{\mathbf{u2}\mathrm{T}}}{\mathrm{d}\xi}\frac{\mathrm{d}\mathbf{N}^{\theta 2}}{\mathrm{d}\xi}\mathrm{d}\xi\mathrm{d}z,$$

$$\mathbf{K}^{lr}=\mathbf{K}^{lr(1)}-\mathbf{K}^{lr(2)},$$

$$\mathbf{K}_{22}^{(1)}=\frac{2Gb}{L}\int_{-\frac{h}{2}}^{\frac{h}{2}}\int_{-1}^{\frac{2l_1}{L}}\frac{\mathrm{d}\mathbf{N}^{\mathbf{w1}\mathrm{T}}}{\mathrm{d}\xi}\frac{\mathrm{d}\mathbf{N}^{\mathbf{w1}}}{\mathrm{d}\xi}\mathrm{d}\xi\mathrm{d}z+\frac{2Gb}{L}\int_{2l_1/L}^{(2l_1+x_c)/L}\int_{-h/2}^{\frac{h}{2}-h_1}\frac{\mathrm{d}\mathbf{N}^{\mathbf{w1}\mathrm{T}}}{\mathrm{d}\xi}\frac{\mathrm{d}\mathbf{N}^{\mathbf{w1}}}{\mathrm{d}\xi}\mathrm{d}\xi\mathrm{d}z$$

$$+\frac{2Gb}{L}\int_{\frac{(2l_1+x_c)}{L}}^{\frac{2l_2}{L}}\int_{-\frac{h}{2}}^{\frac{h}{2}-h_1}\frac{\mathrm{d}\mathbf{N}^{\mathbf{w2}\mathrm{T}}}{\mathrm{d}\xi}\frac{\mathrm{d}\mathbf{N}^{\mathbf{w2}}}{\mathrm{d}\xi}\mathrm{d}\xi\mathrm{d}z+\frac{2Gb}{L}\int_{2l_2/L}^{1}\int_{-h/2}^{h/2}\frac{\mathrm{d}\mathbf{N}^{\mathbf{w2}\mathrm{T}}}{\mathrm{d}\xi}\frac{\mathrm{d}\mathbf{N}^{\mathbf{w2}}}{\mathrm{d}\xi}\mathrm{d}\xi\mathrm{d}z,$$

$$\mathbf{K}_{22}^{(2)}=\frac{4Gb\beta}{L}\int_{-\frac{h}{2}}^{\frac{h}{2}}\int_{-1}^{\frac{2l_1}{L}}\mathrm{z}^2\frac{\mathrm{d}\mathbf{N}^{\mathbf{w1}\mathrm{T}}}{\mathrm{d}\xi}\frac{\mathrm{d}\mathbf{N}^{\mathbf{w1}}}{\mathrm{d}\xi}\mathrm{d}\xi\mathrm{d}z+\frac{4Gb\beta}{L}\int_{2l_1/L}^{(2l_1+x_c)/L}\int_{-h/2}^{\frac{h}{2}-h_1}\mathrm{z}^2\frac{\mathrm{d}\mathbf{N}^{\mathbf{w1}\mathrm{T}}}{\mathrm{d}\xi}\frac{\mathrm{d}\mathbf{N}^{\mathbf{w1}}}{\mathrm{d}\xi}\mathrm{d}\xi\mathrm{d}z$$

$$+\frac{4Gb\beta}{L}\int_{\frac{(2l_1+x_c)}{L}}^{\frac{2l_2}{L}}\int_{-\frac{h}{2}}^{\frac{h}{2}-h_1} z^2\frac{d\mathbf{N}^{w2T}}{d\xi}\frac{d\mathbf{N}^{w2}}{d\xi}d\xi dz+\frac{4Gb\beta}{L}\int_{2l_2/L}^{1}\int_{-h/2}^{h/2} z^2\frac{d\mathbf{N}^{w2T}}{d\xi}\frac{d\mathbf{N}^{w2}}{d\xi}d\xi dz,$$

$$\mathbf{K}_{22}^{(3)}=\frac{2Gb\beta^2}{L}\int_{-\frac{h}{2}}^{\frac{h}{2}}\int_{-1}^{\frac{2l_1}{L}} z^4\frac{d\mathbf{N}^{w1T}}{d\xi}\frac{d\mathbf{N}^{w1}}{d\xi}d\xi dz+\frac{2Gb\beta^2}{L}\int_{2l_1/L}^{(2l_1+x_c)/L}\int_{-h/2}^{\frac{h}{2}-h_1} z^4\frac{d\mathbf{N}^{w1T}}{d\xi}\frac{d\mathbf{N}^{w1}}{d\xi}d\xi dz$$

$$+\frac{2Gb\beta^2}{L}\int_{\frac{(2l_1+x_c)}{L}}^{\frac{2l_2}{L}}\int_{-\frac{h}{2}}^{\frac{h}{2}-h_1} z^4\frac{d\mathbf{N}^{w2T}}{d\xi}\frac{d\mathbf{N}^{w2}}{d\xi}d\xi dz+\frac{2Gb\beta^2}{L}\int_{\frac{2l_2}{L}}^{1}\int_{-\frac{h}{2}}^{\frac{h}{2}} z^4\frac{d\mathbf{N}^{w2T}}{d\xi}\frac{d\mathbf{N}^{w2}}{d\xi}d\xi dz,$$

$$\mathbf{K}_{22}^{(4)}=\frac{8Eb\alpha^2}{L^3}\int_{-\frac{h}{2}}^{\frac{h}{2}}\int_{-1}^{\frac{2l_1}{L}} z^6\frac{d^2\mathbf{N}^{w1T}}{d\xi^2}\frac{d^2\mathbf{N}^{w1}}{d\xi^2}d\xi dz+\frac{8Eb\alpha^2}{L^3}\int_{2l_1/L}^{(2l_1+x_c)/L}\int_{-h/2}^{\frac{h}{2}-h_1} z^6\frac{d^2\mathbf{N}^{w1T}}{d\xi^2}\frac{d^2\mathbf{N}^{w1}}{d\xi^2}d\xi dz$$

$$+\frac{8Eb\alpha^2}{L^3}\int_{\frac{(2l_1+x_c)}{L}}^{\frac{2l_2}{L}}\int_{-\frac{h}{2}}^{\frac{h}{2}-h_1} z^6\frac{d^2\mathbf{N}^{w2T}}{d\xi^2}\frac{d^2\mathbf{N}^{w2}}{d\xi^2}d\xi dz+\frac{8Eb\alpha^2}{L^3}\int_{2l_2/L}^{1}\int_{-h/2}^{h/2} z^6\frac{d^2\mathbf{N}^{w2T}}{d\xi^2}\frac{d^2\mathbf{N}^{w2}}{d\xi^2}d\xi dz,$$

$$\mathbf{K}_{22}=\mathbf{K}_{22}^{(1)}-\mathbf{K}_{22}^{(2)}+\mathbf{K}_{22}^{(3)}+\mathbf{K}_{22}^{(4)},$$

$$\mathbf{K}_{23}^{(1)}=Gb\int_{-\frac{h}{2}}^{\frac{h}{2}}\int_{-1}^{\frac{2l_1}{L}}\frac{d\mathbf{N}^{w1T}}{d\xi}\mathbf{N}^{\theta1}d\xi dz+Gb\int_{2l_1/L}^{(2l_1+x_c)/L}\int_{-h/2}^{\frac{h}{2}-h_1}\frac{d\mathbf{N}^{w1T}}{d\xi}\mathbf{N}^{\theta1}d\xi dz$$

$$+Gb\int_{\frac{(2l_1+x_c)}{L}}^{\frac{2l_2}{L}}\int_{-\frac{h}{2}}^{\frac{h}{2}-h_1}\frac{d\mathbf{N}^{w2T}}{d\xi}\mathbf{N}^{\theta2}d\xi dz+Gb\int_{2l_2/L}^{1}\int_{-h/2}^{h/2}\frac{d\mathbf{N}^{w2T}}{d\xi}\mathbf{N}^{\theta2}d\xi dz,$$

$$\mathbf{K}_{23}^{(2)}=2bG\beta\int_{-\frac{h}{2}}^{\frac{h}{2}}\int_{-1}^{\frac{2l_1}{L}} z^2\frac{d\mathbf{N}^{w1T}}{d\xi}\mathbf{N}^{\theta1}d\xi dz+2bG\beta\int_{2l_1/L}^{(2l_1+x_c)/L}\int_{-h/2}^{\frac{h}{2}-h_1} z^2\frac{d\mathbf{N}^{w1T}}{d\xi}\mathbf{N}^{\theta1}d\xi dz$$

$$+2bG\beta\int_{\frac{(2l_1+x_c)}{L}}^{\frac{2l_2}{L}}\int_{-\frac{h}{2}}^{\frac{h}{2}-h_1} z^2\frac{d\mathbf{N}^{w2T}}{d\xi}\mathbf{N}^{\theta2}d\xi dz+2bG\beta\int_{2l_2/L}^{1}\int_{-h/2}^{h/2} z^2\frac{d\mathbf{N}^{w2T}}{d\xi}\mathbf{N}^{\theta2}d\xi dz,$$

$$\mathbf{K}_{23}^{(3)} = bG\beta^2 \int_{-\frac{h}{2}}^{\frac{h}{2}} \int_{-1}^{\frac{2l_1}{L}} z^4 \frac{d\mathbf{N}^{w1T}}{d\xi} \mathbf{N}^{\theta 1} d\xi dz + bG\beta^2 \int_{2l_1/L}^{(2l_1+x_c)/L} \int_{-h/2}^{\frac{h}{2}-h_1} z^4 \frac{d\mathbf{N}^{w1T}}{d\xi} \mathbf{N}^{\theta 1} d\xi dz$$

$$+ bG\beta^2 \int_{\frac{(2l_1+x_c)}{L}}^{\frac{2l_2}{L}} \int_{-\frac{h}{2}}^{\frac{h}{2}-h_1} z^4 \frac{d\mathbf{N}^{w2T}}{d\xi} \mathbf{N}^{\theta 2} d\xi dz + bG\beta^2 \int_{2l_2/L}^{1} \int_{-h/2}^{h/2} z^4 \frac{d\mathbf{N}^{w2T}}{d\xi} \mathbf{N}^{\theta 2} d\xi dz,$$

$$\mathbf{K}_{23}^{(4)} = \frac{4Eb\alpha}{L^2} \int_{-\frac{h}{2}}^{\frac{h}{2}} \int_{-1}^{\frac{2l_1}{L}} z^4 \frac{d^2\mathbf{N}^{w1T}}{d\xi^2} \frac{d\mathbf{N}^{\theta 1}}{d\xi} d\xi dz + \frac{4Eb\alpha}{L^2} \int_{2l_1/L}^{(2l_1+x_c)/L} \int_{-h/2}^{\frac{h}{2}-h_1} z^4 \frac{d^2\mathbf{N}^{w1T}}{d\xi^2} \frac{d\mathbf{N}^{\theta 1}}{d\xi} d\xi dz$$

$$+ \frac{4Eb\alpha}{L^2} \int_{\frac{(2l_1+x_c)}{L}}^{\frac{2l_2}{L}} \int_{-\frac{h}{2}}^{\frac{h}{2}-h_1} z^4 \frac{d^2\mathbf{N}^{w2T}}{d\xi^2} \frac{d\mathbf{N}^{\theta 2}}{d\xi} d\xi dz + \frac{4Eb\alpha}{L^2} \int_{2l_2/L}^{1} \int_{-h/2}^{h/2} z^4 \frac{d^2\mathbf{N}^{w2T}}{d\xi^2} \frac{d\mathbf{N}^{\theta 2}}{d\xi} d\xi dz,$$

$$\mathbf{K}_{23}^{(5)} = \frac{4Eb\alpha^2}{L^2} \int_{-\frac{h}{2}}^{\frac{h}{2}} \int_{-1}^{\frac{2l_1}{L}} z^6 \frac{d^2\mathbf{N}^{w1T}}{d\xi^2} \frac{d\mathbf{N}^{\theta 1}}{d\xi} d\xi dz + \frac{4Eb\alpha^2}{L^2} \int_{2l_1/L}^{(2l_1+x_c)/L} \int_{-h/2}^{\frac{h}{2}-h_1} z^6 \frac{d^2\mathbf{N}^{w1T}}{d\xi^2} \frac{d\mathbf{N}^{\theta 1}}{d\xi} d\xi dz$$

$$+ \frac{4Eb\alpha^2}{L^2} \int_{\frac{(2l_1+x_c)}{L}}^{\frac{2l_2}{L}} \int_{-\frac{h}{2}}^{\frac{h}{2}-h_1} z^6 \frac{d^2\mathbf{N}^{w2T}}{d\xi^2} \frac{d\mathbf{N}^{\theta 2}}{d\xi} d\xi dz + \frac{4Eb\alpha^2}{L^2} \int_{2l_2/L}^{1} \int_{-h/2}^{h/2} z^6 \frac{d^2\mathbf{N}^{w2T}}{d\xi^2} \frac{d\mathbf{N}^{\theta 2}}{d\xi} d\xi dz,$$

$$\mathbf{K}_{23} = \mathbf{K}_{23}^{(1)} - \mathbf{K}_{23}^{(2)} + \mathbf{K}_{23}^{(3)} - \mathbf{K}_{23}^{(4)} + \mathbf{K}_{23}^{(5)},$$

$$\mathbf{K}_{33}^{(1)} = \frac{LGb}{2} \int_{-\frac{h}{2}}^{\frac{h}{2}} \int_{-1}^{\frac{2l_1}{L}} \mathbf{N}^{\theta 1T} \mathbf{N}^{\theta 1} d\xi dz + \frac{LGb}{2} \int_{2l_1/L}^{(2l_1+x_c)/L} \int_{-h/2}^{\frac{h}{2}-h_1} \mathbf{N}^{\theta 1T} \mathbf{N}^{\theta 1} d\xi dz$$

$$+ \frac{LGb}{2} \int_{\frac{(2l_1+x_c)}{L}}^{\frac{2l_2}{L}} \int_{-\frac{h}{2}}^{\frac{h}{2}-h_1} \mathbf{N}^{\theta 2T} \mathbf{N}^{\theta 2} d\xi dz + \frac{LGb}{2} \int_{2l_2/L}^{1} \int_{-h/2}^{h/2} \mathbf{N}^{\theta 2T} \mathbf{N}^{\theta 2} d\xi dz,$$

$$\mathbf{K}_{33}^{(2)} = LGb\beta \int_{-\frac{h}{2}}^{\frac{h}{2}} \int_{-1}^{\frac{2l_1}{L}} z^2 \mathbf{N}^{\theta 1T} \mathbf{N}^{\theta 1} d\xi dz + LGb\beta \int_{2l_1/L}^{(2l_1+x_c)/L} \int_{-h/2}^{\frac{h}{2}-h_1} z^2 \mathbf{N}^{\theta 1T} \mathbf{N}^{\theta 1} d\xi dz$$

$$+ LGb\beta \int_{\frac{(2l_1+x_c)}{L}}^{\frac{2l_2}{L}} \int_{-\frac{h}{2}}^{\frac{h}{2}-h_1} z^2 \mathbf{N}^{\theta 2T} \mathbf{N}^{\theta 2} d\xi dz + LGb\beta \int_{2l_2/L}^{1} \int_{-h/2}^{h/2} z^2 \mathbf{N}^{\theta 2T} \mathbf{N}^{\theta 2} d\xi dz,$$

$$\mathbf{K}_{33}^{(3)} = \frac{LGb\beta^2}{2}\int_{-\frac{h}{2}}^{\frac{h}{2}}\int_{-1}^{\frac{2l_1}{L}} z^4\mathbf{N}^{\theta 1\mathrm{T}}\mathbf{N}^{\theta 1}\mathrm{d}\xi\mathrm{d}z + \frac{LGb\beta^2}{2}\int_{2l_1/L}^{(2l_1+x_c)/L}\int_{-h/2}^{\frac{h}{2}-h_1} z^4\mathbf{N}^{\theta 1\mathrm{T}}\mathbf{N}^{\theta 1}\mathrm{d}\xi\mathrm{d}z$$

$$+ \frac{LGb\beta^2}{2}\int_{\frac{(2l_1+x_c)}{L}}^{\frac{2l_2}{L}}\int_{-\frac{h}{2}}^{\frac{h}{2}-h_1} z^4\mathbf{N}^{\theta 2\mathrm{T}}\mathbf{N}^{\theta 2}\mathrm{d}\xi\mathrm{d}z + \frac{LGb\beta^2}{2}\int_{2l_2/L}^{1}\int_{-h/2}^{h/2} z^4\mathbf{N}^{\theta 2\mathrm{T}}\mathbf{N}^{\theta 2}\mathrm{d}\xi\mathrm{d}z,$$

$$\mathbf{K}_{33}^{(4)} = \frac{2Eb}{L}\int_{-\frac{h}{2}}^{\frac{h}{2}}\int_{-1}^{\frac{2l_1}{L}} z^2\frac{\mathrm{d}\mathbf{N}^{\theta 1\mathrm{T}}}{\mathrm{d}\xi}\frac{\mathrm{d}\mathbf{N}^{\theta 1}}{\mathrm{d}\xi}\mathrm{d}\xi\mathrm{d}z + \frac{2Eb}{L}\int_{2l_1/L}^{(2l_1+x_c)/L}\int_{-h/2}^{\frac{h}{2}-h_1} z^2\frac{\mathrm{d}\mathbf{N}^{\theta 1\mathrm{T}}}{\mathrm{d}\xi}\frac{\mathrm{d}\mathbf{N}^{\theta 1}}{\mathrm{d}\xi}\mathrm{d}\xi\mathrm{d}z$$

$$+ \frac{2Eb}{L}\int_{\frac{(2l_1+x_c)}{L}}^{\frac{2l_2}{L}}\int_{-\frac{h}{2}}^{\frac{h}{2}-h_1} z^2\frac{\mathrm{d}\mathbf{N}^{\theta 2\mathrm{T}}}{\mathrm{d}\xi}\frac{\mathrm{d}\mathbf{N}^{\theta 2}}{\mathrm{d}\xi}\mathrm{d}\xi\mathrm{d}z + \frac{2Eb}{L}\int_{\frac{2l_2}{L}}^{1}\int_{-\frac{h}{2}}^{\frac{h}{2}} z^2\frac{\mathrm{d}\mathbf{N}^{\theta 2\mathrm{T}}}{\mathrm{d}\xi}\frac{\mathrm{d}\mathbf{N}^{\theta 2}}{\mathrm{d}\xi}\mathrm{d}\xi\mathrm{d}z,$$

$$\mathbf{K}_{33}^{(5)} = \frac{4Eb\alpha}{L}\int_{-\frac{h}{2}}^{\frac{h}{2}}\int_{-1}^{\frac{2l_1}{L}} z^4\frac{\mathrm{d}\mathbf{N}^{\theta 1\mathrm{T}}}{\mathrm{d}\xi}\frac{\mathrm{d}\mathbf{N}^{\theta 1}}{\mathrm{d}\xi}\mathrm{d}\xi\mathrm{d}z + \frac{4Eb\alpha}{L}\int_{2l_1/L}^{(2l_1+x_c)/L}\int_{-h/2}^{\frac{h}{2}-h_1} z^4\frac{\mathrm{d}\mathbf{N}^{\theta 1\mathrm{T}}}{\mathrm{d}\xi}\frac{\mathrm{d}\mathbf{N}^{\theta 1}}{\mathrm{d}\xi}\mathrm{d}\xi\mathrm{d}z$$

$$+ \frac{4Eb\alpha}{L}\int_{\frac{(2l_1+x_c)}{L}}^{\frac{2l_2}{L}}\int_{-\frac{h}{2}}^{\frac{h}{2}-h_1} z^4\frac{\mathrm{d}\mathbf{N}^{\theta 2\mathrm{T}}}{\mathrm{d}\xi}\frac{\mathrm{d}\mathbf{N}^{\theta 2}}{\mathrm{d}\xi}\mathrm{d}\xi\mathrm{d}z + \frac{4Eb\alpha}{L}\int_{2l_2/L}^{1}\int_{-h/2}^{h/2} z^4\frac{\mathrm{d}\mathbf{N}^{\theta 2\mathrm{T}}}{\mathrm{d}\xi}\frac{\mathrm{d}\mathbf{N}^{\theta 2}}{\mathrm{d}\xi}\mathrm{d}\xi\mathrm{d}z,$$

$$\mathbf{K}_{33}^{(6)} = \frac{2Eb\alpha^2}{L}\int_{-\frac{h}{2}}^{\frac{h}{2}}\int_{-1}^{\frac{2l_1}{L}} z^6\frac{\mathrm{d}\mathbf{N}^{\theta 1\mathrm{T}}}{\mathrm{d}\xi}\frac{\mathrm{d}\mathbf{N}^{\theta 1}}{\mathrm{d}\xi}\mathrm{d}\xi\mathrm{d}z + \frac{2Eb\alpha^2}{L}\int_{2l_1/L}^{(2l_1+x_c)/L}\int_{-h/2}^{\frac{h}{2}-h_1} z^6\frac{\mathrm{d}\mathbf{N}^{\theta 1\mathrm{T}}}{\mathrm{d}\xi}\frac{\mathrm{d}\mathbf{N}^{\theta 1}}{\mathrm{d}\xi}\mathrm{d}\xi\mathrm{d}z$$

$$+ \frac{2Eb\alpha^2}{L}\int_{\frac{(2l_1+x_c)}{L}}^{\frac{2l_2}{L}}\int_{-\frac{h}{2}}^{\frac{h}{2}-h_1} z^6\frac{\mathrm{d}\mathbf{N}^{\theta 2\mathrm{T}}}{\mathrm{d}\xi}\frac{\mathrm{d}\mathbf{N}^{\theta 2}}{\mathrm{d}\xi}\mathrm{d}\xi\mathrm{d}z + \frac{2Eb\alpha^2}{L}\int_{2l_2/L}^{1}\int_{-h/2}^{h/2} z^6\frac{\mathrm{d}\mathbf{N}^{\theta 2\mathrm{T}}}{\mathrm{d}\xi}\frac{\mathrm{d}\mathbf{N}^{\theta 2}}{\mathrm{d}\xi}\mathrm{d}\xi\mathrm{d}z,$$

$$\mathbf{K}_{33} = \mathbf{K}_{33}^{(1)} - \mathbf{K}_{33}^{(2)} + \mathbf{K}_{33}^{(3)} + \mathbf{K}_{33}^{(4)} - \mathbf{K}_{33}^{(5)} + \mathbf{K}_{33}^{(6)}$$

$$\mathbf{K}_{n\ell}^{2}(\mathbf{q_w}) = \frac{4Ebh}{L^2}\int_{-1}^{\frac{2l_1}{L}}\frac{\mathrm{d}w_0}{\mathrm{d}\xi}\frac{\mathrm{d}\mathbf{N}^{u1\mathrm{T}}}{\mathrm{d}\xi}\frac{\mathrm{d}\mathbf{N}^{w1}}{\mathrm{d}\xi}\mathrm{d}\xi + \frac{4Eb}{L^2}\int_{-\frac{h}{2}}^{\frac{h}{2}-h_1}\int_{\frac{2l_1}{L}}^{\frac{2l_1+x_c}{L}}\frac{\mathrm{d}w_0}{\mathrm{d}\xi}\frac{\mathrm{d}\mathbf{N}^{u1\mathrm{T}}}{\mathrm{d}\xi}\frac{\mathrm{d}\mathbf{N}^{w1}}{\mathrm{d}\xi}\mathrm{d}\xi\mathrm{d}z$$

$$+\frac{4Eb}{L^2}\int_{-\frac{h}{2}}^{\frac{h}{2}-h_1}\int_{\frac{2l_1+x_c}{L}}^{\frac{2l2}{L}}\frac{\mathrm{d}w_0}{\mathrm{d}\xi}\frac{\mathrm{d}\mathbf{N}^{u2\mathrm{T}}}{\mathrm{d}\xi}\frac{\mathrm{d}\mathbf{N}^{w2}}{\mathrm{d}\xi}\mathrm{d}\xi\mathrm{d}z+\frac{4Ebh}{L^2}\int_{\frac{2l2}{L}}^{1}\frac{\mathrm{d}w_0}{\mathrm{d}\xi}\frac{\mathrm{d}\mathbf{N}^{u2\mathrm{T}}}{\mathrm{d}\xi}\frac{\mathrm{d}\mathbf{N}^{w2}}{\mathrm{d}\xi}\mathrm{d}\xi,$$

$$\mathbf{K}_{n\ell}^{3}(\mathbf{q_w})=2\mathbf{K}_{n\ell}^{2\mathrm{T}}(\mathbf{q_w}),$$

$$\mathbf{K}_{n\ell}^{4(1)}(\mathbf{q_w})=\frac{4Ebh}{L^3}\int_{-1}^{\frac{2l_1}{L}}\left(\frac{\mathrm{d}w_0}{\mathrm{d}\xi}\right)^2\frac{\mathrm{d}\mathbf{N}^{w1\mathrm{T}}}{\mathrm{d}\xi}\frac{\mathrm{d}\mathbf{N}^{w1}}{\mathrm{d}\xi}\mathrm{d}\xi$$

$$+\frac{4Eb}{L^3}\int_{-\frac{h}{2}}^{\frac{h}{2}-h_1}\int_{\frac{2l_1}{L}}^{\frac{2l_1+x_c}{L}}\left(\frac{\mathrm{d}w_0}{\mathrm{d}\xi}\right)^2\frac{\mathrm{d}\mathbf{N}^{w1\mathrm{T}}}{\mathrm{d}\xi}\frac{\mathrm{d}\mathbf{N}^{w1}}{\mathrm{d}\xi}\mathrm{d}\xi\mathrm{d}z$$

$$+\frac{4Eb}{L^3}\int_{-\frac{h}{2}}^{\frac{h}{2}-h_1}\int_{\frac{2l_1+x_c}{L}}^{\frac{2l_2}{L}}\left(\frac{\mathrm{d}w_0}{\mathrm{d}\xi}\right)^2\frac{\mathrm{d}\mathbf{N}^{w2\mathrm{T}}}{\mathrm{d}\xi}\frac{\mathrm{d}\mathbf{N}^{w2}}{\mathrm{d}\xi}\mathrm{d}\xi\mathrm{d}z+\frac{4Ebh}{L^3}\int_{\frac{2l_2}{L}}^{1}\left(\frac{\mathrm{d}w_0}{\mathrm{d}\xi}\right)^2\frac{\mathrm{d}\mathbf{N}^{w2\mathrm{T}}}{\mathrm{d}\xi}\frac{\mathrm{d}\mathbf{N}^{w2}}{\mathrm{d}\xi}\mathrm{d}\xi,$$

$$\mathbf{K}_{n\ell}^{4(2)}(\mathbf{q_w})=\frac{8E\alpha b}{L^3}\int_{-1}^{\frac{2l_1}{L}}\int_{-\frac{h}{2}}^{\frac{h}{2}}z^3\frac{\mathrm{d}w_0}{\mathrm{d}\xi}\frac{\mathrm{d}\mathbf{N}^{w1\mathrm{T}}}{\mathrm{d}\xi}\frac{\mathrm{d}^2\mathbf{N}^{w1}}{\mathrm{d}\xi^2}\mathrm{d}z\mathrm{d}\xi$$

$$+\frac{8E\alpha b}{L^3}\int_{-\frac{h}{2}}^{\frac{h}{2}-h_1}\int_{\frac{2l_1}{L}}^{\frac{2l_1+x_c}{L}}z^3\frac{\mathrm{d}w_0}{\mathrm{d}\xi}\frac{\mathrm{d}\mathbf{N}^{w1\mathrm{T}}}{\mathrm{d}\xi}\frac{\mathrm{d}^2\mathbf{N}^{w1}}{\mathrm{d}\xi^2}\mathrm{d}\xi\mathrm{d}z$$

$$+\frac{8E\alpha b}{L^3}\int_{-\frac{h}{2}}^{\frac{h}{2}-h_1}\int_{\frac{2l_1+x_c}{L}}^{\frac{2l_2}{L}}z^3\frac{\mathrm{d}w_0}{\mathrm{d}\xi}\frac{\mathrm{d}\mathbf{N}^{w1\mathrm{T}}}{\mathrm{d}\xi}\frac{\mathrm{d}^2\mathbf{N}^{w1}}{\mathrm{d}\xi^2}\mathrm{d}\xi\mathrm{d}z$$

$$+\frac{8E\alpha b}{L^3}\int_{-1}^{\frac{2l_1}{L}}\int_{-\frac{h}{2}}^{\frac{h}{2}}z^3\frac{\mathrm{d}w_0}{\mathrm{d}\xi}\frac{\mathrm{d}\mathbf{N}^{w1\mathrm{T}}}{\mathrm{d}\xi}\frac{\mathrm{d}^2\mathbf{N}^{w1}}{\mathrm{d}\xi^2}\mathrm{d}z\mathrm{d}\xi,$$

$$\mathbf{K}_{n\ell}^{4}(\mathbf{q_w})=\mathbf{K}_{n\ell}^{4(1)}(\mathbf{q_w})+\mathbf{K}_{n}^{4(2)}(\mathbf{q_w}),$$

$$\mathbf{K}_{n\ell}^{br(1)}(\mathbf{q_w})=\frac{2Eb}{L^2}\int_{-1}^{\frac{2l_1}{L}}\int_{-\frac{h}{2}}^{\frac{h}{2}}z\frac{\mathrm{d}w_0}{\mathrm{d}\xi}\frac{\mathrm{d}\mathbf{N}^{w1\mathrm{T}}}{\mathrm{d}\xi}\frac{\mathrm{d}\mathbf{N}^{\theta 1}}{\mathrm{d}\xi}\mathrm{d}z\mathrm{d}\xi$$

$$+\frac{2Eb}{L^2}\int_{-\frac{h}{2}}^{\frac{h}{2}-h_1}\int_{\frac{2l_1}{L}}^{\frac{2l_1+x_c}{L}} z\frac{dw_0}{d\xi}\frac{d\mathbf{N}^{w1\mathrm{T}}}{d\xi}\frac{d\mathbf{N}^{\theta 1}}{d\xi}d\xi dz$$

$$+\frac{2Eb}{L^2}\int_{-\frac{h}{2}}^{\frac{h}{2}-h_1}\int_{\frac{2l_1+x_c}{L}}^{\frac{2l_2}{L}} z\frac{dw_0}{d\xi}\frac{d\mathbf{N}^{w2\mathrm{T}}}{d\xi}\frac{d\mathbf{N}^{\theta 2}}{d\xi}d\xi dz$$
$$+\frac{2Eb}{L^2}\int_{-\frac{h}{2}}^{\frac{h}{2}-h_1}\int_{\frac{2l_2}{L}}^{1} z\frac{dw_0}{d\xi}\frac{d\mathbf{N}^{w2\mathrm{T}}}{d\xi}\frac{d\mathbf{N}^{\theta 2}}{d\xi}d\xi dz,$$

$$\mathbf{K}_{n\ell}^{br(2)}(\mathbf{q_w})=\frac{2Eb\alpha}{L^2}\int_{-1}^{\frac{2l_1}{L}}\int_{-\frac{h}{2}}^{\frac{h}{2}} z^3\frac{dw_0}{d\xi}\frac{d\mathbf{N}^{w1\mathrm{T}}}{d\xi}\frac{d\mathbf{N}^{\theta 1}}{d\xi}dz d\xi$$

$$+\frac{2Eb\alpha}{L^2}\int_{-\frac{h}{2}}^{\frac{h}{2}-h_1}\int_{\frac{2l_1}{L}}^{\frac{2l_1+x_c}{L}} z^3\frac{dw_0}{d\xi}\frac{d\mathbf{N}^{w1\mathrm{T}}}{d\xi}\frac{d\mathbf{N}^{\theta 1}}{d\xi}d\xi dz$$

$$+\frac{2Eb\alpha}{L^2}\int_{-\frac{h}{2}}^{\frac{h}{2}-h_1}\int_{\frac{2l_1+x_c}{L}}^{\frac{2l_2}{L}} z^3\frac{dw_0}{d\xi}\frac{d\mathbf{N}^{w2\mathrm{T}}}{d\xi}\frac{d\mathbf{N}^{\theta 2}}{d\xi}d\xi dz$$
$$+\frac{2Eb\alpha}{L^2}\int_{-\frac{h}{2}}^{\frac{h}{2}-h_1}\int_{\frac{2l_2}{L}}^{1} z^3\frac{dw_0}{d\xi}\frac{d\mathbf{N}^{w2\mathrm{T}}}{d\xi}\frac{d\mathbf{N}^{\theta 2}}{d\xi}d\xi dz,$$

$$\mathbf{K}_{n\ell}^{br}(\mathbf{q_w})=\mathbf{K}_{n\ell}^{br(1)}(\mathbf{q_w})-\mathbf{K}_{n}^{br(2)}(\mathbf{q_w}),$$

$$\mathbf{K}_{\ell}^{W}=\frac{L}{2}k_l\int_{-1}^{\frac{2l_1+x_c}{L}}\mathbf{N}^{w1\mathrm{T}}\mathbf{N}^{w1}d\xi+\frac{L}{2}k_l\int_{\frac{2l_1+x_c}{L}}^{1}\mathbf{N}^{w2\mathrm{T}}\mathbf{N}^{w2}d\xi,$$

$$\mathbf{K}_{\ell}^{P}=\frac{8}{L^3}k_p\int_{-1}^{\frac{2l_1+x_c}{L}}\frac{d^2\mathbf{N}^{w1\mathrm{T}}}{d\xi^2}\frac{d^2\mathbf{N}^{w1}}{d\xi^2}d\xi+\frac{8}{L^3}k_p\int_{\frac{2l_1+x_c}{L}}^{1}\frac{d^2\mathbf{N}^{w2\mathrm{T}}}{d\xi^2}\frac{d^2\mathbf{N}^{w1}}{d\xi^2}d\xi,$$

$$\mathbf{K}_{n\ell}^{W}(\mathbf{q_w})=\frac{L}{2}k_{nl}\int_{-1}^{\frac{2l_1+x_c}{L}}(w_0)^3\mathbf{N}^{w1\mathrm{T}}\mathbf{N}^{w1}d\xi+\frac{L}{2}k_{nl}\int_{\frac{2l_1+x_c}{L}}^{1}(w_0)^3\mathbf{N}^{w2\mathrm{T}}\mathbf{N}^{w2}d\xi,$$

where $\mathbf{N}^{u1}$, $\mathbf{N}^{w1}$, and $\mathbf{N}^{\theta 1}$ belong to the domain $-1\le\xi\le 2l_c/L$, and $\mathbf{N}^{u2}$, $\mathbf{N}^{w2}$, and $\mathbf{N}^{\theta 2}$ belong to the domain $\frac{2l_c}{L}\le\xi\le 1,\ l_c=\frac{l_1+l_2}{L}$.

Appendix 2

$$\mathbf{KNL}1_{11} = \frac{1}{8}\mathrm{KNL1}\left(Q_{w_{c0}}, Q_{w_{c0}}\right) + \frac{1}{4}\mathrm{KNL1}\left(Q_{w_{c1}}, Q_{w_{c1}}\right) + \frac{1}{4}\mathrm{KNL1}\left(Q_{w_{c2}}, Q_{w_{c2}}\right) + \frac{1}{4}\mathrm{KNL1}\left(Q_{w_{c3}}, Q_{w_{c3}}\right),$$

$$\mathbf{KNL}1_{12} = \frac{1}{4}\mathrm{KNL1}\left(Q_{w_{c0}}, Q_{w_{c1}}\right) + \frac{1}{4}\mathrm{KNL1}\left(Q_{w_{c1}}, Q_{w_{c0}}\right) + \frac{1}{4}\mathrm{KNL1}\left(Q_{w_{c1}}, Q_{w_{c2}}\right) + \frac{1}{4}\mathrm{KNL1}\left(Q_{w_{c2}}, Q_{w_{c1}}\right) + \frac{1}{4}\mathrm{KNL1}\left(Q_{w_{c2}}, Q_{w_{c3}}\right) + \frac{1}{4}\mathrm{KNL1}\left(Q_{w_{c3}}, Q_{w_{c2}}\right),$$

$$\mathbf{KNL}1_{13} = \frac{1}{4}\mathrm{KNL1}\left(Q_{w_{c0}}, Q_{w_{c2}}\right) + \frac{1}{4}\mathrm{KNL1}\left(Q_{w_{c2}}, Q_{w_{c0}}\right) + \frac{1}{4}\mathrm{KNL1}\left(Q_{w_{c1}}, Q_{w_{c3}}\right) + \frac{1}{4}\mathrm{KNL1}\left(Q_{w_{c3}}, Q_{w_{c1}}\right) + \frac{1}{4}\mathrm{KNL1}\left(Q_{w_{c1}}, Q_{w_{c1}}\right),$$

$$\mathbf{KNL}1_{14} = \frac{1}{4}\mathrm{KNL1}\left(Q_{w_{c0}}, Q_{w_{c3}}\right) + \frac{1}{4}\mathrm{KNL1}\left(Q_{w_{c3}}, Q_{w_{c0}}\right) + \frac{1}{4}\mathrm{KNL1}\left(Q_{w_{c1}}, Q_{w_{c2}}\right) + \frac{1}{4}\mathrm{KNL1}\left(Q_{w_{c2}}, Q_{w_{c1}}\right),$$

$$\mathbf{KNL}1_{21} = \frac{1}{4}\left(\mathrm{KNL1}\left(Q_{w_{c0}}, Q_{w_{c1}}\right) + \mathrm{KNL1}\left(Q_{w_{c1}}, Q_{w_{c0}}\right) + \mathrm{KNL1}\left(Q_{w_{c1}}, Q_{w_{c2}}\right) + \mathrm{KNL1}\left(Q_{w_{c2}}, Q_{w_{c1}}\right) + \mathrm{KNL1}\left(Q_{w_{c2}}, Q_{w_{c3}}\right) + \mathrm{KNL1}\left(Q_{w_{c3}}, Q_{w_{c2}}\right)\right),$$

$$\mathbf{KNL}1_{22} = \frac{1}{4}\left(\mathrm{KNL1}\left(Q_{w_{c0}}, Q_{w_{c2}}\right) + \mathrm{KNL1}\left(Q_{w_{c2}}, Q_{w_{c0}}\right) + \mathrm{KNL1}\left(Q_{w_{c0}}, Q_{w_{c0}}\right) + \mathrm{KNL1}\left(Q_{w_{c1}}, Q_{w_{c3}}\right) + \mathrm{KNL1}\left(Q_{w_{c3}}, Q_{w_{c1}}\right) + 3\mathrm{KNL1}\left(Q_{w_{c1}}, Q_{w_{c1}}\right) + 2\mathrm{KNL1}\left(Q_{w_{c2}}, Q_{w_{c2}}\right) + 2\mathrm{KNL1}\left(Q_{w_{c3}}, Q_{w_{c3}}\right)\right),$$

$$\mathbf{KNL}1_{23} = \frac{1}{4}\left(\mathrm{KNL1}\left(\mathrm{Q}_{\mathrm{W}_{c0}}, \mathrm{Q}_{\mathrm{W}_{c1}}\right) + \mathrm{KNL1}\left(\mathrm{Q}_{\mathrm{W}_{c1}}, \mathrm{Q}_{\mathrm{W}_{c0}}\right) + \mathrm{KNL1}\left(\mathrm{Q}_{\mathrm{W}_{c0}}, \mathrm{Q}_{\mathrm{W}_{c3}}\right) + \mathrm{KNL1}\left(\mathrm{Q}_{\mathrm{W}_{c3}}, \mathrm{Q}_{\mathrm{W}_{c0}}\right) + 2\mathrm{KNL1}\left(\mathrm{Q}_{\mathrm{W}_{c1}}, \mathrm{Q}_{\mathrm{W}_{c2}}\right) + 2\mathrm{KNL1}\left(\mathrm{Q}_{\mathrm{W}_{c2}}, \mathrm{Q}_{\mathrm{W}_{c1}}\right) + \mathrm{KNL1}\left(\mathrm{Q}_{\mathrm{W}_{c2}}, \mathrm{Q}_{\mathrm{W}_{c3}}\right) + \mathrm{KNL1}\left(\mathrm{Q}_{\mathrm{W}_{c3}}, \mathrm{Q}_{\mathrm{W}_{c2}}\right)\right),$$

$$\mathbf{KNL1}_{24} = \frac{1}{4}\Big(\mathrm{KNL1}\big(Q_{w_{c0}}, Q_{w_{c2}}\big) + \mathrm{KNL1}\big(Q_{w_{c2}}, Q_{w_{c0}}\big) + 2\mathrm{KNL1}\big(Q_{w_{c1}}, Q_{w_{c3}}\big) + 2\mathrm{KNL1}\big(Q_{w_{c3}}, Q_{w_{c1}}\big) + \mathrm{KNL1}\big(Q_{w_{c1}}, Q_{w_{c1}}\big) + \mathrm{KNL1}\big(Q_{w_{c2}}, Q_{w_{c2}}\big)\Big),$$

$$\mathbf{KNL1}_{31} = \frac{1}{4}\Big(\mathrm{KNL1}\big(Q_{w_{c0}}, Q_{w_{c2}}\big) + \mathrm{KNL1}\big(Q_{w_{c2}}, Q_{w_{c0}}\big) + \mathrm{KNL1}\big(Q_{w_{c1}}, Q_{w_{c3}}\big) + \mathrm{KNL1}\big(Q_{w_{c3}}, Q_{w_{c1}}\big) + \mathrm{KNL1}\big(Q_{w_{c1}}, Q_{w_{c1}}\big)\Big),$$

$$\mathbf{KNL1}_{32} = \frac{1}{4}\Big(\mathrm{KNL1}\big(Q_{w_{c0}}, Q_{w_{c1}}\big) + \mathrm{KNL1}\big(Q_{w_{c1}}, Q_{w_{c0}}\big) + \mathrm{KNL1}\big(Q_{w_{c0}}, Q_{w_{c3}}\big) + \mathrm{KNL1}\big(Q_{w_{c3}}, Q_{w_{c0}}\big) + 2\mathrm{KNL1}\big(Q_{w_{c1}}, Q_{w_{c2}}\big) + 2\mathrm{KNL1}\big(Q_{w_{c2}}, Q_{w_{c1}}\big) + \mathrm{KNL1}\big(Q_{w_{c2}}, Q_{w_{c3}}\big) + \mathrm{KNL1}\big(Q_{w_{c3}}, Q_{w_{c2}}\big)\Big),$$

$$\mathbf{KNL1}_{33} = \frac{1}{4}\Big(\mathrm{KNL1}\big(Q_{w_{c0}}, Q_{w_{c0}}\big) + \mathrm{KNL1}\big(Q_{w_{c1}}, Q_{w_{c3}}\big) + \mathrm{KNL1}\big(Q_{w_{c3}}, Q_{w_{c1}}\big) + 2\mathrm{KNL1}\big(Q_{w_{c1}}, Q_{w_{c1}}\big) + 3\mathrm{KNL1}\big(Q_{w_{c2}}, Q_{w_{c2}}\big) + 2\mathrm{KNL1}\big(Q_{w_{c3}}, Q_{w_{c3}}\big)\Big),$$

$$\mathbf{KNL1}_{34} = \frac{1}{4}\Big(\mathrm{KNL1}\big(Q_{w_{c0}}, Q_{w_{c1}}\big) + \mathrm{KNL1}\big(Q_{w_{c1}}, Q_{w_{c0}}\big) + \mathrm{KNL1}\big(Q_{w_{c1}}, Q_{w_{c2}}\big) + \mathrm{KNL1}\big(Q_{w_{c2}}, Q_{w_{c1}}\big) + 2\mathrm{KNL1}\big(Q_{w_{c2}}, Q_{w_{c3}}\big) + 2\mathrm{KNL1}\big(Q_{w_{c3}}, Q_{w_{c2}}\big)\Big),$$

$$\mathbf{KNL1}_{41} = \frac{1}{4}\Big(\mathrm{KNL1}\big(Q_{w_{c0}}, Q_{w_{c3}}\big) + \mathrm{KNL1}\big(Q_{w_{c3}}, Q_{w_{c0}}\big) + \mathrm{KNL1}\big(Q_{w_{c1}}, Q_{w_{c2}}\big) + \mathrm{KNL1}\big(Q_{w_{c2}}, Q_{w_{c1}}\big)\Big),$$

$$\mathbf{KNL1}_{42} = \frac{1}{4}\Big(\mathrm{KNL1}\big(Q_{w_{c0}}, Q_{w_{c2}}\big) + \mathrm{KNL1}\big(Q_{w_{c2}}, Q_{w_{c0}}\big) + 2\mathrm{KNL1}\big(Q_{w_{c1}}, Q_{w_{c3}}\big) + 2\mathrm{KNL1}\big(Q_{w_{c3}}, Q_{w_{c1}}\big) + \mathrm{KNL1}\big(Q_{w_{c1}}, Q_{w_{c1}}\big) + \mathrm{KNL1}\big(Q_{w_{c2}}, Q_{w_{c2}}\big)\Big),$$

$$\mathbf{KNL1}_{43} = \frac{1}{4}\Big(\mathrm{KNL1}\big(Q_{w_{c0}}, Q_{w_{c1}}\big) + \mathrm{KNL1}\big(Q_{w_{c1}}, Q_{w_{c0}}\big) + \mathrm{KNL1}\big(Q_{w_{c1}}, Q_{w_{c2}}\big) + \mathrm{KNL1}\big(Q_{w_{c2}}, Q_{w_{c1}}\big) + 2\mathrm{KNL1}\big(Q_{w_{c2}}, Q_{w_{c3}}\big) + 2\mathrm{KNL1}\big(Q_{w_{c3}}, Q_{w_{c2}}\big)\Big),$$

$$\mathbf{KNL1}_{44} = \frac{1}{4}\big(\mathrm{KNL1}(Q_{w_{c0}}, Q_{w_{c0}}) + 2\mathrm{KNL1}(Q_{w_{c1}}, Q_{w_{c1}}) + 2\mathrm{KNL1}(Q_{w_{c2}}, Q_{w_{c2}}) + 3\mathrm{KNL1}(Q_{w_{c3}}, Q_{w_{c3}})\big),$$

$$\mathbf{KNL2}_{11} = \frac{1}{4}\mathrm{KNL2}(Q_{w_{c0}}), \quad \mathbf{KNL2}_{12} = \frac{1}{2}\mathrm{KNL2}(Q_{w_{c1}}),$$

$$\mathbf{KNL2}_{13} = \frac{1}{2}\mathrm{KNL2}(Q_{w_{c2}}), \quad \mathbf{KNL2}_{14} = \frac{1}{2}\mathrm{KNL2}(Q_{w_{c3}}),$$

$$\mathbf{KNL2}_{21} = \frac{1}{2}\mathrm{KNL2}(Q_{w_{c1}}), \quad \mathbf{KNL2}_{22} = \frac{1}{2}\mathrm{KNL2}(Q_{w_{c0}}),$$

$$\mathbf{KNL2}_{23} = \frac{1}{2}\mathrm{KNL2}(Q_{w_{c1}}) + \frac{1}{2}\mathrm{KNL2}(Q_{w_{c3}}), \quad \mathbf{KNL2}_{24} = \frac{1}{2}\mathrm{KNL2}(Q_{w_{c2}}).$$

$$\mathbf{KNL2}_{31} = \frac{1}{2}\mathrm{KNL2}(Q_{w_{c2}}), \quad \mathbf{KNL2}_{32} = \frac{1}{2}\mathrm{KNL2}(Q_{w_{c1}}) + \frac{1}{2}\mathrm{KNL2}(Q_{w_{c3}}),$$

$$\mathbf{KNL2}_{33} = \frac{1}{2}\mathrm{KNL2}(Q_{w_{c0}}), \quad \mathbf{KNL2}_{34} = \frac{1}{2}\mathrm{KNL2}(Q_{w_{c1}}),$$

$$\mathbf{KNL2}_{41} = \frac{1}{2}\mathrm{KNL2}(Q_{w_{c3}}), \quad \mathbf{KNL2}_{44} = \frac{1}{2}\mathrm{KNL2}(Q_{w_{c0}}),$$

$$\mathbf{KNL3}_{11} = \frac{1}{4}\mathrm{KNL3}(Q_{w_{c0}}, Q_{w_{c0}}), \quad \mathbf{KNL3}_{12} = \frac{1}{2}\mathrm{KNL3}(Q_{w_{c1}}, Q_{w_{c1}}),$$

$$\mathbf{KNL3}_{21} = \frac{1}{2}\mathrm{KNL3}(Q_{w_{c1}}, Q_{w_{c1}}),$$

$$\mathbf{KNL3}_{22} = \frac{1}{2}\big(\mathrm{KNL3}(Q_{w_{c0}}, Q_{w_{c0}}) + \mathrm{KNL3}(Q_{w_{c2}}, Q_{w_{c2}})\big),$$

$$\mathbf{KNL3}_{23} = \frac{1}{2}\left(\text{KNL3}\left(Q_{W_{c1}}, Q_{W_{c1}}\right) + \text{KNL3}\left(Q_{W_{c3}}, Q_{W_{c3}}\right)\right),$$

$$\mathbf{KNL3}_{24} = \frac{1}{2}\text{KNL3}\left(Q_{W_{c2}}, Q_{W_{c2}}\right),$$

$$\mathbf{KNL3}_{31} = \frac{1}{2}\text{KNL3}(Q_{W_{c2}}, Q_{W_{c2}}),$$

$$\mathbf{KNL3}_{32} = \frac{1}{2}(\text{KNL3}(Q_{W_{c1}}, Q_{W_{c1}}) + \text{KNL3}(Q_{W_{c3}}, Q_{W_{c3}})),$$

$$\mathbf{KNL3}_{33} = \frac{1}{2}\text{KNL3}(Q_{W_{c0}}, Q_{W_{c0}}), \quad \mathbf{KNL3}_{34} = \frac{1}{2}\text{KNL3}(Q_{W_{c1}}, Q_{W_{c1}}),$$

$$\mathbf{KNL3}_{41} = \frac{1}{2}\text{KNL3}(Q_{W_{c3}}, Q_{W_{c3}}), \quad \mathbf{KNL3}_{42} = \frac{1}{2}\text{KNL3}(Q_{W_{c2}}, Q_{W_{c2}}),$$

$$\mathbf{KNL3}_{43} = \frac{1}{2}\text{KNL3}(Q_{W_{c1}}, Q_{W_{c1}}), \quad \mathbf{KNL3}_{44} = \frac{1}{2}\text{KNL3}(Q_{W_{c0}}, Q_{W_{c0}}).$$

References

[1] A.D. Dimarogonas, Vibration of cracked structures: a state of the art review, Engineering Fracture Mechanics 55 (1996) 831–857.

[2] P. Cawley, R.D. Adams, The locations of defects in structures from measurements of natural frequencies, Journal of Strain Analysis 14 (1979) 49–57.

[3] M.I. Friswell, J.E.T. Penny, D.A.L. Wilson, Using vibration data and statistical measures to locate damage in structures. Modal analysis, The International Journal of Analytical and Experimental Modal Analysis 9 (1994) 239–254.

[4] Y. Narkis, Identification of crack location in vibrating simply supported beams, Journal of Sound and Vibration 172 (1994) 549–558.

[5] J.K. Sinha, M.I. Friswell, S. Edwards, Simplified models for the location of cracks in beam structures using measured vibration data, Journal of Sound and vibration 251 (2002) 13–38.

[6] H. Nahvi, M. Jabbari, Crack detection in beams using experimental modal data and finite element model, International Journal of Mechanical Sciences 47 (2005) 1477–1497.

[7] J.A. Loya, L. Rubio, J. Fernandez-Saez, Natural frequencies for bending vibrations of Timoshenko cracked beams, Journal of Sound and Vibration 290 (2006) 640–653.

[8] A.K. Pandey, M. Biswas, M.M. Samman, Damage detection from changes in curvature mode shapes, Journal of Sound and Vibration 145 (1991) 321–332.

[9] C.P. Ratcliffe, Damage detection using a modified Laplacian operator on mode shape data, Journal of Sound and Vibration 204 (1997) 505–517.
[10] J.T. Kim, Y.S. Ryu, H.M. Cho, N. Stubbs, Damage identification in beam-type structures: frequency-based method vs mode-shape-based method, Engineering Structures 25 (2003) 57–67.
[11] A.K. Pandey, M. Biswas, Damage detection in structures using changes in flexibility, Journal of Sound and Vibration 169 (1994) 3–17.
[12] S.D. Panteliou, T.G. Chondros, V.C. Argyrakis, A.D. Dimarogonas, Damping factors as an indicator of crack severity, Journal of Sound and Vibration 241 (2001) 235–245.
[13] J. Fernandez-Saez, L. Rubio, C. Navarro, Approximate calculation of the fundamental frequency for bending vibrations of cracked beams, Journal of Sound and Vibration 225 (1999) 345–352.
[14] Y.S. Lee, M.J. Chung, A study on crack detection using eigenfrequency test data, Computers & Structures 77 (2000) 327–342.
[15] S. Christides, A.D.S. Barr, One-dimensional theory of cracked Bernoulli-Euler beams, International Journal of Mechanical Science 26 (1984) 639–648.
[16] V. Stojanović, P. Ribeiro, S. Stoykov, Non-linear vibration of Timoshenko damaged beams by a new p-version finite element method, Computers & Structures 120 (2013) 107–119.
[17] İ. Coşkun, The response of a finite beam on a tensionless Pasternak foundation subjected to a harmonic load, European Journal of Mechanics—A/Solids 22 (1) (2003) 151–161.
[18] X. Ma, J.W. Butterworth, G.C. Clifton, Static analysis of an infinite beam resting on a tensionless Pasternak foundation, European Journal of Mechanics—A/Solids 28 (4) (2009) 697–703.
[19] X. Ma, J. Butterworth, C. Clifton, Shear buckling of infinite plates resting on tensionless elastic foundations, European Journal of Mechanics—A/Solids 30 (6) (2011) 1024–1027.
[20] A. Nayfeh, D. Mook, Nonlinear Oscillations, Wiley-VCH, Weinheim, 1995.
[21] S. Shaw, C. Pierre, Normal modes of vibration for nonlinear continuous systems, Journal of Sound and Vibration 169 (1994) 319–347.
[22] K. Nagaya, S. Kato, Transient response of a multi-span beam on non-symmetric non-linear supports, Journal of Sound and Vibration 138 (1990) 59–71.
[23] R. Kanaka, R. Venkateswara, Effect of a non-linear elastic foundation on the mode shapes, instability, and vibration problems of uniform columns/beams, Journal of Sound and Vibration 160 (1993) 369–371.
[24] T.-P. Chang, Y.-N. Liu, Dynamic finite element analysis of a nonlinear beam subjected to a moving load, International Journal of Solids and Structures 33 (1996) 1673–1688.
[25] E.J. Sapountzakis, A.E. Kampitsis, Nonlinear response of shear deformable beams on tensionless nonlinear viscoelastic foundation under moving loads, Journal of Sound and Vibration 330 (2011) 5410–5426.
[26] P. Malekzadeh, A.R. Vosoughi, DQM large amplitude vibration of composite beams on nonlinear elastic foundations with restrained edges, Communications in Nonlinear Science and Numerical Simulation 14 (2009) 906–915.
[27] L. Demeio, S. Lenci, Nonlinear resonances of a semi-infinite cable on a nonlinear elastic foundation, Communications in Nonlinear Science and Numerical Simulation 18 (2013) 785–798.
[28] P. Ribeiro, A p-version, first-order shear deformation, finite element for geometrically non-linear vibration of curved beams, International Journal for Numerical Methods in Engineering 61 (2004) 2696–2715.

[29] P. Ribeiro, Free periodic vibrations of beams with large displacements and initial plastic strains, International Journal of Mechanical Sciences 52 (2010) 1407–1418.
[30] P. Ribeiro, Hierarchical finite element analyses of geometrically non-linear vibration of beams and plane frames, Journal of Sound and Vibration 246 (2001) 225–244.
[31] P. Ribeiro, Non-linear forced vibrations of thin/thick beams and plates by the finite element and shooting methods, Computers & Structures 82 (2004) 1413–1423.
[32] W. Han, M. Petyt, Linear vibration analysis of laminated rectangular plates using the hierarchical finite element method, Part 1: free vibration analysis, Computers & Structures 61 (1996) 705–712.
[33] N.S. Bardell, The application of symbolic computing to the hierarchical finite element method, International Journal for Numerical Methods in Engineering 28 (1989) 1181–1204.
[34] N.S. Bardell, R.S. Langley, J.M. Dunsdon, G.S. Aglietti, An h-p finite element vibration analysis of open conical sandwich panels and conical sandwich frusta, Journal of Sound and Vibration 226 (1999) 345–377.
[35] P. Ribeiro, A hierarchical finite element for geometrically non-linear vibration of doubly curved, moderately thick isotropic shallow shells, International Journal for Numerical Methods in Engineering 56 (2003) 715–738.
[36] P. Ribeiro, First-order shear deformation, p-version, finite element for laminated plate nonlinear vibrations, AIAA Journal 43 (2005) 1371–1379.
[37] A. Houmat, Free vibration analysis of membranes using the h-p version of the finite element method, Journal of Sound and Vibration 282 (2005) 401–410.
[38] B. Szabó, I. Babuska, Finite Element Analysis, John Wiley & Sons, New York, 1991.
[39] Z. Yu, F. Chu, Identification of crack in functionally graded material beams using the p-version of finite element method, Journal of Sound and Vibration 325 (2009) 69–84.
[40] V. Stojanović, Geometrically nonlinear vibrations of beams supported by a nonlinear elastic foundation with variable discontinuity, Communications in Nonlinear Science and Numerical Simulation 28 (2015) 66–80.
[41] A. Worku, Development of a calibrated Pasternak foundation model for practical use, International Journal of Geotechnical Engineering 8 (2013) 26–33.
[42] C.M. Wang, J.N. Reddy, K.H. Lee, Shear Deformable Beams and Plates, Relationships with Classical Solutions, Elsevier, Amsterdam, 2000.
[43] V. Stojanović, M. Petković, Nonlinear dynamic analysis of damaged Reddy–Bickford beams supported on anelastic Pasternak foundation, Journal of Sound and Vibration 385 (2016) 239–266.
[44] N.S. Bardell, Free vibration analysis of a flat plate using the hierarchical finite element method, Journal of Sound and Vibration 151 (2) (1991) 263–289.
[45] H. Wolfe, An experimental investigation of nonlinear behaviour of beams and plates excited to high levels of dynamic response, Ph.D. thesis, University of Southampton, 1995.
[46] C. Chiang, C. Mei, C. Gray, Finite element large-amplitude free and forced vibrations of rectangular thin composite plates, Transactions of the ASME, Journal of Vibration and Acoustics 113 (1991) 309–315.

The purpose of an arch in the stability of nonlinear vibrations of coupled structures

12

12.1 Introduction

The one-dimensional two-layered and multilayered continuous systems can often be found in technical practice. The comprehension of their dynamic behavior has large significance because of these three important structural performances: weight reduction, strength and stiffness increase, and vibration absorption. In the mentioned systems, different layers can be made of different materials, and each layer can have a specific function—not only to increase structural strength and stiffness but also thermal insulation, dissipation, etc. A proper combination of layers can fulfill specific requirements. For these reasons and other characteristics, these systems have applications in aerospace, automotive industry, naval engineering, microelectronic engineering, civil engineering, etc.

The dynamic response of a mechanical system can be changed by changing the geometry of curvatures of certain constituent parts. Characteristic examples of such structures are curved beams and arches. Petyt and Fleischer [1] performed one of the first studies on the linear vibration of arches using the finite element analysis (FEM). To study the radial vibrations of curved beams, they used three FEM models, which had different boundary conditions and subtended angles. Lee et al. [2] found a formulation for solving a problem of shear and membrane locking, and provided three examples; however, none takes into account the dynamics. Eisenberger and Efraim [3] formulated a dynamic stiffness matrix for a circular beam and determined the natural frequencies of vibration. Bi and Dai [4] analyzed the dynamic behavior of shallow arches subjected to periodic excitations. They used a two-mode expansion, and the ensuing two nonlinear motion equations were integrated numerically. Chen et al. [5] examined nonlinear steady-state vibrations of beams, frames, and shallow arches by applying the FEM and incremental harmonic balance methods. By adopting a straight beam element for h-FEM, the researchers calculated fundamental, superharmonic, subharmonic, and combination resonances. The efficiency of the h-FEM in the nonlinear analysis was shown in Ref. [6]. An analysis of internal resonances in undamped and free vibrations of curved Timoshenko beams can be seen in Ref. [7]. There, to present the nonlinear differential equations of motion, the Galerkin method was used in the space and time domain. The vibration analysis of complex structures—comprising a beam and an arch—is difficult because of the viscoelastic nonlinear interactions. A physical model of a bridge is a characteristic example of a dynamic model of such a structure. In addition, there are double-beam cranes, double-beam spectrometers, double-beam interferometers, etc. Double-beam systems are a significant technological extension of the single-beam concept. Being a

Modeling of Complex Dynamic Systems. DOI: https://doi.org/10.1016/B978-0-443-23942-7.00012-7

complex continuous system (two one-dimensional solids coupled by an elastic medium), the double-beam system has drawn attention; different aspects of dynamics have been studied, however, only in the linear vibrational regime. Seelig and Hoppmann II [8] in their study of an elastically connected double-beam system subjected to an impulsive load presented the development and solution for differential equations of motion. Rao [9] investigated the natural vibrations of Timoshenko beam systems. Oniszczuk [10] studied the free transverse vibrations of two simply supported Euler–Bernoulli beams coupled by a Winkler layer. Bochicchio et al. [11] investigated the damped transverse nonlinear vibrations of an elastically coupled straight double-beam system, where the beams were subjected to compressive axial loading. All of that showed that the stationary problem could use both—the unimodal, and the bimodal buckled solutions—and that the number depended on both the structural parameters and the applied axial loads. Many authors have investigated the problems of nonlinear curved beams. Huang et al. [12] studied a curved beam subjected to a uniform base harmonic excitation with both quadratic and cubic nonlinearities; a high-dimensional model was derived that can include nonlinear coupling. Mohamed et al. [13] defined a modified numerical procedure for the analysis of nonlinear vibrations of a curved clamped-clamped beam near the buckling point (after the buckling has unfolded) where von Kármán's large deformation and small strain are taken into account. Huang et al. [14] presented multifarious bifurcation phenomena for a single curved nonlinear beam subjected to a base harmonic excitation. Using the frequency responses of the system without and with an excited antisymmetric mode, the bifurcation phenomena were explained and discussed. Another notable effect is the stretching of rotating beams that results from centrifugal forces. Chen et al. [15] analyzed the stretching effect of a curved beam; the results (obtained from the radial point interpolation method) showed that the configurations in steady and transient states were close. Medina et al. [16] analyzed the problem of nonlinear curved beams at the microlevel. In that study, they analyzed an interesting phenomenon that occurs in latched curved bistable beams when their deformed shape is their buckled state in the absence of the force. The nonlinear dynamic analysis of 3D curved beams is of importance since the advances in the design of microstructures have enabled the construction of various free forms of curved beams. To that end, Weeger et al. [17] found an isogeometric method for shape optimization of curved 3D beams and beam structures. They used geometrically exact Cosserat rod theory to model nonlinear 3D beams subjected to large deformations and rotations.

This chapter is about the study of nonlinear vibrations of a viscoelastically coupled beam–arch system; it is founded on the study by Stojanović et al. [18]. Complex beam structures often operate in convoluted environmental conditions, and they are regularly exposed to several dynamic excitations. Knowing their natural frequencies is important, so resonance phenomena can be prevented. Hence, the focus of this study was first on the determination of natural frequencies in the linear regime, afterwards the study continued in the nonlinear regime under varying excitations. The main aim of this study is to indicate the advantage of the beam–arch system over the classical double-beam system, along with presenting the discovered phenomena concerning the coupled structures and their wide applications in situations where vibrations dominate. The curvature of the lower beam leads to improvements in the two main vibrational aspects that were not detectable in the linear study and without the nonlinear analysis could not be predicted. The first aspect (if the curvature is present) is that the upper beam in the beam–arch system possesses the amplitude–symmetric vibration behavior in the steady-state response regimes. This behavior is present in single-beam models but not in a classical double-beam model for the same excitation parameters. The second advantageous aspect is lower amplitudes of the upper beam or even complete nonlinear dynamic absorption by the lower curved beam. The upper beam's higher nonsymmetric amplitude deviations in the steady-state vibrational regime are transferred to the lower one, only if the lower beam is curved. Thus the lower curved beam has a stabilizing feature. The same phenomenon caused the formation of the vibrational symmetry of the beam in terms of velocity versus deflection. Within the steady-state regime of vibrations, the maximal deflection and the shapes of the beam and the arch were found and presented. Finally, by applying the Richardson extrapolation method within the standard Newmark procedure, an improvement is also presented. The frequency-domain analysis was performed using the harmonic balance and continuation methods.

12.2 Problem statement and modeling of a geometrically nonlinear beam–arch coupled system using the p-FEM

A close study of the coupled beam–arch system with uniform thickness h is presented in this section; the result of the study is the equations of motion that are derived. When the beam is not deformed, the points of the centroidal axis are at the distance w_i from the reference straight line (Fig. 12.1). Within the Oxz rectangular coordinate system, the quadratic centroidal axis of the beam can be presented as:

$$w^i(x) = -\frac{1}{2}\left(\frac{x^2}{R}\right) \tag{12.1}$$

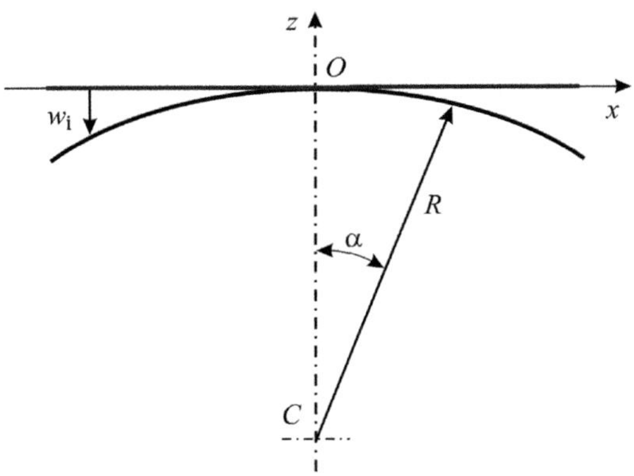

Figure 12.1 Curved beam. The points of the centroidal axis are at the distance w_i from the reference straight line.

where R stands for the principal radius of the curvature ($R \gg x$). The beams with a curvature radius that is greatly larger than their lengths were analyzed in Ref. [19].

By developing the final part in the Taylor series around $x = 0$, the following is obtained:

$$\sqrt{1-\left(\frac{x}{R}\right)^2} = 1 - \frac{1}{2R^2}x^2 + \text{h. o. t} \tag{12.2}$$

In this study, shear effects and rotary inertia are considered in the description of the coupled beam–arch system (Fig. 12.2). Aside from the Timoshenko beam assumption, the beam analog of von Kármán's plate assumption is also used; in this way, the geometric nonlinearity for large deflections can be taken into account. A nonlinear Winkler layer represents a viscoelastic coupling between the beam and the arch.

A viscoelastic layer [20] is defined with the linear stiffness modulus k_L, nonlinear stiffness modulus k_{NL}, and damping coefficient β. The beam and the arch are assumed to be both elastic and isotropic. The axial displacements u_1, u_2 and the transverse displacements w_1, w_2 at points of the arch and the beam are hence given by

$$u_1(x, z, t) = u_1^0(x, t) + z\theta_1^0(x, t), \quad w_1(x, z, t) = w_1^0(x, t), \tag{12.3}$$

$$u_2(x, z, t) = u_2^0(x, t) + z\theta_2^0(x, t), \quad w_2(x, z, t) = w_2^0(x, t), \tag{12.4}$$

When the radius of the curvature, defined by R, is greatly larger than L, ($R \gg L$) the following longitudinal strain at $z = 0$, ε_{x1}^0 can be presented approximately by

$$\varepsilon_{x1}^0 = \frac{\partial u_1^0(x, t)}{\partial x} + \frac{w_1^0(x, t)}{R}, \tag{12.5}$$

where the second term on the right-hand side of Eq. (12.5) is explained in detail for cylindrical shallow shells in Appendix A12.1 [21]. The essential part is the same in a curved beam. Since a von Kármán approach is adopted, the dominant geometric

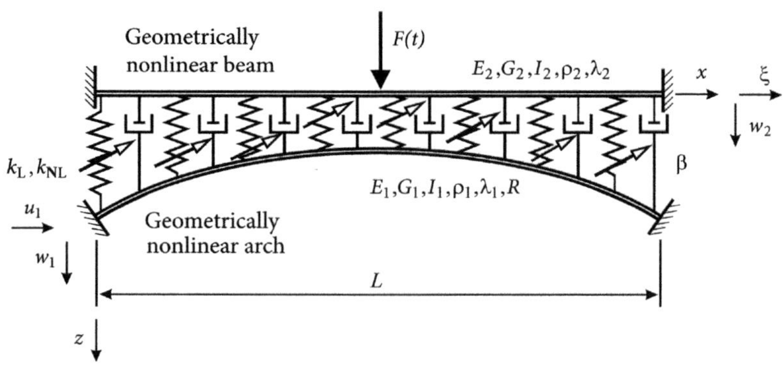

Figure 12.2 Coupled beam–arch system. A viscoelastic layer performs the coupling.

nonlinearity is considered, and the equations for the longitudinal and shear strains have the following form:

$$\varepsilon_{x1} = \varepsilon_{x1}^0 + \frac{1}{2}\left(\frac{\partial w_1^0}{\partial x}\right)^2 + z\frac{\partial \theta_1^0}{\partial x}, \quad \varepsilon_{x2} = \frac{\partial u_2^0(x,t)}{\partial x} + \frac{1}{2}\left(\frac{\partial w_2^0}{\partial x}\right)^2 + z\frac{\partial \theta_2^0}{\partial x}. \tag{12.6}$$

$$\gamma_{xz1} = \frac{\partial w_1^0}{\partial x} + \theta_1^0, \quad \gamma_{xz2} = \frac{\partial w_2^0}{\partial x} + \theta_2^0. \tag{12.7}$$

The fundamental equations for the linear elastic isotropic materials are as follows:

$$\boldsymbol{\sigma} = \mathbf{D}\boldsymbol{\varepsilon} \qquad \begin{Bmatrix} \sigma_{x1} \\ \tau_{xz1} \\ \sigma_{x2} \\ \tau_{xz2} \end{Bmatrix} = \begin{bmatrix} E_1 & 0 & 0 & 0 \\ 0 & \lambda_1 G_1 & 0 & 0 \\ 0 & 0 & E_2 & 0 \\ 0 & 0 & 0 & \lambda_2 G_2 \end{bmatrix} \begin{Bmatrix} \varepsilon_{x1} \\ \gamma_{xz1} \\ \varepsilon_{x2} \\ \gamma_{xz2} \end{Bmatrix}, \tag{12.8}$$

in which E_1 and E_2 are the Young moduli of the arch and the straight beam, respectively; G_1 and G_2 are the shear moduli; λ_1 and λ_2 are the shear correction factors (in line with the first-order shear deformation theory); $\mathbf{D}$ is the matrix of elastic constants. The shear modulus can be determined from $E_i/(2(1+\nu_i))$, where ν_i represents the Poisson ratio. In p-FEM, the middle surface displacements, for each element, are expressed in the form:

$$\begin{Bmatrix} u_1^0(\xi,t) \\ w_1^0(\xi,t) \\ \theta_1^0(\xi,t) \\ u_2^0(\xi,t) \\ w_2^0(\xi,t) \\ \theta_2^0(\xi,t) \end{Bmatrix} = \operatorname{diag}\{[N^{u_1}]^T, [N^{w_1}]^T, [N^{\theta_1}]^T, [N^{u_2}]^T, [N^{w_2}]^T, [N^{\theta_2}]^T\} \begin{Bmatrix} \mathbf{q}_{u1}(t) \\ \mathbf{q}_{w1}(t) \\ \mathbf{q}_{\theta 1}(t) \\ \mathbf{q}_{u2}(t) \\ \mathbf{q}_{w2}(t) \\ \mathbf{q}_{\theta 2}(t) \end{Bmatrix}, \tag{12.9}$$

in which $\mathbf{q}_{ui}(t)$ are the vectors of generalized longitudinal displacements, $\mathbf{q}_{wi}(t)$ are the vectors of generalized transverse displacements, and $\mathbf{q}_{\theta i}(t)$ are the vectors of generalized rotations about the y axis; $\mathbf{N}(\xi)$ is the matrix of shape functions formed by the row vectors of longitudinal, transverse, and rotation shape functions. These row vectors—of longitudinal, transverse, and rotation-shape functions—are represented in the form:

$$\{N^{u1}(\xi)\}^T = \{N^{u2}(\xi)\}^T = \lfloor g_1(\xi) \; g_2(\xi) \; g_3(\xi) \; \cdots \; g_{p(gi)}(\xi) \rfloor,$$

$$\{N^{w1}(\xi)\}^T = \{N^{w2}(\xi)\}^T = \lfloor f_1(\xi) \; f_2(\xi) \; f_3(\xi) \; \cdots \; f_{p(fi)}(\xi) \rfloor,$$

$$\{N^{\theta 1}(\xi)\}^T = \{N^{\theta 2}(\xi)\}^T = \lfloor \theta_1(\xi) \; \theta_2(\xi) \; \theta_3(\xi) \; \cdots \; \theta_{p(\theta i)}(\xi) \rfloor, \tag{12.10}$$

in which $p(gi)$, $p(fi)$, and $p(\theta i)$ are the numbers of longitudinal, transverse, and rotation-shape functions, respectively; while ξ is the local horizontal coordinate. The longitudinal strains Eq. (12.6) can be written as:

$$\varepsilon_{x1} = \lfloor 1 \quad z \rfloor \left(\begin{Bmatrix} \varepsilon_{01}^{p} \\ \varepsilon_{01}^{b} \end{Bmatrix} + \begin{Bmatrix} \varepsilon_{I} \\ 0 \end{Bmatrix} + \begin{Bmatrix} \varepsilon_{L1}^{p} \\ 0 \end{Bmatrix} \right), \quad \varepsilon_{x2} = \lfloor 1 \quad z \rfloor \left(\begin{Bmatrix} \varepsilon_{02}^{p} \\ \varepsilon_{02}^{b} \end{Bmatrix} + \begin{Bmatrix} \varepsilon_{L2}^{p} \\ 0 \end{Bmatrix} \right), \tag{12.11}$$

in which $\{\varepsilon_{0i}^{p}\}$, $\{\varepsilon_{0i}^{b}\}$, and $\{\varepsilon_{Li}^{p}\}$ present the linear longitudinal strains, bending strains, and geometrically nonlinear longitudinal strains, respectively. These strains and the contribution of the initial deformation to the longitudinal strain $\{\varepsilon_I\}$ are defined as:

$$\begin{Bmatrix} \varepsilon_{01}^{p} \\ \varepsilon_{02}^{p} \end{Bmatrix} = \begin{Bmatrix} \dfrac{\partial u_1^0}{\partial x} \\ \dfrac{\partial u_2^0}{\partial x} \end{Bmatrix}, \quad \begin{Bmatrix} \varepsilon_{01}^{b} \\ \varepsilon_{02}^{b} \end{Bmatrix} = \begin{Bmatrix} \dfrac{\partial \theta_1^0}{\partial x} \\ \dfrac{\partial \theta_2^0}{\partial x} \end{Bmatrix}, \quad \begin{Bmatrix} \varepsilon_{L1}^{p} \\ \varepsilon_{L2}^{p} \end{Bmatrix} = \begin{Bmatrix} \dfrac{1}{2}\left(\dfrac{\partial w_1^0}{\partial x}\right)^2 \\ \dfrac{1}{2}\left(\dfrac{\partial w_2^0}{\partial x}\right)^2 \end{Bmatrix}, \quad \varepsilon_I = \frac{w_1^0}{R}. \tag{12.12}$$

The stresses, the coupled resultants per unit lengths, and the shear stresses resultants are:

$$T_{xi} = \int_{-h/2}^{h/2} \sigma_{xi}\,\mathrm{d}z, \quad M_{xi} = \int_{-h/2}^{h/2} \sigma_{xi}\, z\,\mathrm{d}z, \quad Q_{xi} = \int_{-h/2}^{h/2} \tau_{xzi}\,\mathrm{d}z, \quad i = 1,\ 2. \tag{12.13}$$

Substituting Eq. (12.8) into Eq. (12.13) gives the following relations for the beam–arch system:

$$\begin{Bmatrix} T_1 \\ M_1 \\ T_2 \\ M_2 \end{Bmatrix} = \begin{bmatrix} A_{\Delta 1} & 0 & 0 & 0 \\ 0 & D_{\Delta 1} & 0 & 0 \\ 0 & 0 & A_{\Delta 2} & 0 \\ 0 & 0 & 0 & D_{\Delta 2} \end{bmatrix} \left(\begin{Bmatrix} \varepsilon_{01}^{p} \\ \varepsilon_{01}^{b} \\ \varepsilon_{02}^{p} \\ \varepsilon_{02}^{b} \end{Bmatrix} + \begin{Bmatrix} \varepsilon_I \\ 0 \\ 0 \\ 0 \end{Bmatrix} + \begin{Bmatrix} \varepsilon_{L1}^{p} \\ 0 \\ \varepsilon_{L2}^{p} \\ 0 \end{Bmatrix} \right),$$

$$\begin{Bmatrix} Q_{x1} \\ Q_{x2} \end{Bmatrix} = \begin{Bmatrix} \dfrac{E_1 h_1 \lambda_1}{2(1+\nu_1)} \gamma_{xz1} \\ \dfrac{E_2 h_2 \lambda_2}{2(1+\nu_2)} \gamma_{xz2} \end{Bmatrix}, \tag{12.14}$$

$$\begin{Bmatrix} A_{\Delta 1},\ D_{\Delta 1} \\ A_{\Delta 2},\ D_{\Delta 2} \end{Bmatrix} = \begin{Bmatrix} \int_z E_1 (1,\ z^2) \\ \int_z E_2 (1,\ z^2) \end{Bmatrix}. \tag{12.15}$$

The equations of motion are derived from the principle of virtual work, according to which $\delta W_{\mathbf{in}} + \delta W_V + \delta W_{\mathbf{ex}} = 0$, with the components $\delta \mathbf{d} = \mathbf{N}\delta \mathbf{q}$. For the virtual work of internal forces, the following relations apply:

$$\delta W_V = -\int_V \delta \boldsymbol{\varepsilon}^{\mathrm{T}} \boldsymbol{\sigma}\, \mathrm{d}V =$$

$$= -b\int_L^z \left(\begin{Bmatrix} \varepsilon_{01}^p \\ \varepsilon_{01}^b \\ \varepsilon_{02}^p \\ \varepsilon_{02}^b \end{Bmatrix}^{\mathrm{T}} + \begin{Bmatrix} \varepsilon_I \\ 0 \\ 0 \\ 0 \end{Bmatrix}^{\mathrm{T}} + \begin{Bmatrix} \varepsilon_{L1}^p \\ 0 \\ \varepsilon_{L2}^p \\ 0 \end{Bmatrix}^{\mathrm{T}} \right) \begin{bmatrix} A_{\Delta 1} & 0 & 0 & 0 \\ 0 & D_{\Delta 1} & 0 & 0 \\ 0 & 0 & A_{\Delta 2} & 0 \\ 0 & 0 & 0 & D_{\Delta 2} \end{bmatrix} \left(\begin{Bmatrix} \varepsilon_{01}^p \\ \varepsilon_{01}^b \\ \varepsilon_{02}^p \\ \varepsilon_{02}^b \end{Bmatrix} \right.$$

$$\left. + \begin{Bmatrix} \varepsilon_I \\ 0 \\ 0 \\ 0 \end{Bmatrix} + \begin{Bmatrix} \varepsilon_{L1}^p \\ 0 \\ \varepsilon_{L2}^p \\ 0 \end{Bmatrix} \right) \mathrm{dx} - b\int_L^z \begin{Bmatrix} \delta\gamma_{xz1} \\ \delta\gamma_{xz2} \end{Bmatrix}^{\mathrm{T}} \begin{Bmatrix} \dfrac{E_1 h_1 \lambda_1}{2(1+\nu_1)}\gamma_{xz1} \\ \dfrac{E_2 h_2 \lambda_2}{2(1+\nu_2)}\gamma_{xz2} \end{Bmatrix} \mathrm{dx}$$

$$= \delta \mathbf{q}^{\mathrm{T}} \mathbf{K}(\mathbf{q})\mathbf{q}, \tag{12.16}$$

where $\delta \boldsymbol{\varepsilon}$ represents the vector composed of virtual strains. The virtual work of the inertia forces is:

$$\delta W_{\mathbf{in}} = -\int_V \rho \delta \mathbf{d}^{\mathrm{T}} \ddot{\mathbf{d}} \mathrm{d}V = -\begin{Bmatrix} \rho_1 b_1 \int_{-h_1/2}^{h_1/2} \int_{-L/2}^{\frac{L}{2}} (\delta u_1 \ddot{u}_1 + \delta w_1 \ddot{w}_1)\mathrm{dx\ dz} \\ \rho_2 b_2 \int_{-h_2/2}^{h_2/2} \int_{-L/2}^{\frac{L}{2}} (\delta u_2 \ddot{u}_2 + \delta w_2 \ddot{w}_2)\mathrm{dx\ dz} \end{Bmatrix} = \delta \mathbf{q}^{\mathrm{T}} \mathbf{M}\ddot{\mathbf{q}}. \tag{12.17}$$

where $\ddot{\mathbf{d}}$ are the components of acceleration of a point of the beam. The principle of virtual work of external forces is applied to describe the general case of external mechanical activities. Considering the transversal point load, $P_{wi}^j(t)$, the distributed transverse load, $P_{wi}^d(x, t)$, the longitudinal point load, $P_{ui}^j(t)$, the distributed longitudinal load, $P_{ui}^d(x, t)$, and in the same manner the point and distributed moment $M_i^j(t)$, $M_i^j(x, t)$, the virtual work of external forces is defined with the following expressions:

$$\delta W_{\mathbf{ex}(1)} = \int_L [P_{w1}^j(t)\delta(x - x_j) + P_{w1}^d(x, t) - k_w(w_1 - w_2)]\delta w_1 \mathrm{dL}$$

$$+ \int_L [P_{u1}^j(t)\delta(x - x_j) + P_{u1}^d(x, t)]\delta u_1 \mathrm{dL} + \int_L [M_1^j(t)\delta(x - x_j) + M_1^d(x, t)]\delta\theta_1 \mathrm{dL}, \tag{12.18}$$

$$\delta W_{\mathbf{ex}(2)} = \int_L [P_{w2}^j(t)\delta(x - x_j) + P_{w2}^d(x, t) - k_w(w_2 - w_1)]\delta w_2 \mathrm{dL}$$

$$+ \int_L [P_{u2}^j(t)\delta(x - x_j) + P_{u2}^d(x, t)]\delta u_2 \mathrm{dL} + \int_L [M_2^j(t)\delta(x - x_j) + M_2^d(x, t)]\delta\theta_2 \mathrm{dL},$$

$$i = 1,\ 2, \tag{12.19}$$

Considering that in Fig. 12.2 only the transversal load exists:

$$P_{w2}^d(x, t) = F(t), \tag{12.20}$$

the other external forces and moments are equal to zero:

$$P_{wi}^j(t) = P_{ui}^j(t) = P_{w1}^d(x, t) = P_{u1}^d(x, t) = P_{u2}^d(x, t) =$$

$$M_i^j(t) = M_i^d(x, t) = 0, \quad i = 1,\ 2. \tag{12.21}$$

In Eqs. (12.17–12.19), $\delta(x - x_j)$ represents a spatial Dirac delta function, ρ_i symbolizes the mass per unit volume, and $\ddot{u}_i = \mathrm{d}^2 w_i/\mathrm{d}t^2$. Applying the virtual work principle $\begin{Bmatrix} \delta W_{\mathbf{in}(1)} + \delta W_{\mathbf{V}(1)} + \delta W_{\mathbf{ex}(1)} \\ \delta W_{\mathbf{in}(2)} + \delta W_{\mathbf{V}(2)} + \delta W_{\mathbf{ex}(2)} \end{Bmatrix} = \begin{Bmatrix} 0 \\ 0 \end{Bmatrix}$, the following equations of motion of the nonlinearly coupled geometrically nonlinear beam–arch system are obtained:

$$\begin{bmatrix} \mathbf{M}_{11} & 0 & 0 & 0 & 0 & 0 \\ 0 & \mathbf{M}_{12} & 0 & 0 & 0 & 0 \\ 0 & 0 & \mathbf{M}_{13} & 0 & 0 & 0 \\ 0 & 0 & 0 & \mathbf{M}_{21} & 0 & 0 \\ 0 & 0 & 0 & 0 & \mathbf{M}_{22} & 0 \\ 0 & 0 & 0 & 0 & 0 & \mathbf{M}_{23} \end{bmatrix} \begin{Bmatrix} \ddot{\mathbf{q}}_{u1} \\ \ddot{\mathbf{q}}_{w1} \\ \ddot{\mathbf{q}}_{\theta 1} \\ \ddot{\mathbf{q}}_{u2} \\ \ddot{\mathbf{q}}_{w2} \\ \ddot{\mathbf{q}}_{\theta 2} \end{Bmatrix}$$

$$+\widetilde{\alpha}\left(\begin{bmatrix} \mathbf{K}_{11} & \mathbf{K}_{12(r)} & 0 & 0 & 0 & 0 \\ \mathbf{K}_{21(r)} & \mathbf{K}_{22(s)} + \mathbf{K}_{22(\gamma 11)} + \mathbf{K}_{22(k_L)} & \mathbf{K}_{23(\gamma 12)} & 0 & -\mathbf{K}_{22(k_L)} & 0 \\ 0 & \mathbf{K}_{32(\gamma 21)} & \mathbf{K}_{33(\gamma 22)} + \mathbf{K}_{33(0)} & 0 & 0 & 0 \\ 0 & 0 & 0 & \mathbf{K}_{44} & 0 & 0 \\ 0 & -\mathbf{K}_{22(k_L)} & 0 & 0 & \mathbf{K}_{55(\gamma 11)} + \mathbf{K}_{22(k_L)} & \mathbf{K}_{56(\gamma 12)} \\ 0 & 0 & 0 & 0 & \mathbf{K}_{65(\gamma 21)} & \mathbf{K}_{66(\gamma 22)} + \mathbf{K}_{66(0)} \end{bmatrix}\right)$$

$$\cdot\begin{bmatrix} \dot{\mathbf{q}}_{u1} & \dot{\mathbf{q}}_{w1} & \dot{\mathbf{q}}_{\theta 1} & \dot{\mathbf{q}}_{u2} & \dot{\mathbf{q}}_{w2} & \dot{\mathbf{q}}_{\theta 2} \end{bmatrix}^{\mathrm{T}}$$

$$+\widetilde{\beta}\left(\begin{bmatrix} \mathbf{M}_{11} & 0 & 0 & 0 & 0 & 0 \\ 0 & \mathbf{M}_{12} & 0 & 0 & 0 & 0 \\ 0 & 0 & \mathbf{M}_{13} & 0 & 0 & 0 \\ 0 & 0 & 0 & \mathbf{M}_{21} & 0 & 0 \\ 0 & 0 & 0 & 0 & \mathbf{M}_{22} & 0 \\ 0 & 0 & 0 & 0 & 0 & \mathbf{M}_{23} \end{bmatrix}\right)\begin{Bmatrix} \dot{\mathbf{q}}_{u1} \\ \dot{\mathbf{q}}_{w1} \\ \dot{\mathbf{q}}_{\theta 1} \\ \dot{\mathbf{q}}_{u2} \\ \dot{\mathbf{q}}_{w2} \\ \dot{\mathbf{q}}_{\theta 2} \end{Bmatrix}$$

$$+\begin{bmatrix} \mathbf{K}_{11} & \mathbf{K}_{12(r)} & 0 & 0 & 0 & 0 \\ \mathbf{K}_{21(r)} & \mathbf{K}_{22(s)}+\mathbf{K}_{22(\gamma 11)}+\mathbf{K}_{22(k_L)} & \mathbf{K}_{23(\gamma 12)} & 0 & -\mathbf{K}_{22(k_L)} & 0 \\ 0 & \mathbf{K}_{32(\gamma 21)} & \mathbf{K}_{33(\gamma 22)}+\mathbf{K}_{33(0)} & 0 & 0 & 0 \\ 0 & 0 & 0 & \mathbf{K}_{44} & 0 & 0 \\ 0 & -\mathbf{K}_{22(k_L)} & 0 & 0 & \mathbf{K}_{55(\gamma 11)}+\mathbf{K}_{22(k_L)} & \mathbf{K}_{56(\gamma 12)} \\ 0 & 0 & 0 & 0 & \mathbf{K}_{65(\gamma 21)} & \mathbf{K}_{66(\gamma 22)}+\mathbf{K}_{66(0)} \end{bmatrix}$$

$$\cdot\left[\mathbf{q}_{u1}\ \ \mathbf{q}_{w1}\ \ \mathbf{q}_{\theta 1}\ \ \mathbf{q}_{u2}\ \ \mathbf{q}_{u2}\ \ \mathbf{q}_{\theta 2}\right]^{\mathbf{T}}$$

$$+\begin{bmatrix} 0 & \mathbf{K}_{12(NL)} & 0 & 0 & 0 & 0 \\ \mathbf{K}_{21(NL)} & \mathbf{K}_{22(NL)}+\mathbf{K}_{22(r_{NL})}+2\left[\mathbf{K}_{22(r_{NL})}\right]^{\mathrm{T}}+\mathbf{K}_{22(k_{NL})} & 0 & 0 & -\mathbf{K}_{24(k_{NL})} & 0 \\ 0 & 0 & 0 & 0 & 0 & 0 \\ 0 & 0 & 0 & 0 & \mathbf{K}_{45(NL)} & 0 \\ 0 & -\mathbf{K}_{22(k_{NL})} & 0 & \mathbf{K}_{54(NL)} & \mathbf{K}_{55(NL)}+\mathbf{K}_{24(k_{NL})} & 0 \\ 0 & 0 & 0 & 0 & 0 & 0 \end{bmatrix}\begin{Bmatrix} \mathbf{q}_{u1} \\ \mathbf{q}_{w1} \\ \mathbf{q}_{\theta 1} \\ \mathbf{q}_{u2} \\ \mathbf{q}_{w2} \\ \mathbf{q}_{\theta 2} \end{Bmatrix}$$

$$=\left[\mathbf{0}\ \ \mathbf{0}\ \ \mathbf{0}\ \ \mathbf{0}\ \ \mathbf{F}_{w2}\ \ \mathbf{0}\right]^{\mathbf{T}}, \tag{12.22}$$

in which $\{0\ \ 0\ \ 0\ \ 0\ \ \mathbf{F}_{w2}\ \ 0\}^{\mathbf{T}}$ represents the vector of generalized external forces. Constant matrices can be identified by $\mathbf{M}_{**}$, $\mathbf{K}_{**}$, and $\mathbf{K}_{**(r,\gamma,k_L)}$, they lead to linear terms in the equations of motion. Matrices $\mathbf{K}_{22(r_{NL})}$, $\mathbf{K}_{12(NL)}$, $\mathbf{K}_{21(NL)}$, $\mathbf{K}_{45(NL)}$, and $\mathbf{K}_{54(NL)}$ depend linearly on the solution, leading to quadratic nonlinear terms. Matrices $\mathbf{K}_{22(k_{NL})}$ and $\mathbf{K}_{55(k_{NL})}$ depend quadratically on the solution, thereby rendering the cubic nonlinearity. Matrices $\mathbf{K}_{22(k_{NL})}$ and $\mathbf{K}_{24(k_{NL})}$ depend cubically on the solution. The coefficients $\widetilde{\alpha}$ and $\widetilde{\beta}$ describe the Rayleigh proportional type of damping. Expressing the previous equations in a more condensed form will give the following result:

$$\mathbf{M}\ddot{\mathbf{q}}(\mathrm{t}) + \widetilde{\alpha}(\mathbf{K}_{\ell})\dot{\mathbf{q}}(\mathrm{t}) + \widetilde{\beta}(\mathbf{M})\dot{\mathbf{q}}(\mathrm{t}) + (\mathbf{K}_{\ell} + \mathbf{K}_{\mathrm{n}\ell})(\mathbf{q}(t))\mathbf{q}(\mathrm{t}) = \mathbf{F}(\mathrm{t}) \tag{12.23}$$

The linear mass, the linear stiffness, and the nonlinear stiffness matrices are given in detail in Appendix A12.1. Each of the displacement components is associated with a particular set of shape functions. The transverse displacement is specified as a function of the Legendre polynomials in Rodrigues form, plus the four Hermite cubics [22]. At the end of each element, only one of the Hermite cubics has a displacement or rotation not equal to zero. All of the Legendre polynomials, which are higher-order hierarchical functions, have zero amplitudes and slopes at $\xi = \pm 1$. The g set is a set of polynomials applied in conjunction with linear functions for the longitudinal displacements and rotations. These g functions have a value of zero, but at the boundaries, their slope differs from zero. In the numerical experiment that comprises both the time domain and the amplitude–frequency analysis, the vector of generalized external forces $\mathbf{F}(t) = \{0,\ 0,\ 0,\ 0,\ \mathbf{F}_{w2},\ 0\}^{\mathrm{T}}$ is an explicit function of time. Other than that, this function represents the concentrated harmonic excitation exerted on the upper beam in the transverse direction. The beam–arch case with clamped-clamped supports is considered and discussed in detail.

12.3 Convergence study

For verification and comparison, the ANSYS® commercial software package [23] is chosen; and within it is the h-version element BEAM189, which considers rotary inertia and shear effects. This FEM element has 3 nodes with 6 degrees of freedom at each node; which makes it more than required, because only 3 degrees of freedom per node are required for vibrational analysis in a plane. The BEAM189 is an element based on the Timoshenko beam theory, which makes it suitable for the analysis of slender to moderately thick beam structures. The FEM results from ANSYS® are obtained with 300 elements (beam, arch, and elastic continuous layer have 100 elements each). The convergence of transverse natural frequencies can be achieved with circa 30 elements. Here, however, a finer mesh is used so that the *p*-FEM model can be validated against a proper reference model. A similar validation procedure was performed with an intricate layer that has discontinuity [20]. In Tables 12.1–12.4, it is shown that the *p*-FEM formulation introduced here gives accurate results and measures up with the *h*-FEM models. In comparing two cases—one utilizing seven shape functions (with the p-FEM method and a minimum of 42 degrees of freedom (DOF)) and another using 300 DOF with the commercial software ANSYS®—the error is approximately 1%. Further increasing the number of shape functions within the p-version of FEM reduces this error to below 1%. The shear correction factor, which depends on the beam's cross-section, was proposed in Refs. [24–26]. Kaneko [26] concluded that the value implied by Timoshenko [24] is closest to the experimental results. For rectangular beams, this value is $\lambda = (5 + 5\nu)/(6 + 5\nu)$. Therefore in this study, the value $\lambda = (5 + 5\nu)/(6 + 5\nu)$ is used. The values of the present approach (Tables 12.1–12.4) show that by taking the 90 DOF ($p(g, f, \theta) = 15,\ 6{\cdot}15 = 90$) over one element per beam or arch model (the feature of the *p*-FEM) as a reference (used in the further numerical time-domain analysis), the error of the model with 42 DOF ($p(g, f, \theta) = 7,\ 6{\cdot}7 = 42$), also over one element per beam or arch, is lower than 1 % in the first two modes. In Tables 12.1–12.4, the number of shape functions is presented with $p(g, f, \theta)$; the natural frequencies ω_{n1} and ω_{n2} are associated with vibrations in the transverse direction of the coupled beam–arch system.

Table 12.1 Natural frequencies (n=1) (rad/s) of a clamped-clamped beam–arch coupled system.

The curvature effect	r[1/m]	0	0.2	0.4	0.6	DOF
$p(g, f, \theta) = 7$	ω_{11}	44.463	62.693	62.864	62.892	42
	ω_{12}	77.286	112.49	90.373	71.206	
$p(g, f, \theta) = 10$	ω_{11}	44.460	62.691	62.861	62.889	60
	ω_{12}	77.284	112.44	90.317	71.151	
$p(g, f, \theta) = 15$	ω_{11}	44.457	62.688	62.859	62.886	90
	ω_{12}	77.282	112.42	90.296	71.131	
Ansys Beam189	ω_{11}	43.751	61.871	62.183	62.196	300
	ω_{12}	76.193	111.372	89.196	70.029	

Table 12.2 Natural frequencies (n=2) (rad/s) of a clamped-clamped beam–arch coupled system.

The curvature effect	r[1/m]	0	0.2	0.4	0.6	DOF
$p(g, f, \theta) = 7$	ω_{21}	121.58	132.44	130.38	129.90	42
	ω_{22}	137.01	206.67	194.51	173.45	
$p(g, f, \theta) = 10$	ω_{21}	121.54	132.40	130.34	129.86	60
	ω_{22}	136.96	206.06	193.16	171.82	
$p(g, f, \theta) = 15$	ω_{21}	121.53	132.39	130.33	129.84	90
	ω_{22}	136.95	205.94	193.03	171.68	
Ansys Beam189	ω_{21}	120.84	131.68	129.62	129.15	300
	ω_{22}	135.85	204.68	191.91	170.57	

Table 12.3 Natural frequencies (n=3) (rad/s) of a clamped-clamped beam–arch coupled system.

The curvature effect	r[1/m]	0	0.2	0.4	0.6	DOF
$p(g, f, \theta) = 7$	ω_{31}	236.21	240.71	240.63	240.52	42
	ω_{32}	244.47	373.08	347.12	320.56	
$p(g, f, \theta) = 10$	ω_{31}	235.73	240.24	240.15	240.04	60
	ω_{32}	244.01	369.75	342.31	314.55	
$p(g, f, \theta) = 15$	ω_{31}	235.64	240.15	240.06	239.95	90
	ω_{32}	243.92	369.50	341.99	314.18	
Ansys Beam189	ω_{31}	234.83	239.30	239.25	239.11	300
	ω_{32}	242.62	368.21	340.58	312.75	

Table 12.4 Natural frequencies (n=4) (rad/s) of a clamped-clamped beam–arch coupled system.

The curvature effect	r[1/m]	0	0.2	0.4	0.6	DOF
$p(g, f, \theta) = 7$	ω_{41}	387.17	390.06	389.83	389.77	42
	ω_{42}	392.24	413.25	616.99	659.44	
$p(g, f, \theta) = 10$	ω_{41}	384.49	387.40	387.17	387.11	60
	ω_{42}	389.60	407.80	508.55	488.67	
$p(g, f, \theta) = 15$	ω_{41}	384.29	387.20	386.97	386.91	90
	ω_{42}	389.40	407.69	507.44	487.18	
Ansys Beam189	ω_{41}	382.53	385.90	385.63	385.55	300
	ω_{42}	387.60	405.87	505.61	485.38	

The starting parameters for the obtained results that are presented in Tables 12.1–12.4 are:

$$L_i = 10\,\text{m},\ h_i = \frac{2}{5}\sqrt{\frac{3}{5}}\ \text{m},\ b_i = \frac{1}{8}\sqrt{\frac{5}{3}}\ \text{m},\ A_i = 5 \times 10^{-2}\,\text{m}^2,\ \rho_i = 2 \times 10^3\,\text{kgm}^{-3},$$
$$k = 2 \times 10^5\,\text{Pa},\ E_i = 1 \times 10^{10}\,\text{Pa},\ I_i = 4 \times 10^{-4}\,\text{m}^4,\ \lambda_i = \frac{5+5\nu}{6+5\nu},\ \nu_i = 0.34.$$

12.4 Time-domain analysis and improvement of the Newmark method

In this section, all the above mentioned is summed up, by using the obtained results. The following presents several time-domain results on nonlinear vibrations. By applying Newmark's method to the derived system of nonlinear equations of motion, a solution for displacement at time t can be obtained. The accuracy can be moderately increased by modifying the standard Newmark procedure. Accordingly, the improvement related to the accuracy of the results is here presented. The first step in the procedure of the method is to divide time t into smaller intervals Δt and to assume that the second derivatives of displacement (accelerations) are linear functions of time during any time interval Δt. Using the initial conditions, the displacements and velocities at the time Δt can be calculated by Newmark's method. Introducing the calculated displacements and velocities at the time Δt as the initial conditions for the second time interval—from Δt to $2 \times \Delta t$ —will bring about the results for the time $2 \times \Delta t$. This procedure continues until the desired instant has been achieved. With the assumption that acceleration is changing linearly between two successive moments j and $j + 1$ (so-called linear acceleration method), the acceleration between these two moments is:

$$\ddot{q}(\tau) = \ddot{q}_j + \frac{1}{\Delta t}(-\ddot{q}_j + \ddot{q}_{j+1})\tau, \quad 0 \le \tau \le \Delta t. \tag{12.24}$$

Integration with respect to time (t), brings about the velocity:

$$\dot{q}(\tau) = \dot{q}_j + \ddot{q}_j\tau + \frac{1}{2\Delta t}(-\ddot{q}_j + \ddot{q}_{j+1})\tau^2, \tag{12.25}$$

and another integration with respect to time results in the displacements as:

$$q(\tau) = q_j + \dot{q}_j\tau + \frac{1}{2}\ddot{q}_j\tau^2 + \frac{1}{6\Delta t}(-\ddot{q}_j + \ddot{q}_{j+1})\tau^3, \tag{12.26}$$

at time $\tau = \Delta t$ this will be:

$$\dot{q}_{j+1} = \dot{q}_j + \Delta t\left(\frac{1}{2}\ddot{q}_j + \frac{1}{2}\ddot{q}_{j+1}\right), \tag{12.27}$$

$$q_{j+1} = q_j + \dot{q}_j\Delta t + (\Delta t)^2\left(\frac{2}{6}\ddot{q}_j + \frac{1}{6}\ddot{q}_{j+1}\right). \tag{12.28}$$

Accordingly, a generalization of the linear acceleration method is thus the Newmark method, hence

$$\dot{q}_{j+1} = \dot{q}_j + \Delta t[(1-\gamma)\ddot{q}_j + \gamma\ddot{q}_{j+1}], \tag{12.29}$$

$$q_{j+1} = q_j + \dot{q}_j\Delta t + (\Delta t)^2\left[\left(\frac{1}{2} - \beta\right)\ddot{q}_j + \beta\ddot{q}_{j+1}\right]. \tag{12.30}$$

Putting $\gamma = 1/2$ and $\beta = 1/6$ will make the Newmark method equal to the linear acceleration method. Alternatively, it may be assumed that the acceleration is equal to an average value in each time interval:

$$\bar{\ddot{q}} = \left(\frac{\ddot{q}_j + \ddot{q}_{j+1}}{2}\right), \tag{12.31}$$

therefore

$$\dot{q}_{j+1} = \dot{q}_j + \Delta t\left(\frac{\ddot{q}_j + \ddot{q}_{j+1}}{2}\right). \tag{12.32}$$

$$q_{j+1} = q_j + \dot{q}_j\Delta t + \frac{1}{2}(\Delta t)^2\left(\frac{\ddot{q}_j + \ddot{q}_{j+1}}{2}\right). \tag{12.33}$$

Making $\gamma = 1/2$ and $\beta = 1/4$ will bring the Newmark method equal to the constant average acceleration method. From Eqs. (12.29–12.30) the following is obtained:

$$\ddot{q}_{j+1} = \frac{1}{\beta(\Delta t)^2}(q_{j+1} - q_j) - \frac{1}{\beta\Delta t}\dot{q}_j - \left(\frac{1}{2\beta} - 1\right)\ddot{q}_j. \tag{12.34}$$

$$\dot{q}_{j+1} = \frac{\gamma}{\beta\Delta t}(q_{j+1} - q_j) + \left(1 - \frac{\gamma}{\beta}\right)\dot{q}_j + \Delta t\left(1 - \frac{\gamma}{2\beta}\right)\ddot{q}_j. \tag{12.35}$$

or in a general form:

$$\ddot{q}_{j+1} = a_1(q_{j+1} - q_j) - a_3\dot{q}_j - a_5\ddot{q}_j. \tag{12.36}$$

$$\dot{q}_{j+1} = a_2(q_{j+1} - q_j) + a_4\dot{q}_j + a_6\ddot{q}_j, \tag{12.37}$$

where

$$a_1 = \frac{1}{\beta(\Delta t)^2}, \quad a_2 = \frac{\gamma}{\beta \Delta t}, \quad a_3 = \frac{1}{\beta \Delta t},$$

$$a_4 = \left(1 - \frac{\gamma}{\beta}\right), \quad a_5 = \left(\frac{1}{2\beta} - 1\right), \quad a_6 = \Delta t\left(1 - \frac{\gamma}{2\beta}\right).$$

Using the presented substitutions, Eq. (12.22) can be written in the following new form:

$$[a_1 M + a_2 C + K]\{q_{j+1}\} =$$

$$\{f_{j+1}\} + [a_1 M + a_2 C]\{q_j\} + [a_3 M - a_4 C]\{\dot{q}_j\} + [a_5 M - a_6 C]\{\ddot{q}_j\}, \tag{12.38}$$

where $[a_1 M + a_2 C + K]$ presents the effective stiffness and the right-hand side of Eq. (12.38) presents the effective force. For calculating the deformations in time $j + 1$, $\{q_{j+1}\}$, it is necessary to know the displacement, velocity, and acceleration in time j, which is $\{q_j\}$, $\{\dot{q}_j\}$, and $\{\ddot{q}_j\}$. The procedure is initiated by introducing the displacement and velocity in time 0, $\{q_0\}$, and $\{\dot{q}_0\}$ as the initial conditions of the problem. Afterward acceleration in time 0, $\{\ddot{q}_0\}$ should be calculated by

$$[M]\{\ddot{q}_0\} = \{f_0\} - [C]\{q_0\}. \tag{12.39}$$

With knowing $\{q_0\}$, $\{\dot{q}_0\}$, and $\{\ddot{q}_0\}$, the procedure will continue until the targeted time is reached. The next displacements will be calculated using Eq. (12.38). The velocity and acceleration in the new time will be obtained through Eqs. (12.36–12.37). With time, the deformations can converge to a constant value (with static loading) or change (with dynamic loading), which depends on the damping factor. It is noteworthy to mention that, in linear systems, Newmark's method is unconditionally stable when $\gamma \geq \frac{1}{2}$ and $\beta \geq \frac{1}{4}\left(\gamma + \frac{1}{2}\right)^2$; also, the constant average acceleration method is unconditionally stable—stability does not depend on the time step. However, the linear acceleration method is conditionally stable, with respect to the time step. For reasons of accuracy, the constant average acceleration method, with the time step given by $\frac{\omega_0 \Delta t}{2\pi} = 0.01$ or $\Delta t = 0.01\tau_0$, may be used, where τ_0 is the period of vibration.

Following the review of the applied method, it can be observed how an additional improvement in accuracy has been gained. This is done by applying the method for the time step $\Delta t/2$ and then using the extrapolation for $\Delta t = 0$ by the Richardson extrapolation method [27]. This improvement produces the results divulged in this study. The verification is made by comparing two examples with five and six-shape functions. As it is known, a larger number of shape functions provide better accuracy; with the Richardson extrapolation method and five shape functions, the result is closer to the solution obtained with the classical procedure and six shape functions. The result, reached with the classical procedure and five shape functions, has lower accuracy—as it can be concluded by comparison. The maximal deflection in the steady-state regime of vibration is chosen for comparison: beam (standard 0.427551;

improved 0.427634), arch (standard 0.0358066; improved 0.0358155), where the effect of the curvature exists. It is important to note that the standard Newmark procedure has already been verified and compared with the experimental results for the Timoshenko beam model [28] and the Timoshenko curved beam model [29].

12.5 Harmonic balance and continuation methods in frequency domain analysis

The harmonic balance method is used in frequency vibration analysis. The continuation method is applied to solve the system of nonlinear algebraic equations with frequency variation [6,22,30]. The standard routine of the continuation method is exercised with two main loops. In the external loop, a predictor of the solution is defined. The last two determined vectors of the curve, $\{\mathbf{w}\}_i$ and $\{\mathbf{w}\}_{i-1}$, are used for the prediction of the vector of generalized coordinates. The approximated solution is rectified in an internal loop. In system Eq. (12.22), the longitudinal inertia can be omitted, for it represents a small quantity and does not affect the solution in a nonlinear model [31,32]; afterward, the system will be

$$
\begin{aligned}
&\begin{bmatrix} \mathbf{M}_{12} & 0 & 0 & 0 \\ 0 & \mathbf{M}_{13} & 0 & 0 \\ 0 & 0 & \mathbf{M}_{22} & 0 \\ 0 & 0 & 0 & \mathbf{M}_{23} \end{bmatrix} \begin{Bmatrix} \ddot{\mathbf{q}}_{w1} \\ \ddot{\mathbf{q}}_{\theta 1} \\ \ddot{\mathbf{q}}_{w2} \\ \ddot{\mathbf{q}}_{\theta 2} \end{Bmatrix} + \frac{\tilde{\tilde{\beta}}}{\omega} \left(\begin{bmatrix} \mathbf{M}_{12} & 0 & 0 & 0 \\ 0 & \mathbf{M}_{13} & 0 & 0 \\ 0 & 0 & \mathbf{M}_{22} & 0 \\ 0 & 0 & 0 & \mathbf{M}_{23} \end{bmatrix} \right) \begin{Bmatrix} \dot{\mathbf{q}}_{w1} \\ \dot{\mathbf{q}}_{\theta 1} \\ \dot{\mathbf{q}}_{w2} \\ \dot{\mathbf{q}}_{\theta 2} \end{Bmatrix} \\
&+ \begin{bmatrix} \mathbf{Kr}_{11} & \mathbf{Kr}_{12} & \mathbf{Kr}_{13} & 0 \\ \mathbf{Kr}_{21} & \mathbf{Kr}_{22} & 0 & 0 \\ \mathbf{Kr}_{31} & 0 & \mathbf{Kr}_{33} & \mathbf{Kr}_{34} \\ 0 & 0 & \mathbf{Kr}_{43} & \mathbf{Kr}_{44} \end{bmatrix} \begin{Bmatrix} \mathbf{q}_{w1} \\ \mathbf{q}_{\theta 1} \\ \mathbf{q}_{w2} \\ \mathbf{q}_{\theta 2} \end{Bmatrix} + \begin{bmatrix} \mathbf{K1}_{n\ell}(\mathbf{q}_{w1}) & 0 & 0 & 0 \\ 0 & 0 & 0 & 0 \\ 0 & 0 & 0 & 0 \\ 0 & 0 & 0 & 0 \end{bmatrix} \begin{Bmatrix} \mathbf{q}_{w1} \\ \mathbf{q}_{\theta 1} \\ \mathbf{q}_{w2} \\ \mathbf{q}_{\theta 2} \end{Bmatrix} \\
&+ \begin{bmatrix} \mathbf{K1}_{n\ell}(\mathbf{q}_{w1}, \mathbf{q}_{w1}^{\otimes}) & 0 & 0 & 0 \\ 0 & 0 & 0 & 0 \\ 0 & 0 & \mathbf{K3}_{n\ell}(\mathbf{q}_{w2}, \mathbf{q}_{w2}^{\otimes}) & 0 \\ 0 & 0 & 0 & 0 \end{bmatrix} \begin{Bmatrix} \mathbf{q}_{w1} \\ \mathbf{q}_{\theta 1} \\ \mathbf{q}_{w2} \\ \mathbf{q}_{\theta 2} \end{Bmatrix} \\
&+ \begin{bmatrix} \mathbf{K1}_{n\ell}(\mathbf{q}_{w1}, \mathbf{q}_{w1}^{\otimes}, \mathbf{q}_{w1}^{\otimes\otimes}) & 0 & \mathbf{K1}_{n\ell}(\mathbf{q}_{w2}, \mathbf{q}_{w2}^{\otimes}, \mathbf{q}_{w2}^{\otimes\otimes}) & 0 \\ 0 & 0 & 0 & 0 \\ \mathbf{K3}_{n\ell}(\mathbf{q}_{w1}, \mathbf{q}_{w1}^{\otimes}, \mathbf{q}_{w1}^{\otimes\otimes}) & 0 & \mathbf{K3}_{n\ell}(\mathbf{q}_{w2}, \mathbf{q}_{w2}^{\otimes}, \mathbf{q}_{w2}^{\otimes\otimes}) & 0 \\ 0 & 0 & 0 & 0 \end{bmatrix} \begin{Bmatrix} \mathbf{q}_{w1} \\ \mathbf{q}_{\theta 1} \\ \mathbf{q}_{w2} \\ \mathbf{q}_{\theta 2} \end{Bmatrix} = \begin{Bmatrix} \mathbf{0} \\ \mathbf{0} \\ \mathbf{F}_{w2} \\ \mathbf{0} \end{Bmatrix},
\end{aligned}
\tag{12.40}
$$

where $\mathbf{K1}_{n\ell}(\mathbf{q}_{wi}, \mathbf{q}_{wi}^{\otimes})$ and $\mathbf{K3}_{n\ell}(\mathbf{q}_{wi}, \mathbf{q}_{wi}^{\otimes})$ indicates that matrices $\mathbf{K1}_{n\ell}$ and $\mathbf{K3}_{n\ell}$ are a quadratic function of $\mathbf{q}_{wi}$; $\mathbf{K1}_{n\ell}(\mathbf{q}_{wi}, \mathbf{q}_{wi}^{\otimes}, \mathbf{q}_{wi}^{\otimes\otimes})$ and $\mathbf{K3}_{n\ell}(\mathbf{q}_{wi}, \mathbf{q}_{wi}^{\otimes}, \mathbf{q}_{wi}^{\otimes\otimes})$ indicates that matrices $\mathbf{K1}_{n\ell}$ and $\mathbf{K3}_{n\ell}$ are a cubic function of $\mathbf{q}_{wi}$; $\tilde{\tilde{\beta}} = \omega_{l1}^2 \times \tilde{\tilde{\alpha}}$, where $\tilde{\tilde{\alpha}}$

presents the loss factor [32,33]. The reduced form of the general equations results in the following matrices:

$$\mathbf{Kr}_{11} = \mathbf{K}_{22(s)} + \mathbf{K}_{22(\gamma 11)} + \mathbf{K}_{22(k_L)} - (\mathbf{K}_{12(r)})^{\mathrm{T}}(\mathbf{K}_{11})^{-1}\mathbf{K}_{12(r)},$$

$$\mathbf{Kr}_{12} = \mathbf{K}_{23(\gamma 12)},\ \mathbf{Kr}_{21} = \mathbf{K}_{32(\gamma 21)},\ \mathbf{Kr}_{22} = \mathbf{K}_{33(\gamma 22)} + \mathbf{K}_{33(0)},\ \mathbf{Kr}_{13} = \mathbf{Kr}_{31} = -\mathbf{K}_{22(k_L)},$$

$$\mathbf{Kr}_{33} = \mathbf{K}_{55(\gamma 11)} + \mathbf{K}_{22(k_L)},\quad \mathbf{Kr}_{34} = \mathbf{K}_{56(\gamma 12)},\quad \mathbf{Kr}_{43} = \mathbf{K}_{65(\gamma 21)},\ \mathbf{Kr}_{44} = \mathbf{K}_{66(\gamma 22)} + \mathbf{K}_{66(0)},$$

$$\mathbf{K1}_{n\ell}(\mathbf{q}_{w1}) = \mathbf{K}_{22(r_{NL})} + 2[\mathbf{K}_{22(r_{NL})}]^{\mathrm{T}} - \mathbf{K}_{21(NL)}(\mathbf{K}_{11})^{-1}\mathbf{K}_{12(r)} - \left(\mathbf{K}_{12(r)}\right)^{\mathrm{T}}(\mathbf{K}_{11})^{-1}\mathbf{K}_{12(NL)}$$

$$\mathbf{K1}_{n\ell}(\mathbf{q}_{w1}, \mathbf{q}_{w1}^{\otimes}) = \mathbf{K}_{22(NL)} - \mathbf{K}_{21(NL)}(\mathbf{K}_{11})^{-1}\mathbf{K}_{12(NL)},$$

$$\mathbf{K3}_{n\ell}(\mathbf{q}_{w2}, \mathbf{q}_{w2}^{\otimes}) = \mathbf{K}_{55(NL)} - \mathbf{K}_{54(NL)}(\mathbf{K}_{11})^{-1}\mathbf{K}_{45(NL)},$$

$$\mathbf{K1}_{n\ell}(\mathbf{q}_{w1}, \mathbf{q}_{w1}^{\otimes}, \mathbf{q}_{w1}^{\otimes\otimes}) = \mathbf{K}_{22(k_{NL})},\ \mathbf{K1}_{n\ell}(\mathbf{q}_{w2}, \mathbf{q}_{w2}^{\otimes}, \mathbf{q}_{w2}^{\otimes\otimes}) = -\mathbf{K}_{24(k_{NL})}$$

$$\mathbf{K3}_{n\ell}(\mathbf{q}_{w1}, \mathbf{q}_{w1}^{\otimes}, \mathbf{q}_{w1}^{\otimes\otimes}) = -\mathbf{K}_{22(k_{NL})},\ \mathbf{K3}_{n\ell}(\mathbf{q}_{w2}, \mathbf{q}_{w2}^{\otimes}, \mathbf{q}_{w2}^{\otimes\otimes}) = \mathbf{K}_{24(k_{NL})}.$$

The vector of generalized coordinates can be shown as a sum of the first three members of a trigonometric series in the following form:

$$\begin{Bmatrix} \mathbf{q}_{w1}(t) \\ \mathbf{q}_{\theta 1}(t) \end{Bmatrix} = \begin{Bmatrix} \Sigma_{i=1}^{3} Q_{w1_{ci}} \\ \Sigma_{i=1}^{3} Q_{\theta 1_{ci}} \end{Bmatrix} \cos(i\omega t) + \begin{Bmatrix} \Sigma_{i=1}^{3} Q_{w1_{si}} \\ \Sigma_{i=1}^{3} Q_{\theta 1_{si}} \end{Bmatrix} \sin(i\omega t), \tag{12.41}$$

$$\begin{Bmatrix} \mathbf{q}_{w2}(t) \\ \mathbf{q}_{\theta 2}(t) \end{Bmatrix} = \begin{Bmatrix} \Sigma_{i=1}^{3} Q_{w2_{ci}} \\ \Sigma_{i=1}^{3} Q_{\theta 2_{ci}} \end{Bmatrix} \cos(i\omega t) + \begin{Bmatrix} \Sigma_{i=1}^{3} Q_{w2_{si}} \\ \Sigma_{i=1}^{3} Q_{\theta 2_{si}} \end{Bmatrix} \sin(i\omega t), \tag{12.42}$$

where ω represents the natural frequency of the system. The vector of new unknowns is now

$$\{\mathbf{Q}^{\mathrm{HBM}}\} = \begin{Bmatrix} \mathbf{Q1}^{\mathrm{HBM}} \\ \mathbf{Q2}^{\mathrm{HBM}} \end{Bmatrix},\quad \mathbf{Q1}^{\mathrm{HBM}} = \begin{Bmatrix} Q_{w1_{c1}} \\ Q_{w1_{s1}} \\ Q_{w1_{c2}} \\ Q_{w1_{s2}} \\ Q_{w1_{c3}} \\ Q_{w1_{s3}} \\ Q_{\theta 1_{c1}} \\ Q_{\theta 1_{s1}} \\ Q_{\theta 1_{c2}} \\ Q_{\theta 1_{s2}} \\ Q_{\theta 1_{c3}} \\ Q_{\theta 1_{s3}} \end{Bmatrix},\quad \mathbf{Q2}^{\mathrm{HBM}} = \begin{Bmatrix} Q_{w2_{c1}} \\ Q_{w2_{s1}} \\ Q_{w2_{c2}} \\ Q_{w2_{s2}} \\ Q_{w2_{c3}} \\ Q_{w2_{s3}} \\ Q_{\theta 2_{c1}} \\ Q_{\theta 2_{s1}} \\ Q_{\theta 2_{c2}} \\ Q_{\theta 2_{s2}} \\ Q_{\theta 2_{c3}} \\ Q_{\theta 2_{s3}} \end{Bmatrix}. \tag{12.43}$$

Substituting expressions Eq. (12.43) into the initial equations of motion Eq. (12.40) begets the following:

$$[-\omega^2\mathbf{M}^{\mathrm{HBM}} + \mathbf{C}^{\mathrm{HBM}} + \mathbf{KL}^{\mathrm{HBM}} + \mathbf{KNL}^{\mathrm{HBM}}]\{\mathbf{Q}^{\mathrm{HBM}}\} = \{\mathbf{F}\}, \tag{12.44}$$

12.44 where $\mathbf{M}^{\mathrm{HBM}}$ is the mass matrix, which is the function of $\mathbf{M}$; $\mathbf{C}^{\mathrm{HBM}}$ is the damping matrix, which is the function of $\tilde{\tilde{\beta}}\,\mathbf{M}$; $\mathbf{KL}^{\mathrm{HBM}}$ is the linear stiffness matrix, which is the function of $\mathbf{Kr}$ matrix. Lastly, $\mathbf{KNL}^{\mathrm{HBM}}$ is the nonlinear stiffness matrix, which is the function of: $\mathbf{K1}_{n\ell}(\mathbf{q}_{w1})$, which is linearly dependent on the generalized displacements; $\mathbf{K1}_{n\ell}(\mathbf{q}_{w1}, \mathbf{q}_{w1}^{\otimes})$, $\mathbf{K3}_{n\ell}(\mathbf{q}_{w2}, \mathbf{q}_{w2}^{\otimes})$, which is quadratically dependent on the generalized displacements; and $\mathbf{K1}_{n\ell}(\mathbf{q}_{w1,2}, \mathbf{q}_{w1,2}^{\otimes}, \mathbf{q}_{w1,2}^{\otimes\otimes})$, $\mathbf{K3}_{n\ell}(\mathbf{q}_{w1,2}, \mathbf{q}_{w1,2}^{\otimes}, \mathbf{q}_{w1,2}^{\otimes\otimes})$, which is cubically dependent on the generalized displacements. The harmonic balance and continuation methods provide a productive procedure for obtaining the amplitude–frequency responses in the additional analysis.

12.6 Applications

The results from various examples (of the beam–arch nonlinearly coupled system) that are shown previously are summarized in this section. Figs. 12.3–12.7 show both the amplitude–frequency dependence and the response to the transverse harmonic point excitation with the varying curvature effect $r = (0,\ 0.\ 1,\ 0.\ 2,\ 0.\ 3,\ 0.\ 4)\ \mathrm{m}^{-1}$, according to the obtained linear frequencies given in Tables 12.1–12.4 with amplitude $F = 50 \times 10^3$ N and excitation frequency $\omega = 0.9\omega_{\ell 1}$ applied at mid-span. The geometric and material properties (of the beam–arch system) that are used in the calculations are provided in Table 12.1. These figures show the steady-state response, that is, transients have died out. The response presented in Fig. 12.3A is invariably periodic, with the first harmonic predominant. One more important occurrence, in this case, occurs in the example without the curvature. It is apparent that the displacement of the upper beam (w_2) is much more pronounced outwards than inwards (negative transverse displacements). With an increase in the arch reciprocal curvature radius, this phenomenon disappears.

The phase plots presented in Fig. 12.3B show that in each cycle of vibration, zero-velocity appears four times, which is not clear in all cases where the curvature exists. The most interesting effect, by which the new physical insight becomes clear, can be seen by comparing the amplitude–frequency responses depicted in Fig. 12.3C and D with the responses depicted in Figs. 12.4–12.7C and D, where the curvature radius exists. The stabilization effect (dynamic absorption, particularly) externalized through lower amplitudes is now manifested in most subsegments of the frequency range $\{1.05\omega/\omega_{\ell 1},\ 3\omega/\omega_{\ell 1}\}$ when the curvature effect exists. The classical double-beam system possesses two resonances in the frequency range $\{1.05\omega/\omega_{\ell 1},\ 2\omega/\omega_{\ell 1}\}$. There, the two bifurcation points from the main branches are consequences of 1:5 internal resonances, where the first (ω_{11}) and the third mode (ω_{31}, ω_{32}) become coupled. The curvature effect is beneficial in the frequency range that is after the first resonance, with the second resonance emerging at a higher frequency, which is

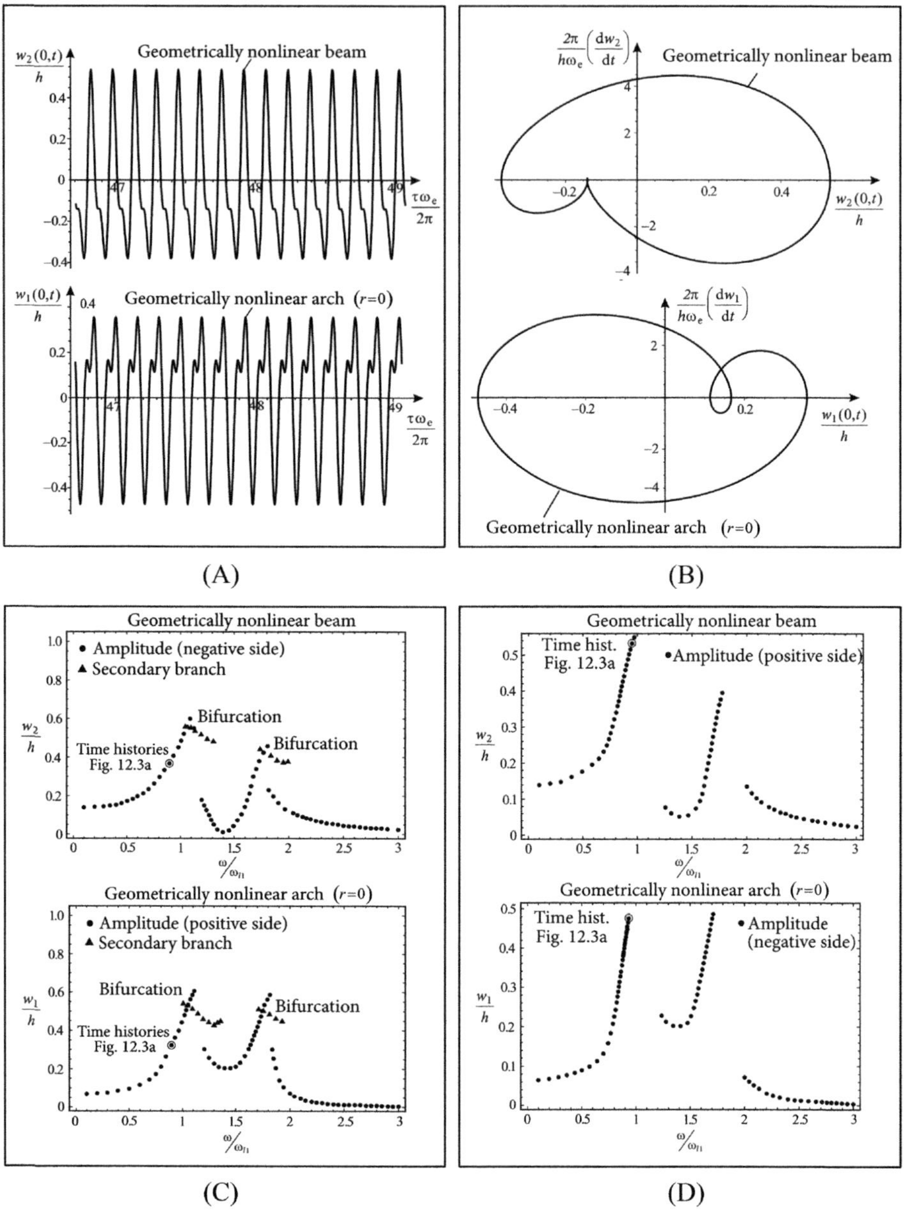

Figure 12.3 Time and frequency domains for beam–arch system. Parameters: r=0, F=50×10^3 N, ω=0.9ω_{11}. (A) Time histories at the middle of the beam and the middle of the arch; (B) Phase plots at the middle of the beam and the middle of the arch; (C) Maximum transverse displacement at x=0 in the function of excitation frequency (presented sides with lower amplitudes near the first two resonance frequencies); (D) Maximum transverse displacement at x=0 in the function of excitation frequency (presented sides with higher amplitudes near the first two resonance frequencies).

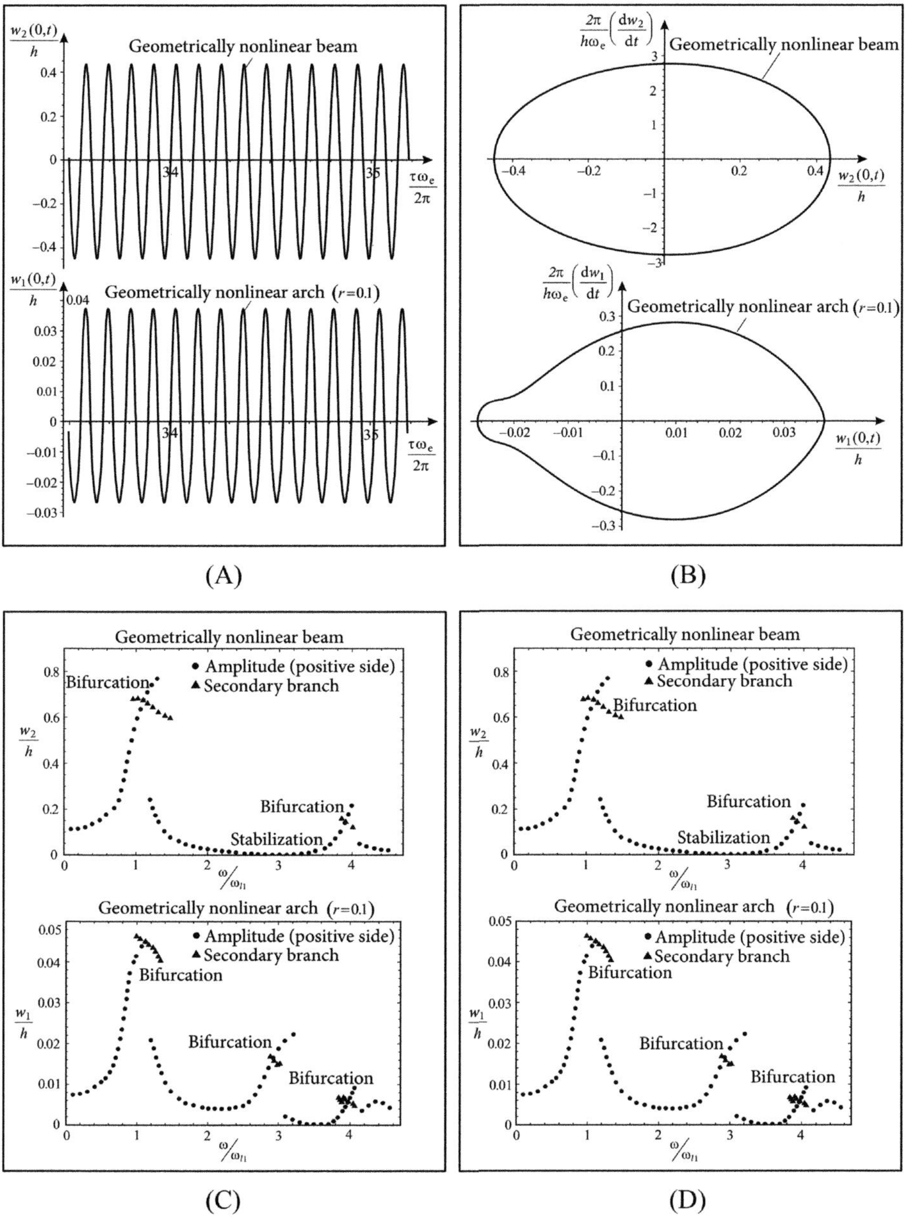

Figure 12.4 Time and frequency domains for beam–arch system. Parameters: r=0.1 m^{-1}, F=50×10^3 N, ω=0.9ω_{11}. (A) Time histories at the middle of the beam and the middle of the arch; (B) Phase plots at the middle of the beam and the middle of the arch; (C) Maximum transverse displacement at x=0 in the function of excitation frequency (presented sides with lower amplitudes near the first two resonance frequencies); (D) Maximum transverse displacement at x=0 in the function of excitation frequency (presented sides with higher amplitudes near the first two resonance frequencies).

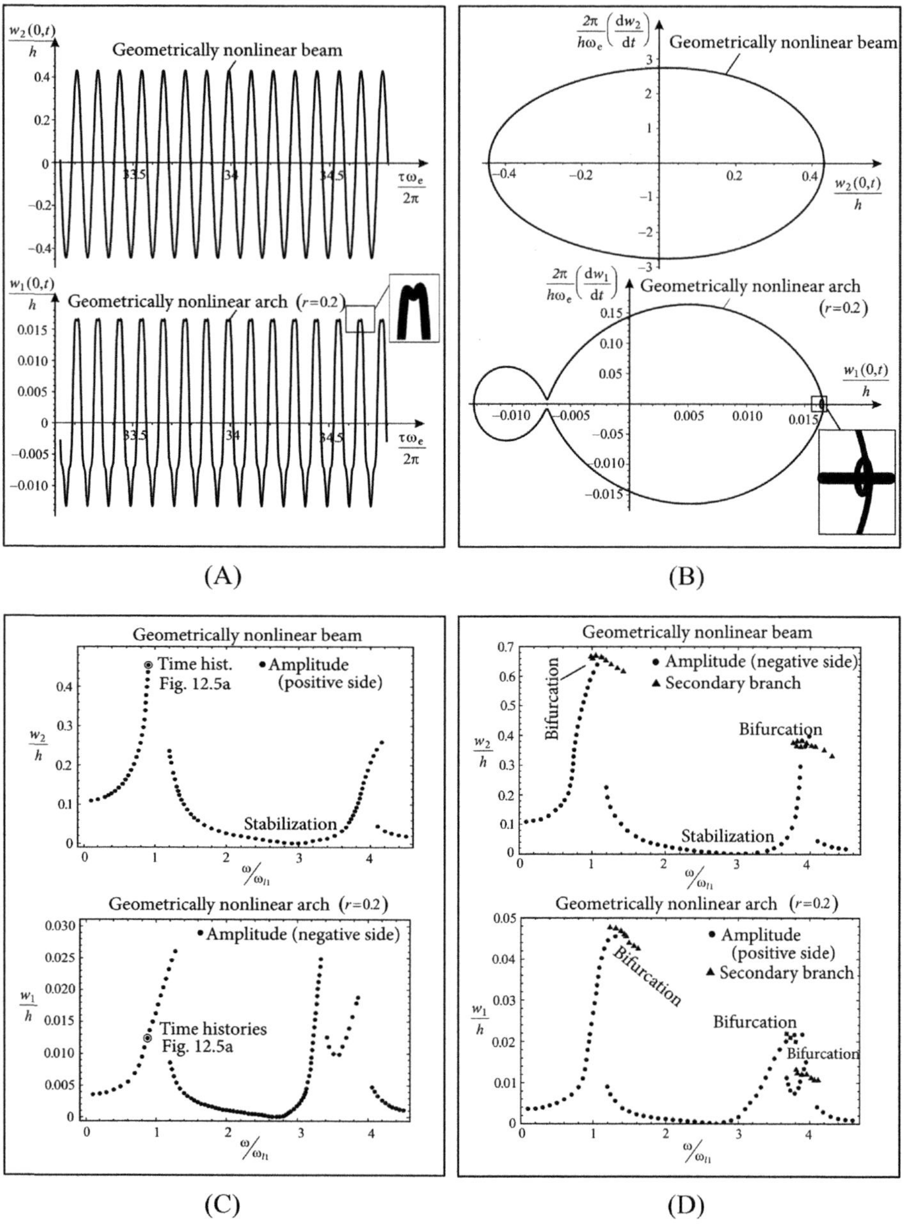

Figure 12.5 Time and frequency domains for beam–arch system. Parameters: r=0.2 m^{-1}, F=50×10^3 N, ω=0.9ω_{11}. (A) Time histories at the middle of the beam and the middle of the arch; (B) Phase plots at the middle of the beam and the middle of the arch; (B) Maximum transverse displacement at x=0 in the function of excitation frequency (presented sides with lower amplitudes near the first two resonance frequencies); (D) Maximum transverse displacement at x=0 in the function of excitation frequency (presented sides with higher amplitudes near the first two resonance frequencies).

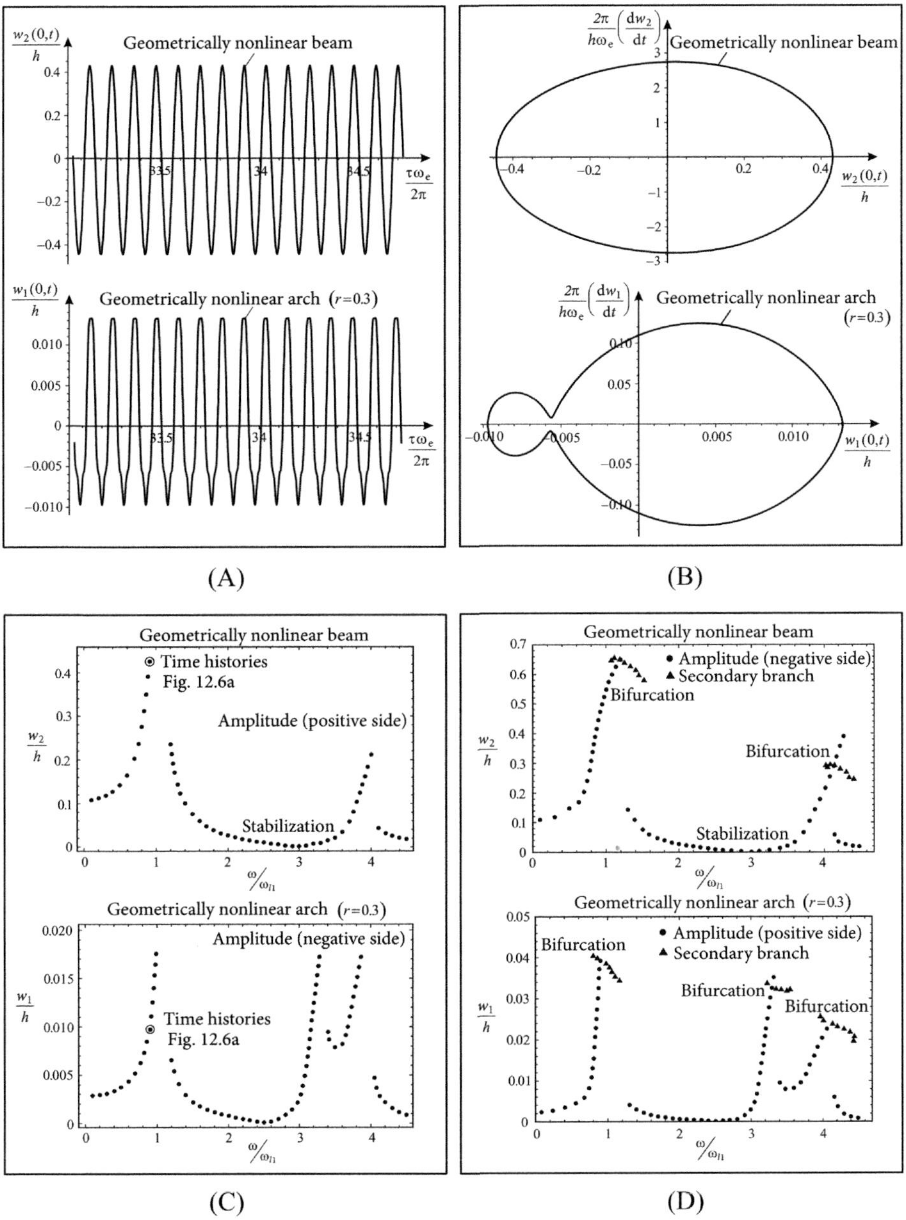

Figure 12.6 Time and frequency domains for beam–arch system. Parameters: $r = 0.3\ \text{m}^{-1}$, F=50×10^3 N, ω=0.9ω_{11}. (A) Time histories at the middle of the beam and the middle of the arch; (B) Phase plots at the middle of the beam and the middle of the arch; (C) Maximum transverse displacement at x=0 in the function of excitation frequency (presented sides with lower amplitudes near the first two resonance frequencies); (D) Maximum transverse displacement at x=0 in the function of excitation frequency (presented sides with higher amplitudes near the first two resonance frequencies).

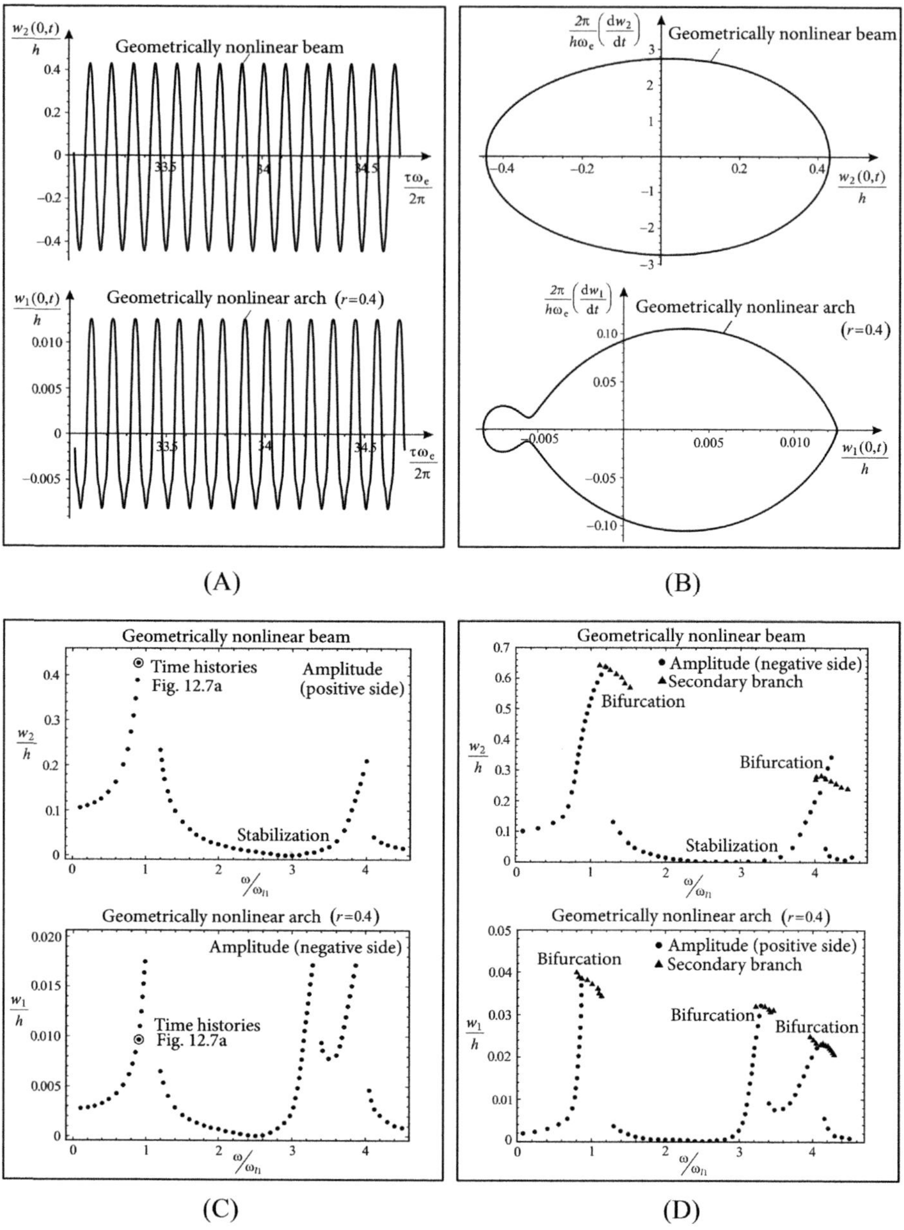

Figure 12.7 Time and frequency domains for beam–arch system. Parameters: r=0.4 m^{-1}, F=50 $\times 10^3$ N, ω=0.9ω_{11}. (A) Time histories at the middle of the beam and the midd le of the arch; (B) Phase plots at the middle of the beam and the middle of the arch; (C) Maximum transverse displacement at x=0 in the function of excitation frequency (presented sides with lower amplitudes near the first two resonance frequencies); (D) Maximum transverse displacement at x=0 in the function of excitation frequency (presented sides with higher amplitudes near the first two resonance frequencies).

not the case in a classical double-beam system. This appears again in the 1:5 internal resonance case; however, the distribution of the frequencies is different when the curvature radius exists. In the lower frequency range, the beam–arch system externalizes lower amplitudes in comparison with the double-beam system. The dynamic absorption emerges in the case when the curvature effect is lower (the first case, $r = 0.1\ m^{-1}$), and it is preserved with almost the same influence for other higher values of r. Fig. 12.3C shows the lower amplitudes (maximal deflections) on the positive or negative side; the same rule is used for Figs. 12.4–12.7 (tag C). Fig. 12.3D shows the highest amplitudes (maximal deflections) on the positive or negative side; again, the same rule is used for Figs. 12.4–12.7 (tag D). An amplitude–frequency response, delineated in Fig. 12.4D, shows that the system possesses the dynamic absorption of the upper beam at around $2.9\omega/\omega_{\ell 1}$. The lower curved beam has a bifurcation point in that spot. In comparison with the classical double-beam system, the interaction between the modes is considerably different; 1:3 internal resonances are observed between the first and the third modes. The dynamic absorption phenomenon is conspicuous in Fig. 12.5C and D; identical second and third resonances are at around $3.6\omega/\omega_{\ell 1}$ and $3.8\omega/\omega_{\ell 1}$. Figs. 12.6C and D and 12.7C and D show the dynamic behavior that is close between the two cases ($r = 0.3\ m^{-1}$ and $r = 0.4\ m^{-1}$), and in comparison, with Fig. 12.5C and D, the greater distance between the second and the third resonance. An interesting occurrence in Figs. 12.6D and 12.7D is that the lower beam (arch) has the first resonance at $0.88\omega/\omega_{\ell 1}$ which is a consequence of 1:3 internal resonances, where the first (ω_{12}) and the third mode (ω_{31}) become coupled.

The stabilizing feature is most pronounced in the case depicted in Fig. 12.7A when the R radius is the lowest and complete assimilation of an asymmetric response of the beam occurs (physical insight becomes clear again). This effect is caused by the presence of nonlinear couplings in the model, and it is not present in the linear vibration regime (Figs. 12.8–12.11). The linear regimes (Figs. 12.8–12.11) produce only symmetric responses (amplitudes in the positive and negative directions are equal) and greater amplitudes. They are also not associated with the change in excitation because the only change was in the radius of curvature for all the examples (Figs. 12.3–12.7).

The comparison of the phase plots shown in Figs. 12.3–12.7 (tag B) is of interest. If the dynamic behavior and change in the velocity of the central point of the upper beam are observed, the positive effect of the curvature's radius can be noticed. It is recognizable that the curvature radius institutes a more uniform change in the velocity of the upper beam at the midpoint. When the radius is zero, the negative effects—in terms of peculiar changes in the velocity of the upper beam's point—are transmitted to the arch. In that case, the arch behaves differently and the change in the velocity of the middle point of the arch is interesting, as shown in Figs. 12.4–12.7 (tag B). An interesting arch's midpoint velocity change characterized by a partial swirl effect can be seen in Fig. 12.5B. The results—related to the existence of the curvature in the lower beam—can now be briefly summarized. The cases in which curvature exists and has influence are presented in Figs. 12.4–12.7. In these cases, any unexpected deviation or effect, such as an asymmetrical response of the upper beam, is transmitted to the arch. This differs from the case where the curvature of the lower beam vanishes (Fig. 12.3).

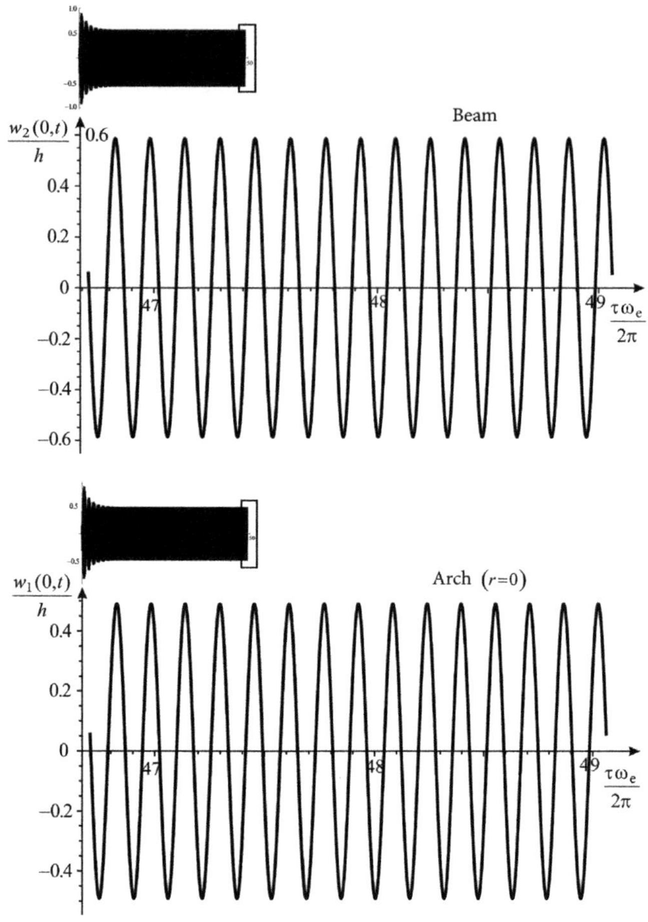

Figure 12.8 Time histories. Linear solution at the middle of the beam and the middle of the arch (r=0.0, F=50×10^3 N, ω=0.9ω_{11}).

The following numerical examples compare the two cases; beam–zero curvature arch system or classical double-beam system and beam–arch system. The influence of the varied amplitude of the harmonic point excitation is shown in Fig. 12.12 and Fig 12.13. The maximal deflections of the beam and arch points under the influence of the amplitude force of the harmonic point excitation are presented in Fig. 12.14. The effect of the location of the excitation point and the amplitude of the harmonic force is shown in Figs. 12.15–12.18. Ultimately, the influence of the varying frequency excitations is shown in Figs. 12.19 and 12.20. All these examples are considered in a comparative relation between the beam–zero arch system and the beam–arch system (r = 0.4). Visible deflections of the beam and arch and the considerable influence of the curvature effect are apparent in Fig. 12.12. Furthermore, for smaller excitation amplitudes, the displacement is almost symmetric with

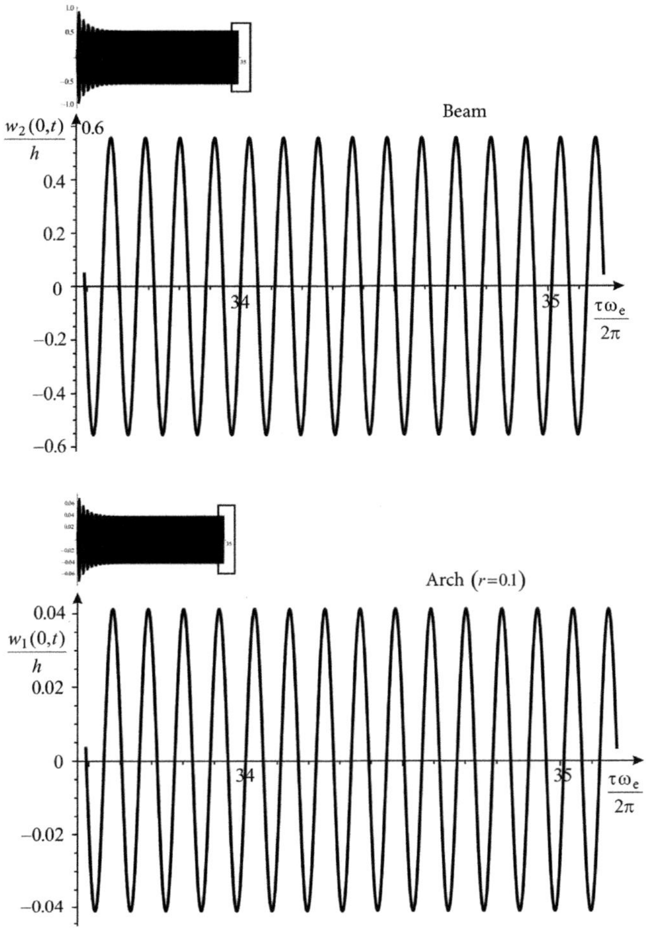

Figure 12.9 Time histories. Linear solution at the middle of the beam and the middle of the arch (r=0.1 m, F=50 × 10^3 N, ω=0.9ω_{11}).

reference to the middle axis of the beam. By increasing the amplitudes of excitations, the displacement becomes asymmetric in the situation when the curvature does not exist. Regardless of the increase in the amplitude of excitation, in the case where the curvature effect is present, the symmetry of the beam's displacement is preserved. The comparison shows that the absence of the curvature effect causes the opposite asymmetric deflections between the beams; deflections are significantly more pronounced outwards than inwards. An increase in the amplitudes of excitation does not have an effect on the emergence of asymmetry when the curvature exists; the phase plots shown in Fig 12.13 confirm this. Furthermore, it can be verified from these plots that for smaller amplitudes of excitation, the phase plots deviate slightly from an ellipse (F = 5 × 10^3 N, F = 20 × 10^3 N, and F = 30 × 10^3 N). The upper right diagram, in Fig 12.13, shows the absence of the increase in amplitude during excitation because

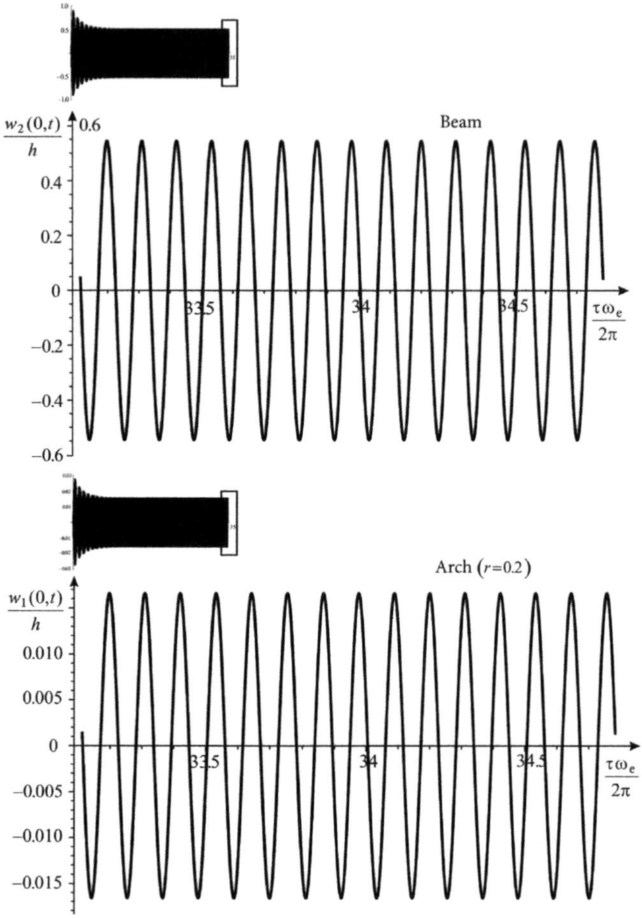

Figure 12.10 Time histories. Linear solution at the middle of the beam and the middle of the arch (r=0.2/m, F=50×10^3 N, ω=0.9ω_{11}).

almost all phase plots are near-perfect ellipses. In the cases where the amplitudes of excitation are the most potent ($F = 50 \times 10^3$ N, $F = 56 \times 10^3$ N), the arch takes unwanted deviations but still keeps the symmetry in relation to the horizontal axis (the lower right diagram in Fig 12.13).

To understand the dynamic behavior of all the continuous systems, it is needed to perform the calculation of the maximal deflections of all points of the beam–arch system; Fig. 12.14 shows these computations graphically. Maximal displacements of all points of both beams are delineated in Fig. 12.14, but in this particular case, attention should be given to the displacements of the points of the arch (the lower right diagram). The arch has a bifurcation point close to the excitation frequency ($0.88\omega/\omega_{\ell 1}$) and presents 1:3 internal resonances, where the first (ω_{12}) and the third mode (ω_{31}) become coupled. Evidently, the analyzed "arch absorber"

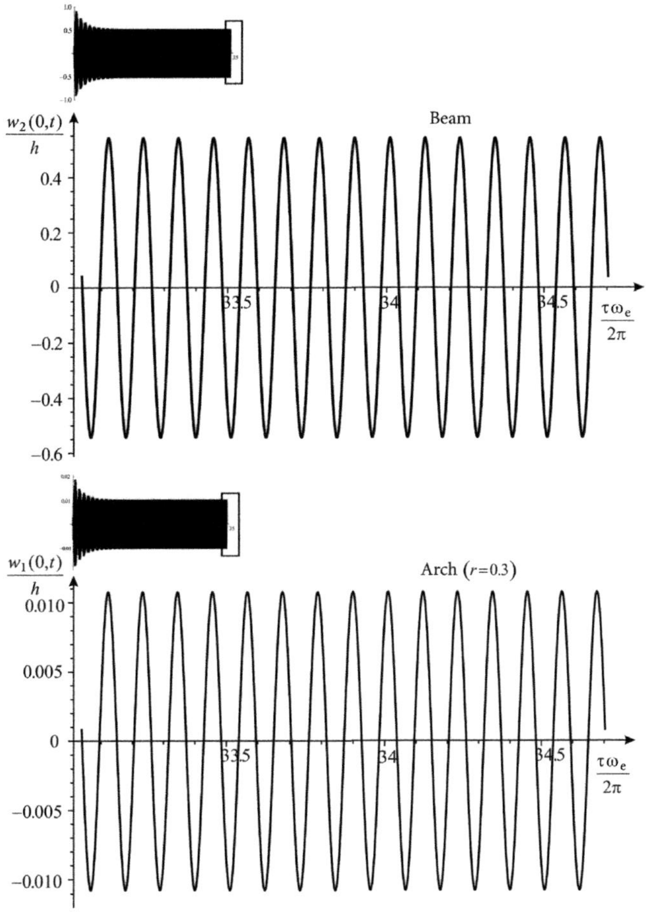

Figure 12.11 Time histories. Linear solution at the middle of the beam and the middle of the arch (r=0.3 m, F=50×10^3 N, ω=0.9ω_{11}).

(Figs. 12.12 and 12.13) possesses stabilization properties that sustain the symmetry of deflections and reduce the amplitudes of the vibrational response of the upper beam. A consequence of this phenomenon is the interaction between the first and the third modes, as exemplified in Fig. 12.14. Another key point, concerning this kind of dynamic behavior of continuous systems, that should be discussed, is the scenario when the excitation is applied at a location that is not at the mid-span of the beam. In accordance with this, Figs. 12.15–12.18, show the effect of the location of the harmonic force. Scenario when the harmonic force excites the system at the location $\xi = 0.5$ is presented in Figs. 12.15 and 12.16; while Figs. 12.17 and 12.18 are for the location at $\xi = 0.7$. Analysis of the maximum deflection points delineated in Figs. 12.15 and 12.17 (beam–arch system) gives one noteworthy observation; four times larger amplitude of excitation leads to a three times larger deflection of the central beam point if the excitation is shifted from $\xi = 0.5$ to $\xi = 0.7$. Thereupon, this

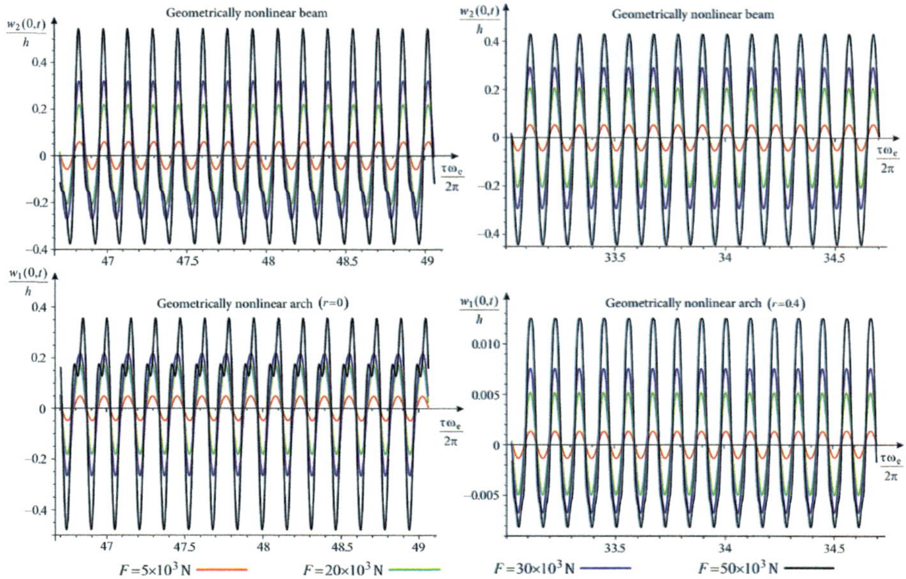

Figure 12.12 Comparison of time histories of the beam–arch systems. Various values of force are used; $F=\{5\times10^3, 2\times10^4, 3\times10^4, 5\times10^4\}$ N, $r=0.4$ m^{-1}, $\omega=0.9\omega_{11}$. The amplitude is proportionate to the force; the force with the highest intensity produces the highest amplitude.

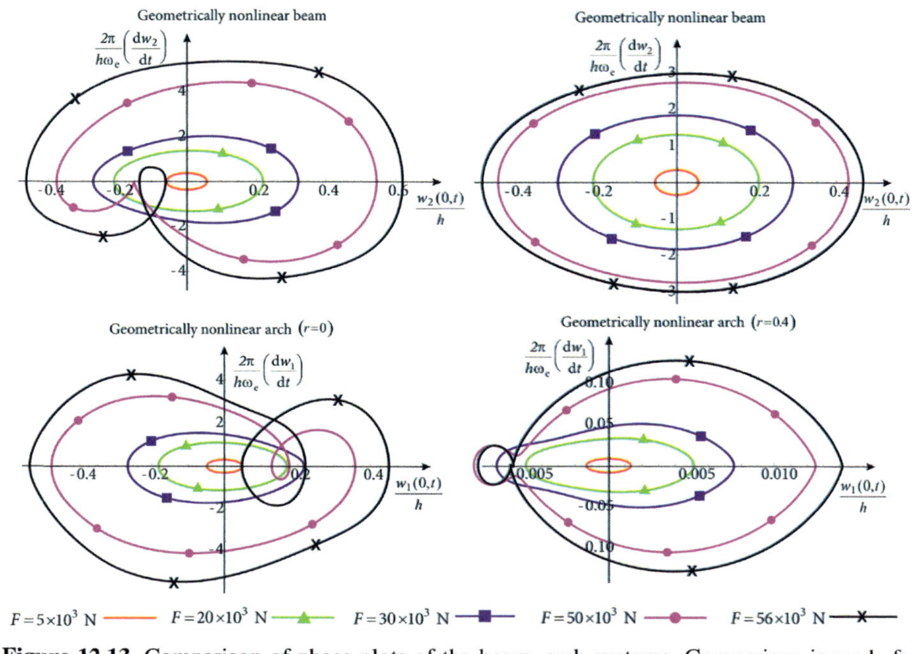

Figure 12.13 Comparison of phase plots of the beam–arch systems. Comparison is made for the various values of force ($r=0.4$ m^{-1}, $\omega=0.9\omega_{11}$).

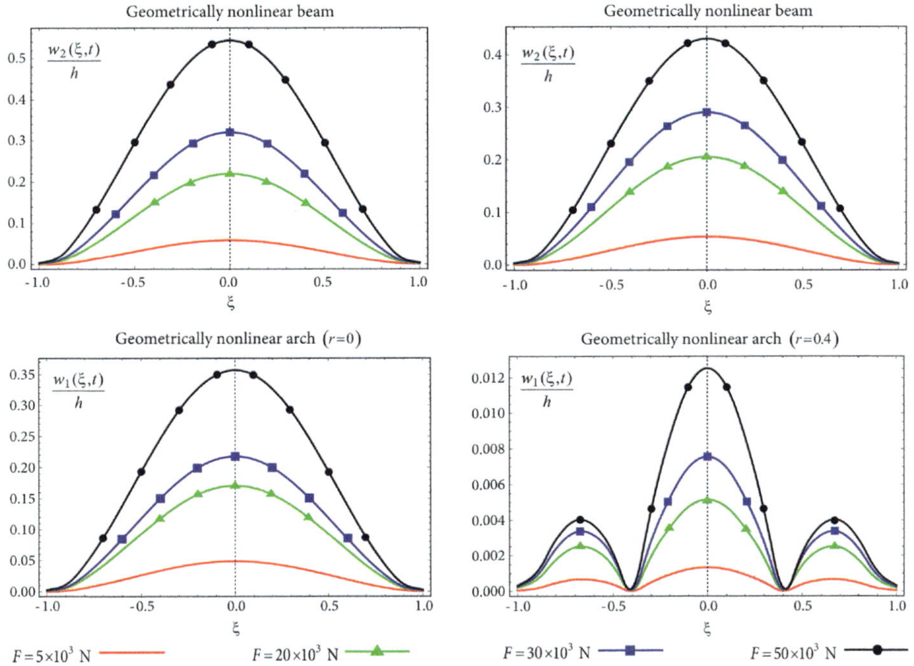

Figure 12.14 Maximal deflections during the forced nonlinear vibration. Comparison is for the nonlinearly coupled beam–arch system. for various values of $F(r{=}0.0,\ r{=}0.4\ \mathrm{m}^{-1},\ \omega{=}0.9\omega_{11})$.

relation may be used to predict the location effect on the deflections of the beam's points, and it is not necessary to calculate any time of deflection for all points of the beam. By nature, this represents just an assumption considering that the nonlinear coupled effects exist. When the results between Figs. 12.16 and 12.18 (without curvature) are compared, it is apparent that the growth in the amplitude of excitation does not have a substantial effect if the location of the excitation point is far from the mid-span. The effect of the varying excitation frequency represented through time histories, phase plots, and maximal deflections is presented in Figs. 12.19 and 12.20. The comparison has been verified by the discovered phenomenon (previously mentioned). However, a new conclusion here expresses that the excitation frequency, which is close to the second linear frequency, leads to the growth of the deflection of the central point of the beam, whether or not the curvature exists. This is contrary to the already mentioned and discussed cases in the comparison between the double-beam system (without curvature) and the beam–arch system. The excitation close to the second linear frequency is $1.7\omega/\omega_{\ell 1}$ for the double-beam system, and $1.4\ \omega/\omega_{\ell 1}$ for the beam–arch system. This example presents a particularly nonstabilizing case, which is unmistakable from the amplitude–frequency responses shown in Figs. 12.3D and 12.7D.

The primary goal of the study was to gain physical insight into the nonlinear stabilizing benefits of the analyzed beam–arch system; with it are determined amplitude–frequency responses, time histories, phase plots, and deformed shapes for each analyzed case. The frequency zones with stabilizing effects (shown with

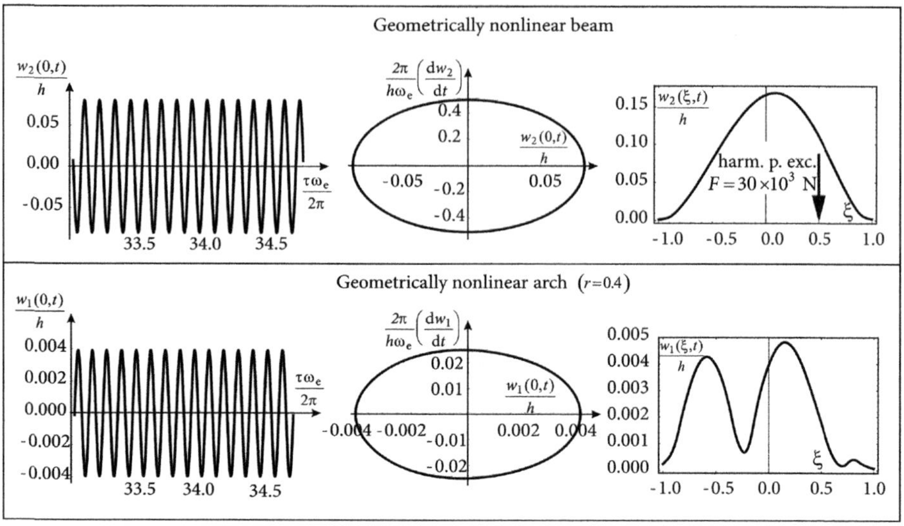

Figure 12.15 Time histories, phase plots, and maximal deflections. This is during the forced steady-state nonlinear vibrations of the nonlinearly coupled beam–arch system for the unsymmetrical transverse harmonic point excitation (r=0.4 m^{-1}, F=30×10^3 N, ω=0.9ω_{11}).

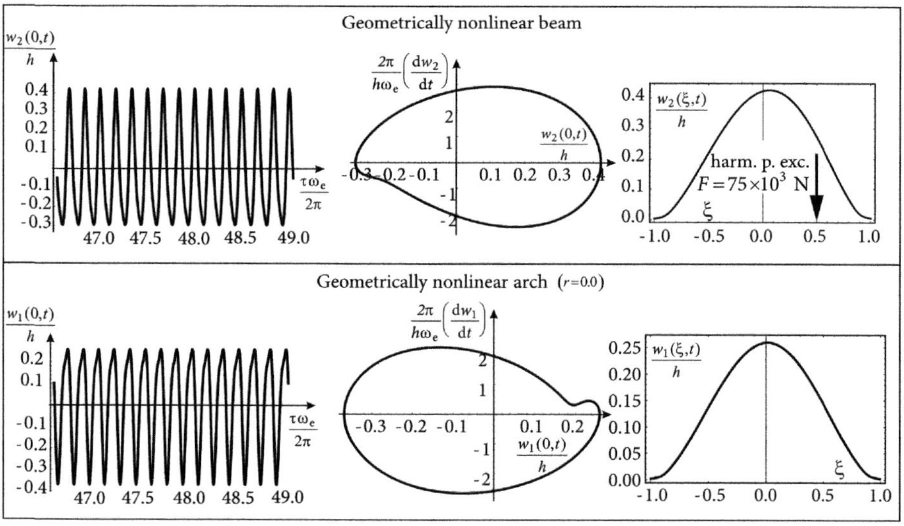

Figure 12.16 Time histories, phase plots, and maximal deflections. This is during the forced steady-state nonlinear vibrations of the nonlinearly coupled system for the unsymmetrical transverse harmonic point excitation (r=0, F=75×10^3 N, ω=0.9ω_{11}).

amplitude–frequency responses) exemplify the complete nonlinear dynamic absorption or, leastwise, a decrease in the amplitudes of the upper excites beam when the radius of the curvature exists in the lower beam. This has been shown in many examples and can be applied in engineering. The disadvantages of the linear solutions, regarding

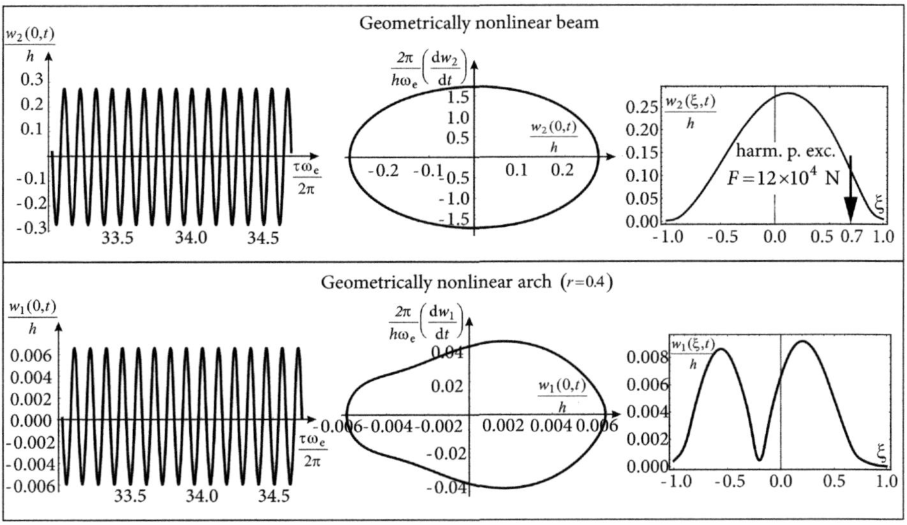

Figure 12.17 Time histories, phase plots, and maximal deflections. This is during the forced steady-state nonlinear vibrations of the nonlinearly coupled system for the unsymmetrical transverse harmonic point excitation (r=0.4 m^{-1}, F=12×10^4 N, ω=0.9ω_{11}).

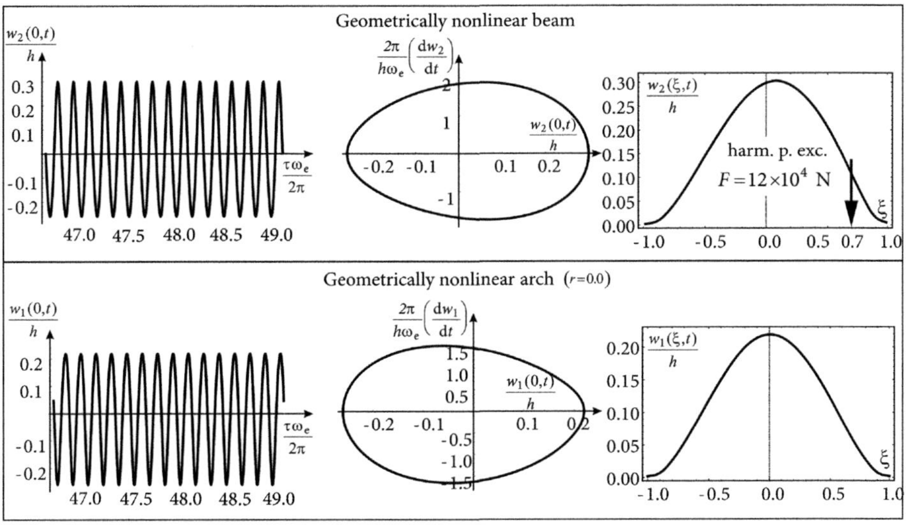

Figure 12.18 Time histories, phase plots, and maximal deflections. This is during the forced steady-state nonlinear vibrations of the nonlinearly coupled system for the unsymmetrical transverse harmonic point excitation (F=12×10^4 N, ω=0.9ω_{11}).

the absence of the asymmetric response with larger obtained amplitudes, show the importance of considering nonlinear couplings in the system, which is yet another advantage discovered in this study.

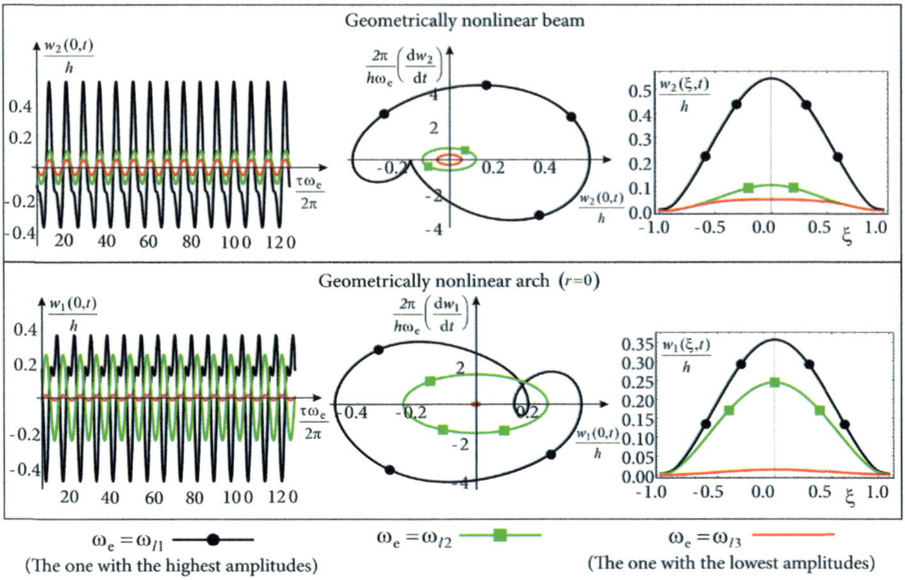

Figure 12.19 Time histories, phase plots, and maximal deflections. This is during the forced steady-state nonlinear vibrations of the nonlinearly coupled system for various frequency excitations (r=0, F=5×10^3 N).

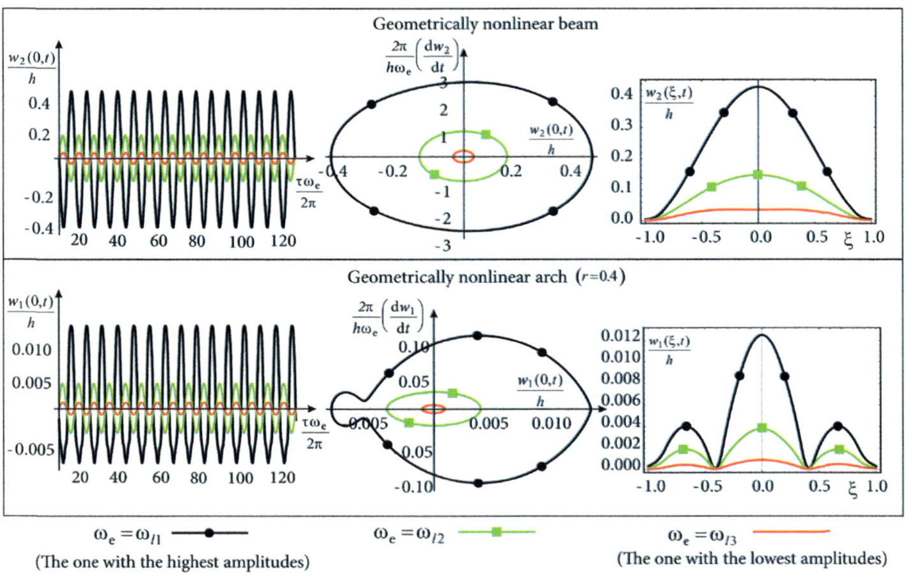

Figure 12.20 Time histories, phase plots, and maximal deflections. This is during the forced steady-state nonlinear vibrations of the nonlinearly coupled system for various frequency excitations (r=0.4 m^{-1}, F=5×10^3 N).

12.7 Conclusion

In this chapter, a viscoelastically nonlinearly coupled beam–arch system was studied. More specifically, the properties of geometrically nonlinear forced transverse vibrations of this kind of system. In order to take into account, the geometric nonlinearity of large deflections, the beam analog of von Kármán's plate theory was applied. The utilized procedures enabled the derivation of response curves of nonlinear structures, including the definition of internal resonances. Secondary branches were also found. The harmonic balance and continuation methods have enabled the features of the curvatures of the arch to be investigated in nonlinear vibration analysis. When the lower beam in the system was curved (which makes it an arch), the nonlinear dynamic absorption or decrease in the amplitudes of the upper excited beam was observed. The time-domain analysis was performed and it showed the stabilizing properties as a consequence of nonlinear couplings. This nonlinear phenomenon was shown in comparison with the linear solutions, where the absence of an asymmetric response with larger amplitudes did not present the real dynamic behavior of the proposed model. Amplitude–frequency responses showed the appearance of 1:3 and 1:5 internal resonances where the first and third modes became coupled. Furthermore, the study shows the effect of the improved steady-state response regimes of vibrations of the beam–arch system in comparison with the conventional double-beam system. It was shown that the upper beam goes back to the stable symmetric regime of vibrations only when it was coupled with a curved beam; as in a single-beam model scenario, but not in a conventional double-beam model (for the same excited scenario). A small improvement of the standard Newmark method in the time domain analysis was also presented. This was accomplished by performing the extrapolation for $\Delta t = 0$ by the Richardson extrapolation method. All of the results are important and have a wider application in the vibrations of coupled structures.

Appendix A 12.1

$$\mathbf{M}_{i1} = \frac{\rho_i h_i b_i L}{2} \int_{-1}^{1} \mathbf{N}^{u\mathrm{T}} \mathbf{N}^{u} \mathrm{d}\xi,$$

$$\mathbf{M}_{i2} = \frac{\rho_i h_i b_i L}{2} \int_{-1}^{1} \mathbf{N}^{w\mathrm{T}} \mathbf{N}^{w} \mathrm{d}\xi, \quad \mathbf{M}_{i3} = \frac{\rho_i b_i h_i^3 L}{24} \int_{-1}^{1} \mathbf{N}^{w\mathrm{T}} \mathbf{N}^{w} \mathrm{d}\xi,$$

$$\mathbf{K}_{11} = \frac{2E_1 h_1 b_1}{L} \int_{-1}^{1} \frac{\mathrm{d}\mathbf{N}^{\mathbf{u}\mathrm{T}}}{\mathrm{d}\xi} \frac{\mathrm{d}\mathbf{N}^{\mathbf{u}}}{\mathrm{d}\xi} \mathrm{d}\xi, \quad \mathbf{K}_{44} = \frac{2E_2 h_2 b_2}{L} \int_{-1}^{1} \frac{\mathrm{d}\mathbf{N}^{\mathbf{u}\mathrm{T}}}{\mathrm{d}\xi} \frac{\mathrm{d}\mathbf{N}^{\mathbf{u}}}{\mathrm{d}\xi} \mathrm{d}\xi,$$

$$\mathbf{K}_{22(\gamma 11)} = \frac{2\lambda_1 G_1 h_1 b_1}{L} \int_{-1}^{1} \frac{\mathrm{d}\mathbf{N}^{\mathbf{w}\mathrm{T}}}{\mathrm{d}\xi} \frac{\mathrm{d}\mathbf{N}^{\mathbf{w}}}{\mathrm{d}\xi} \mathrm{d}\xi,$$

$$\mathbf{K}_{55(\gamma 11)} = \frac{2\lambda_2 G_2 h_2 b_2}{L}\int_{-1}^{1}\frac{d\mathbf{N}^{\mathbf{w}\mathrm{T}}}{d\xi}\frac{d\mathbf{N}^{\mathbf{w}}}{d\xi}d\xi,$$

$$\mathbf{K}_{23(\gamma 12)} = \lambda_1 G_1 h_1 b_1\int_{-1}^{1}\frac{d\mathbf{N}^{\mathbf{w}\mathrm{T}}}{d\xi}\mathbf{N}^{\theta}d\xi,\quad \mathbf{K}_{56(\gamma 12)} = \lambda_2 G_2 h_2 b_2\int_{-1}^{1}\frac{d\mathbf{N}^{\mathbf{w}\mathrm{T}}}{d\xi}\mathbf{N}^{\theta}d\xi,$$

$$\mathbf{K}_{32(\gamma 21)} = \lambda_1 G_1 h_1 b_1\int_{-1}^{1}\mathbf{N}^{\theta\mathrm{T}}\frac{d\mathbf{N}^{\mathbf{w}}}{d\xi}d\xi,\quad \mathbf{K}_{65(\gamma 21)} = \lambda_2 G_2 h_2 b_2\int_{-1}^{1}\mathbf{N}^{\theta\mathrm{T}}\frac{d\mathbf{N}^{\mathbf{w}}}{d\xi}d\xi,$$

$$\mathbf{K}_{33(\gamma 22)} = \frac{L\quad \lambda_1 G_1 h_1 b_1}{2}\int_{-1}^{1}\mathbf{N}^{\theta\mathrm{T}}\mathbf{N}^{\theta}d\xi,\quad \mathbf{K}_{66(\gamma 22)} = \frac{L\quad \lambda_2 G_2 h_2 b_2}{2}\int_{-1}^{1}\mathbf{N}^{\theta\mathrm{T}}\mathbf{N}^{\theta}d\xi,$$

$$\mathbf{K}_{33(0)} = \frac{E_1 b_1 h_1^3}{6L}\int_{-1}^{1}\frac{d\mathbf{N}^{\theta\mathrm{T}}}{d\xi}\frac{d\mathbf{N}^{\theta}}{d\xi}d\xi,\quad \mathbf{K}_{66(0)} = \frac{E_2 b_2 h_2^3}{6L}\int_{-1}^{1}\frac{d\mathbf{N}^{\theta\mathrm{T}}}{d\xi}\frac{d\mathbf{N}^{\theta}}{d\xi}d\xi,$$

$$\mathbf{K}_{12(r)} = rE_1 b_1 h_1\int_{-1}^{1}\frac{d\mathbf{N}^{\mathbf{u}\mathrm{T}}}{d\xi}\mathbf{N}^{\mathbf{w}}d\xi,\quad \mathbf{K}_{21(r)} = [\mathbf{K}_{12(r)}]^{\mathrm{T}},$$

$$\mathbf{K}_{22(s)} = r^2\frac{L\quad E_1 b_1 h_1}{2}\int_{-1}^{1}\mathbf{N}^{\mathbf{w}\mathrm{T}}\mathbf{N}^{\mathbf{w}}d\xi,$$

$$\mathbf{K}_{22(k_L)} = \frac{L}{2}k_L\int_{-1}^{1}\mathbf{N}^{\mathbf{w}\mathrm{T}}\mathbf{N}^{\mathbf{w}}d\xi,$$

$$\mathbf{K}_{12(NL)}(\mathbf{q}_{\mathbf{w1}}) = \frac{2E_1 b_1 h_1}{L^2}\int_{-1}^{1}\frac{dw_{01}}{d\xi}\frac{d\mathbf{N}^{\mathbf{u}\mathrm{T}}}{d\xi}\frac{d\mathbf{N}^{\mathbf{w}}}{d\xi}d\xi,$$

$$\mathbf{K}_{21(NL)}(\mathbf{q}_{\mathbf{w1}}) = \frac{4E_1 b_1 h_1}{L^2}\int_{-1}^{1}\frac{dw_{01}}{d\xi}\frac{d\mathbf{N}^{\mathbf{w}\mathrm{T}}}{d\xi}\frac{d\mathbf{N}^{\mathbf{u}}}{d\xi}d\xi,$$

$$\mathbf{K}_{45(NL)}(\mathbf{q}_{\mathbf{w2}}) = \frac{2E_2 b_2 h_2}{L^2}\int_{-1}^{1}\frac{dw_{02}}{d\xi}\frac{d\mathbf{N}^{\mathbf{u}\mathrm{T}}}{d\xi}\frac{d\mathbf{N}^{\mathbf{w}}}{d\xi}d\xi,$$

$$\mathbf{K}_{54(NL)}(\mathbf{q}_{\mathbf{w2}}) = \frac{4E_2 b_2 h_2}{L^2}\int_{-1}^{1}\frac{dw_{02}}{d\xi}\frac{d\mathbf{N}^{\mathbf{w}\mathrm{T}}}{d\xi}\frac{d\mathbf{N}^{\mathbf{u}}}{d\xi}d\xi,$$

$$\mathbf{K}_{22(NL)}(\mathbf{q_{w1}}) = \frac{4E_1b_1h_1}{L^3}\int_{-1}^{1}\left(\frac{\mathrm{d}w_{01}}{\mathrm{d}\xi}\right)^2\frac{\mathrm{d}\mathbf{N}^{\mathrm{w}\mathrm{T}}}{\mathrm{d}\xi}\frac{\mathrm{d}\mathbf{N}^{\mathrm{w}}}{\mathrm{d}\xi}\mathrm{d}\xi,$$

$$\mathbf{K}_{55(NL)}(\mathbf{q_{w2}}) = \frac{4E_2b_2h_2}{L^3}\int_{-1}^{1}\left(\frac{\mathrm{d}w_{02}}{\mathrm{d}\xi}\right)^2\frac{\mathrm{d}\mathbf{N}^{\mathrm{w}\mathrm{T}}}{\mathrm{d}\xi}\frac{\mathrm{d}\mathbf{N}^{\mathrm{w}}}{\mathrm{d}\xi}\mathrm{d}\xi,$$

$$\mathbf{K}_{22(k_{NL})}(\mathbf{q_{w1}}) = \frac{L}{2}k_{NL}\int_{-1}^{1}(w_{01})^3\mathbf{N}^{\mathrm{w}\mathrm{T}}\mathbf{N}^{\mathrm{w}}\mathrm{d}\xi,$$

$$\mathbf{K}_{24(k_{NL})}(\mathbf{q_{w2}}) = \frac{L}{2}k_{NL}\int_{-1}^{1}(w_{02})^3\mathbf{N}^{\mathrm{w}\mathrm{T}}\mathbf{N}^{\mathrm{w}}\mathrm{d}\xi,$$

$$\mathbf{K}_{22(rNL)}(\mathbf{q_{w1}}) = r\frac{E_1b_1h_1}{L}\int_{-1}^{1}\frac{\mathrm{d}w_{01}}{\mathrm{d}\xi}\mathbf{N}^{\mathrm{w}\mathrm{T}}\frac{\mathrm{d}\mathbf{N}^{\mathrm{w}}}{\mathrm{d}\xi}\mathrm{d}\xi.$$

References

[1] M. Petyt, C. Fleischer, Free vibration of a curved beam, Journal of Sound and Vibration 18 (1971) 17–30.
[2] S.S. Lee, J.S. Koo, J.M. Choi, Development of a new curved beam element with shear effect, Engineering Computations 13 (1996) 9–25.
[3] M. Eisenberger, E. Efraim, In-plane vibrations of shear deformable curved beams, International Journal for Numerical Methods in Engineering 52 (2001) 1221–1234.
[4] Q. Bi, H.H. Dai, Analysis of non-linear dynamics and bifurcations of a shallow arch subjected to periodic excitation with internal resonance, Journal of Sound and Vibration 233 (2000) 557–571.
[5] S.H. Chen, Y.K. Cheung, H.X. Xing, Nonlinear vibration of plane structures by finite element and incremental harmonic balance method, Nonlinear Dynamics 26 (2001) 87–104.
[6] P. Ribeiro, M. Petyt, Non-linear free vibration of isotropic plates with internal resonance, International Journal of Non-Linear Mechanics 35 (2000) 263–278.
[7] J. Luczko, Bifurcations and internal resonances in space-curved rods, Computer Methods in Applied Mechanics and Engineering 191 (2002) 3271–3296.
[8] J.M. Seelig, W.H. Hoppmann, II: Impact on an elastically connected double-beam system, Journal of Applied Mechanics—Transactions of the ASME 31 (1964) 621–626.
[9] S.S. Rao, Natural vibrations of systems of elastically connected Timoshenko beams, Journal of the Acoustical Society of America 55 (1974) 1232–1237.
[10] Z. Oniszczuk, Free transverse vibrations of elastically connected simply supported double-beam complex system, Journal of Sound and Vibration 232 (2) (2000) 387–403.
[11] I. Bochicchio, C. Giorgi, E. Vuk, Buckling and nonlinear dynamics of elastically coupled double-beam systems, International Journal of Non-Linear Mechanics 85 (2016) 161–173.
[12] J.L. Huang, R.K.L. Su, Y.Y. Lee, S.H. Chen, Nonlinear vibration of a curved beam under uniform base harmonic excitation with quadratic and cubic nonlinearities, Journal of Sound and Vibration 330 (21) (2011) 5151–5164.

[13] N. Mohamed, M.A. Eltaher, S.A. Mohamed, L.F. Seddek, Numerical analysis of nonlinear free and forced vibrations of buckled curved beams resting on nonlinear elastic foundations, International Journal of Non-Linear Mechanics 101 (2018) 157–173.
[14] J. Huang, K.L.R. Su, Y.Y.R. Lee, S. Chen, Various bifurcation phenomena in a nonlinear curved beam subjected to base harmonic excitation, International Journal of Bifurcation and Chaos 28 (07) (2018) 1830023.
[15] Y. Chen, D. Zhang, L. Li, Dynamic analysis of rotating curved beams by using Absolute Nodal Coordinate Formulation based on radial point interpolation method, Journal of Sound and Vibration 441 (2019) 63–83.
[16] L. Medina, R. Gilat, S. Krylov, Dynamic release condition in latched curved micro beams, Communications in Nonlinear Science and Numerical Simulation 73 (2019) 291–306.
[17] O. Weeger, B. Narayanan, M.L. Dunn, Isogeometric shape optimization of nonlinear, curved 3D beams and beam structures, Computer Methods in Applied Mechanics and Engineering 345 (2019) 26–51.
[18] V. Stojanović, M.D. Petković, D. Milić, Nonlinear vibrations of a coupled beam–arch bridge system, Journal of Sound and Vibration 464 (2020) 115000.
[19] P. Ribeiro, E. Manoach, The effect of temperature on the large amplitude vibrations of curved beams, Journal of Sound and Vibration 285 (2005) 1093–1107.
[20] V. Stojanović, Geometrically nonlinear vibrations of beams supported by a nonlinear elastic foundation with variable discontinuity, Communications in Nonlinear Science and Numerical Simulation 28 (2015) 66–80.
[21] P. Ribeiro, Forced large amplitude periodic vibrations of cylindrical shallow shells, Finite Elements in Analysis and Design 44 (2008) 657–674.
[22] V. Stojanović, M.D. Petković, Nonlinear dynamic analysis of damaged Reddy–Bickford beams supported on an elastic Pasternak foundation, Journal of Sound and Vibration 385 (2016) 239–266.
[23] ANSYS Workbench User's Guide, 2009, 124p.
[24] S.P. Timoshenko, On the transverse vibrations of bars of uniform cross section, Philosophy Magazine 43 (1922) 125–131.
[25] J.R. Hutchinson, Shear coefficients for Timoshenko beam theory, Journal of Applied Mechanics 68 (2001) 87–92.
[26] T. Kaneko, On Timoshenko's correction for shear in vibrating beams, Journal of Physics D 8 (1975) 1927–1936.
[27] J. Stoer, R. Bulirsch, Introduction to Numerical Analysis, Springer, 2002.
[28] V. Stojanović, P. Ribeiro, S. Stoykov, Non-linear vibration of Timoshenko damaged beams by a new p-version finite element method, Computers & Structures 120 (2013) 107–119.
[29] P. Ribeiro, A p-version, first order shear deformation, finite element for geometrically non-linear vibration of curved beams, International Journal for Numerical Methods in Engineering 61 (2004) 2696–2715.
[30] R. Lewandowski, Non-linear free vibrations of beams by the finite element and continuation methods, Journal of Sound and Vibration 170 (1994) 577–593.
[31] Y.K. Cheung, S.L. Lau, Incremental time-space finite strip method for non-linear structural vibrations, Earthquake Engineering and Structural Dynamics 10 (1982) 239–253.
[32] P. Ribeiro, M. Petyt, Non-linear vibration of beams with internal resonance by the hierarchical finite-element method, Journal of Sound and Vibration 224 (1999) 591–624.
[33] M. Petyt, Introduction to Finite Element Vibration Analysis, Cambridge University Press, Cambridge, 1990.

Quantitative effect of an axial load on the amplitude stability of rotating nano-beams

13

13.1 Introduction

The studies concerning nano-mechanical structures are of great importance for nanotechnology and its future applications. There is a practical interest in knowing the dynamic behavior of a rotating nano-beam, especially when an external load is examined. Such rotating nanostructures have applications in nanomotorshafts, more specifically hybridized nanogenerators with different influences of external load [1]. The hybridized nanogenerator can be introduced as a nanosensor, for which the detected external load represents water or wind. The mechanical behavior of nanostructures greatly depends on their size because their dimensions are minute and comparable to molecular distances. The size effect is not considered in classical continuum mechanics; the classical continuum theory cannot predict such size effects due to the lack of additional material length scale parameters. To overcome this shortcoming—in the study of the mechanical behaviors of microstructures—various higher-order (nonclassical) continuum theories have been proposed. These theories involve additional material length scale parameters, such as couple stress elasticity theories [2–4], surface elasticity theory [5], strain gradient elasticity theory [6], and micro-polar scale-dependent continuum theories [7,8]. One of the more widespread theories—for determining material length scale parameters, which is applied in the here presented study—is the scale-dependent continuum theory proposed by Eringen [7,8]. Omitting small-scale effects and atomic forces in sensitive nano-designs may produce false solutions and, hence, wrong designs. As mentioned, Eringen and his associates have conducted one of the first modifications of the classical continuum theory. They expanded the classical theory of continuum mechanics to nonlocal continuum mechanics by introducing a material parameter that considered size effects and atomic forces. Compared to other methods, a modified continuum theory such as that Refs. [7–10] gives certain advantages in the design of nanoscale devices and has been known as the theory of nonlocal elasticity. According to Eringen [9], the theory of nonlocal elasticity provides a good approximation for a large class of problems in nanosystems in which the influence of length scales is highly pronounced. Carbon nanotubes (CNTs) by Iijima et al. [11,12] have brought attention to scientists and engineers that are in the field of nanotechnology because of their remarkable chemical [13], physical [14], and mechanical properties [15]. Schilhansl [16] published one of the first papers on the vibrations of a rotating cantilever beam; more precisely, the bending frequency of a rotating cantilever beam. Subsequently, Hoa [17] has researched the vibration of a rotating beam with tip mass. Using beam characteristic orthogonal polynomials in the Rayleigh-Ritz method, Baht [18]

Modeling of Complex Dynamic Systems. DOI: https://doi.org/10.1016/B978-0-443-23942-7.00013-9

analyzed natural frequencies and mode shapes of a rotating uniform cantilever beam with tip mass. Ni et al. [19] investigated the free vibration and stability of a cantilever beam attached to an axially moving base that is immersed in a fluid. Diken [20] has examined the vibration control of a rotating Euler–Bernoulli beam. Stojanović and Kozić [21] have analyzed the forced vibration and buckling of a double-beam system that is continuously joined by a Winkler elastic layer and loaded with compressive axial forces, using the Rayleigh and Timoshenko beam assumptions. Furthermore, a detailed analysis of vibrations and stability of complex beam systems has been presented by Stojanović and Kozić [22]. The externally excited single-walled carbon nanotube (SWCNT) system of two, coupled by a Winkler elastic medium, has been examined by Stamenković et al. [23]. Narendar [24] studied a rotating SWCNT using the Euler–Bernoulli beam assumption and the nonlocal/nonclassical continuum mechanics. Babaei et al. [25] have done a free longitudinal vibration analysis of a rotating rod based on Eringen's nonlocal elasticity. Guo et al. [26] have done analyzes based on the theory of nonlocal strain gradients to determine dynamic transverse vibration characteristics and vibrobuckling analyzes of axially moving and rotating nano-beams. Mohammadi et al. [27] have studied the vibration behavior of the rotating viscoelastic nano-beam embedded in the Pasternak foundation. The small-scale effect on the vibrational behavior of a rotary tapered axially functionally graded (AFG) micro-beam has been examined by Shafiei et al. [28]. For the basis, they used Timoshenko and Euler–Bernoulli beam assumptions and modified couple stress theories. Nayak [29] investigated the dynamic stability of a rotating three-layered symmetric sandwich beam with a magneto-rheological elastomer core and conductive skins. Using nonlocal elasticity theory, Murmu and Adhikari [30] analyzed the nonlocal effects in the bending vibration of a prestressed SWCNT. Also, using nonlocal elasticity theory, Narendar and Gopalakrishnan [31] have published their findings on the wave dispersion behavior of a rotating nanotube. Chen and Ho [32] have analyzed the transverse vibration of a rotating twisted Timoshenko beam subjected to axial loading by means of differential transform. Pradhan and Murmu [33] first developed a single nonlocal beam model and then used it to examine the flap-wise bending-vibration features of a rotating nano-cantilever. Younesian and Esmailzadeh [34] presented a novel approach for the vibration suspension of a rotating beam; the solution was to use a time-increasing internal tensile force. Dejin Chen et al. [35] performed a flap-wise vibration analysis of rotating laminated microbeams with geometric imperfections.

The study published by Pradhan and Murmu [33] was the inspiration for the study by Stamenković and Stojanović [36] on which the study in this chapter is based. The novelty of the study lies in the transient responses of the rotating cantilever nanobeams, with the considered nonlocality magnitude effect. The study presents a single rotating nonlocal cantilever nanobeam with external excitations. Two cases of forced vibrations are analyzed: undamped and damped forced vibrations. The governing equation of motion is derived using the Euler–Bernoulli beam assumption and Eringen's nonlocal elasticity theory. This equation of motion cannot be solved by the standard analytical procedure in closed-form, so an appropriate discretization is necessary; this is achieved with the Galerkin method. The further analysis investigates the nonlocal effect, angular velocity effect, and hub radius effect on the vibration of an undamped and damped rotating cantilever nano-beam. The determination of the quantitative values of natural frequencies and dynamic behavior of the rotating nonlocal cantilever nano-beam is done by applying the standard procedure of modal analysis [37], which means using the principal coordinates for uncoupling the differential equations of motion. The values of natural frequencies, which account for various influences of the three mentioned effects, are presented and validated by comparison with the results in Ref. [33]. The agreement between the compared values is very good. In addition, a detailed analysis concerning the dynamic behavior of the described rotating nano-beam is carried out. The dynamic vibratory behavior is discussed in the time domain according to the determined time histories and phase plots. This discussion is based on a few examples that account for various nonlocal effects, angular velocity effects, and hub radius effects. Furthermore, the study discusses the appropriately determined deformed shapes of the nano-beam in the steady-state regime of vibrations. In addition to the influence of transverse external excitations, the influence of the axial extension load on the system is also considered.

13.2 Nonlocal elastic constitutive relations

This section considers the fundamental equations of the nonlocal elasticity theory. The basic hypothesis in this continuum theory states that stress at a point is a function of strains at all points of the body. The constitutive elastic relation for a three-dimensional isotropic homogeneous body in integral form is given by Eringen et al. [2–4] as:

$$\sigma_{ij}(x) = \int \widetilde{\alpha}(|x - x'|, \psi) C_{ijkl}\varepsilon_{kl}(x')\mathrm{d}V(x'), \ \forall\ x \in V \tag{13.1a}$$

$$\sigma_{ij,j} = 0, \tag{13.1b}$$

$$\varepsilon_{ij} = \frac{1}{2}(u_{i,j} + u_{j,i}), \tag{13.1c}$$

where σ_{ij} and ε_{ij} are the stress and strain tensors, respectively; u_i is the displacement vector; C_{ijkl} is the elastic modulus tensor for classical isotropic elasticity; and $\widetilde{\alpha}(|x - x'|, \psi)$ denotes the nonlocal modulus or attenuation function, which incorporates nonlocal effects into the constitutive equation at a reference point x produced by the local strain at a source x'. The absolute value of the difference $|x - x'|$ denotes the Euclidean metric. The parameter ψ is the nonlocal parameter $\psi = (e_0 a)/L$, where L is the external characteristic length (crack length, wavelength), a is the internal characteristic length, and e_0 is a constant appropriate to each material that can be identified from simulations at an atomic scale or by using the dispersive curve of the Born–Karman model of lattice dynamics. Using Eq. (13.1) in practice is difficult; thus, simplified constitutive relations are developed in the differential form. As stated in Ref. [9], these constitutive relations in differential form are given as:

$$\sigma(x) - (e_0 a)^2 \cdot \frac{\mathrm{d}^2\sigma(x)}{\mathrm{d}x^2} = E\varepsilon(x), \tag{13.2}$$

where $\sigma(x)$ is the normal nonlocal stress; $\varepsilon(x) = \partial u/\partial x$ is the axial deformation; $\mu = (e_0 a)^2$ is the nonlocal parameter; E is the modulus of elasticity. The internal characteristic lengths $(e_0 a)$ are often assumed to be in the range 0–2 [nm] for nanomaterials such as carbon nanotubes, zinc oxide, etc. When $e_0 a = 0$, the nonlocal constitutive relation is reduced to the classical constitutive relation of the elastic body. Here will be considered lower values of the nondimensional nonlocal parameter ψ in the range [0, 0.4].

13.3 Mathematical formulations of the equations of motion

The physical model of a rotating nanotube is important for novel miniature devices. Rotating nanotubes can be represented as cantilever nanobeams. This study considers a nano-cantilever of length L that is fixed at point O to a rigid hub. The nano-cantilever is under external excitation (Fig. 13.1). The hub has a radius r, and its rotational speed Ω is constant.

This study was conducted using the elementary Euler–Bernoulli beam assumption. The moment stress resultant relation of the nano-cantilever is specified as:

$$M = \int_A z\sigma_{xx} \mathrm{d}A, \tag{13.3}$$

where x is the coordinate along the axial direction of the cantilever, A is the cross-sectional area, and σ_{xx} is the stress in the x direction. With Eqs. (13.2) and (13.3), the following nonlocal moment relation can be derived:

$$M - (e_0 a)^2 \frac{\partial^2 M}{\partial x^2} = -EI \frac{\partial^2 w}{\partial x^2}. \tag{13.4}$$

The Euler–Bernoulli beam assumption gives the following equations of motion:

$$\frac{\partial^2 M}{\partial x^2} + \frac{\partial}{\partial x}\left(T(x)\frac{\partial w}{\partial x}\right) - \rho A \frac{\partial^2 w}{\partial t^2} + f(x, t) = 0, \tag{13.5}$$

where $T(x)$ is the centrifugal tension force due to the rotation of the cantilever, $f(x, t)$ is the external force on the nano-beam, while ρ presents the mass density and

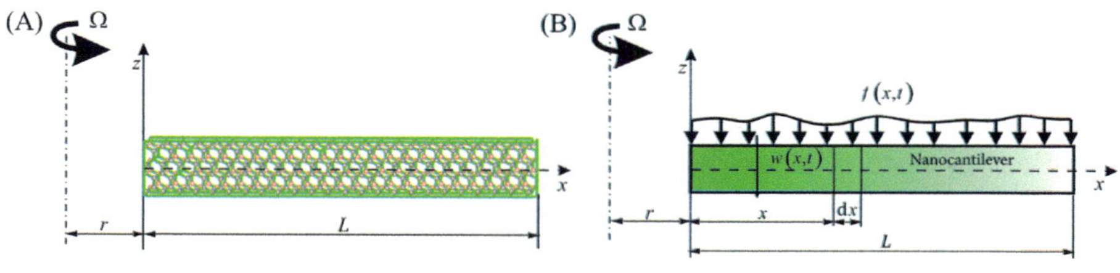

Figure 13.1 Rotating nanocantilever with external excitations; (A) The physical model of the external excited CNT; (B) The equivalent mechanical model.

A is the beam cross-section area. The axial force $T(x)$ due to centrifugal stiffening is given as:

$$T(x) = \int_x^L \rho A \Omega^2 (r + x) \mathrm{d}x, \tag{13.6}$$

where r (known as the hub radius) is the distance from the origin of rotation to the inner end of the cantilever (Fig. 13.1), and Ω is the angular velocity of the cantilever. Collecting the second derivative of M from Eq. (13.5) and substituting it into Eq. (13.4), produces:

$$M = -EI\frac{\partial^2 w}{\partial x^2} + (e_0 a)^2 \left[\rho A \frac{\partial^2 w}{\partial t^2} - \frac{\partial}{\partial x}\left(T\frac{\partial w}{\partial x} \right) - f(x, t) \right], \tag{13.7}$$

where I is the moment of inertia of the nano-beam cross-section. Substituting Eq. (13.7) into Eq. (13.5) leads to the equation in the following form:

$$\begin{aligned} &EI\frac{\partial^4 w}{\partial x^4} + \rho A\frac{\partial^2 w}{\partial t^2} - (e_0 a)^2 \rho A \frac{\partial^4 w}{\partial x^2 \partial t^2} - \frac{\partial}{\partial x}\left(T\frac{\partial w}{\partial x} \right) + (e_0 a)^2 \frac{\partial^2}{\partial x^2}\left[\frac{\partial}{\partial x}\left(T\frac{\partial w}{\partial x} \right) \right] \\ &= f(x, t) - (e_0 a)^2 \frac{\partial^2 f(x, t)}{\partial x^2}. \end{aligned} \tag{13.8}$$

Using the force $T(x)$ from Eq. (13.6) and substituting into Eq. (13.8) will lead to the equation of motion of the rotating nonlocal cantilever nano-beam with an external force.

$$\begin{aligned} &EI\frac{\partial^4 w}{\partial x^4} + \rho A\frac{\partial^2 w}{\partial t^2} - (e_0 a)^2 \rho A \frac{\partial^4 w}{\partial x^2 \partial t^2} - \rho A \Omega^2 \left[(-r - x)\frac{\partial w}{\partial x} + \left(rL + \frac{L^2}{2} - rx - \frac{x^2}{2} \right) \frac{\partial^2 w}{\partial x^2} \right] \\ &+ (e_0 a)^2 \rho A \Omega^2 \left[-3\frac{\partial^2 w}{\partial x^2} + 3(-r - x)\frac{\partial^3 w}{\partial x^3} + \left(rL + \frac{L^2}{2} - rx - \frac{x^2}{2} \right) \frac{\partial^4 w}{\partial x^4} \right] \\ &= f(x, t) - (e_0 a)^2 \frac{\partial^2 f(x, t)}{\partial x^2}. \end{aligned} \tag{13.9}$$

According to Eq. (13.22) from Ref. [38], the boundary conditions for the cantilever nano-beam at two ends are given by

$$\frac{\partial w}{\partial x} = 0 \text{ and } w = 0 \text{ at } x = 0, \tag{13.10a}$$

$$\frac{\partial^2 w}{\partial x^2} = 0 \text{ and } \frac{\partial^3 w}{\partial x^3} - (e_0\ a)^2 \rho A \frac{\partial^3 w}{\partial x \partial t^2} = 0 \text{ at } x = L. \tag{13.10b}$$

13.4 Solution methodology

Eq. (13.9) cannot be solved by the standard analytical procedure in the closed-form. Thus, solving can be done by making an appropriate discretization. One of the useful methods is the Galerkin method. The solution of Eq. (13.9) can be expressed as:

$$w(x, t) = \sum_{i=1}^{n} W_i(x) q_i(t), \tag{13.11}$$

where $W_i(x)$ functions satisfy the given boundary conditions Eqs. (13.10a) and (13.10b) in the following familiar form:

$$W_i(x) = \cosh(\widetilde{\lambda}_i x) - \cos(\widetilde{\lambda}_i x) - \frac{\sinh(\widetilde{\lambda}_i L) - \sin(\widetilde{\lambda}_i L)}{\cosh(\widetilde{\lambda}_i L) + \cos(\widetilde{\lambda}_i L)}[\sinh(\widetilde{\lambda}_i x) - \sin(\widetilde{\lambda}_i x)], \tag{13.12}$$

where $\widetilde{\lambda}_i$, $i = 1, 2, \ldots., n$ are the values that satisfy the characteristic equation $1 + \cos(\lambda)\cosh(\lambda) = 0$.

Substituting the solution of Eq. (13.11) in Eq. (13.9) gives

$$\begin{aligned}
&\sum_{i=1}^{n}\Bigg\{EI\frac{\partial^4 W_i(x)}{\partial x^4}q_i(t) + \rho A W_i(x)\ddot{q}_i(t) - (e_0 a)^2\rho A\frac{\partial^2 W_i(x)}{\partial x^2}\ddot{q}_i(t) \\
&- \rho A\Omega^2\left[(-r - x)\frac{\partial W_i(x)}{\partial x} + \left(rL + \frac{L^2}{2} - rx - \frac{x^2}{2}\right)\frac{\partial^2 W_i(x)}{\partial x^2}\right]q_i(t) \\
&+ (e_0 a)^2\rho A\Omega^2\Bigg[-3\frac{\partial^2 W_i(x)}{\partial x^2} + 3(-r - x)\frac{\partial^3 W_i(x)}{\partial x^3} \\
&+ \left(rL + \frac{L^2}{2} - rx - \frac{x^2}{2}\right)\frac{\partial^4 W_i(x)}{\partial x^4}\Bigg]q_i(t)\Bigg\} \\
&= f(x, t) - (e_0 a)^2\frac{\partial^2 f(x, t)}{\partial x^2},
\end{aligned} \tag{13.13}$$

or in the more condensed form:

$$\sum_{i=1}^{n}\Psi_i(x, t) = \widetilde{f}(x, t). \tag{13.14}$$

Successful multiplication of Eq. (13.14) with trial function Eq. (13.12), for each $j = 1, 2, \ldots n$ that satisfies the boundary conditions Eq. (13.10) and integrating from 0 to L, will give the set of n ordinary differential equations.

$$\int_0^L\left[\sum_{i=1}^{n}\Psi_i(x, t)W_1(x)\right]dx = \int_0^L[\widetilde{f}(x, t)W_1(x)]dx, \tag{13.15a}$$

$$\int_0^L\left[\sum_{i=1}^{n}\Psi_i(x, t)W_2(x)\right]dx = \int_0^L[\widetilde{f}(x, t)W_2(x)]dx, \tag{13.15b}$$

$$\int_0^L\left[\sum_{i=1}^{n}\Psi_i(x, t)W_n(x)\right]dx = \int_0^L[\widetilde{f}(x, t)W_n(x)]dx. \tag{13.15c}$$

Since the integration of Eqs. (13.15a), (13.15b), and (13.15c) depends on the number n, the system of equations of motion can be expressed in the matrix form:

$$\mathbf{M}\ddot{\mathbf{q}} + \mathbf{K}\mathbf{q} = \mathbf{F}, \tag{13.16}$$

where $\mathbf{q}$ is the n-dimensional column vector of generalized coordinates, $\mathbf{M}$ and $\mathbf{K}$ are the symmetric $n \times n$ mass and stiffness matrices, and $\mathbf{F}$ is the system of external loads.

$$\mathbf{F} = \begin{Bmatrix} \int_0^L \left(f(x,t) - (e_0 a)^2 \frac{\partial^2 f(x,t)}{\partial x^2} \right) W_1(x) \mathrm{d}x \\ \int_0^L \left(f(x,t) - (e_0 a)^2 \frac{\partial^2 f(x,t)}{\partial x^2} \right) W_2(x) \mathrm{d}x \\ \vdots \\ \int_0^L \left(f(x,t) - (e_0 a)^2 \frac{\partial^2 f(x,t)}{\partial x^2} \right) W_n(x) \mathrm{d}x \end{Bmatrix}. \tag{13.17}$$

13.5 Natural frequencies of the rotating nonlocal cantilever nano-beam

The natural frequencies of the rotating nonlocal cantilever beam can be determined using Eq. (13.16), but first, the standard form of homogenous differential equations of free vibrations needs to be considered in the form:

$$\mathbf{M}\ddot{\mathbf{q}} + \mathbf{K}\mathbf{q} = 0. \tag{13.18}$$

When there is no damping in the system, dissipative mechanisms are nonexistent, and the free vibrations described by the solution of Eq. (13.18) are periodic. It is assumed that the vibrations are synchronous. Therefore, when free vibrations at a single frequency are initiated for a particular system, the ratio of any two dependent variables is unrelated to time. These assumptions lead to hypothesizing the normal-mode solution of Eq. (13.18) in the form:

$$\mathbf{q}(\mathrm{t}) = \mathbf{Q}e^{i\omega_n t}, \ n = 1.2, \ \ldots, \ \infty, \tag{13.19}$$

where ω_n represents the natural frequencies and $\mathbf{Q}$ is the m-dimensional vector of constants (according to mode shapes). Substituting Eq. (13.19) into Eq. (13.18) leads to

$$(-\omega_n^2 \mathbf{MQ} + \mathbf{KQ})e^{i\omega_n t} = 0. \tag{13.20}$$

Since $e^{i\omega_n t} \neq 0$ for any real value of t,

$$-\omega_n^2 \mathbf{MQ} + \mathbf{KQ} = 0. \tag{13.21}$$

The mass matrix is nonsingular, hence $\mathbf{M}^{-1}$ exists. Multiplying Eq. (13.21) by $\mathbf{M}^{-1}$ and rearranging gives

$$(\mathbf{M}^{-1}\mathbf{K} - \omega_n^2 \mathbf{I})\mathbf{Q} = 0, \tag{13.22}$$

where $\mathbf{I}$ is the $n \times n$ identity matrix. Accordingly, the trivial solution $\mathbf{Q} = 0$ is obtained, unless

$$\det|\mathbf{M}^{-1}\mathbf{K} - \omega_n^2 \mathbf{I}| = 0. \tag{13.23}$$

Based on this, the squares of the eigenvalues of the matrix of $\mathbf{M}^{-1}\mathbf{K}$ present the natural frequencies $(\omega_1, \ \omega_2, \ \omega_3, \ \omega_4, \ \ldots, \omega_n)$ of the system. Pradhan and Murmu

[33] have introduced the nondimensionalization of principal parameters. For this reason, the following parameters are introduced so the results in the present study can be validated:

a. nondimensional hub radius

$$\delta = \frac{r}{L}, \tag{13.24}$$

b. nondimensional angular velocity

$$\gamma = \sqrt{\frac{\rho A \Omega^2 L^4}{EI}}, \tag{13.25}$$

c. nondimensional frequency

$$\lambda = \sqrt{\frac{\rho A \omega_n^2 L^4}{EI}}, \tag{13.26}$$

d. nondimensional nonlocal parameter

$$\psi = \frac{e_0 a}{L}. \tag{13.27}$$

The first four nondimensional frequencies $\lambda_i,\ i = 1,\ 2,\ 3,\ 4$, for the nondimensional angular velocity $\gamma = 1$ and the nondimensional hub radius $\delta = 1$ are presented in Table 13.1. The first four nondimensional frequencies $\lambda_i,\ i = 1,\ 2,\ 3,\ 4$, for the nondimensional angular velocity $\gamma = 1$ and the nondimensional hub radius $\delta = 0$ are presented in Table 13.2. The varying nonlocal parameter ψ is used in the [0, 0.4] range. The value of the first nondimensional frequency λ_1 is the same as in Ref. [33]. In this study, other values of nondimensional frequencies when the hub radius is zero

Table 13.1 Nondimensional frequencies for the nonlocal parameter $\psi = 0; 0.1; 0.2; 0.3; 0.4$, where nondimensional hub radius is $\delta = 1$ and nondimensional angular velocity is $\gamma = 1$.

–		λ_1	λ_2	λ_3	λ_4
$\psi = 0$	Present study,	3.889	22.375	62.043	121.263
	[33]	3.890	22.380	62.050	–
$\psi = 0.1$	Present study,	3.932	21.122	51.603	86.356
	[33]	3.932	21.122	51.603	–
$\psi = 0.2$	Present study,	4.065	18.257	37.744	55.350
	[33]	4.065	18.257	37.744	–
$\psi = 0.3$	Present study,	4.305	15.107	29.519	39.189
	[33]	4.305	15.110	29.520	–
$\psi = 0.4$	Present study,	4.701	12.229	26.018	28.587
	[33]	4.701	12.230	26.0201	–

Table 13.2 Nondimensional frequencies for the varying nonlocal parameter ψ=0; 0.1; 0.2; 0.3; 0.4, where nondimensional hub radius is δ=0 and nondimensional angular velocity is γ=1.

		λ_1	λ_2	λ_3	λ_4
$\psi = 0$	Present study,	3.682	22.181	61.842	121.051
	[33]	3.690	–	–	–
$\psi = 0.1$	Present study,	3.706	20.868	51.295	85.982
	[33]	3.710	–	–	–
$\psi = 0.2$	Present study,	3.783	17.866	37.223	54.704
	[33]	3.783	–	–	–
$\psi = 0.3$	Present study,	3.923	14.589	28.709	38.346
	[33]	3.923	–	–	–
$\psi = 0.4$	Present study,	4.153	11.675	24.468	28.043
	[33]	4.154	–	–	–

Table 13.3 Convergence study of nondimensional frequencies for ψ=0, γ=1, and δ=0.

	λ_1	λ_2	λ_3	λ_4
$n = 2$	3.68167306	22.18102852		
$n = 3$	3.68165070	22.18102833	61.84178166	
$n = 4$	3.68164789	22.18101484	61.84177366	121.05094094
$n = 5$	3.68164718	22.18101281	61.84176519	121.05092897
$n = 6$	3.68164697	22.18101185	61.84176414	121.05092360

($\delta = 0$) are shown for various values of the nonlocal parameter ψ. It can be noticed that an increase in the nonlocal parameter ψ leads to an increase in the first nondimensional frequency λ_1. But, for the other values of nondimensional frequencies λ_2, λ_3, and λ_4, the observation is contrary. An increase in the nonlocal parameter ψ leads to a decrease in the values of the nondimensional frequencies λ_2, λ_3, and λ_4. The highest value of the nondimensional frequency λ_4 is measured for the local model when the nonlocal parameter is $\psi = 0$. The slightly higher values of nondimensional frequencies λ_i, $i = 1, 2, 3, 4$, are obtained when the hub radius r exists ($\delta > 0$). In the vibrational analysis (Tables 13.1 and 13.2), six members of the order in discretization were applied. In this way, a comparison of results with results from Ref. [33] is achieved. Furthermore, using other theories to determine the frequency of the system in this study has led to results that are very close to the results in Refs. [39] and [40]. The convergence is presented in Table 13.3. It is perceptible that an increase in the discretization order leads to a decrease in the value of nondimensional frequencies. The differences between two and six points of the orders in discretization are in the fifth decimal place.

The variation of nondimensional first-mode frequency λ_1 with nondimensional angular velocity γ (for different values of nonlocal parameter ψ and nondimensional hub radius δ) is depicted in Fig. 13.2. An increase in the nonlocal parameter causes an increase in the nondimensional first mode frequency of the rotating nonlocal undamped cantilever nano-beam. The same is observed when the hub radius exists ($\delta > 0$). It can be noticed that the local model has the lowest values of first mode frequency. An increase in the nondimensional angular velocity leads to an increase in the nondimensional first mode frequency. The frequency increases as the angular velocity increases; furthermore, the increase in angular velocity causes the centrifugal force to increase, and with that, the stiffening effect. The opposite observations of the frequency illustration for higher modes are presented in Figs. 13.3–13.5. With an

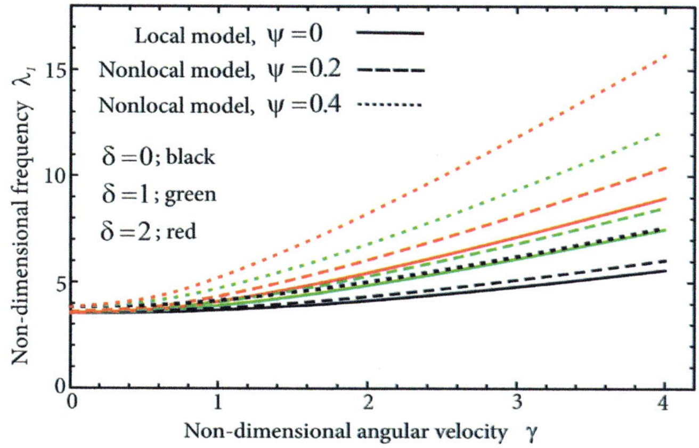

Figure 13.2 Variation of nondimensional first-mode frequency. Over the nondimensional angular velocity for several nonlocal nondimensional parameters and nondimensional hub radii.

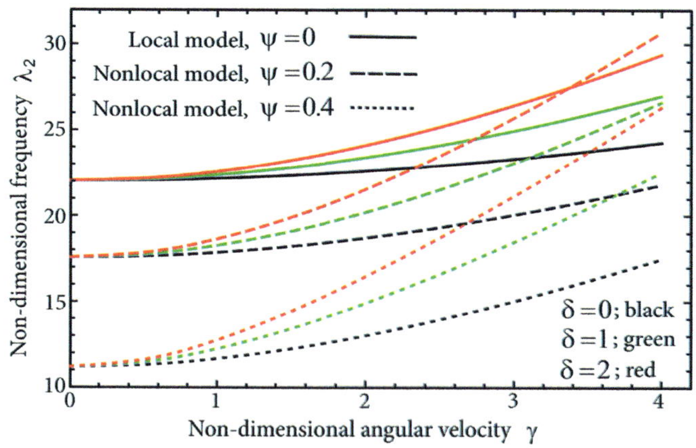

Figure 13.3 Variation of nondimensional second-mode frequency. Over the nondimensional angular velocity for several nonlocal nondimensional parameters and nondimensional hub radii.

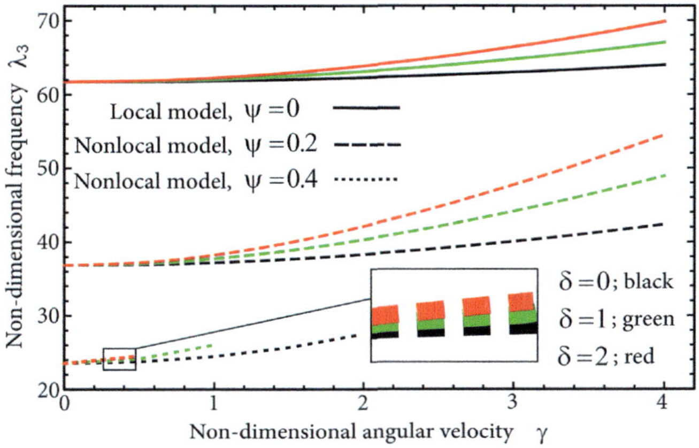

Figure 13.4 Variation of nondimensional third-mode frequency. Over the nondimensional angular velocity for several nonlocal nondimensional parameters and nondimensional hub radii.

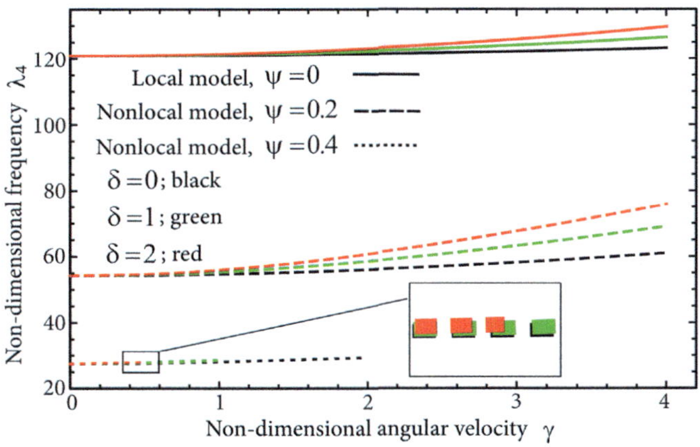

Figure 13.5 Variation of nondimensional fourth-mode frequency. Over the nondimensional angular velocity for several nonlocal nondimensional parameters and nondimensional hub radii.

increase in the nonlocal parameter, the nondimensional second, third, and fourth modes of frequencies, with nondimensional angular velocity, are decreased. The impact of hub radius at higher modes leads to increased frequency values. For this reason, the diagrams of the third and fourth modes are interrupted (Figs. 13.4 and 13.5). Unlike the frequency of the first mode, the frequencies of the higher modes have different starting points among the frequencies of the local and nonlocal models. The present studies are in agreement with those of Ref. [33].

13.6 Forced vibration of the undamped system

The differential equations of motion of the forced vibrations of the undamped system are presented by Eq. (13.16). One of the most common means to determine the response to the forced vibration of the presented system, Eq. (13.16), is the standard procedure of modal analysis based on principal coordinates. In the previous section, the natural frequencies $\omega_1 \leq \omega_2 \leq \omega_3 \leq \omega_4$ presented the natural frequencies of the system whose equations are given by Eq. (13.16). If it is assumed that matrix $\mathbf{P}$ represents the modal matrix of the system, the matrix whose columns are the normalized mode shapes

$$\mathbf{P} = \{\mathbf{Q}_1 \quad \mathbf{Q}_2 \quad \cdots \quad \mathbf{Q}_n\}. \tag{13.28}$$

A mode shape corresponding to a specific natural frequency of an undamped linear n-degree-of-freedom (n-DOF) system chosen to satisfy the normalization constraint is termed a normalized mode shape. All mode shapes must satisfy the same normalization constraint. The constraint should be chosen so that subsequent use of the normalized mode shape is convenient. The system of n-differential equations, which corresponds to the generalized coordinates q(t), can be transformed into a system with the principal coordinates $p_j(t)$, $j = 1, 2, \ldots, n$, by introducing the modal matrix of the system $\mathbf{P}$, whose columns represent the normalized mode shapes. The transformation

$$\mathbf{q}(t) = \sum_{j=1}^{n} p_j(t)\mathbf{Q}_j, \tag{13.29}$$

is equivalent to a linear transformation between the generalized and the principal coordinates of the system

$$\mathbf{q}(t) = \mathbf{P}\mathbf{p}(t). \tag{13.30}$$

Substituting Eq. (13.29) in Eq. (13.16) gives

$$\mathbf{M}\sum_{j=1}^{n} \ddot{p}_j(t)\mathbf{Q}_j + \mathbf{K}\sum_{j=1}^{n} p_j(t)\mathbf{Q}_j = \mathbf{F}. \tag{13.31}$$

Mode-shape vectors are normalized with respect to the kinetic-energy scalar product. Based on this, Eq. (13.31) multiplied by a scalar $\mathbf{Q}_r$, for an arbitrary $r = 1, 2, \ldots, n$, brings on

$$\sum_{j=1}^{n} \ddot{p}_j(t)(\mathbf{Q}_r\mathbf{M}\mathbf{Q}_j) + \sum_{j=1}^{n} p_j(t)(\mathbf{Q}_r\mathbf{K}\mathbf{Q}_j) = (\mathbf{Q}_r\mathbf{F}). \tag{13.32}$$

Using the orthogonality condition of the basic forms of vibrations (mode-shape orthogonality condition) will generate only one member in each of the sums that is different from zero, and that is for $r = j$. The mode shapes are normalized, and Eq. (13.32) is in the form:

$$\sum_{r=1}^{n} \ddot{p}_r(t)(\mathbf{Q}_r\mathbf{Q}_r)_M + \sum_{r=1}^{n} p_r(t)(\mathbf{Q}_r\mathbf{Q}_r)_K = (\mathbf{Q}_r\mathbf{F}), \tag{13.33}$$

or

$$\ddot{p}_r(t) + \omega_r^2 p_r(t) = g_r(t),\ r = 1, 2, 3, \ldots, n \tag{13.34}$$

where

$$\omega_r^2 = \frac{(\mathbf{Q}_r\mathbf{Q}_r)_K}{(\mathbf{Q}_r\mathbf{Q}_r)_M}, \tag{13.35}$$

$$g_r(t) = (\mathbf{Q}_r\mathbf{F}), \tag{13.36}$$

where **F** is the system of external forces presented by Eq. (13.17). For uniformly distributed load, expression Eq. (13.17) can be written in the form:

$$f(x, t) = F_0 \cos(\tilde{\Omega}t). \tag{13.37}$$

If the initial conditions for the principal coordinates are acquired from Eq. (13.30) as $p_r(0) = 0$ and $\dot{p}_r(0) = 0$, the solution of the principal coordinates has the following form:

$$p_r(t) = \frac{1}{\omega_r}\int_0^t g_r(s)\sin[\omega_r(t-s)]\mathrm{d}s\ .\ r = 1, 2, 3, \ldots, n. \tag{13.38}$$

When the solution from Eq. (13.38) is returned to Eq. (13.30), the following is obtained:

$$\mathbf{q}(t) = \begin{Bmatrix} Q_{11}(\omega_1)p_1(t) + Q_{12}(\omega_2)p_2(t)+\ldots+Q_{1n}(\omega_n)p_n(t) \\ Q_{21}(\omega_1)p_1(t) + Q_{22}(\omega_2)p_2(t)+\ldots+Q_{2n}(\omega_n)p_n(t) \\ \vdots \\ Q_{n1}(\omega_1)p_1(t) + Q_{n2}(\omega_2)p_2(t)+\ldots+Q_{nn}(\omega_n)p_n(t) \end{Bmatrix}. \tag{13.39}$$

Finally, the solution of Eq. (13.9) for the undamped system is in the following form:

$$\begin{aligned} w(x,t) &= \sum_{i=1}^{n} W_i(x)q_i(t) \\ &= \{W_1(x)[Q_{11}(\omega_1)p_1(t) + Q_{12}(\omega_2)p_2(t)+\ldots+Q_{1n}(\omega_n)p_n(t)]\} \\ &+ \{W_2(x)[Q_{21}(\omega_1)p_1(t) + Q_{22}(\omega_2)p_2(t)+\ldots+Q_{2n}(\omega_n)p_n(t)]\}+\ldots \\ &+ \{W_n(x)[Q_{n1}(\omega_1)p_1(t) + Q_{n2}(\omega_2)p_2(t)+\ldots+Q_{nn}(\omega_n)p_n(t)]\}, \end{aligned} \tag{13.40}$$

where $W_i(x)$ for the cantilever beam are presented by Eq. (13.12).

13.7 Forced vibration of the damped system

If the proportional damping of Rayleigh type is introduced, the equations of motion that characterize the forced vibration of the damped system will have the form:

$$\mathbf{M}\ddot{\mathbf{q}} + (\alpha\mathbf{K} + \beta\mathbf{M})\dot{\mathbf{q}} + \mathbf{K}\mathbf{q} = \mathbf{F}. \tag{13.41}$$

Eq. (13.41) renders the proportionally damped model, where the damping matrix is a linear combination of the mass matrix $\beta\mathbf{M}$ and the stiffness matrix $\alpha\mathbf{K}$. The modal analysis was mentioned in the previous section; the same modal analysis is implemented here for systems with proportional damping. The same procedure is followed as in the previous section.

Substituting Eq. (13.29) in Eq. (13.41) in this case gives

$$\mathbf{M}\sum_{j=1}^{n}\ddot{p}_j(t)\mathbf{Q}_j + (\alpha\mathbf{K}+\beta\mathbf{M})\sum_{j=1}^{n}\dot{p}_j(t)\mathbf{Q}_j + \mathbf{K}\sum_{j=1}^{n}p_j(t)\mathbf{Q}_j = \mathbf{F}. \tag{13.42}$$

Multiplying Eq. (13.42) by a scalar $\mathbf{Q}_r$, for an arbitrary $r = 1, 2, \ldots, n$, leads to

$$\sum_{j=1}^{n}\ddot{p}_j(t)(\mathbf{Q}_r\mathbf{M}\mathbf{Q}_j) + \sum_{j=1}^{n}\dot{p}_j(t)[\alpha(\mathbf{Q}_r\mathbf{K}\mathbf{Q}_j) + \beta(\mathbf{Q}_r\mathbf{M}\mathbf{Q}_j)] + \sum_{j=1}^{n}p_j(t)(\mathbf{Q}_r\mathbf{K}\mathbf{Q}_j) = (\mathbf{Q}_r\mathbf{F}). \tag{13.43}$$

By employing the orthogonality condition of the basic forms of vibrations (mode-shape orthogonality condition), only one member in each of the sums that is different from zero can be obtained, and that is for $r = j$. The mode shapes are normalized, and Eq. (13.43) is in the form:

$$\sum_{r=1}^{n}\ddot{p}_r(t)(\mathbf{Q}_r\mathbf{Q}_r)_{\mathrm{M}} + \sum_{j=1}^{n}\dot{p}_r(t)[\alpha(\mathbf{Q}_r\mathbf{Q}_r)_K + \beta(\mathbf{Q}_r\mathbf{Q}_r)_{\mathrm{M}}] + \sum_{r=1}^{n}p_r(t)(\mathbf{Q}_r\mathbf{Q}_r)_K = (\mathbf{Q}_r\mathbf{F}), \tag{13.44}$$

or

$$\ddot{p}_r(t) + 2\zeta_r\dot{p}_r(t) + \omega_r^2 p_r(t) = g_r(t),\ r = 1, 2, 3, \ldots, n, \tag{13.45}$$

where

$$\zeta_r = \frac{1}{2}(\alpha\omega_r^2 + \beta), \tag{13.46}$$

The system of external forces $\mathbf{F}$ presented by Eq. (13.17) is applied to the uniformly distributed load in the form:

$$f(x, t) = F_0\cos(\widetilde{\Omega}\mathrm{t}). \tag{13.47}$$

Substituting Eq. (13.47) into Eq. (13.17) gives

$$\mathbf{F} = \mathbf{F}_0\cos(\widetilde{\Omega}\mathrm{t}) = \begin{Bmatrix} \int_0^L F_0 W_1(x)\mathrm{d}x \\ \int_0^L F_0 W_2(x)\mathrm{d}x \\ \vdots \\ \int_0^L F_0 W_n(x)\mathrm{d}x \end{Bmatrix}\cos(\widetilde{\Omega}\mathrm{t}). \tag{13.48}$$

Now, Eq. (13.45) can be written in the form:

$$\ddot{p}_r(t) + 2\zeta_r\dot{p}_r(t) + \omega_r^2 p_r(t) = h_r \cos(\widetilde{\Omega}t),\ r = 1, 2, 3, \ldots, n, \tag{13.49}$$

where

$$h_r = (\mathbf{Q}_r\mathbf{F}_0). \tag{13.50}$$

Next, the attention will be on nonhomogeneous second-order linear equations presented by Eq. (13.49). The general solution can be expressed in the following form:

$$p_r(t) = p_{rh}(t) + p_{rp}(t), \tag{13.51}$$

where $p_{rh}(t)$ and $p_{rp}(t)$ are the homogeneous and the particular solutions of Eq. (13.49). Each nonhomogeneous Eq. (13.49) has a corresponding homogeneous equation.

$$\ddot{p}_r(t) + 2\zeta_r\dot{p}_r(t) + \omega_r^2 p_r(t) = 0,\ r = 1, 2, 3, \ldots, n. \tag{13.52}$$

The homogeneous solution is assumed to have the form:

$$p_{rh} = (A\cos\vartheta t + B\sin\vartheta t)e^{-\zeta_r t}, \tag{13.53}$$

where A and B are the arbitrary constants obtained from the initial conditions and

$$\vartheta = \sqrt{\omega_r^2 - \zeta_r^2}\,. \tag{13.54}$$

Based on the nonhomogeneous term $h_r\cos(\widetilde{\Omega}t)$ given in Eq. (13.49), the type of function can be predicted; the function's particular solution p_{rp} would be

$$p_{rp} = C\cos(\widetilde{\Omega}t) + D\sin(\widetilde{\Omega}t), \tag{13.55}$$

where C and D are the coefficients that are easy to determine. By substituting the $\dot{p}_{rp}$ and $\ddot{p}_{rp}$ in Eq. (13.49), two equations are obtained:

$$(\omega_r^2 - \widetilde{\Omega}^2)C + 2\zeta_r\widetilde{\Omega}D = h_r, \tag{13.56a}$$

$$-2\zeta_r\widetilde{\Omega}C + (\omega_r^2 - \widetilde{\Omega}^2)D = 0. \tag{13.56b}$$

From the systems of two Eqs. (13.56a) and (13.56b), the coefficients are determined:

$$C = \frac{h_r(\omega_r^2 - \widetilde{\Omega}^2)}{(\omega_r^2 - \widetilde{\Omega}^2)^2 + 4\zeta_r^2\widetilde{\Omega}^2} = N\cos\varphi, \tag{13.57}$$

$$D = \frac{2h_r\zeta_r\widetilde{\Omega}}{(\omega_r^2 - \widetilde{\Omega}^2)^2 + 4\zeta_r^2\widetilde{\Omega}^2} = N\sin\varphi, \tag{13.58}$$

where N is the amplitude of the vibration. From Eqs. (13.57) and (13.58) it is simple to obtain newly inserted unknowns, N and φ in the following form:

$$N = \frac{h_r}{\sqrt{(\omega_r^2 - \widetilde{\Omega}^2)^2 + 4\zeta_r^2\widetilde{\Omega}^2}}, \tag{13.59}$$

and

$$\varphi = \text{arctg}\frac{2\zeta_r\widetilde{\Omega}}{(\omega_r^2 - \widetilde{\Omega}^2)}. \tag{13.60}$$

Now the particular and general solutions have the following forms:

$$p_{rp} = N\cos(\widetilde{\Omega}\text{t} - \varphi). \tag{13.61}$$

$$p_r(t) = (A\cos\vartheta t + B\sin\vartheta t)e^{-\zeta_r t} + N\cos(\widetilde{\Omega}\text{t} - \varphi). \tag{13.62}$$

For the arbitrary initial condition $p_r(0) = p_{r0}$ and $\dot{p}_r(0) = \dot{p}_{r0}$, it is possible to obtain the remaining constants A and B:

$$A = p_{r0} - N\cos\varphi, \tag{13.63}$$

$$B = \frac{\dot{p}_{r0} + \zeta_r p_{r0}}{\vartheta} - \frac{N}{\vartheta}(\zeta_r\cos\varphi + \widetilde{\Omega}\sin\varphi). \tag{13.64}$$

After some algebra, the general solution of Eq. (13.49) is obtained:

$$\begin{aligned} p_r(t) &= \left(p_{r0}\cos\vartheta t + \tfrac{\dot{p}_{r0} + \zeta_r p_{r0}}{\vartheta}\sin\vartheta t\right)e^{-\zeta_r t} \\ &- \left[\cos\varphi\cos\vartheta t + \tfrac{(\zeta_r\cos\varphi + \widetilde{\Omega}\sin\varphi)}{\vartheta}\sin\vartheta t\right]Ne^{-\zeta_r t} \\ &+ N\cos(\widetilde{\Omega}\text{t} - \varphi),\ r = 1, 2, \ldots n. \end{aligned} \tag{13.65}$$

Substituting the solutions Eq. (13.65) in Eq. (13.30) gives the same form of expression Eq. (13.39).

At last, after the vector $\mathbf{q}(t)$ is fully determined, the solution of Eq. (13.9) for the damped system Eq. (13.41) has the following form:

$$\begin{aligned} w(x,t) &= \sum_{i=1}^{n} W_i(x)q_i(t) \\ &= \{W_1(x)[Q_{11}(\omega_1)p_1(t) + Q_{12}(\omega_2)p_2(t) + \ldots + Q_{1n}(\omega_n)p_n(t)]\} \\ &+ \{W_2(x)[Q_{21}(\omega_1)p_1(t) + Q_{22}(\omega_2)p_2(t) + \ldots + Q_{2n}(\omega_n)p_n(t)]\} + \ldots \\ &+ \{W_n(x)[Q_{n1}(\omega_1)p_1(t) + Q_{n2}(\omega_2)p_2(t) + \ldots + Q_{nn}(\omega_{nm})p_n(t)]\}, \end{aligned} \tag{13.66}$$

13.8 Results and discussion

The results calculated for both forced cases (damped and undamped vibrations) are presented in this section. The calculated results for the forced vibration of the rotating nanotube are validated by adopting the geometric and material properties of the nano-beam (nanotube) from Ref. [33]. Young's modulus is $E = 971$ GPa, the density is $\rho = 2300\ \mathrm{kg/m^3}$, $L = 100$ nm is the length, and $d = 0.34$ nm is the diameter. Nondimensional angular velocity is expressed with the parameter $\gamma = 1; 2; 3$; the nondimensional hub radius with $\delta = 0; 1; 2$; and the nondimensional nonlocal parameter with $\psi = 0.1; 0.2$. The forced vibrations of the rotating nonlocal model are analyzed using numerical simulations in the time domain. In addition to the influence of the transverse external excitations, the influence of the axial extension load is also considered.

13.9 Numerical discussion of the forced vibration of the undamped system

For the undamped nano system shown in Fig. 13.1, a detailed analysis is performed. For such an undamped system—based on the obtained solution from Eq. (13.40)—the nonlocal effect, the angular velocity effect, and the hub radius effect are considered. The periodic force in the form $F = F_0 \cos(\widetilde{\Omega}t)$, where $F_0 = 0.5$ nN and $\widetilde{\Omega} = 0.9\omega_{l1}$, excites the rotating nonlocal undamped cantilever nano-beam. The nondimensional transverse deflection is presented in the form:

$$\overline{w}(L,\ t) = w(L,\ t)/d. \tag{13.67}$$

The time and longer time histories of the forced undamped vibration of a rotating cantilever nano-beam are shown in Figs. 13.6 and 13.7. When the dumping is not present, the response is periodical. The diagrams are obtained for the initial conditions (deflections and velocities for all cases are zero). The time histories of the forced undamped vibrations of a rotating cantilever nano-beam with nonlocal effects are shown in Fig. 13.6(A). The periodical response describes the transverse deflections of the undamped system. For nondimensional effects of angular velocity and hub radius, values $\gamma = 1$ and $\delta = 1$ are used. The nonlocal effects reduce the deflections that are caused by the vibrations. Fig. 13.6(B) shows the time histories of the forced undamped vibrations of a rotating cantilever nano-beam with angular velocity effects. For nondimensional effects of the nonlocal and hub radius, the values $\psi = 0.1$ and $\delta = 1$ are used. An increase in the angular velocity reduces the transverse deflections of the undamped system. The periodic variation in amplitude for the lower value of angular velocity effect $\gamma = 1$ is significantly higher than for $\gamma = 2$ and $\gamma = 3$ values. Fig. 13.6(C) shows the time histories of the forced undamped vibrations of a rotating cantilever nano-beam with hub radius effects. An increase in the hub radius effect causes a decrease in transverse deflection. The amplitude is lower in this case than in the case of the hub radius absence ($\delta = 0$). In Fig. 13.6(C) the analysis was carried out for nondimensional values of angular velocity $\gamma = 1$ and

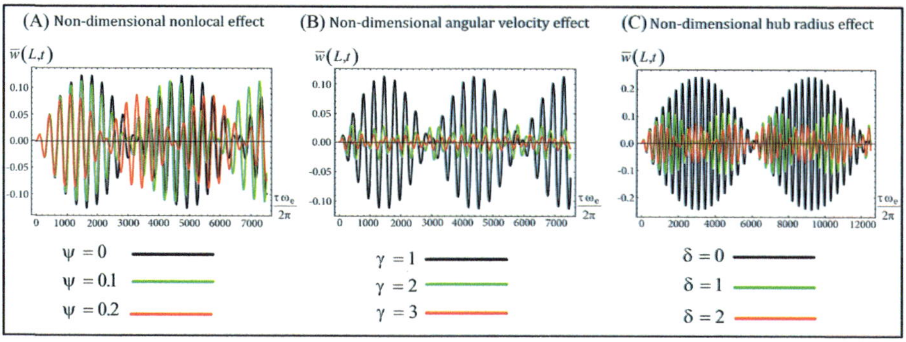

Figure 13.6 Time histories of the forced undamped vibration of the rotating nonlocal cantilever nano-beam. (A) nonlocality effects; (B) angular velocities effects; and (C) hub radius effects.

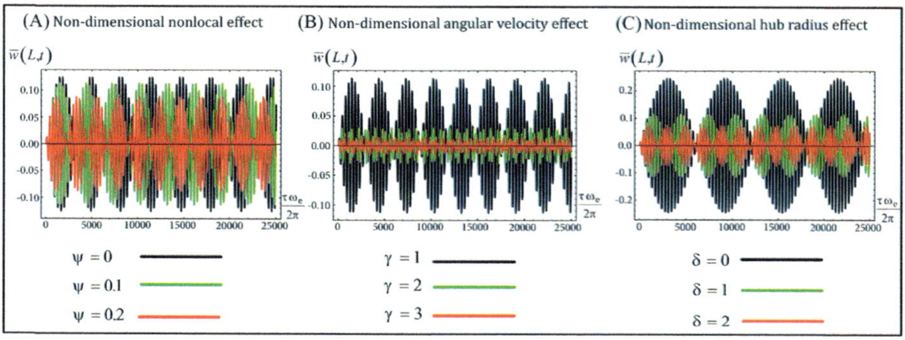

Figure 13.7 Longer time histories of the forced undamped vibration of the rotating nonlocal cantilever nano-beam. (A) nonlocality effects; (B) angular velocities effects; and (C) hub radius effects.

nonlocal parameter $\psi = 0.1$. The longer time history period is presented in Fig. 13.7. The discussion for Fig. 13.7 is the same as for Fig. 13.6; the period response is apparent.

13.10 Numerical discussion of the forced vibration of the damped system

Using the solution from Eq. (13.66), the nonlocal effect, the angular velocity effect, and the radius effect of the hub are analyzed through numerical calculations. The values of the proportional damping coefficients are chosen to be $\alpha = \beta = 10$ nNs/m. The periodic force in the form $F = F_0 \cos(\widetilde{\Omega}t)$, where $F_0 = 1$ nN and $\widetilde{\Omega} = 0.9\omega_{l1}$, excites the rotating nonlocal damped cantilever nano-beam. The analysis of the obtained results leads to the conclusion that an increase in all three effects (individually) leads to a decrease in the transverse deflections and that the system possesses a steady-state regime of vibrations. The general tendencies are shown in Figs. 13.8–13.10. The analysis, which results are shown in Fig. 13.8, was done for nondimensional values of angular velocity $\gamma = 1$ and hub radius $\delta = 1$. The time histories of vibration in the steady-state regime of vibrations are presented in Fig. 13.8(A), and this figure also shows that an increase in the nonlocal effect leads to a decrease in the transverse deflections. The appropriate phase plots are given in Fig. 13.8(B). An interesting phenomenon of the maximal deformed shapes of the rotating clamped-free model beam is shown in Fig. 13.8(C). It can be seen that the nonlocality effect slightly increases with an increase in the distance from the beam's ends. This phenomenon does not exist in the other examples where the hub radius and

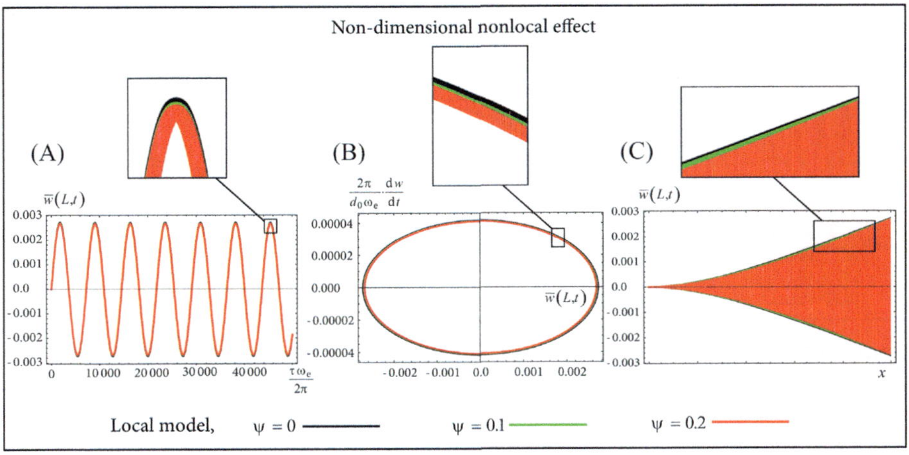

Figure 13.8 Time-domain of the damped vibration of the rotating nonlocal cantilever nano-beam with the nonlocality effect. (A) time histories; (B) phase plots; and (C) deformed shapes.

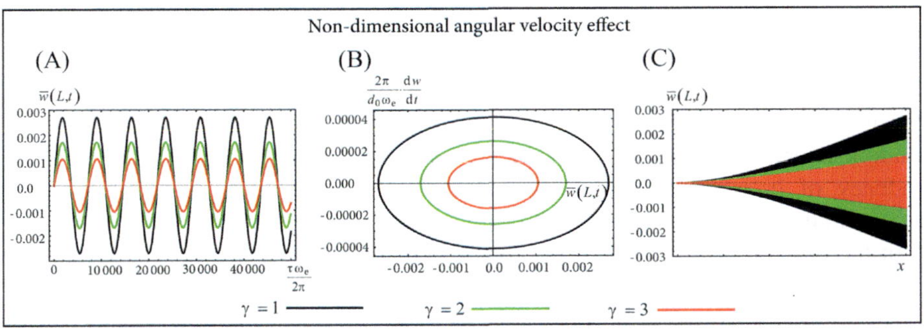

Figure 13.9 Time-domain of the damped vibration of the rotating nonlocal cantilever nano-beam with the effect of angular velocity. (A) time histories; (B) phase plots; and (C) deformed shapes.

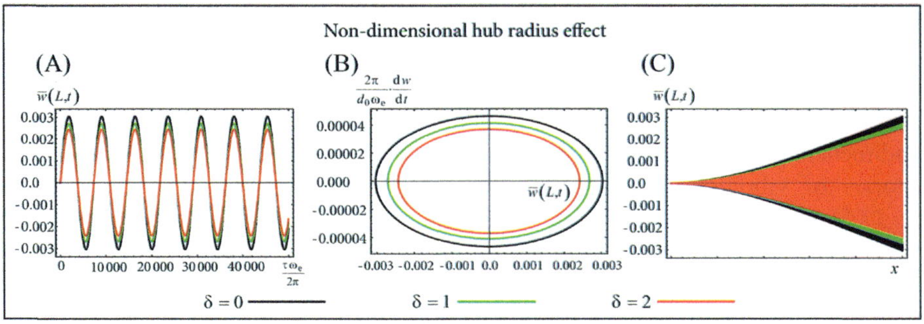

Figure 13.10 Time-domain of the damped vibration of the rotating nonlocal cantilever nanobeam with the hub radius effect. (A) time histories; (B) phase plots; and (C) deformed shapes.

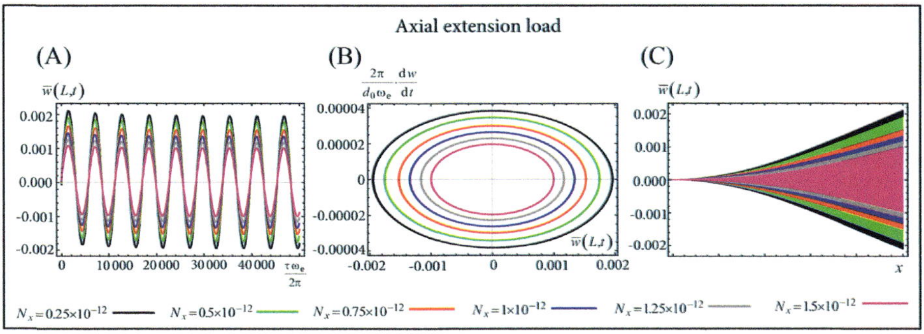

Figure 13.11 Time-domain of the damped vibration of the rotating nonlocal cantilever nanobeam with the axial extension load. (A) time histories; (B) phase plots; and (C) deformed shapes.

velocity effects are analyzed. The influence of angular velocity is depicted in Fig. 13.9. This analysis was performed for nondimensional values of $\psi = 0.1$ for the nonlocal effect and $\delta = 1$ for the hub radius effect. Particularly shown in Fig. 13.9(A and B) is the decrease in transverse deflections that is caused by an increase in angular velocity. The maximal deformed shapes during vibrations are depicted in Fig. 13.9(C). The points of the beam that are farthest from the clamped end are exposed to the highest influence of angular velocity.

The time histories, phase plots, and deformed shapes for various influences from the hub radius can be seen in Fig. 13.10. In Fig. 13.10 the analysis was realized for nondimensional values of angular velocity $\gamma = 1$ and nonlocal parameter $\psi = 0.1$. The tendency of transversal vibrations to decrease, together with the appearance of maximum deflections (at the farthest beam points from the clamped end), is the same as in the previous case (an increase in the radius of the hub leads to a decrease in transversal deflections).

The influence of the axial extension load of the transverse external excitation system is depicted in Fig. 13.11. The axial load effect N_x on the rotating nanobeam without transverse external excitations is analyzed in Ref. [33] and described through

Eq. (13.11) of the mentioned paper. The analysis was performed for the nonlocal parameter $\psi = 0.1$, the hub radius parameter $\delta = 1$, and the angular velocity parameter $\gamma = 1$. Fig. 13.11(A and B) shows that an increase in the axial extension load leads to a decrease in the transverse deflections. The maximal deformed shapes with the impact of axial extension load are depicted in Fig. 13.11(C). Even with the known qualitative impact of the external axial load, a quantitative effect is shown here.

13.11 Conclusion

The study presented in this chapter examines a single rotating nonlocal cantilever nano-beam with external excitations. The governing equation of motion of the forced vibration of a rotating nonlocal cantilever nano-beam is based on the Euler–Bernoulli beam assumption and Eringen's nonlocal elasticity theory. Afterward, the Galerkin method is used to discretize the governing equation. The further determination of the dynamic behavior is performed using the standard modal analysis procedure. Next, the forced damped and undamped vibrations are analyzed for the studied model. The numerical values of the natural frequencies of the rotating nano-model are calculated and verified. The values for the natural frequencies with various effects of nonlocality, angular velocities, and hub radius are presented and discussed—the values for the first four nondimensional frequencies are presented. When the nondimensional hub radius exists, slightly higher values of the natural frequencies are detected. The obtained values of the first nondimensional frequency for various values of the nonlocal parameter are in good agreement with the results available in the literature. Comparison between the values has led to the conclusion that an increase in the nonlocal parameter leads to an increase only in the first natural frequency (for the higher frequencies, this tendency is contrary). A detailed analysis is conducted for the variation of nondimensional first, second, third, and fourth modes frequency with nondimensional angular velocity for different values of nonlocal parameters and nondimensional hub radius. An increase of nonlocal and hub radius parameters leads to an increase of the first mode frequency with the angular velocity of the rotating nano-beam. The opposite is observed for the frequencies of higher modes; the increase in the nonlocal parameter leads to a decrease in higher mode frequencies, while the impact of hub radius leads to an increase in higher mode values of frequencies. The convergence study is also presented. This study indicates that as the discretization increases, the values of nondimensional frequencies will decrease.

From the detailed analysis of the undamped system and presented time and longer time histories, the influence of the analyzed effects is shown to generate a periodic response. The deflections caused by the vibrations are reduced when nonlocal effects are considered. A noteworthy phenomenon was observed with higher values of the angular velocity effect; an increase in the angular velocity decreased the transverse deflections. The hub radius effect also generates a periodic variation of the transverse deflections. The increase in the hub radius effect led to a decrease in transverse deflections.

The detailed analysis of the damped system showed that the increase in the effect of nonlocality, angular velocity, and hub radius leads to a decrease in transverse deflections. The quantitative values associated with the change of each of the aforementioned parameters are also determined and presented. An interesting phenomenon is noticed when the maximally deformed shapes of the nano-beam are analyzed under the nonlocality effect. It was observed that the increase in the nonlocality effect slightly decreases the maximal deflections of the points that are close to the middle of the beam. This means that the nonlocality effect is highest in the middle of the beam. The observed phenomenon does not exist in the other examples of the analyzed hub radius effect and velocity effect. An increase in the hub radius and velocity leads to a decrease in maximal deflections, and their effect is the highest at the free end of the beam. The influence of axial extension load is also considered for the damped system. An increase in the axial extension load leads to a decrease in the transverse deflections.

The model in this study considers the forced vibration of a rotating cantilever nano-beam with the nonlocality magnitude effect. The novelty of the study lies in the transient responses of the model. The presented dynamic behavior of this kind of model possesses practical interest. The functional properties of rotating structures with various effects of external excitations can be improved to better satisfy the motion of the shafts in hybridized nano-generators.

References

[1] B. Zhang, J. Chen, L. Jin, W. Deng, L. Zhang, H. Zhang, et al., Rotating-disk-based hybridized electromagnetic–triboelectric nanogenerator for sustainably powering wireless traffic volume sensors, ACS Nano 10 (6) (2016) 6241–6247.

[2] R. Toupin, Elastic materials with couple-stresses, Archive for Rational Mechanics and Analysis 11 (1) (1962) 385–414.

[3] R.D. Mindlin, H.F. Tiersten, Effects of couple-stresses in linear elasticity, Archive for Rational Mechanics and Analysis 11 (1) (1962) 415–448.

[4] W. Koiter, Couple stresses in the theory of elasticity, I and II. Koninklijke Nederlandse Akademie van Wetenschappen, Proceedings of the Royal Society B 67 (1) (1964) 17–44.

[5] M.E. Gurtin, J. Weissmüller, F. Larche, A general theory of curved deformable interfaces in solids at equilibrium, Philosophical Magazine A 78 (5) (1998) 1093–1109.

[6] D.C. Lam, F. Yang, A.C.M. Chong, J. Wang, P. Tong, Experiments and theory in strain gradient elasticity, Journal of the Mechanics and Physics of Solids 51 (8) (2003) 1477–1508.

[7] A.C. Eringen, Linear theory of nonlocal elasticity and dispersion of plane waves, International Journal of Engineering Science 10 (5) (1972) 425–435.

[8] A.C. Eringen, On differential equations of nonlocal elasticity and solutions of screw dislocation and surface waves, Journal of Applied Physics 54 (9) (1983) 4703–4710.

[9] A.C. Eringen, Nonlocal Continuum Field Theories, Springer Science & Business, 2002.

[10] A.C. Eringen, D.G.B. Edelen, On nonlocal elasticity, International Journal of Engineering Science 10 (3) (1972) 233–248.

[11] S. Iijima, Helical microtubules of graphitic carbon, Nature 354 (1991) 56–58.

[12] S. Iijima, C. Brabec, A. Maiti, J. Bernholc, Structural flexibility of carbon nanotubes, The Journal of Chemical Physics 104 (5) (1996) 2089–2092.

[13] J. Hu, T.W. Odom, C.M. Lieber, Chemistry and physics in one dimension: synthesis and properties of nanowires and nanotubes, Accounts of Chemical Research 32 (5) (1999) 435–445.
[14] S. Reich, C. Thomsen, J. Maultzsch, Carbon Nanotubes: Basic Concepts and Physical Properties, John Wiley & Sons, 2008.
[15] J.P. Lu, Elastic properties of carbon nanotubes and nanoropes, Physical Review Letters 79 (7) (1997) 1297–1300.
[16] M. Schilhansl, Bending frequency of a rotating cantilever beam, Journal of Applied Mechanics 25 (1) (1958).
[17] S.V. Hoa, Vibration of a rotating beam with tip mass, Journal of Sound and Vibration 67 (3) (1979) 369–381.
[18] R.B. Bhat, Transverse vibrations of a rotating uniform cantilever beam with tip mass as predicted by using beam characteristic orthogonal polynomials in the Rayleigh-Ritz method, Journal of Sound and Vibration 105 (2) (1986) 199–210.
[19] Q. Ni, M. Li, M. Tang, L. Wang, Free vibration and stability of a cantilever beam attached to an axially moving base immersed in fluid, Journal of Sound and Vibration 333 (9) (2014) 2543–2555.
[20] H. Diken, Vibration control of a rotating Euler–Bernoulli beam, Journal of Sound and Vibration 232 (3) (2000) 541–551.
[21] V. Stojanović, P. Kozić, Forced transverse vibration of Rayleigh and Timoshenko double-beam system with effect of compressive axial load, International Journal of Mechanical Sciences 60 (1) (2012) 59–71.
[22] V. Stojanović, P. Kozić, Vibrations and stability of complex beam systems, Springer Tracts in Mechanical Engineering, Springer, 2015 ISBN 978-3-319-13766-7.
[23] M. Stamenković, D. Karličić, J. Goran, P. Kozić, Nonlocal forced vibration of a double single-walled carbon nanotube system under the influence of an axial magnetic field, Journal of Mechanics of Materials and Structures 11 (3) (2016) 279–307.
[24] S. Narendar, S. Gopalakrishnan, Nonlocal wave propagation in rotating nanotube, Results in Physics 1 (1) (2011) 17–25.
[25] A. Babaei, C.X. Yang, Vibration analysis of rotating rods based on the nonlocal elasticity theory and coupled displacement field, Microsystem Technologies (2018) 1–9.
[26] S. Guo, Y. He, D. Liu, J. Lei, Z. Li, Dynamic transverse vibration characteristics and vibro-buckling analyses of axially moving and rotating nanobeams based on nonlocal strain gradient theory, Microsystem Technologies 24 (2) (2018) 963–977.
[27] M. Mohammadi, M. Safarabadi, A. Rastgoo, A. Farajpour, Hygro-mechanical vibration analysis of a rotating viscoelastic nanobeam embedded in a visco-Pasternak elastic medium and in a nonlinear thermal environment, Acta Mechanica 227 (8) (2016) 2207–2232.
[28] N. Shafiei, M. Kazemi, M. Ghadiri, Comparison of modeling of the rotating tapered axially functionally graded Timoshenko and Euler–Bernoulli microbeams, Physica E: Low-dimensional Systems and Nanostructures 83 (2016) 74–87.
[29] B. Nayak, S.K. Dwivedy, K.S.R.K. Murthy, Dynamic stability of a rotating sandwich beam with magnetorheological elastomer core, European Journal of Mechanics—A/Solids 47 (2014) 143–155.
[30] T. Murmu, S. Adhikari, Scale-dependent vibration analysis of prestressed carbon nanotubes undergoing rotation, Journal of Applied Physics 108 (12) (2010) 123507.
[31] S. Narendar, Mathematical modelling of rotating single-walled carbon nanotubes used in nanoscale rotational actuators, Defence Science Journal 61 (4) (2011) 317.

[32] C.O.K. Chen, S.H. Ho, Transverse vibration of a rotating twisted Timoshenko beams under axial loading using differential transform, International Journal of Mechanical Sciences 41 (11) (1999) 1339–1356.

[33] S.C. Pradhan, T. Murmu, Application of nonlocal elasticity and DQM in the flapwise bending vibration of a rotating nanocantilever, Physica E: Low-dimensional Systems and Nanostructures 42 (7) (2010) 1944–1949.

[34] D. Younesian, E. Esmailzadeh, Vibration suppression of rotating beams using time-varying internal tensile force, Journal of Sound and Vibration 330 (2) (2011) 308–320.

[35] C. Dejin, F. Kai, Z. Shijie, Flapwise vibration analysis of rotating composite laminated Timoshenko microbeams with geometric imperfection based on a re-modified couple stress theory and isogeometric analysis, European Journal of Mechanics—A/Solids 76 (2019) 25–35.

[36] M.S. Atanasov, V. Stojanović, Nonlocal forced vibrations of rotating cantilever nano-beams, European Journal of Mechanics/A Solids 79 (2020) 103850.

[37] S.G. Kelly, Mechanical vibrations: theory and applications, Cengage learning (2012).

[38] S.M. Han, H. Benaroya, T. Wei, Dynamics of transversely vibrating beams using four engineering theories, Journal of Sound and Vibration 225 (5) (1999) 935–988.

[39] D.Y. Hodges, M.Y. Rutkowski, Free-vibration analysis of rotating beams by a variable-order finite-element method, AIAA Journal 19 (11) (1981) 1459–1466.

[40] N. Shafiei, A. Mousavi, M. Ghadiri, Vibration behavior of a rotating non-uniform FG microbeam based on the modified couple stress theory and GDQEM, Composite Structures 149 (2016) 157–169.

Coupled multiple plate systems and their stability characteristics

14

14.1 Introduction

Plate systems comprising multiple parallel elastically connected plates have their place in theoretical research. In the fields of structural engineering, these systems also have practical importance, for example, lightweight structures in aeronautical and mechanical engineering applications. Solutions obtained from theoretical research of this type of structure may be used in civil engineering to create an approximate procedure used to determine the behavior of multistory buildings. The need for better approximations of results always exists, especially if the models are more complex. The precision of the solutions is often related to the number of accounted effects. Here, various plate theories can produce differences in approximations of solutions—thus comparison studies between them are of interest.

The simplest plate theory that can be used to describe plate vibrations, the classical plate theory (CPT), does not take into account the effects of shear deformation and rotary inertia, that is, the plate is considered infinitely stiff in transverse shear. This theory only considers flexural deformations. Therefore CPT has sufficient accuracy only when it comes to thin plates ($h/L < 0.1$). When the thickness ratio of a plate is fairly large ($h/L > 0.1$), CPT generates a significant overprediction of natural frequencies and buckling loads and underprediction of deflections and stresses. To obtain more accurate results related to plate vibrations, deflections, and stresses, the first-order shear deformation theories (FSDTs) were developed. The most commonly used among them is the Mindlin–Reissner theory of plates [1]. However, FSDTs take the transverse shear stress to be constant through the plate thickness, which is their weakness. In reality, the transverse shear stress has parabolic distribution through the plate thickness, and its value on the top and the bottom surface of the plate is zero. This discrepancy is corrected by inserting the shear correction factor into the equations. The selection of the value for the shear correction factor can cause overprediction or underprediction of the value of natural frequencies. To improve the accuracy of shear deformation plate theories (SDPTs), higher-order shear deformation theories were developed. The most known among them is the high-order shear deformation theory (HSDT) developed by Reddy and Phan [2]. In addition, Reddy and Wang [3] summarized the relationships between CPT and SDPTs.

The vibrations of a double-plate rectangular system have been studied in many papers. Using the CPT, Oniszczuk [4] studied the vibration of a rectangular double-plate system. Zhang et al. [5] investigated the influences of axial forces on the vibration of a double-beam system, through the use of the Euler beam assumption. Stojanović and Kozić [6] in their study of a double-beam system, which is loaded with axial forces, included the effects of shear and rotary inertia. Aided by Green's function, Kukla [7] analyzed the free vibrations of two elastically connected plates.

Modeling of Complex Dynamic Systems. DOI: https://doi.org/10.1016/B978-0-443-23942-7.00014-0

Often, in analytical research concerning linear vibrations of plates, the Navier method is applied. The Navier method [8–14], which is a particular case of the Lévy method, is adjusted for analyzing a plate with all four lateral sides being simply supported. In the study presented here, the Navier procedure is employed to formulate the exact theoretical solutions for the free transverse vibrations of a system that comprises multiple simple supported rectangular plates.

The relevance of more accurate results of natural frequencies, amplitudes, and buckling load rises with complex models; for example, the system of stacked elastically connected plates. The presented study, provides an analytical solution and results for further vibration analysis of systems of stacked elastically connected thin and thick isotropic plates.

Some studies can be of interest for further investigation and expansion of this work. For example, a study [15] concerning coupled plate systems, in which the free vibrations of a bundle of identical rectangular plates that are in full contact with an ideal liquid are analyzed and the effect of the liquid gap between the plates on the natural frequencies is discussed. In another research, Shyong and Chang [16] analyzed vibrations of a multistep beam where different sections of the beam have discretely different characteristics; exact equations that describe vibrations of the entire system are quite useful and can be important for future consideration of the present study. With multiple plate systems, there is an assembly of different parts that interact with each other [17]; finding a method that can mathematically describe the behavior of the entire system of this kind can be notable.

14.2 Theoretical formulation

Theoretical analysis of vibrations is done on a model that comprises multiple rectangular plates stacked on top of each other and connected via elastic layers of the Winkler type. The entire system rests on an elastic foundation (Fig. 14.1). The assumptions are that the plates are isotropic, homogenous, and of constant thickness. The considered case is when all four lateral sides on each plate are simply supported. In the general case, it is assumed that the plates are subjected to continuous axial compressive loads on the lateral sides. Small undamped vibrations of the system are analyzed. The determination of the system's natural frequencies and the critical buckling loads is achieved using the trigonometric method, hitherto employed in the paper by Stojanović et al. [18], in which the vibration analysis of the multibeam system was represented. The major problem in this kind of multiplate system was eliminating the variables (rotations) when the FSDT or HSDT is used, due to the implemented trigonometric method. The authors have created a program for the elimination of variables [19].

The dimensions of the plates are $a \times b \times h$. All of the plates can be subjected to axial compressive loading in the direction of x-axis and/or y-axis. For the multiplate system, the transverse vibrations are described by the following differential equations obtained from the principle of virtual displacements and zero value of total virtual work. These principles are applied to the system of p plates in which each plate is

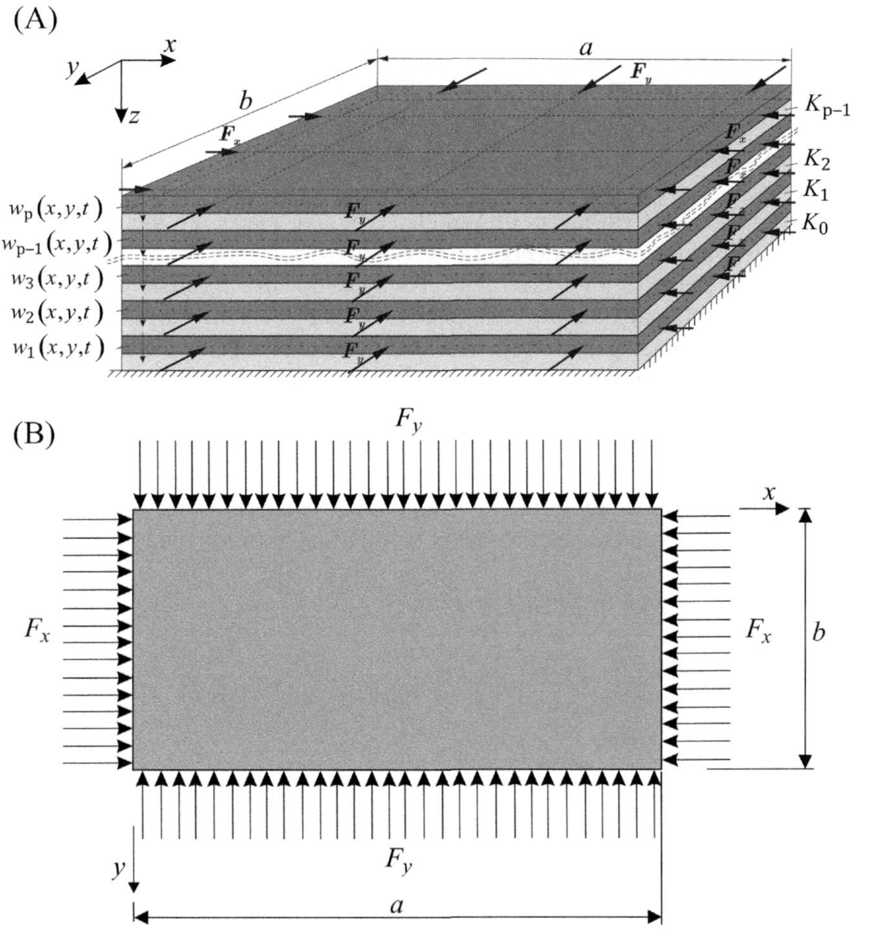

Figure 14.1 System of plates. (A) A set of p elastically connected plates on a Winkler type foundation; (B) Uniformly compressed complex plate model (all plates are uniformly compressed).

loaded with the axial forces F_x and F_y. Analogously to the equations of motion with neglected rotary inertia and shear effects of multibeam systems [20], when the CPT is used for the multiplate system, the equations of motion have the following form:

$$D_1\nabla^4 w_1 + h_1\rho_1\frac{\partial^2 w_1}{\partial t^2} + F_x\frac{\partial^2 w_1}{\partial x^2} + F_y\frac{\partial^2 w_1}{\partial y^2} + K_0 w_1 - K_1(w_2 - w_1) = 0, \tag{14.1a}$$

$$\begin{aligned} &D_i\nabla^4 w_i + h_i\rho_i\frac{\partial^2 w_i}{\partial t^2} + F_x\frac{\partial^2 w_i}{\partial x^2} + F_y\frac{\partial^2 w_i}{\partial y^2} \\ &+ K_{i-1}(w_i - w_{i-1}) - K_i(w_{i+1} - w_i) = 0, \end{aligned} \tag{14.1b}$$

$$D_p\nabla^4 w_p + h_p\rho_p\frac{\partial^2 w_p}{\partial t^2} + F_x\frac{\partial^2 w_p}{\partial x^2} + F_y\frac{\partial^2 w_p}{\partial y^2} + K_{p-1}(w_p - w_{p-1}) = 0. \tag{14.1c}$$

where

$$D_i = \frac{E_i h_i^3}{12(1-\nu_i^2)}, \quad i = 2,\ 3,\ \ldots, p-1\ . \tag{14.2}$$

Eqs. (14.1a–14.1c) present a system of equations that describes transversal vibrations of a set of p plates, elastically connected (the bottom plate is connected with the rigid ground) via layers of Winkler type. All plates are identically axially loaded on all four lateral sides (Fig. 14.1).

Each plate is made of a material with Young's modulus E_i, mass density ρ_i, and uniform cross-sectional area A_i. All plates are subjected to the same uniformly compressive axial loading F_x and F_y. The first plate is connected to the rigid ground via a Winkler foundation that has stiffness modulus K_0. The other plates are connected also via a continuous linear elastic layer of Winkler type with stiffness modulus K_i. The transverse displacements of the plates are $w_i = w_i(x, y, t)$.

The transverse vibrations of a rectangular multiplate system (using FSDT) are described by the following differential equations obtained using the principle of virtual displacements and the zero value of total virtual work, applied to the system of p plates.

Coupled set of equations for the first plate:

$$\rho h \frac{\partial^2 w_1}{\partial t^2} - kGh\left(\frac{\partial^2 w_1}{\partial x^2} + \frac{\partial^2 w_1}{\partial y^2} + \frac{\partial \psi_{x1}}{\partial x} + \frac{\partial \psi_{y1}}{\partial y}\right) + F_x \frac{\partial^2 w_1}{\partial x^2} + F_y \frac{\partial^2 w_1}{\partial y^2} + K_0 w_1 - K_1(w_2 - w_1) = 0, \tag{14.3a}$$

$$\frac{D}{2}\left[(1-\nu)\left(\frac{\partial^2 \psi_{x1}}{\partial x^2} + \frac{\partial^2 \psi_{x1}}{\partial y^2}\right) + (1+\nu)\frac{\partial}{\partial x}\left(\frac{\partial \psi_{x1}}{\partial x} + \frac{\partial \psi_{y1}}{\partial y}\right)\right] - kGh\left(\frac{\partial w_1}{\partial x} + \psi_{x1}\right) - \frac{\rho h^3}{12}\frac{\partial^2 \psi_{x1}}{\partial t^2} = 0, \tag{14.3b}$$

$$\frac{D}{2}\left[(1-\nu)\left(\frac{\partial^2 \psi_{y1}}{\partial x^2} + \frac{\partial^2 \psi_{y1}}{\partial y^2}\right) + (1+\nu)\frac{\partial}{\partial y}\left(\frac{\partial \psi_{x1}}{\partial x} + \frac{\partial \psi_{y1}}{\partial y}\right)\right] - kGh\left(\frac{\partial w_1}{\partial y} + \psi_{y1}\right) - \frac{\rho h^3}{12}\frac{\partial^2 \psi_{y1}}{\partial t^2} = 0, \tag{14.3c}$$

Coupled set of equations for the i-th *plate*:

$$\rho h \frac{\partial^2 w_i}{\partial t^2} - kGh\left(\frac{\partial^2 w_i}{\partial x^2} + \frac{\partial^2 w_i}{\partial y^2} + \frac{\partial \psi_{xi}}{\partial x} + \frac{\partial \psi_{yi}}{\partial y}\right) + F_x \frac{\partial^2 w_1}{\partial x^2} + F_y \frac{\partial^2 w_1}{\partial y^2} + K_{i-1}(w_i - w_{i-1}) - K_i(w_{i+1} - w_i) = 0, \tag{14.4a}$$

$$\frac{D}{2}\left[(1-\nu)\left(\frac{\partial^2\psi_{xi}}{\partial x^2}+\frac{\partial^2\psi_{xi}}{\partial y^2}\right)+(1+\nu)\frac{\partial}{\partial x}\left(\frac{\partial\psi_{xi}}{\partial x}+\frac{\partial\psi_{yi}}{\partial y}\right)\right]$$
$$-kGh\left(\frac{\partial w_i}{\partial x}+\psi_{xi}\right)-\frac{\rho h^3}{12}\frac{\partial^2\psi_{xi}}{\partial t^2}=0, \tag{14.4b}$$

$$\frac{D}{2}\left[(1-\nu)\left(\frac{\partial^2\psi_{yi}}{\partial x^2}+\frac{\partial^2\psi_{yi}}{\partial y^2}\right)+(1+\nu)\frac{\partial}{\partial y}\left(\frac{\partial\psi_{xi}}{\partial x}+\frac{\partial\psi_{yi}}{\partial y}\right)\right]$$
$$-kGh\left(\frac{\partial w_i}{\partial y}+\psi_{yi}\right)-\frac{\rho h^3}{12}\frac{\partial^2\psi_{yi}}{\partial t^2}=0, \tag{14.4c}$$

Coupled set of equations for the p-th plate:

$$\rho h\frac{\partial^2 w_p}{\partial t^2}-kGh\left(\frac{\partial^2 w_p}{\partial x^2}+\frac{\partial^2 w_p}{\partial y^2}+\frac{\partial\psi_{xp}}{\partial x}+\frac{\partial\psi_{yp}}{\partial y}\right)$$
$$+F_x\frac{\partial^2 w_1}{\partial x^2}+F_y\frac{\partial^2 w_1}{\partial y^2}+K_{p-1}(w_p-w_{p-1})=0, \tag{14.5a}$$

$$\frac{D}{2}\left[(1-\nu)\left(\frac{\partial^2\psi_{xp}}{\partial x^2}+\frac{\partial^2\psi_{xp}}{\partial y^2}\right)+(1+\nu)\frac{\partial}{\partial x}\left(\frac{\partial\psi_{xp}}{\partial x}+\frac{\partial\psi_{yp}}{\partial y}\right)\right]$$
$$-kGh\left(\frac{\partial w_p}{\partial x}+\psi_{xp}\right)-\frac{\rho h^3}{12}\frac{\partial^2\psi_{xp}}{\partial t^2}=0, \tag{14.5b}$$

$$\frac{D}{2}\left[(1-\nu)\left(\frac{\partial^2\psi_{yp}}{\partial x^2}+\frac{\partial^2\psi_{yp}}{\partial y^2}\right)+(1+\nu)\frac{\partial}{\partial y}\left(\frac{\partial\psi_{xp}}{\partial x}+\frac{\partial\psi_{yp}}{\partial y}\right)\right]$$
$$-kGh\left(\frac{\partial w_p}{\partial y}+\psi_{yp}\right)-\frac{\rho h^3}{12}\frac{\partial^2\psi_{yp}}{\partial t^2}=0. \tag{14.5c}$$

Reddy [2], HSDT is based on the displacement field:

$$u_i(x,y,z,t)=u_{0i}(x,y,t)+$$
$$+z\left[\psi_{xi}(x,y,t)-\frac{4}{3}\left(\frac{z}{h}\right)^2\left(\psi_{xi}(x,y,t)+\frac{\partial w_i(x,y,t)}{\partial x}\right)\right], \tag{14.6a}$$

$$v_i(x,y,z,t)=v_{0i}(x,y,t)+$$
$$+z\left[\psi_{yi}(x,y,t)-\frac{4}{3}\left(\frac{z}{h}\right)^2\left(\psi_{0y}(x,y,t)+\frac{\partial w_i(x,y,t)}{\partial y}\right)\right], \tag{14.6b}$$

$$w_i(x,y,z,t)=w_i^0(x,y,t),\quad i=1,\ 2,\ \ldots,p. \tag{14.6c}$$

The transverse vibrations of the rectangular multiplate system using HSDT are obtained from the principle of virtual work applied to the system of p plates. Virtual works of inertia and internal forces have a well-known form. The virtual work of external forces is:

$$\delta W_{ex1} = \int_A \left\{ \delta w_1^0(x,y,t)[-K_0 w_1^0(x,y,t) + K_1(w_2^0(x,y,t) - w_1^0(x,y,t))] + F_x \frac{\partial w_1^0(x,y,t)}{\partial x}\frac{\partial w_1^0(x,y,t)}{\partial x} + F_y \frac{\partial w_1^0(x,y,t)}{\partial y}\frac{\partial \delta w_1^0(x,y,t)}{\partial y} \right\} dA, \tag{14.7a}$$

$$\delta W_{exi} = \int_A \left\{ \delta w_i^0(x,y,t)[-K_{i-1}(w_i^0(x,y,t) - w_{i-1}^0(x,y,t)) + K_i(w_{i+1}^0(x,y,t) - w_i^0(x,y,t))] + F_x \frac{\partial w_i^0(x,y,t)}{\partial x}\frac{\partial w_i^0(x,y,t)}{\partial x} + F_y \frac{\partial w_i^0(x,y,t)}{\partial y}\frac{\partial \delta w_i^0(x,y,t)}{\partial y} \right\} dA, \quad i = 2, 3, \ldots, p-1. \tag{14.7b}$$

$$\delta W_{exp} = \int_A \left\{ \delta w_p^0(x,y,t)\left[-K_{p-1}\left(w_p^0(x,y,t) - w_{p-1}^0(x,y,t)\right)\right] + F_x \frac{\partial w_i^0(x,y,t)}{\partial x}\frac{\partial w_i^0(x,y,t)}{\partial x} + F_y \frac{\partial w_i^0(x,y,t)}{\partial y}\frac{\partial \delta w_i^0(x,y,t)}{\partial y} \right\} dA. \tag{14.7c}$$

Applying the virtual work principle $\delta W_{INi} + \delta W_{Vi} + \delta W_{exi} = 0, \ i = 1, 2, \ldots p$, equations for Reddy HSDT are obtained but are omitted here because of their length.

14.3 Natural frequencies

14.3.1 Natural frequencies of the multiplate system obtained using classical plate theory

Boundary conditions for plates (dimensions: $a \times b \times h$) with simply supported lateral sides are assumed as follows:

$$w_i(0,y,t) = w_i(a,y,t) = w_i(x,0,t) = w_i(x,b,t) = 0, \tag{14.8}$$

$$\left.\frac{\partial^2 w_i}{\partial x^2}\right|_{(0,y,t)} = \left.\frac{\partial^2 w_i}{\partial x^2}\right|_{(a,y,t)} = \left.\frac{\partial^2 w_i}{\partial x^2}\right|_{(x,0,t)} = \left.\frac{\partial^2 w_i}{\partial x^2}\right|_{(x,b,t)} = 0, \quad i = 1, 2, \ldots, p. \tag{14.9}$$

The initial conditions in the general form may be written as:

$$w_i(x,y,0) = w_{i0}(x,y), \quad \dot{w}_i|_{(x,y,0)} = v_{i0}(x,y), \quad i = 1, 2, \ldots, p. \tag{14.10}$$

Assuming the harmonic motion, and using the separation of variables, the system of Eqs. (14.1a–14.1c) can be solved. Solutions may be assumed to be in the form:

$$w_i(x, y, t) = \sum_{m,n=1}^{\infty} W_{mn}(x, y)\, T_{imn}(t), \quad i = 1, 2, \ldots, p, \tag{14.11}$$

where $T_{imn}(t) = 1, 2, 3, \ldots$ is the unknown time function, and $W_{mn}(x, y)$ is the known mode shape function for simply supported plate, and it is defined as:

$$W_{mn}(x, y) = X_m(x)\, Y_n(y) = \sin(a_m x)\sin(b_n y), \tag{14.12}$$

where

$$X_m(x) = \sin(a_m x), \quad Y_n(y) = \sin(b_n y), \quad a_m = \frac{m\pi}{a}, \quad b_n = \frac{n\pi}{b}, \quad m, n = 1, 2, 3\ldots$$

By introducing the general solution (14.11) into the Eqs. (14.1a–14.1c), the following system of ordinary differential equations (ODEs) is obtained:

$$h_1\rho_1\frac{d^2T_{1mn}}{dt^2} + \left[D_1\pi^4H^2 - \left(\frac{\pi m}{a}\right)^2F_x - \left(\frac{\pi n}{b}\right)^2F_y + K_0 + K_1\right]T_{1mn} - K_1T_{2mn} = 0, \tag{14.13a}$$

$$h_i\rho_i\frac{d^2T_{i\,mn}}{dt^2} - K_{(i-1)}T_{(i-1)mn} + \left[D_i\pi^4H^2 - \left(\frac{\pi m}{a}\right)^2F_x - \left(\frac{\pi n}{b}\right)^2F_y + K_{(i-1)} + K_i\right]T_{imn} - K_iT_{(i+1)mn} = 0, \tag{14.13b}$$

$$h_p\rho_p\frac{d^2T_{pmn}}{dt^2} + \left[D_p\pi^4H^2 - \left(\frac{\pi m}{a}\right)^2F_x - \left(\frac{\pi n}{b}\right)^2F_y + K_{(p-1)}\right]T_{pmn} - K_{(p-1)}T_{(p-1)mn} = 0. \tag{14.13c}$$

where

$$H = \left(\frac{m}{a}\right)^2 + \left(\frac{n}{b}\right)^2.$$

The solutions of ODE can be assumed to have the following forms:

$$T_{imn} = A_{imn}e^{j\omega_{mn}t}, \quad i = 1, 2, \ldots p, \tag{14.14}$$

where ω_{mn} denotes the natural frequency of the system. Substituting Eq. (14.12) into the ODEs results in the following system of homogenous algebraic equations for the unknown constants $A_{1mn}, A_{2mn}, \ldots, A_{pmn}$

$$\left\{-h_1\rho_1\omega_{mn}^2 + \left[D_1\pi^4H^2 - \left(\frac{\pi m}{a}\right)^2F_x - \left(\frac{\pi n}{b}\right)^2F_y + K_0 + K_1\right]\right\}A_{1mn} - K_1A_{2mn} = 0, \tag{14.15a}$$

$$
\begin{aligned}
&- K_{(i-1)}A_{(i-1)mn} \\
&+ \left\{ -h_i\rho_i\omega_{mn}^2 + \left[D_i\pi^4H^2 - \left(\tfrac{\pi m}{a}\right)^2 F_x - \left(\tfrac{\pi n}{b}\right)^2 F_y + K_{(i-1)} + K_i \right] \right\} A_{imn} \\
&- K_iA_{(i+1)mn} = 0,
\end{aligned}
\tag{14.15b}
$$

$$
\begin{aligned}
&- K_{(p-1)}A_{(p-1)mn} \\
&+ \left\{ -h_p\rho_p\omega_{mn}^2 + \left[D_p\pi^4H^2 - \left(\tfrac{\pi m}{a}\right)^2 F_x - \left(\tfrac{\pi n}{b}\right)^2 F_y + K_{(p-1)} \right] \right\} A_{pmn} = 0.
\end{aligned}
\tag{14.15c}
$$

When the multiplate system comprises identical plates and identical elastic layers between the plates; $h_i = h, E_i = E, D_i = D, K_{i-1} = K_i = K,\ K_p = 0$, and $\rho_i = \rho$, where $i = 1, 2, 3, \ldots p$, then the system of homogenous algebraic equations can be rewritten in the matrix form. The result of this is the following system of p homogenous algebraic equations:

$$
\begin{bmatrix}
S_{mn} & -u_{mn} & 0 & \cdots & 0 & 0 & 0 & \cdots & 0 & 0 & 0 \\
-u_{mn} & S_{mn} & -u_{mn} & \cdots & 0 & 0 & 0 & \cdots & 0 & 0 & 0 \\
0 & 0 & 0 & \cdots & 0 & 0 & 0 & \cdots & 0 & 0 & 0 \\
\cdots & \cdots & \cdots & \cdots & \cdots & \cdots & \cdots & \cdots & \cdots & \cdots & \cdots \\
0 & 0 & 0 & \cdots & S_{mn} & -u_{mn} & 0 & \cdots & 0 & 0 & 0 \\
0 & 0 & 0 & \cdots & -u_{mn} & S_{mn} & -u_{mn} & \cdots & 0 & 0 & 0 \\
0 & 0 & 0 & \cdots & 0 & -u_{mn} & S_{mn} & \cdots & 0 & 0 & 0 \\
\cdots & \cdots & \cdots & \cdots & \cdots & \cdots & \cdots & \cdots & \cdots & \cdots & \cdots \\
0 & 0 & 0 & \cdots & 0 & 0 & 0 & \cdots & 0 & 0 & 0 \\
0 & 0 & 0 & \cdots & 0 & 0 & 0 & \cdots & -u_{mn} & S_{mn} & -u_{mn} \\
0 & 0 & 0 & \cdots & 0 & 0 & 0 & \cdots & 0 & -u_{mn} & S_{mn} - u_{mn}
\end{bmatrix}
\begin{Bmatrix}
A_{1mn} \\ A_{2mn} \\ A_{3mn} \\ \cdots \\ A_{(i-1)mn} \\ A_{imn} \\ A_{(i+1)mn} \\ \cdots \\ A_{(p-2)mn} \\ A_{(p-1)mn} \\ A_{pmn}
\end{Bmatrix}
=
\begin{Bmatrix}
0 \\ 0 \\ 0 \\ \cdots \\ 0 \\ 0 \\ 0 \\ \cdots \\ 0 \\ 0 \\ 0
\end{Bmatrix},
\tag{14.16}
$$

where

$$
S_{mn} = -h\rho\omega_{mn}^2 + D\pi^4H^2 - \left(\frac{\pi m}{a}\right)^2 F_x - \left(\frac{\pi n}{b}\right)^2 F_y + 2K, \quad u_{mn} = K.
$$

The system of algebraic Eq. (14.16) encompasses the equations of free vibrations and the stability equations. These equations are for a set of identical plates that are simply supported on all four lateral sides and linked to each other via Winkler layers of the same stiffness modulus. The next step would be to single out the determinant of the matrix $(p \times p)$ and equate it to zero. This will yield an ordinary polynomial of the $2p$-th order, with the undamped natural frequencies ω_{mn} as the unknowns. Finding the analytical solution for a system of this kind that comprises three plates can be a problem. For systems with more than three plates, the values of natural frequencies can only be determined numerically and for specific system parameters. Be that as it may, if a homogenous system of algebraic Eq. (14.11) is utilized, analytical formulations for natural frequencies of the plate system can be determined by employing the trigonometric method [10]. Values of the unknown constants $A_{i\,mn}$ are assumed to alternate in an oscillatory fashion:

$$A_{imn} = N\sin(i\varphi), \quad i = 1,\ 2,\ 3,\ \ldots p. \tag{14.17}$$

If close attention is paid to the first and last equations of the system (14.16), a different structure of these two equations can be noticed, in comparison to all the other equations. Substituting the postulated solution (14.17) into the i-th equation of the system (14.16) for $i = 2,\ 3,\ \ldots,\ i - 1$ produces:

$$\begin{aligned} &-u_{mn}N\sin(i-1)\varphi + S_{mn}N\sin(i\varphi) \\ &-u_{mn}N\sin(i+1)\varphi = 0, \quad i = 2,\ 3,\ \ldots p-1. \end{aligned} \tag{14.18}$$

After algebraic rearranging, the following is obtained:

$$(S_{mn} - 2u_{mn}\cos\varphi)N\sin(i\varphi) = 0, \quad i = 2,\ 3,\ \ldots p-1. \tag{14.19}$$

Eq. (14.19) needs to have $N \neq 0$, and $\sin(i\varphi) \neq 0$, because then $A_{imn} \neq 0$, for $i = 2,\ 3,\ \ldots p-1$. The only solution left is:

$$S_{mn} - 2u_{mn}\cos\varphi = 0 \Rightarrow S_{mn} = 2u_{mn}\cos\varphi. \tag{14.20}$$

Eq. (14.20) is a frequency equation, which is a second-order polynomial in ω_{mn}. The unknown φ can be determined based on the condition that the assumed solutions (14.17) satisfy the first and the last equation of the system (14.16).

In the case when $A_{1mn} = N\sin\varphi$ and $A_{2mn} = N\sin 2\varphi$ are substituted into the first equation of system (14.16), the following is obtained:

$$S_{mn}N\sin\varphi - u_{mn}N\sin 2\varphi = 0 \Rightarrow (S_{mn} - 2u_{mn}\cos\varphi)N\sin\varphi \equiv 0. \tag{14.21}$$

Eq. (14.21) is identical since it is determined that $S_{mn} - 2u_{mn}\cos\varphi = 0$, where $N \neq 0$, and $\sin\varphi \neq 0$. When the assumed solutions for $A_{(p-1)mn}$ and A_{imn} are substituted into the p-th equation of system (14.16), the following is obtained:

$$-u_{mn}N\sin[(p-1)\varphi] + (S_{mn} - u_{mn})N\sin(p\varphi) = 0. \tag{14.22}$$

After algebraic rearranging, the following is obtained:

$$Nu_{mn}\sin\left(\frac{\varphi}{2}\right)\cos\left(\frac{2p+1}{2}\varphi\right) = 0. \tag{14.23}$$

Solutions of Eq. (14.23) are:

$$\cos\left(\frac{2p+1}{2}\varphi\right) = 0 \Rightarrow \varphi_s = \frac{2s-1}{2p+1}\pi, \quad s = 1,\ 2,\ 3,\ \ldots p, \tag{14.24}$$

since $N \neq 0$, $u_{mn} \neq 0$ and $\sin(\varphi/2) \neq 0$. By substituting solutions (14.24) into (14.20), this frequency equation follows:

$$\begin{aligned} &-h\rho\omega_{mn}^2 + D\pi^4 H^2 - \left(\frac{\pi m}{a}\right)^2 F_x - \left(\frac{\pi n}{b}\right)^2 F_y + 2K = \\ &= 2K\cos\left(\frac{2s-1}{2p+1}\pi\right), \quad s = 1,\ 2,\ 3,\ \ldots p. \end{aligned} \tag{14.25}$$

From frequency Eq. (14.25), the following expression for the determination of natural frequencies of individual plates is obtained:

$$\omega_{mn,s}^2 = \frac{1}{h\rho}\left\{D\pi^4 H^2 - \left(\frac{\pi m}{a}\right)^2 F_x - \left(\frac{\pi n}{b}\right)^2 F_y + 2K\left[1 - \cos\left(\frac{2s-1}{2p+1}\pi\right)\right]\right\}. \tag{14.26}$$

14.3.2 Natural frequencies of the multiplate system obtained using first-order shear deformation theories and high-order shear deformation theory

Solving multiplate problems with FSDTs and HSDTs can also be done by employing the procedure shown for the CPT. The CPT has one variable in the equation of motion: the transverse displacement. As a result, with CPT, only one equation is needed to describe the vibration of the plate. In the FSDTs and HSDTs, there are two more variables besides transverse displacement. Those additional variables describe angular rotations of the normal to the midplane. This leads to three equations of motion for each plate. By eliminating angular rotations [18,19], the number of equations of motion can be reduced to one for each plate. After that, a system of equations comparable to the system Eqs. (14.1a–14.1c) can be defined. Because of the lengthy procedure and lengthy equations as the result, that will not be shown here. Merely, for example, here are the final equations for determining natural frequencies using the FSDT:

$$\begin{aligned} \begin{Bmatrix} [\omega_{mn,s}^2]_l \\ [\omega_{mn,s}^2]_h \end{Bmatrix} &= B_{F24}\cos\left(\frac{2s-1}{2p+1}\pi\right) - \frac{1}{2}A_{F24} \\ &\mp \frac{1}{2}\sqrt{\left[A_{F24} - 2\cos\left(\frac{2s-1}{2p+1}\pi\right)B_{F24}\right]^2 - 4\left[A_{F04} - 2\cos\left(\frac{2s-1}{2p+1}\pi\right)B_{F04}\right]}, \end{aligned} \tag{14.27}$$

where

$$A_{F24} = -\frac{12Gk}{\rho h^3}\left[h + H\pi^2\left(\frac{h^3}{12} + \frac{D}{Gk}\right) + \frac{h^2}{12Gk}\left(2K - \left(\frac{\pi m}{a}\right)^2 F_x - \left(\frac{\pi n}{b}\right)^2 F_y\right)\right],$$

$$A_{F04} = \frac{12Gk}{\rho^2 h^3}\left[DH^2\pi^4 + \left(1 + \frac{DH\pi^2}{Ghk}\right)\left(2K - \left(\frac{\pi m}{a}\right)^2 F_x - \left(\frac{\pi n}{b}\right)^2 F_y\right)\right],$$

$$B_{F24} = -\frac{K}{h\rho}, \quad B_{F04} = \frac{12K}{\rho^2 h^4}(Ghk + DH\pi^2).$$

Unlike expression (14.26), with expression (14.27) beside lower (fundamental) natural frequencies, higher frequencies can additionally be obtained. The lowest frequency, by value, obtained from Eq. (14.27) is associated with the mode for which the transverse component is dominant, while the higher frequencies are associated with the modes for which displacements in the tangent plane are dominant [21].

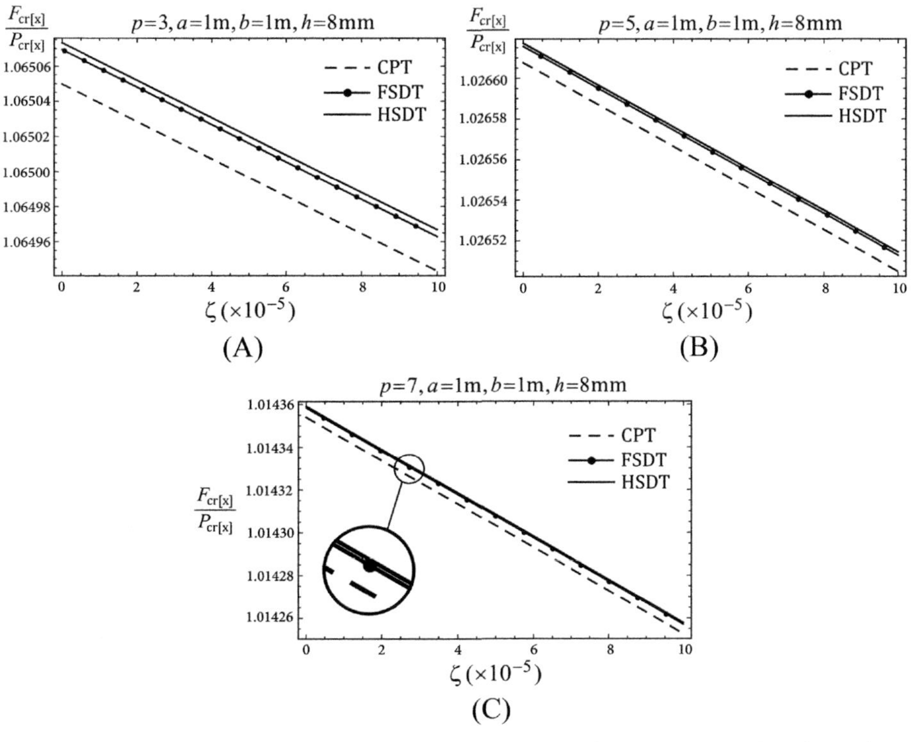

Figure 14.2 Critical buckling force for different numbers of plates; (A) $p=3$, (B) $p=5$, (C) $p=7$.

14.4 Stability analysis of the multiplate system using CPT, FSDT, and HSDT

The buckling of the system occurs when the axial loading increases until the natural frequency becomes zero. Introduction of $\omega_{mn,s} = 0$ and $F_y = 0$, into Eq. (14.25) for the case of the CPT leads to

$$D\pi^4 H^2 + 2K\left[1 - \cos\left(\frac{2s-1}{2p+1}\pi\right)\right] - \left(\frac{\pi m}{a}\right)^2 F_x = 0, \tag{14.28}$$

Further from Eq. (14.28), the buckling force—corresponding to vibration modes m and n of p identically elastically connected plates—is determined in the following form:

$$F_{cr[x]} = \min_{\substack{m=1,2,3\cdots \\ n=1,2,3\cdots \\ s=1,2,3\cdots p}} \left\{ \left(\frac{a}{\pi m}\right)^2 \left[D\pi^4 H^2 + 2K\left(1 - \cos\left(\frac{2s-1}{2p+1}\pi\right)\right)\right]\right\}. \tag{14.29}$$

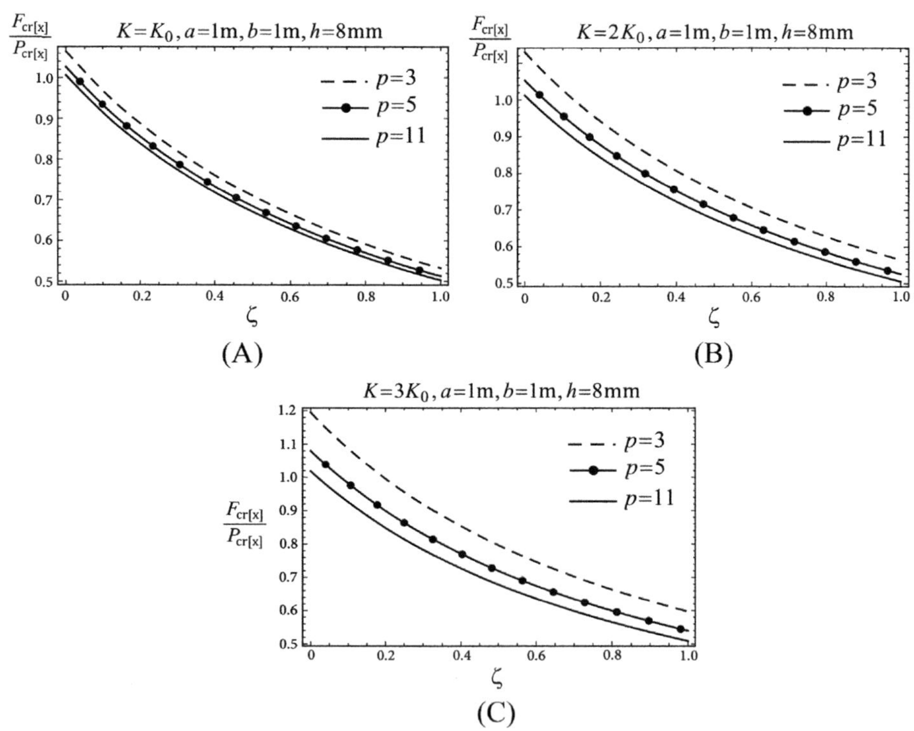

Figure 14.3 Critical buckling force for different stiffness moduli K ($h = 8$ mm); (A) $K = K_0$, (B) $K = 2K_0$, (C) $K = 3K_0$.

Following the same procedure, the critical buckling force is obtained for the case of elastically connected plates using FSDT:

$$F_{cr\,[x]} = \min_{\substack{m=1,2,3\cdots \\ n=1,2,3\cdots \\ s=1,2,3\cdots p}} \frac{1}{Ghk + DH\pi^2}\left\{DGhk\pi^2\left[H + \left(\frac{n}{b}\right)^2\left(1 + \left(\frac{an}{bm}\right)^2\right)\right]\right.$$

$$\left. + 2K\left[Ghk\left(\frac{a}{\pi m}\right)^2 + D\left(1 + \left(\frac{an}{bm}\right)^2\right)\right]\left(1 - \cos\left(\frac{2s-1}{2p+1}\pi\right)\right)\right\}. \tag{14.30}$$

By going further, the same procedure can lead to the analytical formulation of the critical buckling force for the case when HSDT is used. For further analysis in this chapter, that analytical form is determined, but is left out from showing here because of its length. Without losing the generalization it can be taken that $F_y = \zeta F_x$, $0 \le \zeta \le 1$, the critical buckling force of the system $F_{cr\,[x]}$ will depend on the compression ratio ζ. The changes in the critical buckling force of the multiplate system are illustrated in Figs. 14.2–14.4, this is done for various stiffness moduli values of elastic layers, and various numbers of plates. In the diagrams, the $P_{cr\,[x]}$ represents the critical buckling force for a multiplate system without elastic layers and foundation. The stability

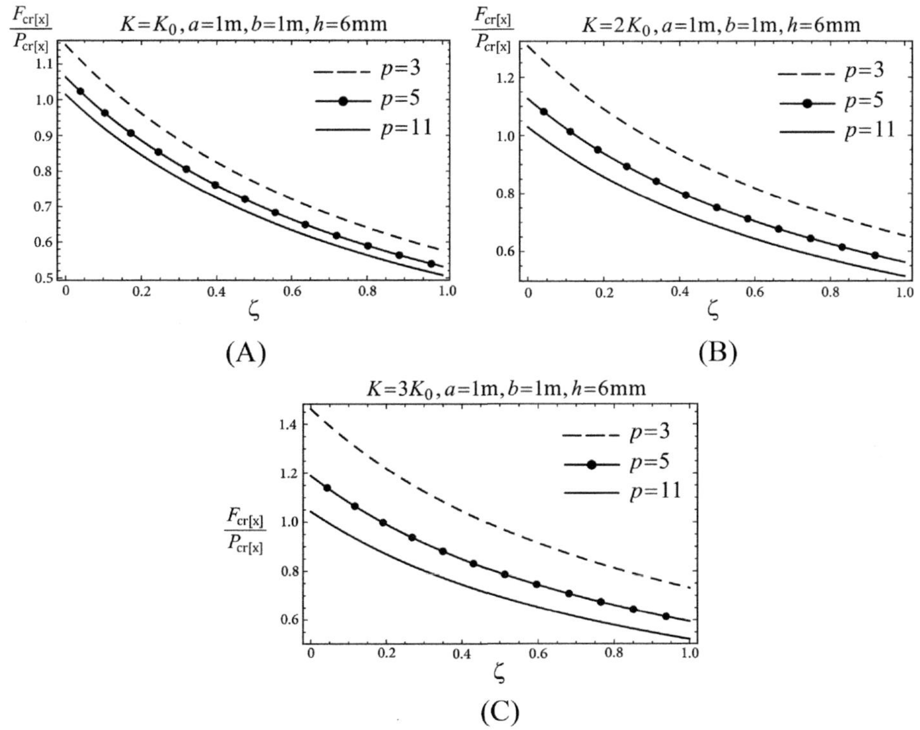

Figure 14.4 Critical buckling force for different stiffness moduli K ($h = 6$ mm); (A) $K = K_0$, (B) $K = 2K_0$, (C) $K = 3K_0$.

of the system is presented for the plates and layers with the following material properties: $K = K_0 = 6 \times 10^4\ \mathrm{N/m^3}$, $E = 1 \times 10^{10}\ \mathrm{N/m^2}$, $\rho = 5 \times 10^3\ \mathrm{kg/m^3}$, $\nu = 0.3$. By analyzing the diagrams, it can be deduced that the stability region decreases when the number of plates increases. Generally, the increase of compression ratio ζ leads to a decrease in critical buckling force. In the thin plate instance (Fig. 14.4), the stiffness moduli of elastic layers are greater than the stiffness of the plates, and the nondimensional critical buckling force is bigger.

14.5 General natural mode shapes of multiplate systems

Observation of the system of homogenous Eq. (14.16) shows that the coefficients A_{pimn} depend on natural frequencies. These coefficients can be determined analytically if one coefficient is fixed and the corresponding functions are used

$$W_{1imn}(x, y) = X_m(x) Y_n(y) = \sin(a_m x) \sin(b_n y), \tag{14.31}$$

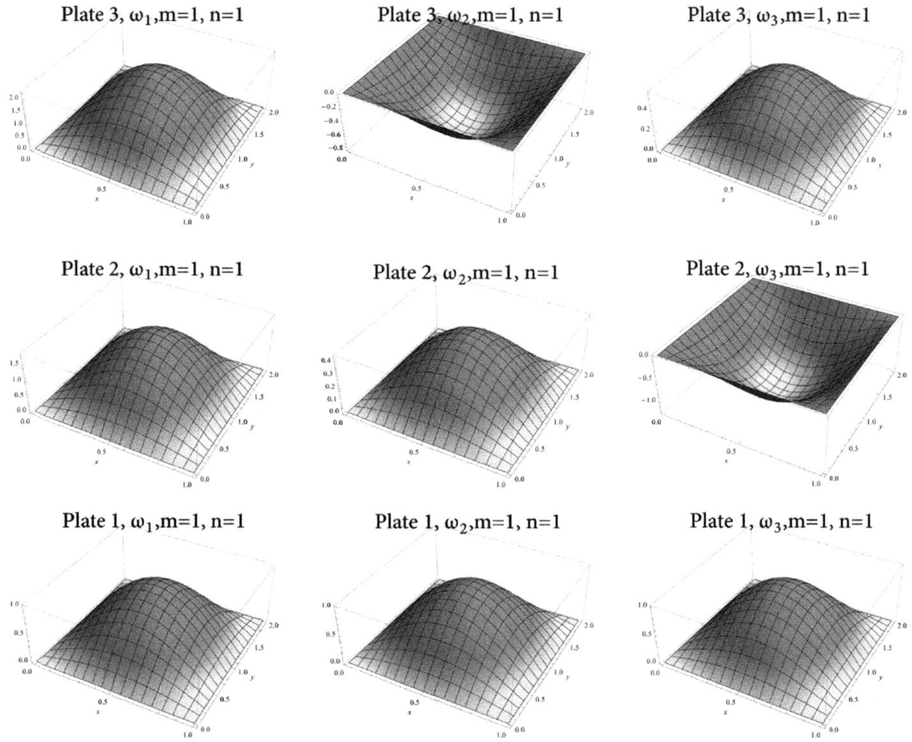

(A) (B) (C)

Figure 14.5 General mode shapes of vibration of a rectangular three-plate complex system. The plates are elastically connected, with all four lateral sides simply supported. The bottom plate is the system's first plate, and it is elastically connected to the rigid foundation. The system corresponds to three sequences of natural frequencies (ω_{imn}, $i = 1, 2, 3$, $m = 1$, $n = 1$); (A) ω_{111}, (B) ω_{211}, (C) ω_{311}.

$$W_{pimn}(x, y) = X_m(x) Y_n(y) = A_{p\ imn} \sin(a_m x) \sin(b_n y), \tag{14.32}$$

where

$$X_m(x) = \sin(a_m x), \quad Y_n(y) = \sin(b_n y), \quad a_m = \frac{m\pi}{a}, \quad b_n = \frac{n\pi}{b}, \quad m,\ n = 1,\ 2,\ 3\ldots$$

Here, Eqs. (14.31) and (14.32) are used for acquiring the general mode shape functions (W_{1imn}, W_{pimn}) for the chosen multiplate system and mode of vibration. Frequencies for the chosen mode of vibration need to be obtained beforehand. The general mode shapes for multiplate systems comprising three, four, and five plates are shown in Figs. 14.5–14.9. In all figures, the bottom plate is the bottom plate of the system—the plate linked to the foundation or the first plate. The multiplate system can vibrate at several natural frequencies. The number of frequencies at which the system can vibrate is equal to the number of plates in the system. In Figs. 14.5–14.9, the plates that vibrate at the lowest natural frequencies are the plates on the far-left

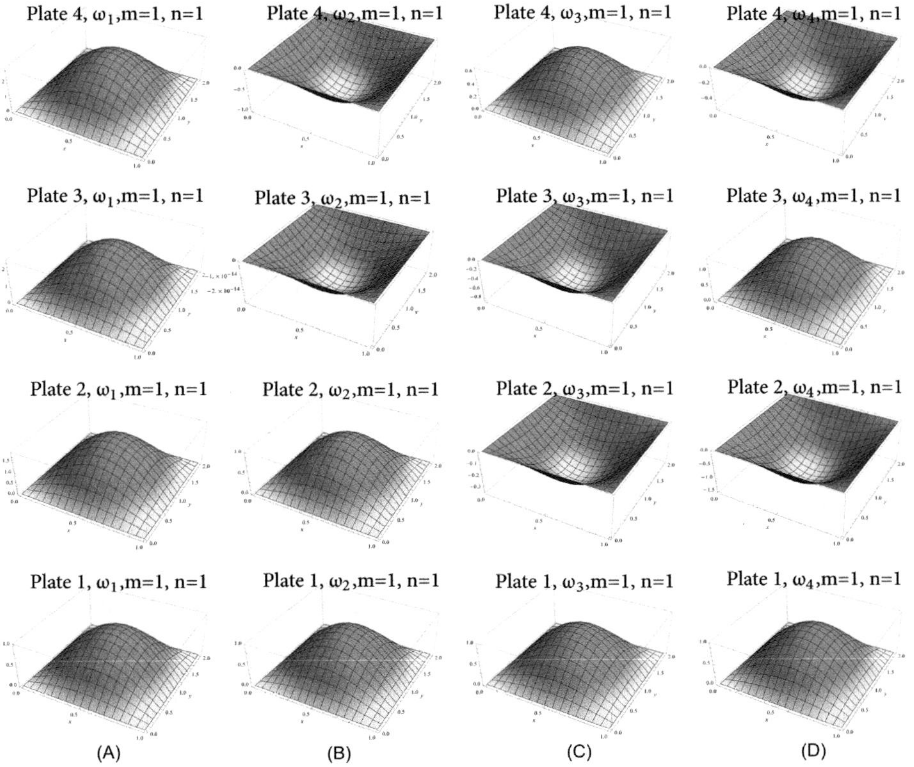

Figure 14.6 General mode shapes of vibration of a rectangular four-plate complex system. The plates are elastically connected with all four lateral sides simply supported. The bottom plate is the system's first plate, and it is elastically connected to the rigid foundation. The system corresponds to four sequences of natural frequencies (ω_{imn}, $i = 1, 2, 3, 4$, $m = 1$, $n = 1$); (A) ω_{111}, (B) ω_{211}, (C) ω_{311}, (D) ω_{411}.

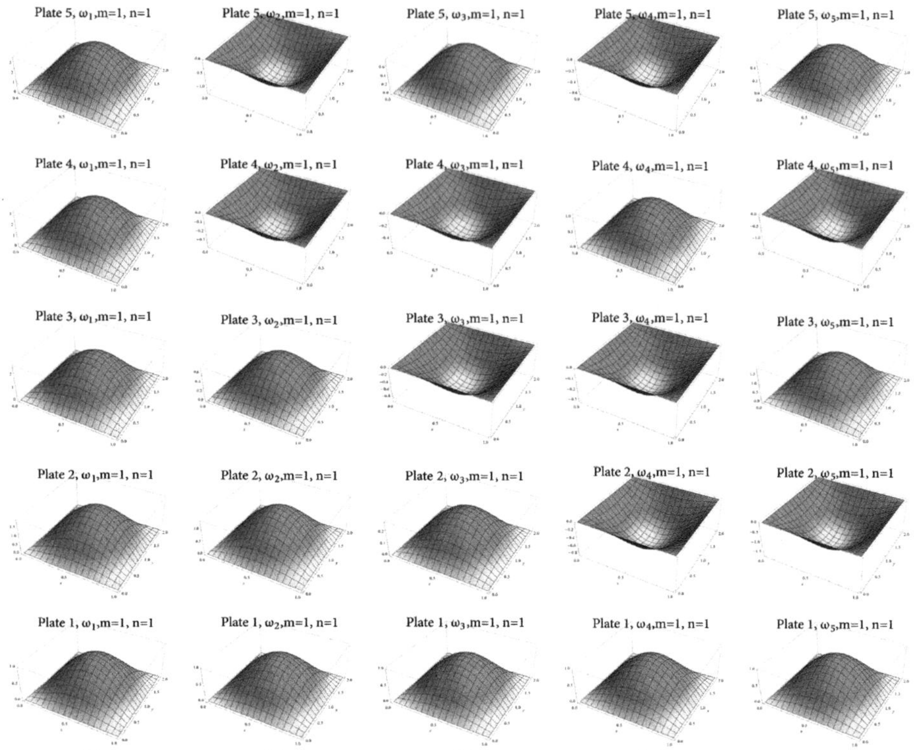

(A) (B) (C) (D) (E)

Figure 14.7 General mode shapes of vibration of a rectangular five-plate complex system. The plates are elastically connected, with all four lateral sides simply supported. The bottom plate is the system's first plate, and it is elastically connected to the rigid foundation. The system corresponds to five sequences of natural frequencies (ω_{imn}, $i = 1, 2, 3, 4, 5$, $m = 1$, $n = 1$); (A) ω_{111}, (B) ω_{211}, (C) ω_{311}, (D) ω_{411}, (E) ω_{511}.

side. The plates that vibrate at the highest natural frequency are on the far-right side. If the bottom plate is selected as the referent plate, then when the system vibrates at the lowest natural frequency, all plates are in synchronous mode with the bottom plate. At the second natural frequency, the plates in the upper chunk are vibrating asynchronously in relation to the bottom chunk and as well as to the bottom plate. As the migration to the higher frequency progresses, the plates that vibrate asynchronously move closer to the bottom plate. Lastly, at the highest natural frequency, the vibration mode shapes alternate between every adjacent plate.

14.6 Numerical analysis

For the selected physical parameters ($\nu = 0.3$, $k = 5/6$) natural frequencies for multiplate systems are shown in Tables 14.1–14.3. It can be stated that an increase in the number of plates in the system causes a decrease in the value of the lowest natural

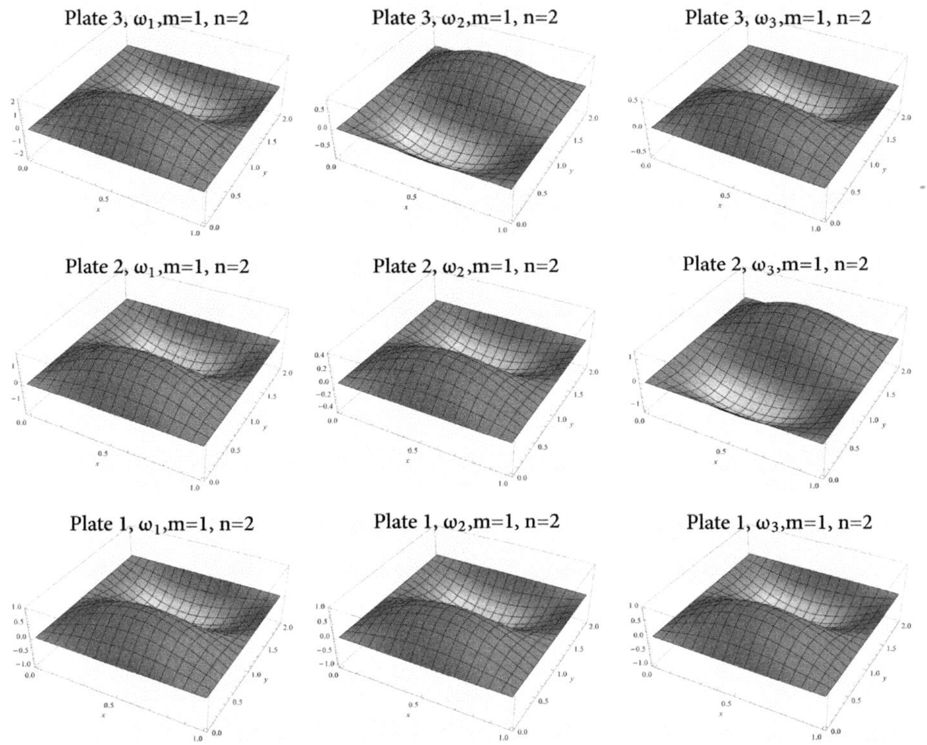

(A) (B) (C)

Figure 14.8 General mode shapes of vibration of a rectangular three-plate complex system. The plates are elastically connected, with all four lateral sides simply supported. The bottom plate is the system's first plate, and it is elastically connected to the rigid foundation. The system corresponds to three sequences of natural frequencies (ω_{imn}, $i = 1, 2, 3$, $m = 1$, $n = 2$); (A) ω_{112}, (B) ω_{212}, (C) ω_{312}.

frequency. The natural frequency has the highest value in the special case when there is only one plate resting on an elastic layer that is connected to the foundation. Adding more plates to the system will cause the lowest natural frequency to decrease. The lowest natural frequency of the multiplate system—in the case where the number of plates is increasing—gravitates towards the natural frequency of a single plate that is not laid on an elastic layer [6] and is presented in Table 14.3, for the case of 50 or 100 elastically connected plates. For the situation with thick plates, better approximations of natural frequencies are obtained if FSDT and HSDT are used, as seen in Table 14.2. The validation of the derived equations and the frequency equation is performed by comparison with the results obtained for the case of three elastically connected plates using commercial ANSYS® software.

Mutual system parameters for the obtained values of natural frequencies presented in Tables 14.1–14.3 are $m = 1$, $n = 1$, $a = 2$ m, $b = 1$ m, $F_x = 0$, $F_y = 0$, $K = K_0 = 6 \times 10^4$ N/m^3, $E = 10^{10}$ Pa, $\rho = 5 \times 10^3$ kg/m^3, $\nu = 0.3$.

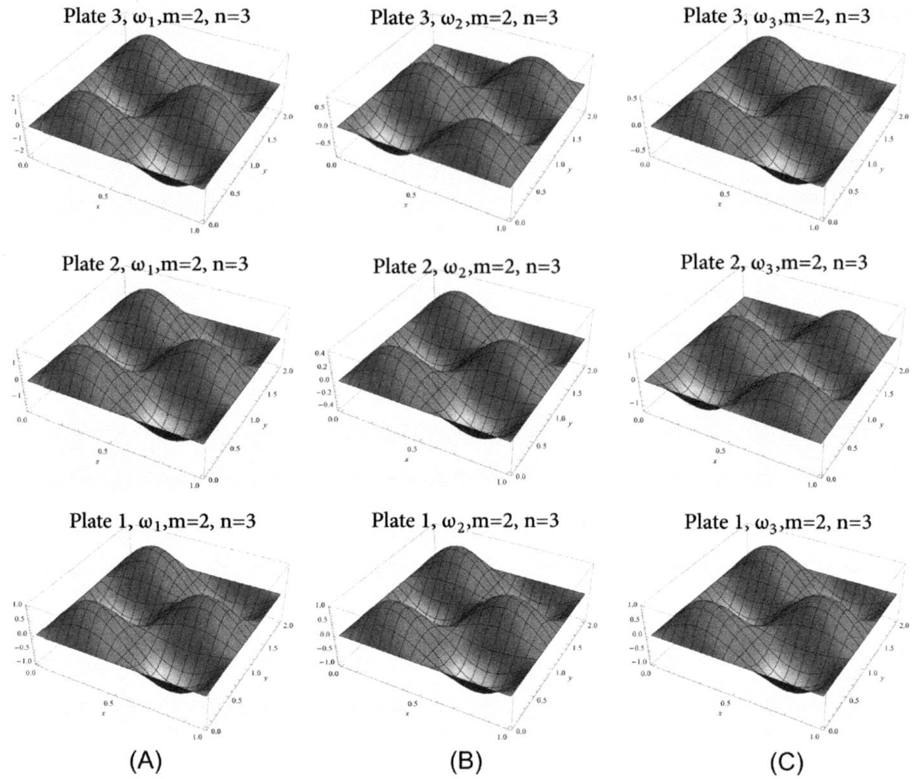

Figure 14.9 General mode shapes of vibration of a rectangular three-plate complex system. The plates are elastically connected, with all four lateral sides simply supported. The bottom plate is the system's first plate, and it is elastically connected to the rigid foundation. The system corresponds to three sequences of natural frequencies (ω_{imn}, $i=1, 2, 3$, $m=2$, $n=3$); (A) ω_{123}, (B) ω_{223}, (C) ω_{323}.

Table 14.1 Natural frequencies (Hz) calculated using CPT, FSDT, and Reddy HSDT for systems with $p=3, 5, 7$ plates of thickness $h=10$ mm.

Number of plates—p	Frequency number—s	CPT (14.26)	FSDT (14.27)	HSDT Reddy
3	1	55.00229	54.99053	54.9777
	2	68.21676	68.20606	68.1905
	3	81.75544	81.74524	81.7267
5	1	53.71028	53.69837	53.6858
	2	60.13245	60.12119	60.1073
	3	69.61337	69.60274	69.5868
	4	78.64206	78.63177	78.6139
	5	84.89161	84.88146	84.8623

(*Continued*)

7	1	53.29186	53.27991	53.2674
	2	56.97312	56.96157	56.9483
	3	63.14726	63.13624	63.1217
	4	70.26172	70.25112	70.2351
	5	77.00141	76.99108	76.9736
	6	82.42263	82.41244	82.3938
	7	85.90743	85.89730	85.8779

CPT, Classical plate theory; *FSDT*, first-order shear deformation theories; *HSDT*, high-order shear deformation theory.

Table 14.2 Natural frequencies (Hz) calculated using CPT, FSDT, and Reddy HSDT for systems with $p=3$, 5, 7 plates of thickness $h=100$ mm.

Number of plates—p	Frequency number—s	CPT (14.26)	FSDT (14.27)	HSDT Reddy
3	1	527.99760	516.45977	515.7681
	2	528.15177	516.61588	515.9245
	3	528.34396	516.81049	516.1193
5	1	527.98430	516.44630	515.7547
	2	528.05353	516.51640	515.8249
	3	528.17000	516.63434	515.9429
	4	528.29700	516.76263	516.0714
	5	528.39342	516.86057	516.1694
7	1	527.98006	516.44201	515.7504
	2	528.01850	516.48093	515.7893
	3	528.08872	516.55204	515.8605
	4	528.17858	516.64303	515.9516
	5	528.27253	516.73816	516.0469
	6	528.35432	516.82099	516.1298
	7	528.40983	516.87720	516.1861

CPT, Classical plate theory; *FSDT*, first-order shear deformation theories; *HSDT*, high-order shear deformation theory.

Table 14.3 Natural frequencies (Hz) calculated using CPT, FSDT, and Reddy HSDT for systems with $p=10$, 20, 50, and 100 plates of thickness $h=10$ mm.

Number of plates—p	Frequency number—s	CPT (14.26)	FSDT (14.27)	HSDT Reddy
1	1	63.1472	63.1362	63.1217
20	1	52.8641	52.8521	52.8398
	20	86.9450	86.9349	86.9154
50	1	52.8085	52.7964	52.7841
	50	87.0800	87.0699	87.0503
100	1	52.8002	52.7882	52.7759
	100	87.0999	87.0898	87.0702

CPT, Classical plate theory; *FSDT*, first-order shear deformation theories; *HSDT*, high-order shear deformation theory.

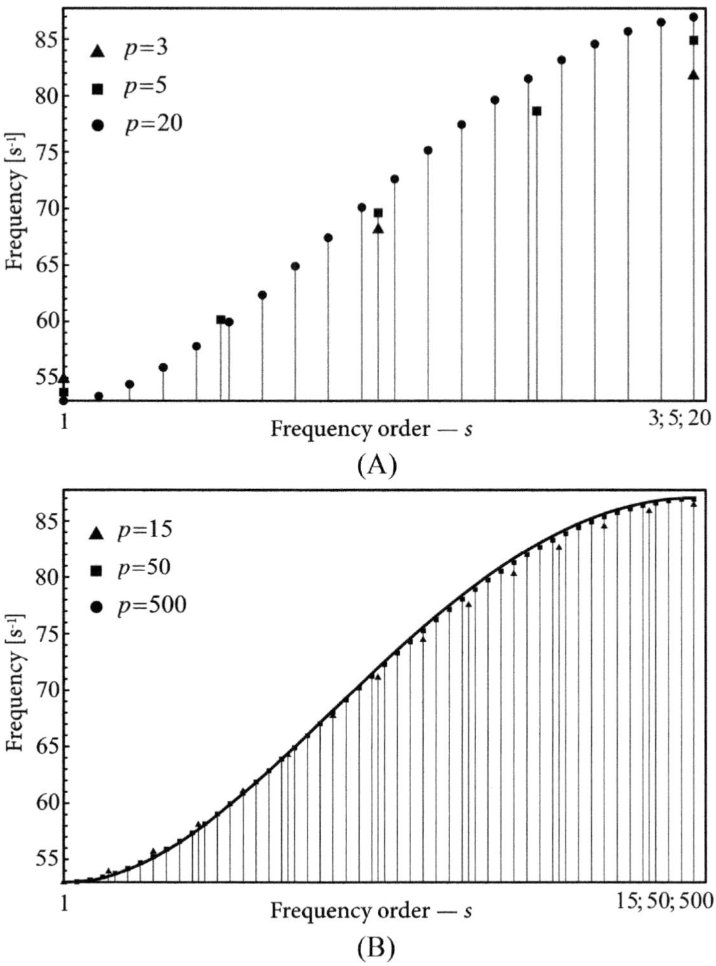

Figure 14.10 Schematic presentation of the range and tendency of natural frequencies. The first-mode; $m = n = 1$; (A) $p = 3$, 5, 20; (B) $p = 15$, 50, 500.

In the finite element analysis (performed in ANSYS®) of a system with three elastically connected plates, the results were close to those calculated using FSDT (Table 14.1) 54.98, 68.19, and 81.65 Hz. In Table 14.3, it is shown that the range between the lowest and the highest natural frequencies depends on the number of plates in the system. All frequencies in the first mode 11 and the range are given in Fig. 14.10. A conclusion can be drawn that the value of the highest natural frequency of the multiplate system will increase as the number of plates increases. Changes of natural frequencies in the function of thickness h, based on the properties given in Table 14.1 are shown in Fig. 14.11. These functions are for modes 11 and 13 and a system consisting of 7 plates.

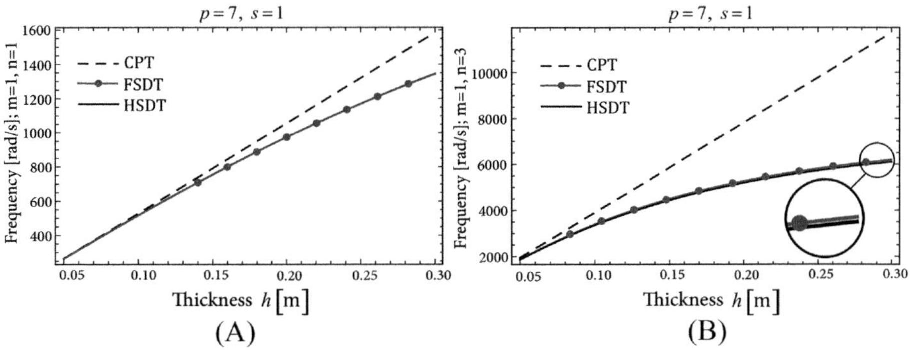

Figure 14.11 Natural frequency of an elastically connected rectangular plates. Different modes for a 7 plate system; (A) $m=1$, $n=1$, (B) $m=1$, $n=3$.

14.7 Conclusion

The analysis done in this chapter was on the properties of free transverse vibrations of multiplate systems, in which constituent parts, that is, plates, are elastically connected between each other and are simply supported on all four lateral sides. In the analysis, three plate theories were used: CPT, FSDT (Mindlin–Reissner), and HSDT (Reddy). Analytical solutions for natural frequencies and critical buckling loads are obtained with the trigonometric method. The natural mode shapes are shown for systems comprising 3, 4, and 5 elastically connected plates. A vibratory system comprising elastically connected thick plates was studied using FSDT and HSDT because these theories describe deformations of thick plates more precisely. It can be concluded that as the number of elastically connected plates in the system increases, the value of the first fundamental natural frequency will get closer to the value of the natural frequency of one foundationless plate. The critical buckling force for the multiplate system is explicitly determined and with it the stability regions. The influence of the stiffness moduli of the elastic layers on the stability is presented. The convergence of the present solution is checked and the accuracy is validated by comparing with numerical results using different theories. By using different theories, the accuracy of the results is validated. The study of the elastically connected multiplate system is expanded by using not only the CPT but also the FSDT and the HSDT. That has led to the conclusion that in multiplate systems (depending on the stiffness of elastic layers), natural frequencies can be in proximity to each other; accordingly, the need for better approximations is important, especially with thick plates. Analytical study of the natural frequencies and critical buckling loads can be used for application in real multistory buildings.

The formulation described in this study possesses several advantages. Primarily, the presented formulation is general since a different number of plates, various geometric dimensions (wide range of thickness), and different mechanical systems can be easily adopted in the formulation. Additionally, the proposed method offers a unified operation for the change of mathematical model, which describes a multisystem, from one case to another as easy as changing structural parameters without changing the solving procedure. Last, of all, it can be readily applied to shells with complex boundaries such as point supports, partial supports, nonuniform elastic restraints, and their combinations.

References

[1] R.D. Mindlin, Influence of rotatory inertia and shear on flexural motions of isotropic elastic plates, Journal of Applied Mechanics 73 (1951) 31–38. American Society of Mechanical Engineers.

[2] J.N. Reddy, N.D. Phan, Stability and vibration of isotropic, orthotropic and laminated plates according to a higher-order shear deformation theory, Journal of Sound and Vibration 98 (1985) 157–170.

[3] J.N. Reddy, C.M. Wang, An overview of the relationships between solutions of the classical and shear deformation plate theories, Composites Science and Technology 60 (2000) 2327–2335.

[4] Z. Oniszczuk, Free transverse vibrations of an elastically connected rectangular simply supported double-plate complex system, Journal of Sound and Vibration 236 (2000) 595–608.

[5] Y.Q. Zhang, Y. Lu, G.W. Ma, Effect of compressive axial load on forced transverse vibrations of a double-beam system, International Journal of Mechanical Sciences 50 (2008) 299–305.

[6] V. Stojanović, P. Kozić, Forced transverse vibration of Rayleigh and Timoshenko double-beam system with effect of compressive axial load, International Journal of Mechanical Sciences 60 (2012) 59–71.

[7] S. Kukla, Application of Green's functions in free vibration analysis of plate systems, Zeitschrift fur Angewandte Mathematik und Mechanik 76 (1996) 279–280.

[8] A.W. Leissa, The free vibration of rectangular plates, Journal of Sound and Vibration 31 (1973) 257–293.

[9] R.R. Craig JR, Structural Dynamics, John Wiley, New York, 1981.

[10] M. Levinson, D.W. Cooke, Thick rectangular plates-I. The generalized Navier solution, International Journal of Mechanical Sciences 25 (1983) 199–205.

[11] I.E. Harik, X. Liu, N. Balakrishnan, Analytic solution to free vibration of rectangular plates, Journal of Sound and Vibration 153 (1992) 51–62.

[12] S.Y. Lee, S.M. Lin, Free vibrations of elastically restrained non-uniform plates, Journal of Sound and Vibration 158 (1992) 121–131.

[13] S.Y. Lee, S.M. Lin, Lévy-type solution for the analysis of nonuniform plates, Computers and Structures 49 (1993) 931–939.

[14] U. Lee, J. Lee, Spectral-element method for Lévy-type plates subject to dynamic loads, Journal of Engineering Mechanics 125 (1999) 243–247.

[15] K.-H. Jeong, H.-S. Kang, Free vibration of multiple rectangular plates coupled with a liquid, International Journal of Mechanical Sciences 74 (2013) 161–172.

[16] J.W. Shyong, B.H. Chang, Free vibration of axial-loaded multi-step Timoshenko beam carrying arbitrary concentrated elements using continuous-mass transfer matrix method, European Journal of Mechanics—A/Solids 38 (2013) 20–37.

[17] V. Stojanović, P. Kozić, M. Ristić, Vibrations and stability analysis of multiple rectangular plates coupled with elastic layers based on different plate theories, International Journal of Mechanical Sciences 92 (2015) 233–244.

[18] V. Stojanović, P. Kozić, G. Janevski, Exact closed-form solutions for the natural frequencies and stability of elastically connected multiple beam system using Timoshenko and high-order shear deformation theory, Journal of Sound and Vibration 332 (2013) 563–576.

[19] V. Stojanović, P. Kozić, Vibrations and Stability of Complex Beam Systems, Springer Tracts in Mechanical Engineering, 2015 Springer ISBN 978-3-319-13766-7.

[20] S.G. Kelly, S. Srinivas, Free vibrations of elastically connected stretched beams, Journal of Sound and Vibration 326 (2009) 883–893.

[21] W. Soedel, Vibrations of Shells and Plates, Marcel Dekker Inc., 2004, pp. 97–98.

Part IV

Stochastic stability of structures and mechanical systems: Methodology and examples

Moment Lyapunov exponents and stochastic stability of the vibrationally isolated laminated plates

15

15.1 Introduction

In technical practice, composite materials are typically assembled in coupled structures together with elements composed of classical materials; dynamical analysis of such structures can only be done as a whole. These kinds of problems can be analyzed, if methods related to multi-degrees-of-freedom (MDOF) mechanical systems [1,2] are used. Dynamical analysis of coupled structures (with two or multi-degree-of-freedom) has a special case—the analysis of their stochastic stability, i.e., when they are exposed to random disturbances of various types [3,4]. Dynamic systems that are multibody parts—comprising a combination of composite and classical materials—are habitually exposed to such random disturbances. Such systems are common when there is a need to achieve appropriate thermal, mechanical, acoustical, or other kinds of isolation, protection, or change of some system properties depending on construction constraints or other needs. Sandwich and multilayered structures [5–7] are examples of continuous mechanical systems that comprise classical and composite materials. The considered model has a significant place in practical applications, which has motivated the authors to perform a detailed analytical analysis of the dynamic stability problem. Therefore, here are some of the technical applications: a) basic structure of most satellites include sandwich panels that comprise aluminum core and Carbon Fiber Reinforced Polymer face sheets; b) some of the larger antennas on telecommunication satellites are sandwich structures or viscoelastically connected laminated and classical plates which are not merely flat; c) often the payload adapter on which a satellite is connected to the launch vehicle is a combination of coupled thin composite plates and protection-isolation plates; d) the payload fairing (the cylindrical-dome shaped structure on top of most launch vehicles). These examples represent isolated systems of composite structures.

By reviewing the literature, it has been determined that until now, the problem of dynamic stability (at the level of determining the moment Lyapunov exponents and Lyapunov exponents) of systems with 3-DOF and MDOF has not been studied for cases in which complex coupling is present, and when a special procedure is needed for uncoupling the stochastic differential equations (SDEs). However, comprehending the dynamic stability conditions of such systems (as a whole, when all coupling effects are accounted for) is important for their application. A better part of the papers pertains to the analysis of stochastic

Modeling of Complex Dynamic Systems. DOI: https://doi.org/10.1016/B978-0-443-23942-7.00015-2

systems with 2-DOF, which are special cases of models of two coupled oscillators where it is unnecessary to implement the decoupling of SDEs [3] and [8–11]. The above papers consider the problems associated with systems of discrete oscillators or are expanded to two coupled one-dimensional continuous systems. Problems associated with the stochastic stability of classical or laminated composite plates have been regarded as 1-DOF or 2-DOF systems or they are reduced to them. There are research papers where the dynamic stability of laminated plates (boundaries of almost-sure instability) is analyzed via the direct Lyapunov method. Thus, Tylikowski et al. [12] systematically analyzed the dynamic stability of symmetric cross-ply laminated plates, including the shear deformation effect. In [13], studies of dynamic stability are addressed to the analyses of anti-symmetrically-laminated cross-ply rectangular plates, where the coupling of in-plane and transverse motion is taken into account. Studies like that show the correlations among the number of layers and classes of parametric excitation on the stability domains. Tylikowski and Frischmuth [14] studied the stability of parametric vibrations of a circular plate subjected to forces in the plane using the direct Lyapunov method. The result of their study is a method for the stabilization of stochastic vibrations, which consists of using a sufficiently large active damping coefficient. Besides the aforementioned standard direct Lyapunov method, the stability analysis—of stochastic systems comprising composite structures—can be performed by numerical methods. One of them is the moving least square method that is based on FEM and its implementation is given in [15] by Dey et al. They examined the numerical stochastic stability of composite curved panels; the obtained curvature effects and effects of fiber parameters were discussed, aside from non-uniform load parameters on the stability boundaries. Within the context of the study presented here, it is important to mention another study associated with the stability of hybrid composite plates [16]; the authors used the Bolotin method [17] to identify the regions of dynamic stability of a laminated hybrid composite plate. In it [16], the determination of the stability boundaries was done through a comprehensive analytical procedure that included deriving equations of motion.

In the field of stochastic dynamic analysis, the study here presented is one of the first that tackles the problem of coupled plates affected by in-plane time-depended forces that are analyzed by determining the moment of Lyapunov exponents and Lyapunov exponents. The model of the plate system is a multibody part, comprising a symmetric cross-ply laminated plate that is viscoelastically connected to two thin plates, with the purpose of achieving adequate structural requirements (of the whole system), or different structurally external material properties, or to provide isolation (mechanical, thermal, acoustical, etc.). The viscoelastic connection between the composite laminated plate and the external thin plate-isolators is provided with layers of Kelvin–Voigt type. The bases for the derived stochastic partial differential equations (SPDEs) are the first-order shear deformation theory (FSDT; used on laminated plates), and the classical plate theory (CPT; used on thin plate-isolators). The stochastic stability is considered analytically through the moment Lyapunov exponents and Lyapunov exponents when the system is under parametric excitation of white noise. The system of SPDEs is reduced to the system of stochastic differential equations (SDEs) using the Galerkin method. Then the contact analysis method [4,18] is used to decouple equations of motion and convert them to the system of Stratonovitch differential equations. The adequate transformations are used to obtain the Itô differential equation system, as is shown in the special case of coupled beam systems [19]. The trigonometric transformations that are applied are pivotal because the classical stochastic averaging is unnecessary in that case [20]. The moment Lyapunov exponents are derived as explicit asymptotic expressions by the perturbation method. Local and global stochastic stability benefits of plate-isolators are shown and discussed. The present study is based on the study by Stojanović et al. [21]. The primary objective of this study is to determine stochastic stable regions of an isolated plate and to validate the mechanical benefits that the thin isolators have—which is the enlargement of stable regions. The analyzed model is a complex coupled plate system that is not considered a compact entity and is loaded with in-plane stochastic forces. The novelty of the study comes from the way in which the parametric conditions of stability of the structure are determined. The novelty also presents the applicability of the considered model, which can satisfy varying isolation (mechanical, thermal, acoustical, etc.), or structural requirements in the technical practice. The complex plate system, which is given as an isolated model, belongs to the six-dimensional class of problems and presents a novelty in the dynamic stability analysis of coupled plates.

15.2 Problem formulation

Fig. 15.1 shows an isolated symmetric cross-ply laminated plate that is subjected to uniformly distributed stochastic in-plane time-depended loading. The core plate comprises unidirectional layers (0° or 90°) and is described through FSDT [22]. The plate-isolators are described through Kirchhoff–Love (CPT) plate theory.

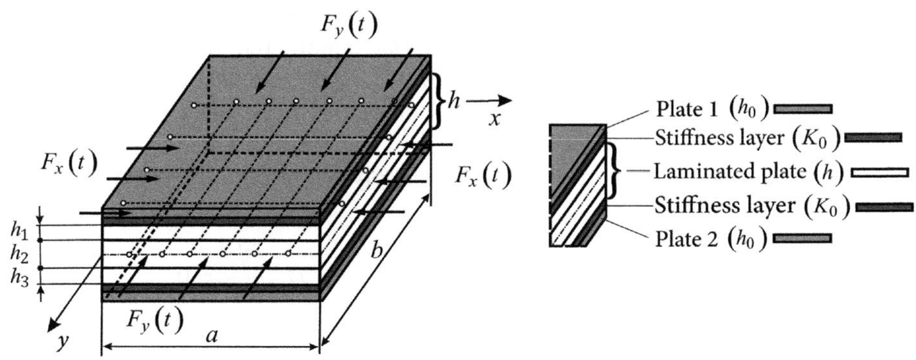

Figure 15.1 Laminated plate. Geometry of an isolated cross-ply symmetric plate system.

The connection between the core laminated plate and the plate-isolators is achieved with elastic layers (which have stiffness modulus K_0 and damping coefficient β). The form of the PDEs of motion is

$$\rho_0 h_0 \frac{\partial^2 w_{01}}{\partial t^2} + \beta\left(\frac{\partial w_{01}}{\partial t} - \frac{\partial w}{\partial t}\right) + \frac{E_0 h_0^3}{12(1-\mu_0^2)}\left(\frac{\partial^4 w_{01}}{\partial x^4} + 2\frac{\partial^4 w_{01}}{\partial x^2 \partial y^2} + \frac{\partial^4 w_{01}}{\partial y^4}\right)$$

$$+ F_x(t)\frac{\partial^2 w_{01}}{\partial x^2} + F_y(t)\frac{\partial^2 w_{01}}{\partial y^2} + K_0(w_{01} - w) = 0, \tag{15.1}$$

$$D_{11}\frac{\partial^2 \psi_x}{\partial x^2} + D_{66}\frac{\partial^2 \psi_x}{\partial y^2} + (D_{12} + D_{66})\frac{\partial^2 \psi_y}{\partial x \partial y} - k_1 A_{55}\left(\psi_x + \frac{\partial w}{\partial x}\right) = 0, \tag{15.2a}$$

$$(D_{12} + D_{66})\frac{\partial^2 \psi_x}{\partial x \partial y} + D_{66}\frac{\partial^2 \psi_y}{\partial x^2} + D_{22}\frac{\partial^2 \psi_y}{\partial y^2} - k_2 A_{44}\left(\psi_y + \frac{\partial w}{\partial y}\right) = 0, \tag{15.2b}$$

$$\rho h \frac{\partial^2 w}{\partial t^2} + \beta\left(2\frac{\partial w}{\partial t} - \frac{\partial w_{01}}{\partial t} - \frac{\partial w_{02}}{\partial t}\right) + F_x(t)\frac{\partial^2 w}{\partial x^2} + F_y(t)\frac{\partial^2 w}{\partial y^2} - k_1 A_{55}\left(\frac{\partial \psi_x}{\partial x} + \frac{\partial^2 w}{\partial x^2}\right)$$

$$- k_2 A_{44}\left(\frac{\partial \psi_y}{\partial y} + \frac{\partial^2 w}{\partial y^2}\right) + K_0(2w - w_{01} - w_{02}) = 0, \tag{15.2c}$$

$$\rho_0 h_0 \frac{\partial^2 w_{02}}{\partial t^2} + \beta\left(\frac{\partial w_{02}}{\partial t} - \frac{\partial w}{\partial t}\right) + \frac{E_0 h_0^3}{12(1-\mu_0^2)}\left(\frac{\partial^4 w_{02}}{\partial x^4} + 2\frac{\partial^4 w_{02}}{\partial x^2 \partial y^2} + \frac{\partial^4 w_{02}}{\partial y^4}\right)$$

$$+ F_x(t)\frac{\partial^2 w_{02}}{\partial x^2} + F_y(t)\frac{\partial^2 w_{02}}{\partial y^2} + K_0(w_{02} - w) = 0, \tag{15.3}$$

where A_{44}, A_{55}, D_{11}, D_{12}, D_{22}, and D_{66} are the stiffness moduli of the laminate; w, w_{01}, and w_{02} are the transverse displacements of laminated and thin plates, respectively; ψ_x and ψ_y represent the bending slopes of the laminated plate; k_1 and k_2 are the shear correction factors associated with the shear stiffness moduli A_{55} and A_{44}, respectively; ρ and ρ_0 denote the mass densities of the laminated plate and the thin isolators, respectively; h and h_0 denote the thicknesses of the laminated plate and thin isolators; $F_x(t)$ and $F_y(t)$ are the stochastic forces. In Eqs. (15.1) and (15.3), E_0 is the Young modulus and μ_0 is the Poisson ratio; both are associated with thin plate-isolators.

Boundary conditions that correspond to the simply supported lateral sides have the following form

$$\left.\begin{array}{l} x = 0 \\ x = a \end{array}\right\} w = 0, \quad \psi_y = 0, \quad M_x = D_{11}\frac{\partial \psi_x}{\partial x} + D_{12}\frac{\partial \psi_y}{\partial y} = 0,$$

$$w_{01} = w_{02} = \frac{\partial^2 w_{01}}{\partial x^2} = \frac{\partial^2 w_{02}}{\partial x^2} = 0,$$

$$\left.\begin{array}{l} y=0 \\ y=b \end{array}\right\} w=0, \quad \psi_x=0, \quad M_y=D_{12}\frac{\partial \psi_x}{\partial x}+D_{22}\frac{\partial \psi_y}{\partial y}=0,$$

$$w_{01}=w_{02}=\frac{\partial^2 w_{01}}{\partial y^2}=\frac{\partial^2 w_{02}}{\partial y^2}=0. \tag{15.4}$$

An assumption is that the axial loads $F_x(t)$ and $F_y(t)$ are stationary stochastic processes having a mean value P_0 and a fluctuation part $\zeta(t)$. In that scenario, the stochastic forces could be defined with $F_x(t)=P_0+\zeta(t)$ and $F_y(t)=\chi F_x(t)$, $0\le\chi\le 1$. After some algebra, it is possible to eliminate the bending slopes ψ_x and ψ_y in Eq. (15.2) and according to that the new system of three PDEs of motion is

$$\rho_0 h_0\frac{\partial^2 w_{01}}{\partial t^2}+\beta\left(\frac{\partial w_{01}}{\partial t}-\frac{\partial w}{\partial t}\right)+\frac{E_0 h_0^3}{12(1-\mu_0^2)}\left(\frac{\partial^4 w_{01}}{\partial x^4}+2\frac{\partial^4 w_{01}}{\partial x^2\partial y^2}+\frac{\partial^4 w_{01}}{\partial y^4}\right)$$
$$+(P_0+\zeta(t))\frac{\partial^2 w_{01}}{\partial x^2}+\chi(P_0+\zeta(t))\frac{\partial^2 w_{01}}{\partial y^2}+K_0(w_{01}-w)=0, \tag{15.5}$$

$$C_{002}\frac{\partial^2 w}{\partial t^2}+C_{202}\frac{\partial^4 w}{\partial x^2\partial t^2}+C_{022}\frac{\partial^4 w}{\partial y^2\partial t^2}+C_{222}\frac{\partial^6 w}{\partial x^2\partial y^2\partial t^2}+C_{402}\frac{\partial^6 w}{\partial x^4\partial t^2}$$
$$+C_{042}\frac{\partial^6 w}{\partial y^4\partial t^2}+C_{240}[\zeta(t)]\frac{\partial^6 w}{\partial x^2\partial y^4}+C_{220}[\zeta(t)]\frac{\partial^4 w}{\partial x^2\partial y^2}$$
$$+C_{420}[\zeta(t)]\frac{\partial^6 w}{\partial x^4\partial y^2}+C_{400}[\zeta(t)]\frac{\partial^4 w}{\partial x^4}+C_{040}[\zeta(t)]\frac{\partial^4 w}{\partial y^4}$$
$$+C_{200}[\zeta(t)]\frac{\partial^2 w}{\partial x^2}+C_{020}[\zeta(t)]\frac{\partial^2 w}{\partial y^2}+C_{600}[\zeta(t)]\frac{\partial^6 w}{\partial x^6}+C_{060}[\zeta(t)]\frac{\partial^6 w}{\partial y^6}$$
$$+\beta\mathcal{L}_\beta[w,w_{01},w_{02}]+K_0\mathcal{L}_{K_0}[w,w_{01},w_{02}]=0, \tag{15.6}$$

$$\rho_0 h_0\frac{\partial^2 w_{02}}{\partial t^2}+\beta\left(\frac{\partial w_{02}}{\partial t}-\frac{\partial w}{\partial t}\right)+\frac{E_0 h_0^3}{12(1-\mu_0^2)}\left(\frac{\partial^4 w_{02}}{\partial x^4}+2\frac{\partial^4 w_{02}}{\partial x^2\partial y^2}+\frac{\partial^4 w_{02}}{\partial y^4}\right)$$
$$+(P_0+\zeta(t))\frac{\partial^2 w_{02}}{\partial x^2}+\chi(P_0+\zeta(t))\frac{\partial^2 w_{02}}{\partial y^2}+K_0(w_{02}-w)=0. \tag{15.7}$$

15.3 Stochastic stability analysis

The system of Eqs. (15.5–15.7) presents the system of SPDEs. With the given boundary conditions (15.4) and with the use of the Galerkin method, an approximate solution can be obtained. Substituting

$$w_{01}(x,y,t)=x_1(t)\sin\frac{m\pi x}{a}\sin\frac{n\pi y}{b},$$

$$w(x, y, t) = x_2(t)\sin\frac{m\pi x}{a}\sin\frac{n\pi x}{b}, \quad w_{02}(x, y, t) = x_3(t)\sin\frac{m\pi x}{a}\cos\frac{n\pi y}{b}, \tag{15.8}$$

into Eqs. (15.5–15.7) and multiplying by $\sin\frac{m\pi x}{a}\sin\frac{n\pi x}{b}$ and integrating from 0 to a and from 0 to b, yields the matrix equation for the m-th and n-th considered mode

$$\begin{bmatrix} m_{11} & 0 & 0 \\ 0 & m_{22} & 0 \\ 0 & 0 & m_{33} \end{bmatrix}\ddot{\mathbf{x}} + 2\varepsilon\begin{bmatrix} \beta & 0 & 0 \\ \beta_{21} & \beta_{22}+\beta & \beta_{23} \\ 0 & 0 & \beta \end{bmatrix}\dot{\mathbf{x}} + 2\beta\begin{bmatrix} 0 & -1 & 0 \\ 1 & 0 & 1 \\ 0 & -1 & 0 \end{bmatrix}\dot{\mathbf{x}}$$

$$+\begin{bmatrix} k_{11} & k_{12} & 0 \\ k_{21} & k_{22} & k_{23} \\ 0 & k_{32} & k_{33} \end{bmatrix}\mathbf{x} + \sqrt{\varepsilon}\xi(t)\begin{bmatrix} \widetilde{k}_{11} & 0 & 0 \\ 0 & \widetilde{k}_{22} & 0 \\ 0 & 0 & \widetilde{k}_{33} \end{bmatrix}\mathbf{x} = 0. \tag{15.9}$$

In the matrix Eq. (15.9), the small parameter $\varepsilon(0 < \varepsilon \ll 1)$ is introduced to show that damping and parametric excitations are small, where $\xi(t)$ represents the excitation modeled as a Gaussian white noise process in the fluctuation part $\zeta(t) = \sqrt{\varepsilon}\xi(t)$. The constants are given in [21]. The condensed form of the system of Eq. (15.9) is

$$\mathbf{M}\ddot{\mathbf{x}} + (\mathbf{G} + \varepsilon\mathbf{D})\dot{\mathbf{x}} + \mathbf{K}\mathbf{x} + \sqrt{\varepsilon}\xi(t)\mathbf{S}\mathbf{x} = \mathbf{0}, \tag{15.10}$$

where $\mathbf{M}$, $\mathbf{G}$, $\mathbf{D}$, and $\mathbf{K}$ are the 3×3 mass, gyroscopic, damping, and stiffness matrices, respectively, with an additional skew-symmetric property of the matrix $\mathbf{G}$ given by $\mathbf{G}^T = -\mathbf{G}$. Unlike a non-gyroscopic system (examined later), a point transformation $\mathbf{x} = \mathbf{T}\mathbf{q}$ when $\varepsilon = 0$ cannot be applied to convert Eq. (15.10) to the form

$$\mathbf{I}\ddot{\mathbf{q}} - \mathbf{\Omega}^2\mathbf{q} = \mathbf{0}, \quad \mathbf{\Omega} = \mathrm{diag}\{\omega_1, \ \omega_2, \ \ldots., \omega_n\}. \tag{15.11}$$

For this, a method known as contact analysis [4] can be used. This method is a transformation that preserves the Hamiltonian structure of the system. Taking the state vector $\mathbf{x}$ as the generalized coordinate vector $\mathbf{q}$, it can be noticed that the kinetic energy and the potential energy of the system can be written in the form of

$$\mathrm{T} = \frac{1}{2}(\dot{\mathbf{q}}^{\mathrm{T}}\mathbf{M}\dot{\mathbf{q}} + \dot{\mathbf{q}}^{\mathrm{T}}\mathbf{G}\mathbf{q}), \quad V = \frac{1}{2}\mathbf{q}^{\mathrm{T}}\mathbf{K}\mathbf{q}. \tag{15.12}$$

The Lagrangian, defined as $L = T - V$, is

$$L(\mathbf{q}, \dot{\mathbf{q}}) = \frac{1}{2}(\dot{\mathbf{q}}^{\mathrm{T}}\mathbf{M}\dot{\mathbf{q}} + \dot{\mathbf{q}}^{\mathrm{T}}\mathbf{G}\mathbf{q} - \mathbf{q}^{\mathrm{T}}\mathbf{K}\mathbf{q}). \tag{15.13}$$

The Lagrange's equations are

$$\frac{\mathrm{d}}{\mathrm{dt}}\left(\frac{\partial L}{\partial \dot{\mathrm{q}}_r}\right) - \frac{\partial L}{\partial \mathrm{q}_r} = 0, \quad r = 1, \ 2, \ \ldots, n, \tag{15.14}$$

Which are identical to (15.10). The momentum conjugate with respect to $\mathbf{q}$ is defined as

$$\mathbf{p} = \frac{\partial L}{\partial \dot{\mathbf{q}}} = \mathbf{M}\dot{\mathbf{q}} + \frac{1}{2}\mathbf{G}\mathbf{q}. \tag{15.15}$$

Given that $\mathbf{M}$ is a positive definite matrix, its inverse exists. Thereby, using Eq. (15.15) leads to the generalized velocity as

$$\dot{\mathbf{q}} = \mathbf{M}^{-1}\left(\mathbf{p} - \frac{1}{2}\mathbf{G}\mathbf{q}\right). \tag{15.16}$$

The Hamiltonian function is defined as

$$h(\mathbf{q}, \mathbf{p}) = \dot{\mathbf{q}}^{\mathrm{T}}\mathbf{p} - L(\mathbf{q}, \mathbf{p}). \tag{15.17}$$

From Eqs. (15.13–15.16), the Hamiltonian function can be written as

$$h(\mathbf{q}, \mathbf{p}) = \frac{1}{2}\mathbf{p}^{\mathrm{T}}\mathbf{M}^{-1}\mathbf{p} - \frac{1}{2}\mathbf{p}^{\mathrm{T}}\mathbf{M}^{-1}\mathbf{G}\mathbf{q} - \frac{1}{8}\mathbf{q}^{\mathrm{T}}\mathbf{G}\mathbf{M}^{-1}\mathbf{G}\mathbf{q} + \frac{1}{2}\mathbf{q}^{\mathrm{T}}\mathbf{K}\mathbf{q} = \frac{1}{2}\mathbf{u}^{\mathrm{T}}\mathbf{H}\mathbf{u}, \tag{15.18}$$

where $\mathbf{u}$ is a state vector of dimension $2n$ and $\mathbf{H}$ is a matrix of dimension $2n$ x $2n$ given by

$$\mathbf{u} = \begin{pmatrix} \mathbf{q} \\ \mathbf{p} \end{pmatrix}, \quad \mathbf{H} = \begin{bmatrix} -\frac{1}{4}\mathbf{G}\mathbf{M}^{-1}\mathbf{G} + \mathbf{K} & \frac{1}{2}\mathbf{G}\mathbf{M}^{-1} \\ \frac{1}{2}\mathbf{M}^{-1}\mathbf{G} & \mathbf{M}^{-1} \end{bmatrix}.$$

The equations of motion in the form of Hamiltonian equations are

$$\dot{\mathbf{q}} = \frac{\partial h}{\partial \mathbf{p}}, \quad \dot{\mathbf{p}} = -\frac{\partial h}{\partial \mathbf{q}}. \tag{15.19}$$

Expressing Eq. (15.9) in terms of the state vector $\mathbf{u}$ leads to

$$\dot{\mathbf{u}} = \mathbf{J}\mathbf{H}\mathbf{u}, \tag{15.20}$$

where

$$\mathbf{J} = \begin{bmatrix} \mathbf{0} & \mathbf{1} \\ -\mathbf{1} & \mathbf{0} \end{bmatrix}_{2n \text{ x } 2n}.$$

Verification shows that $\mathbf{J}^{\mathrm{T}} = \mathbf{J}^{-1} = -\mathbf{J}$, which means that $\mathbf{J}$ is skew-symmetric. Since $\mathbf{G}$ is skew-symmetric, the matrix $\mathbf{H}$ can be proved to be symmetric. When $\varepsilon \neq 0$, Eq. (15.10) is of the form

$$\dot{\mathbf{u}} = \mathbf{J}\mathbf{H}\mathbf{u} - \varepsilon[\widetilde{\mathbf{D}}\mathbf{J}\mathbf{H} + f(t)\widetilde{\mathbf{S}}]\mathbf{u}, \tag{15.21}$$

where

$$\widetilde{\mathbf{D}} = \begin{bmatrix} 0 & 0 \\ \mathrm{D} & 0 \end{bmatrix}, \quad \widetilde{\mathbf{S}} = \begin{bmatrix} 0 & 0 \\ \mathrm{S} & 0 \end{bmatrix}.$$

Looking for the solution for Eq. (15.18) in the form $\mathbf{u} = \hat{\mathbf{u}}e^{\lambda t}$ leads to the eigenvalue problem of the form

$$(\mathbf{JH} - \lambda\mathbf{I})\hat{\boldsymbol{\mu}} = 0 \text{ or } (\mathbf{H} + \lambda\mathbf{j})\hat{\boldsymbol{\mu}} = 0. \tag{15.22}$$

It can be proven that the eigenvalues λ are purely imaginary and occur in conjugate pairs, i.e. $\lambda_r = \pm i\omega_r$, where ω_r is real. The eigenvectors can be expressed as $\hat{\mathbf{u}}_r = \mathbf{x}_r + i\mathbf{y}_r$, where $\mathbf{x}_r$ and $\mathbf{y}_r$ satisfy the following relationships

$$\mathbf{Hx}_r = \omega_r\mathbf{Jy}_r, \quad \mathbf{Hy}_r = -\omega_r\mathbf{Jx}_r, \quad \text{for } \lambda = i\omega_r, \tag{15.23}$$

$$\mathbf{Hx}_r = -\omega_r\mathbf{Jy}_r, \quad \mathbf{Hy}_r = \omega_r\mathbf{Jx}_r, \quad \text{for } \lambda = i\omega_r. \tag{15.24}$$

Eliminating $\mathbf{y}_r$ in Eq. (15.23) results in

$$\mathbf{HJHx}_r = \omega_r^2\mathbf{Jx}_r. \tag{15.25a}$$

Similarly, eliminating $\mathbf{x}_r$ from Eq. (15.24) leads to

$$\mathbf{HJHy}_r = \omega_r{}^2\mathbf{Jy}_r. \tag{15.25b}$$

Eq. (15.22) is a complex eigenvalue problem of dimension $2n$. By identifying that eigenvalues have the form $\lambda_r = \pm i\omega_r$, and expressing the eigenvectors as $\hat{\mathbf{u}}_r = \mathbf{x}_r + i\mathbf{y}_r$, then the complex eigenvalue problem (15.22) can be rendered by two real eigenvalue problems of dimensions n represented by Eq. (15.25). The eigenvectors $\mathbf{x}_r$ and $\mathbf{y}_r$, obtained from Eq. (15.25), have the orthogonality properties in relation to matrices $\mathbf{J}$ and $\mathbf{H}$. The orthogonal correlations can be obtained by considering Eq. (15.23) as follows; multiplying $\mathbf{y}_s{}^T$ and $\mathbf{x}_s{}^T$ from the left to both sides of Eq. (15.23), respectively, leads to

$$\mathbf{y}_s{}^T\mathbf{Hx}_r = \omega_r\mathbf{y}_s{}^T\mathbf{Jy}_r, \quad \mathbf{x}_s{}^T\mathbf{Hy}_r = -\omega_r\mathbf{x}_s{}^T\mathbf{Jx}_r. \tag{15.26}$$

Since $\mathbf{J}$ is skew-symmetric, simple verification shows that

$$\mathbf{y}_r{}^T\mathbf{Jy}_r = 0, \quad \mathbf{x}_r{}^T\mathbf{Jx}_r = 0. \tag{15.27}$$

By using Eq. (15.27) and setting $r = s$ in Eq. (15.26), the bi-orthogonal correlation will be obtained

$$\mathbf{y}_r{}^T\mathbf{Hy}_r = 0, \quad \mathbf{x}_r{}^T\mathbf{Hx}_r = 0. \tag{15.28}$$

Likewise, multiplying $\mathbf{x}_s{}^T$ and $\mathbf{y}_s{}^T$ from the left on both sides of Eq. (15.23), respectively, yields

$$\mathbf{x}_s{}^T\mathbf{Hx}_r = \omega_r\mathbf{x}_s{}^T\mathbf{Jy}_r, \quad \mathbf{y}_s{}^T\mathbf{Hy}_r = -\omega_r\mathbf{y}_s{}^T\mathbf{Jx}_r. \tag{15.29}$$

Using the fact that $\mathbf{J}$ is a skew-symmetric and setting $r = s$, will lead to the following correlation

$$\mathbf{x}_r{}^T\mathbf{Hx}_r = \mathbf{y}_r{}^T\mathbf{Hy}_r = \omega_r\mathbf{x}_r{}^T\mathbf{Jy}_r. \tag{15.30}$$

If $\mathbf{x}_r{}^T\mathbf{Jy}_r$ is normalized such that

$$\mathbf{x}_r{}^T\mathbf{Jy}_r = 1, \tag{15.31}$$

the following orthogonal correlations are obtained from Eq. (15.30)

$$\mathbf{x}_r^{\mathrm{T}}\mathbf{H}\mathbf{x}_r = \mathbf{y}_r^{\mathrm{T}}\mathbf{H}\mathbf{y}_r = \omega_r. \tag{15.32}$$

For the case when $r \neq s$, it may be verified that

$$\mathbf{x}_s^{\mathrm{T}}\mathbf{J}\mathbf{x}_r = 0, \quad \mathbf{y}_s^{\mathrm{T}}\mathbf{J}\mathbf{y}_r = 0, \quad \mathbf{x}_s^{\mathrm{T}}\mathbf{J}\mathbf{y}_r = 0, \quad \mathbf{y}_s^{\mathrm{T}}\mathbf{J}\mathbf{x}_r = 0. \tag{15.33}$$

From Eqs. (15.26), (15.27), and (15.29) additional orthogonal relations may be obtained as

$$\mathbf{y}_s^{\mathrm{T}}\mathbf{H}\mathbf{x}_r = 0, \quad \mathbf{x}_s^{\mathrm{T}}\mathbf{H}\mathbf{y}_r = 0, \quad \mathbf{x}_s^{\mathrm{T}}\mathbf{H}\mathbf{x}_r = 0, \quad \mathbf{y}_s^{\mathrm{T}}\mathbf{H}\mathbf{y}_r = 0. \tag{15.34}$$

Denoting

$$\mathbf{X} = [\mathbf{x}_1\mathbf{x}_2\ldots\mathbf{x}_n]_{2n \times n}, \quad \mathbf{Y} = [\mathbf{y}_1\mathbf{y}_2\ldots\mathbf{y}_n]_{2n \times n}$$

the orthogonality properties (15.27), (15.28), (15.31–15.34) become

$$\begin{aligned}\mathbf{X}^{\mathrm{T}}\mathbf{J}\mathbf{X} = \mathbf{Y}^{\mathrm{T}}\mathbf{J}\mathbf{Y} = \mathbf{Y}^{\mathrm{T}}\mathbf{H}\mathbf{X} = \mathbf{X}^{\mathrm{T}}\mathbf{H}\mathbf{Y} = \mathbf{0}_{n \times n},\\ \mathbf{X}^{\mathrm{T}}\mathbf{H}\mathbf{X} = \mathbf{Y}^{\mathrm{T}}\mathbf{H}\mathbf{Y} = \boldsymbol{\Omega} = \mathrm{diag}\{\omega_1, \omega_2, \ldots., \omega_n\},\\ \mathbf{X}^{\mathrm{T}}\mathbf{J}\mathbf{Y} = -\mathbf{Y}^{\mathrm{T}}\mathbf{J}\mathbf{X} = \mathbf{I}_{n \times n}.\end{aligned} \tag{15.35}$$

A $2n$ x $2n$ transformation matrix is formed as $\mathbf{T} = [\mathbf{XY}]$. Orthogonal relations (15.35) can prove that

$$\mathbf{T}^{\mathrm{T}}\mathbf{J}\mathbf{T} = \mathbf{J}. \tag{15.36}$$

A matrix $\mathbf{T}$, which satisfies the condition (15.36), is known as a simplectic matrix. In addition, the matrix $\mathbf{T}$ also satisfies the condition

$$\mathbf{T}^{\mathrm{T}}\mathbf{H}\mathbf{T} = \begin{bmatrix} \boldsymbol{\Omega} & \boldsymbol{0} \\ \boldsymbol{0} & \boldsymbol{\Omega} \end{bmatrix}. \tag{15.37}$$

Considering the transformation

$$\mathbf{u} = \mathbf{T}\mathbf{v}, \quad \mathbf{v} = \begin{Bmatrix} \mathbf{Q} \\ \mathbf{P} \end{Bmatrix}. \tag{15.38}$$

The Hamiltonian function defined in Eq. (15.18) can be written as

$$\hat{h}(\mathbf{v}) = \frac{1}{2}\mathbf{u}^{\mathrm{T}}(\mathbf{v})\mathbf{H}, \quad \mathbf{u}(\mathbf{v}) = \frac{1}{2}\mathbf{v}^{\mathrm{T}}\mathbf{T}^{\mathrm{T}}\mathbf{H}\mathbf{T}, \quad \mathbf{v} = \frac{1}{2}\mathbf{v}^{\mathrm{T}}\hat{\mathbf{H}}\mathbf{v}, \tag{15.39}$$

where $\hat{\mathbf{H}} = \mathbf{T}^{\mathrm{T}}\mathbf{H}\mathbf{T}$ and is given by Eq. (15.37). Because the Hamiltonian structure of the system is preserved, such transformation is canonical and is termed a contact transformation. The quantity $\hat{h}(\mathbf{v})$ is the new Hamiltonian in the new variables ($\mathbf{Q}$, $\mathbf{P}$). When $\varepsilon \neq 0$, the equations of motion (15.12) are of the form Eq. (15.21). Applying the transformation $\mathbf{u} = \mathbf{T}\mathbf{v}$ to Eq. (15.21) and multiplying the results of the equation from the left by $\mathbf{T}^{-1}$ yields

$$\dot{\mathbf{v}} = \mathbf{T}^{-1}\mathbf{J}\mathbf{H}\mathbf{T}\mathbf{v} - \varepsilon\mathbf{T}^{-1}[\widetilde{\mathbf{D}}\mathbf{J}\mathbf{H} + f(t)\widetilde{\mathbf{S}}]\mathbf{T}\mathbf{v}. \tag{15.40}$$

Using the orthogonal relations from Eq. (15.35), the matrix $\mathbf{T}^{-1}\mathbf{JHT}$ may be rewritten as

$$\mathbf{T}^{-1}\mathbf{JHT} = \mathbf{T}^{-1}\mathbf{J}(\mathbf{T}^{\mathrm{T}})^{-1}(\mathbf{T}^{T}\mathbf{HT}) = \mathbf{J}\hat{\mathbf{H}},$$

and Eq. (15.40) is reduced to

$$\dot{\mathbf{v}} = \mathbf{J}\hat{\mathbf{H}}\mathbf{v} - \varepsilon[\widetilde{\mathbf{D}} + f(t)\widetilde{\mathbf{S}}]\mathbf{v}, \tag{15.41}$$

where

$$\widetilde{\mathbf{D}} = \mathbf{T}^{-1}\widetilde{\mathbf{D}}\mathbf{JHT}, \quad \widetilde{\mathbf{S}} = \mathbf{T}^{-1}\widetilde{\mathbf{S}}\mathbf{T}.$$

Eq. (15.41) may be rewritten in the matrix form

$$\begin{bmatrix} \dot{\mathbf{Q}} \\ \dot{\mathbf{P}} \end{bmatrix} = \begin{bmatrix} \mathbf{0} & \mathbf{1} \\ -\mathbf{1} & \mathbf{0} \end{bmatrix}\begin{bmatrix} \mathbf{\Omega} & \mathbf{0} \\ \mathbf{0} & \mathbf{\Omega} \end{bmatrix} - \varepsilon\left\{\begin{bmatrix} \overline{\mathbf{D}}^1 & \overline{\mathbf{D}}^2 \\ \overline{\mathbf{D}}^3 & \overline{\mathbf{D}}^4 \end{bmatrix} + f(t)\begin{bmatrix} \overline{\mathbf{S}}^1 & \overline{\mathbf{S}}^2 \\ \overline{\mathbf{S}}^3 & \overline{\mathbf{S}}^4 \end{bmatrix}\right\}\begin{bmatrix} \mathbf{Q} \\ \mathbf{P} \end{bmatrix}, \tag{15.42}$$

or

$$\begin{aligned} \dot{\mathbf{Q}} &= +\mathbf{\Omega}\mathbf{P} - \varepsilon[\overline{\mathbf{D}}^1\mathbf{Q} + \overline{\mathbf{D}}^2\mathbf{P} + f(t)(\overline{\mathbf{S}}^1\mathbf{Q} + \overline{\mathbf{S}}^2\mathbf{P})], \\ \dot{\mathbf{P}} &= -\mathbf{\Omega}\mathbf{Q} - \varepsilon[\overline{\mathbf{D}}^3\mathbf{Q} + \overline{\mathbf{D}}^4\mathbf{P} + f(t)(\overline{\mathbf{S}}^3\mathbf{Q} + \overline{\mathbf{S}}^4\mathbf{P})]. \end{aligned} \tag{15.43}$$

Eq. (15.43) can be rewritten as a system of $2n$ first-order ordinary equations, for $r = 1, 2, \ldots, n$. In the particular case when $n = 3$, the expanded form is a system of six first-order Stratonovich differential equations

$$\begin{aligned} \mathrm{d}\{\mathbf{x}\} &= \begin{bmatrix} \mathbf{0} & \mathbf{1} \\ -\mathbf{1} & \mathbf{0} \end{bmatrix}[\mathrm{diag}\{\omega_1, \omega_2, \ldots, \omega_n\}]\{\mathbf{x}\}\mathrm{d}t \\ &- \varepsilon\begin{bmatrix} \boldsymbol{d}_{1(.)} & \boldsymbol{d}_{2(.)} \\ \boldsymbol{d}_{3(.)} & \boldsymbol{d}_{4(.)} \end{bmatrix}\{\mathbf{x}\}\mathrm{d}t - \sqrt{\varepsilon}\begin{bmatrix} s_{1(.)} & s_{2(.)} \\ s_{3(.)} & s_{4(.)} \end{bmatrix}\{\mathbf{x}\}\mathrm{d}\xi(t), \end{aligned} \tag{15.44}$$

where white-noise process $\xi(t)$ is given with zero mean and autocorrelation function

$$R_{\xi\xi}(t_1, t_2) = E[\xi(t_1)\xi(t_2)] = \sigma^2\delta(t_2 - t_1), \tag{15.45}$$

the Dirac delta function is denoted with $\delta(\,)$, σ is the intensity of the random process $\xi(t)$, and $E[\,]$ is expectation. The dynamic stability of an oscillatory system can be known if the maximal Lyapunov exponent and the p=th moment Lyapunov exponent are determined. The Lyapunov exponent is

$$\lambda_q = \lim_{t\to\infty}\frac{1}{t}\log\|\boldsymbol{q}(t; q_0)\|, \tag{15.46}$$

where $\boldsymbol{q}(t; q_0)$ is the solution process of a linear dynamic system. The almost-sure stability depends on the sign of the maximal Lyapunov exponent, which is an exponential growth rate of the solution of the perturbed dynamic system. A negative sign of the maximal Lyapunov exponent indicates almost-sure stability, while a non-negative value means instability. The exponential growth rate $E[\|\boldsymbol{q}(t; q_{0,}\, \dot{q}_0)\|^p]$ is provided by the moment Lyapunov exponent defined as

$$\Lambda_q(p) = \lim_{t\to\infty} \frac{1}{t} \log E[\,\|\boldsymbol{q}(t; q_0)\|^p]. \tag{15.47}$$

If $\Lambda_T(p) < 0$, then by definition $E[\|\boldsymbol{q}(t; q_0, \dot{q}_0)\|^p] \to 0$ as $t \to 0$ and this is referred to as the p-th moment stability. Even though the moment Lyapunov exponents are vital in the study of the dynamic stability of stochastic systems, the actual evaluations of the moment Lyapunov exponents are difficult. With the corresponding transformations, which can be derived through direct substitution for 3DOF systems from the general transformations presented in Chapter 3, Subsection 3.9, we obtain the following coordinates in the form

$$x_1 = a\cos\varphi_1\cos\varphi_2\cos\theta_1,\quad x_2 = -a\cos\varphi_1\cos\varphi_2\sin\theta_1,\quad x_3 = a\cos\varphi_1\sin\varphi_2\cos\theta_2,$$

$$x_4 = -a\cos\varphi_1\sin\varphi_2\sin\theta_2,\quad x_5 = a\sin\varphi_1\cos\theta_3,\quad x_6 = -a\sin\varphi_1\sin\theta_3,$$

$$P = a^p = (x_1^2 + x_2^2 + x_3^2 + x_4^2 + x_5^2 + x_6^2)^{\frac{p}{2}},$$

$$-\infty < p < \infty,\quad 0 \le \theta_i \le 2\pi,\quad 0 \le \varphi_j \le \frac{\pi}{2},\quad i = 1,\ 2,\ 3,\quad j = 1,\ 2,$$

Itô's rule brings the set of equations for the p-th power of the norm of the response and phase variables. The trigonometric transformation represents a, as a norm of the response, where θ_1, θ_2, and θ_3 are the angles of the three oscillators and φ_1 and φ_2 describe the couplings or exchange of energy between the oscillators, where

$$\mathrm{d}\theta_1 = m_1(\theta_1, \theta_2, \theta_3, \varphi_1, \varphi_2)\mathrm{d}t + \sigma_{11}(\theta_1, \theta_2, \theta_3, \varphi_1, \varphi_2)\mathrm{d}W(t),$$

$$\mathrm{d}\theta_2 = m_2(\theta_1, \theta_2, \theta_3, \varphi_1, \varphi_2)\mathrm{d}t + \sigma_{21}(\theta_1, \theta_2, \theta_3, \varphi_1, \varphi_2)\mathrm{d}W(t),$$

$$\mathrm{d}\theta_3 = m_3(\theta_1, \theta_2, \theta_3, \varphi_1, \varphi_2)\mathrm{d}t + \sigma_{31}(\theta_1, \theta_2, \theta_3, \varphi_1, \varphi_2)\mathrm{d}W(t),$$

$$\mathrm{d}\varphi_1 = m_4(\theta_1, \theta_2, \theta_3, \varphi_1, \varphi_2)\mathrm{d}t + \sigma_{41}(\theta_1, \theta_2, \theta_3, \varphi_1, \varphi_2)\mathrm{d}W(t),$$

$$\mathrm{d}\varphi_2 = m_5(\theta_1, \theta_2, \theta_3, \varphi_1, \varphi_2)\mathrm{d}t + \sigma_{51}(\theta_1, \theta_2, \theta_3, \varphi_1, \varphi_2)\mathrm{d}W(t),$$

$$\mathrm{d}P = Pm_6(\theta_1, \theta_2, \theta_3, \varphi_1, \varphi_2)\mathrm{d}t + P\sigma_{61}(\theta_1, \theta_2, \theta_3, \varphi_1, \varphi_2)\mathrm{d}W(t), \tag{15.48}$$

and $W(t)$ presents the standard Weiner process. Applying a linear stochastic transformation given by Wedig [23]

$$S = T(\theta_1, \theta_2, \theta_3, \varphi_1, \varphi_2)P,\quad P = \frac{S}{T(\theta_1, \theta_2, \theta_3, \varphi_1, \varphi_2)}, \tag{15.49}$$

introducing the new norm process S, by means of the scalar function $T(\theta_1, \theta_2, \theta_3, \varphi_1, \varphi_2)$, which is defined in the stationary phase process φ_j, the Itô equation for the new p=th norm process S can be obtained from Itô‘s Lemma

$$\mathrm{d}S = P\left[\frac{1}{2}\left(\sigma_{11}^2\frac{\partial^2 T}{\partial\theta_1^2} + \sigma_{21}^2\frac{\partial^2 T}{\partial\theta_2^2} + \sigma_{31}^2\frac{\partial^2 T}{\partial\theta_3^2} + \sigma_{41}^2\frac{\partial^2 T}{\partial\varphi_1^2} + \sigma_{51}^2\frac{\partial^2 T}{\partial\varphi_2^2}\right)\right.$$

$$+ \sigma_{11}\sigma_{21}\frac{\partial^2 T}{\partial\theta_1\partial\theta_2} + \sigma_{11}\sigma_{31}\frac{\partial^2 T}{\partial\theta_1\partial\theta_3} + \sigma_{11}\sigma_{41}\frac{\partial^2 T}{\partial\theta_1\partial\varphi_1} + \sigma_{31}\sigma_{51}\frac{\partial^2 T}{\partial\theta_3\partial\varphi_2}$$

$$+ \sigma_{41}\sigma_{51}\frac{\partial^2 T}{\partial\varphi_1\partial\varphi_2} + m_1\frac{\partial T}{\partial\theta_1} + m_2\frac{\partial T}{\partial\theta_2} + m_3\frac{\partial T}{\partial\theta_3} + m_4\frac{\partial T}{\partial\varphi_1} + m_5\frac{\partial T}{\partial\varphi_2}$$

$$\left. + m_6 T + \sigma_{11}\sigma_{61}\frac{\partial T}{\partial\theta_1} + \sigma_{21}\sigma_{61}\frac{\partial T}{\partial\theta_2} + \sigma_{31}\sigma_{61}\frac{\partial T}{\partial\theta_3} + \sigma_{41}\sigma_{61}\frac{\partial T}{\partial\varphi_1} + \sigma_{51}\sigma_{61}\frac{\partial T}{\partial\varphi_2}\right] \mathrm{d}T$$

$$+ P\left(\sigma_{11}\frac{\partial T}{\partial\theta_1} + \sigma_{21}\frac{\partial T}{\partial\theta_2} + \sigma_{31}\frac{\partial T}{\partial\theta_3} + \sigma_{41}\frac{\partial T}{\partial\varphi_1} + \sigma_{51}\frac{\partial T}{\partial\varphi_2} + \sigma_{61}T\right)\mathrm{d}W(t). \tag{15.50}$$

In the case when the transformation function $T(\varphi_j)$ is bounded and non-singular, both processes P and S possess the same stability. Therefore, the transformation function $T(\varphi_j)$ is selected so that the drift term, of the Itô differential Eq. (15.50), does not depend on the phase processes φ_j, θ_i, so that

$$\mathrm{d}S = \Lambda(p)S\mathrm{d}t + \frac{S}{T}\left(\sigma_{11}\frac{\partial T}{\partial\theta_1} + \sigma_{21}\frac{\partial T}{\partial\theta_2} + \sigma_{31}\frac{\partial T}{\partial\theta_3} + \sigma_{41}\frac{\partial T}{\partial\varphi_1} + \sigma_{51}\frac{\partial T}{\partial\varphi_2} + \sigma_{61}T\right)\mathrm{d}W(t). \tag{15.51}$$

Comparing Eqs. (15.50) and (15.51), the transformation function $T(\theta_1, \theta_2, \theta_3, \varphi_1, \varphi_2)$ is given by the following equation

$$[L_0 + \varepsilon L_1]T(\theta_1, \theta_2, \theta_3, \varphi_1, \varphi_2) = \Lambda(p)T(\theta_1, \theta_2, \theta_3, \varphi_1, \varphi_2). \tag{15.52}$$

Where L_0 and L_1 are the differential operators in the forms

$$L_0 = \omega_1\frac{\partial}{\partial\theta_1} + \omega_2\frac{\partial}{\partial\theta_2} + \omega_3\frac{\partial}{\partial\theta_3},$$

$$L_1 = a_1\frac{\partial^2}{\partial\theta_1^2} + a_2\frac{\partial^2}{\partial\theta_2^2} + a_3\frac{\partial^2}{\partial\theta_3^2} + a_4\frac{\partial^2}{\partial\varphi_1^2} + a_5\frac{\partial^2}{\partial\varphi_2^2}$$

$$+ a_6\frac{\partial^2}{\partial\theta_1\partial\theta_2} + a_7\frac{\partial^2}{\partial\theta_1\partial\theta_3} + a_8\frac{\partial^2}{\partial\theta_1\partial\varphi_1} + a_9\frac{\partial^2}{\partial\theta_1\partial\varphi_2} + a_{10}\frac{\partial^2}{\partial\theta_2\partial\theta_3}$$

$$+ a_{11}\frac{\partial^2}{\partial\theta_2\partial\varphi_1} + a_{12}\frac{\partial^2}{\partial\theta_2\partial\varphi_2} + a_{13}\frac{\partial^2}{\partial\theta_3\partial\varphi_1} + a_{14}\frac{\partial^2}{\partial\theta_3\partial\varphi_2} + a_{15}\frac{\partial^2}{\partial\varphi_1\partial\varphi_2}$$

$$+ b_1\frac{\partial}{\partial\theta_1} + b_2\frac{\partial}{\partial\theta_2} + b_3\frac{\partial}{\partial\theta_3} + b_4\frac{\partial}{\partial\varphi_1} + b_5\frac{\partial}{\partial\varphi_2} + c, \tag{15.53}$$

with functional coefficients $a_{1\ldots15}$, $b_{1\ldots5}$ and $c = c(\theta_1, \theta_2, \theta_3, \varphi_1, \varphi_2)$.

15.4 Averaged equations in the general case of a three-degree-of-freedom-stochastic-system

The general case of stochastic stability of a three-degree-of-freedom stochastic system can be rendered as a non-gyroscopic system. In contrast to some specific models (for example, the viscoelastic rotating shaft [24]) the procedure shown here is needed to obtain the averaged equations that are related to the general three-degree-of-freedom stochastic system. In a case like that, after discretization, the excited stochastic system with three degrees of freedom can be written in the following form

$$\ddot{q}_i(t) + 2\varepsilon \sum_{j=1}^{3} \beta_{ij}\dot{q}_j(t) + \omega_i^2 q_i(t) + \varepsilon^{\frac{1}{2}}\xi(t)\omega_i \sum_{j=1}^{3} k_{ij}q_j(t) = 0, \quad i = 1,\ 2,\ 3, \tag{15.54}$$

where excitation $\xi(t)$ is considered to be an ergodic stochastic process with zero mean value. As is known, when parametric excitation is a deterministic function of time, i.e., $\xi(t) = \mu \cos \nu t$, subharmonic resonance occurs when $\nu = 2\omega_i$, in which only the particular mode is excited into motion. Combination resonance occurs when

$$\nu = \begin{cases} \omega_i + \omega_j, & i \neq j, \text{ if } k_{ij}k_{ji} > 0, \\ |\omega_i - \omega_j|, & i \neq j, \text{ if } k_{ij}k_{ji} < 0, \end{cases} \tag{15.55}$$

in which a pair of modes are excited into motion while the remaining modes are at rest. In the general Eq. (15.54), the generalized coordinates $q_i(t)$ and the velocities $\dot{q}_i(t)$ are transformed into polar coordinates, for $i = 1, 2, 3$,

$$q_i(t) = a_i(t)\cos\Theta_i(t),\ \dot{q}_i(t) = -a_i(t)\omega_i \sin\Theta_i(t),\ \Theta_i(t) = \omega_i t + \theta_i(t). \tag{15.56}$$

Differentiating q_i with respect to t and comparing with the $\dot{q}_i(t)$ equation yields

$$\dot{a}_i \cos\Theta_i - a_i\dot{\theta}_i \sin\Theta_i = 0. \tag{15.57a}$$

Substituting Eq. (15.56) into (15.54) leads to

$$\dot{a}_i\omega_i \sin\Theta_i + a_i\dot{\theta}_i\omega_i \cos\Theta_i = F_i(a, \Theta), \tag{15.57b}$$

where

$$F_i(a, \Theta) = -2\varepsilon \sum_{j=1}^{n} \beta_{ij}a_j\omega_j \sin\Theta_j + \sqrt{\varepsilon}\xi(t)\omega_i \sum_{j=1}^{n} k_{ij}a_j \cos\Theta_j.$$

Solving for a_i and θ_i between Eq. (15.57) results in $2n$ first-order ODEs, for $i = 1, 2, 3$,

$$\dot{a}_i = \varepsilon^{1/2}F_{a,i}^{(0)} + \varepsilon F_{a,i}^{(1)},\quad \dot{\theta}_i = \varepsilon^{1/2}F_{\theta,i}^{(0)} + \varepsilon F_{\theta,i}^{(1)}, \tag{15.58}$$

where

$$F_{a,i}^{(0)} = \xi(t) \sum_{j=1}^{n} k_{ij}a_j \sin\Theta_i \cos\Theta_j,\quad F_{a,i}^{(1)} = -\frac{2}{\omega_i} \sum_{j=1}^{n} \beta_{ij}a_j\omega_j \sin\Theta_i \sin\Theta_j,$$

$$F_{\theta,i}^{(0)} = \xi(t)\frac{1}{a_i} \sum_{j=1}^{n} k_{ij}a_j \cos\Theta_i \cos\Theta_j,\quad F_{\theta,i}^{(1)} = -\frac{2}{a_i\omega_i} \sum_{j=1}^{n} \beta_{ij}a_j\omega_j \cos\Theta_i \sin\Theta_j.$$

For small values of ε, the relaxation times of the $a_i(t)$ and $\theta_i(t)$ processes are much larger than the correlation time of the excitation process $\xi(t)$. The stochastic averaging method may be applied to any non-gyroscopic system of equations to produce the Itô equations for the averaged amplitudes $\bar{a}_i(t)$ and phase angles $\bar{\theta}_i(t)$, whose solutions provide a valid first-order approximation to the exact values. It can be proved that both the averaged amplitude and the phase angle equations do not entail the phase angles and the amplitude equations that are decoupled from the phase angle equations. Therefore, the averaged equations can be written as

$$\mathrm{d}a_i = \varepsilon m_i^a \mathrm{d}t + \sqrt{\varepsilon}\sum_{j=1}^{3} \sigma_{ij}^a \mathrm{d}W_j^a, \tag{15.59a}$$

$$\mathrm{d}\theta_i = \varepsilon m_i^\theta \mathrm{d}t + \sqrt{\varepsilon}\sum_{j=1}^{3} \sigma_{ij}^\theta \mathrm{d}W_j^\theta, \tag{15.59b}$$

where $\mathbf{W}^a = \{W_1^a,\ W_2^a,\ \ldots,\ W_n^a\}^{\mathrm{T}}$, $\mathbf{W}^\theta = \{W_1^\theta,\ W_2^\theta,\ \ldots,\ W_n^\theta\}^{\mathrm{T}}$ are *three*-dimensional vectors of the independent standard Wiener processes. The drift coefficients εm_i^a, εm_i^θ, $i = 1,\ 2,\ 3,$ and the 3×3 diffusion matrices $\varepsilon b^a = \varepsilon\sigma^a(\sigma^a)^T$, $\varepsilon b^\theta = \varepsilon\sigma^\theta(\sigma^\theta)^T$, in which $\sigma^a = [\sigma_{ij}^a]$, $\sigma^\theta = [\sigma_{ij}^a]$, are given by

$$m_i^a = \underset{t}{\mathcal{M}}\left\{F_{a,i}^{(1)} + \int_{-\infty}^{0} \mathrm{E}\left[\sum_{m=1}^{3}\left(\frac{\partial F_{a,i}^{(0)}}{\partial a_m}F_{a,m\tau}^{(0)} + \frac{\partial F_{a,i}^{(0)}}{\partial \theta_m}F_{\theta,m\tau}^{(0)}\right)\right]\mathrm{d}\tau\right\}$$

$$= \left[-\beta_{ii} + \frac{3}{16}k_{ii}^2 S(a\omega_i) + \frac{1}{8}\sum_{j=1}^{3} k_{ij}k_{ji}S_{ij}^-\right]a_i + \frac{1}{16}\sum_{\substack{j=1\\ j\neq 1}}^{3} k_{ij}^2 S_{ij}^+ \frac{a_j^2}{a_i},$$

$$m_i^\theta = \underset{t}{\mathcal{M}}\left\{F_{\theta,i}^{(1)} + \int_{-\infty}^{0} \mathrm{E}\left[\sum_{m=1}^{3}\left(\frac{\partial F_{\theta,i}^{(0)}}{\partial a_m}F_{a,m\tau}^{(0)} + \frac{\partial F_{\theta,i}^{(0)}}{\partial \theta_m}F_{\theta,m\tau}^{(0)}\right)\right]\mathrm{d}\tau\right\} = -\frac{1}{8}\sum_{j=1}^{3} k_{ij}k_{ji}\psi_{ij}^-,$$

and

$$b_{ii}^a = \underset{t}{\mathcal{M}}\left\{\int_{-\infty}^{+\infty} \mathrm{E}[F_{a,i}^{(0)}F_{a,i\tau}^{(0)}]\mathrm{d}\tau\right\} = \frac{1}{8}k_{ii}^2 S(2\omega_i)a_i^2 + \frac{1}{8}\sum_{\substack{j=1\\ j\neq 1}}^{n} k_{ij}^2 S_{ij}^+ a_j^2,$$

$$b_{ij}^a = \underset{t}{\mathcal{M}}\left\{\int_{-\infty}^{+\infty} \mathrm{E}[F_{a,i}^{(0)}F_{a,j\tau}^{(0)}]\mathrm{d}\tau\right\} = \frac{1}{8}k_{ij}k_{ji}S_{ij}^- a_i a_j,\ i \neq j,$$

$$b_{ii}^\theta = \underset{t}{\mathcal{M}}\left\{\int_{-\infty}^{+\infty} \mathrm{E}[F_{\theta,i}^{(0)}F_{\theta,i\tau}^{(0)}]\mathrm{d}\tau\right\} = \frac{1}{8}k_{ii}^2[2S(0) + S(2\omega_i)] + \frac{1}{8}\sum_{\substack{j=1\\ j\neq 1}}^{3} k_{ij}^2 S_{ij}^+ \frac{a_j^2}{a_i^2},$$

$$b_{ij}^\theta = \underset{t}{\mathcal{M}}\left\{\int_{-\infty}^{+\infty} E[F_{\theta,i}^{(0)}F_{\theta,j\tau}^{(0)}]\mathrm{d}\tau\right\} = \frac{1}{4}k_{ii}k_{jj}S(0) + \frac{1}{8}k_{ij}k_{ji}S_{ij}^+,\ i \neq j.$$

In the above expressions, the quantities S^+, S^-, ψ^+, and ψ^- are defined by

$$S_{ij}^{\pm} = S(\omega_i + \omega_j) \pm S(\omega_i - \omega_j), \quad \psi_{ij}^{\pm} = \psi(\omega_i + \omega_j) \pm \psi(\omega_i - \omega_j), \tag{15.60}$$

And $S(\omega)$, $\psi(\omega)$ are the cosine and sine power spectral density functions of $\xi(t)$.

15.5 Moment Lyapunov exponents

After the Itô equations (for gyroscopic and non-gyroscopic systems) are derived, the procedure can be continued with the determination of the moment Lyapunov exponents and Lyapunov exponents. The eigenvalue problem for a differential operator of five independent variables is identified from Eq. (15.52) in which $\Lambda(p)$ is the eigenvalue and $T(\theta_1, \theta_2, \theta_3, \varphi_1, \varphi_2)$ is the associated eigenfunction. Eq. (15.52) defines that the eigenvalue $\Lambda(p)$ is the Lyapunov exponent of the p-th moment. Wedig [23] was the first to employ this method, to derive the eigenvalue problem for the moment Lyapunov exponent of a two-dimensional linear Itô stochastic system. When the method of regular perturbation is applied, the moment Lyapunov exponent $\Lambda(p)$ and the eigenfunction $T(\theta_1, \theta_2, \theta_3, \varphi_1, \varphi_2)$ are both expanded in power series of ε; the purpose of this is to obtain a weak noise expansion of the Lyapunov exponent of a system with stochastic excitation described by six-dimensional components as

$$\begin{aligned}
\Lambda(p) &= \Lambda_0(p) + \varepsilon\Lambda_1(p) + \varepsilon^2\Lambda_2(p) + \ldots + \varepsilon^n\Lambda_n(p) + \ldots + T(\theta_1, \theta_2, \theta_3, \varphi_1, \varphi_2) \\
&= T_0(\theta_1, \theta_2, \theta_3, \varphi_1, \varphi_2) + \varepsilon T_1(\theta_1, \theta_2, \theta_3, \varphi_1, \varphi_2) + \varepsilon^2 T_2(\theta_1, \theta_2, \theta_3, \varphi_1, \varphi_2) + \ldots \\
&\quad + \varepsilon^2 T_2(\theta_1, \theta_2, \theta_3, \varphi_1, \varphi_2) + \ldots + \varepsilon^n T_n(\theta_1, \theta_2, \theta_3, \varphi_1, \varphi_2) + \ldots.
\end{aligned} \tag{15.61}$$

Equating the terms of equal powers of ε from Eqs. (15.52) and (15.61) the following equations are obtained

$$\begin{aligned}
&\varepsilon^0 : \; L_0 T_0(\theta_1, \theta_2, \theta_3, \varphi_1, \varphi_2) = \Lambda_0(p) T_0(\theta_1, \theta_2, \theta_3, \varphi_1, \varphi_2), \\
&\varepsilon^1 : \; L_0 T_1(\theta_1, \theta_2, \theta_3, \varphi_1, \varphi_2) + L_1 T_0(\theta_1, \theta_2, \theta_3, \varphi_1, \varphi_2) \\
&= \Lambda_0(p) T_1(\theta_1, \theta_2, \theta_3, \varphi_1, \varphi_2) + \Lambda_1(p) T_0(\theta_1, \theta_2, \theta_3, \varphi_1, \varphi_2), \\
&\varepsilon^2 : \; L_0 T_2(\theta_1, \theta_2, \theta_3, \varphi_1, \varphi_2) + L_1 T_1(\theta_1, \theta_2, \theta_3, \varphi_1, \varphi_2) \\
&= \Lambda_0(p) T_2(\theta_1, \theta_2, \theta_3, \varphi_1, \varphi_2) + \Lambda_1(p) T_1(\theta_1, \theta_2, \theta_3, \varphi_1, \varphi_2) \\
&\quad + \Lambda_2(p) T_0(\theta_1, \theta_2, \theta_3, \varphi_1, \varphi_2), \\
&\varepsilon^3 : \; L_0 T_3(\theta_1, \theta_2, \theta_3, \varphi_1, \varphi_2) + L_1 T_2(\theta_1, \theta_2, \theta_3, \varphi_1, \varphi_2) \\
&= \Lambda_0(p) T_3(\theta_1, \theta_2, \theta_3, \varphi_1, \varphi_2) + \Lambda_1(p) T_2(\theta_1, \theta_2, \theta_3, \varphi_1, \varphi_2) \\
&\quad + \Lambda_2(p) T_1(\theta_1, \theta_2, \theta_3, \varphi_1, \varphi_2) + \Lambda_3(p) T_0(\theta_1, \theta_2, \theta_3, \varphi_1, \varphi_2),
\end{aligned}$$

…

$$\varepsilon^n: \; L_0 T_n(\theta_1, \theta_2, \theta_3, \varphi_1, \varphi_2) + L_1 T_{n-1}(\theta_1, \theta_2, \theta_3, \varphi_1, \varphi_2)$$
$$= \Lambda_0(p) T_n(\theta_1, \theta_2, \theta_3, \varphi_1, \varphi_2) + \Lambda_1(p) T_{n-1}(\theta_1, \theta_2, \theta_3, \varphi_1, \varphi_2)$$
$$+ \ldots + \Lambda_n(p) T_0(\theta_1, \theta_2, \theta_3, \varphi_1, \varphi_2), \tag{15.62}$$

where each function $T_s(\theta_1, \theta_2, \theta_3, \varphi_1, \varphi_2)$, $s = 1, 2, 3, \ldots$ must be positive and periodic in the range $0 \le \theta_i \le 2\pi$.

15.6 Zeroth order perturbation

From Eqs. (15.52) and (15.62) the *zero*-th order perturbation equation is $L_0 T_0 = \Lambda_0(p) T_0$ and can be written as

$$\omega_1 \frac{\partial T_0(\theta_1, \theta_2, \theta_3, \varphi_1, \varphi_2)}{\partial \theta_1} + \omega_2 \frac{\partial T_0(\theta_1, \theta_2, \theta_3, \varphi_1, \varphi_2)}{\partial \theta_2} + \omega_3 \frac{\partial T_0(\theta_1, \theta_2, \theta_3, \varphi_1, \varphi_2)}{\partial \theta_3}$$
$$= \Lambda_0(p) T_0(\theta_1, \theta_2, \theta_3, \varphi_1, \varphi_2) \tag{15.63}$$

Eq. (15.63) can be easily solved from the moment Lyapunov exponent characteristic, which results in $\Lambda_n(0) = 0$ for $n = 0, 1, 2, 3, \ldots$ because of $\Lambda(0) = \Lambda_0(0) + \varepsilon\Lambda_1(0) + \varepsilon^2\Lambda_2(0) + \ldots + \varepsilon^n\Lambda_n(0) = 0$. The eigenvalue $\Lambda_0(p)$ is independent of p. Hence $\Lambda_0(0) = 0$ leads to $\Lambda_0(p) = 0$. The solutions of Eq. (15.63) have a periodic solution if and only if

$$\Lambda_0(p) = 0, \quad q_0(\theta_1, \theta_2, \theta_3, \varphi_1, \varphi_2) = 1. \tag{15.64}$$

15.7 First-order perturbation

The equation obtained from the first-order perturbation is

$$L_0 T_1(\theta_1, \theta_2, \theta_3, \varphi_1, \varphi_2) + L_1 T_0(\theta_1, \theta_2, \theta_3, \varphi_1, \varphi_2)$$
$$= \Lambda_0(p) T_1(\theta_1, \theta_2, \theta_3, \varphi_1, \varphi_2) + \Lambda_1(p) T_0(\theta_1, \theta_2, \theta_3, \varphi_1, \varphi_2), \tag{15.65}$$

and Eq. (15.65) has a periodic solution if and only if

$$\int_0^{2\pi}\int_0^{2\pi}\int_0^{2\pi}\int_0^{\pi/2}\int_0^{\pi/2} [L_1 1 - \Lambda_1(p)] d\varphi_1 d\varphi_2 d\theta_1 d\theta_2 d\theta_3 = 0, \tag{15.66}$$

which yields

$$\Lambda_1(p) = \frac{1}{2\pi^5}\int_0^{2\pi}\int_0^{2\pi}\int_0^{2\pi}\int_0^{\pi/2}\int_0^{\pi/2} [c(\theta_1, \theta_2, \theta_3, \varphi_1, \varphi_2)] d\varphi_1 d\varphi_2 d\theta_1 d\theta_2 d\theta_3$$
$$= \frac{1}{1024} p\sigma^2 [f_1(s_{ijk}) + p f_2(s_{ijk})]$$
$$- \frac{p}{8}[d_{111} + d_{122} + d_{133} + d_{411} + 2(d_{422} + d_{433})]. \tag{15.67}$$

The first-order perturbation equation can be rewritten as

$$\omega_1 \frac{\partial T_1(\theta_1, \theta_2, \theta_3, \varphi_1, \varphi_2)}{\partial \theta_1} + \omega_2 \frac{\partial T_1(\theta_1, \theta_2, \theta_3, \varphi_1, \varphi_2)}{\partial \theta_2} + \omega_3 \frac{\partial T_1(\theta_1, \theta_2, \theta_3, \varphi_1, \varphi_2)}{\partial \theta_3} +$$

$$c(\theta_1, \theta_2, \theta_3, \varphi_1, \varphi_2) = \Lambda_1(p). \tag{15.68}$$

It is important to consider the commensurable frequencies, in which exists a relation of the form of $\omega_1 = \frac{\omega_2}{\mu_1} = \frac{\omega_3}{\mu_2}$. The function $c(\theta_1, \theta_2, \theta_3, \varphi_1, \varphi_2)$ in Eq. (15.68) can be written in the form

$$c(\theta_1, \theta_2, \theta_3, \varphi_1, \varphi_2)=$$

$$\Lambda_1(p) + f_0(\theta_1, \theta_2, \theta_3) + \mathcal{L}_g(\theta_1, \theta_2, \theta_3)\mathfrak{R}_g(\varphi_1, \varphi_2), \quad g = 1, 2, \ldots 35 \tag{15.69}$$

Functions $\mathcal{L}_g$ are periodic in θ_1, θ_2, and θ_3 with $\mathfrak{R}_g$ trigonometric combinations of (φ_1, φ_2). Eq. (15.68) cannot be obtained explicitly for $T_1(\theta_1, \theta_2, \theta_3, \varphi_1, \varphi_2)$. The combination of coefficients φ_1 and φ_2 in Eq. (15.69) suggests that the function $T_1(\theta_1, \theta_2, \theta_3, \varphi_1, \varphi_2)$ can be written as

$$T_1(\theta_1, \theta_2, \theta_3, \varphi_1, \varphi_2) = T_{10}(\theta_1, \theta_2, \theta_3) + T_{1g}(\theta_1, \theta_2, \theta_3)\mathfrak{R}_g(\varphi_1, \varphi_2). \tag{15.70}$$

Substituting Eq. (15.70) into Eq. (15.68), and equating terms of the equal trigonometry function gives a set of PDEs

$$\omega_1 \frac{\partial T_{1r}(\theta_1, \theta_2, \theta_3)}{\partial \theta_1} + \omega_2 \frac{\partial T_{1r}(\theta_1, \theta_2, \theta_3)}{\partial \theta_2}$$

$$+ \omega_3 \frac{\partial T_{1r}(\theta_1, \theta_2, \theta_3)}{\partial \theta_3} + f_r(\theta_1, \theta_2, \theta_3) = 0, \quad r = 0, 1, \ldots, 35. \tag{15.71}$$

The functions $T_{1r}(\theta_1, \theta_2, \theta_3)$ can be written as

$$T_{1r}(\theta_1, \theta_2, \theta_3) = \mho + C_r(\theta_2 - \mu_1\theta_1, \theta_3 - \mu_2\theta_1), \quad r = 0, 1, \ldots, 35. \tag{15.72}$$

Here $C_r(\theta_2 - \mu_1\theta_1, \theta_3 - \mu_2\theta_1)$ are the same arbitrary functions of three variables for which the following assumptions can be made

$$C_r(\theta_2 - l_1\theta_1, \theta_3 - l_2\theta_1) = A_{1r} + B_{1r}\sin(2\theta_2 - 2\mu_1\theta_1) + C_{1r}\sin(4\theta_2 - 4\mu_1\theta_1)$$

$$+ D_{1r}\sin(2\theta_3 - 2\mu_2\theta_1) + E_{1r}\sin(4\theta_3 - 4\mu_2\theta_1), \quad r = 0, 1, \ldots, 35. \tag{15.73}$$

As each function $T_{1r}(\theta_1, \theta_2, \theta_3)$, $r = 0, 1, \ldots, 35$ must be positive and periodic, unknown constants $A_{1r}, B_{1r}, C_{1r}, D_{1r}, E_{1r}$ can be determined using conditions

$$T_{1r}(0, 0, 0) = T_{1r}(0, 0, 2\pi) = T_{1r}(0, 2\pi, 0) = T_{1r}(2\pi, 0, 0)$$

$$= T_{1r}(2\pi, 2\pi, 0) = T_{1r}(0, 2\pi, 2\pi) = T_{1r}(2\pi, 0, 2\pi) = T_{1r}(2\pi, 2\pi, 2\pi) = 0, \tag{15.74}$$

$$\frac{\partial T_{1r}(0, 0, 0)}{\partial \theta_1} = \frac{\partial T_{1r}(2\pi, 0, 0)}{\partial \theta_1}, \quad \frac{\partial T_{1r}(0, 0, 0)}{\partial \theta_2} = \frac{\partial T_{1r}(0, 2\pi, 0)}{\partial \theta_2},$$

$$\frac{\partial T_{1r}(0,\ 0,\ 0)}{\partial \theta_3} = \frac{\partial T_{1r}(0,\ 0,\ 2\pi)}{\partial \theta_3},\quad r = 0,\ 1,\ \ldots,\ 35. \tag{15.75}$$

15.8 Second-order perturbation

The second-order perturbation equation must satisfy the condition of the periodic function $T_2(\theta_1, \theta_2, \theta_3, \varphi_1, \varphi_2)$ in θ_1, θ_2, and θ_3 given as

$$L_0 T_2(\theta_1, \theta_2, \theta_3, \varphi_1, \varphi_2) + L_1 T_2(\theta_1, \theta_2, \theta_3, \varphi_1, \varphi_2)$$
$$= \Lambda_1(p) T_1(\theta_1, \theta_2, \theta_3, \varphi_1, \varphi_2) + \Lambda_2(p). \tag{15.76}$$

Which gives

$$\Lambda_2(p) = \frac{1}{2\pi^5} \int_0^{2\pi} \int_0^{2\pi} \int_0^{2\pi} \int_0^{\frac{\pi}{2}} \int_0^{\frac{\pi}{2}} \left\{ a_1 \frac{\partial^2 T_1}{\partial \theta_1^2} + a_2 \frac{\partial^2 T_1}{\partial \theta_2^2} + a_3 \frac{\partial^2 T_1}{\partial \theta_3^2} + a_4 \frac{\partial^2 T_1}{\partial \varphi_1^2} \right.$$
$$+ a_5 \frac{\partial^2 T_1}{\partial \varphi_2^2} + a_6 \frac{\partial^2 T_1}{\partial \theta_1 \partial \theta_2} + a_7 \frac{\partial^2 T_1}{\partial \theta_1 \partial \theta_3} + a_8 \frac{\partial^2 T_1}{\partial \theta_1 \partial \varphi_1} + a_9 \frac{\partial^2 T_1}{\partial \theta_1 \partial \varphi_2}$$
$$+ a_{10} \frac{\partial^2 T_1}{\partial \theta_2 \partial \theta_3} + a_{11} \frac{\partial^2 T_1}{\partial \theta_2 \partial \varphi_1} + a_{12} \frac{\partial^2 T_1}{\partial \theta_2 \partial \varphi_2} + a_{13} \frac{\partial^2 T_1}{\partial \theta_3 \partial \varphi_1}$$
$$+ a_{14} \frac{\partial^2 T_1}{\partial \theta_3 \partial \varphi_2} + a_{15} \frac{\partial^2 T_1}{\partial \varphi_1 \partial \varphi_2} + b_1 \frac{\partial T_1}{\partial \theta_1} + b_2 \frac{\partial T_1}{\partial \theta_2} b_3 \frac{\partial T_1}{\partial \theta_3} + b_4 \frac{\partial T_1}{\partial \varphi_1}$$
$$\left. + b_5 \frac{\partial T_1(\theta_1, \theta_2, \theta_3, \varphi_1, \varphi_2)}{\partial \varphi_2} + [c - \Lambda_1(p)] T_1 \right\} d\varphi_1 d\varphi_2 d\theta_1 d\theta_2 d\theta_3. \tag{15.77}$$

The weak noise expansion of the moment Lyapunov exponent in the second-order perturbation is determined in the form

$$\Lambda(p) = \varepsilon \Lambda_1(p) + \varepsilon^2 \Lambda_2(p) + O(\varepsilon^3). \tag{15.78}$$

The Lyapunov exponent can now be obtained from Eq. (15.78) by using a property of the moment Lyapunov exponent

$$\lambda = \left. \frac{d\Lambda(p)}{dp} \right|_{p=0} = \varepsilon \lambda_1 + \varepsilon^2 \lambda_2 + O(\varepsilon^3)$$

$$= \varepsilon \left\{ \left[\frac{1}{1024} \Psi \right] \sigma^2 - \frac{p}{8} [d_{111} + d_{122} + d_{133} + d_{411} + 2(d_{422} + d_{433})] \right\} + \varepsilon^2 \mathcal{F} + O(\varepsilon^3), \tag{15.79}$$

where Ψ is given in [21].

15.9 Stochastic stability conditions and discussion

The applied values for the parameters of the stochastic system (in the Moment Lyapunov exponent and Lyapunov exponent calculations) are given as follows for a graphite/epoxy regular cross-ply laminated plate, glass/epoxy cross-ply laminated plate, thin isolators, and stiffness of the K_0 layer, respectively

$$
\begin{gathered}
F = E_2/E_1 = 1/25, \quad G_{12}/E_2 = G_{13}/E_2 = 0.5, \quad G_{23}/E_2 = 0.2, \\
\nu_{12} = 0.25, \quad \rho = 1.55 \times 10^3\,\text{kgm}^{-3}, \quad k_1 = 0.7031, \quad k_2 = 0.8676, \\
F = E_2/E_1 = 1/4.6, \quad G_{12}/E_2 = G_{23}/E_2 = 0.47, \quad G_{13}/E_2 = 0.4, \\
\nu_{12} = 0.31, \quad \rho = 1.90 \times 10^3\,\text{kgm}^{-3}, \quad k_1 = 5/6, \quad k_2 = 2/3, \\
\rho_0 = 2 \times 10^3\,\text{kgm}^{-3}, \quad E = 1 \times 10^{10}\,\text{Pa}, \quad \nu = 0.3, \\
K = 2 \times 10^3\,\text{kgm}^{-3}, \quad K_0 = \eta K.
\end{gathered}
\tag{15.80}
$$

When selecting these two sets of parameters for different composite materials, it is important to highlight their distinguishing differences. Graphite/epoxy composites are known for their superior strength and stiffness compared to glass/epoxy composites. This makes them the material of choice for high-performance scenarios that demand maximum structural integrity. One of the key advantages of graphite/epoxy composites is their lighter weight, stemming from the lower density of graphite fibers. This attribute is particularly valuable in industries such as aerospace and high-end sports manufacturing, where minimizing weight is crucial. When it comes to cost, glass/epoxy composites are typically more economical than their graphite/epoxy counterparts. For projects with tight budget constraints, glass/epoxy may be the more viable option, despite offering lower performance in some aspects. Graphite/epoxy composites excel in thermal stability and electrical conductivity, making them suitable for environments with high thermal demands or where electrical conductivity is beneficial. Conversely, glass/epoxy composites are excellent electrical insulators, which can be advantageous in applications that require robust electrical insulation. In summary, both graphite/epoxy and glass/epoxy composites present distinct benefits tailored to different uses based on their physical properties. The choice between them should be guided by specific performance needs, weight requirements, budgetary limitations, and the necessary thermal and electrical characteristics for the application at hand.

An analytical solution is extensive and some of the determined constants used in numerical examples are hefty so the authors decided to provide them on request. The stiffness moduli of the laminate are described in [21]. The p-th moment stability boundary in the first-order perturbation for $p = 1$ is determined analytically; moreover, using the obtained analytical form, this can be calculated for any numerical value of p with the definition of the moment stability $\Lambda(p) < 0$. The $\Lambda_1(1)$ is obtained from the results for the moment Lyapunov exponent from the first-order perturbation; $\Lambda(p) = \varepsilon\Lambda_1(p) + O(\varepsilon^2)$. The oscillatory stochastic system is asymptotically stable only if the Lyapunov exponent is $\lambda < 0$; this is used for determining the almost-sure stability boundary of the oscillatory system in the first-order perturbation. The condition $\lambda < 0$ is obtained from the expression $\lambda = \varepsilon\lambda_1 + O(\varepsilon^2)$. Variations of the moment Lyapunov exponent are presented in Figs. 15.2–15.7. The crux of the reported results is that the stochastic stability possesses a growth in the stable region with the rise in the stiffness of the K_0 layer (Fig. 15.2), rise in damping (Fig. 15.3), rise in the thickness of the isolators (Fig. 15.4), and rise in the number of layers of the laminated plate (Fig. 15.5).

The influence of these qualitative characters, of some parameters, on dynamic stability could be known even without the analysis. However, without these results,

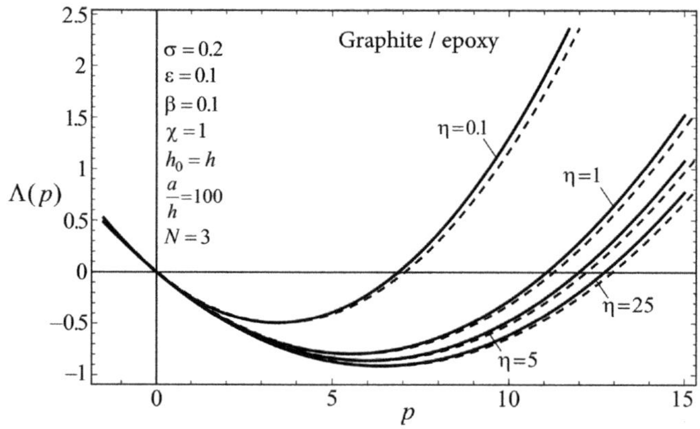

Figure 15.2 Moment Lyapunov exponent $\Lambda(p)$. The variable is the stiffness modulus of the K_0 layer (graphite/epoxy).

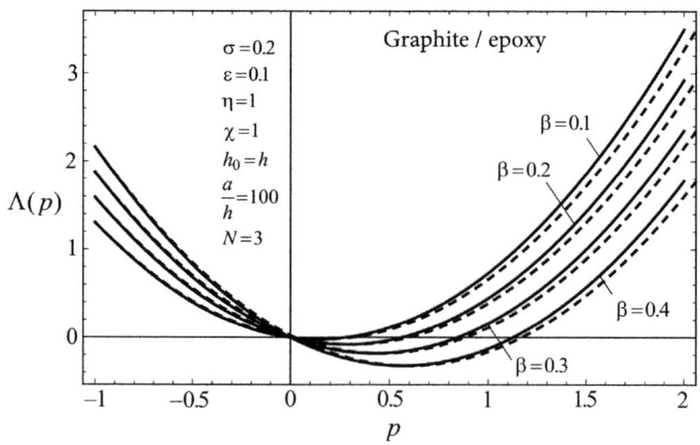

Figure 15.3 Moment Lyapunov exponent $\Lambda(p)$. The variable is the damping coefficient (graphite/epoxy).

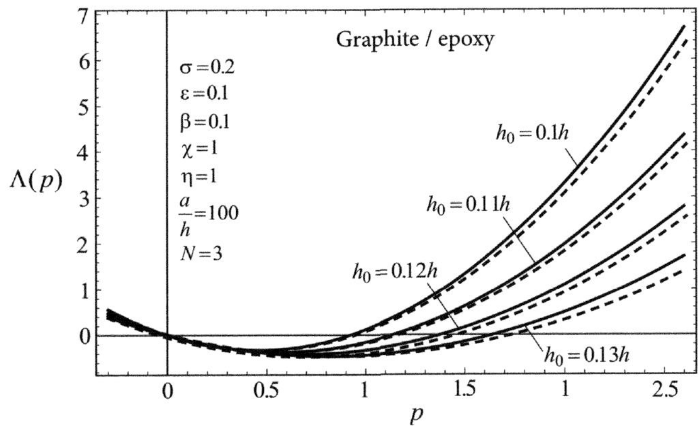

Figure 15.4 Moment Lyapunov exponent $\Lambda(p)$. The variable is the thickness of the core-plates (graphite/epoxy).

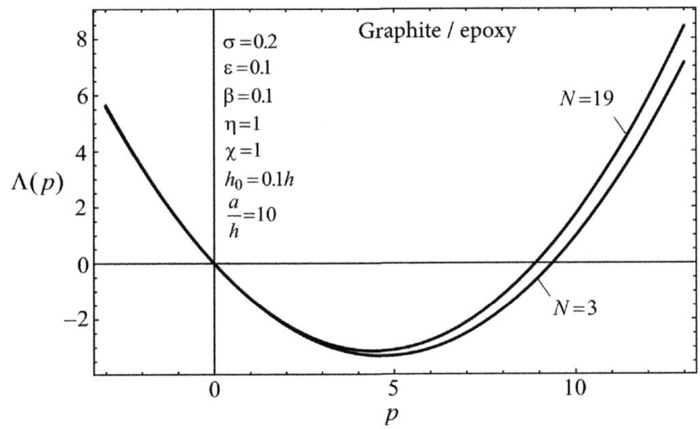

Figure 15.5 Moment Lyapunov exponent (Λp). The variable is the number of layers in LP (graphite/epoxy).

what would be unknown would be the change in the dynamic stability conditions based on the quantitative change of a chosen parameter. Due to this, it is possible to analyze the dynamic stability effects of the proposed isolated structure while gratifying some specific engineering conditions; a decrease in the weight of the structure, an increase in its volume, or an improvement in the shielding of the laminated plate. Figs. 15.6 and 15.7 depict the influences of stiffness and damping on the dynamic stability of the glass/epoxy laminated plate. The effect of the thickness of the isolators and the effect of the number of layers of the glass/epoxy laminated plate are presented in Figs. 15.8 and 15.9. All figures are displayed for the normalized values of p and $\Lambda(p)$. A simple single laminated plate has a substantially large difference between stable regions; for that reason, its dynamic stability analysis is not shown here. The proposed complex isolated plate model is always considerably more

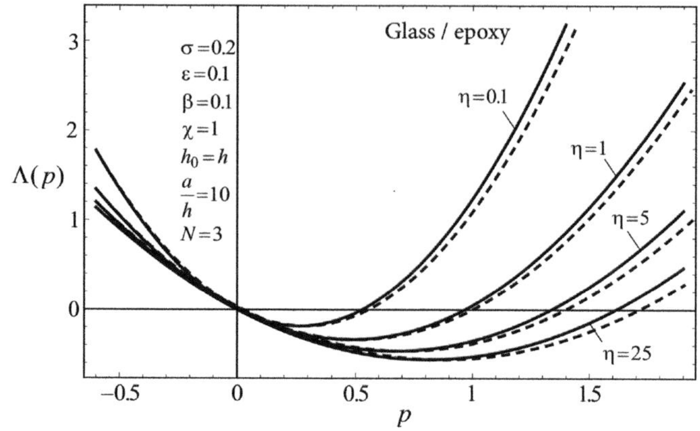

Figure 15.6 Moment Lyapunov exponent $\Lambda(p)$. The variable is the stiffness modulus of the K_0 layer (glass/epoxy).

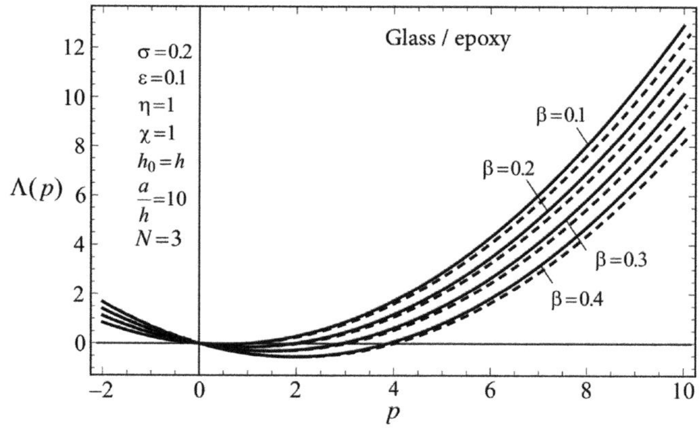

Figure 15.7 Moment Lyapunov exponent $\Lambda(p)$. The variable is the damping coefficient (glass/epoxy).

stable. It is important to note that this kind of structure naturally incorporates more space for fluctuating influences, which should result in the fulfillment of specific required conditions. This means that the proposed type of structure opens a new space for research that could be applied in engineering; concerning the fulfillment of required dynamic behavior or ensuring the existence of stable regions under more intensive stochastic disturbances. It is also important to mention that the simpler case of coupled one laminated and one classical plate is not considered in stochastic stability, so the application of this kind of structure could be found in technical practice, and, a simple reduction of one generalized coordinate from the present study will directly give the stability scenario and results of the moment Lyapunov exponents of the simpler model.

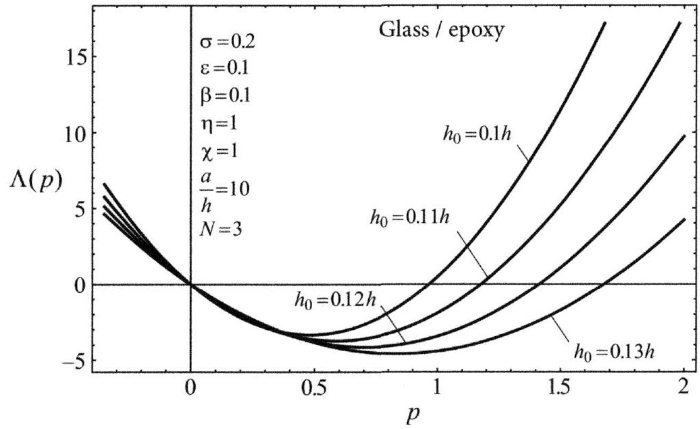

Figure 15.8 Moment Lyapunov exponent *Λ(p)*. The variable is the thickness of the core-plates (glass/epoxy).

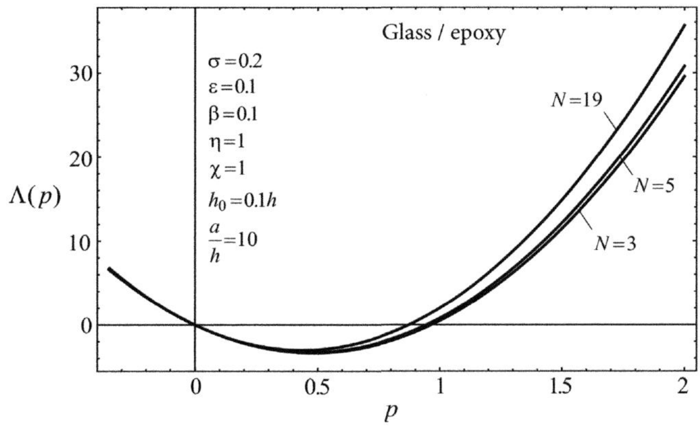

Figure 15.9 Moment Lyapunov exponent *Λ(p)*. The variable is the number of layers in LP (glass/epoxy).

15.10 Conclusions

The discussion in this chapter was about the dynamic instability of a complexly coupled structure that in the middle has a cross-ply symmetrically laminated plate, which is viscoelastically coupled to two thin outer plates. The effect of the transverse shear on the laminated plate is considered. Positive effects on dynamic stability are elaborated. The stochastic dynamic stability is determined analytically, for the case when the system is under white noise excitations; that is achieved by determining the moment Lyapunov exponents and the Lyapunov exponents. An eigenvalue problem for the moment Lyapunov exponent is established using the theory of stochastic dynamic systems. The second-order expansions of the moment Lyapunov exponent (when weak noise excitations are at play) are obtained by employing a singular perturbation method. Subsequently, the Lyapunov exponent has been obtained via the correlation between the moment Lyapunov exponent and the Lyapunov exponent. The almost-sure and p-th moment stability boundaries are determined in correlation to the damping coefficients. The comparison among the stochastic stabilities (of the proposed model with varying parameters) reveals the expansion of the stable regions for various changes in the values of the parameter given with; a) an increase in the stiffness of the elastic layers that are connecting the central laminated plate with the outer plate-isolators; b) an increase in the damping coefficient; c) an increase in the thickness of the plate-isolators; d) an increase in the number of layers in the laminated plate. The system is analyzed scrupulously from the analytical side; the findings are significant for the overall stability of complex structures. Additionally, the study contains a short procedure for the stochastic analysis of non-gyroscopic coupled systems (which could be applied to the gyroscopic systems) if any kind of coupled stochastic oscillator has non-gyroscopic couplings.

References

[1] S.G. Kelly, Advanced Vibration Analaysis, CRC Press, New York, 2006.

[2] V. Stojanović, P. Kozić, Vibrations and Stability of Complex Beam Systems, ISBN 978-3-319-13766-7, Springer Tracts in Mechanical Engineering, Springer, 2015.

[3] S.T. Ariaratnam, W.C. Xie, Effect of correlation on the almost-sure asymptotic stability of second-order linear stochastic systems. ASME, J Appl Mech 56 (8) (1989) 685–690.

[4] W.C. Xie, Dynamic Stability of Structures, Cambridge University Press, Cambridge, 2006.

[5] A.J.M. Ferreira, A.L. Araújo, A.M.A. Neves, J.D. Rodrigues, E. Carrera, M. Cinefra, et al., A finite element model using a unified formulation for the analysis of viscoelastic sandwich laminates, Composites Part B Engineering 45 (2013) 1258–1264.

[6] W. He, W. Chen, H. Qiao, Two-scale analytical solutions of multilayered composite rectangular plates with in-plane small periodic structure, European Journal of Mechanics—A/Solids 40 (2013) 123–130.

[7] D.K. Biswal, S.C. Mohanty, Free vibration and damping characteristics study of doubly curved sandwich shell panels with viscoelastic core and isotropic/laminated constraining layer, European Journal of Mechanics—A/Solids 72 (2018) 424–439.

[8] N.S. Namachchivaya, H.J. Van Roessel, Maximal Lyapunov exponent and rotation numbers for two coupled oscillators driven by real noise, Journal of Statistical Physics 71 (1993) 549–567.

[9] L. Arnold, M.M. Doyle, N.S. Namachchivaya, Small noise expansion of moment Lyapunov exponents for two-dimensional systems, Dynamics and Stability of Systems 12 (1997) 187–211.

[10] N.S. Namachchivaya, H.J. Van Roessel, Moment Lyapunov exponent and stochastic stability of two coupled oscillators driven by real noise. ASME, J Appl Mech 68 (2001) 903–914.

[11] P. Kozić, G. Janevski, R. Pavlović, Moment Lyapunov exponents and stochastic stability of a double-beam system under compressive axial loading, International Journal of Solids and Structures 47 (2010) 1435–1442.

[12] A. Tylikowski, R. Pavlović, P. Kozić, Influence of transverse shear on stochastic instability of symmetric cross-ply laminated plates, Probabilistic Engineering Mechanics 26 (2011) 454–460.
[13] A. Tylikowski, Dynamic Stability of Nonlinear Antisymmetrically-Laminated Cross-Ply Rectangular Plates, Journal of Applied Mechanics 56 (1989) 375–381.
[14] A. Tylikowski, K. Frischmuth, Stability and stabilization of circular plate parametric vibrations, International Journal of Solids and Structures 40 (2003) 5187–5196.
[15] S. Dey, T. Mukhopadhyay, S.K. Sahu, S. Adhikari, Stochastic dynamic stability analysis of composite curved panels subjected to non-uniform partial edge loading, European Journal of Mechanics—A/Solids 67 (2018) 108–122.
[16] C.S. Chen, W.R. Chen, R.D. Chien, Stability of parametric vibrations of hybrid composite plates, European Journal of Mechanics—A/Solids 28 (2009) 329–337.
[17] V.V. Bolotin, The Dynamic Stability of Elastic Systems, Holden-Day, San Francisco, 1964.
[18] V. Stojanović, P. Kozić, Stochastic stability of a thick beams using contact transformation method, Probabilistic Engineering Mechanics 34 (2013) 110–113.
[19] V. Stojanović, M. Petković, Moment Lyapunov Exponents and Stochastic Stability of a Three-Dimensional System on Elastic Foundation Using a Perturbation Approach, Journal of Applied Mechanics 80 (2013) 051009.
[20] J. Deng, W.C. Xie, M.D. Pandey, Moment Lyapunov exponent and stochastic stability of coupled viscoelastic systems under white noise excitation, J Mech Mater Struct 9 (1) (2014) 27–50.
[21] V. Stojanović, M. Petković, J. Deng, Stability of parametric vibrations of an isolated symmetric cross-ply laminated plate, Composites Part B 167 (2019) 631–642.
[22] P. Pai, A new look at shear correction factors and warping functions of anisotropic laminates, International Journal of Solids and Structures 32 (6) (1995) 2295–2313.
[23] W. Wedig, Lyapunov exponent of stochastic systems and related bifurcation problem, in: S.T. Ariaratnam, G.I. Schuëller, I. Elishakoff (Eds.), Stochastic Structural Dynamics - Progress in Theory and Applications, Elsevier Applied Science, London, 1998, pp. 315–327.
[24] J. Deng, Dynamic stability of a viscoelastic rotating shaft under parametric random excitation, International Journal of Non-Linear Mechanics 84 (2016) 56–67.

Higher-order stochastic averaging method in fractional stochastic dynamics

16

16.1 Introduction

A mechanical behavior that combines liquid-like and solid-like characteristics is called viscoelasticity. Manifestation of viscoelasticity is present in many materials, like polymers, metals at elevated temperatures, concrete, and rocks; these materials have been amply used in engineering [1]. The majority of theories on viscoelasticity are dedicated to traditional binary viscoelastic models that permit only two states for their components: pure elasticity and pure viscosity. In them, pure elasticity is represented via Hoek's spring, while pure viscosity is represented via Newton's dashpot. Several conventional mechanical models for viscoelasticity were contemplated. For example, the Maxwell model is characterized by a dash-pot and a spring connected in series, and the Kelvin–Voigt model comprises a dash-pot and a spring in parallel.

However, real-world materials often comprise multistate components with various functional levels between elasticity and viscosity; leading to diverse effects on the performance of the entire system. The theory of fractional calculus (Appendix A1) is used to define a component with the characteristics between spring and dashpot (Appendix A2). Application of fractional calculus in viscoelasticity [2] and stability analysis of fractional-order systems is available in theoretical and practical studies [3]. Kelvin-Voigt constitutive relation is represented by a purely viscous damper and a purely elastic spring connected in parallel. The Kelvin-Voigt model exhibits exponential strain creep due to integer-order damping. However, there is a theoretical basis for using fractional calculus in viscoelasticity. In this research, a fractional Kelvin-Voigt constitutive relation is chosen to better capture the complex viscoelastic behavior of materials. Loadings from wind, earthquakes, ocean waves, and explosions can be described satisfactorily in a probabilistic manner. Hence, the equation of motion of the viscoelastic system under such excitations is usually governed by the stochastic differential equation (SDE); getting the exact response and stability of the system is problematic. The method of stochastic averaging has been frequently used to solve approximately SDEs that contain small parameters. Under certain conditions, stochastic averaging can reduce the dimension of problems and then considerably simplify the solution [4]. Several of the papers [5,6] are a testimony to the popularity of stochastic averaging.

The basic concept of the stochastic averaging method is to approximate the original Stratonovich stochastic system by an averaged Itô stochastic system (diffusive Markov process). The Itô system is presumably simpler for analysis; the properties of the dynamics of the original system can be deduced from the dynamics of the

Modeling of Complex Dynamic Systems. DOI: https://doi.org/10.1016/B978-0-443-23942-7.00016-4

averaged system. Ordinary stochastic averaging is a first-order approximation method [4]. Using a first-order stochastic averaging method, Deng et al. [7] analyzed the stochastic stability of a fractional viscoelastic column under bounded noise excitation. A second-order stochastic averaging based on the Stratonovich–Khasminskii limit theorem in a systematic and unified approach can be found in Ref. [8]. Huang and Xie found that the second-order averaging results are in better concurrence with the results estimated by Monte Carlo simulation than the first-order averaging [9]. Using the first-order and the second-order method of stochastic averaging, the moment Lyapunov exponents of the Hill equation were studied in Ref. [10].

For engineering applications, stochastic loadings have been modeled as Gaussian white noise processes, real noise processes, or bounded noise processes. The power spectral density of a white noise process is constant over the entire frequency range, which is obviously an idealization. Real noise is often characterized by an Ornstein–Uhlenbeck process, which is not bounded and may take arbitrarily large values with small probabilities, making it less realistic for many engineering applications. In contrast, bounded noise is a more realistic and versatile model of stochastic fluctuation in engineering applications. Under extreme conditions, bounded noise may reduce to either a deterministic sinusoidal function or a white noise of constant spectral density. The purpose of the study presented in this chapter is to analyze the stability of fractional viscoelastic systems under bounded noise excitation by using the method of higher-order stochastic averaging. The study extends [7] from stochastic averaging of first-order to higher-order. Furthermore, it extends [9–11] from white noises to bounded noises.

16.2 Stochastic averaging

16.2.1 Formulation

The analysis involves the stability problem of a uniform cross-section column under a dynamic axial compressive load $F(t)$. A study concerning a vibrating column can provide valuable acumen about the behavior of a wider class of structures [12]. The equation of motion is derived as [4]

$$\frac{\partial^2 M(x,t)}{\partial x^2} = \rho A \frac{\partial^2 v(x,t)}{\partial t^2} + \beta_0 \frac{\partial v(x,t)}{\partial t} + F(t)\frac{\partial^2 v(x,t)}{\partial x^2}, \tag{16.1}$$

where A is the cross-sectional area, ρ is the mass density per unit volume of the column, L is the length of the column, $v(x, t)$ is the transverse displacement of the central axis, β_0 is the damping constant. The moment $M(x, t)$ at the cross-section x and the geometry relation are:

$$M(x,t) = \int_A \sigma(x,t) z \mathrm{d}A, \quad \varepsilon(x,t) = -\frac{\partial^2 v(x,t)}{\partial x^2} z. \tag{16.2}$$

The viscoelastic material is supposed to follow the fractional constitutive model in Eq. (A10), which can be recast as $\sigma(t) = [E + \eta\, {}^{\mathrm{RL}}_{0}\boldsymbol{D}_t^{\mu}]\varepsilon(t)$. Here ${}^{\mathrm{RL}}_{0}\boldsymbol{D}_t^{\mu}$ is an operator for the Riemann–Liouville (RL) fractional derivative of order μ, which is defined in Eq. (A8). Substituting into Eq. (16.2) and then to Eq. (16.1) yields:

$$\rho A\frac{\partial^2 v(x,t)}{\partial t^2}+\beta_0\frac{\partial v(x,t)}{\partial t}+EI\frac{\partial^4 v(x,t)}{\partial x^4}$$
$$+I\eta\,{}^{\mathrm{RL}}_{0}\boldsymbol{D}_t^{\mu}\left[\frac{\partial^4 v(x,t)}{\partial x^4}\right]+F(t)\frac{\partial^2 v(x,t)}{\partial x^2}=0, \tag{16.3}$$

where I is the minimum moment of inertia of the column's cross-sectional area. If the column is simply supported, the solution of Eq. (16.3) can be written in the following form:

$$v(x,t)=q(t)\sin\frac{\pi x}{L}. \tag{16.4}$$

When Eq. (16.4) is substituted into Eq. (16.3) and it is assumed that the amplitude of load, damping, and viscoelastic effect are all diminutive, and a small parameter is introduced $0<\varepsilon\ll 1$, then the equation of motion—if the function $F(t)$ is taken to be a stochastic process—is described by

$$\ddot{q}(t)+2\varepsilon\beta\dot{q}(t)+\omega^2[1-\varepsilon\xi(t)+\varepsilon\tau_\varepsilon\,{}^{\mathrm{RL}}_{0}\boldsymbol{D}_t^{\mu}]q(t)=0, \tag{16.5}$$

where

$$\beta=\frac{\beta_0}{2\rho A},\quad \omega^2=\frac{EI}{\rho A}\left(\frac{\pi}{L}\right)^4,\quad P=EI\left(\frac{\pi}{L}\right)^2,\quad \xi(t)=\frac{F(t)}{P},\quad \tau_\varepsilon=\frac{\eta}{E}. \tag{16.6}$$

Random excitations can be described in mathematical form as stochastic processes, either wide-band noise processes or narrow-band noise processes. A representative wide-band noise process is white noise, which cannot be physically achieved. A bounded noise $\eta(t)$ is an achievable and versatile narrow-band stochastic fluctuation; in engineering, it is usually depicted in the form:

$$\eta(t)=\zeta\cos[\nu t+\sqrt{\varepsilon}\sigma W(t)+\theta], \tag{16.7}$$

where ζ is the noise amplitude, σ is the noise intensity, $W(t)$ is the standard Wiener process, and θ is a random variable uniformly distributed in the interval $[0, 2\pi]$. The bounded noise surmounts the deficiency of the Ornstein–Uhlenbeck process, which is not bounded. The insertion of the phase angle θ forms the bounded noise $\eta(t)$ into a stationary process. Eq. (16.7) may be written as:

$$\eta(t)=\zeta\cos Z(t),\quad \mathrm{d}Z(t)=\nu t+\sigma\mathrm{d}W(t), \tag{16.8}$$

where the initial condition of $Z(t)$ is $Z(0)=\theta$. This process is bounded between $-\zeta$ and $+\zeta$ for all time t and therefore is a bounded stochastic process. The autocorrection function of $\eta(t)$ is given by

$$R(\tau)=\mathbf{E}[\eta(t)\eta(t+\tau)]=\frac{1}{2}\zeta^2\cos\nu\tau\exp\left(-\frac{\sigma^2}{2}|\tau|\right), \tag{16.9}$$

and the spectral density function of $\eta(t)$ is:

$$S(\omega) = \int_{-\infty}^{+\infty} R(\tau)e^{-i\omega\tau}\mathrm{d}\tau = \frac{\zeta^2\sigma^2\left(\omega^2 + \nu^2 + \frac{1}{4}\sigma^4\right)}{2\left[(\omega+\nu)^2 + \frac{1}{4}\sigma^4\right]\left[(\omega-\nu)^2 + \frac{1}{4}\sigma^4\right]}. \quad (16.10)$$

When the noise intensity σ is minute, the bounded noise can be used to model a process about frequency ν. Therefore bounded noise is a splendid realistic model of narrow-band noise and is used extensively in engineering applications [4], such as wind turbulence and earthquake load. As σ approaches zero in the limit, the bounded noise reduces to a deterministic sinusoidal function. In contrast, when σ approaches infinity in the limit, the bounded noise becomes a white noise with a constant spectral density. However, since the mean-square value is fixed at 1/2, this constant spectral density level reduces to zero in the limit.

The moment stability of a stochastic dynamical system with state vector $\mathbf{x}(t)$ is determined by moment Lyapunov exponents, which are defined by Xie [4]

$$\Lambda(p) = \lim_{t\to\infty} \frac{1}{t} \log \mathbf{E}[\,\|\mathbf{x}(t)\|^p\,], \quad (16.11)$$

where $\mathbf{E}[\cdot]$ designates the expected value and $\|\mathbf{x}(t)\|$ describes a suitable vector norm. If $\Lambda(p) < 0$, then the p-th moments are asymptotically stable. Moreover, $\Lambda(p)$ is a convex function of p and $\Lambda'(p)$ is equal to the largest Lyapunov exponent λ, which is defined by

$$\lambda = \lim_{t\to\infty} \frac{1}{t} \log \|\mathbf{x}(t)\|, \quad (16.12)$$

and portrays the almost-sure or sample stability of the system. As shown in Fig. 16.1, the almost-sure stability or instability cannot always ensure the moment stability, which can be explained by the theory of large deviation [4]. Therefore the complete properties of

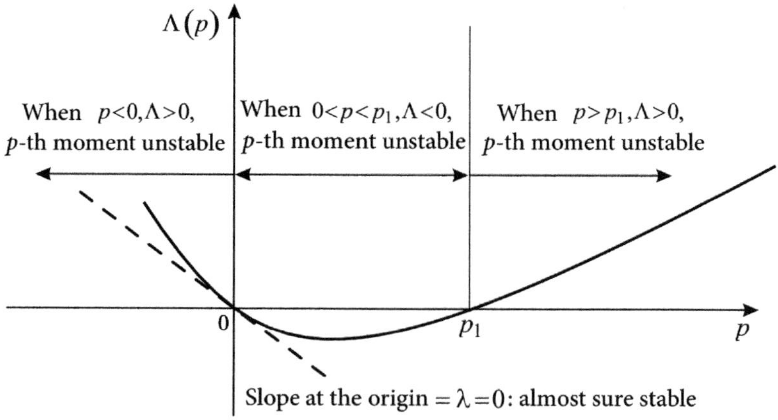

Figure 16.1 Almost sure stability and moment stability. The almost-sure stability or instability cannot always ensure the moment stability. Regardless of the potential stability or instability of the system shown in the segmented regions, the moment stability can vary.

dynamic stability can be described only when the moment Lyapunov exponents of stochastic systems are obtained.

However, getting the exact moment Lyapunov exponents of the stochastic system described in Eq. (16.5) is difficult. In this study, the differential equations governing the *p-th* moment are solved by using the method of higher-order stochastic averaging. Afterward, the moment stability of the viscoelastic system Eq. (16.5) is determined by solving the averaged equations.

16.2.2 First-order stochastic averaging

For thoroughness, this subsection describes the first-order stochastic averaging, which was used in Ref. [7], because higher-order stochastic averaging is based on the lower order. To apply the averaging method, the transformation is made to the amplitude and phase variables a and φ by means of the relation

$$q(t) = a(t)\cos\Phi(t),\ \dot{q}(t) = -\omega a(t)\sin\Phi(t),\quad \Phi(t) = \frac{\nu}{2}t + \varphi(t). \tag{16.13}$$

Substituting Eq. (16.13) into Eq. (16.5) yields:

$$\begin{aligned} &\dot{a}\cos\Phi(t) - a\dot{\varphi}\sin\Phi(t) = -\varepsilon\Delta a\sin\Phi(t), \\ &\dot{a}\sin\Phi(t) + a\dot{\varphi}\cos\Phi(t) = \varepsilon\Delta a\cos\Phi(t) \\ &- 2\varepsilon a\beta\sin\Phi(t) + \varepsilon\omega\xi(t)a\cos\Phi(t) + \varepsilon\omega\tau_\varepsilon\, {}^{\mathrm{RL}}_{\ 0}\boldsymbol{D}_t^{\mu} q(s), \end{aligned} \tag{16.14}$$

where $\varepsilon\Delta = \omega - \nu/2$. When it is assumed $P = a^P$, and Eq. (16.14) is solved, the p-th norm of the system Eq. (16.5) will be

$$\begin{aligned} \dot{P}(t) &= \varepsilon pP\left[-2\beta\sin^2\Phi(t) + \frac{1}{2}\omega\xi(t)\sin 2\Phi(t) + \omega\tau_\varepsilon\mathscr{U}^{ss}\right], \\ \dot{\varphi}(t) &= \varepsilon[\Delta - \beta\sin 2\Phi(t) + \omega\cos^2\Phi(t)\xi(t) + \omega\tau_\varepsilon\mathscr{U}^{cs}], \end{aligned} \tag{16.15}$$

where

$$\mathscr{U}^{ss} = -\frac{\omega\sin\Phi(t)}{\Gamma(1-\mu)}\int_0^t\left[\frac{P(s)}{P(t)}\right]^{1/p}\frac{\sin\Phi(t)}{(t-s)^{\mu}}\mathrm{d}s,$$

$$\mathscr{U}^{cs} = -\frac{\omega\cos\Phi(t)}{\Gamma(1-\mu)}\int_0^t\left[\frac{P(s)}{P(t)}\right]^{1/p}\frac{\sin\Phi(t)}{(t-s)^{\mu}}\mathrm{d}s,$$

and for ease of presentation, the fractional derivate in Eq. (A8) is rewritten as:

$${}^{\mathrm{RL}}_{\ 0}\boldsymbol{D}_t^{\mu}[f(\tau)g(t)] = \frac{g(t)}{\Gamma(1-\mu)}\int_0^t\frac{f'(\tau)}{(t-\tau)^{\mu}}\mathrm{d}\tau. \tag{16.16}$$

By assuming that the magnitude is small and then introducing a small parameter $\sqrt{\varepsilon}$, the bounded noise is rewritten as:

$$\xi(t) = \zeta\cos[\nu t + \psi(t)],\quad \psi(t) = \sqrt{\varepsilon}\vartheta W(t) + \theta.$$

Substituting $\xi(t)$ into Eq. (16.15) yields:

$$\dot{P}(t) = \varepsilon p P\left\{-\beta[1 - \cos 2\Phi] + \frac{\zeta\omega}{2}\cos[\nu t + \psi(t)]\sin 2\Phi + \omega\tau_\varepsilon \mathscr{U}^{ss}\right\},$$

$$\dot{\varphi}(t) = \varepsilon\{\Delta - \beta\sin 2\Phi + \omega\zeta\cos[\nu t + \psi(t)]\cos^2\Phi + \omega\tau_\varepsilon \mathscr{U}^{cs}\},$$

$$\dot{\psi}(t) = \sqrt{\varepsilon}\vartheta\dot{W}(t). \tag{16.17}$$

Eq. (16.17) are precisely equivalent to Eq. (16.5) and still cannot be solved accurately. However, both P and φ change slowly due to the presence of a small parameter ε. Therefore reasonably accurate results can be obtained by averaging the response over one period. This can be done by applying the averaging operator given by

$$\underset{\tau}{\mathscr{M}}(\cdot) = \lim_{T\to\infty}\frac{1}{T}\int_{\tau}^{\tau+T}(\cdot)\mathrm{d}\tau,$$

where $(\cdot)$ represents the equation of motion. When the averaging operator is applied, the integration is performed over explicitly appearing τ only. The averaging method can be applied to obtain the averaged equations, without distinction between the averaged and the original nonaveraged variables P and φ. When the averaging operation is exercised, variables $P(t)$ and $\varphi(t)$ are treated as unchanged, that is, they are directly replaced by the corresponding averaged variables.

$$\dot{P}(t) = \varepsilon p P\left[-\beta + \frac{1}{4}\zeta\omega\sin(2\varphi - \psi) + \omega\tau_\varepsilon \underset{t}{\mathscr{M}}\{\mathscr{U}^{ss}\}\right],$$

$$\dot{\varphi}(t) = \varepsilon\left[\Delta + \frac{1}{4}\zeta\omega\cos(2\varphi - \psi) + \omega\tau_\varepsilon \underset{t}{\mathscr{M}}\{\mathscr{U}^{cs}\}\right], \tag{16.18}$$

where some averaged identities have been used,

$$\underset{t}{\mathscr{M}}(\cos 2\Phi) = \underset{t}{\mathscr{M}}(\sin 2\Phi) = 0,$$

$$\underset{t}{\mathscr{M}}\{\cos[\nu t + \psi(t)]\sin 2\Phi\} = \frac{1}{2}\sin(2\varphi - \psi),$$

$$\underset{t}{\mathscr{M}}\{\cos[\nu t + \psi(t)]\cos^2\Phi\} = \frac{1}{4}\cos(2\varphi - \psi). \tag{16.19}$$

Applying the transformation $\tau = t - s$ and changing the order of integration produce:

$$\underset{t}{\mathscr{M}}\{\mathscr{U}^{ss}\} = -\frac{\omega}{2}\mathscr{H}^c\left(\frac{\nu}{2}\right), \quad \underset{t}{\mathscr{M}}\{\mathscr{U}^{cs}\} = \frac{\omega}{2}\mathscr{H}^s\left(\frac{\nu}{2}\right), \tag{16.20}$$

where

$$\mathscr{H}^c\left(\frac{\nu}{2}\right) = \frac{1}{\Gamma(1-\mu)}\int_0^\infty \tau^{-\mu}\cos\frac{\nu\tau}{2}\mathrm{d}\tau = \left(\frac{\nu}{2}\right)^{\mu-1}\sin\frac{\mu\pi}{2},$$

$$\mathscr{H}^s\left(\frac{\nu}{2}\right) = \frac{1}{\Gamma(1-\mu)}\int_0^\infty \tau^{-\mu}\sin\frac{\nu\tau}{2}\mathrm{d}\tau = \left(\frac{\nu}{2}\right)^{\mu-1}\cos\frac{\mu\pi}{2}. \tag{16.21}$$

Substituting Eq. (16.20) into Eq. (16.18) yields:

$$\dot{P}(t) = \varepsilon p P\left[-\hat{\beta} + \frac{1}{4}\zeta\omega \sin(2\varphi - \psi)\right],$$

$$\dot{\varphi}(t) = \varepsilon\left[\hat{\Delta} + \frac{1}{4}\zeta\omega \cos(2\varphi - \psi)\right], \tag{16.22}$$

where

$$\hat{\beta} = \beta + \frac{1}{2}\omega\tau_\varepsilon \mathscr{H}^c\left(\frac{\nu}{2}\right), \quad \hat{\Delta} = \Delta + \frac{1}{2}\omega\tau_\varepsilon \mathscr{H}^s\left(\frac{\nu}{2}\right). \tag{16.23}$$

Introducing the transformation $\Theta = \varphi - \psi/2$ and Itô Lemma in three Itô stochastic differential equations:

$$\mathrm{d}P(t) = m_P P \mathrm{d}t,$$

$$\mathrm{d}\Theta(t) = m_\Theta \mathrm{d}t - \sqrt{\varepsilon}\frac{\vartheta}{2}\mathrm{d}W(t),$$

$$\mathrm{d}\psi(t) = \sqrt{\varepsilon}\vartheta \mathrm{d}W(t), \tag{16.24}$$

where

$$m_P = \varepsilon p\left[-\hat{\beta} + \frac{1}{4}\zeta\omega \sin(2\Theta)\right], \quad m_\Theta = \varepsilon\left[\hat{\Delta} + \frac{1}{4}\zeta\omega \cos(2\Theta)\right]. \tag{16.25}$$

The eigenvalue problem governing the p-th moment Lyapunov exponent can be formulated from these stochastic differential equations. Applying a linear stochastic transformation [13]

$$S = T(\Theta)P, \quad P = T^{-1}(\Theta)S, \quad 0 \le \Theta \le \pi, \tag{16.26}$$

the Itô equation for the transformed p-th norm process S is given by

$$\mathrm{d}S = \left(m_P T + m_\Theta T_\Theta + \frac{\varepsilon\vartheta^2}{8}T_{\Theta\Theta}\right)P\mathrm{d}t - \frac{\sqrt{\varepsilon}\vartheta}{2}T_\Theta P \mathrm{d}W(t)$$

$$= \frac{1}{T}\left(m_P T + m_\Theta T_\Theta + \frac{\varepsilon\vartheta^2}{8}T_{\Theta\Theta}\right)S\mathrm{d}t - \frac{\sqrt{\varepsilon}\vartheta}{2}\frac{T_\Theta}{T}S\mathrm{d}W(t), \tag{16.27}$$

where T_Θ and $T_{\Theta\Theta}$ are the first and second derivatives of T with respect to Θ, respectively. For bounded and nonsingular transformation $T(\Theta)$, both processes P and S are expected to have the same stable behavior, as Eq. (16.26) is a linear transformation. Therefore $T(\Theta)$ is chosen so that the drift term of the Itô Eq. (16.27) is independent of Θ, that is,

$$\frac{1}{T}\left(m_P T + m_\Theta T_\Theta + \frac{\varepsilon\vartheta^2}{8}T_{\Theta\Theta}\right) = \Lambda \text{ and } \mathrm{d}S = \Lambda S\mathrm{d}t - \frac{\sqrt{\varepsilon}\vartheta}{2}\frac{T_\Theta}{T}S\mathrm{d}W(t). \tag{16.28}$$

Comparing the drift terms in Eqs. (16.27) and (16.28), such a transformation $T(\Theta)$ is given by the following differential equation:

$$\frac{\vartheta^2}{8}T_{\Theta\Theta} + m_{\Theta 1}T_\Theta + m_{P1}T = \hat{\Lambda}T, \tag{16.29}$$

where

$$m_{\Theta 1} = \frac{m_\Theta}{\varepsilon} = \hat{\Delta} + \frac{1}{4}\zeta\omega\cos(2\Theta), \quad m_{P1} = \frac{m_P}{\varepsilon} = p\left[-\hat{\beta} + \frac{1}{4}\zeta\omega\sin(2\Theta)\right]. \tag{16.30}$$

$T(\Theta)$ is a periodic function in terms of Θ of period π, and $\Lambda_{q(t)}(p) = \Lambda(p) = \varepsilon\hat{\Lambda}(p)$. Eq. (16.29) defines an eigenvalue problem for a second-order differential operator with $\hat{\Lambda}(p)$ being the eigenvalue and $T(\Theta)$ the associated eigenfunction.

Taking the expected value of Eq. (16.28) leads to $\mathrm{d}\mathbf{E}[S] = \Lambda\mathbf{E}[S]\mathrm{d}t$, from which it can be seen that Λ is the Lyapunov exponent of the transformed *p-th* moment $\mathbf{E}[S]$ of Eqs. (16.24) or (16.5). Since both processes P and S have the same stability behavior, Λ is the Lyapunov exponent of p-th moment $\mathbf{E}[P]$. Solving the eigenvalue problem is the remaining task needed for the moment Lyapunov exponents determination.

16.2.3 Determination of moment Lyapunov exponents

Since the coefficients in Eq. (16.29) are periodic with period π, a Fourier series expansion of the eigenfunction $T(\Theta)$ can be considered as:

$$T(\Theta) = C_0 + \sum_{k=1}^{K} (C_k\cos 2k\Theta + S_k\sin 2k\Theta). \tag{16.31}$$

Substituting this expansion into the eigenvalue problem (16.29) and equating the coefficients of similar trigonometric terms $\sin 2k\Theta$ and $\cos 2k\Theta$, $k = 0,1, \ldots, K$ produces a set of homogeneous linear algebraic equations with immeasurably many equations for the unknown coefficients C_0, C_k, and S_k,

$$\text{Constant: } -(c_0 + \hat{\Lambda})C_0 + b_0S_1 = 0,$$

$$\cos 2\Theta\text{: } -(c_1 + \hat{\Lambda})C_1 + d_1S_1 + b_1S_2 = 0,$$

$$\sin 2\Theta\text{: } 2a_1C_0 + d_1C_1 + b_1C_2 + (c_1 + \hat{\Lambda})S_1 = 0,$$

$$\cos 2k\Theta\text{: } a_kS_{k-1} + d_kS_k + b_kS_{k+1} - (c_k + \hat{\Lambda})C_k = 0,$$

$$\sin 2k\Theta\text{: } a_kC_{k-1} + d_kC_k + b_kC_{k+1} + (c_k + \hat{\Lambda})S_k = 0,$$

$$\cos 2K\Theta\text{: } a_KS_{K-1} + d_KS_K - (c_K + \hat{\Lambda})C_K = 0,$$

$$\sin 2K\Theta\text{: } a_KC_{K-1} + d_KC_K + (c_K + \hat{\Lambda})S_K = 0, \tag{16.32}$$

where

$$k = 2,\ 3,\ \ldots, K - 1,$$

$$a_n = \frac{1}{8}\zeta\omega(2n - 2 - p), \quad b_n = \frac{1}{8}\zeta\omega(2n + 2 + p),$$

$$c_n = \frac{1}{2}\sigma^2 n^2 + p\hat{\beta}, \quad d_n = 2n\hat{\Delta}, \quad n = 0,\ 1,\ 2,\ \ldots,\ K.$$

A nontrivial solution of the C_0, C_k, and S_k will be acquired when the determinant of the coefficient matrix of Eq. (16.32) equal to zero yields a polynomial equation of degree $2K + 1$ for $\hat{\Lambda}^{(K)}(p)$,

$$e_{2K+1}^{(K)}[\hat{\Lambda}^{(K)}]^{2K+1} + e_{2K}^{(K)}[\hat{\Lambda}^{(K)}]^{2K} + \cdots + e_1^{(K)}\hat{\Lambda}^{(K)} + e_0^{(K)} = 0, \tag{16.33}$$

where $\hat{\Lambda}^{(K)}$ denotes the approximate moment Lyapunov exponent. Then, the moment Lyapunov exponent of the system can be approximated with

$$\Lambda_{q(t)}(p) \approx \varepsilon\hat{\Lambda}^{(K)}(p). \tag{16.34}$$

For $K = 1$, Eq. (16.33) reduces to a cubic equation with

$$e_3^{(1)} = 1, \quad e_2^{(1)} = \sigma^2 + 3p\hat{\beta},$$

$$e_1^{(1)} = 4\hat{\Delta}^2 + \frac{\sigma^4}{4} - \frac{1}{32}\zeta^2\omega^2 p(p+2) + 3\hat{\beta}^2 p^2 + 2\sigma^2\hat{\beta}p,$$

$$e_0^{(1)} = \hat{\beta}^3 p^3 + \sigma^2\hat{\beta}^2 p^2 + p\hat{\beta}\left[4\hat{\Delta}^2 + \frac{\sigma^4}{4} - \frac{1}{32}\zeta^2\omega^2 p(p+2)\right] - \frac{1}{64}\zeta^2\omega^2\sigma^2 p(p+2),$$

and an approximate expression of the moment Lyapunov exponent is given by

$$\hat{\Lambda}^{(1)}(p) = \frac{A_2}{6} - \frac{2(3e_1 - e_2^2)}{3A_2} - \frac{e_2}{3},$$

$$A_2 = (12A_1 - 108e_0 + 36e_1e_2 - 8e_2^3)^{1/3},$$

$$A_1 = (81e_0^2 + 12e_1^3 - 54e_0e_1e_2 + 12e_0e_2^3 - 3e_1^2e_2^2)^{1/2}, \tag{16.35}$$

where the superscript "(1)" in the coefficients e's is dropped for clarity of presentation. When $K > 1$, there are no analytical solutions for the polynomial Eq. (16.33), so the solutions can be found only using numerical approaches.

16.2.4 Second-order stochastic averaging

The transformed equation of motion in Eq. (16.17) can be rewritten in the form:

$$\dot{P}(t) = \varepsilon f_1(P, \varphi, t),\ \dot{\varphi}(t) = \varepsilon f_2(P, \varphi, t),\ \dot{\psi}(t) = \sqrt{\varepsilon}\sigma\dot{W}(t), \tag{16.36}$$

where

$$f_1(P, \varphi, t) = pP\left\{-\beta[1 - \cos 2\Phi(t)] + \frac{\zeta\omega}{2}\cos[\nu t + \psi(t)]\sin 2\Phi + \omega\tau_\varepsilon\mathscr{U}^{ss}\right\},$$

$$f_2(P, \varphi, t) = \Delta - \beta\sin 2\Phi + \omega\zeta\cos[\nu t + \psi(t)]\cos^2\Phi + \omega\tau_\varepsilon\mathscr{U}^{cs}. \tag{16.37}$$

The near-identity transformation is introduced [8] with the purpose of formulating a second-order stochastic averaging

$$P(t) = \overline{P}(t) + \varepsilon P_1(\overline{P}, \overline{\varphi}, t), \quad \varphi(t) = \overline{\varphi}(t) + \varepsilon \varphi_1(\overline{P}, \overline{\varphi}, t), \tag{16.38}$$

where $\overline{P}(t)$ and $\overline{\varphi}(t)$ stand for nonoscillatory amplitude and phase angle, respectively. Differentiating Eq. (16.38) and equating each result with the corresponding functions from Eq. (16.36) make:

$$\left\{\begin{matrix} \dfrac{d\overline{P}}{dt} \\ \dfrac{d\overline{\varphi}}{dt} \end{matrix}\right\} + \varepsilon \left\{\begin{matrix} \dfrac{\partial P_1}{\partial \overline{P}}\dfrac{d\overline{P}}{dt} + \dfrac{\partial P_1}{\partial \overline{\varphi}}\dfrac{d\overline{\varphi}}{dt} + \dfrac{\partial P_1}{\partial t} \\ \dfrac{\partial \varphi_1}{\partial \overline{P}}\dfrac{d\overline{P}}{dt} + \dfrac{\partial \varphi_1}{\partial \overline{\varphi}}\dfrac{d\overline{\varphi}}{dt} + \dfrac{\partial \varphi_1}{\partial t} \end{matrix}\right\} = \varepsilon \left\{\begin{matrix} f_1(\overline{P} + \varepsilon P_1, \overline{\varphi} + \varepsilon \varphi_1, t) \\ f_2(\overline{P} + \varepsilon P_1, \overline{\varphi} + \varepsilon \varphi_1, t) \end{matrix}\right\}. \tag{16.39}$$

Substituting the Taylor expansion

$$f_j(\overline{P} + \varepsilon P_1, \overline{\varphi} + \varepsilon \varphi_1, t) = f_j(\overline{P}, \overline{\varphi}, t) + \varepsilon P_1 \frac{\partial f_j}{\partial \overline{P}} + \varepsilon \varphi_1 \frac{\partial f_j}{\partial \overline{\varphi}} + o(\varepsilon), \quad j = 1, \ 2, \tag{16.40}$$

and the inverse matrix

$$\mathbf{A} = \begin{bmatrix} 1 + \varepsilon\dfrac{\partial P_1}{\partial \overline{P}} & \varepsilon\dfrac{\partial P_1}{\partial \overline{\varphi}} \\ \varepsilon\dfrac{\partial \varphi_1}{\partial \overline{P}} & 1 + \varepsilon\dfrac{\partial \varphi_1}{\partial \overline{\varphi}} \end{bmatrix}, \quad \mathbf{A}^{-1} = \begin{bmatrix} 1 - \varepsilon\dfrac{\partial P_1}{\partial \overline{P}} & -\varepsilon\dfrac{\partial P_1}{\partial \overline{\varphi}} \\ -\varepsilon\dfrac{\partial \varphi_1}{\partial \overline{P}} & 1 - \varepsilon\dfrac{\partial \varphi_1}{\partial \overline{\varphi}} \end{bmatrix} + o(\varepsilon), \tag{16.41}$$

into Eq. (16.39) yields:

$$\frac{d\overline{P}}{dt} = \varepsilon \overline{u}_1 + \varepsilon^2 \overline{u}_2 + o(\varepsilon^2), \quad \frac{d\overline{\varphi}}{dt} = \varepsilon \overline{v}_1 + \varepsilon^2 \overline{v}_2 + o(\varepsilon^2), \tag{16.42}$$

where

$$\begin{aligned} \overline{u}_1 &= f_1 - \frac{\partial P_1}{\partial t}, \quad \overline{u}_2 = \frac{\partial f_1}{\partial \overline{P}} P_1 + \frac{\partial f_1}{\partial \overline{\varphi}} \varphi_1 - \frac{\partial P_1}{\partial \overline{P}} f_1 - \frac{\partial P_1}{\partial \overline{\varphi}} f_2 + \frac{\partial P_1}{\partial t}\frac{\partial P_1}{\partial \overline{P}} + \frac{\partial \varphi_1}{\partial t}\frac{\partial P_1}{\partial \overline{\varphi}}, \\ \overline{v}_1 &= f_2 - \frac{\partial \varphi_1}{\partial t}, \quad \overline{v}_2 = \frac{\partial f_2}{\partial \overline{P}} P_1 + \frac{\partial f_2}{\partial \overline{\varphi}} \varphi_1 - \frac{\partial \varphi_1}{\partial \overline{P}} f_1 - \frac{\partial \varphi_1}{\partial \overline{\varphi}} f_2 + \frac{\partial P_1}{\partial t}\frac{\partial \varphi_1}{\partial \overline{P}} + \frac{\partial \varphi_1}{\partial t}\frac{\partial \varphi_1}{\partial \overline{\varphi}}, \end{aligned} \tag{16.43}$$

A further focus is on obtaining $\overline{u}_1$, $\overline{u}_2$, $\overline{v}_1$, and $\overline{v}_2$, and their averaged values. Firstly, obtaining $\overline{u}_1$ and $\overline{v}_1$ can be done by using the assumption:

$$\frac{\partial P_1}{\partial t} = f_1 - \underset{t}{\mathcal{M}}\{f_1\}, \quad \frac{\partial \varphi_1}{\partial t} = f_2 - \underset{t}{\mathcal{M}}\{f_2\}, \tag{16.44}$$

then the terms $f_1 - (\partial P_1/\partial t)$ and $f_2 - (\partial \varphi_1/\partial t)$ contain only nonoscillatory terms. It also suggests that P_1 and φ_1 are oscillatory functions. Substituting Eq. (16.20) into the averaged Eq. (16.37) yields:

$$\bar{u}_1 = \underset{t}{\mathscr{M}}\{f_1\} = pP\left[-\hat{\beta} + \frac{1}{4}\zeta\omega\sin(2\varphi - \psi)\right],$$

$$\bar{v}_1 = \underset{t}{\mathscr{M}}\{f_2\} = \hat{\Delta} + \frac{1}{4}\zeta\omega\cos(2\varphi - \psi), \tag{16.45}$$

where P and φ are treated as a constant under the averaging operation. Substituting Eqs. (16.37) and (16.45) into Eq. (16.44) makes:

$$\frac{\partial P_1}{\partial t} = \frac{1}{4}p\bar{P}\left[4\beta\cos(vt + 2\bar{\varphi}) + \omega\zeta\sin(\psi - 2\bar{\varphi}) + 2\omega\tau_\varepsilon\mathscr{H}^s\left(\frac{v}{2}\right)\sin(vt + 2\bar{\varphi})\right.$$

$$\left. + 2\omega\tau_\varepsilon\mathscr{H}^c\left(\frac{v}{2}\right)\cos(vt + 2\bar{\varphi}) + 2\omega\zeta\cos(vt + \psi)\sin(vt + 2\bar{\varphi})\right],$$

$$\frac{\partial \varphi_1}{\partial t} = -\beta\sin(vt + 2\bar{\varphi}) + \frac{1}{2}\omega\tau_\varepsilon\mathscr{H}^s\left(\frac{v}{2}\right)\cos(vt + 2\bar{\varphi}) - \frac{1}{2}\omega\tau_\varepsilon\mathscr{H}^c\left(\frac{v}{2}\right)\sin(vt + 2\bar{\varphi})$$

$$- \frac{1}{4}\omega\zeta\cos(\psi - 2\bar{\varphi}) + \frac{1}{2}\omega\zeta\cos(vt + \psi) + \frac{1}{2}\omega\zeta\cos(vt + \psi)\cos(vt + 2\bar{\varphi}). \tag{16.46}$$

During the derivation of Eq. (16.46), the following equations must be used:

$$\begin{aligned}
\mathscr{U}^{sc} &= -\frac{v}{2\Gamma(1-\mu)}\int_0^t \frac{\sin\left(\frac{v}{2}t + \bar{\varphi}\right)\sin\left(\frac{v}{2}s + \bar{\varphi}\right)}{(t-s)^\mu}\mathrm{d}s \\
&= -\frac{v}{2\Gamma(1-\mu)}\left[\int_0^\infty \frac{\sin\left(\frac{vt}{2} + \bar{\varphi}\right)\sin\left(\frac{vt}{2} - \frac{v\tau}{2} + \bar{\varphi}\right)}{\tau^\mu}\mathrm{d}\tau\right. \\
&\quad \left. - \int_0^\infty \frac{\sin\left(\frac{vt}{2} + \bar{\varphi}\right)\sin\left(\frac{vt}{2} - \frac{v\tau}{2} + \bar{\varphi}\right)}{\tau^\mu}\mathrm{d}\tau\right] \\
&= -\frac{v}{4\Gamma(1-\mu)}\int_0^\infty \frac{\cos\left(vt - \frac{v}{2}\tau + 2\bar{\varphi}\right)}{\tau^\mu}\mathrm{d}\tau - \frac{1}{2}\mathscr{H}^c\left(\frac{v}{2}\right) \\
&= -\frac{1}{2}\mathscr{H}^c\left(\frac{v}{2}\right) + \frac{1}{2}\left[\cos(vt + 2\bar{\varphi})\mathscr{H}^c\left(\frac{v}{2}\right) + \sin(vt + 2\bar{\varphi})\mathscr{H}^s\left(\frac{v}{2}\right)\right].
\end{aligned} \tag{16.47}$$

The third equation exists when $t \to \infty$. The following equations are obtained in a similar way:

$$\mathscr{U}^{cc} = \frac{1}{2}\mathscr{H}^{s}\left(\frac{\nu}{2}\right) - \frac{1}{2}\left[\sin(\nu t + 2\overline{\varphi})\mathscr{H}^{c}\left(\frac{\nu}{2}\right) - \cos(\nu t + 2\overline{\varphi})\mathscr{H}^{s}\left(\frac{\nu}{2}\right)\right],$$

$$\mathscr{U}^{ss} = \frac{1}{2}\mathscr{H}^{s}\left(\frac{\nu}{2}\right) + \frac{1}{2}\left[\sin(\nu t + 2\overline{\varphi})\mathscr{H}^{c}\left(\frac{\nu}{2}\right) - \cos(\nu t + 2\overline{\varphi})\mathscr{H}^{s}\left(\frac{\nu}{2}\right)\right],$$

$$\mathscr{U}^{cs} = \frac{1}{2}\mathscr{H}^{c}\left(\frac{\nu}{2}\right) + \frac{1}{2}\left[\cos(\nu t + 2\overline{\varphi})\mathscr{H}^{c}\left(\frac{\nu}{2}\right) + \sin(\nu t + 2\overline{\varphi})\mathscr{H}^{s}\left(\frac{\nu}{2}\right)\right]. \tag{16.48}$$

Solving the ODEs in Eq. (16.46) results in:

$$P_1 = \frac{p\overline{P}}{8\nu}\left\{4\left[2\beta + \omega\tau_\varepsilon\mathscr{H}^{s}\left(\frac{\nu}{2}\right)\right]\sin(\nu t + 2\overline{\varphi}) - 4\omega\tau_\varepsilon\mathscr{H}^{s}\left(\frac{\nu}{2}\right)\cos(\nu t + 2\overline{\varphi})\right.$$
$$\left. -\omega\zeta\cos(2\nu t + \psi + 2\overline{\varphi})\right\},$$

$$\varphi_1 = \frac{1}{8\nu}\left\{4\left[2\beta + \omega\tau_\varepsilon\mathscr{H}^{c}\left(\frac{\nu}{2}\right)\right]\cos(\nu t + 2\overline{\varphi}) + 4\omega\zeta\sin(\nu t + \psi)\right.$$
$$\left. +4\omega\tau_\varepsilon\mathscr{H}^{s}\left(\frac{\nu}{2}\right)\sin(\nu t + 2\overline{\varphi}) + \omega\zeta\sin(2\nu t + \psi + 2\overline{\varphi})\right\}. \tag{16.49}$$

Secondly, to obtain $\overline{u}_2$ and $\overline{v}_2$, the following derivatives of f_1 and f_2 in Eq. (16.37) are required:

$$\frac{\partial f_1}{\partial \overline{P}} = -2\beta p\sin^2\Phi + \frac{p}{2}\omega\zeta\cos(\nu t + \psi)\sin 2\Phi + \omega\tau_\varepsilon(p\mathscr{U}^{sc} + \overline{P}\mathscr{V}^{sc}),$$

$$\frac{\partial f_1}{\partial \overline{\varphi}} = p\overline{P}\left[-2\beta\sin 2\Phi + \omega\zeta\cos(\nu t + \psi)\cos 2\Phi + \omega\tau_\varepsilon(\mathscr{U}^{cc} - \mathscr{U}^{ss})\right],$$

$$\frac{\partial f_2}{\partial \overline{P}} = \frac{\omega\tau_\varepsilon\mathscr{V}^{cc}}{p},$$

$$\frac{\partial f_2}{\partial \overline{\varphi}} = -2\beta\cos 2\Phi - \omega\zeta\cos(\nu t + \psi)\sin 2\Phi - \omega\tau_\varepsilon(\mathscr{U}^{cs} - \mathscr{U}^{sc}), \tag{16.50}$$

where

$$\mathscr{U}^{cc} = {}_{0}^{\mathrm{RL}}\boldsymbol{D}_t^{\mu}\left\{\left[\frac{\overline{P}(s)}{\overline{P}(t)}\right]^{1/p}\cos(\omega t + \overline{\varphi})\cos(\omega s + \overline{\varphi})\right\},$$

$$\mathscr{U}^{ss} = {}_{0}^{\mathrm{RL}}\boldsymbol{D}_t^{\mu}\left\{\left[\frac{\overline{P}(s)}{\overline{P}(t)}\right]^{1/p}\sin(\omega t + \overline{\varphi})\sin(\omega s + \overline{\varphi})\right\},$$

$$\mathscr{U}^{cs} = {}^{\mathrm{RL}}_{0}\boldsymbol{D}^{\mu}_{t}\left\{\left[\frac{\overline{P}(s)}{\overline{P}(t)}\right]^{1/p}\cos(\omega t+\overline{\varphi})\sin(\omega s+\overline{\varphi})\right\},$$

$$\mathscr{U}^{sc} = {}^{\mathrm{RL}}_{0}\boldsymbol{D}^{\mu}_{t}\left\{\left[\frac{\overline{P}(s)}{\overline{P}(t)}\right]^{1/p}\sin(\omega t+\overline{\varphi})\cos(\omega s+\overline{\varphi})\right\},$$

$$\mathscr{V}^{cc} = {}^{\mathrm{RL}}_{0}\boldsymbol{D}^{\mu}_{t}\left\{\left[\frac{\overline{P}(s)}{\overline{P}(t)}\right]^{1/p}\left[\frac{1}{\overline{P}(s)}-\frac{1}{\overline{P}(t)}\right]\cos(\omega t+\overline{\varphi})\cos(\omega s+\overline{\varphi})\right\},$$

$$\mathscr{V}^{ss} = {}^{\mathrm{RL}}_{0}\boldsymbol{D}^{\mu}_{t}\left\{\left[\frac{\overline{P}(s)}{\overline{P}(t)}\right]^{1/p}\left[\frac{1}{\overline{P}(s)}-\frac{1}{\overline{P}(t)}\right]\sin(\omega t+\overline{\varphi})\sin(\omega s+\overline{\varphi})\right\},$$

$$\mathscr{V}^{cs} = {}^{\mathrm{RL}}_{0}\boldsymbol{D}^{\mu}_{t}\left\{\left[\frac{\overline{P}(s)}{\overline{P}(t)}\right]^{1/p}\left[\frac{1}{\overline{P}(s)}-\frac{1}{\overline{P}(t)}\right]\cos(\omega t+\overline{\varphi})\sin(\omega s+\overline{\varphi})\right\},$$

$$\mathscr{V}^{sc} = {}^{\mathrm{RL}}_{0}\boldsymbol{D}^{\mu}_{t}\left\{\left[\frac{\overline{P}(s)}{\overline{P}(t)}\right]^{1/p}\left[\frac{1}{\overline{P}(s)}-\frac{1}{\overline{P}(t)}\right]\sin(\omega t+\overline{\varphi})\cos(\omega s+\overline{\varphi})\right\}, \tag{16.51}$$

and,

$$\frac{\partial\mathscr{U}^{sc}}{\partial\overline{P}} = \frac{\mathscr{V}^{sc}}{p},\quad \frac{\partial\mathscr{U}^{cc}}{\partial\overline{P}} = \frac{\mathscr{V}^{cc}}{p},\quad \frac{\partial\mathscr{U}^{cc}}{\partial\overline{\varphi}} = -\mathscr{U}^{sc}-\mathscr{U}^{cs},\quad \frac{\partial\mathscr{U}^{sc}}{\partial\overline{\varphi}} = \mathscr{U}^{cc}-\mathscr{U}^{ss}. \tag{16.52}$$

Substituting P_1 and φ_1 into Eq. (16.43) leads to the expressions of $\overline{u}_2$ and $\overline{v}_2$, but they are too wearisome to be displayed here; the averaged values are of main interest. Performing averaging on them gives:

$$\underset{t}{\mathscr{M}}\{\overline{u}_2\} = \underset{t}{\mathscr{M}}\left\{\frac{\partial f_1}{\partial\overline{P}}P_1+\frac{\partial f_1}{\partial\overline{\varphi}}\varphi_1-\frac{\partial P_1}{\partial\overline{P}}\underset{t}{\mathscr{M}}\{f_1\}-\frac{\partial P_1}{\partial\overline{\varphi}}\underset{t}{\mathscr{M}}\{f_2\}\right\}$$

$$= -\frac{p\overline{P}\omega\zeta}{4v}\left\{\left[2\beta+\omega\tau_\varepsilon\mathscr{H}^c\left(\frac{v}{2}\right)\right]\cos(\psi-2\overline{\varphi})-\omega\tau_\varepsilon\mathscr{H}^s\left(\frac{v}{2}\right)\sin(\psi-2\overline{\varphi})\right\}$$

$$= -\frac{p\overline{P}\omega\zeta}{2v}[\hat{\beta}\cos(\psi-2\overline{\varphi})+\hat{\alpha}\sin(\psi-2\overline{\varphi})],$$

$$\underset{t}{\mathscr{M}}\{\overline{v}_2\} = \underset{t}{\mathscr{M}}\left\{\frac{\partial f_2}{\partial\overline{P}}P_1+\frac{\partial f_2}{\partial\overline{\varphi}}\varphi_1-\frac{\partial \varphi_1}{\partial\overline{P}}\underset{t}{\mathscr{M}}\{f_1\}-\frac{\partial \varphi_1}{\partial\overline{\varphi}}\underset{t}{\mathscr{M}}\{f_2\}\right\}$$

$$= -\frac{1}{32v}\left\{8\zeta\omega^2\tau_\varepsilon\mathscr{H}^s\left(\frac{v}{2}\right)\cos(\psi-2\overline{\varphi})+8\omega\zeta\left[2\beta+\omega\tau_\varepsilon\mathscr{H}^c\left(\frac{v}{2}\right)\right]\sin(\psi-2\overline{\varphi})\right.$$

$$+\omega^2\zeta^2 + 8\left[\omega\tau_\varepsilon \mathscr{H}^s\left(\frac{v}{2}\right)\right]^2 + 8\left[\omega\tau_\varepsilon \mathscr{H}^c\left(\frac{v}{2}\right)\right]^2 + 32\omega\beta\tau_\varepsilon \mathscr{H}^c\left(\frac{v}{2}\right) + 32\beta^2\Bigg\}$$

$$= \frac{1}{2v}[\zeta\omega\hat{\alpha}\cos(\psi - 2\overline{\varphi}) - \hat{\beta}\omega\zeta\sin(\psi - 2\overline{\varphi}) + \hat{\delta}], \tag{16.53}$$

where

$$\hat{\alpha} = -\frac{1}{2}\omega\tau_\varepsilon \mathscr{H}^s\left(\frac{v}{2}\right),$$

$$\hat{\delta} = -\left\{\frac{1}{16}\omega^2\zeta^2 + \frac{1}{2}\left[\omega\tau_\varepsilon \mathscr{H}^s\left(\frac{v}{2}\right)\right]^2 + \frac{1}{2}\left[\omega\tau_\varepsilon \mathscr{H}^c\left(\frac{v}{2}\right)\right]^2 + 2\omega\beta\tau_\varepsilon \mathscr{H}^c\left(\frac{v}{2}\right) + 2\beta^2\right\}.$$

Applying the transformation $\Theta = \varphi - \psi/2$ and Itô Lemma results in three Itô stochastic differential equations:

$$\begin{aligned} dP(t) &= (\varepsilon m_{P1} + \varepsilon^2 m_{P2})P\,dt, \\ d\Theta(t) &= (\varepsilon m_{\Theta 1} + \varepsilon^2 m_{\Theta 2})dt - \sqrt{\varepsilon}\frac{\sigma}{2}dW(t), \\ d\psi(t) &= \sqrt{\varepsilon}\sigma dW(t), \end{aligned} \tag{16.54}$$

where

$$\begin{aligned} m_{P1} &= \frac{1}{P}\ \underset{t}{\mathscr{M}}\{\bar{u}_1\} = p\left[-\hat{\beta} + \frac{\zeta\omega}{4}\sin(2\Theta)\right], \\ m_{P2} &= \frac{1}{P}\ \underset{t}{\mathscr{M}}\{\bar{u}_2\} = -\frac{p\zeta\omega}{2v}[\hat{\beta}\cos(2\Theta) - \hat{\alpha}\sin(2\Theta)], \\ m_{\Theta 1} &= \underset{t}{\mathscr{M}}\{\bar{v}_1\} = \hat{\Delta} + \frac{\zeta\omega}{4}\cos(2\Theta), \\ m_{\Theta 2} &= \underset{t}{\mathscr{M}}\{\bar{v}_2\} = \frac{1}{2v}[\zeta\omega\hat{\alpha}\cos(2\Theta) + \hat{\beta}\omega\zeta\sin(2\Theta) + \hat{\delta}]. \end{aligned} \tag{16.55}$$

It can be noticed that Eq. (16.54) are also coupled. By introducing a linear stochastic transformation:

$$S = T(\Theta)P, \quad P = T^{-1}(\Theta)S, \quad 0 \le \Theta \le \pi,$$

Itô equation for the transformed p-th norm process S is given by, again using Itô Lemma in the vector case:

$$\begin{aligned} dS &= \left[(\varepsilon m_{P1} + \varepsilon^2 m_{P2})T + (\varepsilon m_{\Theta 1} + \varepsilon^2 m_{\Theta 2})T_\Theta + \frac{\varepsilon\sigma^2}{8}T_{\Theta\Theta}\right]P\,dt - \frac{\varepsilon^{1/2}\sigma}{2}T_\Theta P\,dW(t) \\ &= \frac{1}{T}\left[(\varepsilon m_{P1} + \varepsilon^2 m_{P2})T + (\varepsilon m_{\Theta 1} + \varepsilon^2 m_{\Theta 2})T_\Theta + \frac{\varepsilon\sigma^2}{8}T_{\Theta\Theta}\right]S\,dt \\ &\quad - \frac{\varepsilon^{1/2}\sigma}{2}\frac{T_\Theta}{T}S\,dW(t). \end{aligned} \tag{16.56}$$

The eigenvalue problem governing the *p-th* moment Lyapunov exponent is:

$$\frac{\sigma^2}{8}T_{\Theta\Theta} + (m_{\Theta 1} + \varepsilon m_{\Theta 2})T_\Theta + (m_{P1} + \varepsilon m_{P2})T = \hat{\Lambda}T, \tag{16.57}$$

where $T(\Theta)$ is a periodic function in Θ of period π, and $\Lambda_{q(t)}(p) = \Lambda(p) = \varepsilon\hat{\Lambda}(p)$.

Substituting the expansion in Eq. (16.31) into an eigenvalue problem (16.57) and equating the coefficients of similar trigonometric terms $\sin 2k\Theta$ and $\cos 2k\Theta$, $k = 0,1, \ldots$. render a set of homogeneous linear algebraic equations with immeasurably many equations for the unknown coefficients C_0, C_k, and S_k, $k = 1,2, \ldots$.

Constant terms:

$$(c_0 + \hat{\Lambda})C_0 - b_0 S_1 + \varepsilon\frac{2b_0}{v}(\hat{\beta}C_1 - \hat{\alpha}S_1) = 0,$$

coefficients for $\cos 2\Theta$:

$$(c_1 + \hat{\Lambda})C_1 - d_1 S_1 - b_1 S_2 + \varepsilon\frac{1}{v}(-e_1 S_1 - 2\hat{\alpha}b_1 S_2 - 4a_1\hat{\beta}C_0 + 2\hat{\beta}b_1 C_2) = 0,$$

coefficients for $\sin 2\Theta$:

$$2a_1 C_0 + d_1 C_1 + b_1 C_2 + (c_1 + \hat{\Lambda})S_1 + \varepsilon\frac{1}{v}(e_1 C_1 - 2\hat{\alpha}b_1 C_2 - 4a_1\hat{\beta}C_0 + 2\hat{\beta}b_1 S_2) = 0,$$

coefficients for $\cos 2k\Theta$:

$$a_k S_{k-1} - (c_k + \hat{\Lambda})C_k + d_k S_k + b_k S_{k+1}$$
$$+ \varepsilon\frac{1}{v}(2a_k\hat{\alpha}S_{k-1} + e_k S_k + 2b_k\hat{\alpha}S_{k+1} + 2a_k\hat{\alpha}C_{k-1} - 2b_k\hat{\beta}C_{k+1}) = 0,$$
$$k = 2,\ 3,\ \ldots, K - 1,$$

coefficients for $\sin 2k\Theta$:

$$a_k C_{k-1} + d_k C_k + b_k C_{k+1} - (c_k + \hat{\Lambda})S_k$$
$$+ \varepsilon\frac{1}{v}(2a_k\hat{\alpha}C_{k-1} + e_k C_k + 2b_k\hat{\alpha}C_{k+1} - 2a_k\hat{\alpha}S_{k-1} + 2b_k\hat{\beta}S_{k+1}) = 0,$$
$$k = 2,\ 3,\ \ldots, K - 1,$$

coefficients for $\cos 2K\Theta$:

$$a_k S_{k-1} - (c_k + \hat{\Lambda})C_k + d_k S_k + \varepsilon\frac{1}{v}(2a_k\hat{\alpha}S_{k-1} + e_k S_k + 2a_k\hat{\beta}C_{k+1}) = 0,$$

coefficients for $\sin 2K\Theta$:

$$a_k C_{k-1} + d_k C_k + (c_k + \hat{\Lambda})S_k + \varepsilon\frac{1}{v}(2a_k\hat{\alpha}C_{k-1} + e_k C_k + 2a_k\hat{\beta}S_{k+1}) = 0, \tag{16.58}$$

where

$$a_k = \frac{1}{8}\zeta\omega(2k - 2 - p),\ b_k = \frac{1}{8}\zeta\omega(2k + 2 + p),$$

$$c_k = \frac{1}{2}\sigma^2 k^2 + p\hat{\beta},\ d_k = 2k\hat{\Delta},\quad e_k = k\hat{\delta},\quad k = 0,\ 1,\ 2,\ \ldots \tag{16.59}$$

As in the first-order stochastic averaging, a set of Eq. (16.58) is reduced to a set of $2K + 1$ homogenous linear equations for C_0, C_k, and $S_k, k = 1,2, \ldots, K$. The determinant is equated to zero to yield a polynomial equation of degree $2K + 1$ for $\hat{\Lambda}^{(K)}(p)$. Solving the polynomial equation for the largest root yields the approximate moment Lyapunov exponents. The terms $\mathscr{H}^s\left(\frac{\nu}{2}\right)$ and $\mathscr{H}^c\left(\frac{\nu}{2}\right)$ depend on the constitutive model of the material.

16.2.5 Higher-order stochastic averaging

Higher-order stochastic averaging analysis can be done by following the procedure shown in Section 16.2.4. As an example of a higher-order stochastic method, the three-order stochastic averaging is illustrated here. A near-identity transformation is introduced, as in Eq. (16.38), but up to order ε^2

$$P(t) = \overline{P}(t) + \varepsilon P_1 + \varepsilon^2 P_2,\quad \varphi(t) = \overline{\varphi}(t) + \varepsilon\varphi_1 + \varepsilon^2\varphi_2, \tag{16.60}$$

where $P_1 = P_1(\overline{P}, \overline{\varphi}, t), P_2 = P_2(\overline{P}, \overline{\varphi}, t), \varphi_1 = \varphi_1(\overline{P}, \overline{\varphi}, t)$, and $\varphi_2 = \varphi_2(\overline{P}, \overline{\varphi}, t)$ are unknown functions. When $P_2 = 0$ and $\varphi_2 = 0$, it is reduced to second-order averaging.

Differentiating Eq. (16.60) and equating each result with the corresponding functions from Eq. (16.36) gives:

$$\begin{Bmatrix} \frac{d\overline{P}}{dt} \\ \frac{d\overline{\varphi}}{dt} \end{Bmatrix} + \varepsilon \begin{Bmatrix} \frac{\partial P_1}{\partial \overline{P}}\frac{d\overline{P}}{dt} + \frac{\partial P_1}{\partial \overline{\varphi}}\frac{d\overline{\varphi}}{dt} + \frac{\partial P_1}{\partial t} \\ \frac{\partial \varphi_1}{\partial \overline{P}}\frac{d\overline{P}}{dt} + \frac{\partial \varphi_1}{\partial \overline{\varphi}}\frac{d\overline{\varphi}}{dt} + \frac{\partial \varphi_1}{\partial t} \end{Bmatrix} + \varepsilon^2 \begin{Bmatrix} \frac{\partial P_2}{\partial \overline{P}}\frac{d\overline{P}}{dt} + \frac{\partial P_2}{\partial \overline{\varphi}}\frac{d\overline{\varphi}}{dt} + \frac{\partial P_2}{\partial t} \\ \frac{\partial \varphi_2}{\partial \overline{P}}\frac{d\overline{P}}{dt} + \frac{\partial \varphi_2}{\partial \overline{\varphi}}\frac{d\overline{\varphi}}{dt} + \frac{\partial \varphi_2}{\partial t} \end{Bmatrix}$$

$$= \varepsilon \begin{Bmatrix} f_1(\overline{P} + \varepsilon P_1 + \varepsilon^2 P_2,\ \overline{\varphi} + \varepsilon\varphi_1 + \varepsilon^2\varphi_2,\ t) \\ f_2(\overline{P} + \varepsilon P_1 + \varepsilon^2 P_2,\ \overline{\varphi} + \varepsilon\varphi_1 + \varepsilon^2\varphi_2,\ t) \end{Bmatrix}, \tag{16.61}$$

or in a matrix form:

$$\mathbf{A}\begin{Bmatrix} \frac{d\overline{P}}{dt} \\ \frac{d\overline{\varphi}}{dt} \end{Bmatrix} = \varepsilon \begin{Bmatrix} f_1(\overline{P} + \varepsilon P_1 + \varepsilon^2 P_2,\ \overline{\varphi} + \varepsilon\varphi_1 + \varepsilon^2\varphi_2,\ t) - \frac{\partial P_1}{\partial t} \\ f_2(\overline{P} + \varepsilon P_1 + \varepsilon^2 P_2,\ \overline{\varphi} + \varepsilon\varphi_1 + \varepsilon^2\varphi_2,\ t) - \frac{\partial \varphi_1}{\partial t} \end{Bmatrix} - \varepsilon^2 \begin{Bmatrix} \frac{\partial P_2}{\partial t} \\ \frac{\partial \varphi_2}{\partial t} \end{Bmatrix}, \tag{16.62}$$

where

$$\mathbf{A} = \begin{bmatrix} 1 + \varepsilon\frac{\partial P_1}{\partial \overline{P}} + \varepsilon^2\frac{\partial P_2}{\partial \overline{P}} & \varepsilon\frac{\partial P_1}{\partial \overline{\varphi}} + \varepsilon^2\frac{\partial P_2}{\partial \overline{\varphi}} \\ \varepsilon\frac{\partial \varphi_1}{\partial \overline{P}} + \varepsilon^2\frac{\partial \varphi_2}{\partial \overline{P}} & 1 + \varepsilon\frac{\partial \varphi_1}{\partial \overline{\varphi}} + \varepsilon^2\frac{\partial \varphi_2}{\partial \overline{\varphi}} \end{bmatrix}. \tag{16.63}$$

Using the expansion of $1/|\mathbf{A}|$ to the order of ε^2, the inverse matrix of $\mathbf{A}$ is:

$$\mathbf{A}^{-1} = \frac{1}{|\mathbf{A}|} = \begin{bmatrix} 1 + \varepsilon\frac{\partial \varphi_1}{\partial \overline{\varphi}} + \varepsilon^2\frac{\partial \varphi_2}{\partial \overline{\varphi}} & -\varepsilon\frac{\partial P_1}{\partial \overline{\varphi}} - \varepsilon^2\frac{\partial P_2}{\partial \overline{\varphi}} \\ -\varepsilon\frac{\partial \varphi_1}{\partial \overline{P}} - \varepsilon^2\frac{\partial \varphi_2}{\partial \overline{P}} & 1 + \varepsilon\frac{\partial P_1}{\partial \overline{P}} + \varepsilon^2\frac{\partial P_2}{\partial \overline{P}} \end{bmatrix} = \begin{bmatrix} A11 & A12 \\ A21 & A22 \end{bmatrix}, \tag{16.64}$$

where

$$A11 = 1 - \varepsilon\frac{\partial P_1}{\partial \overline{P}} + \varepsilon^2\left[\frac{\partial P_1}{\partial \overline{\varphi}}\frac{\partial \varphi_1}{\partial \overline{P}} - \frac{\partial P_2}{\partial \overline{P}} + \left(\frac{\partial P_1}{\partial \overline{P}}\right)^2\right],$$

$$A12 = -\varepsilon\frac{\partial P_1}{\partial \overline{P}} + \varepsilon^2\left[\frac{\partial P_1}{\partial \overline{\varphi}}\frac{\partial \varphi_1}{\partial \overline{\varphi}} - \frac{\partial P_2}{\partial \overline{\varphi}} + \frac{\partial P_1}{\partial \overline{P}}\frac{\partial P_1}{\partial \overline{\varphi}}\right],$$

$$A21 = -\varepsilon\frac{\partial \varphi_1}{\partial \overline{P}} + \varepsilon^2\left[\frac{\partial \varphi_1}{\partial \overline{P}}\frac{\partial \varphi_1}{\partial \overline{\varphi}} - \frac{\partial \varphi_2}{\partial \overline{P}} + \frac{\partial P_1}{\partial \overline{P}}\frac{\partial \varphi_1}{\partial \overline{P}}\right],$$

$$A22 = 1 - \varepsilon\frac{\partial \varphi_1}{\partial \overline{\varphi}} + \varepsilon^2\left[\frac{\partial P_1}{\partial \overline{\varphi}}\frac{\partial \varphi_1}{\partial \overline{P}} - \frac{\partial \varphi_2}{\partial \overline{\varphi}} + \left(\frac{\partial \varphi_1}{\partial \overline{\varphi}}\right)^2\right].$$

Taylor expansion of a function with multivariables is considered:

$$\begin{aligned}
&f_j\left(\overline{P} + \varepsilon P_1 + \varepsilon^2 P_2,\ \overline{\varphi} + \varepsilon\varphi_1 + \varepsilon^2\varphi_2,\ t\right) \\
&= f_j(\overline{P}, \overline{\varphi}, t) + \varepsilon\left\{P_1\frac{\partial f_j(\overline{P}, \overline{\varphi}, t)}{\partial \overline{P}} + \varphi_1\frac{\partial f_j(\overline{P}, \overline{\varphi}, t)}{\partial \overline{\varphi}}\right\} \\
&+ \varepsilon^2\left\{P_2\frac{\partial f_j(\overline{P}, \overline{\varphi}, t)}{\partial \overline{P}} + \varphi_2\frac{\partial f_j(\overline{P}, \overline{\varphi}, t)}{\partial \overline{\varphi}} + \frac{1}{2}P_1^2\frac{\partial^2 f_j(\overline{P}, \overline{\varphi}, t)}{\partial \overline{P}^2}\right. \\
&\left.+ \frac{1}{2}\varphi_1^2\frac{\partial^2 f_j(\overline{P}, \overline{\varphi}, t)}{\partial \varphi^2} + P_1\varphi_1\frac{\partial^2 f_j(\overline{P}, \overline{\varphi}, t)}{\partial \overline{P}\partial \overline{\varphi}}\right\} + o(\varepsilon^2), \quad j = 1,\ 2.
\end{aligned} \tag{16.65}$$

Substituting Eqs. (16.64) and (16.65) into Eq. (16.62) makes:

$$\frac{d\overline{P}}{dt} = \varepsilon\overline{\mu}_1 + \varepsilon^2\overline{\mu}_2 + \varepsilon^3\overline{\mu}_3 + o(\varepsilon^3), \quad \frac{d\overline{\varphi}}{dt} = \varepsilon\overline{\nu}_1 + \varepsilon^2\overline{\nu}_2 + \varepsilon^3\overline{\nu}_3 + o(\varepsilon^3), \tag{16.66}$$

where

$$\overline{\mu}_1 = f_1 - \frac{\partial P_1}{\partial t},$$

$$\overline{\mu}_2 = \frac{\partial f_1}{\partial \overline{P}}P_1 + \frac{\partial f_1}{\partial \overline{\varphi}}\varphi_1 - \frac{\partial P_1}{\partial \overline{P}}\left(f_1 - \frac{\partial P_1}{\partial t}\right) - \frac{\partial P_1}{\partial \overline{\varphi}}\left(f_2 - \frac{\partial \varphi_1}{\partial t}\right) - \frac{\partial P_2}{\partial t},$$

$$\begin{aligned}
\overline{\mu}_3 &= \frac{\partial f_1}{\partial \overline{P}}P_2 + \frac{\partial f_1}{\partial \overline{\varphi}}\varphi_2 + \frac{P_1^2}{2}\frac{\partial^2 f_1}{\partial \overline{P}^2} + \frac{\varphi_1^2}{2}\frac{\partial^2 f_1}{\partial \overline{\varphi}^2} + P_1\varphi_1\frac{\partial^2 f_1}{\partial \overline{P}\partial \overline{\varphi}} \\
&- \frac{\partial P_1}{\partial \overline{P}}\left(P_1\frac{\partial f_1}{\partial \overline{P}} + \varphi_1\frac{\partial f_1}{\partial \overline{\varphi}} - \frac{\partial P_2}{\partial t}\right) - \frac{\partial P_1}{\partial \overline{\varphi}}\left(P_1\frac{\partial f_2}{\partial \overline{P}} + \varphi_1\frac{\partial f_2}{\partial \overline{\varphi}} - \frac{\partial \varphi_2}{\partial t}\right)
\end{aligned}$$

$$+\left(f_1-\frac{\partial P_1}{\partial t}\right)\left[\frac{\partial P_1}{\partial \overline{\varphi}}\frac{\partial \varphi_1}{\partial \overline{P}}-\frac{\partial P_2}{\partial \overline{P}}+\left(\frac{\partial P_1}{\partial \overline{P}}\right)^2\right]$$

$$+\left(f_2-\frac{\partial \varphi_1}{\partial t}\right)\left[\frac{\partial P_1}{\partial \overline{\varphi}}\frac{\partial \varphi_1}{\partial \overline{\varphi}}-\frac{\partial P_2}{\partial \overline{\varphi}}+\frac{\partial P_1}{\partial \overline{P}}\frac{\partial P_1}{\partial \overline{\varphi}}\right], \tag{16.67}$$

$$\overline{\nu}_1=f_2-\frac{\partial \varphi_1}{\partial t},$$

$$\overline{\nu}_2=\frac{\partial f_2}{\partial \overline{P}}P_1+\frac{\partial f_2}{\partial \overline{\varphi}}\varphi_1-\frac{\partial \varphi_1}{\partial \overline{P}}\left(f_1-\frac{\partial P_1}{\partial t}\right)-\frac{\partial \varphi_1}{\partial \overline{\varphi}}\left(f_2-\frac{\partial \varphi_1}{\partial t}\right)-\frac{\partial \varphi_2}{\partial t},$$

$$\overline{\nu}_3=\frac{\partial f_2}{\partial \overline{P}}P_2+\frac{\partial f_2}{\partial \overline{\varphi}}\varphi_2+\frac{P_1^2}{2}\frac{\partial^2 f_2}{\partial \overline{P}^2}+\frac{\varphi_1^2}{2}\frac{\partial^2 f_2}{\partial \overline{\varphi}^2}+P_1\varphi_1\frac{\partial^2 f_2}{\partial \overline{P}\partial \overline{\varphi}}$$

$$-\frac{\partial \varphi_1}{\partial \overline{P}}\left(P_1\frac{\partial f_1}{\partial \overline{P}}+\varphi_1\frac{\partial f_1}{\partial \overline{\varphi}}-\frac{\partial P_2}{\partial t}\right)-\frac{\partial \varphi_1}{\partial \overline{\varphi}}\left(P_1\frac{\partial f_2}{\partial \overline{P}}+\varphi_1\frac{\partial f_2}{\partial \overline{\varphi}}-\frac{\partial \varphi_2}{\partial t}\right)$$

$$+\left(f_2-\frac{\partial \varphi_1}{\partial t}\right)\left[\frac{\partial P_1}{\partial \overline{\varphi}}\frac{\partial \varphi_1}{\partial \overline{P}}-\frac{\partial \varphi_2}{\partial \overline{\varphi}}+\left(\frac{\partial \varphi_1}{\partial \overline{\varphi}}\right)^2\right]$$

$$+\left(f_1-\frac{\partial P_1}{\partial t}\right)\left[\frac{\partial P_1}{\partial \overline{P}}\frac{\partial \varphi_1}{\partial \overline{P}}-\frac{\partial \varphi_2}{\partial \overline{P}}+\frac{\partial \varphi_1}{\partial \overline{P}}\frac{\partial \varphi_1}{\partial \overline{\varphi}}\right]. \tag{16.68}$$

Since third-order averaging can be reduced to second-order averaging, it can be assumed that P_1 and φ_1 have the same forms as in Eq. (16.49). Using the same concept used for deriving Eq. (16.49), $\overline{\mu}_2$ can be integrated with respect to t, the result can be solved for P_2, any terms from P_2 that are linear in $\overline{\varphi}$ should be removed—in order for the near-identity transformation (16.60) to be uniformly valid in the large t limit. It can be assumed that:

$$\frac{\partial P_2}{\partial t}=\overline{u}_2-\underset{t}{\mathscr{M}}\{\overline{u}_2\},\quad \frac{\partial \varphi_2}{\partial t}=\overline{\nu}_2-\underset{t}{\mathscr{M}}\{\overline{\nu}_2\}. \tag{16.69}$$

Substituting $\overline{u}_2$ and $\overline{\nu}_2$ in Eq. (16.43) and their averaging values in Eq. (16.53) into Eq. (16.69) and solving the differential equations yields the functions P_2 and φ_2, however, these functions are too lengthy to be shown here.

Finding $\overline{\mu}_3$ and $\overline{\nu}_3$ form in Eq. (16.66) requires derivatives of f_1 and f_2 in Eq. (16.50) and some second-order derivatives:

$$\frac{\partial^2 f_1}{\partial \overline{P}^2}=\frac{\partial}{\partial \overline{P}}\left[-2\beta p\sin^2\Phi+\frac{p}{2}\omega\zeta\cos(\nu t+\psi)\sin 2\Phi+\omega\tau_\varepsilon(p\mathscr{U}^{sc}+\overline{P}\mathscr{V}^{sc})\right]$$

$$=\omega\tau_\varepsilon\mathscr{V}^{sc}+\omega\tau_\varepsilon\overline{P}\frac{\partial \mathscr{V}^{sc}}{\partial \overline{P}},$$

$$\frac{\partial^2 f_1}{\partial \overline{P} \partial \overline{\varphi}} = \frac{\partial}{\partial \overline{\varphi}} \left[-2\beta p \sin^2 \Phi + \frac{p}{2} \omega \zeta \cos(\nu t + \psi) \sin 2\Phi + \omega \tau_\varepsilon (p \mathscr{U}^{sc} + \overline{P} \mathscr{V}^{sc}) \right]$$

$$= -2\beta p \sin 2\Phi + p\omega\zeta \cos(\nu t + \psi) \cos 2\Phi + \omega\tau_\varepsilon p (\mathscr{U}^{cc} - \mathscr{U}^{ss}) + \omega \tau_\varepsilon \overline{P} (\mathscr{V}^{cc} - \mathscr{V}^{ss}),$$

$$\frac{\partial^2 f_1}{\partial \overline{\varphi}^2} = \frac{\partial}{\partial \overline{\varphi}} p \overline{P} [-2\beta \sin 2\Phi + \omega\zeta \cos(\nu t + \psi) \cos 2\Phi + \omega \tau_\varepsilon (\mathscr{U}^{cc} - \mathscr{U}^{ss})]$$

$$= -2p\overline{P} [2\beta \cos 2\Phi + \omega\zeta \cos(\nu t + \psi) \sin 2\Phi + \omega\tau_\varepsilon (\mathscr{U}^{sc} + \mathscr{U}^{cs})],$$

$$\frac{\partial^2 f_2}{\partial \overline{P}^2} = \frac{\partial}{\partial \overline{P}} \left[\frac{\omega \tau_\varepsilon \mathscr{V}^{cc}}{p} \right] = \frac{\omega \tau_\varepsilon}{p} {}^{\mathrm{RL}}_{0}\boldsymbol{D}_t^{\mu} \left\{ \left[\frac{\overline{P}(s)}{\overline{P}(t)} \right]^{1/p} \left[\left(\frac{1}{p} - 1 \right) \frac{1}{\overline{P}(s)^2} - \frac{2}{p} \frac{1}{\overline{P}(t) \overline{P}(s)} \right.\right.$$

$$\left.\left. + \left(\frac{1}{p} + 1 \right) \frac{1}{\overline{P}(t)^2} \right] \cos(\omega t + \overline{\varphi}) \cos(\omega s + \overline{\varphi}) \right\},$$

$$\frac{\partial^2 f_2}{\partial \overline{P} \partial \overline{\varphi}} = \frac{\partial}{\partial \overline{\varphi}} \left[\frac{\omega \tau_\varepsilon \mathscr{V}^{cc}}{p} \right] = -\frac{\omega \tau_\varepsilon}{p} (\mathscr{V}^{sc} - \mathscr{V}^{cs}),$$

$$\frac{\partial^2 f_2}{\partial \overline{\varphi}^2} = \frac{\partial}{\partial \overline{\varphi}} [-2\beta \cos 2\Phi - \omega\zeta \cos(\nu t + \psi) \sin 2\Phi - \omega \tau_\varepsilon (\mathscr{U}^{cs} - \mathscr{U}^{sc})]$$

$$= 4\beta \sin 2\Phi - 2\omega\zeta \cos(\nu t + \psi) \cos 2\Phi - 2\omega\tau_\varepsilon (\mathscr{U}^{cc} - \mathscr{U}^{ss}), \tag{16.70}$$

where Eq. (16.52) and

$$\frac{\partial \mathscr{U}^{ss}}{\partial \overline{\varphi}} = \mathscr{U}^{cs} + \mathscr{U}^{sc}, \quad \frac{\partial \mathscr{U}^{cs}}{\partial \overline{\varphi}} = -\mathscr{U}^{ss} + \mathscr{U}^{cc},$$

$$\frac{\partial \mathscr{V}^{sc}}{\partial \overline{\varphi}} = \mathscr{V}^{cc} - \mathscr{V}^{ss}, \quad \frac{\partial \mathscr{V}^{cc}}{\partial \overline{\varphi}} = -\mathscr{V}^{sc} - \mathscr{V}^{cs}, \tag{16.71}$$

are used, and $\mathscr{U}^{sc}$, $\mathscr{U}^{cc}$, $\mathscr{U}^{cs}$, $\mathscr{U}^{ss}$, $\mathscr{V}^{ss}$, $\mathscr{V}^{cc}$, $\mathscr{V}^{sc}$, and $\mathscr{V}^{cs}$ are defined in Eq. (16.51).

Substituting these derivatives and P_1, φ_1, P_2, and φ_2 into Eqs. (16.67) and (16.68) yields the values of $\overline{\mu}_3$ and $\overline{\nu}_3$. Their average values are:

$$\underset{t}{\mathscr{M}} \{\overline{\mu}_3\} = \frac{p\overline{P}\zeta\omega}{256\nu^2} \{\hat{\gamma} \sin(\psi - 2\overline{\varphi}) - 128\hat{\alpha}\hat{\beta} \cos(\psi - 2\overline{\varphi})\},$$

$$\underset{t}{\mathscr{M}} \{\overline{\nu}_3\} = \frac{-\zeta\omega}{256\nu^2} \{\hat{\gamma} \cos(\psi - 2\overline{\varphi}) + 128\hat{\alpha}\hat{\beta} \sin(\psi - 2\overline{\varphi}) - \hat{\lambda}\}, \tag{16.72}$$

where

$$\hat{\gamma} = 7\omega^2\zeta^2 - 128\hat{\alpha}^2, \quad \hat{\alpha} = -\frac{1}{2}\omega\tau_\varepsilon \mathscr{H}^s\left(\frac{\nu}{2}\right),$$

$$\hat{\lambda} = 128\omega\tau_\varepsilon\hat{\Delta}\left\{\omega\tau_\varepsilon\left[\mathscr{H}^c\left(\frac{\nu}{2}\right)\right]^2 + 4\beta\mathscr{H}^c\left(\frac{\nu}{2}\right) + \omega\tau_\varepsilon\left[\mathscr{H}^s\left(\frac{\nu}{2}\right)\right]^2\right\}$$
$$- 8\hat{\alpha}(64\beta^2 - 3\omega^2\zeta^2) + 8\Delta(64\beta^2 + \omega^2\zeta^2).$$

Applying the transformation $\Theta = \varphi - \psi/2$ and Itô Lemma results in three Itô stochastic differential equations:

$$\mathrm{d}P(t) = (\varepsilon m_{P1} + \varepsilon^2 m_{P2} + \varepsilon^3 m_{P3})P\mathrm{d}t,$$
$$\mathrm{d}\Theta(t) = (\varepsilon m_{\Theta 1} + \varepsilon^2 m_{\Theta 2} + \varepsilon^3 m_{\Theta 3})\mathrm{d}t - \sqrt{\varepsilon}\frac{\sigma}{2}\mathrm{d}W(t),$$
$$\mathrm{d}\psi(t) = \sqrt{\varepsilon}\sigma\mathrm{d}W(t), \tag{16.73}$$

where

$$m_{P1} = \frac{1}{P}\underset{t}{\mathscr{M}}\{\bar{u}_1\} = p\left[-\hat{\beta} + \frac{\zeta\omega}{4}\sin(2\Theta)\right],$$
$$m_{P2} = \frac{1}{P}\underset{t}{\mathscr{M}}\{\bar{u}_2\} = -\frac{p\zeta\omega}{2\nu}[\hat{\beta}\cos(2\Theta) - \hat{\alpha}\sin(2\Theta)],$$
$$m_{P3} = \frac{1}{P}\underset{t}{\mathscr{M}}\{\bar{\mu}_2\} = \frac{p\zeta\omega}{256\nu^2}[\hat{\gamma}\sin(\psi - 2\overline{\varphi}) - 128\hat{\alpha}\hat{\beta}\cos(\psi - 2\overline{\varphi})],$$
$$m_{\Theta 1} = \underset{t}{\mathscr{M}}\{\bar{v}_1\} = \hat{\Delta} + \frac{\zeta\omega}{4}\cos(2\Theta),$$
$$m_{\Theta 2} = \underset{t}{\mathscr{M}}\{\bar{v}_2\} = \frac{1}{2\nu}[\zeta\omega\hat{\alpha}\cos(2\Theta) + \hat{\beta}\omega\zeta\sin(2\Theta) + \hat{\delta}],$$
$$m_{\Theta 3} = \underset{t}{\mathscr{M}}\{\bar{v}_3\} = \frac{-\zeta\omega}{256\nu^2}[\hat{\gamma}\cos(\psi - 2\overline{\varphi}) + 128\hat{\alpha}\hat{\beta}\sin(\psi - 2\overline{\varphi}) + \hat{\lambda}]. \tag{16.74}$$

Eq. (16.73) are coupled. Introducing a linear stochastic transformation:

$$S = T(\Theta)P, \quad P = T^{-1}(\Theta)S, \quad 0 \leq \Theta \leq \pi,$$

the Itô equation for the transformed p-th norm process S is given by, again using Itô Lemma in the vector case:

$$\mathrm{d}S = \left[(\varepsilon m_{P1} + \varepsilon^2 m_{P2} + \varepsilon^3 m_{P3})T + (\varepsilon m_{\Theta 1} + \varepsilon^2 m_{\Theta 2} + \varepsilon^3 m_{\Theta 3})T_\Theta + \frac{\varepsilon\sigma^2}{8}T_{\Theta\Theta}\right]P\mathrm{d}t$$
$$- \frac{\sqrt{\varepsilon}\sigma}{2}T_\Theta P\mathrm{d}W(t)$$
$$= \frac{1}{T}\left[(\varepsilon m_{P1} + \varepsilon^2 m_{P2} + \varepsilon^3 m_{P3})T + (\varepsilon m_{\Theta 1} + \varepsilon^2 m_{\Theta 2} + \varepsilon^3 m_{\Theta 3})T_\Theta + \frac{\varepsilon\sigma^2}{8}T_{\Theta\Theta}\right]S\mathrm{d}t$$
$$- \frac{\sqrt{\varepsilon}\sigma}{2}\frac{T_\Theta}{T}S\mathrm{d}W(t). \tag{16.75}$$

The eigenvalue problem governing the p-th moment Lyapunov exponent is:

$$\frac{\sigma^2}{8}T_{\Theta\Theta} + (m_{\Theta 1} + \varepsilon m_{\Theta 2} + \varepsilon^3 m_{\Theta 3})T_\Theta + (m_{P1} + \varepsilon m_{P2} + \varepsilon^3 m_{P3})T = \hat{\Lambda}T, \tag{16.76}$$

where $T(\Theta)$ is a periodic function in Θ of period π, and $\Lambda_{q(t)}(p) = \Lambda(p) = \varepsilon\hat{\Lambda}(p)$.

Substituting the expansion in Eq. (16.31) into an eigenvalue problem (16.76) and equating the coefficients of similar trigonometric terms $\sin 2k\Theta$ and $\cos 2k\Theta$, $k = 0,1, \ldots$. render a set of homogeneous linear algebraic equations with immeasurably many equations for the unknown coefficients C_0, C_k, and S_k, $k = 1,2, \ldots$.

Constant terms:

$$(c_0 + \hat{\Lambda})C_0 - b_0 S_1 + \varepsilon\frac{2b_0}{v}(\hat{\beta}C_1 - \hat{\alpha}S_1) + \varepsilon^2\frac{a_1}{8v^2}(\hat{\lambda}C_1 - \hat{\gamma}S_1) = 0,$$

coefficients for $\cos 2\Theta$:

$$(c_1 + \hat{\Lambda})C_1 - d_1 S_1 - b_1 S_2 + \varepsilon\frac{1}{v}(-e_1 S_1 - 2\hat{\alpha}b_1 S_2 - 4a_1\hat{\beta}C_0 + 2\hat{\beta}b_1 C_2)$$

$$+ \varepsilon^2\frac{1}{32v^2}(-4a_3\hat{\gamma}S_2 - p\hat{\lambda}C_0 + 4a_3\hat{\lambda}C_2) = 0,$$

coefficients for $\sin 2\Theta$:

$$2a_1 C_0 + d_1 C_1 + b_1 C_2 + (c_1 + \hat{\Lambda})S_1 + \varepsilon\frac{1}{v}(e_1 C_1 + 2\hat{\alpha}b_1 C_2 + 4a_1\hat{\beta}C_0 + 2\hat{\beta}b_1 S_2)$$

$$+ \varepsilon^2\frac{1}{32v^2}(4a_3\hat{\lambda}S_2 + p\hat{\gamma}C_0 + 4a_3\hat{\gamma}C_2) = 0,$$

coefficients for $\cos 2k\Theta$:

$$a_k S_{k-1} - (c_k + \hat{\Lambda})C_k + d_k S_k + b_k S_{k+1}$$

$$+ \varepsilon\frac{1}{v}(2a_k\hat{\alpha}S_{k-1} + e_k S_k + 2b_k\hat{\alpha}S_{k+1} + 2a_k\hat{\alpha}C_{k-1} - 2b_k\hat{\beta}C_{k+1})$$

$$+ \varepsilon^2\frac{1}{8v^2}(-b_{k-2}\hat{\gamma}S_{k-1} - a_{k+2}\hat{\gamma}S_{k+1} - b_{k-2}\hat{\lambda}C_{k-1} + a_{k+2}\hat{\lambda}C_{k+1}) = 0,$$

$$k = 2,\ 3,\ \ldots, K-1,$$

coefficients for $\sin 2k\Theta$:

$$a_k C_{k-1} + d_k C_k + b_k C_{k+1} + (c_k + \hat{\Lambda})S_k$$

$$+ \varepsilon\frac{1}{v}(2a_k\hat{\alpha}C_{k-1} + e_k C_k + 2b_k\hat{\alpha}C_{k+1} - 2a_k\hat{\alpha}S_{k-1} + 2b_k\hat{\beta}S_{k+1})$$

$$+ \varepsilon^2\frac{1}{8v^2}(-b_{k-2}\hat{\lambda}S_{k-1} - a_{k+2}\hat{\lambda}S_{k+1} + b_{k-2}\hat{\gamma}C_{k-1} + a_{k+2}\hat{\gamma}C_{k+1}) = 0,$$

$$k = 2,\ 3,\ \ldots, K-1,$$

coefficients for $\cos 2K\Theta$:

$$a_k S_{k-1} - (c_k + \hat{\Lambda})C_k + d_k S_k + \varepsilon\frac{1}{\nu}(2a_k\hat{\alpha}S_{k-1} + e_k S_k + 2a_k\hat{\beta}C_{k+1}) + \varepsilon\frac{1}{8\nu^2}(b_{k-2}\hat{\gamma}S_{k-1} + b_{k-2}\hat{\lambda}C_{k-1}) = 0,$$

coefficients for $\sin 2K\Theta$:

$$a_k C_{k-1} + d_k C_k + (c_k + \hat{\Lambda})S_k + \varepsilon\frac{1}{\nu}(2a_k\hat{\alpha}C_{k-1} + e_k C_k + 2a_k\hat{\beta}S_{k+1}) + \varepsilon\frac{1}{8\nu^2}(b_{k-2}\hat{\gamma}C_{k-1} - b_{k-2}\hat{\lambda}S_{k-1}) = 0, \tag{16.77}$$

where

$$a_k = \frac{1}{8}\zeta\omega(2k - 2 - p),\ b_k = \frac{1}{8}\zeta\omega(2k + 2 + p),$$

$$c_k = \frac{1}{2}\sigma^2 k^2 + p\hat{\beta},\quad d_k = 2k\hat{\Delta},\quad e_k = k\hat{\delta},\quad k = 0,\ 1,\ 2,\ \ldots$$

As in the first-order stochastic averaging, a set of Eq. (16.77) is reduced to a set of $2K + 1$ homogenous linear equations for C_0, C_k, and S_k, $k = 1,2, \ldots, K$. The determinant is equated to zero to yield a polynomial equation of degree $2K + 1$ for $\hat{\Lambda}^{(K)}(p)$. Solving the polynomial equation for the largest root yields the approximate moment Lyapunov exponents. The terms $\mathscr{H}^s\left(\frac{\nu}{2}\right)$ and $\mathscr{H}^c\left(\frac{\nu}{2}\right)$ depend on the constitutive model of the material.

16.3 Results and discussions

In Eq. (16.33), the Fourier series expansion for $T(\Theta)$ is truncated to the K-th order; the nontrivial solution of this equation is the moment Lyapunov exponent. Theoretically, this procedure renders the set of approximate moment Lyapunov exponents that converge to the corresponding true exponent values as $K \to \infty$. Practically, the convergence of the moment Lyapunov exponents with the increase of Fourier expansion is fast, and the results of order $K = 3$ are accurate enough, which can be seen from Fig. 16.2. Still, this convergence also depends on the width of the power spectral density of the bounded noise process. Fig. 16.3 shows when $\zeta = 5$, even $K = 12$ is not sufficient. This result differs from Ref. [14], where real noise was considered. Fig. 16.4 shows that for small σ or large ζ, the bounded noise is a narrow-band process; whereas when σ is increased or ζ is decreased, the power spectral density function will flatten. Therefore if the power spectral density function of the bounded noise is relatively sharp, which implies that it is a narrow-band process, a

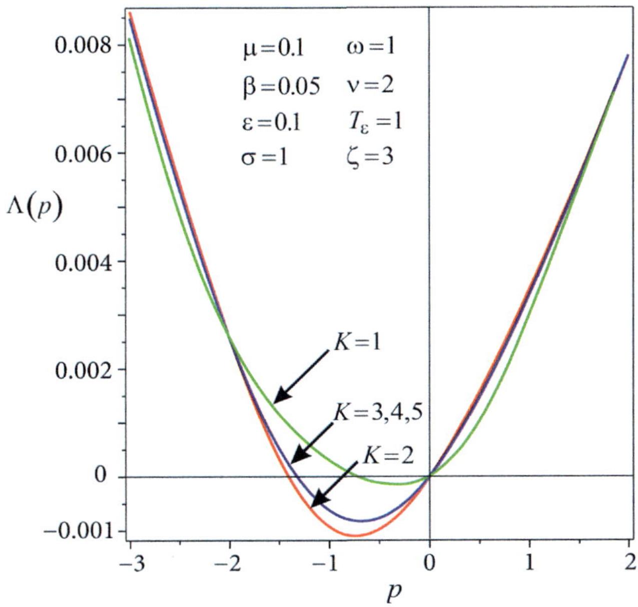

Figure 16.2 Effect of Fourier series expansion on moment Lyapunov exponents. A series of three terms is sufficient to achieve convergence when the noise amplitude is low.

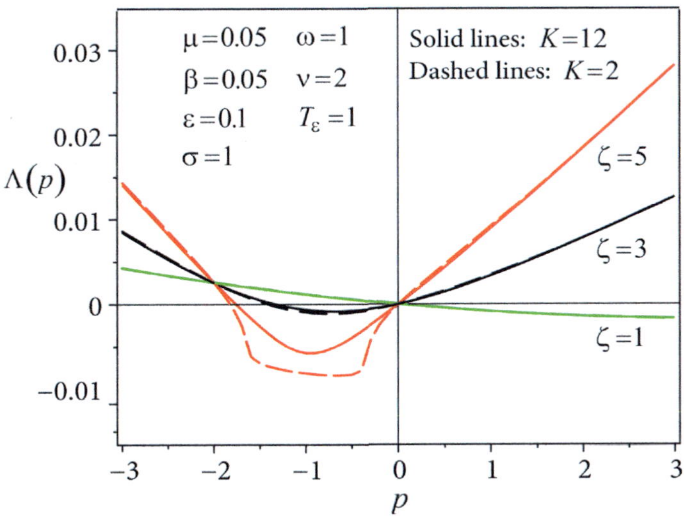

Figure 16.3 Fourier series expansion on moment Lyapunov exponents. The figure shows the convergence of the moment Lyapunov exponents for different values of noise amplitude and illustrates that more terms in the series are required for an accurate approximation when the noise amplitude is high.

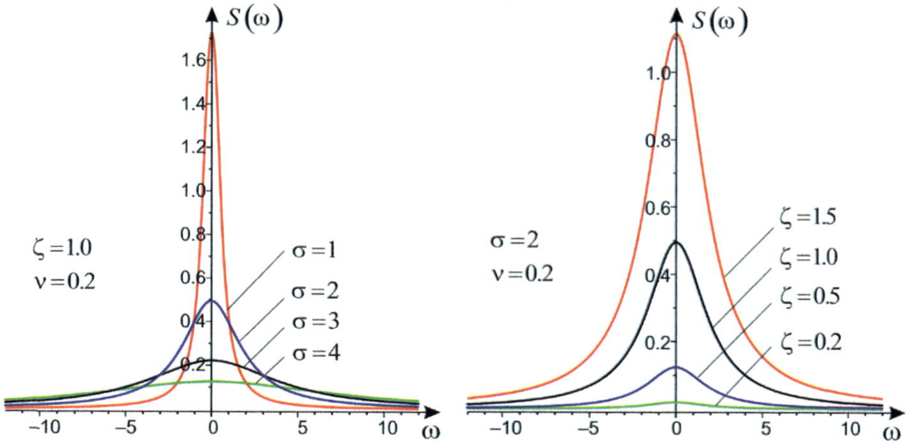

Figure 16.4 Power spectral density of a bounded noise process. The figure illustrates the power spectral density for different values of noise amplitude and noise intensity.

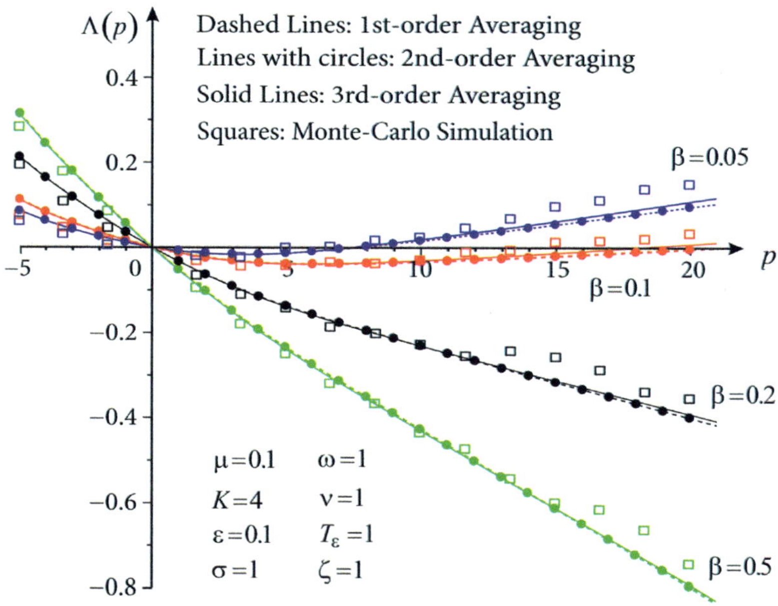

Figure 16.5 Effect of damping β on moment Lyapunov exponents. The figure shows the effect of damping on the moment Lyapunov exponents for 1st-, 2nd-, and 3rd-order averaging, as well as on Monte Carlo simulations.

large number of sinusoidal terms in the Fourier series Eq. (16.33) must be adopted to obtain a good approximation of the moment Lyapunov exponent $\Lambda(p)$. On the contrary, if the power spectral density function of the bounded noise is relatively flat,

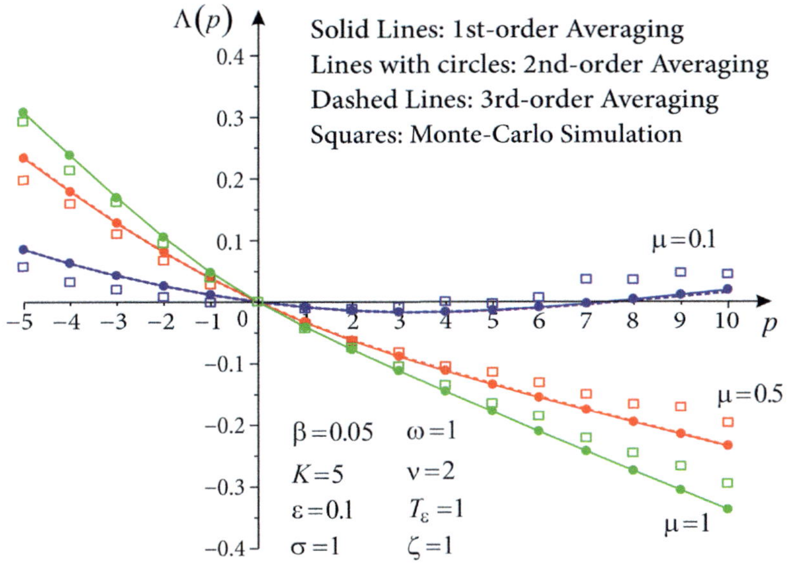

Figure 16.6 Effect of fractional orders μ on moment Lyapunov exponents. The figure shows the variation in moment Lyapunov exponents for 1st-, 2nd-, and 3rd-order averaging, as well as for Monte Carlo simulations across varying fractional orders.

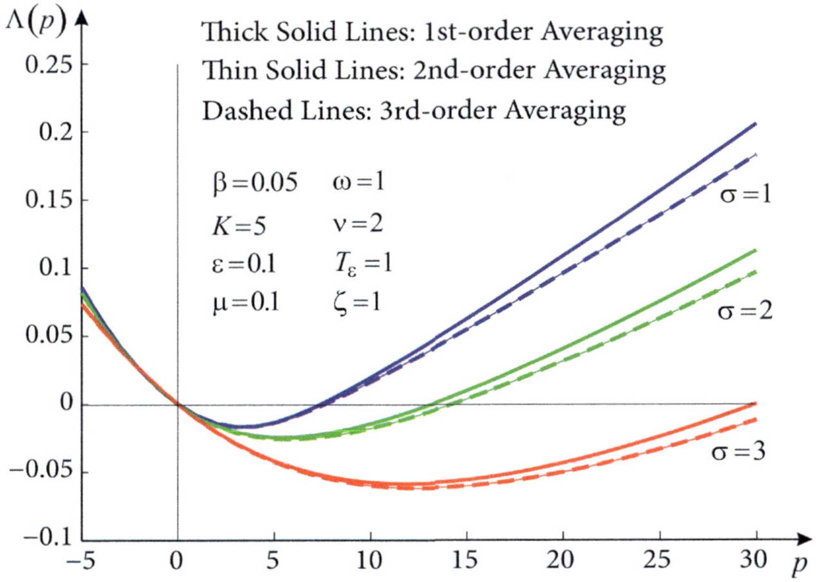

Figure 16.7 Effect of σ of bounded noise on moment Lyapunov exponents. The figure shows the variation in the moment Lyapunov exponents for 1st-, 2nd-, and 3rd-order averaging, as well as for Monte Carlo simulations, across different levels of noise intensity.

which implies that it is a wide-band process, fewer sinusoidal terms in the Fourier series Eq. (16.33) can provide a good enough approximation of $\Lambda(p)$.

The effect of damping β on moment Lyapunov exponents is depicted in Fig. 16.5. It shows that damping has a stabilizing effect on the stability of the system. The approximate moment Lyapunov exponents are compared with the results from the Monte Carlo simulation [7]. This comparison shows that the two results match well, although deviation occurs when p is large.

The effect of fractional orders μ on moment Lyapunov exponents is depicted in Fig. 16.6. The illustration shows that the parameter μ is substantial for the stability of the system. As the fractional-order μ increases, the stability of the system improves. When μ changes from 0 to 1, the system gradually becomes stabilized. As noted in Appendix A2, when μ varies from 0 to 1, the fractional Newton element in a fractional Kelvin–Voigt model changes from

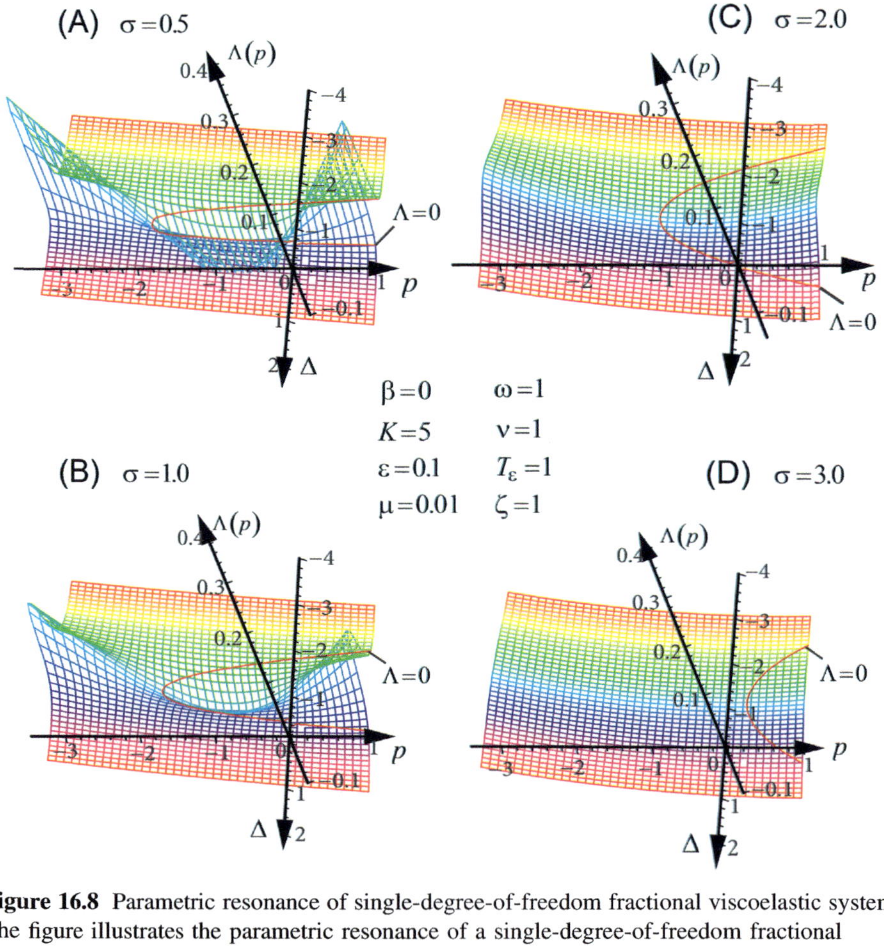

Figure 16.8 Parametric resonance of single-degree-of-freedom fractional viscoelastic system. The figure illustrates the parametric resonance of a single-degree-of-freedom fractional viscoelastic system under varying levels of noise intensity. (A) $\sigma = 0.5$, (A) $\sigma = 0.5$, (B) $\sigma = 2.0$, (D) $\sigma = 3.0$.

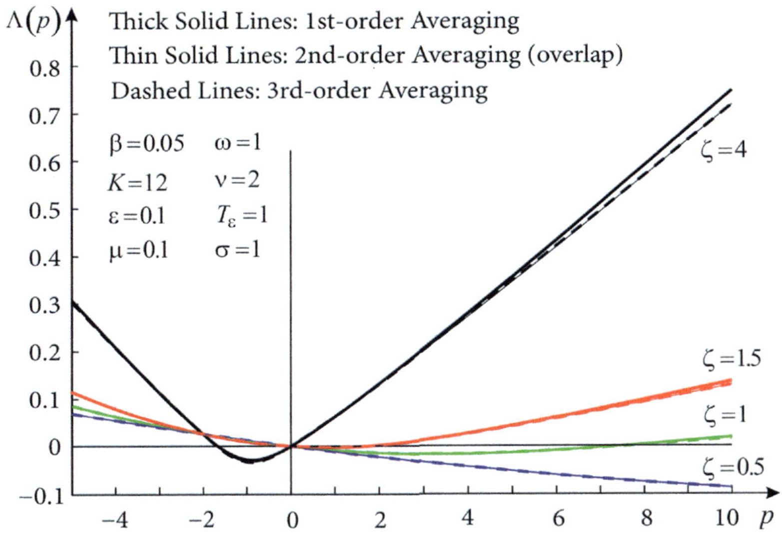

Figure 16.9 Effect of noise amplitude ζ on moment Lyapunov exponents. The figure illustrates the effect of noise amplitude on the moment Lyapunov exponents for 1st-, 2nd-, and 3rd-order averaging.

a Hooke element to an integer-value Newton element, that is, from pure elasticity to pure viscosity. Hence, the fractional-order μ enables a fluent transition from elasticity to viscosity in one element. This abandons the conventional binary theory, by which an element can either function purely elastic (a spring) or administer perfect viscosity (a dashpot). This multistate fractional component has multifarious levels of performance and many viscoelastic modes. As it is shown in Fig. 16.6, fractional components have major effects on the performance of the entire system.

It is illustrated in Fig. 16.7 that with the increase of the noise intensity parameter σ, the system's stability increases. Three-dimensional diagrams in Fig. 16.8A–D show that as the noise intensity σ gets smaller, the more significant the parametric resonance becomes. The effect of noise amplitude ζ on moment Lyapunov exponents is illustrated in Fig. 16.9, which implies the destabilizing effect of noise amplitude on the stability of the system. The effect of ζ on parametric resonance is observed in Fig. 16.10A–D. A larger noise amplitude would produce a more prominent parametric resonance. One plausible explanation is from the power spectrum density function of bounded noise in Eq. (16.10); in Fig. 16.4 as σ is larger or as ζ is smaller, the frequency band of the power spectrum will be wider. This implies that the power of the noise is not concentrated in the locality of the central frequency ν, which reduces the effect of the primary parametric resonance.

The system retardation factor τ_ε plays a stabilizing role in moment stability, as illustrated in Fig. 16.11. As can be seen, the stability index increases with the retardation factor. The stability index is the nontrivial zero δ of the moment stability $\Lambda(p)$, that is, $\Lambda(\delta) = 0$. This indicates the probability with which the response of an almost-surely system exceeds a threshold.

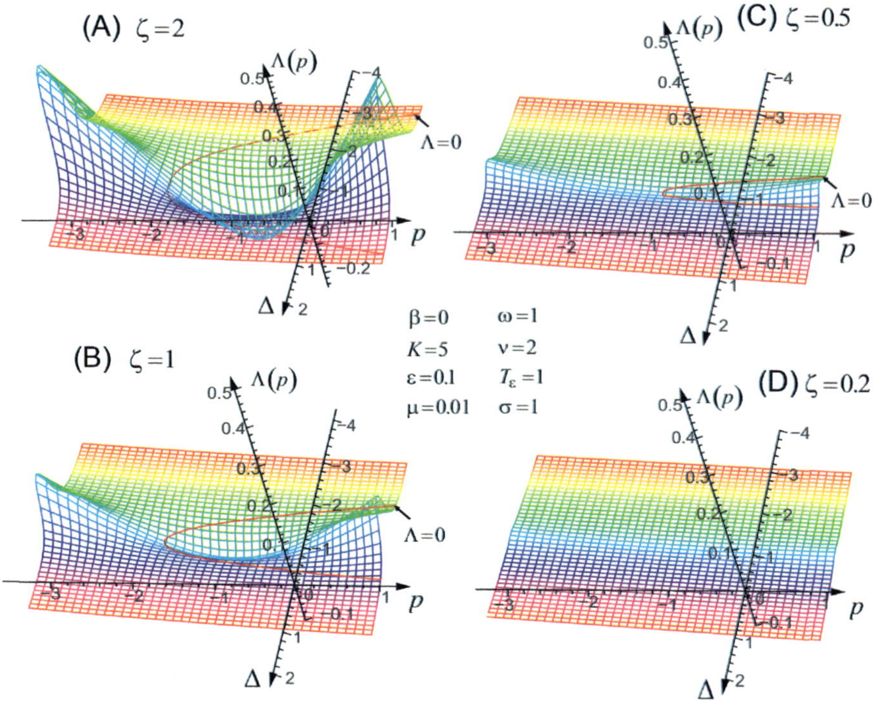

Figure 16.10 Parametric resonance of an single-degree-of-freedom fractional viscoelastic system. The figure illustrates the parametric resonance of a single-degree-of-freedom fractional viscoelastic system under varying values of noise amplitude. (A) $\zeta = 2$, (B) $\zeta = 1$, (C) $\zeta = 0.5$, (D) $\zeta = 2$.

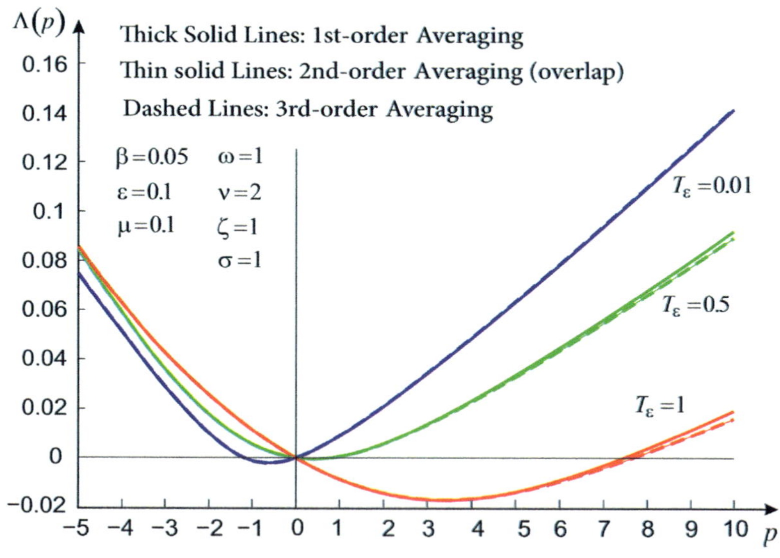

Figure 16.11 Effect of retardation τ_e on moment Lyapunov exponents. The figure shows the effect of retardation on the moment Lyapunov exponents for 1st-, 2nd-, and 3rd-order averaging.

The above figures suggest that there is some difference between first-order and higher-order averaging. Nonetheless, the third-order averaging is approximately identical to the second-order averaging, which implies that, for viscoelastic structures, the second-order averaging analysis is adequate.

16.4 Conclusions

In this chapter, the method of higher-order stochastic averaging is proposed and used for studying the stochastic stability of a fractional viscoelastic column under the excitation of bounded noise. The higher-order stochastic averaging is founded on lower-order stochastic averaging and is formulated by introducing a near-identity transformation. The stochastic stability of the system is analyzed by means of the moment Lyapunov exponent, which is an excellent approach for the characterization of moment stability. The moment Lyapunov exponent is determined numerically by solving a set of homogeneous linear equations. These equations are acquired from an eigenvalue problem by using the linear stochastic transformation and the Itô Lemma. A minute difference has been found between the results from first-order and higher-order averaging. However, results acquired from the third-order averaging are nearly equal to those of the second-order averaging, which suggests that for this viscoelastic structure, second-order averaging analysis is sufficient. The approximate results are in good agreement with the Monte Carlo simulation, which corroborates the higher-order stochastic averaging of this study. The convergence issue and the effects of various parameters on the stochastic stability of the system are discussed in detail and possible explanations are explored. Damping has a stabilizing effect on the system. As the fractional-order increases, the system's stability improves. With an increase in the bounded noise intensity parameter, the system's stability also increases. The system retardation factor plays a stabilizing role, with the stability index increasing as the retardation factor increases. However, the noise amplitude has a destabilizing effect on the system's stability, with larger noise amplitudes leading to more prominent parametric resonance. These results pave the way for the utilization or control of parametric resonances in engineering applications.

Appendix A1 Fractional calculus

The first three repeated integrations should be reduced to a single integral by using the usual change of variables formula

$$f^{(-1)}(t) = \int_0^t f(\tau)\mathrm{d}\tau,$$

$$f^{(-2)}(t) = \int_0^t \mathrm{d}\tau_2 \int_0^{\tau_2} f(\tau_1)\mathrm{d}\tau_1 = \int_0^t f(\tau_1)\mathrm{d}\tau_1 \int_{\tau_1}^t \mathrm{d}\tau_2 = \int_0^t (t-\tau_1) f(\tau_1)\mathrm{d}\tau_1,$$

$$f^{(-3)}(t) = \int_0^t \mathrm{d}\tau_3 \int_0^{\tau_3} \mathrm{d}\tau_2 \int_0^{\tau_2} f(\tau_1)\mathrm{d}\tau_1 = \int_0^t \mathrm{d}\tau_3 \int_0^{t_3} (t_3-\tau_1) f(\tau_1)\mathrm{d}\tau_1$$

$$= \int_0^t f(\tau_1)\mathrm{d}\tau_1 \int_{\tau_1}^t (t_3-\tau_1)\mathrm{d}\tau_3 = \frac{1}{2}\int_0^t (t-\tau_1) f(\tau_1)\mathrm{d}\tau_1. \tag{A1}$$

This can be continued until it becomes clear that

$$f^{(-n)}(t) = \frac{\mathrm{d}^{-n} f(t)}{\mathrm{d}t^{-n}} = \frac{1}{(n-1)!}\int_0^t (t-\tau)^{n-1} f(\tau)\mathrm{d}\tau, \quad n \in \mathscr{N}^+, \tag{A2}$$

for almost all x with $-\infty \le 0 < t \le \infty$, and $\mathscr{N}^+$ is a set of positive integers.

By replacing the fractional expression with its gamma function equivalent, the formula for repeated integration in Eq. (A2) for all $\mathscr{R}^+$ can be generated:

$$f^{(-\mu)}(t) = {}_0^{\mathrm{RL}}\boldsymbol{D}_t^{-\mu}[f(t)] = \frac{1}{\Gamma(\mu)}\int_0^t (t-\tau)^{n-1} f(\tau)\mathrm{d}\tau, \quad \mu \in \mathscr{R}^+, \tag{A3}$$

where μ is the fractional coefficient, $\mathscr{R}^+$ is a set of positive real numbers, ${}_0^{\mathrm{RL}}\boldsymbol{D}_t^{-\mu}[\cdot]$ is an operator for the Reimann–Liouville (RL) fractional integral of order μ, $\Gamma(\mu)$ is the gamma function:

$$\Gamma(\mu) = \int_0^\infty e^{-t} t^{\mu-1}\mathrm{d}t. \tag{A4}$$

if μ is a positive integer, then $\Gamma(\mu) = (\mu - 1)!$ so this definition of $f^{(-\mu)}(t)$ agrees with Eq. (A2) when μ is a positive integer. Eq. (A3) is the definition of RL fractional integral [15].

In a similar fashion, a fractional derivative can be defined from a repeated derivative of a function. Although, it is more fitting to use the fractional integral and, via it, to directly formulate a definition of fractional derivative. For a fixed $n \ge 1$ and integer $m \ge n$, the $(m - n)$-th derivative of the function $f(t)$ can be written as:

$$f^{(m-n)}(t) = f^{(m)}[f^{(-n)}(t)] = \frac{1}{\Gamma(\mu)}\boldsymbol{D}^m \int_0^t (t-\tau)^{n-1} f(\tau)\mathrm{d}\tau, \quad m \in \mathscr{N}^+, \tag{A5}$$

where the symbol $\boldsymbol{D}^m (m \ge 0)$ denotes m-th iterated differentiations. Substituting Eq. (A3) into Eq. (A5) makes:

$$\begin{gathered} {}_0^{\mathrm{RL}}\boldsymbol{D}_t^{m-\alpha}[f(t)] = {}_0^{\mathrm{RL}}\boldsymbol{D}_t^{m}[{}_0^{\mathrm{RL}}\boldsymbol{D}_t^{-\alpha} f(t)] = \frac{1}{\Gamma(\alpha)}\frac{\mathrm{d}^m}{\mathrm{d}t^m}\int_0^t (t-\tau)^{\alpha-1} f(\tau)\mathrm{d}\tau, \\ 0 < \alpha \le 1, \quad m \in \mathscr{N}^+. \end{gathered} \tag{A6}$$

by expressing $\mu = m - \alpha$, fractional derivatives can be written as:

$$ {}_0^{\mathrm{RL}}\boldsymbol{D}_t^{\mu}[f(t)] = \frac{1}{\Gamma(m-\mu)}\frac{\mathrm{d}^m}{\mathrm{d}t^m}\int_0^t \frac{f(\tau)}{(t-\tau)^{-(m-\mu)+1}}\mathrm{d}\tau, \quad m-1 \le \mu < m, \tag{A7}$$

where ${}_a^{\mathrm{RL}}\boldsymbol{D}_t^{\mu}[\cdot]$ is an operator for the RL fractional derivative of order μ. It can be observed that ${}_0^{\mathrm{RL}}\boldsymbol{D}_t^{0}[f(t)] = f(t)$ and ${}_0^{\mathrm{RL}}\boldsymbol{D}_t^{m}[f(t)] = f^{(m)}(t)$. Specifically, for

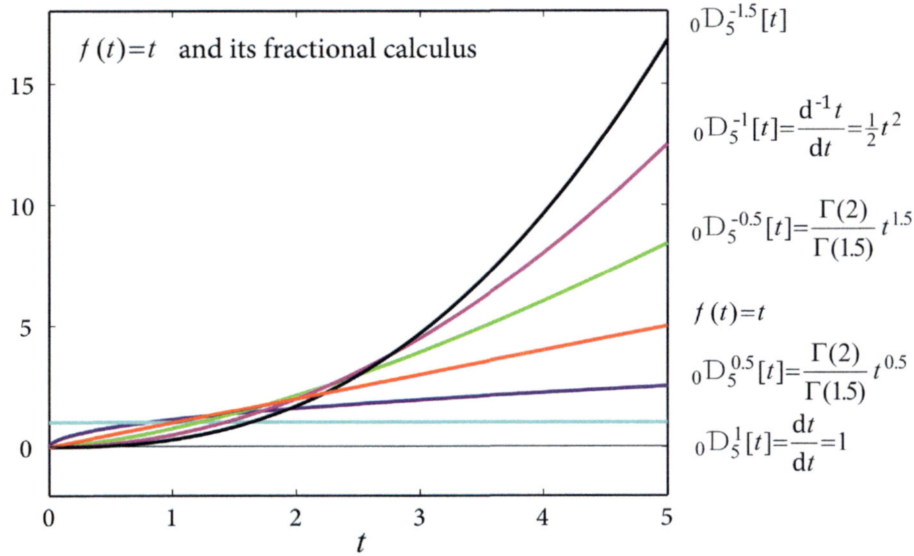

Figure 16.12 Fractional calculus of a power function. The figure illustrates the fractional calculus of a power function for varying orders of the fractional derivative.

$0 < \mu \leq 1$, assuming $f(t)$ is continuous for $t > 0$, it turns out (see Eq. [2.1.28] of Ref. [15]):

$$ {}^{RL}_{0}\boldsymbol{D}_{t}^{\mu}[f(t)] = \frac{1}{\Gamma(1-\mu)}\frac{\mathrm{d}}{\mathrm{d}t}\int_0^t \frac{f(\tau)}{(t-\tau)^{\mu}}\mathrm{d}\tau = \frac{1}{\Gamma(1-\mu)}\left[\frac{f(0)}{t^{\mu}} + \int_0^t \frac{f'(\tau)}{(t-\tau)^{\mu}}\mathrm{d}\tau\right]. \tag{A8} $$

Further assuming $f(0) = 0$ gives the fractional derivative in Caputo (C) form:

$$ {}^{C}_{0}\boldsymbol{D}_{t}^{\mu}[f(t)] = \frac{1}{\Gamma(1-\mu)}\int_0^t \frac{f'(\tau)}{(t-\tau)^{\mu}}\mathrm{d}\tau = \frac{1}{\Gamma(1-\mu)}\int_0^t \frac{f'(t-\tau)}{\tau^{\mu}}\mathrm{d}\tau. \tag{A9} $$

A diagram for a simple function $f(t) = t$ and its fractional calculus as a showcase ${}^{RL}_{0}\boldsymbol{D}_{t}^{\mu}[t] = t^{1-\mu}\Gamma(2)/\Gamma(2-\mu)$ is delineated in Fig. 16.12. The fractional calculus of arbitrary real order μ can be considered as an interpolation of the sequence of integer-order calculus.

Appendix A2 Fractional viscoelasticity

It is feasible to mathematically construct a fractional viscoelastic component, shown in Fig. 16.13, by using fractional calculus. The property of the fractional element can be changed from elastic to viscous by changing the value of μ from 0 to 1 (various in-between values can be used).

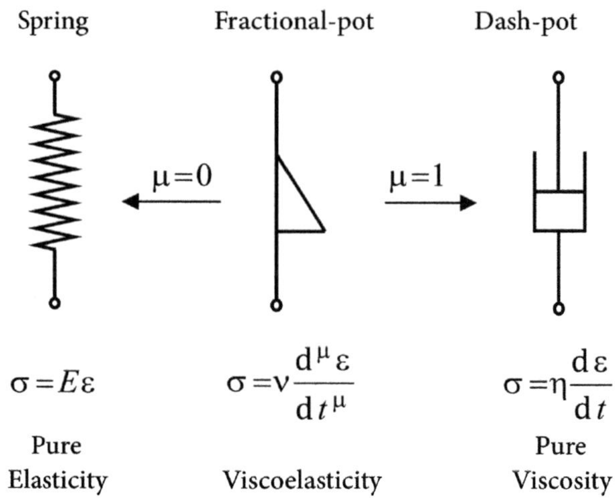

Figure 16.13 Schematic of a fractional element. The fractional element is between a spring ($\mu = 0$) and a dashpot ($\mu = 1$).

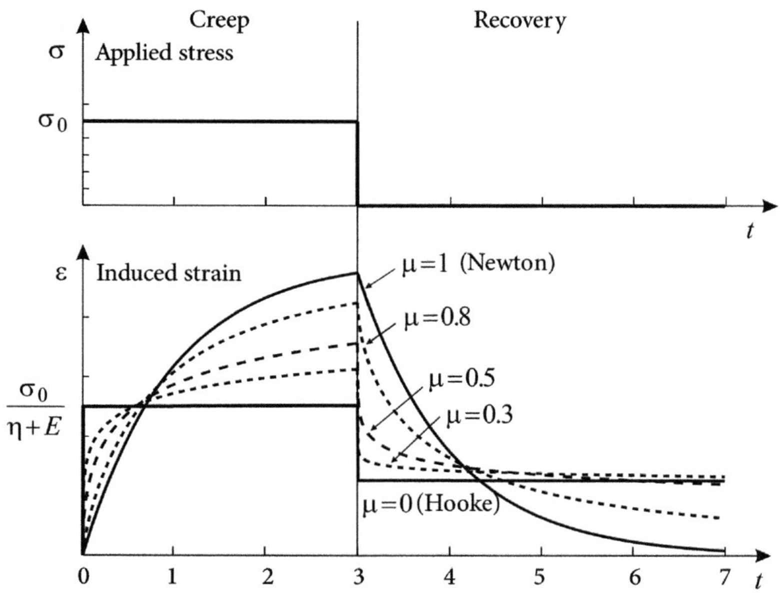

Figure 16.14 Creep and recovery for a fractional viscoelastic material. The figure shows the creep and recovery behavior of a fractional viscoelastic material for varying levels of elasticity, viscoelasticity, and up to the characteristics of pure viscosity.

When Hook springs, fractional pots, and dash-pots are combined in parallel or series, it is possible to formulate fractional viscoelasticity. The fractional Kelvin–Voigt viscoelastic model comprises a fractional Newton dash-pot and a Hooke spring in parallel, which is mathematically defined as:

$$\sigma(t) = E\varepsilon(t) + \eta \frac{d^{\mu}\varepsilon(t)}{dt^{\mu}} = E\varepsilon(t) + \eta\, {}^{RL}_{0}\boldsymbol{D}_{t}^{\mu}\left[\varepsilon(t)\right], \tag{A10}$$

where E is Young's elastic modulus and η is the viscosity modulus or coefficient of viscosity.

When $\mu = 0$, the fractional Newton part turns into a constant η, which is the case for the Hooke model. Furthermore, only when $\mu = 0$, this model manifests transient elasticity at $t = 0$. When $\mu = 1$, the relaxation modulus of the fractional Newton part turns into a Dirac delta function, that is, $\sigma(t) = \eta\delta(t)$, which is the case of the integer Newton model.

Ergo, the fractional Newton element is a continuum between the spring model and the Newton model (Fig. 16.13). A characteristic creep and recovery correlation for a viscoelastic material is depicted in Fig. 16.14, which indubitably shows that the extra degree-of-freedom from the fractional order can enhance the performance of traditional viscoelastic elements. In particular, when $\mu = 1$, this fractional model reduces to the ordinary integer Kelvin–Voigt model.

References

[1] R. Mainardi, Fractional Calculus and Waves in Linear Viscoelasticity: An Introduction to Mathematical Models, Imperial College Press, London, 2010.

[2] H. Xu, X. Jiang, Creep constitutive models for viscoelastic materials based on fractional derivatives, Computers & Mathematics with Applications 73 (6) (2017) 1377–1384.

[3] Q. Song, X. Yang, C. Li, T. Huang, X. Chen, Stability analysis of nonlinear fractional-order systems with variable-time impulses, Journal of the Franklin Institute 354 (2017) 2959–2978.

[4] W.-C. Xie, Dynamic Stability of Structures, Cambridge University Press, Cambridge, 2006.

[5] Q. Ling, X.L. Jin, Z.L. Huang, Response and stability of SDOF viscoelastic system under wideband noise excitations, Journal of the Franklin Institute 348 (2011) 2026–2043.

[6] Y. Xu, B. Pei, J.L. Wu, Stochastic averaging principle for differential equations with non-Lipschitz coefficients driven by fractional Brownian motion, Stochastics and Dynamics 17 (2017) 1750013.

[7] J. Deng, W.-C. Xie, M.D. Pandey, Stochastic stability of a fractional viscoelastic column under bounded noise excitation, Journal of Sound and Vibration 333 (6) (2014) 1629–1643.

[8] M. Hijawi, N. Moshchuk, R.A. Ibrahim, Unified second-order stochastic averaging approach, Journal of Applied Mechanics 64 (2) (1997) 281–291.

[9] Q. Huang, W.-C. Xie, Stability of SDOF linear viscoelastic system under the excitation of wideband noise, Journal of Applied Mechanics 75 (2) (2008) 021012.

[10] G. Janevski, P. Kozic, I. Pavlovic, Moment Lyapunov exponents of the parametrical Hill's equation under the excitation of two correlated wideband noises, Structural Engineering and Mechanics 52 (3) (2014) 525–540.

[11] J. Deng, W.-C. Xie, M.D. Pandey, Higher-order stochastic averaging to study stability of a fractional viscoelastic column, Journal of Sound and Vibration 333 (23) (2014) 6121–6139.
[12] C. Floris, Stochastic stability of a viscoelastic column axially loaded by a white noise force, Mechanics Research Communications 38 (2011) 57–61.
[13] W. Wedig, Analysis and simulation of nonlinear stochastic systems, in: W. Schiehlen (Ed.), Nonlinear Dynamics in Engineering Systems, Springer-Verlag, Berlin, 1989, pp. 337–344.
[14] N.S. Namachchivaya, H.J.V. Roessel, Moment Lyapunov exponent and stochastic stability of two coupled oscillators driven by real noise, Journal of Applied Mechanics 68 (6) (2001) 903–914.
[15] A.A. Kilbas, H.M. Srivastava, J.J. Trujillo, Theory and Applications of Fractional Differential Equations, Elsevier, New York, 2006.

Parametric stochastic stability of viscoelastic rotating shafts

17

17.1 Introduction

The dynamic stability of rotating shafts has gained respectful attention. This is because the stability and characterization of vibrations are imperatives to the analysis and optimum design of diverse mechanical systems that have a rotating body as one of their components. Tylikowski studied uniform stochastic stability of rotating shafts, where the axial force is defined as a wide-band Gaussian process [1]. Namachchivaya analyzed the mean square stability of a rotating shaft subjected to combined harmonic and stochastic excitations [2]. Young et al. [3,4] studied the stability of a pretwisted cantilever beam subjected to axial random forces and constant and nonconstant spin rates. Viscoelasticity is observed in a number of materials that are important in engineering applications, such as polymers and plastics, composite materials, and metals and alloys at elevated temperatures. These materials exhibit both elastic and viscous characteristics, which are crucial for their performance in various engineering contexts. By using the direct Lyapunov method, Pavlovic et al. [5] have obtained the almost-sure asymptotic stability conditions for a viscoelastic Voigt–Kelvin rotating shaft subjected to axial forces.

In modern dynamic stability theory, Lyapunov exponents signal almost-sure instability. Nevertheless, this sample stability cannot guarantee moment stability due to the theory of large deviation [6]. The moment Lyapunov exponents, however, can give information on both—the moment stability and almost-sure stability [7]. The moment Lyapunov exponents are important for the dynamic stability analysis, but the determination of the moment Lyapunov exponents is difficult, and the exact evaluation is almost impossible. Thereupon, some perturbation approximations [8] and asymptotic expansion series [9] for the moment Lyapunov exponents of a two-dimensional linear stochastic system have been defined. Namachchivaya and Van Roessel [10] conducted a study on the moment Lyapunov exponents of coupled elastic oscillators with two-degrees-of-freedom (2-DOF), subjected to real noise excitations. Vedula completed his Ph.D. thesis on the stability of parametrically excited linear and nonlinear gyroscopic systems [11]; however, he did not consider viscoelasticity. Kozić et al. [12] determined weak noise expansions of the maximum Lyapunov exponents and moment Lyapunov exponents for stochastically coupled 2-DOF systems.

The study presented in this chapter is the investigation of the moment stability and almost-sure stability of viscoelastic rotating shafts subjected to white noise excitation. A full overview of the dynamic stability can be gained by investigating both of the aforementioned quantities.

Modeling of Complex Dynamic Systems. DOI: https://doi.org/10.1016/B978-0-443-23942-7.00017-6

17.2 Vibration of a rotating shaft

A rotor-bearing system shown in Fig. 17.1 is used for this analysis. The rotor is mounted on two identical rigid bearings; it rotates about the centerline Oz with a constant angular velocity W and is subjected to a dynamic axial compressive load $P(t)$. The motion of the system is described via the nonrotating Cartesian coordinates xyz and the rotating coordinates uwz, which rotates with the shaft about the z-axis with angular velocity W. The parameters of the shaft are length L, mass per unit length m, coefficient of external damping D_e, and flexural rigidities EI_u and EI_w (in the Ou and Ow directions, respectively). The internal damping and the torsional rigidities of the shaft are minute and thus without meaningful effect.

For this elastic shaft, the transverse flexural equations of motion in the rotating coordinate system are given by Refs. [7,13]:

$$EI_u\frac{\partial^4 u}{\partial z^4} + P(t)\frac{\partial^2 u}{\partial z^2} + m\left(\frac{\partial^2 u}{\partial t^2} - 2W\frac{\partial w}{\partial t} - W^2 u\right) + D_e\left(\frac{\partial u}{\partial t} - Ww\right) = 0,$$
$$I_w\frac{\partial^4 w}{\partial z^4} + P(t)\frac{\partial^2 w}{\partial z^2} + m\left(\frac{\partial^2 w}{\partial t^2} + 2W\frac{\partial u}{\partial t} - W^2 w\right) + D_e\left(\frac{\partial w}{\partial t} + Wu\right) = 0, \quad (17.1)$$

with boundary conditions

$$u(0, t) = u(L, t) = \frac{\partial^2 u(0, t)}{\partial z^2} = \frac{\partial^2 u(L, t)}{\partial z^2} = 0,$$

$$w(0, t) = w(L, t) = \frac{\partial^2 w(0, t)}{\partial z^2} = \frac{\partial^2 w(L, t)}{\partial z^2} = 0, \quad (17.2)$$

for the case of simply supported ends for small bearings.

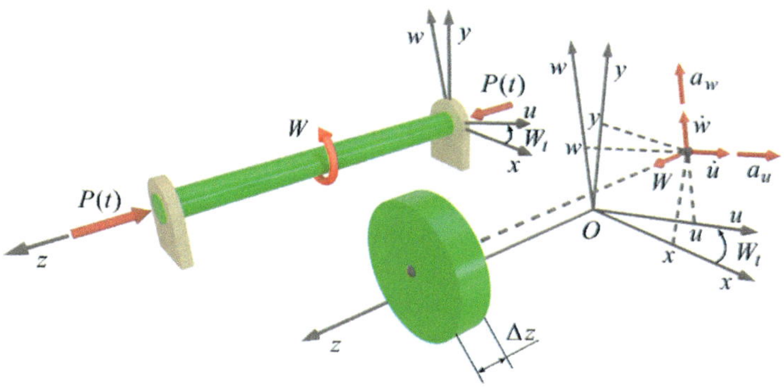

Figure 17.1 Vibration of a rotating shaft. Model for this can be a rotor-bearing system.

In the problems with viscoelastic materials, the elastic modulus E (in the elastic problem) needs to be replaced by the Voltera operators $E(1-\mathbb{H})$ [14], where $\mathbb{H}$ is a linear viscoelastic (relaxation) operator defined by

$$\mathbb{H}[\psi(t)] = \int_0^t \mathscr{H}(t-\tau)\psi(t)\mathrm{d}\tau, \quad 0 \le \int_0^t \mathscr{H}(\theta)\mathrm{d}\theta < 1, \tag{17.3}$$

where $\mathscr{H}(\theta)$ is the relaxation kernel.

The governing equations for a viscoelastic rotating shaft can be derived from Eq. (17.1) as:

$$E(1-\mathbb{H})I_u\frac{\partial^4 u}{\partial z^4} + P(t)\frac{\partial^2 u}{\partial z^2} + m\left(\frac{\partial^2 u}{\partial t^2} - 2W\frac{\partial w}{\partial t} - W^2 u\right) + D_e\left(\frac{\partial u}{\partial t} - Ww\right) = 0,$$

$$E(1-\mathbb{H})I_w\frac{\partial^4 w}{\partial z^4} + P(t)\frac{\partial^2 w}{\partial z^2} + m\left(\frac{\partial^2 w}{\partial t^2} + 2W\frac{\partial u}{\partial t} - W^2 w\right) + D_e\left(\frac{\partial w}{\partial t} + Wu\right) = 0. \tag{17.4}$$

These are partial differential equations (PDEs). By applying the Galerkin method and taking

$$u(z,t) = q_1(t)\sin\frac{\pi z}{L}, \quad w(z,t) = q_2(t)\sin\frac{\pi z}{L}, \tag{17.5}$$

the PDEs in Eq. (17.4) can be converted into two ordinary differential equations (ODEs), that is, multiplying Eq. (17.4) by $\sin\pi z/L$ and integrating from 0 to L,

$$q_1'' + \frac{D_e}{m}q_1' - 2Wq_2' + \left\{\frac{\pi^2}{mL^2}\left[\frac{\pi^2 EI_u}{L^2}(1-\mathbb{H}) - P(t)\right] - W^2\right\}q_1 - W\frac{D_e}{m}q_2 = 0,$$

$$q_2'' + \frac{D_e}{m}q_2' + 2Wq_1' + \left\{\frac{\pi^2}{mL^2}\left[\frac{\pi^2 EI_w}{L^2}(1-\mathbb{H}) - P(t)\right] - W^2\right\}q_2 + W\frac{D_e}{m}q_1 = 0. \tag{17.6}$$

It should be noted that Eq. (17.6) considers only the first mode because in most cases the first mode vibration dominates dynamic stability. The axial force $P(t)$ is assumed to be a stationary stochastic process having a mean value P_0 and a fluctuating part $\zeta(t)$.

$$P(t) = P_0 + \zeta(t), \quad \mathrm{E}[\zeta(t)] = 0, \tag{17.7}$$

where $\mathrm{E}[\cdot]$ denotes the expected value. When the nondimensional time t_1 is introduced by

$$t_1 = \tilde{\omega}_1 t = \frac{\pi^2}{L^2}\sqrt{\frac{EI_u}{m}}\,t, \tag{17.8}$$

and the dot is used to denote differentiation with respect to the new time t_1, Eq. (17.6) becomes:

$$\ddot{q}_1 + 2\beta\dot{q}_1 - 2\Omega\dot{q}_2 + [\overline{\omega}_1^2 - \Omega^2 - \mathbb{H} - \tilde{p}(t)]q_1 - 2\beta\Omega q_2 = 0,$$

$$\ddot{q}_2 + 2\beta\dot{q}_2 + 2\Omega\dot{q}_1 + [\overline{\omega}_2^2 - \Omega^2 - g_2\mathbb{H} - \widetilde{p}(t)]q_2 + 2\beta\Omega q_1 = 0, \tag{17.9}$$

or in a matrix form:

$$\ddot{\mathbf{q}} + 2\beta\dot{\mathbf{q}} + 2\Omega\begin{bmatrix} 0 & -1 \\ 1 & 0 \end{bmatrix}\dot{\mathbf{q}} + \begin{bmatrix} \overline{\omega}_1^2 - \Omega^2 - \mathbb{H} & -2\beta\Omega \\ 2\beta\Omega & \overline{\omega}_2^2 - \Omega^2 - g_2\mathbb{H} \end{bmatrix}\mathbf{q} - \widetilde{p}(t)\mathbf{q} = 0, \tag{17.10}$$

where $\mathbf{q} = \{q_1, q_2\}^T$ and

$$\overline{\omega}_1^2 = \frac{\pi^2}{mL^2}\frac{\pi^2 EI_u}{L^2}, \quad \overline{\omega}_2^2 = \frac{\pi^2}{mL^2}\frac{\pi^2 EI_w}{L^2},$$

$$\beta = \frac{D_e}{2m\widetilde{\omega}_1}, \quad \widetilde{p}(t) = \frac{\pi^2}{mL^2\widetilde{\omega}_1^2}\zeta(t), \quad \Omega = \frac{W}{\widetilde{\omega}_1}, \quad g_2 = \frac{\widetilde{\omega}_2}{\widetilde{\omega}_1},$$

$$\overline{\omega}_1^2 = 1 - \frac{\pi^2}{mL^2\widetilde{\omega}_1^2}P_0, \quad \overline{\omega}_2^2 = g_2 - \frac{\pi^2}{mL^2\widetilde{\omega}_1^2}P_0. \tag{17.11}$$

17.3 Stochastic averaging

To show that damping, viscoelasticity, and parametric excitation are small, a small parameter $0 < \varepsilon \ll 1$ is introduced; then the gyroscopic system in Eq. (17.10) becomes:

$$\ddot{\mathbf{q}} + 2\varepsilon\beta\dot{\mathbf{q}} + 2\Omega\begin{bmatrix} 0 & -1 \\ 1 & 0 \end{bmatrix}\dot{\mathbf{q}}$$

$$+ \begin{bmatrix} \overline{\omega}_1^2 - \Omega^2 - \varepsilon\mathbb{H} & -2\varepsilon\beta\Omega \\ 2\varepsilon\beta\Omega & \overline{\omega}_2^2 - \Omega^2 - \varepsilon g_2\mathbb{H} \end{bmatrix}\mathbf{q} - \sqrt{\varepsilon}\xi(t)\mathbf{q} = 0, \tag{17.12}$$

where $\mathbf{q} = \{q_1, q_2\}^{\mathrm{T}}$ are state coordinates, $\overline{\omega}_{1,2}$ are two quantities with $\overline{\omega}_1 < \overline{\omega}_2$, β is the damping coefficient, Ω is the axial angular velocity, $\mathbb{H}$ is the linear viscoelastic operator in Eq. (17.3), and $\xi(t)$ is the excitation, which is modeled as the Gaussian white noise process. Considering that the excitation is in the parameter of the equations, the system is an example of a parametric random excitation system.

Eq. (17.12) accepts the trivial solution $q_1 = q_2 = 0$. To implement the averaging method, first, the unperturbed system should be considered, that is, $\varepsilon = 0$ and $\xi(t) = 0$, which reduces to:

$$\ddot{q}_1 + (\overline{\omega}_1^2 - \Omega^2)q_1 - 2\Omega\dot{q}_2 = 0, \quad \ddot{q}_2 + (\overline{\omega}_2^2 - \Omega^2)q_2 + 2\Omega\dot{q}_1 = 0. \tag{17.13}$$

This is a system of two linear second-order ODEs, which can be solved by the method of the operator. Employing the $\mathfrak{D}$ – operator, $\mathfrak{D}(\cdot) \equiv \mathrm{d}(\cdot)/\mathrm{d}t$, the differential equations turn into:

$$[\mathfrak{D}^2 + (\overline{\omega}_1^2 - \Omega^2)]q_1 - 2\Omega\mathfrak{D}q_2 = 0, \quad 2\Omega\mathfrak{D}q_1 + [\mathfrak{D}^2 + (\overline{\omega}_2^2 - \Omega^2)]q_2 = 0. \tag{17.14}$$

The determinant of the coefficient matrix is:

$$\phi(\mathfrak{D}) = \begin{vmatrix} \mathfrak{D}^2 + (\overline{\omega}_1^2 - \Omega^2) & -2\Omega\mathfrak{D} \\ 2\Omega\mathfrak{D} & \mathfrak{D}^2 + (\overline{\omega}_2^2 - \Omega^2) \end{vmatrix} =$$

$$= \mathfrak{D}^4 + (\overline{\omega}_1^2 + \overline{\omega}_2^2 + 2\Omega^2)\mathfrak{D}^2 + (\overline{\omega}_1^2 - \Omega^2)(\overline{\omega}_2^2 - \Omega^2). \tag{17.15}$$

Thus, the characteristic equation is as follows:

$$\phi(\chi) = \chi^4 + (\overline{\omega}_1^2 + \overline{\omega}_2^2 + 2\Omega^2)\chi^2 + (\overline{\omega}_1^2 - \Omega^2)(\overline{\omega}_2^2 - \Omega^2),$$

and its solutions are:

$$\chi_{1,2}^2 = -\omega_{1,2}^2 =$$

$$= \frac{1}{2}[-(\overline{\omega}_1^2 + \overline{\omega}_2^2 + 2\Omega^2) \pm \sqrt{(\overline{\omega}_1^2 + \overline{\omega}_2^2 + 2\Omega^2)^2 - 4(\overline{\omega}_1^2 - \Omega^2)(\overline{\omega}_2^2 - \Omega^2)}\,], \tag{17.16}$$

where $\omega_{1,2}^2$ can be rewritten as:

$$\omega_{1,2}^2 = \frac{1}{2}[(\overline{\omega}_1^2 + \overline{\omega}_2^2 + 2\Omega^2) \mp \sqrt{(\overline{\omega}_1^2 - \overline{\omega}_2^2)^2 + 8\Omega^2(\overline{\omega}_1^2 + \overline{\omega}_2^2)}\,]. \tag{17.17}$$

The undamped and unperturbed system will be stable if the eigenfrequencies $\chi_{1,2}$ are pure imaginary numbers, which also means that $\omega_{1,2}$ are real numbers. This is possible if $(\overline{\omega}_1^2 - \Omega^2)(\overline{\omega}_2^2 - \Omega^2) > 0$, that is, $\Omega < \overline{\omega}_1$, and $\Omega > \overline{\omega}_2$, which implies that $0 \le \Omega \le \overline{\omega}_1$ and $\Omega \ge \overline{\omega}_2$ are the stability regions. The system is unstable when $\overline{\omega}_1 \le \Omega \le \overline{\omega}_2$. Therefore, $\overline{\omega}_1$ and $\overline{\omega}_2$ are the two critical velocities of the system.

The complementary solutions of q_1 and q_2 in Eq. (17.13) are defined by

$$q_1 = A_{11}\cos\omega_1 t + A_{12}\sin\omega_1 t + A_{13}\cos\omega_2 t + A_{14}\sin\omega_2 t,$$

$$q_2 = A_{21}\cos\omega_1 t + A_{22}\sin\omega_1 t + A_{23}\cos\omega_2 t + A_{24}\sin\omega_2 t, \tag{17.18}$$

which contain eight arbitrary constants. Yet, since $\phi(\mathfrak{D})$ is a polynomial of degree 4 in $\mathfrak{D}$, the complementary solutions should comprise just four arbitrary constants. The four extra constants can be eliminated by substituting Eq. (17.18) into the first equation of Eq. (17.13), which gives:

$$\frac{A_{22}}{A_{11}} = -\frac{\omega_1^2 - (\overline{\omega}_1^2 - \Omega^2)}{2\Omega\omega_1}, \quad \frac{A_{21}}{A_{12}} = \frac{\omega_1^2 - (\overline{\omega}_1^2 - \Omega^2)}{2\Omega\omega_1},$$

$$\frac{A_{24}}{A_{13}} = -\frac{\omega_2^2 - (\overline{\omega}_1^2 - \Omega^2)}{2\Omega\omega_2}, \quad \frac{A_{23}}{A_{14}} = \frac{\omega_2^2 - (\overline{\omega}_1^2 - \Omega^2)}{2\Omega\omega_2}. \tag{17.19}$$

Likewise, substituting Eq. (17.18) into the second equation of Eq. (17.13) yields:

$$\frac{A_{22}}{A_{11}} = -\frac{2\Omega\omega_1}{\omega_1^2 - (\overline{\omega}_2^2 - \Omega^2)}, \quad \frac{A_{21}}{A_{12}} = \frac{2\Omega\omega_1}{\omega_1^2 - (\overline{\omega}_2^2 - \Omega^2)},$$

$$\frac{A_{24}}{A_{13}} = -\frac{2\Omega\omega_2}{\omega_2^2 - (\overline{\omega}_2^2 - \Omega^2)}, \quad \frac{A_{23}}{A_{14}} = \frac{2\Omega\omega_2}{\omega_2^2 - (\overline{\omega}_2^2 - \Omega^2)}. \tag{17.20}$$

Comparing Eqs. (17.19) and (17.20) yields:

$$\frac{\omega_i^2 - (\overline{\omega}_1^2 - \Omega^2)}{2\Omega\omega_i} = \frac{2\Omega\omega_i}{\omega_i^2 - (\overline{\omega}_2^2 - \Omega^2)} = \alpha_i, \quad i = 1,\ 2. \tag{17.21}$$

Substituting Eq. (17.21) into (17.19) and then into (17.18) gives the solution of Eq. (17.13)

$$q_1 = \alpha_{11}a_1 \sin\Phi_1 + \alpha_{22}a_2 \sin\Phi_2,$$

$$q_2 = \alpha_{11}a_1\alpha_1 \cos\Phi_1 + \alpha_{22}a_2\alpha_2 \sin\Phi_2, \tag{17.22}$$

where $\Phi_1 = \omega_1 t + \phi_1$ and $\Phi_2 = \omega_2 t + \phi_2$. The four constants are a_1, a_2, ϕ_1, and ϕ_2. To make the evaluation of the moment Lyapunov exponent somewhat simpler, α_{11} and α_{22} are added as coordinate scaling parameters.

To determine the solutions of the original perturbed system in Eq. (17.12), the method of variation of parameters is applied. Varying the parameters a_1, a_2, ϕ_1, and ϕ_2 in Eq. (17.22) to make them the functions of t gives the solutions in the form:

$$q_1 = \alpha_{11}a_1(t) \sin\Phi_1 + \alpha_{22}a_2(t) \sin\Phi_2,$$

$$q_2 = \alpha_{11}a_1(t)\alpha_1 \cos\Phi_1 + \alpha_{22}a_2(t)\alpha_2 \sin\Phi_2, \tag{17.23}$$

where a_r and Φ_r, $r = 1,\ 2$, are solutions of the following four first-order equations:

$$\dot{a}_1 = \frac{A}{\alpha_{11}}(+B_1 F \cos\Phi_1 - G \sin\Phi_1),$$

$$\dot{a}_1\dot{\Phi}_1 = \frac{A}{\alpha_{11}}(-B_1 F \sin\Phi_1 - G \cos\Phi_1),$$

$$\dot{a}_2 = \frac{A}{\alpha_{22}}(-B_2 F \cos\Phi_2 - G \sin\Phi_2),$$

$$\dot{a}_2\dot{\Phi}_2 = \frac{A}{\alpha_{22}}(+B_2 F \sin\Phi_2 + G \cos\Phi_2), \tag{17.24}$$

where

$$A = \frac{1}{\alpha_1\omega_1 - \alpha_2\omega_2} = \frac{2\Omega}{\omega_1^2 - \omega_2^2},$$

$$B_1 = -\frac{\alpha_2(\alpha_1\omega_1 - \alpha_2\omega_2)}{\alpha_1\omega_2 - \alpha_2\omega_1} = -\frac{\alpha_2\omega_1\omega_2}{\overline{\omega}_1^2 - \Omega^2}$$

$$= -\frac{\omega_1}{\overline{\omega}_1^2 - \Omega^2}\,\frac{\omega_2^2 - (\overline{\omega}_1^2 - \Omega^2)}{2\Omega}$$

$$= -\frac{\omega_1}{\overline{\omega}_1^2 - \Omega^2} \frac{\frac{(\overline{\omega}_1^2 - \Omega^2)(\overline{\omega}_2^2 - \Omega^2)}{\omega_1^2} - (\overline{\omega}_1^2 - \Omega^2)}{2\Omega} = \frac{1}{\alpha_1},$$

$$B_2 = -\frac{\alpha_1(\alpha_1\omega_1 - \alpha_2\omega_2)}{\alpha_1\omega_2 - \alpha_2\omega_1} = -\frac{\alpha_1\omega_1\omega_2}{\overline{\omega}_1^2 - \Omega^2}$$

$$= -\frac{\omega_1}{\overline{\omega}_1^2 - \Omega^2} \frac{\omega_1^2 - (\overline{\omega}_1^2 - \Omega^2)}{2\Omega}$$

$$= -\frac{\omega_1}{\overline{\omega}_1^2 - \Omega^2} \frac{\frac{(\overline{\omega}_1^2 - \Omega^2)(\overline{\omega}_2^2 - \Omega^2)}{\omega_2^2} - (\overline{\omega}_1^2 - \Omega^2)}{2\Omega} = \frac{1}{\alpha_2},$$

$$F = -\varepsilon\{-\mathbb{H}[q_1] + 2\beta\dot{q}_1 - 2\beta\Omega q_2\} + \sqrt{\varepsilon}\xi(t)q_1,$$

$$G = -\varepsilon\{-g_2\mathbb{H}[q_2] + 2\beta\dot{q}_2 + 2\beta\Omega q_1\} + \sqrt{\varepsilon}\xi(t)q_2.$$

On deriving Eq. (17.24), Eq. (17.16) and the relations $\omega_1^2 + \omega_2^2 = \overline{\omega}_1^2 + \overline{\omega}_2^2 - 2\Omega^2$ and $\omega_1^2\omega_2^2 = (\overline{\omega}_1^2 - \Omega^2)(\overline{\omega}_2^2 - \Omega^2)$ are used. Eq. (17.24) can be expanded as:

$$\dot{a}_r = \sqrt{\varepsilon}F_{a,r}^{(0)} + \varepsilon F_{a,r}^{(1)}, \quad \dot{\phi}_r = \sqrt{\varepsilon}F_{\phi,r}^{(0)} + \varepsilon F_{\phi,r}^{(1)}, \quad r = 1,\ 2, \tag{17.25}$$

where

$$F_{a,r}^{(0)} = \frac{(-1)^r\, 2\Omega\, \xi(t)}{\alpha_r\alpha_{rr}(\omega_1^2 - \omega_2^2)}\left\{-\cos\Phi_r \sum_{i=1}^{2} \alpha_{ii}a_i \sin\Phi_i + a_r \sin\Phi_r \sum_{i=1}^{2} \alpha_{ii}\alpha_i a_i \cos\Phi_i\right\},$$

$$F_{a,r}^{(1)} = \frac{(-1)^r\, 2\Omega}{\alpha_r\alpha_{rr}(\omega_1^2 - \omega_2^2)}\Bigg\{-\cos\Phi_r\Bigg[2\beta \sum_{i=1}^{2} \alpha_{ii}a_i\omega_i\cos\Phi_i$$

$$-2\beta\Omega \sum_{i=1}^{2} \alpha_{ii}\alpha_i a_i \cos\Phi_i - \sum_{i=1}^{2} \alpha_{ii}a_i\mathbb{H}[\sin\Phi_i]\Bigg]\Bigg]$$

$$-\alpha_r \sin\Phi_r\Bigg[2\beta \sum_{i=1}^{2} \alpha_{ii}\alpha_i a_i\omega_i \sin\Phi_i - 2\beta\Omega \sum_{i=1}^{2} \alpha_{ii}a_i \sin\Phi_i + \sum_{i=1}^{2} \alpha_i\alpha_{ii}a_i e_i\mathbb{H}[\cos\Phi_i]\Bigg]\Bigg\},$$

$$F_{\phi,r}^{(0)} = \frac{(-1)^r\, 2\Omega\, \xi(t)}{a_r\alpha_r\alpha_{rr}(\omega_1^2 - \omega_2^2)}\left\{\sin\Phi_r \sum_{i=1}^{2} \alpha_{ii}a_i \sin\Phi_i + \alpha_r \cos\Phi_r \sum_{i=1}^{2} \alpha_{ii}\alpha_i a_i \cos\Phi_i\right\},$$

$$F_{\phi,r}^{(1)} = \frac{(-1)^r\, 2\Omega}{a_r\alpha_r\alpha_{rr}(\omega_1^2 - \omega_2^2)}\Bigg\{-\sin\Phi_r\Bigg[-2\beta \sum_{i=1}^{2} \alpha_{ii}a_i\omega_i\cos\Phi_i$$

$$+2\beta\Omega \sum_{i=1}^{2} \alpha_{ii}\alpha_i a_i \cos\Phi_i + \sum_{i=1}^{2} \alpha_{ii}a_i\mathbb{H}[\sin\Phi_i]\Bigg]\Bigg]$$

$$+ \alpha_r \cos\Phi_r \left[-2\beta \sum_{i=1}^{2} \alpha_{ii}\alpha_i a_i \omega_i \sin\Phi_i + 2\beta \sum_{i=1}^{2} \alpha_{ii} a_i \sin\Phi_i \right.$$

$$\left. \left. - \sum_{i=1}^{2} \alpha_i \alpha_{ii} a_i e_i \mathbb{H}[\cos\Phi_i] \right] \right\}. \tag{17.26}$$

If the compressive axial force is modeled as white noise, then diffusive processes for the averaged amplitudes a_r and phase angles Φ_r can be obtained by applying the method of stochastic averaging to Eq. (17.25).

$$\mathrm{d}a_r = \varepsilon m_r^a \mathrm{d}t + \sqrt{\varepsilon} \sum_{s=1}^{2} \sigma_{rs}^a \mathrm{d}W_s^a, \tag{17.27a}$$

$$\mathrm{d}\phi_r = \varepsilon m_r^\phi \mathrm{d}t + \sqrt{\varepsilon} \sum_{s=1}^{2} \sigma_{rs}^\phi \mathrm{d}W_s^\phi, \quad r = 1,\ 2, \tag{17.27b}$$

where $W^a = \{W_1^a,\ W_2^a\}^T$ and $W^\phi = \{W_1^\phi,\ W_2^\phi\}^T$ are two-dimensional vectors of independent standard Wiener processes with the drift coefficients εm_r^a, εm_r^ϕ, and the 2×2 diffusion matrices $\mathbf{b}^a = \sigma^a(\sigma^a)^T$, $\mathbf{b}^\phi = \sigma^\phi(\sigma^\phi)^T$, where $\sigma^a = [\sigma_{rs}^a]$, $\sigma^\phi = [\sigma_{rs}^\phi]$, $\mathbf{b}^a = [b_{rs}^a]$, and $\mathbf{b}^\phi = [b_{rs}^\phi]$. The drift and diffusion coefficients of the amplitude are defined by

$$m_1^a = \mathop{\mathscr{M}}_t \left\{ F_{a,1}^{(1)} + \int_{-\infty}^{0} \mathrm{E}\left[\sum_{j=1}^{2} \left(\frac{\partial F_{a,1}^{(0)}}{\partial a_j} F_{a,j\tau}^{(0)} + \frac{\partial F_{a,1}^{(0)}}{\partial \phi_j} F_{\phi,j\tau}^{(0)} \right) \right] \mathrm{d}\tau \right\}$$

$$= a_1 E_1 + V_0 \left[(3V_1 + 2V_3) a_1 + \frac{\alpha_{21}^2 D_2 a_2^2}{a_1} \right],$$

$$m_2^a = \mathop{\mathscr{M}}_t \left\{ F_{a,2}^{(1)} + \int_{-\infty}^{0} \mathrm{E}\left[\sum_{j=1}^{2} \left(\frac{\partial F_{a,2}^{(0)}}{\partial a_j} F_{a,j\tau}^{(0)} + \frac{\partial F_{\phi,2}^{(0)}}{\partial \phi_j} F_{\phi,j\tau}^{(0)} \right) \right] \mathrm{d}\tau \right\}$$

$$= a_2 E_2 + V_0 \left[(3V_2 + 2V_3) a_2 + \frac{\alpha_{12}^2 D_1 a_1^2}{a_2} \right],$$

$$b_{11}^a = \mathop{\mathscr{M}}_t \left\{ \int_{-\infty}^{\infty} \mathrm{E}[F_{a,1}^{(0)} F_{a,1\tau}^{(0)}] \mathrm{d}\tau \right\} = 2V_0 (V_1 a_1^2 + \alpha_{21}^2 D_2 a_2^2),$$

$$b_{12}^a = \mathop{\mathscr{M}}_t \left\{ \int_{-\infty}^{\infty} \mathrm{E}[F_{a,1}^{(0)} F_{a,2\tau}^{(0)}] \mathrm{d}\tau \right\} = 2V_0 V_3 a_1 a_2,$$

$$b_{22}^a = \mathop{\mathscr{M}}_t \left\{ \int_{-\infty}^{\infty} \mathrm{E}[F_{a,2}^{(0)} F_{a,2\tau}^{(0)}] \mathrm{d}\tau \right\} = 2V_0 (V_2 a_2^2 + \alpha_{12}^2 D_1 a_1^2), \tag{17.28}$$

where the constants are given by

$$\alpha_{12} = \frac{\alpha_{11}}{\alpha_{22}}, \quad \alpha_{21} = \frac{\alpha_{22}}{\alpha_{11}}, \quad V_0 = \frac{\Omega^2}{4(\omega_1^2 - \omega_2^2)^2}, \quad V_1 = \left(\alpha_1 - \frac{1}{\alpha_1}\right)^2 S(2\omega_1),$$

$$V_2 = \left(\alpha_2 - \frac{1}{\alpha_2}\right)^2 S(2\omega_2), \quad V_3 = 2S^+ - \left(\alpha_1\alpha_2 + \frac{1}{\alpha_1\alpha_2}\right)^2 S^-,$$

$$D_1 = \left(\alpha_1^2 + \frac{1}{\alpha_2^2}\right)S^+ - 2\frac{\alpha_1}{\alpha_2}S^-, \quad D_2 = \left(\alpha_2^2 + \frac{1}{\alpha_1^2}\right)S^+ - 2\frac{\alpha_2}{\alpha_1}S^-, \tag{17.29}$$

$$S^{\pm} = S(\omega_1 + \omega_2) \pm S(\omega_1 - \omega_2),$$

$$E_1 = -\beta + \frac{2\Omega}{\alpha_{11}(\omega_1^2 - \omega_2^2)}\left[\frac{1}{\alpha_1} \underset{t}{\mathcal{M}}\{\cos\Phi_1\mathbb{H}[q_1]\} - g_2 \underset{t}{\mathcal{M}}\{\sin\Phi_1\mathbb{H}[q_2]\}\right],$$

$$E_2 = -\beta + \frac{2\Omega}{\alpha_{22}(\omega_1^2 - \omega_2^2)}\left[-\frac{1}{\alpha_2} \underset{t}{\mathcal{M}}\{\cos\Phi_2\mathbb{H}[q_1]\} + g_2 \underset{t}{\mathcal{M}}\{\sin\Phi_2\mathbb{H}[q_2]\}\right], \tag{17.30}$$

where $S(\omega)$ is the cosine power spectral density of the process $\xi(t)$. For the Gaussian white noise process, the cosine power spectral density is constant, $S(\omega) = \vartheta^2$, across the whole frequency range. The sine power spectral density is $\Psi(\omega) = 0$.

The averaging operator is defined as:

$$\underset{t}{\mathcal{M}}(\cdot) = \lim_{T\to\infty} \frac{1}{T}\int_t^{t+T} (\cdot)\mathrm{d}t.$$

The integration in the averaging operator is performed over explicitly appearing t only.

If it is assumed that the kernel functions $\mathcal{H}(t)$ and $t\cdot\mathcal{H}(t)$ are integrable across $[0, \infty)$, then the transformation $s = t - \tau$ and the change of order of integration leads to:

$$\underset{t}{\mathcal{M}}[I_r^{CS}] = \underset{t}{\mathcal{M}}\{\cos\Phi_r\mathbb{H}[\sin\Phi_r]\} = \underset{t}{\mathcal{M}}\left\{\cos\Phi_r \int_0^t \mathcal{H}(t-s)\sin\Phi_r(s)\mathrm{d}s\right\}$$

$$= -\frac{1}{2}\int_0^\infty \mathcal{H}(t)\sin\omega_r\tau\mathrm{d}\tau = -\frac{1}{2}\mathcal{H}^S(\omega_r), \quad r = 1, 2.$$

$$\underset{t}{\mathcal{M}}\{\cos\Phi_r\mathbb{H}[\sin\Phi_i]\} = 0, \quad r \neq i,$$

$$\underset{t}{\mathcal{M}}\{I_r^{CC}\} = \underset{t}{\mathcal{M}}\{\cos\Phi_r\mathbb{H}[\cos\Phi_r]\} = \frac{1}{2}\mathcal{H}^C(\omega_r),$$

$$\underset{t}{\mathcal{M}}\{\cos\Phi_r\mathbb{H}[\cos\Phi_i]\} = 0,$$

$$\underset{t}{\mathcal{M}}\{I_r^{SC}\} = \underset{t}{\mathcal{M}}\{\sin\Phi_r\mathbb{H}[\cos\Phi_r]\} = \frac{1}{2}\mathcal{H}^S(\omega_1),$$

$$\underset{t}{\mathscr{M}}\{\sin\Phi_r\mathbb{H}[\cos\Phi_i]\} = 0,$$

$$\underset{t}{\mathscr{M}}\{l_r^{SS}\} = \underset{t}{\mathscr{M}}\{\sin\Phi_r\mathbb{H}[\sin\Phi_r]\} = \frac{1}{2}\mathscr{H}^C(\omega_1),$$

$$\underset{t}{\mathscr{M}}\{\sin\Phi_r\mathbb{H}[\sin\Phi_i]\} = 0, \tag{17.31}$$

where

$$\mathscr{H}^S(\omega) = \int_0^\infty \mathscr{H}(\tau)\sin\omega\tau\mathrm{d}\tau, \quad \mathscr{H}^C(\omega) = \int_0^\infty \mathscr{H}(\tau)\cos\omega\tau\mathrm{d}\tau, \tag{17.32}$$

are the sine and cosine transformations of the viscoelastic kernel function $\mathscr{H}(\tau)$, respectively. Combining Eqs. (17.23) and (17.31) produces

$$\underset{t}{\mathscr{M}}\{\cos\Phi_r\mathbb{H}[q_1]\} = -\frac{1}{2}\alpha_{rr}a_r\mathscr{H}^S(\omega_r), \qquad \underset{t}{\mathscr{M}}\{\sin\Phi_r\mathbb{H}[q_1]\} = \frac{1}{2}\alpha_{rr}a_r\mathscr{H}^C(\omega_r),$$

$$\underset{t}{\mathscr{M}}\{\sin\Phi_r\mathbb{H}[q_2]\} = \frac{1}{2}\alpha_{rr}\alpha_r a_r\mathscr{H}^S(\omega_r), \qquad \underset{t}{\mathscr{M}}\{\cos\Phi_r\mathbb{H}[q_2]\} = \frac{1}{2}\alpha_{rr}\alpha_r a_r\mathscr{H}^C(\omega_r). \tag{17.33}$$

Substituting Eq. (17.33) into Eq. (17.30) yields:

$$E_1 = -\beta - \frac{\Omega\mathscr{H}^S(\omega_1)}{\omega_1^2 - \omega_2^2}\left(\frac{1}{\alpha_1} + \alpha_1\right),$$

$$E_2 = -\beta + \frac{\Omega\mathscr{H}^S(\omega_2)}{\omega_1^2 - \omega_2^2}\left(\frac{1}{\alpha_2} + g_2\alpha_2\right), \tag{17.34}$$

which may be named pseudo-damping coefficients. In this study, the viscoelastic kernel function is hypothesized to follow the ordinary Maxwell model.

$$\mathscr{H}(t) = \gamma \mathrm{e}^{-\eta t}, \tag{17.35}$$

where γ is the intensity coefficient and η is the viscosity coefficient. This model is applied because it can make an approximation that is as close as possible to linear viscoelastic behavior if enough Maxwell units are arranged in parallel. Its sine and cosine transformations in Eq. (17.32) are given by

$$\mathscr{H}^S(\omega) = \frac{\omega\gamma}{\eta^2 + \omega^2}, \quad \mathscr{H}^C(\omega) = \frac{\gamma\eta}{\eta^2 + \omega^2}. \tag{17.36}$$

It is of essence to mention that both the averaged amplitude a_i in Eq. (17.27a) and the phase angle equations ϕ_i in Eq. (17.27b) do not involve the phase angles; furthermore, the amplitude equations are decoupled from the phase angle equations. Thus, the averaged amplitude vector (a_1, a_2) is a two-dimensional diffusion process.

17.4 Moment Lyapunov exponents

The almost-sure stability of the stochastic system (Eq. 17.12) is governed by the largest Lyapunov exponent defined as:

$$\lambda_{\mathbf{q}} = \lim_{t\to\infty} \frac{1}{t} \log \|\mathbf{q}\|, \tag{17.37}$$

where $\|\mathbf{q}\| = (\mathbf{q}^{\mathrm{T}}\mathbf{q})^{1/2}$ is the Euclidean norm. The trivial solution of system (Eq. 17.12) is stable (with probability 1) if the largest Lyapunov exponent is negative; otherwise, it is unstable almost surely. On another note, the stability of the p-th moment $\mathrm{E}[\|\mathbf{q}\|^p]$ of the solution of system (17.12) is governed by the p-th moment Lyapunov exponents defined by

$$\Lambda_{\mathbf{q}}(p) = \lim_{t\to\infty} \frac{1}{t} \log \mathrm{E}[\|\mathbf{q}\|^p]. \tag{17.38}$$

If $\Lambda_{\mathbf{q}}(p)$ is negative, then the p-th moment is stable; otherwise, it is unstable.

Moment Lyapunov exponents for a system with two-dimensional amplitudes in Eq. (17.27a) can be obtained with Khasminskii's transformation and Wedig's transformation. Complex mechanical systems can also be directly considered using the MDOF transformations presented in Chapter 3, where Stojanovic et al. have determined general transformations for an arbitrary number of MDOF stochastic systems.

Firstly, via Khasminskii's transformation [9], the Itô stochastic differential equations for the amplitudes are transformed:

$$r = \sqrt{a_1^2 + a_2^2}, \quad \varphi = \tan^{-1}\frac{a_2}{a_1},$$

$$a_1 = r\cos\varphi, \quad a_2 = r\sin\varphi, \quad P = r^p, \tag{17.39}$$

and then the Moment Lyapunov exponent is given by

$$\Lambda = \lim_{t\to\infty} \frac{1}{t} \log \mathrm{E}[P]. \tag{17.40}$$

The Itô equation for P and φ can be obtained using Itô Lemma,

$$\mathrm{d}P = m_P(P,\varphi)\mathrm{d}t + \sigma_{P1}\mathrm{d}W_1 + \sigma_{P2}\mathrm{d}W_2, \tag{17.41}$$

where

$$\begin{aligned} m_P(P,\varphi) &= \varepsilon\left\{m_1^a\frac{\partial P}{\partial a_1} + m_2^a\frac{\partial P}{\partial a_2} + \frac{1}{2}\left[b_{11}^a\frac{\partial^2 P}{\partial a_1^2} + (b_{12}^a + b_{21}^a)\frac{\partial^2 P}{\partial a_1\partial a_2} + b_{22}^a\frac{\partial^2 P}{\partial a_2^2}\right]\right\} \\ &= \varepsilon pP\left\{\frac{m_1^a}{\sqrt{a_1^2+a_2^2}}\cos\varphi + \frac{m_2^a}{\sqrt{a_1^2+a_2^2}}\sin\varphi + \frac{1}{2}\left[\frac{b_{11}^a}{a_1^2+a_2^2}[(p-2)\cos^2\varphi + 1]\right.\right. \\ &\quad \left.\left. + \frac{b_{12}^a + b_{21}^a}{2(a_1^2+a_2^2)}(p-2)\sin 2\varphi + \frac{b_{22}^a}{a_1^2+a_2^2}[(p-2)\sin^2\varphi + 1]\right]\right\}, \end{aligned} \tag{17.42}$$

and

$$\Sigma_P^2 = \sigma_{P1}^2 + \sigma_{P2}^2$$

$$= \varepsilon\left\{\sigma_{11}^{a}\frac{\partial P}{\partial a_1}+\sigma_{21}^{a}\frac{\partial P}{\partial a_2}+\sigma_{12}^{a}\frac{\partial P}{\partial a_1}+\sigma_{22}^{a}\frac{\partial P}{\partial a_2}\right\}\left\{\begin{matrix}\sigma_{11}^{a}\frac{\partial P}{\partial a_1}+\sigma_{21}^{a}\frac{\partial P}{\partial a_2}\\ \sigma_{12}^{a}\frac{\partial P}{\partial a_1}+\sigma_{22}^{a}\frac{\partial P}{\partial a_2}\end{matrix}\right\}$$

$$= \varepsilon\left[((\sigma_{11}^{a})^2+(\sigma_{12}^{a})^2)\left(\frac{\partial P}{\partial a_1}\right)^2+2(\sigma_{11}^{a}\sigma_{21}^{a}+\sigma_{12}^{a}\sigma_{22}^{a})\frac{\partial P}{\partial a_1}\frac{\partial P}{\partial a_2}\right.$$

$$\left.+((\sigma_{21}^{a})^2+(\sigma_{22}^{a})^2)\left(\frac{\partial P}{\partial a_2}\right)^2\right]$$

$$= \varepsilon\left[b_{11}^{a}\left(\frac{\partial P}{\partial a_2}\right)^2+(b_{12}^{a}+b_{21}^{a})\frac{\partial P}{\partial a_1}\frac{\partial P}{\partial a_2}+b_{22}^{a}\left(\frac{\partial P}{\partial a_2}\right)^2\right]. \tag{17.43}$$

Hence,

$$dP = m_P(P,\varphi)dt+\sigma_{P1}dW_1+\sigma_{P2}dW_2 = m_P(P,\varphi)dt+\Sigma_P(P,\varphi)dW. \tag{17.44}$$

Similarly, the Itô equation for φ can be obtained as:

$$d\varphi = m_\varphi(\varphi)dt+\sigma_{\varphi 1}dW_1+\sigma_{\varphi 2}dW_2 = m_\varphi(\varphi)dt+\Sigma_\varphi(\varphi)dW, \tag{17.45}$$

where

$$m_\varphi(\varphi) = \varepsilon\left\{m_1^{a}\frac{\partial\varphi}{\partial a_1}+m_2^{a}\frac{\partial\varphi}{\partial a_2}+\frac{1}{2}\left[b_{11}^{a}\frac{\partial^2\varphi}{\partial a_1^2}+(b_{12}^{a}+b_{21}^{a})\frac{\partial^2\varphi}{\partial a_1\partial a_2}+b_{22}^{a}\frac{\partial^2\varphi}{\partial a_2^2}\right]\right\}$$

$$= \varepsilon\left\{-\frac{m_1^{a}}{\sqrt{a_1^2+a_2^2}}\sin\varphi+\frac{m_2^{a}}{\sqrt{a_1^2+a_2^2}}\cos\varphi+\frac{1}{2}\left[\frac{b_{11}^{a}}{a_1^2+a_2^2}\sin 2\varphi\right.\right.$$

$$\left.\left.-\frac{b_{12}^{a}+b_{21}^{a}}{2(a_1^2+a_2^2)}(2\cos^2\varphi-1)-\frac{b_{22}^{a}}{a_1^2+a_2^2}\sin 2\varphi\right]\right\}, \tag{17.46}$$

and

$$\Sigma_\varphi^2 = \sigma_{\varphi 1}^2+\sigma_{\varphi 2}^2$$

$$= \varepsilon\left[b_{11}^{a}\left(\frac{\partial\varphi}{\partial a_1}\right)^2+(b_{12}^{a}+b_{21}^{a})\frac{\partial\varphi}{\partial a_1}\frac{\partial\varphi}{\partial a_2}+b_{22}^{a}\left(\frac{\partial\varphi}{\partial a_2}\right)^2\right]$$

$$= \varepsilon\left[\frac{b_{11}^{a}}{a_1^2+a_2^2}\sin^2\varphi-\frac{b_{12}^{a}+b_{21}^{a}}{a_1^2+a_2^2}\sin\varphi\cos\varphi+\frac{b_{22}^{a}}{a_1^2+a_2^2}\cos^2\varphi\right]. \tag{17.47}$$

It is indicated that the coefficients of the right-side terms of amplitude equations in Eq. (17.27a), like $m_i^{a},\ i = 1,\ 2$, are homogenous of degree one in a_1 and a_2, which causes the diffusion term $\mathbf{b}^{a}$ in Eq. (17.28) to be homogenous of degree two in a_1 and a_2. Therefore, submitting $a_1 = r\cos\varphi$ and $a_2 = r\sin\varphi$ in Eq. (17.39) into Eq. (17.42) will reveal that the drift $m_P(P, \varphi)$ and diffusion term $\Sigma_P(P, \varphi)$ are functions of P of degree one and φ. Nevertheless, the drift $m_\varphi(\varphi)$ in Eq. (17.46) and diffusion term $\Sigma_\varphi(\varphi)$ in Eq. (17.47) are functions of φ only, which shows $P(t)$ in Eq. (17.44) and $\varphi(t)$ in Eq. (17.45) are coupled, although $\varphi(t)$ is itself a diffusion process.

Secondly, a linear stochastic transformation is adopted [15]

$$S = T(\varphi)P, \quad P = T^{-1}(\varphi)S, \quad 0 \le \varphi < \pi,$$

which gives the following equations:

$$\frac{\partial S}{\partial P} = T(\varphi), \quad \frac{\partial S}{\partial \varphi} = T'_\varphi P, \quad \frac{\partial^2 S}{\partial P^2} = 0, \quad \frac{\partial^2 S}{\partial P \partial \varphi} = T'_\varphi, \quad \frac{\partial^2 S}{\partial \varphi^2} = PT''_{\varphi\varphi}, \tag{17.48}$$

where T'_φ and $T''_{\varphi\varphi}$ denote the first and the second derivative of $T(\varphi)$ with respect to φ.

The Itô equation for the transformed p-th norm process S may as well be derived using Itô's Lemma.

$$\mathrm{d}S = m_S \mathrm{d}t + \left(\sigma_{P1} \frac{\partial S}{\partial P} \mathrm{d}W_1 + \sigma_{P2} \frac{\partial S}{\partial P} \mathrm{d}W_2 + \sigma_{\varphi 1} \frac{\partial S}{\partial \varphi} \mathrm{d}W_1 + \sigma_{\varphi 2} \frac{\partial S}{\partial \varphi} \mathrm{d}W_2 \right). \tag{17.49}$$

The drift coefficient is defined by

$$\begin{aligned} m_S &= m_P \frac{\partial S}{\partial P} + m_\varphi \frac{\partial S}{\partial \varphi} + \frac{1}{2} \left[b^S_{11} \frac{\partial^2 S}{\partial P^2} + (b^S_{12} + b^S_{21}) \frac{\partial^2 S}{\partial P \partial \varphi} + b^S_{22} \frac{\partial^2 S}{\partial \varphi^2} \right] \\ &= \frac{1}{2} P(\sigma^2_{\varphi 1} + \sigma^2_{\varphi 2}) T''_{\varphi\varphi} + [m_\varphi P + \sigma_{P1}\sigma_{\varphi 1} + \sigma_{P2}\sigma_{\varphi 2}] T'_\varphi + m_P T, \end{aligned} \tag{17.50}$$

and the diffusion terms have the relation:

$$\mathbf{b}^S = [\boldsymbol{\sigma}^S (\boldsymbol{\sigma}^S)^{\mathrm{T}}], \quad \boldsymbol{\sigma}^S = \begin{bmatrix} \sigma_{P1} & \sigma_{P2} \\ \sigma_{\varphi 1} & \sigma_{\varphi 2} \end{bmatrix}. \tag{17.51}$$

Substituting σ_{P1}, σ_{P2}, $\sigma_{\varphi 1}$, and $\sigma_{\varphi 2}$ into Eq. (17.51) leads to

$$b^2_{11} = \sigma^2_{P1} + \sigma^2_{P2} = \Sigma^2_P, \quad b^2_{22} = \sigma^2_{\varphi 1} + \sigma^2_{\varphi 2} = \Sigma^2_\varphi,$$

$$\begin{aligned} b^S_{12} = b^S_{21} &= \sigma_{P1}\sigma_{\varphi 1} + \sigma_{P2}\sigma_{\varphi 2} = \varepsilon[(\sigma^a_{11})^2 + (\sigma^a_{12})^2] \frac{\partial P}{\partial a_1} \frac{\partial \varphi}{\partial a_1} \\ &+ \varepsilon(\sigma^a_{11}\sigma^a_{21} + \sigma^a_{12}\sigma^a_{22}) \left[\frac{\partial P}{\partial a_1} \frac{\partial \varphi}{\partial a_2} + \frac{\partial P}{\partial a_2} \frac{\partial \varphi}{\partial a_1} \right] + \varepsilon[(\sigma^a_{21})^2 + (\sigma^a_{22})^2] \frac{\partial P}{\partial a_2} \frac{\partial \varphi}{\partial a_2} \\ &= \varepsilon \left\{ b^a_{11} \frac{\partial P}{\partial a_1} \frac{\partial \varphi}{\partial a_1} + b^a_{12} \left[\frac{\partial P}{\partial a_1} \frac{\partial \varphi}{\partial a_2} + \frac{\partial P}{\partial a_2} \frac{\partial \varphi}{\partial a_1} \right] + b^a_{22} \frac{\partial P}{\partial a_2} \frac{\partial \varphi}{\partial a_2} \right\} \\ &= \varepsilon p P \left[-\frac{b^a_{11}}{2(a_1^2 + a_2^2)} \sin 2\varphi + \frac{b^a_{12}}{a_1^2 + a_2^2} \cos 2\varphi + \frac{b^a_{22}}{2(a_1^2 + a_2^2)} \sin 2\varphi \right]. \end{aligned} \tag{17.52}$$

For bounded and nonsingular transformation $T(\varphi)$, both processes P and S are expected to have the same stability behavior. For this reason, $T(\varphi)$ is chosen so that the drift term of the Itô differential Eq. (17.49) is independent of the phase process φ and

$$\mathrm{d}S = \varepsilon\Lambda S\mathrm{d}t + \sqrt{\varepsilon}\,(\sigma_{S1}\mathrm{d}W_1 + \sigma_{S2}\mathrm{d}W_2). \tag{17.53}$$

Comparison of the drift terms of Eqs. (17.49) and (17.53), show that a transformation $T(\varphi)$, is defined by the following equation:

$$\frac{1}{2}P(\sigma_{\varphi 1}^2 + \sigma_{\varphi 2}^2)T''_{\varphi\varphi} + [m_\varphi P + \sigma_{P1}\sigma_{\varphi 1} + \sigma_{P2}\sigma_{\varphi 2}]T'_\varphi + m_P T = \varepsilon\Lambda S, \quad 0 \le \varphi < \pi, \tag{17.54}$$

where $T(\varphi)$ is a periodic function in φ of period π. Eq. (17.54) defines an eigenvalue problem of a second-order differential operator with Λ being the eigenvalue and $T(\varphi)$ the associated eigenfunction. Taking the expected value of both sides of Eq. (17.53) leads to $\mathrm{E}[\sigma_{S1}\mathrm{d}W_1 + \sigma_{S2}\mathrm{d}W_2] = 0$ [7]. Accordingly, $\mathrm{E}[S] = \varepsilon\Lambda\mathrm{E}[S]$ is obtained; this portrays the eigenvalue Λ as the Lyapunov exponent of the p-th moment of the system (Eq. 17.12), that is, $\Lambda = \Lambda_{y(t)}(p)$.

Substituting Eq. (17.52) into Eq. (17.54) yields:

$$\mathscr{L}(p)[T] = \frac{1}{p}\left\{\frac{1}{2}\Sigma_\varphi^2 PT''_{\varphi\varphi} + [m_\varphi P + b_{12}{}^S]T'_\varphi + m_P T\right\} = \varepsilon\Lambda T, \quad 0 \le \varphi < \pi, \tag{17.55}$$

where $\mathscr{L}(p)[T]$ denotes an operator on T, Σ_φ^2, m_φ, b_{12}^S, and m_P are given in Eqs. (17.47), (17.46), (17.52), and (17.42), respectively. Substituting these equations into Eq. (17.55) yields:

$$\mathscr{L}(p)[T] = \lambda_2 T''_{\varphi\varphi} + \lambda_1 T'_\varphi + \lambda_0 T, \quad 0 \le \varphi < \pi, \tag{17.56}$$

where

$$\lambda_2 = \frac{1}{2}\left[\frac{b_{11}^a}{a_1^2 + a_2^2}\sin^2\varphi - \frac{b_{12}^a + b_{21}^a}{2(a_1^2 + a_2^2)}\sin 2\varphi + \frac{b_{22}^a}{a_1^2 + a_2^2}\cos^2\varphi\right],$$

$$\lambda_1 = \left\{\left[\frac{m_2^a\cos\varphi}{\sqrt{a_1^2 + a_2^2}} - \frac{m_1^a\sin\varphi}{\sqrt{a_1^2 + a_2^2}} + \frac{b_{11}^a\sin 2\varphi}{2(a_1^2 + a_2^2)} - \frac{(b_{12}^a + b_{21}^a)\cos 2\varphi}{2(a_1^2 + a_2^2)}\right.\right.$$

$$\left.- \frac{b_{22}^a}{a_1^2 + a_2^2}\sin 2\varphi\right] + p\left[-\frac{b_{11}^a}{2(a_1^2 + a_2^2)}\sin 2\varphi + \frac{b_{12}^a}{a_1^2 + a_2^2}\cos 2\varphi\right.$$

$$\left.\left. + \frac{b_{22}^a}{2(a_1^2 + a_2^2)}\sin 2\varphi\right]\right\},$$

$$\lambda_0 = p\left\{\frac{m_1^a}{\sqrt{a_1^2+a_2^2}}\cos\varphi + \frac{m_2^a}{\sqrt{a_1^2+a_2^2}}\sin\varphi + \frac{1}{2}\left[\frac{b_{11}^a}{a_1^2+a_2^2}[(p-2)\cos^2\varphi+1]\right.\right.$$

$$\left.\left.+\frac{b_{12}^a+b_{21}^a}{2(a_1^2+a_2^2)}(p-2)\sin 2\varphi + \frac{b_{22}^a}{a_1^2+a_2^2}[(p-2)\sin^2\varphi+1]\right]\right\}. \tag{17.57}$$

Substitution of $a_1 = r\cos\varphi$, $a_2 = r\sin\varphi$ into Eq. (17.57) would show that the coefficients λ_0, λ_1, and λ_2 are functions of φ and p only. Substituting Eqs. (17.28) and $a_{17.1} = r\cos\varphi$, $a_2 = r\sin\varphi$ into Eq. (17.56) renders the eigenvalue problem, from which moment Lyapunov exponents can be determined by solving Eq. (17.66).

Simplification of the eigenvalue problem can be achieved by deliberately choosing the parameters α_{11} and α_{22} to make

$$\alpha_{21}^2 = \frac{\alpha_{22}^2}{\alpha_{11}^2} = \begin{cases} 1, & \text{if } D_1D_2 = 0, \\ \sqrt{D_1/D_2}, & \text{if } D_1D_2 \neq 0. \end{cases} \tag{17.58}$$

From Eq. (17.29), it is seen that D_1 and D_2 can be rewritten as:

$$D_1 = \left(\alpha_1 - \frac{1}{\alpha_2}\right)^2 S(\omega_1+\omega_2) + \left(\alpha_1 + \frac{1}{\alpha_2}\right)^2 S(\omega_1-\omega_2),$$

$$D_2 = \left(\alpha_2 - \frac{1}{\alpha_1}\right)^2 S(\omega_1+\omega_2) + \left(\alpha_2 + \frac{1}{\alpha_1}\right)^2 S(\omega_1-\omega_2),$$

which indicate that D_1 and D_2 are nonnegative and D_1 is zero if and only if D_2 equals to zero.

Employing the notations:

$$D = \alpha_{21}^2 D_2 = \alpha_{12}^2 D_1 = \sqrt{D_1D_2} = \left|\left(\alpha_1\alpha_2 + \frac{1}{\alpha_1\alpha_2}\right)S^+ - 8S^-\right|,$$

$$\widetilde{a} = \frac{V_0}{4}(V_1+V_2-2V_3+2D), \quad \widetilde{b} = \frac{V_0}{4}(V_1+V_2-2V_3-2D),$$

$$\Delta_1 = E_1 + 2V_0(V_1+V_3), \quad \Delta_2 = E_2 + 2V_0(V_2+V_3), \tag{17.59}$$

and substituting Eqs. (17.28) and $a_{17.1} = r\cos\varphi$, $a_2 = r\sin\varphi$ into Eq. (17.56) lead to the eigenvalue problem:

$$\mathscr{L}(p) = (D - \widetilde{b}\cos^2 2\varphi)T'' + \left\{p\left[\frac{1}{2}V_0(V_2-V_1)\sin 2\varphi - \widetilde{b}\sin 4\varphi\right]\right.$$

$$\left.+\frac{1}{\sin 2\varphi}\left[\frac{1}{2}(\Delta_2-\Delta_1)\sin^2 2\varphi + 2(\widetilde{a} - \widetilde{b}\cos^2 2\varphi)\cos 2\varphi\right]\right\}T'$$

$$+p\left\{\frac{1}{2}(\Delta_1+\Delta_2) + \frac{1}{2}(\Delta_1-\Delta_2)\cos 2\varphi + 2(D - \widetilde{b}\cos^2 2\varphi)\right.$$

$$\left.+p(V_1\cos^2\varphi + V_2\sin^2\varphi + \widetilde{b}\cos^2 2\varphi)\right\}T = \Lambda T, \quad 0 \le \varphi < \pi, \tag{17.60}$$

where T' and T'' are the first-order and second-order derivatives of $T(\varphi)$ with respect to φ, respectively.

Determination of moment Lyapunov exponents: The coefficients in Eq. (17.56) are periodic with period π, reasonable consideration would be a Fourier cosine series expansion of the eigenfunction $T(\Theta)$ in the form:

$$T(\varphi) = \sum_{i=0}^{K} C_i \cos(2i\varphi). \tag{17.61}$$

The eigenvalue problems in Eq. (17.56) sometimes contain $1/\sin(2\varphi)$; for this reason, only cosine functions are adopted. In actuality, this Fourier expansion method is a method for solving PDEs.

Substituting this expansion and $a_1 = r\cos\varphi$, $a_2 = r\sin\varphi$ into the eigenvalue problem (17.56), multiplying both sides by $\cos(2i\varphi)$, $j = 0,\ 1,\ \ldots,$ and performing integration with respect to φ from 0 to $\pi/2$ produces an infinite set of equations for the unknown coefficients C_i, $i = 0,\ 1,\ \ldots,\ K$:

$$\sum_{i=0}^{\infty} a_{ij} C_i = \Lambda C_j, \quad j = 0,\ 1,\ \ldots,\ K, \tag{17.62}$$

where

$$a_{ij} = \frac{4}{\pi}\int_0^{\pi/2} \mathscr{L}(p)[\cos(2i\varphi)]\cos(2j\varphi)\mathrm{d}\varphi,$$

and

$$\int_0^{\pi/2} \cos(2i\varphi)\cos(2j\varphi)\mathrm{d}\varphi = \begin{matrix} \pi/4, & i = j, \\ 0, & i \neq j. \end{matrix}$$

It is seen that the moment Lyapunov exponent is the eigenvalue of the matrix $A = (a_{mk})$

$$A = \begin{bmatrix} a_{00} & a_{01} & a_{02} & a_{03} & \cdots \\ a_{10} & a_{11} & a_{12} & a_{13} & \cdots \\ a_{20} & a_{21} & a_{22} & a_{23} & \cdots \\ a_{30} & a_{31} & a_{32} & a_{33} & \cdots \\ \cdots & \cdots & \cdots & \cdots & \ddots \end{bmatrix}. \tag{17.63}$$

The eigenvalue can also be obtained by the solution of a polynomial equation as follows. Rearranging Eq. (17.62) leads to:

$$\begin{bmatrix} a_{00} - \hat{\Lambda} & a_{01} & a_{02} & a_{03} & \cdots \\ a_{10} & a_{11} - \hat{\Lambda} & a_{12} & a_{13} & \cdots \\ a_{20} & a_{21} & a_{22} - \hat{\Lambda} & a_{23} & \cdots \\ a_{30} & a_{31} & a_{32} & a_{33} - \hat{\Lambda} & \cdots \\ \cdots & \cdots & \cdots & \cdots & \ddots \end{bmatrix} \begin{Bmatrix} C_0 \\ C_1 \\ C_2 \\ C_3 \\ \cdots \end{Bmatrix} = \mathbf{0}. \tag{17.64}$$

The third-order submatrix is listed here. For convenience, Λ is inserted in a_{ij}:

$$a_{00} = \frac{V_0}{4}(3V_1 + 3V_2 + 2V_3 + 2D)p^2 + \left[\frac{V_0}{2}(5V_1 + 5V_2 + 6V_3 + 6D) - E_1 - E_2\right]p - 2\Lambda,$$

$$a_{01} = \frac{V_0}{2}(V_1 - V_2)p^2 + \left[(V_1 - V_2)V_0 - \frac{1}{2}(E_1 - E_2)\right]p,$$

$$a_{02} = \frac{V_0}{8}(V_1 + V_2 - 2V_3 - 2D)p^2 - \frac{V_0}{4}(V_1 + V_2 - 2V_3 - 2D)p,$$

$$a_{10} = \frac{V_0}{2}(V_1 - V_2)p^2 + \left[2(V_1 - V_2)V_0 - \frac{1}{2}(E_1 - E_2)\right]p + 2(V_1 - V_2)V_0 - (E_1 - E_2),$$

$$a_{11} = \frac{V_0}{16}(7V_1 + 7V_2 + 2V_3 + 2D)p^2 + \left[\frac{V_0}{8}(11V_1 + 11V_2 + 10V_3 + 10D) - \frac{1}{2}(E_1 + E_2)\right]p - \frac{1}{2}(V_1 + V_2 - 2V_3 + 7D)V_0 - \Lambda,$$

$$a_{12} = \frac{V_0}{4}(V_1 - V_2)p^2 - \frac{1}{4}(E_1 - E_2)p - (V_1 - V_2)V_0 + \frac{1}{2}(E_1 - E_2),$$

$$a_{20} = \frac{V_0}{8}(V_1 + V_2 - 2V_3 - 2D)p^2 + \frac{3}{4}V_0(V_1 + V_2 - 2V_3 - 2D)p + (V_1 + V_2 - 2V_3 - 18D)V_0,$$

$$a_{21} = \frac{V_0}{4}(V_1 - V_2)p^2 + \left[\frac{3}{2}(V_1 - V_2)V_0 - \frac{1}{4}(E_1 - E_2)\right]p + 2(V_1 - V_2)V_0 - (E_1 - E_2),$$

$$a_{22} = \frac{V_0}{8}(3V_1 + 3V_2 + 2V_3 + 2D)p^2 - 2(V_1 + V_2 - 2V_3 + 10D)V_0 - \Lambda + \left[\frac{V_0}{4}(5V_1 + 5V_2 + 6V_3 + 6D) - \frac{1}{2}(E_1 + E_2)\right]p, \tag{17.65}$$

where E_i is the pseudo-damping defined in Eq. (17.34).

The requirement for a nontrivial solution of the C_k is the determinant of the coefficient matrix of Eq. (17.64) that is equal to zero; the eigenvalue $\hat{\Lambda}(p)$ can be obtained from this requirement. System Eq. (17.55) is truncated to a set of $K + 1$ homogenous linear equations for C_i, $i = 0, 1, 2, \ldots, K$. A nontrivial solution is obtained when the determinant is set to zero, which yields a polynomial equation of degree $K + 1$ for $\hat{\Lambda}^{(K)}(p)$,

$$e_{K+1}^{(K)}[\hat{\Lambda}^{(K)}]^{K+1} + e_K^{(K)}[\hat{\Lambda}^{(K)}]^K + \cdots + e_1^{(K)}\hat{\Lambda}^{(K)} + e_0^{(K)} = 0, \tag{17.66}$$

where $e_K^{(K)}$ $(k = 0,\ 1,\ \ldots,\ K+1)$ are coefficients and are functions of a_{ij}, and $\hat{\Lambda}^{(K)}$ represents the approximate moment Lyapunov exponent in accordance with the assumption that the expansion of the eigenfunction $T(\varphi)$ is up to K-th order Fourier cosine series, as shown in Eq. (17.61). This procedure gives a set of approximate eigenvalues that converge to the corresponding true eigenvalues as $K \to \infty$. It can be demonstrated that the approximate eigenvalues converge so rapidly that the approximations almost coincide after the order $K \geq 1$.

The moment Lyapunov exponent of the system can be approximated by

$$\Lambda_{q(t)}(p) \approx \hat{\Lambda}^{(K)}(p). \tag{17.67}$$

Determination of Lyapunov exponents: The p-th moment Lyapunov exponent $\lambda_{q(t)}(p)$ is a convex analytic function in p that passes through the origin, and the slope at the origin is equal to the largest Lyapunov exponent $\lambda_{q(t)}$,that is,

$$\lambda_{q(t)}(p) = \lim_{p \to 0} \frac{\hat{\Lambda}^{(K)}(p)}{p}, \tag{17.68}$$

which is obtained directly from Eq. (17.67).

For comparison, the largest Lyapunov exponent for the system (Eq. 17.27a) can be directly derived from invariant probability density by solving a Fokker–Plank equation [16] (Eq. 17.30), for the case of $\hat{a}_1 \neq \hat{a}_2$ and $\hat{a}_1 \neq 0$:

1. If $\hat{a}_2 > 0$,

$$\lambda = \frac{\varepsilon}{2}\left[(\hat{\lambda}_1 + \hat{\lambda}_2) + (\hat{\lambda}_1 - \hat{\lambda}_2)\cosh\left(\frac{\hat{\lambda}_1 - \hat{\lambda}_2}{\sqrt{\Delta_0}}\cosh^{-1}\left(\frac{\mathscr{K}}{D}\right)\right)\right]. \tag{17.69}$$

2. If $\hat{a}_2 < 0$,

$$\lambda = \frac{\varepsilon}{2}\left[(\hat{\lambda}_1 + \hat{\lambda}_2) + (\hat{\lambda}_1 - \hat{\lambda}_2)\cosh\left(\frac{\hat{\lambda}_1 - \hat{\lambda}_2}{\sqrt{-\Delta_0}}\cos^{-1}\left(\frac{\mathscr{K}}{D}\right)\right)\right]. \tag{17.70}$$

3. If $\hat{a}_2 = 0$,

$$\lambda = \frac{\varepsilon}{2}\left[(\hat{\lambda}_1 + \hat{\lambda}_2) + (\hat{\lambda}_1 - \hat{\lambda}_2)\cosh\left(\frac{\hat{\lambda}_1 - \hat{\lambda}_2}{2\hat{a}_1}\right)\right]. \tag{17.71}$$

The constants $\mathscr{K}$ and Δ_0 are defined as:

$$\hat{a}_r = \frac{V_0}{2}[V_1 + V_2 - 2V_3 - (-1)^r 2D]$$

$$= \frac{\Omega^2}{32(\omega_1^2 - \omega_1^2)^2}\left[\left(\alpha_1 - \frac{1}{\alpha_1}\right)^2 S(2\omega_1) + \left(\alpha_2 - \frac{1}{\alpha_2}\right)^2 S(2\omega_2) - 4S^+\right.$$

$$+ 2\left(\alpha_1\alpha_2 + \frac{1}{\alpha_1\alpha_2}\right)S^- - (-1)^r 2\left|\left(\alpha_1\alpha_2 + \frac{1}{\alpha_1\alpha_2}\right)S^+ - 8S^-\right|\Bigg],$$

$$\mathscr{K} = \frac{\hat{a}_1 + \hat{a}_2}{2V_0}, \quad \Delta_0 = 16\hat{a}_1\hat{a}_2,$$

$$\hat{\lambda}_r = -E_r + 2V_0(V_r + V_3)$$

$$= \frac{\Omega^2}{2(\omega_1^2 - \omega_1^2)^2}\left[\left(\alpha_r - \frac{1}{\alpha_r}\right)^2 S(2\omega_r) + 2S^+ - \left(\alpha_1\alpha_2 + \frac{1}{\alpha_1\alpha_2}\right)S^-\right]$$

$$- \beta + (-1)^r \frac{\Omega \mathscr{H}^S(\omega_r)}{\omega_1^2 - \omega_1^2}\left(\frac{1}{\alpha_r} + \alpha_r\right), \quad r = 1,\ 2.$$

17.4.1 Stability boundary

Although Eq. (17.60) provides numerical results for the moment Lyapunov exponents, sometimes analytical expressions are required, which can be obtained by setting the Fourier series order $K = 1,\ 2$ in Eq. (17.61) as follows.

Case 1:. *Fourier cosine series expansion order* $K = 1$. When $K = 1$, the eigenfunction is $T(\varphi) = C_0 + C_1 \cos 2\varphi$, the moment Lyapunov exponent can be solved from:

$$\begin{vmatrix} a_{00} - \hat{\Lambda}^{(1)} & a_{01} \\ a_{10} & a_{11} - \hat{\Lambda}^{(1)} \end{vmatrix} = 0, \quad \text{or} \quad (\hat{\Lambda}^{(1)})^2 + e_1^{(1)}\hat{\Lambda}^{(1)} + e_0^{(1)} = 0. \tag{17.72}$$

This is a quadratic equation; its analytical solution is the moment Lyapunov exponent. The coefficients for this quadratic equation are given by

$$e_1^{(1)} = -\frac{1}{8}(13V_1 + 13V_2 + 6V_3 + 6D)V_{0p}{}^2$$

$$- \left[\frac{1}{4}(21V_1 + 21V_2 + 22V_3 + 22D)V_0 + 2(E_1 + E_2)\right]p$$

$$+ (V_1 + V_2 - 2V_3 + 14D)V_0,$$

$$e_0^{(1)} = \left[\frac{5}{64}(V_1^2 + V_2^2) + \frac{37}{32}V_1V_2 + \frac{5}{16}(V_3 + D)(V_1 + V_2) + \frac{1}{16}(V_3 + D)^2\right]V_0^2 p^4$$

$$+ \left\{\left[\frac{5}{8}(V_1^2 + V_2^2) + \frac{29}{4}V_1V_2 + \frac{13}{4}(V_3 + D)(V_1 + V_2) + (V_3 + D)^2\right]V_0\right.$$

$$\left. - \frac{3}{8}(E_1 + E_2)(D + V_3) - \left(\frac{5}{16}E_1 + \frac{21}{16}E_2\right)V_1 - \left(\frac{5}{16}E_2 + \frac{21}{16}E_1\right)V_2\right\}V_0 p^3$$

$$+\left\{\left[\frac{1}{16}(V_1^2+V_2^2)+\frac{97}{8}V_1V_2+\left(\frac{31}{4}V_3+\frac{7}{4}D\right)(V_1+V_2)\right.\right.$$
$$\left.+\frac{1}{4}(17V_3+D)(V_3+D)\right]V_0^2$$
$$-\left[\frac{11}{4}(E_1+E_2)(D+V_3)+\left(\frac{5}{8}E_1+\frac{37}{8}E_2\right)V_1+\left(\frac{5}{8}E_2+\frac{37}{8}E_1\right)V_2\right]V_0$$
$$\left.+\frac{1}{4}E_1^2+\frac{1}{4}E_2^2+\frac{3}{2}E_1E_2\right\}p^2$$
$$+\left\{\left[-\frac{13}{4}(V_1^2+V_2^2)+V_1V_3+\left(\frac{3}{2}V_2-19D\right)V_1+(V_3-19D)V_2\right.\right.$$
$$\left.-18DV_3-21D^2+V_3^2\right]V_0^2$$
$$\left.+\left[7(E_1+E_2)D+\left(\frac{5}{2}E_1-\frac{3}{2}E_2\right)V_1+\left(\frac{5}{2}E_2-\frac{2}{2}E_1\right)V_2-(E_1+E_2)V_3\right]V_0\right\}p. \tag{17.73}$$

The Lyapunov exponent is given by

$$\lambda^{(1)}=-\lim_{p\to 0}\frac{e_0^{(K)}}{pe_1^{(K)}}=\frac{1}{4}\frac{A^{(1)}}{V_1+V_2-2V_3+14D}, \tag{17.74}$$

where

$$A^{(1)}=[13(V_1^2+V_2^2)-6V_1V_2+(76D-4V_3)(V_1+V_2)+(84D-12V_3)(D+V_3)]V_0$$
$$+(E_1+E_2)(4V_3-28D)-(10E_1-6E_2)V_1-(10E_2-6E_1)V_2+2(E_1+E_2)^2.$$

For white noise, there exists $S^+=2S_0$, $S^-=0$, and

$$V_1=\left(\alpha_1-\frac{1}{\alpha_1}\right)^2S_0,\quad V_2=\left(\alpha_2-\frac{1}{\alpha_2}\right)^2S_0,\quad V_3=4S_0,$$
$$D_1=\left(\alpha_1^2+\frac{1}{\alpha_2^2}\right)2S_0,\quad D_2=\left(\alpha_2^2+\frac{1}{\alpha_1^2}\right)2S_0. \tag{17.75}$$

The moment Lyapunov exponent is obtained by solving the quadric equation in Eq. (17.72) with coefficients in Eq. (17.73), and the corresponding largest Lyapunov exponent is given in Eq. (17.74).

Case 2:. *Fourier cosine series order* $K=2$. When $K=2$, the eigenfunction is $T(\varphi)=C_0+C_1\cos 2\varphi+C_2\cos 4\varphi$, the moment Lyapunov exponent can be solved from

$$\begin{vmatrix} a_{00} - \hat{\Lambda}^{(2)} & a_{01} & a_{02} \\ a_{10} & a_{11} - \hat{\Lambda}^{(2)} & a_{12} \\ a_{20} & a_{21} & a_{22} - \hat{\Lambda}^{(2)} \end{vmatrix} = 0,$$

$$\text{or} \quad (\hat{\Lambda}^{(2)})^3 + e_2^{(2)}(\hat{\Lambda}^{(2)})^2 + e_1^{(2)}\hat{\Lambda}^{(2)} + e_0^{(2)} = 0. \tag{17.76}$$

This is a cubic equation; its coefficients are too lengthy to be listed here. For $K \geq 3$, no explicit expressions for the moment Lyapunov exponent can be presented.

17.4.2 Stability index

The stability index is the nontrivial zeros of the moment Lyapunov exponent, which makes its determination as a root-finding problem such that $\lambda_{q(t)}(\delta_{q(t)}) = 0$. When $K = 1$, the stability index is the nontrivial root of Eq. (17.72). The characteristic results of the stability index are shown in Fig. 17.2. It can be noticed that the stability index decreases from positive to negative values with the increase of the amplitude of the power spectrum, which implies that the system is destabilized by the noise. Additionally, as the damping coefficient β increases, the stability index will increase too, which makes the system more stable.

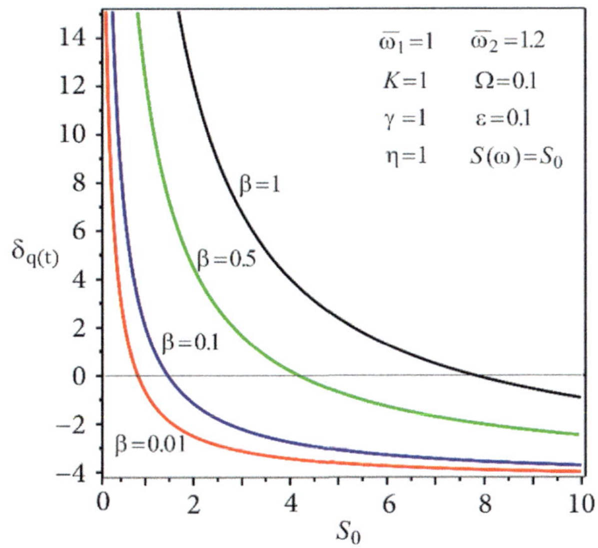

Figure 17.2 Stability index for gyroscopic system under white noise excitation. The stability index is for varying normalized damping coefficients.

17.4.3 Simulation

If the equation of motion Eq. (17.12) is considered and the excitation is approximated by a Gaussian white noise—with spectral density $S(\omega) = \vartheta^2 = constant\ \forall\ \omega$ —then $\xi(t)\mathrm{d}t = \vartheta\,\mathrm{d}W(t)$. Let,

$$x_1(t) = q_1(t), \quad x_2(t) = \dot{q}_1(t), \quad x_3(t) = q_2(t), \quad x_4(t) = \dot{q}_2(t)$$

$$x_5(t) = \int_0^t \gamma e^{-\eta(t-s)} q_1(s)\mathrm{d}s, \quad x_6(t) = \int_0^t \gamma e^{-\eta(t-s)} q_2(s)\mathrm{d}s. \tag{17.77}$$

Eq. (17.12) can be written as a six-dimensional system of Itô differential equations:

$$\mathbf{dx} = \mathbf{Ax}\mathrm{d}t + \mathbf{Bx}\mathrm{d}W, \tag{17.78}$$

where $\mathbf{x} = \{x_1, \quad x_2, \quad x_3, \quad x_4, \quad x_5, \quad x_6\}^{\mathbf{T}}$, and

$$\mathbf{A} = \begin{bmatrix} 0 & 1 & 0 & 0 & 0 & 0 \\ -(\overline{\omega}_1^2 - \Omega^2) & -2\varepsilon\beta & 2\varepsilon\beta\Omega & 2\Omega & \varepsilon & 0 \\ 0 & 0 & 0 & 1 & 0 & 0 \\ -2\varepsilon\beta\Omega & -2\Omega & -(\overline{\omega}_2^2 - \Omega^2) & -2\varepsilon\beta & 0 & \varepsilon \\ \gamma & 0 & 0 & 0 & -\eta & 0 \\ 0 & 0 & \gamma & 0 & 0 & -\eta \end{bmatrix},$$

$$\mathbf{B} = \begin{bmatrix} 0 & 0 & 0 & 0 & 0 & 0 \\ -\sqrt{\varepsilon} & 0 & 0 & 0 & 0 & 0 \\ 0 & 0 & 0 & 0 & 0 & 0 \\ 0 & 0 & -\sqrt{\varepsilon} & 0 & 0 & 0 \\ 0 & 0 & 0 & 0 & 0 & 0 \\ 0 & 0 & 0 & 0 & 0 & 0 \end{bmatrix}. \tag{17.79}$$

Eq. (17.78) is linear homogeneous. The simulation of the moment Lyapunov exponent is done by applying the algorithm introduced in Ref. [7]. The moment the Lyapunov exponent is evaluated by the norm $\|\mathbf{x}(t)\| = \sqrt{x_1^2 + x_2^2 + x_3^2 + x_4^2}$, for x_1, x_2, x_3, and x_4 are related to the state variables of the original system. Using the explicit Euler scheme, the iteration equations are given by

$$x_1^{k+1} = x_1^k + x_2^k \cdot \Delta t,$$

$$x_2^{k+1} = x_2^k + [-(\overline{\omega}_1^2 - \Omega^2)x_1^k - 2\varepsilon\beta x_2^k + 2\varepsilon\beta\Omega x_3^k + 2\Omega x_4^k + \varepsilon x_5^k]\Delta t - \sqrt{\varepsilon}\vartheta x_1^k \cdot \Delta W^k,$$

$$x_3^{k+1} = x_3^k + x_4^k \cdot \Delta t,$$

$$x_4^{k+1} = x_4^k + [-2\varepsilon\beta\Omega x_1^k - 2\Omega x_2^k - (\overline{\omega}_2^2 - \Omega^2)x_3^k - 2\varepsilon\beta x_4^k + \varepsilon x_6^k]\Delta t - \sqrt{\varepsilon}\vartheta x_3^k \cdot \Delta W^k,$$

$$x_5^{k+1} = x_5^k + (\gamma x_1^k - \eta x_5^k) \cdot \Delta t,$$

$$x_6^{k+1} = x_6^k + (\gamma x_3^k - \eta x_6^k) \cdot \Delta t,$$

with Δt being the time step and k denoting the k-th iteration.

17.4.4 Results and discussions

A quite perfect overlapping in Fig. 17.3 between the analytical results from Eq. (17.69) and Fourier series expansion numerical results from Eq. (17.60) proves that the long and tedious derivations in Sections 17.3 and 17.4 are correct. The simulations in this figure confirm the analytical and numerical results. Furthermore, this figure shows that for small values of $S(\omega)$, the numerical results are consistent with the simulation results. The analytical and numerical Lyapunov exponents are tangent lines of Monte Carlo simulation results, which confirms that the method of stochastic averaging is a valid first-order approximation method. Numerical results may become unstable due to round-off errors when $S(\omega) \to 0$; that is why there is a need for caution in this region. For these simulations, the sample size for estimating the expected value is $N = 50000$, the time step $\Delta t = 0.005$, and the total time for simulation is $T = 5000$, the number of iterations is 10^6.

With the increase of β, as shown in Fig. 17.4, the slope of the moment curves at the original point switches from negative to positive; thereby the status of the shaft changes from unstable to stable. Thus, damping plays a stabilizing role in the rotating shaft under white noise excitation.

When $0 \le \Omega \le \bar{\omega}_1$ and near $\bar{\omega}_1$, the instability of the system will increase with the increase of the angular velocity Ω; this is demonstrated in Figs. 17.5 and 17.6. When $\bar{\omega}_1 \le \Omega \le \bar{\omega}_2$, the system is already unstable. When $\Omega \ge \bar{\omega}_2$ and near $\bar{\omega}_2$, the stability area increases with the increase of parameter Ω. After a certain minimum point, with the increase of parameter Ω, the stability area decreases. The boundary, in Fig. 17.6, is V-shaped, which indicates that the axial angular velocity has a critical value at which the shaft is most stable. Before or after that point, the shaft is unstable. As an

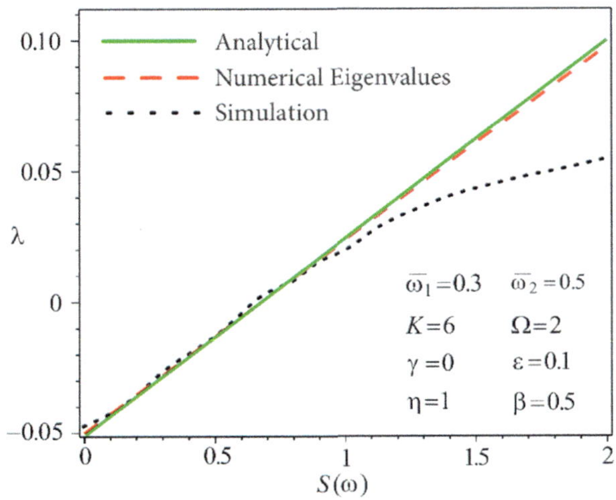

Figure 17.3 Largest Lyapunov exponents of gyroscopic system under white noise excitation. The figure illustrates the largest Lyapunov exponents of a gyroscopic system under white noise excitation.

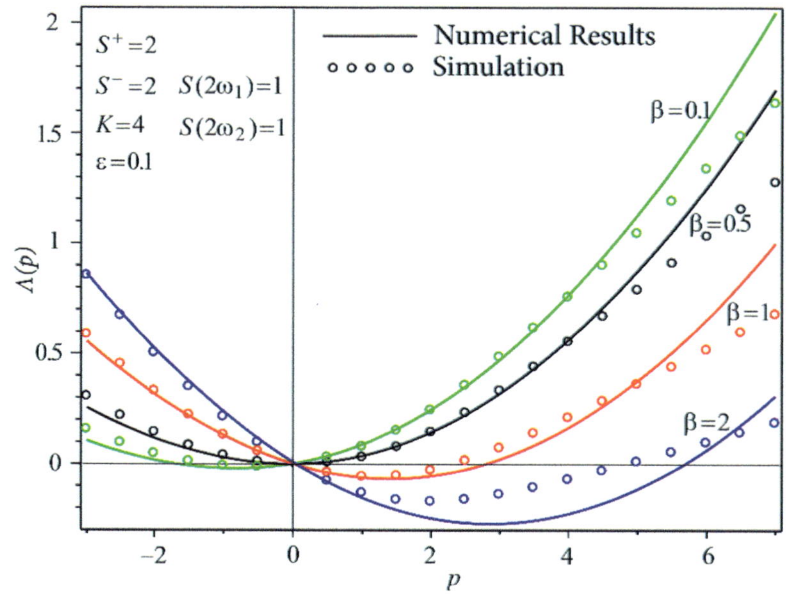

Figure 17.4 Effect of damping on moment Lyapunov exponents under white noise excitation. The figure shows the effect of damping on the moment Lyapunov exponents under white noise excitation.

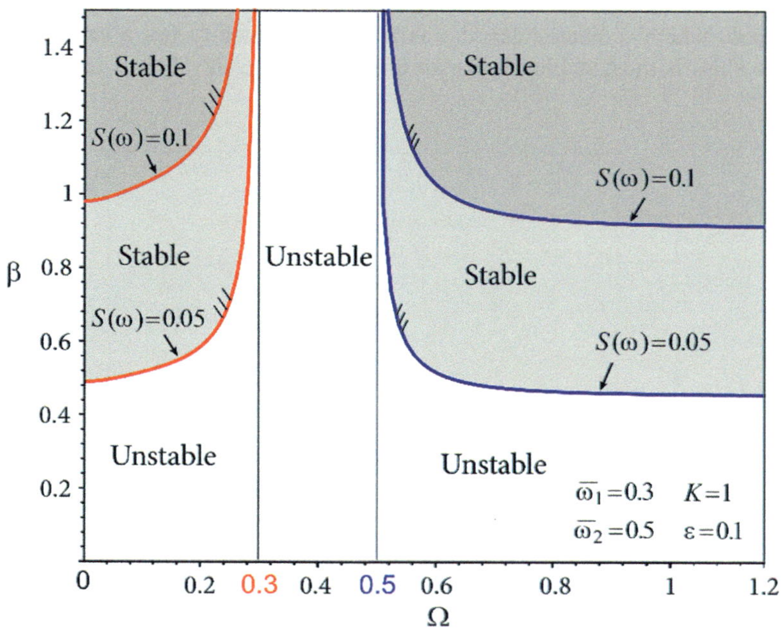

Figure 17.5 Stability boundaries of a gyroscopic elastic system with white noise excitation. The figure depicts the stability boundaries of a gyroscopic purely elastic system with white noise excitation for varying spectral densities.

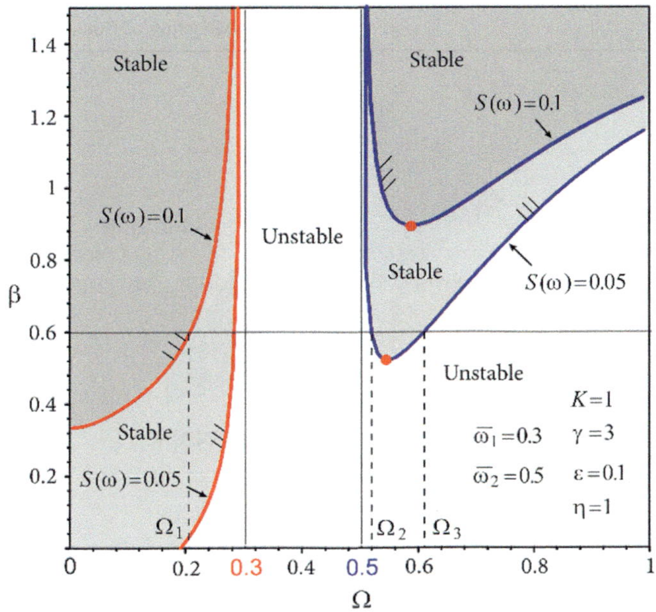

Figure 17.6 Stability boundaries of a gyroscopic viscoelastic system with white noise excitation. The figure shows the stability boundaries of a gyroscopic viscoelastic system with white noise excitation for varying spectral densities.

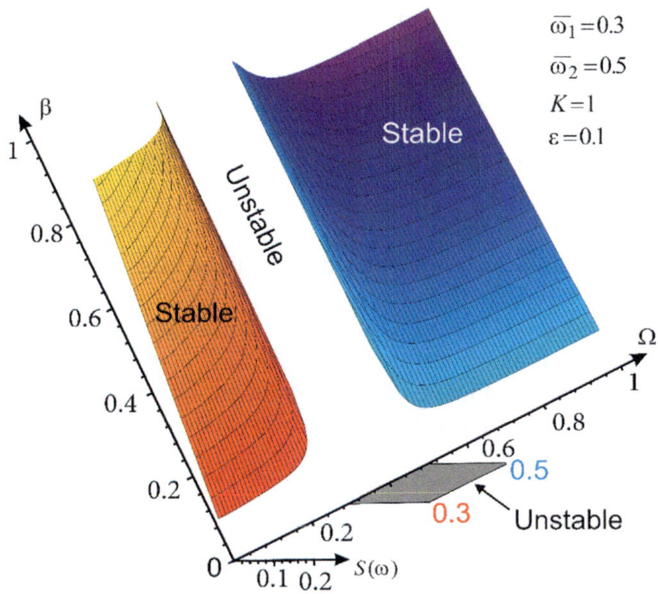

Figure 17.7 Stability boundaries of a gyroscopic elastic system with white noise excitation (3D). The figure illustrates the three-dimensional representation of the stability boundaries of a gyroscopic purely elastic system with white noise excitation, where, in addition to the normalized damping parameter and frequency, the third dimension corresponds to the continuous variation of spectral density.

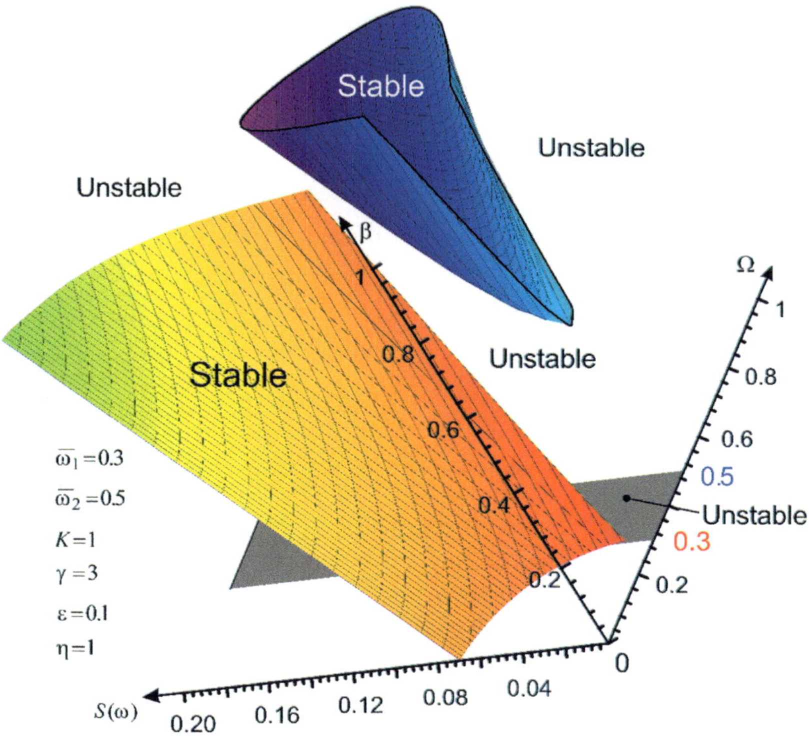

Figure 17.8 Stability boundaries of a gyroscopic viscoelastic system with white noise excitation (3D). The figure shows the three-dimensional representation of the stability boundaries of a gyroscopic viscoelastic system with white noise excitation, where, in addition to the normalized damping parameter and frequency, the third dimension corresponds to the continuous variation of spectral density.

example, when $\beta = 0.6$, which is annotated in the figure, the stability range includes both $\Omega \leq \Omega_1$ and $\Omega_2 \leq \Omega \leq \Omega_3$.

Furthermore, Figs. 17.5 and 17.6 show that noise destabilizes the system; the increase in $S(\omega)$ will diminish the stable region. This is explicitly expressed in the three-dimensional graphs in Figs. 17.7 and 17.8 for elastic and viscoelastic structures, respectively. Fig. 17.8 shows that if the velocity Ω is beyond a certain point, the shaft becomes unstable.

17.5 Conclusion

The dynamic stability of a viscoelastic rotating shaft is the focus of the study presented in this chapter. The shaft is under parametric random excitations. The Galerkin truncation was used to derive and discretize the governing equation of motion for the shaft. The truncated system was a system of gyroscopic stochastic differential equations, which were then decoupled via stochastic averaging into the Itô differential equations.

The largest Lyapunov exponents identify almost-sure stability. Yet, this sample stability cannot assure the moment stability. Moment Lyapunov exponents render the information of both—the moment stability and the almost-sure stability. Therefore, the stability analysis was done by applying both—the largest and the moment Lyapunov exponents. An orderly scheme for determining the moment Lyapunov exponents was developed; it comprises formulating an eigenvalue problem by using mathematical transformations from the decoupled Itô equations. For lower-order (<3) eigenvalue problems, analytical expressions were derived for the moment Lyapunov exponents. For higher-order eigenvalue problems, a numerical method was developed using the Fourier series expansion. The stability boundaries and the stability index were obtained by both analytical approaches and Fourier series expansion numerical methods. The largest Lyapunov exponent was determined through its relation with the moment Lyapunov exponents.

Monte Carlo simulation was used to evaluate the validity and to determine the ranges of the applicability of the stochastic averaging results. Iteration equations are defined via the Euler scheme for Monte Carlo simulation of the moment Lyapunov exponents.

Comparison between the analytical results and Fourier series expansion numerical results for the Lyapunov exponents has shown excellent agreement and near overlap between them. On the other hand, the analytical and numerical Lyapunov exponents are tangent lines of Monte Carlo simulation results, which imply that the method of stochastic averaging is a valid first-order approximation method.

The results for Lyapunov exponents and moment Lyapunov exponents have shown good similarity with the Monte Carlo simulation results and other analytical expressions. Intricate curves have appeared for the stability boundaries in terms of the axial angular velocity Ω and the two natural velocities $\bar{\omega}_1$ and $\bar{\omega}_2$; this has provided insights into the dynamic instability mechanism of rotating shafts. It was also found the noise destabilized the system in terms of the power spectral density of the white noise $S(\omega)$.

The methodology in this study can be readily extended to rotating shafts under real noise excitation or bounded noise excitation. It can also be applied to other gyroscopic systems.

References

[1] A. Tylikowski, Dynamic stability of rotating shafts, Ingenicur-Archiv 50 (1981) 41–48.

[2] N. Sri Namachchivaya, Mean square stability of a rotating shaft under combined harmonic and stochastic excitations, Journal of Sound and Vibration 133 (2) (1989) 323–336.

[3] T.H. Young, C.Y. Gau, Dynamic stability of spinning pre-twisted beams subjected to axial random forces, Journal of Sound and Vibration 268 (2003) 149–165.

[4] T.H. Young, C.Y. Gau, Dynamic stability of pre-twisted beams with non-constant spin rates under axial random forces, International Journal of Solids and Structures 40 (18) (2003) 4675–4698.

[5] R. Pavlovic, P. Rajkovic, I. Pavlovic, Dynamic stability of the viscoelastic rotating shaft subjected to random excitation, International Journal of Mechanical Sciences 50 (2008) 359–364.

[6] P. Baxendale, D. Stroock, Large deviations and stochastic flows of diffeomorphisms, Probability Theory and Related Fields 80 (1988) 169–215.

[7] W.-C. Xie, Dynamic Stability of Structures, Cambridge University Press, Cambridge, 2006.

[8] L. Arnold, M.M. Doyle, N. Sri Namachchivaya, Small noise expansion of moment Lyapunov exponents for two-dimensional systems, Dynamics and Stability of Systems 12 (3) (1997) 122–187.

[9] R.Z. Khasminskii, N. Moshchuk, Moment Lyapunov exponent and stability index for linear conservative system with small random perturbation, SIAM Journal on Applied Mathematics 58 (1) (1998) 245–256.

[10] N. Sri Namachchivaya, H.J. Van Roessel, Moment Lyapunov exponent and stochastic stability of two coupled oscillators driven by real noise, Journal of Applied Mechanics 68 (6) (2001) 903–914.

[11] N.L. Vedula, Dynamics and Stability of Parametrically Excited Gyroscopic Systems (Ph.D. Thesis), University of Illinois at Urbana-Champaign, 2005.

[12] P. Kozic, G. Janevski, R. Pavlovic, Moment Lyapunov exponents and stochastic stability for two coupled oscillators, Journal of Mechanics of Materials and Structures 4 (10) (2009) 1689–1701.

[13] F.M. Dimentberg, Flexural Vibrations of Rotating Shafts, Butterworths, London, 1961.

[14] J. Deng, W.-C. Xie, M.D. Pandey, Moment Lyapunov exponent and stochastic stability of coupled viscoelastic systems under white noise excitation, Journal of Mechanics of Materials and Structures 9 (1) (2014) 27–50.

[15] W. Wedig, Lyapunov exponent of stochastic systems and related bifurcation problems, in: S.T. Ariaratnam, G.I. Schuëller, I. Elishakoff (Eds.), Stochastic Structural Dynamics—Progress in Theory and Applications, Elsevier Applied Science, London, 1988, pp. 315–317.

[16] M.G. Yin, Lyapunov Exponents and Stochastic Stability of Linear Gyroscopic Systems (Master Thesis), University of Waterloo, Waterloo, Canada, 1991.

Stochastic stability of circular cylindrical shells

18

18.1 Introduction

The stochastic instability of plates and shells and their vibrations have an important place in many fields of foundation engineering and construction, and also in aerospace, mechanical engineering, and applications of solids and structures in technical practice. This work contains mathematical procedures that deal with the determination of stochastic stability conditions as well as almost certain stability conditions for coupled stochastic structures. The overview of past studies could give a clear insight into the dynamic behavior of uncoupled and coupled mechanical systems that are exposed to axial compression time-dependent forces. Such kind of considered dynamic models contain a good introduction to the importance of stochastic analysis of solids and structures as well as its applications. The dynamic stability of continuous systems under time-dependent deterministic or stochastic loading has been studied by many authors, but most of the previous studies deal directly with already uncoupled dynamic systems. Contrary to the mentioned research, in most cases of realistic dynamic systems, the couplings exist in the mathematical modeling so that more than two generalized coordinates are necessary for describing the dynamic behavior. An example of one such structure where at least three generalized coordinates can describe its dynamic behavior is the class of cylindrical shells. Thus among the first to publish papers dedicated to the stochastic problem of cylindrical shells were Elishakoff and Arbocz [1] and Kurnik and Tylikowski [2]; many years later the problem of stochastic dynamics of shell structures exposed to buckling represents an interesting problem in mechanics and encourages researchers to study them. Bessa and Pellegrino [3] recently developed a computational framework for the design of a complex shell structure in the stochastic postbuckling range by combining Bayesian regression for imperfection-sensitive quantities of interest. The work is highly stochastic but useful for limited and specific shell problems. It largely depends on computer and automatic technology which can barely be controlled. Some recent research has been devoted to the experimental analysis of a cylindrical shell under axial compression [4]. The authors found that compared to other existing methods, the worst multiple-perturbation load approach (WMPLA) is better for predicting lower buckling load limits for unstiffened cylindrical shells. The authors presented the first attempt to investigate the envelope of WMPLA over random imperfection production.

All dynamic stability problems that have been considered so far at the level of determining moments of Lyapunov exponents belonged to the class with a maximum of two- or three-degrees-of-freedom of vibration, regardless of whether it is a mathematical formulation [5] or the application of procedures to a concrete dynamic model [6,7] or take into consideration only a problem of dynamic stability of coupled

Modeling of Complex Dynamic Systems. DOI: https://doi.org/10.1016/B978-0-443-23942-7.00018-8

structures [8]. Li and Liu [9] presented a mathematical formulation for a three-dimensional system under real noise excitation. Contrary to the three-dimensional system mentioned, on the single-degree-of-freedom problem of the plate model, Jiancheng and Xianbin [10] analyzed the nonGaussian colored noise form of excitation and determined the moments of Lyapunov exponents. In a group of dynamical systems with two-degree-of-freedom, Deng et al. [11] proposed a new scheme for determining the moment Lyapunov exponents of a coupled viscoelastic system by using only analytical mathematical transformations. In addition he recently studied the dynamic stochastic stability of the gyroscopic system under parametric excitation of bounded noise [12]. Depending on the numerical relationship between the natural frequency and the excitation frequencies, four types of parametric resonance were found for a gyroscopic system: no resonance, subharmonic resonance, combination additive resonance, and combination differential resonance.

In addition to the aforementioned research in the field of stochastic stability of dynamic systems, which is based on the determination of the moments of Lyapunov exponents as a very effective known procedure, it is important to highlight the studies of other research groups that deal with the problems of cylindrical shells such as Refs. [13–16]. In addition to the aforementioned published results of various research groups, it is important to highlight recent research in the field of shell stability and the improvement of numerical methods for their analysis, which shows the importance of their application in many real complex mechanical systems. Thus Ansari [17] considered the effects of the inclusion of a bend/twist coupling on the compression buckling of cylindrical shells. They considered the critical buckling load for cylindrical shells on elastic foundations with varying levels of bend/twist anisotropy. In the results, it is shown that the inherent bend/twist anisotropy could lead to a significant reduction in the buckling load, which should initiate further detailed studies in this area. In order to better understand some other influences on the stability of cylindrical shells, Zhang [18] considered an example from technical practice related to the analysis of the stability of cylindrical shells in axial flow. They presented a numerical approach to applying the differential quadrature method to the stability of cylindrical shells subjected to axial flow and found the relationship between the shell frequencies and critical flow speeds to further determine a formula for predicting the critical flow speed. In the analysis of stochastic stability, frequent improvement of the applied mathematical methods is necessary. Thus the application of perturbation methods is very intensive and always current in stochastic and nonlinear dynamics, especially due to the rapid development of computer resources. Intense interest in their application is shown through recently consolidated published studies that will enable the analysis of even more complex mechanical systems or enable an even more detailed description of dynamic phenomena in solids and structures [19].

The objective of this paper is to study the moment and almost-sure stability of a 3DOF coupled system driven by white-noise using the contact transformation method and perturbation approach with application to circular cylindrical shells assuming that the system possesses low damping. A scheme for determining the moment Lyapunov exponents is presented for 3DOF coupled systems. The generalized transformations for arbitrary multi degrees of freedom (MDOF) stochastic systems, as established by Stojanovic et al. and outlined in Chapter 3, allow for the direct consideration of complex mechanical systems. This research is motivated by problems in the dynamic stability of elastic systems subjected to stochastically fluctuating loads. A set of transformations is applied for obtaining Itô differential equations for a 3DOF system as a form of six Stratonovich differential equations. The noise is assumed to be white-noise of small intensity that can obtain an asymptotic growth rate. Stochastic stability boundaries are determined analytically and presented for various parameters of this mechanical system for the first time. By reviewing the literature in the area of stability analysis of stochastic mechanical systems, in terms of determining the moment of Lyapunov exponents, so far, no case of determining stability at the level of the second perturbation for a 3DOF stochastic system is presented. This also represents one of the major contributions presented in the chapter.

18.2 Problem formulation and deterministic vibrations

Let us consider a subcase of the circular conical shell that presents an example of an elastic circular cylindrical shell compressed by time-dependent stochastic membrane forces, as shown in Fig. 18.1. The shell dynamic equilibrium equations are derived from general Love equations [20] in the present form:

$$\frac{\partial N_{xx}}{\partial x} + \frac{1}{a}\frac{\partial N_{x\theta}}{\partial \theta} - P_x(t)\frac{\partial^2 u_x}{\partial x^2} - \rho h\frac{\partial^2 u_x}{\partial t^2} = 0, \tag{18.1}$$

$$\frac{\partial N_{x\theta}}{\partial x} + \frac{1}{a}\frac{\partial N_{\theta\theta}}{\partial \theta} + \frac{Q_{\theta 3}}{a} - \rho h\frac{\partial^2 u_\theta}{\partial t^2} = 0, \tag{18.2}$$

$$\frac{\partial Q_{x3}}{\partial x} + \frac{1}{a}\frac{\partial Q_{\theta 3}}{\partial \theta} + \frac{N_{\theta\theta}}{a} - \rho h\frac{\partial^2 u_3}{\partial t^2} = 0, \tag{18.3}$$

where a is the radius of the circular cylindrical shell, h is the thickness, L is the length, θ is the angle with respect to the vertical axis, x is the axis along its length, and finally u_x, u_θ, and u_3 are the displacements in the x, θ, and z directions,

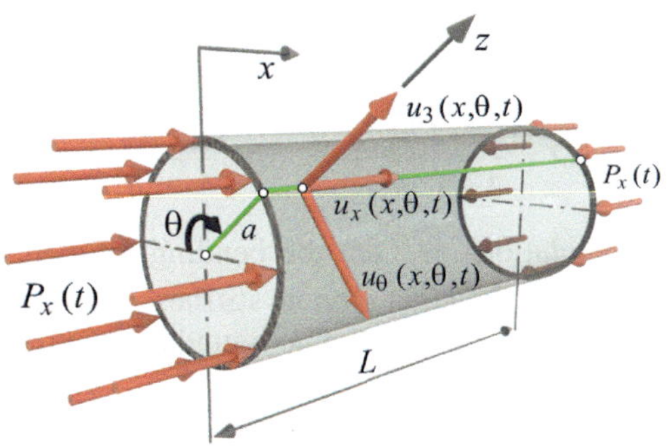

Figure 18.1 Geometry of a circular cylindrical shell. Cylindrical shell subjected to stochastic excitations.

respectively, where z is the distance from the middle surface to an arbitrary point on the shell, $N_{()}$ and $Q_{()}$ are forces and shear forces per unit length. Determining natural frequencies and modes while neglecting time-dependent axial forces at a natural frequency:

$$u_x(x, \theta, t) = U_x(x, \theta)e^{j\omega t}, \tag{18.4}$$

$$u_\theta(x, \theta, t) = U_\theta(x, \theta)e^{j\omega t}, \tag{18.5}$$

$$u_3(x, \theta, t) = U_3(x, \theta)e^{j\omega t}, \tag{18.6}$$

gives:

$$\frac{\partial N'_{xx}}{\partial x} + \frac{1}{a}\frac{\partial N'_{x\theta}}{\partial \theta} - \rho h\omega^2 U_x = 0, \tag{18.7}$$

$$\frac{\partial N'_{x\theta}}{\partial x} + \frac{1}{a}\frac{\partial N'_{\theta\theta}}{\partial \theta} + \frac{1}{a}Q'_{\theta 3} - \rho h\omega^2 U_\theta = 0, \tag{18.8}$$

$$\frac{\partial Q'_{x3}}{\partial x} + \frac{1}{a}\frac{\partial Q'_{\theta 3}}{\partial \theta} - \frac{N'_{\theta\theta}}{a} - \rho h\omega^2 U_3 = 0, \tag{18.9}$$

with

$$Q'_{x3} = \frac{\partial M'_{xx}}{\partial x} + \frac{1}{a}\frac{\partial M'_{x\theta}}{\partial \theta}, \quad Q'_{\theta 3} = \frac{\partial M'_{x\theta}}{\partial x} + \frac{1}{a}\frac{\partial M'_{\theta\theta}}{\partial \theta}, \quad N'_{xx} = K(\varepsilon'^{o}_{xx} + \mu\varepsilon'^{o}_{\theta\theta}),$$

$$N'_{\theta\theta} = K(\varepsilon'^{o}_{\theta\theta} + \mu\varepsilon'^{o}_{xx}), \quad N'_{x\theta} = \frac{K(1-\mu)}{2}\varepsilon'^{o}_{x\theta}, \tag{18.10}$$

$$M'_{xx} = D(k'_{xx} - \mu k'_{\theta\theta}), \quad M'_{\theta\theta} = D(k'_{\theta\theta} - \mu k'_{xx}), \quad M'_{x\theta} = \frac{D(1-\mu)}{2}k'_{x\theta},$$

$$\varepsilon'^{o}_{xx} = \frac{\partial U_x}{\partial x}, \quad \varepsilon'^{o}_{\theta\theta} = \frac{1}{a}\frac{\partial U_\theta}{\partial \theta} + \frac{U_3}{a}, \quad \varepsilon'^{o}_{x\theta} = \frac{\partial U_\theta}{\partial x} + \frac{1}{a}\frac{\partial U_x}{\partial \theta}, \tag{18.11}$$

$$k'_{xx} = \frac{\partial \beta'_x}{\partial x}, \quad k'_{\theta\theta} = \frac{1}{a}\frac{\partial \beta'_\theta}{\partial \theta}, \quad k'_{x\theta} = \frac{\partial \beta'_\theta}{\partial x} + \frac{1}{a}\frac{\partial \beta'_x}{\partial \theta}, \quad \beta'_x = -\frac{\partial U_3}{\partial x},$$

$$D = \frac{Eh^3}{12(1-\mu^2)}, \quad K = \frac{Eh}{1-\mu^2}, \tag{18.12}$$

where from constitutive relations $M'_{()}$ are the bending moments per unit length, D presents the bending stiffness, K the membrane stiffness, μ is the Poisson ratio, and E denotes the Young modulus of elasticity. The boundary conditions become:

$$\begin{aligned} U_3(0, \theta) &= U_\theta(0, \theta) = M'_{xx}(0, \theta) = N'_{xx}(0, \theta) = U_3(L, \theta) \\ &= U_\theta(L, \theta) = M'_{xx}(L, \theta) = N'_{xx}(L, \theta) = 0. \end{aligned} \tag{18.13}$$

Based on the experience with simply supported beams, we assume the following solutions:

$$U_x(x,\theta) = A\cos\frac{m\pi x}{L}\cos n(\theta-\varphi),$$

$$U_\theta(x,\theta) = B\sin\frac{m\pi x}{L}\sin n(\theta-\varphi),$$

$$U_3(x,\theta) = C\sin\frac{m\pi x}{L}\cos n(\theta-\varphi). \tag{18.14}$$

While the assumptions for $U_\theta(x,\theta)$ and $U_3(x,\theta)$ are fairly obvious, the assumption for $U_x(x,\theta)$ needs some explanation. First, the term $n(\theta-\varphi)$ was chosen since it is to be expected that a longitudinal node line will not experience deflections in the x direction. Next, the term $\cos(m\pi x/L)$ is based on the boundary condition requirement that:

$$N'_{xx}(0,\theta) = 0,\quad N'_{xx}(L,\theta) = 0 \tag{18.15}$$

Substituting Eq. (18.14) in all boundary conditions shows that everything is satisfied. Next, these assumed solutions are substituted into Eqs. (18.7–18.9). This leads to:

$$\beta'_\theta = \frac{1}{a}(B+nC)\sin\frac{m\pi x}{L}\sin n(\theta-\varphi),$$

$$\beta'_x = -\frac{m\pi}{L}C\cos\frac{m\pi x}{L}\cos n(\theta-\varphi),$$

$$k'_{x\theta} = \frac{m\pi}{La}(B+2nC)\cos\frac{m\pi x}{L}\sin n(\theta-\varphi),$$

$$k'_{\theta\theta} = \frac{n}{a^2}(B+nC)\sin\frac{m\pi x}{L}\cos n(\theta-\varphi),$$

$$k'_{xx} = \left(\frac{m\pi}{L}\right)^2 C\sin\frac{m\pi x}{L}\cos n(\theta-\varphi),$$

$$k'_{xx} = \left(\frac{m\pi}{L}\right)^2 C\sin\frac{m\pi x}{L}\cos n(\theta-\varphi),$$

$$\varepsilon'^{o}_{x\theta} = \left(\frac{m\pi}{L}B - \frac{n}{a}A\right)\cos\frac{m\pi x}{L}\sin n(\theta-\varphi),$$

$$\varepsilon'^{o}_{\theta\theta} = \frac{1}{a}(Bn+C)\sin\frac{m\pi x}{L}\cos n(\theta-\varphi),$$

$$\varepsilon'^{o}_{xx} = -A\frac{m\pi}{L}\sin\frac{m\pi x}{L}\cos n(\theta-\varphi),$$

$$M'_{x\theta} = \frac{D(1-\mu)}{2}\frac{m\pi}{La}(B+2nC)\cos\frac{m\pi x}{L}\sin n(\theta-\varphi),$$

$$M'_{\theta\theta} = D\left\{\frac{n}{a^2}B + \left[\frac{n^2}{a^2} + \mu\left(\frac{m\pi}{L}\right)^2\right]C\right\}\sin\frac{m\pi x}{L}\cos n(\theta - \varphi),$$

$$M'_{xx} = D\left\{\frac{\mu n}{a^2}B + \left[\mu\frac{n^2}{a^2} + \left(\frac{m\pi}{L}\right)^2\right]C\right\}\sin\frac{m\pi x}{L}\cos n(\theta - \varphi),$$

$$N'_{x\theta} = \frac{K(1-\mu)}{2}\left(\frac{m\pi}{L}B - \frac{n}{a}A\right)\cos\frac{m\pi x}{L}\sin n(\theta - \varphi),$$

$$N'_{\theta\theta} = K\left(\frac{n}{a}B + \frac{1}{a}C - \mu\frac{mn}{L}A\right)\sin\frac{m\pi x}{L}\cos n(\theta - \varphi),$$

$$N'_{xx} = K\left(\frac{\mu n}{a}B + \frac{\mu}{a}C - \frac{mn}{L}A\right)\sin\frac{m\pi x}{L}\cos n(\theta - \varphi),$$

$$Q'_{x3} = D\frac{m\pi}{L}\left\{\frac{n}{a^2}\frac{1+\mu}{2}B + \left[\left(\frac{n}{a}\right)^2 + \left(\frac{mn}{L}\right)^2\right]C\right\}\cos\frac{m\pi x}{L}\cos n(\theta - \varphi),$$

$$Q'_{x3} = D\frac{m\pi}{L}\left\{\frac{n}{a^2}\frac{1+\mu}{2}B + \left[\left(\frac{n}{e}\right)^2 + \left(\frac{mn}{L}\right)^2 + \left(\frac{n}{a}\right)^2\right]C\right\}\sin\frac{m\pi x}{L}\sin n(\theta - \varphi). \tag{18.16}$$

Thus Eqs. (18.7–18.9) become:

$$\left\{\rho h\omega^2 - K\left[\left(\frac{m\pi}{L}\right)^2 + \frac{1-\mu}{2}\left(\frac{n}{a}\right)^2\right]\right\}A + \left(K\frac{1+\mu}{2}\frac{n}{a}\frac{m\pi}{L}\right)B + \left(K\frac{\mu}{a}\frac{m\pi}{L}\right)C = 0, \tag{18.17}$$

$$\left(K\frac{1+\mu}{2}\frac{m\pi}{L}\frac{n}{a}\right)A + \left\{\rho h\omega^2 - \left(K - \frac{D}{a^2}\right)\left[\frac{1-\mu}{2}\left(\frac{m\pi}{L}\right)^2 + \left(\frac{n}{a}\right)^2\right]\right\}B + \left\{-\frac{K}{a}\frac{n}{a} - \frac{D}{a}\frac{n}{a}\left[\left(\frac{m\pi}{L}\right)^2 + \left(\frac{n}{a}\right)^2\right]\right\}C = 0, \tag{18.18}$$

$$\left(\frac{\mu K}{a}\frac{m\pi}{L}\right)A + \left\{-\frac{K}{a}\frac{n}{a} - \frac{D}{a}\frac{n}{a}\left[\left(\frac{m\pi}{L}\right)^2 + \left(\frac{n}{a}\right)^2\right]\right\}B$$
$$+ \left\{\rho h\omega^2 - D\left[\left(\frac{m\pi}{L}\right)^2 + \left(\frac{n}{a}\right)^2\right]^2 - \frac{K}{a^2}\right\}C = 0, \tag{18.19}$$

or

$$\begin{bmatrix} \rho h\omega^2 - k_{11} & k_{12} & k_{13} \\ k_{21} & \rho h\omega^2 - k_{22} & k_{23} \\ k_{31} & k_{32} & \rho h\omega^2 - k_{33} \end{bmatrix}\begin{Bmatrix} A \\ B \\ C \end{Bmatrix} = 0, \tag{18.20}$$

where

$$k_{11} = K\left[\left(\frac{m\pi}{L}\right)^2 + \frac{1-\mu}{2}\left(\frac{n}{a}\right)^2\right],$$
$$k_{12} = k_{21} = K\frac{1+\mu}{2}\frac{m\pi}{L}\frac{n}{a},$$
$$k_{13} = k_{31} = \frac{\mu K}{a}\frac{m\pi}{L},$$
$$k_{22} = \left(K + \frac{D}{a^2}\right)\left[\frac{1-\mu}{2}\left(\frac{m\pi}{L}\right)^2 + \left(\frac{n}{a}\right)^2\right],$$
$$k_{23} = k_{32} = -\frac{K}{a}\frac{n}{a} - \frac{D}{a}\frac{n}{a}\left[\left(\frac{m\pi}{L}\right)^2 + \left(\frac{n}{a}\right)^2\right],$$
$$k_{22} = D\left[\left(\frac{m\pi}{L}\right)^2 + \left(\frac{n}{a}\right)^2\right]^2 + \frac{K}{a^2}. \tag{18.21}$$

For a nontrivial solution, the determinant of Eq. (18.20) has to be zero. Expanding the determinant gives:

$$\omega^6 + a_1\omega^6 + a_2\omega^2 + a_3 = 0, \tag{18.22}$$

where

$$a_1 = \frac{1}{\rho h}(k_{11} + k_{22} + k_{33}),$$
$$a_2 = \frac{1}{(\rho h)^2}(k_{11}k_{33} + k_{22}k_{33} + k_{11}k_{22} - k_{23}{}^2 - k_{12}{}^2 - k_{13}{}^2),$$
$$a_3 = \frac{1}{(\rho h)^3}(k_{11}k_{23}{}^2 + k_{22}k_{13}{}^2 + k_{33}k_{12}{}^2 + 2k_{12}k_{23}k_{13} - k_{11}k_{22}k_{33}). \tag{18.23}$$

Table 18.1 Fundamental natural frequencies (Hz) of a circular cylindrical shell for a varying ratio *a/L*.

***a/L* = 0.3**					
h	***h/5***	***h/2***	***2***	***2h***	***5h***
			a/L = 0.3		
ω_1	22.7504	22.6928	22.4858	21.6333	14.0130
ω_2	80.5471	80.5644	80.6264	80.8719	82.4924
ω_3	192.4984	192.5148	192.5735	192.8095	94.5077
			a/L = 1		
ω_1	31.4934	31.5329	31.6733	32.2283	46.1664
ω_2	73.3110	73.3118	73.3146	73.3261	73.9571
ω_3	120.3101	120.3104	120.3112	120.3146	120.4378
			a/L = 3		
ω_1	11.7360	11.8346	12.1803	13.4744	20.3455
ω_2	69.7355	69.7353	69.7351	69.7344	69.7297
ω_3	117.4682	117.4683	117.4683	117.4685	117.4696

The solutions of this equation are:

$$\omega_{1mn}^2 = -\frac{2}{3}\sqrt{a_1^2 - 3a_2}\cos\frac{\alpha}{3} - \frac{a_1}{3},$$

$$\omega_{2mn}^2 = -\frac{2}{3}\sqrt{a_1^2 - 3a_2}\cos\frac{\alpha + 2\pi}{3} - \frac{a_1}{3},$$

$$\omega_{3mn}^2 = -\frac{2}{3}\sqrt{a_1^2 - 3a_2}\cos\frac{\alpha + 4\pi}{3} - \frac{a_1}{3}, \tag{18.24}$$

where

$$\alpha = \cos^{-1}\frac{27a_3 + 2a_1^3 - 9a_1a_2}{2\sqrt{(a_1^2 - 3a_2)^3}}. \tag{18.25}$$

Thereby, for every *m*, *n* combination there are three frequencies. The lowest is associated with the mode where the transverse component dominates, while the other two are usually significantly higher and are associated with the mode where the displacements in the tangent plane dominate. The list of fundamental natural frequencies is reported in Table 18.1 for a varying thickness h and varying values of the ratio $r_{\varpi} = a/L$, other parameters are: $h = \frac{2}{5}\sqrt{\frac{3}{5}}$ m, $\rho = 2 \times 10^3$ kg/m^3, $E = 1 \times 10^{10}$ N/m^2, $\nu = 0.3$.

Similarly, for every combination of *m*, *n*, there are three different combinations of A, B, and C. Solving A and B in terms of C, renders,

$$\begin{bmatrix} \rho h\omega_{imn}^2 - k_{11} & k_{12} \\ k_{12} & \rho h\omega_{imn}^2 - k_{22} \end{bmatrix} \begin{Bmatrix} A_i \\ B_i \end{Bmatrix} = -C_i \begin{Bmatrix} k_{13} \\ k_{23} \end{Bmatrix}, \tag{18.26}$$

where $i = 1, 2, 3$. Thus

$$\frac{A_i}{C_i} = -\frac{\begin{vmatrix} k_{13} & k_{12} \\ k_{23} & \rho h\omega_{imn}^2 - k_{22} \end{vmatrix}}{D}, \quad \frac{B_i}{C_i} = -\frac{\begin{vmatrix} \rho h\omega_{imn}^2 - k_{11} & k_{13} \\ k_{21} & k_{23} \end{vmatrix}}{D}, \tag{18.27}$$

where

$$D = \begin{vmatrix} \rho h\omega_{imn}^2 - k_{11} & k_{12} \\ k_{21} & \rho h\omega_{imn}^2 - k_{22} \end{vmatrix}, \tag{18.28}$$

or,

$$\frac{A_i}{C_i} = -\frac{k_{13}(\rho h\omega_{imn}^2 - k_{22}) - k_{12}k_{23}}{(\rho h\omega_{imn}^2 - k_{11})(\rho h\omega_{imn}^2 - k_{22}) - k_{12}^2},$$

$$\frac{B_i}{C_i} = -\frac{k_{23}(\rho h\omega_{imn}^2 - k_{11}) - k_{21}k_{13}}{(\rho h\omega_{imn}^2 - k_{11})(\rho h\omega_{imn}^2 - k_{22}) - k_{12}^2}. \tag{18.29}$$

Thus in summary, the three natural modes that are associated with the three natural frequencies ω_{imn} at each m, n combinations are:

$$\begin{Bmatrix} U_x \\ U_\theta \\ U_3 \end{Bmatrix} = C_i \begin{Bmatrix} \frac{A_i}{C_i} \cos \frac{m\pi x}{L} \cos n(\theta - \varphi) \\ \frac{B_i}{C_i} \sin \frac{m\pi x}{L} \sin n(\theta - \varphi) \\ \sin \frac{m\pi x}{L} \sin n(\theta - \varphi) \end{Bmatrix}. \tag{18.30}$$

where C_i are the arbitrary constants.

18.3 Stochastic vibration analysis

It is assumed that the system possesses a small damping which can classify the presented mechanical system in the group of gyroscopic mechanical systems under parametric excitation with the properties of the mass and stiffness matrices: $\mathbf{M} = \mathbf{M}^T > 0$, $\mathbf{K} = \mathbf{K}^T > 0$. In the general case, the system of stochastic partial differential equations emerges as a consequence of the introduced fluctuation part in axial thrust $\zeta(t)$, $(\mathrm{E}[\zeta(t)] = 0)$, where $\mathrm{E}[\cdot]$ denotes the expected value in $P(t) = P_0 + \zeta(t)$ and according to the derived equations of motion:

$$\frac{\partial N_{xx}}{\partial x} + \frac{1}{a}\frac{\partial N_{x\theta}}{\partial \theta} - [P_0 + \zeta(t)]\frac{\partial^2 u_x}{\partial x^2} - \rho h \frac{\partial^2 u_x}{\partial t^2} = 0, \tag{18.31}$$

$$\frac{\partial N_{x\theta}}{\partial x} + \frac{1}{a}\frac{\partial N_{\theta\theta}}{\partial \theta} + \frac{Q_{\theta 3}}{a} - \rho h \frac{\partial^2 u_\theta}{\partial t^2} = 0, \tag{18.32}$$

$$\frac{\partial Q_{x3}}{\partial x} + \frac{1}{a}\frac{\partial Q_{\theta 3}}{\partial \theta} + \frac{N_{\theta\theta}}{a} - \rho h \frac{\partial^2 u_3}{\partial t^2} = 0, \tag{18.33}$$

with boundary conditions detailed in Ref. [20]

$$\begin{aligned} u_3(0, \theta, t) &= U_\theta(0, \theta, t) = M'_{xx}(0, \theta, t) = N'_{xx}(0, \theta, t) = U_3(L, \theta, t) \\ &= U_\theta(L, \theta, t) = M'_{xx}(L, \theta, t) = N'_{xx}(L, \theta, t) = 0. \end{aligned} \tag{18.34}$$

These equations are partial stochastic differential equations where an approximate solution can be obtained by using the Galerkin method. Substituting

$$\begin{aligned} u_x(x, \theta, t) &= x_1(t)\cos\frac{\pi x}{L}\cos(\theta - \varphi), \\ u_\theta(x, \theta, t) &= x_2(t)\sin\frac{\pi x}{L}\sin(\theta - \varphi), \\ u_3(x, \theta, t) &= x_3(t)\sin\frac{\pi x}{L}\cos(\theta - \varphi), \end{aligned} \tag{18.35}$$

into Eqs. (18.31–18.33), multiplying Eq. (18.31) by $\cos\frac{\pi x}{L}\cos(\theta - \varphi)$, Eq. (18.32) by $\sin\frac{\pi x}{L}\sin(\theta - \varphi)$, Eq. (18.33) by $\sin\frac{\pi x}{L}\cos(\theta - \varphi)$ and integrating from 0 to L yields:

$$\begin{aligned} &\rho h\ddot{x}_1 + \left\{K\left[\left(\frac{\pi}{L}\right)^2 - \frac{1-\mu}{2a^2}\right] - \frac{\pi^2}{L^2}[P_0 + \zeta(t)]\right\}x_1 \\ &- \left(K\frac{1+\mu}{2a}\frac{\pi}{L}\right)x_2 - \left(K\frac{\mu}{a}\frac{\pi}{L}\right)x_3 = 0, \end{aligned} \tag{18.36}$$

$$\begin{aligned} &\rho h\ddot{x}_2 + \left\{\left(K - \frac{D}{a^2}\right)\left[\frac{1-\mu}{2}\left(\frac{\pi}{L}\right)^2 + \frac{1}{a^2}\right]\right\}x_2 \\ &- \left(K\frac{1+\mu}{2a}\frac{\pi}{L}\right)x_1 + \left\{\frac{K}{a^2} + \frac{D}{a^2}\left[\left(\frac{\pi}{L}\right)^2 + \frac{1}{a^2}\right]\right\}x_3 = 0, \end{aligned} \tag{18.37}$$

$$\begin{aligned} &\rho h\ddot{x}_3 + \left\{D\left[\left(\frac{\pi}{L}\right)^2 + \left(\frac{1}{a}\right)^2\right]^2 + \frac{K}{a^2}\right\}x_3 \\ &- \left(\frac{\mu K}{a}\frac{\pi}{L}\right)x_1 + \left\{\frac{K}{a^2} + \frac{D}{a^2}\left[\left(\frac{\pi}{L}\right)^2 + \left(\frac{1}{a}\right)^2\right]\right\}x_2 = 0, \end{aligned} \tag{18.38}$$

or in the matrix notation

$$\begin{bmatrix} m_{11} & 0 & 0 \\ 0 & m_{22} & 0 \\ 0 & 0 & m_{33} \end{bmatrix}\ddot{\mathbf{x}} + \begin{bmatrix} k_{11} & k_{12} & k_{13} \\ k_{21} & k_{22} & k_{23} \\ k_{31} & k_{32} & k_{33} \end{bmatrix}\mathbf{x} + \begin{bmatrix} p(t) & 0 & 0 \\ 0 & 0 & 0 \\ 0 & 0 & 0 \end{bmatrix}\mathbf{x} = 0, \tag{18.39}$$

where

$$\mathbf{x} = \{x_1, x_2, x_3\}^{\mathrm{T}}, \quad m_{11} = m_{22} = m_{33} = \rho h,$$

$$k_{11} = K\left[\left(\frac{\pi}{L}\right)^2 - \frac{1-\mu}{2a^2}\right] - \frac{\pi^2 P_0}{L^2},$$

$$k_{12} = k_{21} = -\left(K\frac{1+\mu}{2a}\frac{\pi}{L}\right), \quad k_{13} = k_{31} = -K\frac{\mu}{a}\frac{\pi}{L},$$

$$k_{22} = \left(K - \frac{D}{a^2}\right)\left[\frac{1-\mu}{2}\left(\frac{\pi}{L}\right)^2 + \frac{1}{a^2}\right],$$

$$k_{23} = k_{32} = \frac{K}{a^2} + \frac{D}{a^2}\left[\left(\frac{\pi}{L}\right)^2 + \frac{1}{a^2}\right],$$

$$k_{33} = D\left[\left(\frac{\pi}{L}\right)^2 + \frac{1}{a^2}\right]^2 + \frac{K}{a^2}, \quad p(t) = -\frac{\pi^2}{L^2}\zeta(t).$$

Introducing the small parameter $\varepsilon(0 < \varepsilon \ll 1)$ to show that the damping and parametric excitations are small (under the already introduced assumption that the system has a small damping), the system of Eq. (18.39) becomes:

$$\begin{bmatrix} m_{11} & 0 & 0 \\ 0 & m_{22} & 0 \\ 0 & 0 & m_{33} \end{bmatrix}\ddot{\mathbf{x}} + 2\varepsilon\beta\dot{\mathbf{x}} + 2\beta k\begin{bmatrix} 0 & 1 & 0 \\ -1 & 0 & -1 \\ 0 & 1 & 0 \end{bmatrix}\dot{\mathbf{x}}$$

$$+\begin{bmatrix} k_{11} & k_{12} & k_{13} \\ k_{21} & k_{22} & k_{23} \\ k_{31} & k_{32} & k_{33} \end{bmatrix}\mathbf{x} + \sqrt{\varepsilon}\xi(t)\begin{bmatrix} \widetilde{k}_{11} & 0 & 0 \\ 0 & 0 & 0 \\ 0 & 0 & 0 \end{bmatrix}\mathbf{x} = 0, \tag{18.40}$$

where β is the small damping coefficient and $\xi(t)$ is the excitation which is modeled as a Gaussian white-noise process $(\widetilde{k}_{11} = -\pi^2/L^2)$. The system of stochastic differential Eq. (18.40) can be rewritten in a more condensed form, as follows:

$$\mathbf{M}\ddot{\mathbf{x}} + (\mathbf{G} + \varepsilon\mathbf{D})\dot{\mathbf{x}} + \mathbf{K}\mathbf{x} + \sqrt{\varepsilon}\xi(t)\mathbf{S}\mathbf{x} = \mathbf{0}, \tag{18.41}$$

where **M**, **G**, **D**, and **K** are the 3×3 mass, gyroscopic, damping, and stiffness matrices, respectively, with an additional skew-symmetric property of the matrix **G** given by $\mathbf{G}^T = -\mathbf{G}$. Following the procedure given by Xie [21], in Chapter 3, we obtain an expanded form of a system of six first-order Stratonovich differential equations in the form:

$$\mathrm{d}\{\mathbf{x}\} = \begin{bmatrix} \mathbf{0} & \mathbf{1} \\ -\mathbf{1} & \mathbf{0} \end{bmatrix}[\mathrm{diag}\{\omega_1, \omega_2, \omega_3, \omega_1, \omega_2, \omega_3\}]\{\mathbf{x}\}\mathrm{d}t$$

$$-\varepsilon\begin{bmatrix} \boldsymbol{d}_{1()} & \boldsymbol{d}_{2()} \\ \boldsymbol{d}_{3()} & \boldsymbol{d}_{4()} \end{bmatrix}\{\mathbf{x}\}\mathrm{d}t - \sqrt{\varepsilon}\begin{bmatrix} s_{1()} & s_{2()} \\ s_{3()} & s_{4()} \end{bmatrix}\{\mathbf{x}\}\mathrm{d}\xi(t),$$

$$\{\mathbf{x}\} = \lfloor x_1 x_2 x_3 x_4 x_5 x_6 \rfloor^{\mathrm{T}}, \tag{18.42}$$

where the stochastic term $\sqrt{\varepsilon}\xi(t)$ presents a white-noise process with small intensity. Dynamic stability of an oscillatory system can be known in the case to determine the maximal Lyapunov exponent and the p-th moment Lyapunov exponent which is described as:

$$\lambda_T = \lim_{t\to\infty} \frac{1}{t} \log \|\boldsymbol{q}(t; q_0)\|, \tag{18.43}$$

where $\boldsymbol{q}(t;q_0)$ is the solution process of a linear dynamic system. The almost-sure stability depends upon the sign of the maximal Lyapunov exponent, which is an exponential growth rate of the solution of the randomly perturbed dynamic system. A negative sign of the maximal Lyapunov exponent implies the almost-sure stability whereas a nonnegative value indicates instability. The exponential growth rate $E[\|\boldsymbol{q}(t;q_0,\dot{q}_0)\|^p]$ is provided by the moment Lyapunov exponent defined as:

$$\Lambda_T(p) = \lim_{t\to\infty} \frac{1}{t} \log E[\|\boldsymbol{q}(t; q_0)\|^p],$$

where $E[\,]$ denotes the expectation. If $\Lambda_T(p) < 0$, then by definition $E[\|\boldsymbol{q}(t;q_0,\dot{q}_0)\|^p] \to 0$ as $t \to 0$ and this is referred to as p-th moment stability. Although the moment Lyapunov exponents are important in the study of the dynamic stability of stochastic systems, the actual evaluations of the moment Lyapunov exponents are very difficult, and the almost-sure and moment stability of the equilibrium state $q = \dot{q} = 0$. The stochastic part $\xi(t)$ is taken into account and it represents the white-noise process with zero mean and autocorrelation function:

$$R_{\xi\xi}(t_1, t_2) = E[\xi(t_1)\xi(t_2)] = \sigma^2\delta(t_2 - t_1), \tag{18.44}$$

where σ is the intensity of the random process $\delta(\,)$ is the Dirac delta function. Using the corresponding transformation:

$$x_1 = a\cos\varphi_1\cos\varphi_2\cos\theta_1,\quad x_2 = -a\cos\varphi_1\cos\varphi_2\sin\theta_1,$$

$$x_3 = a\cos\varphi_1\sin\varphi_2\cos\theta_2,\quad x_4 = -a\cos\varphi_1\sin\varphi_2\sin\theta_2,$$

$$x_5 = a\sin\varphi_1\cos\theta_3,\quad x_6 = -a\sin\varphi_1\sin\theta_3,$$

$$P = a^p = (x_1^2 + x_2^2 + x_3^2 + x_4^2 + x_5^2 + x_6^2)^{p/2},$$

$$-\infty < p < \infty,\quad 0 \le \theta_i \le 2\pi,\quad 0 \le \varphi_j \le \frac{\pi}{2},\quad i = 1,\ 2,\ 3,\quad j = 1,\ 2.$$

Itô's rule yields the set of equations for the p-th power of the norm of the response and phase variables. Trigonometric transformation represents a as a norm of the response, where θ_1, θ_2, and θ_3 are the angles of the three oscillators and φ_1 and φ_2 describe the couplings or exchange of energy between the oscillators where

$$d\theta_1 = m_1(\theta_1, \theta_2,\theta_3, \varphi_1, \varphi_2)dt + \sigma_{11}(\theta_1, \theta_2,\theta_3, \varphi_1, \varphi_2)dW(t),$$

$$d\theta_2 = m_2(\theta_1, \theta_2,\theta_3, \varphi_1, \varphi_2)dt + \sigma_{21}(\theta_1, \theta_2,\theta_3, \varphi_1, \varphi_2)dW(t),$$

$$d\theta_3 = m_3(\theta_1, \theta_2,\theta_3, \varphi_1, \varphi_2)dt + \sigma_{31}(\theta_1, \theta_2,\theta_3, \varphi_1, \varphi_2)dW(t),$$

$$d\varphi_1 = m_4(\theta_1, \theta_2, \theta_3, \varphi_1, \varphi_2)dt + \sigma_{41}(\theta_1, \theta_2, \theta_3, \varphi_1, \varphi_2)dW(t), \quad (18.45)$$

$$d\varphi_2 = m_5(\theta_1, \theta_2, \theta_3, \varphi_1, \varphi_2)dt + \sigma_{51}(\theta_1, \theta_2, \theta_3, \varphi_1, \varphi_2)dW(t),$$

$$dP = Pm_6(\theta_1, \theta_2, \theta_3, \varphi_1, \varphi_2)dt + P\sigma_{61}(\theta_1, \theta_2, \theta_3, \varphi_1, \varphi_2)dW(t),$$

and $W(t)$ presents the standard Weiner process. Coefficients m_{ij} are not presented due to cumbersome expressions. Applying a linear stochastic transformation given by Wedig [22]

$$S = T(\theta_1, \theta_2, \theta_3, \varphi_1, \varphi_2)P, \quad P = \frac{S}{T(\theta_1, \theta_2, \theta_3, \varphi_1, \varphi_2)}, \quad (18.46)$$

introducing the new norm process S by means of the scalar function $T(\theta_1, \theta_2, \theta_3, \varphi_1, \varphi_2)$ which is defined on the stationary phase process φ_j, the Itô equation for the new p-th norm process S can be obtained from Itô's Lemma:

$$\begin{aligned} dS = P\Bigg[&\frac{1}{2}\left(\sigma_{11}^2\frac{\partial^2 T}{\partial\theta_1^2} + \sigma_{21}^2\frac{\partial^2 T}{\partial\theta_2^2} + \sigma_{31}^2\frac{\partial^2 T}{\partial\theta_3^2} + \sigma_{41}^2\frac{\partial^2 T}{\partial\varphi_1^2} + \sigma_{51}^2\frac{\partial^2 T}{\partial\varphi_2^2}\right) \\ &+ \sigma_{11}\sigma_{21}\frac{\partial^2 T}{\partial\theta_1\partial\theta_2} + \sigma_{11}\sigma_{31}\frac{\partial^2 T}{\partial\theta_1\partial\theta_3} + \sigma_{11}\sigma_{41}\frac{\partial^2 T}{\partial\theta_1\partial\varphi_1} + \sigma_{31}\sigma_{51}\frac{\partial^2 T}{\partial\theta_3\partial\varphi_2} \\ &+ \sigma_{41}\sigma_{51}\frac{\partial^2 T}{\partial\varphi_1\partial\varphi_2} + m_1\frac{\partial T}{\partial\theta_1} + m_2\frac{\partial T}{\partial\theta_2} + m_3\frac{\partial T}{\partial\theta_3} + m_4\frac{\partial T}{\partial\varphi_1} + m_5\frac{\partial T}{\partial\varphi_2} \\ &+ m_6 T + \sigma_{11}\sigma_{61}\frac{\partial T}{\partial\theta_1} + \sigma_{21}\sigma_{61}\frac{\partial T}{\partial\theta_2} + \sigma_{31}\sigma_{61}\frac{\partial T}{\partial\theta_3} + \sigma_{41}\sigma_{61}\frac{\partial T}{\partial\varphi_1} + \sigma_{51}\sigma_{61}\frac{\partial T}{\partial\varphi_2}\Bigg]dT \\ &+ P\left(\sigma_{11}\frac{\partial T}{\partial\theta_1} + \sigma_{21}\frac{\partial T}{\partial\theta_2} + \sigma_{31}\frac{\partial T}{\partial\theta_3} + \sigma_{41}\frac{\partial T}{\partial\varphi_1} + \sigma_{51}\frac{\partial T}{\partial\varphi_2}\sigma_{61}T)dW(t)\right). \end{aligned} \quad (18.47)$$

In the case where the transformation function $T(\varphi_j)$ is bounded and nonsingular, both processes P and S possess the same stability behavior. Therefore the transformation function $T(\varphi_j)$ is chosen so that the drift term, of the Itô differential Eq. (18.47), does not depend on the phase processes φ_j, θ_i so that

$$dS = \Lambda(p)Sdt$$

$$+ \frac{S}{T}\left(\sigma_{11}\frac{\partial T}{\partial\theta_1} + \sigma_{21}\frac{\partial T}{\partial\theta_2} + \sigma_{31}\frac{\partial T}{\partial\theta_3} + \sigma_{41}\frac{\partial T}{\partial\varphi_1} + \sigma_{51}\frac{\partial T}{\partial\varphi_2} + \sigma_{61}T\right)dW(t). \quad (18.48)$$

Comparing Eqs. (18.47) and (18.48), the transformation function $T(\theta_1, \theta_2, \theta_3, \varphi_1, \varphi_2)$ is given by the following equation:

$$[L_0 + \varepsilon L_1]T(\theta_1, \theta_2, \theta_3, \varphi_1, \varphi_2) = \Lambda(p)T(\theta_1, \theta_2, \theta_3, \varphi_1, \varphi_2), \quad (18.49)$$

where L_0 and L_1 are the differential operators in the forms:

$$L_0 = \omega_1 \frac{\partial}{\partial \theta_1} + \omega_2 \frac{\partial}{\partial \theta_2} + \omega_3 \frac{\partial}{\partial \theta_3},$$

$$L_1 = a_1 \frac{\partial^2}{\partial \theta_1^2} + a_2 \frac{\partial^2}{\partial \theta_2^2} + a_3 \frac{\partial^2}{\partial \theta_3^2} + a_4 \frac{\partial^2}{\partial \varphi_1^2} + a_5 \frac{\partial^2}{\partial \varphi_2^2} + a_6 \frac{\partial^2}{\partial \theta_1 \partial \theta_2}$$

$$+ a_7 \frac{\partial^2}{\partial \theta_1 \partial \theta_3} + a_8 \frac{\partial^2}{\partial \theta_1 \partial \varphi_1} + a_9 \frac{\partial^2}{\partial \theta_1 \partial \varphi_2} + a_{10} \frac{\partial^2}{\partial \theta_2 \partial \theta_3} + a_{11} \frac{\partial^2}{\partial \theta_2 \partial \varphi_1}$$

$$+ a_{12} \frac{\partial^2}{\partial \theta_2 \partial \varphi_2} + a_{13} \frac{\partial^2}{\partial \theta_3 \partial \varphi_1} + a_{14} \frac{\partial^2}{\partial \theta_3 \partial \varphi_2} + a_{15} \frac{\partial^2}{\partial \varphi_1 \partial \varphi_2}$$

$$+ b_1 \frac{\partial}{\partial \theta_1} + b_2 \frac{\partial}{\partial \theta_2} + b_3 \frac{\partial}{\partial \theta_3} + b_4 \frac{\partial}{\partial \varphi_1} + b_5 \frac{\partial}{\partial \varphi_2} + c, \tag{18.50}$$

with functional coefficients $a_{1\ldots15}$, $b_{1\ldots5}$ and $c = c(\theta_1, \theta_2, \theta_3, \varphi_1, \varphi_2)$.

18.4 Moment Lyapunov exponents

The eigenvalue problem for a differential operator of five independent variables is identified from Eq. (18.49) in which $\Lambda(p)$ is the eigenvalue and $T(\theta_1, \theta_2, \theta_3, \varphi_1, \varphi_2)$ is the associated eigenfunction. Eq. (18.49) defines that the eigenvalue $\Lambda(p)$ is the Lyapunov exponent of the p-th moment of the system (18.42). This approach was first applied by Wedig [22] to derive the eigenvalue problem for the moment Lyapunov exponent of a two-dimensional linear Itô stochastic system. Applying the method of regular perturbation, both the moment Lyapunov exponent $\Lambda(p)$ and the eigenfunction $T(\theta_1, \theta_2, \theta_3, \varphi_1, \varphi_2)$ are expanded in power series of ε to obtain a weak noise expansion of the Lyapunov exponent of a system with stochastic excitation described by six-dimensional components as:

$$\Lambda(p) = \Lambda_0(p) + \varepsilon\Lambda_1(p) + \varepsilon^2\Lambda_2(p) + \ldots + \varepsilon^n\Lambda_n(p) + \ldots,$$

$$T(\theta_1, \theta_2, \theta_3, \varphi_1, \varphi_2) = T_0(\theta_1, \theta_2, \theta_3, \varphi_1, \varphi_2) + \varepsilon T_1(\theta_1, \theta_2, \theta_3, \varphi_1, \varphi_2)$$

$$+ \varepsilon^2 T_2(\theta_1, \theta_2, \theta_3, \varphi_1, \varphi_2) + \ldots \tag{18.51}$$

$$+\varepsilon^2 T_2(\theta_1, \theta_2, \theta_3, \varphi_1, \varphi_2) + \ldots + \varepsilon^n T_n(\theta_1, \theta_2, \theta_3, \varphi_1, \varphi_2) + \ldots.$$

Equating the terms of equal powers of ε from Eqs. (18.49) and (18.51), the following equations may be written as:

$$\varepsilon^0: L_0 T_0(\theta_1, \theta_2, \theta_3, \varphi_1, \varphi_2) = \Lambda_0(p) T_0(\theta_1, \theta_2, \theta_3, \varphi_1, \varphi_2),$$

$$\varepsilon^1: L_0 T_1(\theta_1, \theta_2, \theta_3, \varphi_1, \varphi_2) + L_1 T_0(\theta_1, \theta_2, \theta_3, \varphi_1, \varphi_2) = \Lambda_0(p) T_1(\theta_1, \theta_2, \theta_3, \varphi_1, \varphi_2)$$

$$+ \Lambda_1(p) T_0(\theta_1, \theta_2, \theta_3, \varphi_1, \varphi_2),$$

$$\varepsilon^2\text{: } L_0T_2(\theta_1, \theta_2, \theta_3, \varphi_1, \varphi_2) + L_1T_1(\theta_1, \theta_2, \theta_3, \varphi_1, \varphi_2) = \Lambda_0(p)T_2(\theta_1, \theta_2, \theta_3, \varphi_1, \varphi_2)$$
$$+ \Lambda_1(p)T_1(\theta_1, \theta_2, \theta_3, \varphi_1, \varphi_2) + \Lambda_2(p)T_0(\theta_1, \theta_2, \theta_3, \varphi_1, \varphi_2),$$

$$\varepsilon^3\text{: } L_0T_3(\theta_1, \theta_2, \theta_3, \varphi_1, \varphi_2) + L_1T_2(\theta_1, \theta_2, \theta_3, \varphi_1, \varphi_2) = \Lambda_0(p)T_3(\theta_1, \theta_2, \theta_3, \varphi_1, \varphi_2)$$
$$+ \Lambda_1(p)T_2(\theta_1, \theta_2, \theta_3, \varphi_1, \varphi_2)$$
$$+\Lambda_2(p)T_1(\theta_1, \theta_2, \theta_3, \varphi_1, \varphi_2) + \Lambda_3(p)T_0(\theta_1, \theta_2, \theta_3, \varphi_1, \varphi_2),$$
$$\ldots$$
$$\varepsilon^n\text{: } L_0T_n(\theta_1, \theta_2, \theta_3, \varphi_1, \varphi_2) + L_1T_{n-1}(\theta_1, \theta_2, \theta_3, \varphi_1, \varphi_2)$$
$$= \Lambda_0(p)T_n(\theta_1, \theta_2, \theta_3, \varphi_1, \varphi_2) + \Lambda_1(p)T_{n-1}(\theta_1, \theta_2, \theta_3, \varphi_1, \varphi_2)$$
$$+\ldots+\Lambda_n(p)T_0(\theta_1, \theta_2, \theta_3, \varphi_1, \varphi_2), \tag{18.52}$$

where each function $T_s(\theta_1, \theta_2, \theta_3, \varphi_1, \varphi_2)$, $s = 1, 2, 3, \ldots$ must be positive and periodic in the range $0 \le \theta_i \le 2\pi$.

18.4.1 Zeroth order perturbation

From Eqs. (18.49) and (18.52) the zero-*th* order perturbation equation is $L_0T_0 = \Lambda_0(p)T_0$ and can be written as,

$$\omega_1\frac{\partial T_0(\theta_1, \theta_2, \theta_3, \varphi_1, \varphi_2)}{\partial\theta_1} + \omega_2\frac{\partial T_0(\theta_1, \theta_2, \theta_3, \varphi_1, \varphi_2)}{\partial\theta_2}$$
$$+ \omega_3\frac{\partial T_0(\theta_1, \theta_2, \theta_3, \varphi_1, \varphi_2)}{\partial\theta_3} = \Lambda_0(p)T_0(\theta_1, \theta_2, \theta_3, \varphi_1, \varphi_2), \tag{18.53}$$

Eq. (18.53) can be easily solved from the moment Lyapunov exponent characteristic which results $\Lambda_n(0) = 0$ for $n = 0, 1, 2, 3, \ldots$ because of $\Lambda(0) = \Lambda_0(0) + \varepsilon\Lambda_1(0) + \varepsilon^2\Lambda_2(0) + \ldots+\varepsilon^n\Lambda_n(0) = 0$. The eigenvalue $\Lambda_0(p)$ is independent of p. Hence $\Lambda_0(0) = 0$ leads to $\Lambda_0(p) = 0$. The solutions of Eq. (18.53) are periodic on the condition that

$$\Lambda_0(p) = 0, \quad q_0(\theta_1, \theta_2, \theta_3, \varphi_1, \varphi_2) = 1. \tag{18.54}$$

18.4.2 First-order perturbation

The equation given from the first-order perturbation is:

$$L_0T_1(\theta_1, \theta_2, \theta_3, \varphi_1, \varphi_2) + L_1T_0(\theta_1, \theta_2, \theta_3, \varphi_1, \varphi_2)$$
$$= \Lambda_0(p)T_1(\theta_1, \theta_2, \theta_3, \varphi_1, \varphi_2) + \Lambda_1(p)T_0(\theta_1, \theta_2, \theta_3, \varphi_1, \varphi_2), \tag{18.55}$$

and Eq. (18.55) has a periodic solution on the condition that

$$\int_0^{2\pi}\int_0^{2\pi}\int_0^{2\pi}\int_0^{\pi/2}\int_0^{\pi/2}[L_1 1 - \Lambda_1(p)]d\varphi_1 d\varphi_2 d\theta_1 d\theta_2 d\theta_3 = 0, \quad (18.56)$$

which yields

$$\Lambda_1(p) = \frac{1}{2\pi^5}\int_0^{2\pi}\int_0^{2\pi}\int_0^{2\pi}\int_0^{\pi/2}\int_0^{\pi/2}[c(\theta_1, \theta_2, \theta_3, \varphi_1, \varphi_2)]d\varphi_1 d\varphi_2 d\theta_1 d\theta_2 d\theta_3$$

$$= \frac{1}{1024}p\sigma^2[f_1(s_{ijk}) + pf_2(s_{ijk})]$$

$$-\frac{p}{8}[d_{111} + d_{122} + d_{133} + d_{411} + 2(d_{422} + d_{433})], \quad (18.57)$$

where $f_1(s_{ijk})$ and $f_1(s_{ijk})$ are given in Appendix A18.1. The first-order perturbation equation can be rewritten as:

$$\omega_1\frac{\partial T_1(\theta_1, \theta_2, \theta_3, \varphi_1, \varphi_2)}{\partial\theta_1} + \omega_2\frac{\partial T_1(\theta_1, \theta_2, \theta_3, \varphi_1, \varphi_2)}{\partial\theta_2} +$$

$$\omega_3\frac{\partial T_1(\theta_1, \theta_2, \theta_3, \varphi_1, \varphi_2)}{\partial\theta_3} + c(\theta_1, \theta_2, \theta_3, \varphi_1, \varphi_2) = \Lambda_1(p). \quad (18.58)$$

It is important to consider the commensurable frequencies in which a relation of the form of $\omega_1 = \omega_2/\mu_1 = \omega_3/\mu_2$ exists. The function $c(\theta_1, \theta_2, \theta_3, \varphi_1, \varphi_2)$ in Eq. (18.58) can be written in the form:

$$c(\theta_1, \theta_2, \theta_3, \varphi_1, \varphi_2) = \Lambda_1(p) + f_0(\theta_1, \theta_2, \theta_3) + \mathcal{L}_g(\theta_1, \theta_2, \theta_3)\mathfrak{R}_g(\varphi_1, \varphi_2),$$

$$g = 1,\ 2,\ \ldots 35. \quad (18.59)$$

Functions $\mathscr{L}_g$ are periodic in θ_1, θ_2, and θ_3 with $\mathfrak{R}_g$ trigonometric combinations of (φ_1, φ_2) given in Appendix A18.2. Eq. (18.58) cannot be obtained explicitly for $T_1(\theta_1, \theta_2, \theta_3, \varphi_1, \varphi_2)$. The combination of coefficients φ_1 and φ_2 in Eq. (18.59) suggests that the function $T_1(\theta_1, \theta_2, \theta_3, \varphi_1, \varphi_2)$ can be written as:

$$T_1(\theta_1, \theta_2, \theta_3, \varphi_1, \varphi_2) = T_{10}(\theta_1, \theta_2, \theta_3) + T_{1g}(\theta_1, \theta_2, \theta_3)\mathfrak{R}_g(\varphi_1, \varphi_2). \quad (18.60)$$

Substituting Eq. (18.60) into Eq. (18.58), and equating terms of the equal trigonometry function, gives a set of partial differential equations:

$$\omega_1\frac{\partial T_{1r}(\theta_1, \theta_2, \theta_3)}{\partial\theta_1} + \omega_2\frac{\partial T_{1r}(\theta_1, \theta_2, \theta_3)}{\partial\theta_2} + \omega_3\frac{\partial T_{1r}(\theta_1, \theta_2, \theta_3)}{\partial\theta_3}$$

$$+ f_r(\theta_1, \theta_2, \theta_3) = 0, \quad r = 0,\ 1,\ \ldots,\ 35. \quad (18.61)$$

The functions $T_{1r}(\theta_1, \theta_2, \theta_3)$ can be written as:

$$T_{1r}(\theta_1, \theta_2, \theta_3) = \mho + C_r(\theta_2 - \mu_1\theta_1, \theta_3 - \mu_2\theta_1), \quad r = 0,\ 1,\ \ldots,35, \quad (18.62)$$

where the calculated values Ʊ cannot be displayed because of cumbersome and long terms in the analytic form. Here $C_r(\theta_2 - \mu_1\theta_1, \theta_3 - \mu_2\theta_1)$ are the same arbitrary functions of three variables for which the following assumptions are made:

$$C_r(\theta_2 - l_1\theta_1,\ \theta_3 - l_2\theta_1) = A_{1r} + B_{1r}\sin(2\theta_2 - 2\mu_1\theta_1) + C_{1r}\sin(4\theta_2 - 4\mu_1\theta_1)$$
$$+ D_{1r}\sin(2\theta_3 - 2\mu_2\theta_1) + E_{1r}\sin(4\theta_3 - 4\mu_2\theta_1),\quad r = 0,\ 1,\ \ldots, 35. \tag{18.63}$$

As each function $T_{1r}(\theta_1, \theta_2, \theta_3)$, $r = 0, 1, \ldots, 35$ must be positive and periodic, unknown constants A_{1r}, B_{1r}, C_{1r}, D_{1r}, and E_{1r} can be determined using conditions:

$$T_{1r}(0, 0, 0) = T_{1r}(0, 0, 2\pi) = T_{1r}(0, 2\pi, 0) = T_{1r}(2\pi, 0, 0) = T_{1r}(2\pi, 2\pi, 0)$$
$$= T_{1r}(0, 2\pi, 2\pi) = T_{1r}(2\pi, 0, 2\pi) = T_{1r}(2\pi, 2\pi, 2\pi) = 0, \tag{18.64}$$

$$\frac{\partial T_{1r}(0, 0, 0)}{\partial\theta_1} = \frac{\partial T_{1r}(2\pi, 0, 0)}{\partial\theta_1},\quad \frac{\partial T_{1r}(0, 0, 0)}{\partial\theta_2} = \frac{\partial T_{1r}(0, 2\pi, 0)}{\partial\theta_2},$$
$$\frac{\partial T_{1r}(0,\ 0,\ 0)}{\partial\theta_3} = \frac{\partial T_{1r}(0,\ 0,\ 2\pi)}{\partial\theta_3},\quad r = 0,\ 1,\ \ldots, 35. \tag{18.65}$$

18.4.3 Second-order perturbation

The second-order perturbation equation must satisfy the condition of the periodic function $T_2(\theta_1, \theta_2, \theta_3, \varphi_1, \varphi_2)$ in θ_1, θ_2, and θ_3 given as:

$$L_0T_2(\theta_1, \theta_2, \theta_3, \varphi_1, \varphi_2) + L_1T_2(\theta_1, \theta_2, \theta_3, \varphi_1, \varphi_2)$$
$$= \Lambda_1(p)T_1(\theta_1, \theta_2, \theta_3, \varphi_1, \varphi_2) + \Lambda_2(p). \tag{18.66}$$

We have

$$\Lambda_2(p) = \frac{1}{2\pi^5}\int_0^{2\pi}\int_0^{2\pi}\int_0^{2\pi}\int_0^{\frac{\pi}{2}}\int_0^{\frac{\pi}{2}}\left\{a_1\frac{\partial^2T_1}{\partial\theta_1^2} + a_2\frac{\partial^2T_1}{\partial\theta_2^2} + a_3\frac{\partial^2T_1}{\partial\theta_3^2}\right.$$
$$+ a_4\frac{\partial^2T_1}{\partial\varphi_1^2} + a_5\frac{\partial^2T_1}{\partial\varphi_2^2} + a_6\frac{\partial^2T_1}{\partial\theta_1\partial\theta_2} + a_7\frac{\partial^2T_1}{\partial\theta_1\partial\theta_3} + a_8\frac{\partial^2T_1}{\partial\theta_1\partial\varphi_1} + a_9\frac{\partial^2T_1}{\partial\theta_1\partial\varphi_2}$$
$$+ a_{10}\frac{\partial^2T_1}{\partial\theta_2\partial\theta_3} + a_{11}\frac{\partial^2T_1}{\partial\theta_2\partial\varphi_1} + a_{12}\frac{\partial^2T_1}{\partial\theta_2\partial\varphi_2} + a_{13}\frac{\partial^2T_1}{\partial\theta_3\partial\varphi_1}$$
$$+ a_{14}\frac{\partial^2T_1}{\partial\theta_3\partial\varphi_2} + a_{15}\frac{\partial^2T_1}{\partial\varphi_1\partial\varphi_2} + b_1\frac{\partial T_1}{\partial\theta_1} + b_2\frac{\partial T_1}{\partial\theta_2} + b_3\frac{\partial T_1}{\partial\theta_3} + b_4\frac{\partial T_1}{\partial\varphi_1}$$
$$\left.+ b_5\frac{\partial T_1(\theta_1, \theta_2, \theta_3, \varphi_1, \varphi_2)}{\partial\varphi_2} + [c - \Lambda_1(p)]T_1\right\}\mathrm{d}\varphi_1\mathrm{d}\varphi_2\mathrm{d}\theta_1\mathrm{d}\theta_2\mathrm{d}\theta_3. \tag{18.67}$$

After long computation, $\Lambda_2(p)$ can be obtained symbolically. After integrating, the solution cannot be displayed due to long terms but after applying numerical substitutions, the second perturbation has been done for the first time according to the coupled gyroscopic stochastic system with three-degree-of-freedom. The weak noise expansion of the moment Lyapunov exponent in the second-order perturbation is determined in the form:

$$\Lambda(p) = \varepsilon\Lambda_1(p) + \varepsilon^2\Lambda_2(p) + O(\varepsilon^3). \tag{18.68}$$

The Lyapunov exponent can now be obtained from Eq. (18.68) by using the property of the moment Lyapunov exponent.

$$\lambda = \left.\frac{d\Lambda(p)}{dp}\right|_{p=0} = \varepsilon\lambda_1 + \varepsilon^2\lambda_2 + O(\varepsilon^3)$$

$$= \varepsilon\left\{\left[\frac{1}{1024\Psi}\right]\sigma^2\frac{p}{8}[d_{111} + d_{122} + d_{133} + d_{411} + 2(d_{422} + d_{433})]\right\}$$

$$+ \varepsilon^2\mathscr{F} + O(\varepsilon^3), \tag{18.69}$$

where Ψ is given in Appendix A18.3.

18.5 Stochastic stability conditions and discussion

The values used for the parameters of the stochastic system in the calculations to determine the Moment Lyapunov exponent and Lyapunov exponent are given as follows:

$$L = 10\,\text{m},\quad h = \frac{2}{5}\sqrt{\frac{3}{5}}\,\text{m},\quad a = 10\,\text{m},\quad \rho = 2\times10^3\,\text{kg/m}^3,$$

$$E = 1\times10^{10}\,\text{N/m}^2,\quad \nu = 0.3. \tag{18.70}$$

We have analytically determined the p-th moment stability boundary in the first-order perturbation for $p = 1$, but using the obtained analytical form this can be calculated for any concrete numerical value of p with the definition of the moment stability $\Lambda(p) < 0$. The almost-sure stability boundary of the oscillatory system can be determined in the first-order perturbation from knowing that the oscillatory stochastic system is asymptotically stable only if the Lyapunov exponent is $\lambda < 0$. Variations of the moment Lyapunov exponent are presented in Fig. 18.2. The main point of the reported results is that stochastic stability has growth in the stable region with an increase in the radius of the circular shell and the damping ratio of the system, as well as with an increase in the ratio of the length and radius of the circular shell. Fig. 18.2 (C) depicts the increase in the stability region with the decrease in intensity of the random process. Thus the curvature is responsible for the appearance of the more stable stochastic properties for the same conditions in comparison with the same shell with a smaller radius. An interesting conclusion reported in Fig. 18.2 (A) shows that an increase in the radius of the shell leads to an increase in the stable region but only to a certain limit of the radius, which is a consequence of the other

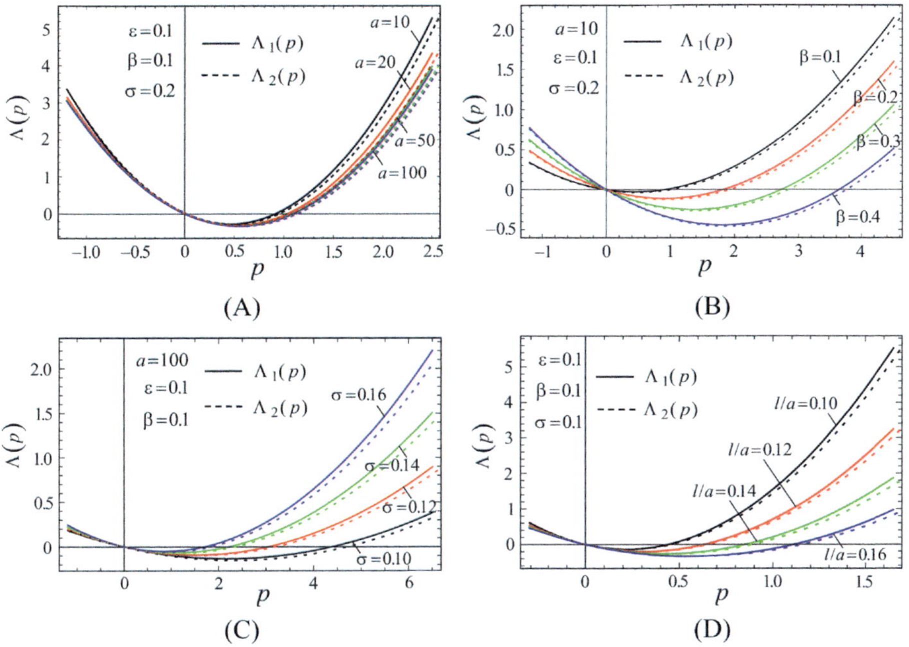

Figure 18.2 Moment Lyapunov exponent $\Lambda(p)$. (A) for varying shell radius a, (B) for varying damping coefficients, (C) for the varying intensity of the random process σ, (D) for varying length to radius ratio l/a.

material and geometric properties of the shell. A further increase in the radius does not lead to an increase in the stochastic stable region, which we can see from the curve obtained for the highest limit case of the chosen radius. Fig. 18.2 (B) shows an increase in the stable region by an increase in the damping coefficient, where the tendency is completely expected. Fig. 18.2 (D) shows the change in stability with the increase of length to radius ratio. It can be seen that the stable region is significantly larger for the curve representing the highest observed ratio compared to the curve representing the lowest ratio.

18.6 Conclusions

The stochastic stability of a cylindrical shell is investigated in this chapter. It is assumed that the system possesses a small damping. The stochastic system of the six-dimensional class under white-noise excitations is studied analytically through the determination of the moment Lyapunov exponents and the Lyapunov exponents. An eigenvalue problem for the moment Lyapunov exponent is established using the theory of stochastic dynamic systems. For weak noise excitations, a singular perturbation method is employed to obtain the second-order expansions of the moment Lyapunov exponent. The Lyapunov exponent is then obtained using the relationship between the moment Lyapunov exponent and the Lyapunov exponent. The almost-sure and p-th moment stability boundaries in the first and second perturbations with respect to the damping coefficients are determined. When the Lyapunov exponent is negative, the system is almost-surely stable with probability 1, otherwise, it is unstable. Comparison of the stochastic stabilities of circular cylindrical shells, chosen with varying parameters of the radius, shows the growth in the stochastic stable region with an increase in the radius. This qualitative character of the stability of the mechanical system of the cylindrical shell is determined quantitatively, and it was analyzed for the first time in the field of stochastic stability using the six-dimensional system of Stratonovich differential equations. The system is analyzed in detail from the analytical side and the reported results are of great importance for the stability of structures. Also, the study contains a short procedure for stochastic analysis of coupled mechanical systems of the class of six-dimensional Stratonovich differential equation systems, and it could be used for any mechanical system with three coupled stochastic oscillators that possess a small damping. It is also necessary to point out a possible further direction in the analysis of the stability of coupled or complex stochastic systems in terms of the determination of higher-order perturbations (in stochastic systems of MDOF), which provides the need for detailed studies in this area. This can be justified by the fact that the procedure is known, so it will be possible to determine even more accurate approximations of the solutions.

Appendix A18.1

$$f_1(s_{ijk}) = 74s_{111}^2 + 46s_{112}^2 + 52s_{113}^2 + 92s_{112}s_{121}$$

$$+ 46s_{121}^2 - 36s_{111}s_{122} + 74s_{122}^2 + 52s_{123}^2 + 104s_{113}s_{131}$$

$$+ 52s_{131}^2 + 104s_{123}s_{132} + 52s_{132}^2 - 24s_{111}s_{133} - 24s_{122}s_{133}$$

$$+ 74s_{133}^2 + 52s_{211}^2 + 112s_{212}^2 + 112s_{213}^2 + 52s_{221}^2 + 112s_{222}^2$$

$$+ 112s_{223}^2 + 46s_{231}^2 + 112s_{232}^2 + 112s_{233}^2 + 104s_{211}s_{311}$$

$$+ 52s_{311}^2 + 104s_{221}s_{312} + 52s_{312}^2 + 92s_{231}s_{313} + 46s_{313}^2 + 160s_{212}s_{321}$$

$$+ 48s_{321}^2 + 160s_{222}s_{322} + 48s_{322}^2 + 160s_{232}s_{323} + 48s_{323}^2$$

$$+ 160s_{213}s_{331} + 48s_{331}^2 + 160s_{223}s_{332} + 48s_{332}^2 + 160s_{233}s_{333}$$

$$+ 48s_{333}^2 - 24s_{111}s_{411} - 24s_{122}s_{411} - 36s_{133}s_{411} + 74s_{411}^2$$

$$+ 112s_{412}^2 + 112s_{413}^2 + 160s_{412}s_{421} + 48s_{421}^2 - 32s_{111}s_{422}$$

$$- 32s_{122}s_{422} - 32s_{133}s_{422} + 160s_{413}s_{431} + 48s_{431}^2 + 160s_{423}s_{432}$$

$$- 32s_{411}s_{422} + 112s_{422}^2 + 80s_{423}^2 + 80s_{432}^2 - 32s_{111}s_{433}$$

$$- 32s_{122}s_{433} - 32s_{133}s_{433} - 32s_{411}s_{433} - 96s_{422}s_{433} + 112s_{433}^2.$$

$$f_2(s_{ijk}) = 27s_{111}^2 + 9s_{112}^2 + 6s_{113}^2 + 18s_{112}s_{121}$$

$$+ 9s_{121}^2 + 18s_{111}s_{122} + 27s_{122}^2 + 6s_{123}^2 + 12s_{113}s_{131}$$

$$+ 6s_{131}^2 + 12s_{123}s_{132} + 6s_{132}^2 + 12s_{111}s_{133} + 12s_{122}s_{133}$$

$$+ 27s_{133}^2 + 6s_{211}^2 + 8s_{212}^2 + 8s_{213}^2 + 6s_{221}^2 + 8s_{222}^2 + 8s_{223}^2$$

$$+ 9s_{231}^2 + 8s_{232}^2 + 8s_{233}^2 + 12s_{211}s_{311} + 6s_{311}^2 + 12s_{221}s_{312}$$

$$+ 6s_{312}^2 + 18s_{231}s_{313} + 9s_{313}^2 + 16s_{212}s_{321} + 8s_{321}^2$$

$$+ 16s_{222}s_{322} + 8s_{322}^2 + 16s_{232}s_{323} + 8s_{323}^2 + 16s_{213}s_{331}$$

$$+ 8s_{331}^2 + 16s_{223}s_{332} + 8s_{332}^2 + 16s_{233}s_{333} + 8s_{333}^2$$

$$+ 12s_{111}s_{411} + 12s_{122}s_{411} + 18s_{133}s_{411} + 27s_{411}^2 + 8s_{412}^2$$

$$+ 8s_{413}^2 + 16s_{412}s_{421} + 8s_{421}^2 + 16s_{111}s_{422} + 16s_{122}s_{422}$$

$$+ 16s_{133}s_{422} + 16s_{411}s_{422} + 72s_{422}^2 + 24s_{423}^2 + 16s_{413}s_{431} + 8s_{431}^2$$

$$+ 48s_{423}s_{432} + 24s_{432}^2 + 16s_{111}s_{433} + 16s_{122}s_{433}$$

$$+ 16s_{133}s_{433} + 16s_{411}s_{433} + 48s_{422}s_{433} + 72s_{433}^2.$$

Appendix A18.2

$$\mathscr{L}_g(\theta_1, \theta_2, \theta_3)\mathfrak{R}_g(\phi_1, \phi_2)=$$

$$+\begin{Bmatrix} f_{25} \\ \vdots \\ f_{31} \end{Bmatrix}^{\mathrm{T}} \begin{bmatrix} \cos\left(2\phi_1\right)\cos\left(2\phi_2\right) \\ \cos\left(4\phi_1\right)\cos\left(2\phi_2\right) \\ \cos\left(2\phi_1\right)\cos\left(4\phi_2\right) \\ \cos\left(4\phi_1\right)\cos\left(4\phi_2\right) \\ \cos\left(\phi_2\right)\sin\left(\phi_2\right) \\ \cos\left(3\phi_2\right)\sin\left(\phi_2\right) \\ \cos\left(\phi_2\right)\sin\left(3\phi_2\right) \end{bmatrix} + \begin{Bmatrix} f_{32} \\ \vdots \\ f_{35} \end{Bmatrix}^{\mathrm{T}} \begin{bmatrix} \cos\left(2\phi_1\right) \\ \cos\left(4\phi_1\right) \\ \cos\left(2\phi_2\right) \\ \cos\left(4\phi_2\right) \end{bmatrix}.$$

Appendix A18.3

$$\Psi = 74s_{111}^2 + 46s_{112}^2 + 52s_{113}^2 + 92s_{112}s_{121} + 46s_{121}^2 - 36s_{111}s_{122}$$

$$+ 74s_{122}^2 + 52s_{123}^2 + 104s_{113}s_{131} + 52s_{131}^2 + 104s_{123}s_{132}$$

$$+ 52s_{132}^2 - 24s_{111}s_{133} - 24s_{122}s_{133} + 74s_{133}^2 + 52s_{211}^2$$

$$+ 112s_{212}^2 + 112s_{213}^2 + 52s_{221}^2 + 112s_{222}^2 + 112s_{223}^2$$

$$+ 46s_{231}^2 + 112s_{232}^2 + 112s_{233}^2 + 104s_{211}s_{311} + 52s_{311}^2$$

$$+ 104s_{221}s_{312} + 52s_{312}^2 + 92s_{231}s_{313} + 46s_{313}^2 + 160s_{212}s_{321}$$

$$+ 48s_{321}^2 + 160s_{222}s_{322} + 48s_{322}^2 + 160s_{232}s_{323} + 48s_{323}^2$$

$$+ 160s_{213}s_{331} + 48s_{331}^2 + 160s_{223}s_{332} + 48s_{332}^2 + 160s_{233}s_{333}$$

$$+ 48s_{333}^2 - 24s_{111}s_{411} - 24s_{122}s_{411} - 36s_{133}s_{411} + 74s_{411}^2$$

$$+ 112s_{412}^2 + 112s_{413}^2 + 160s_{412}s_{421} + 48s_{421}^2 - 32s_{111}s_{422}$$

$$- 32s_{122}s_{422} - 32s_{133}s_{422} - 32s_{411}s_{422} + 112s_{422}^2 + 80s_{423}^2$$

$$+ 160s_{413}s_{431} + 48s_{431}^2 + 160s_{423}s_{432} + 80s_{432}^2 - 32s_{111}s_{433}$$

$$- 32s_{122}s_{433} - 32s_{133}s_{433} - 32s_{411}s_{433} - 96s_{422}s_{433} + 112s_{433}^2.$$

References

[1] I. Elishakoff, J. Arbocz, Reliability of axially compressed cylindrical shells with random axisymmetric imperfections, International Journal of Solids and Structures 18 (1982) 563–585.

[2] W. Kurnik, A. Tylikowski, Stochastic stability and nonstability of a linear cylindrical shell, Ingenieur-Archiv 53 (1983) 363–369.

[3] M.A. Bessa, S. Pellegrino, Design of ultra-thin shell structures in the stochastic post-buckling range using Bayesian machine learning and optimization, International Journal of Solids and Structures 139-140 (2018) 174–188.

[4] B. Wang, K. Du, P. Hao, K. Tian, Y.J. Chao, L. Jiang, S. Xu, X. Zhang, Experimental validation of cylindrical shells under axial compression for improved knockdown factors, International Journal of Solids and Structures 164 (2019) 37–51.

[5] N. Sri Namachchivaya, H.J. Van Roessel, S. Talwar, Maximal Lyapunov exponent and almost-sure stability for coupled two-degree-of-freedom stochastic systems, Journal of Applied Mechanics 61 (2) (1994) 446–452.

[6] W.-C. Xie, Moment Lyapunov exponents of a two-dimensional system under real-noise excitation, Journal of Sound and Vibration 239 (2001) 139–155.

[7] P. Kozić, G. Janevski, R. Pavlović, Moment Lyapunov exponents and stochastic stability of a double-beam system under compressive axial loading, International Journal of Solids and Structures 47 (2010) 1435–1442.

[8] V. Stojanović, P. Kozić, M.D. Petković, Dynamic instability and critical velocity of a mass moving uniformly along a stabilized infinity beam, International Journal of Solids and Structures 108 (2017) 164–174.

[9] S. Li, X. Liu, Moment Lyapunov exponent for three-dimensional system under real noise excitation, Applied Mathematics and Mechanics 34 (2013) 613–626.

[10] W. Jiancheng, L. Xianbin, Moment stability of viscoelastic system influenced by non-Gaussian colored noise, Journal of Sound and Vibration 502 (2021) 116080.

[11] J. Deng, W.-C. Xie, M. Pandey, Moment Lyapunov exponents and stochastic stability of coupled viscoelastic systems driven by white noise, Journal of Mechanics of Materials and Structures 9 (1) (2014) 27–50.

[12] J. Deng, Stochastic stability of gyroscopic systems under bounded noise excitation, International Journal of Structural Stability and Dynamics 18 (2018) 1850022.

[13] D. Haixu, Strength and stability of double cylindrical shell structure subjected to hydrostatic external pressure—II: stability, Marine Structures 16 (5) (2003) 397–415.

[14] V.V. Vasil'ev, To the problem of stability of a cylindrical shell under axial compression, Mechanics of Solids 46 (2) (2011) 161–169.

[15] R.M.J. Groh, A. Pirrera, Orthotropy as a driver for complex stability phenomena in cylindrical shell structures, Composite Structures 198 (2018) 63–72.

[16] Y. Wu, K. Gao, Q. Ren, Hydrothermal effect on Bi-stability of composite cylindrical shell, Composite Structures 232 (2020) 111554.

[17] Q.M. Ansari, G. Zucco, L.C. Trinh, P.M. Weaver, Compression buckling of elastically supported cylindrical shells with bend/twist coupling, Composite Structures 309 (2023) 116691.

[18] D. Zhang, P. Li, J. Lu, C. Zhang, Y. Yang, Stability analysis of cylindrical shell in axial flow: A DQ-based approach and an instability prediction formula, Ocean Engineering 267 (2023) 113198.

[19] À. Jorba, Computational Methods in Perturbation Theory, in: G. Gaeta (Ed.), Perturbation Theory. Encyclopedia of Complexity and Systems Science Series, Springer, New York, 2022.

[20] W. Soedel, Vibrations of Shells and Plates, CRC Press, New York, NY, 2004.

[21] W.-C. Xie, Dynamic Stability of Structures, Cambridge University Press, New York, NY, 2006.

[22] W. Wedig, Lyapunov exponent of stochastic systems and related bifurcation problem, in: S.T. Ariaratnam, G.I. Schuëller, I. Elishakoff (Eds.), Stochastic Structural Dynamics-Progress in Theory and Applications, Elsevier Applied Science, London, 1988, pp. 315–327.

Generalized transformations for MDOF stochastic systems

19

19.1 Introduction

In the field of engineering, dynamic stability and vibration are frequently observed in associated structures, and their understanding is of interest in aerospace, mechanical and civil engineering [1]. This research centers on examining conditions for stochastic stability in complex coupled systems. It presents valuable mathematical methods for identifying both stochastic stability and almost sure stability conditions [2]. Gaining insights into the stability of dynamic systems through parameter analysis is crucial. The analysis of parameters in each analytical-numerical simulation of a mechanical system is imperative to enable and enhance the development, design, and refinement of mechanical systems in technical applications.

To provide an all-encompassing examination, this study investigates the existing literature concerning stochastic systems characterized by random vibrations and stability. Specifically, the focus extends to single-degree-of-freedom, two-degrees-of-freedom, and multi-degrees-of-freedom configurations, emblematic of coupled mechanical systems that are widely encountered in technical applications. Nevertheless, the intricacy of these challenges might emerge due to simplified physical models, potentially yielding outcomes that do not entirely capture the genuine dynamic behavior. As an illustration, Huang and Xie [3] applied stochastic averaging to ascertain moment Lyapunov exponents in the context of a single-degree-of-freedom system subjected to wideband noise. Similarly, Pang et al. [4] used the asymptotic stability of single-degree-of-freedom systems by incorporating fractional operators to articulate the physical models of oscillators, resulting in a novel criterion for asymptotic stability based on the distribution of characteristic roots. Single-degree-of-freedom stochastic systems are commonly explored through specialized methods designed to address particular mechanical system issues, with a continual focus on refinement and enhancement. Zan et al. [5] introduced notable progress in mathematical tools for analyzing single-degree-of-freedom stochastic systems. They presented an improved version of the path integral method specifically designed for assessing the response statistics of single-degree-of-freedom systems under the influence of Lévy noise. The suggested methodology tackles the issues associated with high computational demands and memory needs by streamlining and parallelizing the path integral method. This optimization makes the approach efficient and viable for single-degree-of-freedom systems experiencing Lévy noise. The essential component of the path integral method revolves around deducing and verifying the probability density function for short-time transitions in the single-degree-of-freedom system. This process ensures its adherence to the relevant fractional Fokker-Planck-Kolmogorov equation, which dictates the dynamics of the system. The fractional Fokker-Planck-Kolmogorov (FPK) equation is derived using the characteristic

Modeling of Complex Dynamic Systems. DOI: https://doi.org/10.1016/B978-0-443-23942-7.00019-X

function and the Chapman–Kolmogorov equation. In order to address the challenges related to memory and computational complexity inherent in the path integral method, the authors simplify the one-step iteration formula and execute parallel computations through domain decomposition. The simplification of the one-step iteration formula leads to a reduction in the number of integrations and thereby reduces the memory requirements for the one-step transition matrix. Furthermore, the utilization of parallel computations reduces the total computational time, as evidenced by the outcomes derived from two illustrative examples. To validate the effectiveness of the enhanced path integral method, the authors conduct a comparison with Monte Carlo solutions and analytical solutions, employing them as benchmarks for reference. The outcomes demonstrate the efficacy of the improved method in examining single-degree-of-freedom systems under the influence of Lévy noise.

Furthermore, the analysis of stochastic stability has been broadened to encompass viscoelastic mechanical systems, aiming for a more comprehensive representation of veritable mechanical systems. This expansion includes inquiries such as the examination of single-degree-of-freedom viscoelastic systems exposed to bounded noise excitation using single or higher-order stochastic averaging [6]. Deng et al. [7] investigated stochastic stability in a plate exposed to supersonic flow, characterized as a coupled non-gyroscopic viscoelastic two-degrees-of-freedom stochastic system. These investigations underscore that the intricacy does not stem solely from the inherent nature of the structure but is instead attributed to the level of mathematical analysis needed to attain more accurate approximations. The assessment of moment Lyapunov exponents in three-degrees-of-freedom systems featuring viscoelastic connections, as investigated by Stojanović and Petković [8], affirms the difficulty associated with predicting stochastic stability in systems characterized by variable parameters and susceptibility to external excitations. Notably, stochastic stability predictions are absent for specific instances of technical system models involving systems with more than three degrees of freedom [8]. This particular case of a three-degrees-of-freedom system further expands the challenge of establishing moment Lyapunov exponents for two-dimensional stochastic systems [9–11] and continuous two-dimensional systems [12]. Many mechanical systems, in addition to being categorized and examined based on viscoelastic material properties, can also be classified as aeroelastic mechanical systems. Fixed-interval smoothing, considered a crucial type of state estimation, has attracted substantial attention in diverse practical applications, with a particular emphasis on its utilization in the analysis of flight test data. Nonetheless, traditional sequential filters and smoothers frequently encounter challenges when dealing with nonlinear or high-dimensional systems. Consequently, Liu et al. [13] employed an advanced approach to tackle the fixed-interval smoothing problem concerning a conceptual two-dimensional airfoil model in incompressible flow, through the usage of noisy measurement data. The governing equations of the airfoil model are presumed to be either fully known or partially known. The authors devised a single-objective optimization problem using the classical Runge–Kutta scheme. Through the minimization of this objective function, they concurrently acquire estimations for the system states, measurement noise, and potentially unknown parameters. The efficiency and practicality of this approach are evaluated using simulated measurement data, where the data acquired through measurement is affected by various types of measurement noise. The findings indicate that the proposed algorithm is well-suited for the airfoil model, even when there is cubic or free-play structural nonlinearity, and it produces accurate estimations of both the system states and parameters. Moreover, the algorithm demonstrates notable resilience in the presence of Gaussian white noise and more intricate heavy-tailed measurement noises. Remarkably, the algorithm proves effective even in the context of high-dimensional nonlinear aeroelastic systems. The aforementioned investigations tackle concerns related to the stochastic stability of particular mechanical systems represented in technical applications, predominantly relying on perturbations attributed to white noise fluctuations. A noteworthy array of recent studies is also capturing the interest of numerous researchers engaged in addressing problems involving these types of perturbing stochastic forces [14–17].

Especially notable are certain studies focusing on stochastic differential equations of n-dimensional nature. In their research, Zan et al. [18] tackled a pivotal feature of stochastic differential equations (SDEs) by exploring path integral solutions for n-dimensional SDEs characterized by stable Lévy noise. They conducted a methodical examination, deducing and validating these solutions, with broad implications for various fields like physics and engineering. The researchers initiated their inquiry by formulating the governing equations for the solutions of n-dimensional SDEs subject to stable Lévy noise. In the first stage, they utilized the characteristic function of stochastic processes, establishing a fundamental foundation for their subsequent investigations. Following this, they advanced to deduce the short-time transition probability density function for the path integral solution. This derivation was accomplished by adeptly integrating the Chapman–Kolmogorov–Smoluchowski (CKS) equation with the characteristic function. Furthermore, they took additional steps to validate the precision of their derived solution, ensuring its alignment with the governing equation of the SDE solution, commonly referred to as the Fokker–Planck–Kolmogorov equation. To demonstrate the practical significance of their suggested path integral method, the authors provided illustrative instances supported by numerical simulations. These examples effectively highlighted the viability and relevance of their approach in real-world situations. Furthermore, the authors reinforced their conclusions by computing the relevant Monte Carlo solution, offering additional confirmation of the accuracy and efficacy of their proposed method. In essence, the study not only conducts a thorough exploration of path integral solutions for n-dimensional SDEs with stable Lévy noise but also imparts valuable insights into comprehending and applying these solutions in practical contexts. As we extend the groundwork laid by their foundational research, this study acts as a fundamental building block for the progression in the field of stochastic processes, holding considerable potential for future developments in diverse scientific and industrial domains.

Although shafts are prevalent components in mechanical systems and have undergone extensive examination, the stochastic stability of systems with four or more degrees of freedom has not been documented in the current literature. Shafts, being standard components of mechanical systems, have been the subject of extensive study by numerous scientists over the years. Saeed et al. [19] investigated the nonlinear vibration control of a vertically supported asymmetric nonlinear shaft. Xiao et al. [20] employed the finite element method to examine the bending-torsional-axial vibrations of a variable-section shaft system. Yang and Pei [21] examine a model of a rotating flexible shaft with unstressed initial deformation, exploring its dynamics and vibration that occurs in the suppression, under various parametric excitations. In this research, we have formulated new transformations applicable to n-dimensional systems, facilitating the direct derivation of Itô equations. Additionally, we have introduced a new stochastic model of a coupled shaft within the category of dynamic stochastic systems exhibiting gyroscope effects, characterized by four degrees of freedom. We have analyzed this model both as an example and as an extension of the previously outlined model in Reference [22]. To our understanding, this investigation marks the initial endeavor to explore the stochastic stability of intricate stochastic systems of this nature, thereby filling a notable void in existing scholarly inquiries. The primary goals and novelties of this study are centered on establishing a more comprehensive framework for assessing moment Lyapunov exponents and examining general stochastic Multi-Degree-of-Freedom (MDOF) processes, encompassing both gyroscopic and non-gyroscopic systems. In the second section, generalized transformations designed for decoupling of the equations of motion within MDOF stochastic systems are presented. This facilitates an approximation of moment Lyapunov exponents through perturbation methods, enabling a discourse on the stochastic stability of the system. In the third section, the implementation of stochastic stability analysis on a coupled rotating shaft system is illustrated.

19.2 Stochastic stability of MDOF systems

19.2.1 Formulation

In the broader context of stochastic mechanical models, both non-gyroscopic and gyroscopic systems emerge through suitable discretization. The non-gyroscopic and gyroscopic systems, each characterized by n-DOF, necessitate transformation into 2n-DOF first-order Stratonovich stochastic differential equations before undertaking subsequent stochastic stability analysis.

The stochastic system of differential equations with non-gyroscopic parametric excitation and n degrees of freedom can be expressed as follows:

$$\ddot{q}_i(t) + 2\varepsilon \sum_{j=1}^{n} \beta_{ij}\dot{q}_j(t) + \omega_i^2 q_i(t) + \sqrt{\varepsilon}\xi(t)\omega_i \sum_{j=1}^{n} k_{ij}q_j(t) = 0,\ i = 1,\ 2,\ \ldots,n, \tag{19.1a}$$

where excitation $\xi(t)$ is the ergodic stochastic process with zero mean value, ε represents the small parameter, ω_i is the natural frequency, β_{ij} is the viscous damping coefficient, and $i = 1,\ 2,\ \ldots,n$. By replacing $\dot{q}_i$ by the new variable [23], Eq. (19.1a and b) can be transformed into an equivalent system of $2n$ first-order Stratonovich stochastic differential equations

$$\mathrm{d}\{\mathbf{x}\} = \mathbf{A}\{\mathbf{x}\}\mathrm{d}t + \varepsilon\mathbf{B}\{\mathbf{x}\}\mathrm{d}t + \sqrt{\varepsilon}\mathbf{C}\{\mathbf{x}\}\mathrm{d}\xi(t), \tag{19.1b}$$

where **A**, **B** and **C** are constant coefficient matrices.

For gyroscopic systems, it is required to initially conduct a suitable process to decouple the system of equations. The differential equations describing the gyroscopic system with n degrees of freedom can be compactly expressed as follows

$$\mathbf{M}\ddot{\tilde{\mathbf{x}}} + (\mathbf{G} + \varepsilon\mathbf{D})\dot{\tilde{\mathbf{x}}} + \mathbf{K}\tilde{\mathbf{x}} + \sqrt{\varepsilon}\xi(t)\mathbf{V}\tilde{\mathbf{x}} = 0, \tag{19.2}$$

where **M**, **G**, **D**, and**K** are the $n \times n$ mass, gyroscopic, damping, and stiffness matrices, respectively, with an additional skew symmetric property of the matrix **G** given by $\mathbf{G}^T = \mathbf{G}$. In accordance with the method delineated in Chapter 3 [2], an expanded form of a system of $2n$ first-order Stratonovich differential equations can be obtained, as follows

$$\mathrm{d}\{\mathbf{x}\} = \begin{bmatrix} 0\ 1 \\ -10 \end{bmatrix}[\mathrm{diag}\{\omega_1, \omega_2, \ldots, \omega_n\}]\{\mathbf{x}\}\mathrm{d}t - \varepsilon\begin{bmatrix} \boldsymbol{d}_{1()} & \boldsymbol{d}_{2()} \\ \boldsymbol{d}_{3()} & \boldsymbol{d}_{4()} \end{bmatrix}\{\mathbf{x}\}\mathrm{d}t - \sqrt{\varepsilon}\begin{bmatrix} \boldsymbol{s}_{1()} & \boldsymbol{s}_{2()} \\ \boldsymbol{s}_{3()} & \boldsymbol{s}_{4()} \end{bmatrix}\{\mathbf{x}\}\mathrm{d}\xi(t). \tag{19.3}$$

where $\boldsymbol{d}_{i()}$ and $\boldsymbol{s}_{i()}$ $(i = 1, 2, 3, 4)$ are constant coefficient submatrices. As the generalization is shown in this MDOF case, the submatrices $\boldsymbol{d}_{i()}$ and $\boldsymbol{s}_{i()}$ denote matrices of constants that are obtainable and demonstrable for both a two-degree-of-freedom system and a three-degree-of-freedom system. However, due to the extensive length of the expression, displaying them for mechanical systems with more than three degrees of freedom becomes impractical. The constants are defined for the 4 degrees of freedom system that is analyzed for the first time in this research; this serves as an application of the novel MDOF transformations. However, the complexity of the expressions precludes their analytical presentation.

The stochastic term $\sqrt{\varepsilon}\xi(t)$ in Eqs. (19.2) and (19.3) is a small-intensity white-noise process. Now, both systems (non-gyroscopic and gyroscopic) are transformed into the system of $2n$ first-order Stratonovich differential equations. From this point onward, the same procedure is used for further dynamic analysis of both MDOF systems.

In this scenario, the dynamic stability of an oscillatory system can be ascertained by examining the maximal Lyapunov exponent and the p-th moment Lyapunov exponent, as defined

$$\lambda_T = \lim_{t\to\infty}\frac{1}{t}\log \boldsymbol{q}(t; q_0), \tag{19.4}$$

where $\boldsymbol{q}(t; q_0)$ is the solution process of a linear dynamic system. The almost-sure stability is contingent on the sign of the maximal Lyapunov exponent, which represents the exponential growth rate of the solution in a perturbed dynamic system. Almost-sure stability is indicated by a negative maximal Lyapunov exponent, while instability is suggested by a non-negative value. The exponential growth of $E[\|\boldsymbol{q}(t;\, q_0,\dot{q}_0)\|^p]$ is provided by the moment Lyapunov exponent defined as $\Lambda_T(p) = \lim_{t\to\infty} \frac{1}{t} \log E\left[\|\boldsymbol{q}(t;\, q_0)\|^p\right]$, where E[] denotes the expectation. If $\Lambda_T(p) < 0$, then by definition $E\left[\|\boldsymbol{q}(t;\, q_0,\dot{q}_0)\|^p\right] \to 0$ as $t \to 0$, and this is referred to as the p-th moment stability. Although Lyapunov exponents are vital for the examination of dynamic stability in stochastic systems, their practical assessment presents significant challenges. Furthermore, it ensures the almost-sure and moment stability of the equilibrium state where $q = \dot{q} = 0$ remains a convoluted task. The stochastic part $\xi(t)$, which represents the white-noise process with zero mean and autocorrelation function is considered

$$R_{\xi\xi}(t_1, t_2) = E[\xi(t_1)\xi(t_2)] = \sigma^2\delta(t_2 - t_1), \tag{19.5}$$

where σ is the intensity of the random process $\xi(t)$, $\delta()$ is the Dirac delta function. As noted in the Chapter 3, the general transformations are thoroughly detailed in this chapter. Additionally, Chapter 3 provides all the necessary mathematical tools and additional resources required for handling complex stochastic mechanical systems, ensuring a comprehensive understanding and practical application of these transformations. By employing the newly acquired generalized transformations

$$\begin{aligned}
&\begin{Bmatrix} x_1 \\ x_2 \end{Bmatrix} = a \begin{Bmatrix} \cos\theta_1 \\ -\sin\theta_1 \end{Bmatrix} \prod_{i=1}^{n-1} \cos\varphi_i, \\
&\qquad\vdots \\
&\begin{Bmatrix} x_{2k-1} \\ x_{2k} \end{Bmatrix} = a\sin\varphi_k \begin{Bmatrix} \cos\theta_k \\ -\sin\theta_k \end{Bmatrix} \frac{\prod_{j=1}^{n-1}\cos\varphi_j}{\prod_{i=2}^{k}\cos\varphi_i}, \quad (k = 2, \ldots, n-2), \\
&\qquad\vdots \\
&\begin{Bmatrix} x_{2n-3} \\ x_{2n-2} \end{Bmatrix} = a\cos\varphi_1 \sin\varphi_{n-1} \begin{Bmatrix} \cos\theta_{n-1} \\ -\sin\theta_{n-1} \end{Bmatrix}, \\
&\begin{Bmatrix} x_{2n-1} \\ x_{2n} \end{Bmatrix} = a\sin\varphi_1 \begin{Bmatrix} \cos\theta_n \\ -\sin\theta_n \end{Bmatrix}, \\
&-\infty < p < \infty, \; 0 \le \theta_i \le 2\pi, \; 0 \le \varphi_j \le \frac{\pi}{2}, \\
&(i = 1, \ldots, n), \; (j = 1, \ldots, n-1), \\
&P = a^p = \sum_{i=1}^{2n} (x_i^2)^{\frac{p}{2}}
\end{aligned} \tag{19.6}$$

Itô's rule yields a set of equations for the p-th power of the norm of the response and phase variables according to stochastic systems Eqs. (19.2) or (19.3). The trigonometric transformations represent a as a norm of the response, where $\theta_1, \theta_2, \ldots, \theta_n$ are the angles of the oscillators and $\varphi_1, \varphi_2, \ldots, \varphi_{n-1}$ describe the couplings or transfer of energy among the oscillators where

$$
\begin{aligned}
&\mathrm{d}\theta_i = m_i\left(\theta_1, \theta_2, \ldots, \theta_n, \varphi_1, \varphi_2 \ldots, \varphi_{n-1}\right)\mathrm{d}t\\
&+ \sigma_i\left(\theta_1, \theta_2, \ldots, \theta_n, \varphi_1, \varphi_2, \ldots, \varphi_{n-1}\right)\mathrm{d}W(t),\\
&\mathrm{d}\varphi_j = m_{j+n}\left(\theta_1, \theta_2, \ldots, \theta_n, \varphi_1, \varphi_2, \ldots, \varphi_{n-1}\right)\mathrm{d}t\\
&+ \sigma_{j+n}\left(\theta_1, \theta_2, \ldots, \theta_n, \varphi_1, \varphi_2, \ldots, \varphi_{n-1}\right)\mathrm{d}W(t),\\
&\mathrm{d}P = Pm_{2\times n}\left(\theta_1, \theta_2, \ldots, \theta_n, \varphi_1, \varphi_2, \ldots, \varphi_{n-1}\right)\mathrm{d}t\\
&+ P\sigma_{2n}\left(\theta_1, \theta_2, \ldots, \theta_n, \varphi_1, \varphi_2, \ldots, \varphi_{n-1}\right)\mathrm{d}W(t),\\
&(i = 1, \ldots, n), (j = 1, \ldots, n-1),
\end{aligned} \tag{19.7}
$$

and $W(t)$ presents the standard Weiner process. Then, a linear stochastic transformation given by Wedig [24] is applied

$$
S = T\left(\theta_1, \theta_2, \ldots, \theta_n, \varphi_1, \varphi_2, \ldots, \varphi_{n-1}\right)P, \tag{19.8}
$$

establishing the new norm process S by using the scalar function $T\left(\theta_1, \theta_2, \ldots, \theta_n, \varphi_1, \varphi_2, \ldots, \varphi_{n-1}\right)$, which is defined in the stationary phase process φ_j. The Itô equations for the new p-th norm process S can be obtained from Itô's Lemma

$$
\begin{aligned}
\mathrm{d}S = P\Bigg[&\frac{1}{2}\sum_{i=1}^{n}\sum_{j=1}^{n-1}\left(\frac{\partial^2 T}{\partial\theta_i^2}\sigma_i^2 + \frac{\partial^2 T}{\partial\varphi_j^2}\sigma_{n+j}^2\right) + \sum_{\substack{i,j=1\\ i<j}}^{n}\sigma_i\sigma_j\frac{\partial^2 T}{\partial\theta_i\partial\theta_j}\\
&+ \sum_{i=1}^{n}\sum_{j=1}^{n-1}\sigma_i\sigma_{n+j}\frac{\partial^2 T}{\partial\theta_i\partial\varphi_j} + \sum_{\substack{i,j=1\\ i<j}}^{n-1}\sigma_i\sigma_j\frac{\partial^2 T}{\partial\varphi_i\partial\varphi_j} + \sum_{i=1}^{n}\sigma_i\sigma_{2n}\frac{\partial T}{\partial\theta_i}\\
&+ \sum_{j=1}^{n-1}\sigma_{j+n}\sigma_{2n}\frac{\partial T}{\partial\varphi_j} + \sum_{i=1}^{n}m_i\frac{\partial T}{\partial\theta_i} + \sum_{j=1}^{n-1}m_{j+n}\frac{\partial T}{\partial\varphi_j} + m_{2n}T\Bigg]\\
&+ P\left[\sum_{i=1}^{n}\sigma_i\frac{\partial T}{\partial\theta_i} + \sum_{j=1}^{n-1}\sigma_{j+n}\frac{\partial T}{\partial\varphi_j} + \sigma_{2n}T\right]\mathrm{d}W(t).
\end{aligned} \tag{19.9}
$$

Both processes P and S have the same stability behavior in the case when the transformation function $T\left(\theta_i, \varphi_j\right)$ is bounded and non-singular. Therefore, the transformation function $T\left(\theta_i, \varphi_j\right)$ is chosen so that the drift term, of the Itô differential Eq. (19.9), does not depend on the phase processes φ_j, θ_i so that

$$
\mathrm{d}S = \Lambda(p)S\mathrm{d}t + \frac{S}{T}\left(\sum_{i=1}^{n}\sigma_i\frac{\partial T}{\partial\theta_i} + \sum_{j=1}^{n-1}\sigma_{j+n}\frac{\partial T}{\partial\varphi_j} + \sigma_{2n}T\right)\mathrm{d}W(t). \tag{19.10}
$$

Comparing Eqs. (19.9) and (19.10), the transformation function $T\left(\theta_1, \theta_2, \ldots, \theta_n, \varphi_1, \varphi_2, \ldots, \varphi_{n-1}\right)$ is given by the following eigenvalue problem for moment Lyapunov exponents

$$[L_0 + \varepsilon L_1]T(\theta_1, \theta_2, \ldots, \theta_n, \varphi_1, \varphi_2, \ldots, \varphi_{n-1})$$
$$= \Lambda(p)T(\theta_1, \theta_2, \ldots, \theta_n, \varphi_1, \varphi_2, \ldots, \varphi_{n-1}), \tag{19.11}$$

where L_0 and L_1 are the differential operators in the forms

$$L_0 = \sum_{i=1}^{n} \omega_i \frac{\partial}{\partial \theta_i},$$
$$L_1 = \sum_{i=1}^{n} a_{\theta_i} \frac{\partial^2}{\partial \theta_i^2} + \sum_{j=1}^{n-1} a_{\varphi_j} \frac{\partial^2}{\partial \varphi_j^2} + \sum_{\substack{i,j=1 \\ i<j}}^{n} a_{\theta_i\theta_j} \frac{\partial^2}{\partial \theta_i \partial \theta_j} + \sum_{i=1}^{n}\sum_{j=1}^{n-1} a_{\theta_i\varphi_j} \frac{\partial^2}{\partial \theta_i \partial \varphi_j}$$
$$+ \sum_{\substack{i,j=1 \\ i<j}}^{n-1} a_{\varphi_i\varphi_j} \frac{\partial^2}{\partial \varphi_i \partial \varphi_j} + \sum_{i=1}^{n} b_{\theta_i} \frac{\partial}{\partial \theta_i} + \sum_{j=1}^{n-1} b_{\varphi_j} \frac{\partial}{\partial \varphi_j} + c. \tag{19.12}$$

19.2.2 Moment Lyapunov exponents

The eigenvalue problem for a differential operator of $2n$-1 independent variables is recognized from Eq. (19.11) in which $\Lambda(p)$ is the eigenvalue and $T(\theta_1, \theta_2, \ldots, \theta_n, \varphi_1, \varphi_2, \ldots, \varphi_{n-1})$ is the associated eigenfunction. Eq. (19.11) defines that the eigenvalue $\Lambda(p)$ is the Lyapunov exponent of the p-th moment. Administrating the method of regular perturbation, both the moment Lyapunov exponent $\Lambda(p)$ and the eigenfunction $T(\theta_1, \theta_2, \ldots, \theta_n, \varphi_1, \varphi_2, \ldots, \varphi_{n-1})$ are expanded in power series of ε to obtain a weak noise expansion of the Lyapunov exponent of a system with stochastic excitation described by n-dimensional components as

$$\Lambda(p) = \Lambda_0(p) + \varepsilon\Lambda_1(p) + \varepsilon^2\Lambda_2(p) + \ldots + \varepsilon^n\Lambda_n(p) + \ldots,$$
$$T(\theta_1, \theta_2, \ldots, \theta_n, \varphi_1, \varphi_2, \ldots, \varphi_{n-1})$$
$$= T_0(\theta_1, \theta_2, \ldots, \theta_n, \varphi_1, \varphi_2, \ldots, \varphi_{n-1}) + \varepsilon T_1(\theta_1, \theta_2, \ldots, \theta_n, \varphi_1, \varphi_2, \ldots, \varphi_{n-1})$$
$$+ \varepsilon^2 T_2(\theta_1, \theta_2, \ldots, \theta_n, \varphi_1, \varphi_2, \ldots, \varphi_{n-1}) + \ldots$$
$$+ \varepsilon^3 T_3(\theta_1, \theta_2, \ldots, \theta_n, \varphi_1, \varphi_2, \ldots, \varphi_{n-1}) + \ldots$$
$$+ \varepsilon^n T_n(\theta_1, \theta_2, \ldots, \theta_n, \varphi_1, \varphi_2, \ldots, \varphi_{n-1}) + \ldots. \tag{19.13}$$

Equating the terms of equal powers of ε from Eqs. (19.11) and (19.13), the following perturbation equations may be written as

$$\varepsilon^0: \; L_0 T_0(\theta_1, \theta_2, \ldots, \theta_n, \varphi_1, \varphi_2, \ldots, \varphi_{n-1})$$
$$= \Lambda_0(p) T_0(\theta_1, \theta_2, \ldots, \theta_n, \varphi_1, \varphi_2, \ldots, \varphi_{n-1}),$$

$$\varepsilon^1: \; L_0 T_1(\theta_1, \theta_2, \ldots, \theta_n, \varphi_1, \varphi_2, \ldots, \varphi_{n-1}) + L_1 T_0(\theta_1, \theta_2, \ldots, \theta_n, \varphi_1, \varphi_2, \ldots, \varphi_{n-1})$$
$$= \Lambda_0(p) T_1(\theta_1, \theta_2, \ldots, \theta_n, \varphi_1, \varphi_2, \ldots, \varphi_{n-1}) + \Lambda_1(p) T_0(\theta_1, \theta_2, \ldots, \theta_n, \varphi_1, \varphi_2, \ldots, \varphi_{n-1}),$$

$$\varepsilon^2: L_0T_2(\theta_1, \theta_2, \ldots, \theta_n, \varphi_1, \varphi_2, \ldots, \varphi_{n-1}) + L_1T_1(\theta_1, \theta_2, \ldots, \theta_n, \varphi_1, \varphi_2, \ldots, \varphi_{n-1})$$
$$= \Lambda_0(p)T_2(\theta_1, \theta_2, \ldots, \theta_n, \varphi_1, \varphi_2, \ldots, \varphi_{n-1})$$
$$+ \Lambda_1(p)T_1(\theta_1, \theta_2, \ldots, \theta_n, \varphi_1, \varphi_2, \ldots, \varphi_{n-1}) + \Lambda_2(p)T_0(\theta_1, \theta_2, \ldots, \theta_n, \varphi_1, \varphi_2, \ldots, \varphi_{n-1}),$$

$$\varepsilon^3: L_0T_3(\theta_1, \theta_2, \ldots, \theta_n, \varphi_1, \varphi_2, \ldots, \varphi_{n-1}) + L_1T_2(\theta_1, \theta_2, \ldots, \theta_n, \varphi_1, \varphi_2, \ldots, \varphi_{n-1})$$
$$= \Lambda_0(p)T_3(\theta_1, \theta_2, \ldots, \theta_n, \varphi_1, \varphi_2, \ldots, \varphi_{n-1}) + \Lambda_1(p)T_2(\theta_1, \theta_2, \ldots, \theta_n, \varphi_1, \varphi_2, \ldots, \varphi_{n-1})$$
$$+ \Lambda_2(p)T_1(\theta_1, \theta_2, \ldots, \theta_n, \varphi_1, \varphi_2, \ldots, \varphi_{n-1})$$
$$+ \Lambda_3(p)T_0(\theta_1, \theta_2, \ldots, \theta_n, \varphi_1, \varphi_2, \ldots, \varphi_{n-1}), \ldots$$

$$\varepsilon^n: L_0T_n(\theta_1, \theta_2, \ldots, \theta_n, \varphi_1, \varphi_2, \ldots, \varphi_{n-1}) + L_1T_{n-1}(\theta_1, \theta_2, \ldots, \theta_n, \varphi_1, \varphi_2, \ldots, \varphi_{n-1})$$
$$= \Lambda_0(p)T_n(\theta_1, \theta_2, \ldots, \theta_n, \varphi_1, \varphi_2, \ldots, \varphi_{n-1}) + \Lambda_1(p)T_{n-1}(\theta_1, \theta_2, \ldots, \theta_n, \varphi_1, \varphi_2, \ldots, \varphi_{n-1})$$
$$+ \ldots + \Lambda_n(p)T_0(\theta_1, \theta_2, \ldots, \theta_n, \varphi_1, \varphi_2, \ldots, \varphi_{n-1}), \quad (19.14)$$

where each function $T_s(\theta_1, \theta_2, \ldots, \theta_n, \varphi_1, \varphi_2, \ldots, \varphi_{n-1})$, $s = 1, 2, 3, \ldots$ has to be positive and periodic in the range $0 \le \theta_i \le 2\pi$; as it is elucidated in Refs. [9,12]. The first three order perturbations are commonly considered due to affordable calculation effort.

19.2.2.1 Zeroth-order perturbation

From Eqs. (19.11) and (19.14) the zeroth order perturbation equation is $L_0T_0 = \Lambda_0(p)T_0$ and it can be written as

$$\sum_{i=1}^{n} \omega_i \frac{\partial T_0(\theta_1, \theta_2, \ldots, \theta_n, \varphi_1, \varphi_2, \ldots, \varphi_{n-1})}{\partial \theta_i}$$
$$= \Lambda_0(p)T_0(\theta_1, \theta_2, \ldots, \theta_n, \varphi_1, \varphi_2, \ldots, \varphi_{n-1}). \quad (19.15)$$

Eq. (19.15) can be easily solved from the moment Lyapunov exponent characteristic, which results in $\Lambda_n(0) = 0$ for $n = 0, 1, 2, 3, \ldots$ because of $\Lambda(0) = \Lambda_0(0) + \varepsilon\Lambda_1(0) + \varepsilon^2\Lambda_2(0) + \ldots + \varepsilon^n\Lambda_n(0) = 0$. The eigenvalue $\Lambda_0(p)$ is independent of p. Consequently $\Lambda_0(0) = 0$ leads to $\Lambda_0(p) = 0$. The solutions of Eq. (19.15) have a periodic solution only when

$$\Lambda_0(p) = 0, \quad T_0(\theta_1, \theta_2, \ldots, \theta_n, \varphi_1, \varphi_2, \ldots, \varphi_{n-1}) = 1. \quad (19.16)$$

19.2.2.2 First-order perturbation

The equation obtained from the first-order perturbation is

$$
\begin{aligned}
&L_0T_1(\theta_1, \theta_2, \ldots, \theta_n, \varphi_1, \varphi_2, \ldots, \varphi_{n-1}) + L_1T_0(\theta_1, \theta_2, \ldots, \theta_n, \varphi_1, \varphi_2, \ldots, \varphi_{n-1}) \\
&= \Lambda_0(p)T_1(\theta_1, \theta_2, \ldots, \theta_n, \varphi_1, \varphi_2, \ldots, \varphi_{n-1}) \\
&+ \Lambda_1(p)T_0(\theta_1, \theta_2, \ldots, \theta_n, \varphi_1, \varphi_2, \ldots, \varphi_{n-1}),
\end{aligned} \tag{19.17}
$$

and Eq. (19.17) has a periodic solution only when

$$
\int_0^{2\pi} \cdots \int_0^{2\pi} \int_0^{2\pi} \int_0^{\frac{\pi}{2}} \cdots \int_0^{\frac{\pi}{2}} \int_0^{\frac{\pi}{2}} [L_1 \cdot 1 - \Lambda_1(p)] \mathrm{d}\varphi_1 \mathrm{d}\varphi_2 \ldots \mathrm{d}\varphi_{n-1} \mathrm{d}\theta_1 \mathrm{d}\theta_2 \ldots \mathrm{d}\theta_n = 0, \tag{19.18}
$$

resulting in

$$
\Lambda_1(p) = \frac{1}{2\pi^{2n-1}} \int_0^{2\pi} \cdots \int_0^{2\pi} \int_0^{2\pi} \int_0^{\frac{\pi}{2}} \cdots \int_0^{\frac{\pi}{2}} \int_0^{\frac{\pi}{2}} \left[c(\theta_1, \ldots, \varphi_1, \ldots) \right] \mathrm{d}\varphi_1 \mathrm{d}\varphi_2 \ldots \mathrm{d}\varphi_{n-1} \mathrm{d}\theta_1 \mathrm{d}\theta_2 \ldots \mathrm{d}\theta_n. \tag{19.19}
$$

The first-order perturbation equation can be reformulated as

$$
\begin{aligned}
&\sum_{i=1}^{n} \omega_i \frac{\partial T_1(\theta_1, \theta_2, \ldots, \theta_n, \varphi_1, \varphi_2, \ldots, \varphi_{n-1})}{\partial \theta_i} \\
&+ c(\theta_1, \theta_2, \ldots, \theta_n, \varphi_1, \varphi_2, \ldots, \varphi_{n-1}) = \Lambda_1(p).
\end{aligned} \tag{19.20}
$$

It is important to consider the commensurable frequencies, in which a relation of the form of $\omega_1 = \omega_2/\mu_1 \ldots = \omega_n/\mu_{n-1}$ exists. The function $c(\theta_1, \theta_2, \theta_3, \varphi_1, \varphi_2)$ in Eq. (19.20) can be written in the form

$$
\begin{aligned}
&c(\theta_1, \theta_2, \ldots, \theta_n, \varphi_1, \varphi_2, \ldots, \varphi_{n-1}) \\
&= \Lambda_1(p) + f_0(\theta_1, \ldots) + \mathcal{L}(\theta_1, \ldots)\mathfrak{R}(\varphi_1, \ldots),
\end{aligned} \tag{19.21}
$$

Functions $\mathcal{L}$ are periodic in $(\theta_1, \ldots)$ with $\mathfrak{R}$ trigonometric combinations of $(\varphi_1, \ldots)$. Eq. (19.20) cannot be obtained explicitly for $T_1(\theta_1, \theta_2, \ldots, \theta_n, \varphi_1, \varphi_2, \ldots, \varphi_{n-1})$. The combination of coefficients in Eq. (19.21) reveals that the function $T_1(\theta_1, \theta_2, \ldots, \theta_n, \varphi_1, \varphi_2, \ldots, \varphi_{n-1})$ can be written as

$$
\begin{aligned}
&T_1(\theta_1, \theta_2, \ldots, \theta_n, \varphi_1, \varphi_2, \ldots, \varphi_{n-1}) \\
&= T_{10}(\theta_1, \ldots) + \widetilde{T}_r(\theta_1, \ldots)\mathfrak{R}(\varphi_1, \ldots).
\end{aligned} \tag{19.22}
$$

Substituting Eq. (19.22) into Eq. (19.20), and equating terms of the equal trigonometry function, gives a set of PDEs

$$
\sum_{i=1}^{n} \omega_i \frac{\partial \widetilde{T}_r(\theta_1, \theta_2, \ldots, \theta_n)}{\partial \theta_i} + f_r(\theta_1, \theta_2, \ldots, \theta_n) = 0, \quad r = 0, 1, \ldots \tag{19.23}
$$

The functions $\widetilde{T}_r(\theta_1, \theta_2, \ldots, \theta_n)$ can be written as

$$
\widetilde{T}_r(\theta_1, \theta_2, \ldots, \theta_n) = \mathcal{E} + C_r(\theta_2 - \mu_1\theta_1, \theta_3 - \mu_2\theta_1, \ldots, \theta_n - \mu_n\theta_1). \tag{19.24}
$$

Here $C_r(\theta_2 - \mu_1\theta_1, \theta_3 - \mu_2\theta_1, \ldots, \theta_n - \mu_n\theta_1)$ are the same arbitrary functions of n variables for which the following assumptions can be made

$$\begin{aligned} & C_r\left(\theta_2 - \mu_1\theta_1, \theta_3 - \mu_2\theta_1, \ldots, \theta_n - \mu_n\theta_1\right) \\ & = A_{1r} + B_{1r}\sin\left(2\theta_2 - 2\mu_1\theta_1\right) + \ldots \end{aligned} \tag{19.25}$$

As each function $\widetilde{T}_r(\theta_1, \theta_2, \ldots, \theta_n)$ must be positive and periodic, unknown constants $A_{1r}, B_{1r}, \ldots$ can be determined using conditions

$$\begin{aligned} & \widetilde{T}_r(0, 0, 0, \ldots) = \widetilde{T}_r(0, 0, \ldots, 2\pi) = \widetilde{T}_r(0, \ldots, 2\pi, 0) = \ldots = \widetilde{T}_r(2\pi, \ldots, 0, 0) \\ & = \widetilde{T}_r(2\pi, 2\pi, \ldots, 0) = \ldots = \widetilde{T}_r(0, \ldots, 2\pi, 2\pi) = \ldots \\ & = \widetilde{T}_r(2\pi, \ldots, 0, 2\pi) = \ldots = \widetilde{T}_r(2\pi, \ldots, 2\pi, 2\pi) = 0, \end{aligned} \tag{19.26}$$

$$\begin{aligned} & \frac{\partial \widetilde{T}_r(0, 0, \ldots, 0)}{\partial \theta_1} = \frac{\partial \widetilde{T}_r(2\pi, 0, 0, \ldots, 0)}{\partial \theta_1}, \\ & \frac{\partial \widetilde{T}_r(0, 0, \ldots, 0)}{\partial \theta_2} = \frac{\partial \widetilde{T}_r(0, 2\pi, \ldots, 0)}{\partial \theta_2}, \ldots \end{aligned} \tag{19.27}$$

19.2.2.3 Second-order perturbation

The second-order perturbation equation must satisfy the requirement of periodic function $T_2\left(\theta_1, \theta_2, \ldots, \theta_n, \varphi_1, \varphi_2, \ldots, \varphi_{n-1}\right)$ in $(\theta_1, \theta_2, \ldots, \theta_n)$ given as

$$\begin{aligned} & L_0 T_2\left(\theta_1, \theta_2, \ldots, \theta_n, \varphi_1, \varphi_2, \ldots, \varphi_{n-1}\right) + L_1 T_2\left(\theta_1, \theta_2, \ldots, \theta_n, \varphi_1, \varphi_2, \ldots, \varphi_{n-1}\right) \\ & = \Lambda_1(p) T_1\left(\theta_1, \theta_2, \ldots, \theta_n, \varphi_1, \varphi_2, \ldots, \varphi_{n-1}\right) + \Lambda_2(p). \end{aligned} \tag{19.28}$$

Lastly

$$\begin{aligned} \Lambda_2(p) = & \frac{1}{2\pi^{2n-1}} \int_0^{2\pi} \cdots \int_0^{2\pi} \int_0^{2\pi} \int_0^{\pi/2} \cdots \int_0^{\pi/2} \int_0^{\pi/2} \left[\sum_{i=1}^{n} a_{\theta_i} \frac{\partial^2}{\partial \theta_i^2} + \sum_{j=1}^{n-1} a_{\varphi_j} \frac{\partial^2}{\partial \varphi_j^2} \right. \\ & + \sum_{\substack{i,j=1 \\ i<j}}^{n} a_{\theta_i\theta_j} \frac{\partial^2}{\partial \theta_i \partial \theta_j} + \sum_{i=1}^{n} \sum_{j=1}^{n-1} a_{\theta_i\varphi_j} \frac{\partial^2}{\partial \theta_i \partial \varphi_j} + \sum_{\substack{i,j=1 \\ i<j}}^{n-1} a_{\varphi_i\varphi_j} \frac{\partial^2}{\partial \varphi_i \partial \varphi_j} + \sum_{i=1}^{n} b_{\theta_i} \frac{\partial}{\partial \theta_i} \\ & \left. + \sum_{j=1}^{n-1} b_{\varphi_j} \frac{\partial}{\partial \varphi_j} + [c - \Lambda_1(p)] T_1 \right] d\varphi_1 d\varphi_2 \ldots d\varphi_{n-1} d\theta_1 d\theta_2 \ldots d\theta_n. \end{aligned} \tag{19.29}$$

The weak noise expansion of the moment Lyapunov exponent in the second-order perturbation is defined in the form

$$\Lambda(p) = \varepsilon\Lambda_1(p) + \varepsilon^2\Lambda_2(p) + O(\varepsilon^3). \tag{19.30}$$

The Lyapunov exponent can now be obtained from Eq. (19.30) by use of a property of the moment Lyapunov exponent

$$\lambda = \left. \frac{d\Lambda(p)}{dp} \right|_{p=0} = \varepsilon\lambda_1 + \varepsilon^2\lambda_2 + O(\varepsilon^3). \tag{19.31}$$

19.3 Stochastic stability of a coupled rotating shaft system

In this segment, the configuration of a rotating coupled shaft system, which is depicted in Fig. 19.1, is examined. The system experiences a dynamic axial compressive load denoted as *P(t)*. To minimize weight or meet specific structural requirements, the shaft system is frequently composed of two segments that are interconnected viscoelastically through a stiffness shell layer denoted as k. Both segments of the shaft system revolve around the central axis Oz, which is also coaxial with the bearings, maintaining a consistent angular velocity denoted as Ω. Additionally, the relative angular velocity between the two segments is maintained at zero.

The model under consideration represents an extended iteration of the shaft system outlined in our prior study [22]. The general equations of motion in this context can be expressed in the following manner.

$$EI_{u1}\frac{\partial^4 u_1}{\partial x^4} + \rho A_1\left(\frac{\partial^2 u_1}{\partial t^2} - \Omega^2 u_1 - 2\Omega\frac{\partial w_1}{\partial t}\right) + P(t)\frac{\partial^2 u_1}{\partial z^2} + \beta\left(\frac{\partial u_1}{\partial t} - \Omega w_1\right) + k(u_1 - u_2) = 0, \tag{19.32a}$$

$$EI_{w1}\frac{\partial^4 w_1}{\partial x^4} + \rho A_1\left(\frac{\partial^2 w_1}{\partial t^2} - \Omega^2 w_1 + 2\Omega\frac{\partial u_1}{\partial t}\right) + P(t)\frac{\partial^2 w_1}{\partial z^2} + \beta\left(\frac{\partial w_1}{\partial t} + \Omega w_1\right) + k(w_1 - w_2) = 0, \tag{19.32b}$$

$$EI_{u2}\frac{\partial^4 u_2}{\partial x^4} + \rho A_2\left(\frac{\partial^2 u_2}{\partial t^2} - \Omega^2 u_2 - 2\Omega\frac{\partial w_2}{\partial t}\right) + P(t)\frac{\partial^2 u_2}{\partial z^2} + \beta\left(\frac{\partial u_2}{\partial t} - \Omega w_2\right) + k(u_2 - u_1) = 0, \tag{19.33a}$$

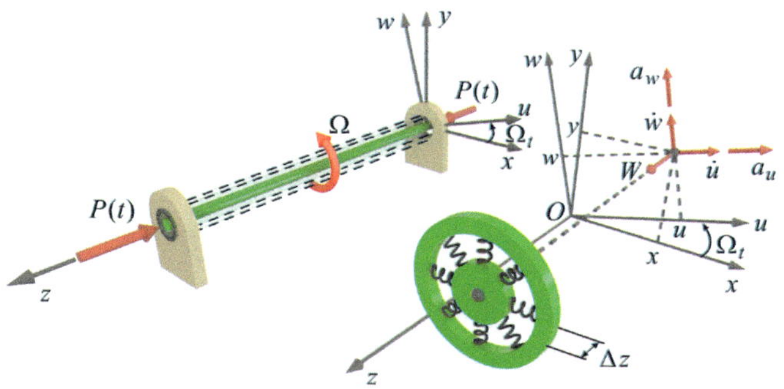

Figure 19.1 Rotating coupled shaft system. The shaft system is composed of two segments that are interconnected viscoelastically.

$$EI_{w2}\frac{\partial^4 w_2}{\partial x^4} + \rho A_2\left(\frac{\partial^2 w_2}{\partial t^2} - \Omega^2 w_2 + 2\Omega\frac{\partial u_2}{\partial t}\right) + P(t)\frac{\partial^2 w_2}{\partial z^2}$$
$$+ \beta\left(\frac{\partial w_2}{\partial t} + \Omega w_2\right) + k(w_2 - w_1) = 0, \tag{19.33b}$$

where A_i is the area of the cross-section, I_i is the second moment of inertia for the cross-sectional area, ρ is the mass per unit volume, E is the Young modulus and $\widetilde{\beta}$ represents the external damping. The shaft system is assumed to be elastic and isotropic. The length of the shaft system is l. The boundary conditions according to simply supported ends are

$$u_i(0, t) = u_i(l, t) = 0,\ w_i(0, t) = w_i(l, t) = 0,$$
$$\frac{\partial^2 u_i(0, t)}{\partial z^2} = \frac{\partial^2 u_i(l, t)}{\partial z^2} = 0,\ \frac{\partial^2 w_i(0, t)}{\partial z^2} = \frac{\partial^2 w_i(l, t)}{\partial z^2} = 0. \tag{19.34}$$

The dynamic axial compressive load is postulated to incorporate a variable component denoted as $\xi(t)$, and its representation is formulated in the form $P(t) = P_0 + \xi(t)$, where P_0 is the constant static load, $\xi(t)$ represents the white-noise process with zero mean and autocorrelation function

$$R_{\xi\xi}(t_1, t_2) = E[\xi(t_1)\xi(t_2)] = \sigma^2\delta(t_2 - t_1), \tag{19.35}$$

and where σ is the intensity of the random process $\xi(t)$, $\delta()$ is the Dirac delta function, and E denotes the expectation. An approximate solution can be derived through the application of the Galerkin method. The intention is to find solutions in the following format

$$u_1(z, t) = q_1(t)\sin\left(\frac{\pi z}{l}\right),\ w_1(z, t) = q_2(t)\sin\left(\frac{\pi z}{l}\right),$$
$$u_2(z, t) = q_3(t)\sin\left(\frac{\pi z}{l}\right),\ w_2(z, t) = q_4(t)\sin\left(\frac{\pi z}{l}\right), \tag{19.36}$$

multiplying Eqs. (19.32a–19.32) and (19.33a–19.33b) by $\sin(\pi z/l)$ and integrating from 0 to l with the introduced small parameter ε to show that damping and parametric excitations are small, yields a new system of coupled differential equations in the form

$$\begin{bmatrix} \rho A_1 & 0 & 0 & 0 \\ 0 & \rho A_1 & 0 & 0 \\ 0 & 0 & \rho A_2 & 0 \\ 0 & 0 & 0 & \rho A_2 \end{bmatrix}\ddot{\widetilde{\mathbf{x}}} + \varepsilon\beta\dot{\widetilde{\mathbf{x}}} + \begin{bmatrix} 0 & -2\rho A_1\Omega & 0 & 0 \\ 2\rho A_1\Omega & 0 & 0 & 0 \\ 0 & 0 & 0 & -2\rho A_2\Omega \\ 0 & 0 & 2\rho A_2\Omega & 0 \end{bmatrix}\dot{\widetilde{\mathbf{x}}}$$
$$+ \begin{bmatrix} k_{11} & k_{12} & k_{13} & 0 \\ k_{21} & k_{22} & 0 & k_{24} \\ k_{31} & 0 & k_{33} & k_{34} \\ 0 & k_{42} & k_{43} & k_{44} \end{bmatrix}\widetilde{\mathbf{x}} + \varepsilon^{\frac{1}{2}}\xi(t)\begin{bmatrix} \widetilde{k} & 0 & 0 & 0 \\ 0 & \widetilde{k} & 0 & 0 \\ 0 & 0 & \widetilde{k} & 0 \\ 0 & 0 & 0 & \widetilde{k} \end{bmatrix}\widetilde{\mathbf{x}} = 0, \tag{19.37}$$

where

$$k_{11} = k - \frac{P_0\pi^2}{l^2} + \frac{EI_{u1}\pi^4}{l^4} - \rho A_1\Omega^2,\ k_{22} = k - \frac{P_0\pi^2}{l^2} + \frac{EI_{w1}\pi^4}{l^4} - \rho A_1\Omega^2,$$
$$k_{33} = k - \frac{P_0\pi^2}{l^2} + \frac{EI_{u2}\pi^4}{l^4} - \rho A_2\Omega^2,\ k_{44} = k - \frac{P_0\pi^2}{l^2} + \frac{EI_{w2}\pi^4}{l^4} - \rho A_2\Omega^2,$$
$$k_{13} = k_{31} = k_{24} = k_{42} = -k,\ k_{21} = -k_{12} = k_{43} = k_{34} = \beta\Omega,\ \widetilde{k} = -\frac{\pi^2}{l^2}.$$

The set of stochastic differential Eq. (19.37) pertains to the gyroscopic category of systems, precisely aligning with the structure of Eq. (19.3). Employing the methodology outlined in Xie [2], a set of first-order Stratonovich differential equations can be derived. When substituting $n = 4$ into Eq. (19.6), a set of associated transformations emerges in the following configuration

$$\begin{aligned}
&x_1 = a\cos\varphi_1\cos\varphi_2\cos\varphi_3\cos\theta_1, \quad x_2 = -a\cos\varphi_1\cos\varphi_2\cos\varphi_3\sin\theta_1,\\
&x_3 = a\cos\varphi_1\sin\varphi_2\cos\varphi_3\cos\theta_2, \quad x_4 = -a\cos\varphi_1\sin\varphi_2\cos\varphi_3\sin\theta_2,\\
&x_5 = a\cos\varphi_1\sin\varphi_3\cos\theta_3, \quad x_6 = -a\cos\varphi_1\sin\varphi_3\sin\theta_3,\\
&x_7 = a\sin\varphi_1\cos\theta_4, \quad x_8 = -a\sin\varphi_1\sin\theta_4,\\
&P = a^p = (x_1^2 + x_2^2 + x_3^2 + x_4^2 + x_5^2 + x_6^2 + x_7^2 + x_8^2)^{\frac{p}{2}},\\
&-\infty < p < \infty, \quad 0 \le \theta_i \le 2\pi,\ 0 \le \varphi_j \le \frac{\pi}{2},\\
&i = 1,\ 2,\ 3,\ 4, \quad j = 1, 2, 3, 4.
\end{aligned} \tag{19.38}$$

Continuing with the outlined approach in Section 19.1, the moment Lyapunov exponent is determined through the first-order perturbation method in the following manner

$$\begin{aligned}
\Lambda(p) = {} & \frac{1}{8192}p\sigma^2\left[f_1(s) + pf_2(s)\right]\\
& - \frac{p}{16}\{d_{111} + d_{122} + d_{133} + d_{144} + 2[d_{411} + d_{422} + 2(d_{433} + d_{444})]\},
\end{aligned} \tag{19.39}$$

The functions $f_1(s)$ and $f_2(s)$ are provided in Appendix A19. The constants, denoted in forms like d_{111} or s_{111} (as listed in App. A), represent elements positioned in the first row and first column within the submatrices $\boldsymbol{d}_{1()}$ and $\boldsymbol{s}_{1()}$, respectively. The identical notation convention has been utilized for additional constant elements within the submatrices. The constants expressed as d_{211} or s_{211}, following the same format, denote elements in the first row and first column of the submatrices $\boldsymbol{d}_{2()}$ and $\boldsymbol{s}_{2()}$, respectively. In a general sense, $d_{i\,i_1\,i_2}$ (employing the same notation rule for $s_{i\,i_1\,i_2}$) signifies the constant element positioned in the i_1-row and i_2-column of the submatrix $\boldsymbol{d}_{i()}$.

An examination of an intricate mechanical system was carried out and stability regions were identified for a specified combination of geometric and material parameters for a modeled coupled shaft viscoelastic system:

$$\begin{aligned}
&E = 10^{10}\text{N/m}^2,\ \rho = 2\times 10^3\text{kg/m}^3,\ k = 2\times 10^4\text{N/m},\ l = 4\text{m},\\
&D_L = 0.35\text{m},\ d_1 = 0.15\text{m},\ d_2 = 0.1\text{m},\\
&A_1 = (D_L^2 - d_1^2)\frac{\pi}{4},\ A_2 = \frac{d_2^2\pi}{4}.
\end{aligned} \tag{19.40}$$

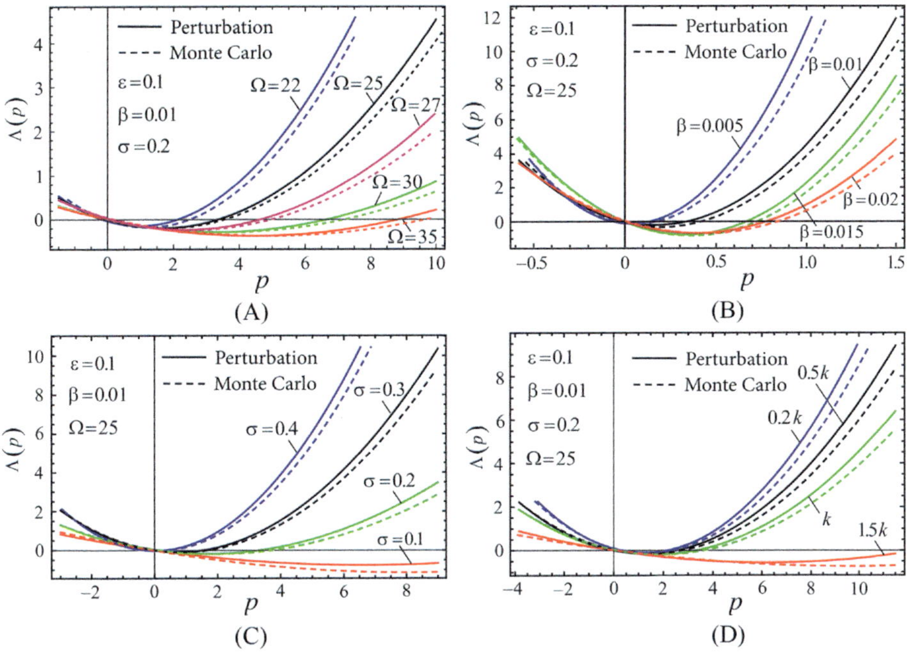

Figure 19.2 Moment Lyapunov exponent $\Lambda(p)$, the stable region is characterized by $\Lambda(p)<0$; (a) for varying angular velocity Ω, (b) for varying damping parameter β, (c) for the intensity of the random process σ, and (d) for varying stiffness of the k layer.

We analytically establish the stability boundary for the p-th moment within the first-order perturbation when $p = 1$. However, utilizing the derived analytical expression, this boundary can be calculated for any specific numerical value of p by applying the definition of the moment stability, where $\Lambda(p) < 0$. Using the results derived for the moment Lyapunov exponent from the first-order perturbation, expressed as $\Lambda(p) = \varepsilon\Lambda_1(p) + O(\varepsilon^2)$, we acquire the value $\Lambda_1(1)$. The almost-sure stability boundary of the oscillatory system can be deduced in the first-order perturbation, with the recognition that the oscillatory stochastic system attains asymptotic stability solely when the Lyapunov exponent $\lambda < 0$. From expression $\lambda = \varepsilon\lambda_1 + O(\varepsilon^2)$ the condition $\lambda < 0$ is obtained. Variations of the normalized values of the moment Lyapunov exponent are presented in Fig. 19.2. A system becomes increasingly unstable as the stability region diminishes for $p > 0$. Within the specified domain, the angular velocity Ω, the damping parameter β, and the stiffness of the k layer all contribute to stabilizing the system. The impact of damping resembles the findings of Deng [22]. In contrast, the intensity of the random process, denoted as σ, has a destabilizing effect. In other words, as σ decreases, the stability region of the p-th moment (for $p > 0$) expands, indicating increased stability in the system, as illustrated in Fig. 19.2(c).

Under a consistent angular velocity, we have identified both the quantitative and qualitative effects of parameter-induced alterations in shaft geometric characteristics.

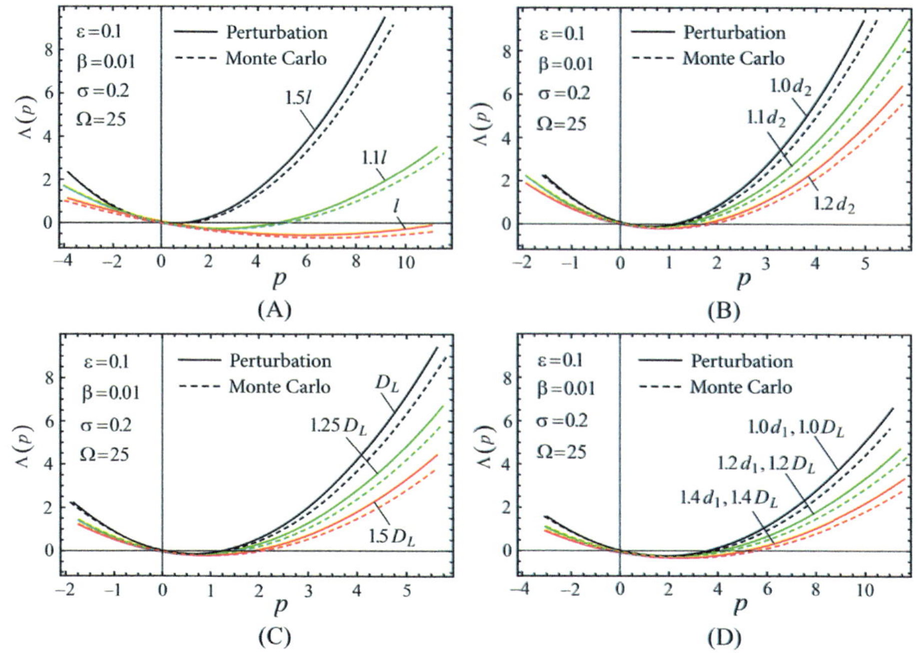

Figure 19.3 Moment Lyapunov exponent $\Lambda(p)$, the stable region is characterized by $\Lambda(p)<0$; (a) for varying length l of the coupled shaft system, (b) for varying inner diameter d_2 of the coupled shaft system, (c) for varying main external diameter D_L of the coupled shaft system, (d) for simultaneously varying diameters D_L and d_1 of the coupled shaft system.

The stability regions are visualized in Fig. 19.3. Extending the length of the coupled shaft system results in a reduction of dynamic stability, whereas adjustments in diameter parameters enhance the overall dynamic stability of the mechanical system, as illustrated in Fig. 19.3(a–d).

19.4 Conclusions

The substance of the study in this chapter was the determination of stochastic stability and moment Lyapunov exponents of multi-degree-of-freedom (MDOF) mechanical systems with both gyroscopic and non-gyroscopic configurations under parametric excitations. Moreover, this study is focused on the generalization of this problem. The primary novelty of this work lies in introducing a broader context of generalized transformations for MDOF systems. These transformations are essential for deriving Itô's equations from Stratonovich form. Additionally, the study outlines a generalized procedure for analytically approximating the moment Lyapunov exponents, by means of the method of perturbation.

To exemplify the extended methodology, a rotating coupled shaft system is employed, necessitating the description of its dynamic behavior through a minimum of four generalized coordinates. Building upon the analysis conducted for MDOF systems, appropriate transformations were initially employed on the model. This model requires a set of four coupled second-order stochastic differential equations for a comprehensive analysis of dynamic stability. The almost-sure and p-th moment stability boundaries are identified across diverse physical parameters. Within this domain, the system is stabilized by the angular velocity Ω, the damping parameter β, and the stiffness of the k layer, while the intensity of the random process σ introduces a destabilizing influence. These findings offer valuable insights into the stability dynamics of rotating shafts. Subsequent research endeavors might explore Monte Carlo simulations for moment Lyapunov exponents of MDOF stochastic systems.

Appendix A19

$$
\begin{aligned}
f_1(s) = {} & 350s_{111}^2 + 202s_{112}^2 + 220s_{113}^2 + 220s_{114}^2 + 404s_{112}s_{121} + 202s_{121}^2 - 108s_{111}s_{122} \\
& + 350s_{122}^2 + 220s_{123}^2 + 220s_{124}^2 + 440s_{113}s_{131} + 220s_{131}^2 + 440s_{123}s_{132} + 220s_{132}^2 \\
& - 72s_{111}s_{133} - 72s_{122}s_{133} + 350s_{133}^2 + 202s_{134}^2 + 440s_{114}s_{141} + 220s_{141}^2 \\
& + 440s_{124}s_{142} + 220s_{142}^2 + 404s_{134}s_{143} + 202s_{143}^2 - 72s_{111}s_{144} - 72s_{122}s_{144} \\
& - 108s_{133}s_{144} + 350s_{144}^2 + 464s_{211}^2 + 464s_{212}^2 + 960s_{213}^2 + 960s_{214}^2 + 464s_{221}^2 \\
& + 464s_{222}^2 + 960s_{223}^2 + 960s_{224}^2 + 464s_{231}^2 + 464s_{232}^2 + 960s_{233}^2 + 960s_{234}^2 \\
& + 464s_{241}^2 + 464s_{242}^2 + 960s_{243}^2 + 960s_{244}^2 + 672s_{211}s_{311} + 208s_{311}^2 \\
& + 672s_{221}s_{312} + 208s_{312}^2 + 672s_{231}s_{313} + 208s_{313}^2 + 672s_{241}s_{314} + 208s_{314}^2 \\
& + 672s_{212}s_{321} + 208s_{321}^2 + 672s_{222}s_{322} + 208s_{322}^2 + 672s_{232}s_{323} + 208s_{323}^2 \\
& + 672s_{242}s_{324} + 208s_{324}^2 + 1152s_{213}s_{331} + 192s_{331}^2 + 1152s_{223}s_{332} + 192s_{332}^2 \\
& + 1152s_{233}s_{333} + 192s_{333}^2 + 1152s_{243}s_{334} + 192s_{334}^2 + 1152s_{214}s_{341} + 192s_{341}^2 \\
& + 1152s_{224}s_{342} + 192s_{342}^2 + 1152s_{234}s_{343} + 192s_{343}^2 + 1152s_{244}s_{344} + 192s_{344}^2 \\
& - 96s_{111}s_{411} - 96s_{122}s_{411} - 96s_{133}s_{411} - 96s_{144}s_{411} + 592s_{411}^2 + 368s_{412}^2 \\
& + 896s_{413}^2 + 896s_{414}^2 + 736s_{412}s_{421} + 368s_{421}^2 - 96s_{111}s_{422} - 96s_{122}s_{422} \\
& - 96s_{133}s_{422} - 96s_{144}s_{422} - 288s_{411}s_{422} + 592s_{422}^2 + 896s_{423}^2 + 896s_{424}^2 \\
& + 1280s_{413}s_{431} + 384s_{431}^2 + 1280s_{423}s_{432} + 384s_{432}^2 - 128s_{111}s_{433} \\
& - 128s_{122}s_{433} - 128s_{133}s_{433} - 128s_{144}s_{433} - 256s_{411}s_{433} - 256s_{422}s_{433} \\
& + 896s_{433}^2 + 640s_{434}^2 + 1280s_{414}s_{441} + 384s_{441}^2 + 1280s_{424}s_{442} + 384s_{442}^2 \\
& + 1280s_{434}s_{443} + 640s_{443}^2 - 128s_{111}s_{444} - 128s_{122}s_{444} - 128s_{133}s_{444} \\
& - 128s_{144}s_{444} - 256s_{411}s_{444} - 256s_{422}s_{444} - 768s_{433}s_{444} + 896s_{444}^2
\end{aligned}
$$

$$
\begin{aligned}
f_2(s) = {} & 81s_{111}^2 + 27s_{112a}^2 + 18s_{113}^2 + 18s_{114}^2 + 54s_{112}s_{121} + 27s_{121}^2 + 54s_{111}s_{122} + 81s_{122}^2 \\
& + 18s_{123}^2 + 18s_{124}^2 + 36s_{113}s_{131} + 18s_{131}^2 + 36s_{123}s_{132} + 18s_{132}^2 + 36s_{111}s_{133} \\
& + 36s_{122}s_{133} + 81s_{133}^2 + 27s_{134}^2 + 36s_{114}s_{141} + 18s_{141}^2 + 36s_{124}s_{142} + 18s_{142}^2 \\
& + 54s_{134}s_{143} + 27s_{143}^2 + 36s_{111}s_{144} + 36s_{122}s_{144} + 54s_{133}s_{144} + 81s_{144}^2 + 24s_{211}^2 \\
& + 24s_{212}^2 + 32s_{213}^2 + 32s_{214}^2 + 24s_{221}^2 + 24s_{222}^2 + 32s_{223}^2 + 32s_{224}^2 + 24s_{231}^2 \\
& + 24s_{232}^2 + 32s_{233}^2 + 32s_{234}^2 + 24s_{241}^2 + 24s_{242}^2 + 32s_{243}^2 + 32s_{244}^2 + 48s_{211}s_{311} \\
& + 24s_{311}^2 + 48s_{221}s_{312} + 24s_{312}^2 + 48s_{231}s_{313} + 24s_{313}^2 + 48s_{241}s_{314} + 24s_{314}^2 \\
& + 48s_{212}s_{321} + 24s_{321}^2 + 48s_{222}s_{322} + 24s_{322}^2 + 48s_{232}s_{323} + 24s_{323}^2 \\
& + 48s_{242}s_{324} + 24s_{324}^2 + 64s_{213}s_{331} + 32s_{331}^2 + 64s_{223}s_{332} + 32s_{332}^2 \\
& + 64s_{233}s_{333} + 32s_{333}^2 + 64s_{243}s_{334} + 32s_{334}^2 + 64s_{214}s_{341} + 32s_{341}^2 \\
& + 64s_{224}s_{342} + 32s_{342}^2 + 64s_{234}s_{343} + 32s_{343}^2 + 64s_{244}s_{344} + 32s_{344}^2 \\
& + 48s_{111}s_{411} + 48s_{122}s_{411} + 48s_{133}s_{411} + 48s_{144}s_{411} + 216s_{411}^2 + 72s_{412}^2 + 64s_{413}^2 \\
& + 64s_{414}^2 + 144s_{412}s_{421} + 72s_{421}^2 + 48s_{111}s_{422} + 48s_{122}s_{422} + 48s_{133}s_{422} \\
& + 48s_{144}s_{422} + 144s_{411}s_{422} + 216s_{422}^2 + 64s_{423}^2 + 64s_{424}^2 + 128s_{413}s_{431} + 64s_{431}^2 \\
& + 128s_{423}s_{432} + 64s_{432}^2 + 64s_{111}s_{433} + 64s_{122}s_{433} + 64s_{133}s_{433} + 64s_{144}s_{433} \\
& + 128s_{411}s_{433} + 128s_{422}s_{433} + 576s_{433}^2 + 192s_{434}^2 + 128s_{414}s_{441} + 64s_{441}^2 \\
& + 128s_{424}s_{442} + 64s_{442}^2 + 384s_{434}s_{443}2s_{443}^2 + 64s_{111}s_{444} + 64s_{122}s_{444} \\
& + 64s_{133}s_{444} + 64s_{144}s_{444} + 128s_{411}s_{444} + 128s_{422}s_{444} + 384s_{433}s_{444} + 576s_{444}^2
\end{aligned}
$$

References

[1] S.G. Kelly, Advanced Vibration Analysis, CRC Press, New York, 2006.

[2] W.-C. Xie, Dynamic Stability of Structures, Cambridge University Press, Cambridge, 2006.

[3] Q. Huang, W.-C. Xie, Stability of SDOF linear viscoelastic system under the excitation of wideband noise, J. Appl. Mech. 75 (2) (2008) 021012.

[4] D. Pang, W. Jiang, S. Liu, D. Jun, Stability analysis for a single degree of freedom fractional oscillator, Phys. A: Stat. Mech. Appl. 523 (2019) 498–506.

[5] W. Zan, W. Jia, Y. Xu, Response Statistics of Single-Degree-of-Freedom Systems with Lévy Noise by Improved Path Integral Method, International Journal of Applied Mechanics 14 (2022), http://doi.org/10.1142/s1758825122500296.

[6] J. Deng, Higher-order stochastic averaging for a SDOF fractional viscoelastic system under bounded noise excitation, Journal of the Franklin Institute 354 (2017) 7917–7945.

[7] J. Deng, Z. Zhong, A. Liu, Stochastic stability of viscoelastic plates under bounded noise excitation, Eur. J. Mech. Solid. 78 (2019) 103849.

[8] V. Stojanović, M. Petković, Moment Lyapunov exponents and stochastic stability of a three-dimensional system on elastic foundation using a perturbation approach, J. Appl. Mech 80 (2013) 051009.
[9] N. Sri Namachchivaya, H.J. Van Roessel, Maximal Lyapunov exponent and rotation numbers for two coupled oscillators driven by real noise, J. Stat. Phys. 71 (1993) 549–567.
[10] R. Khasminskii, N. Moshchuk, Moment Lyapunov exponent and stability index for linear conservative system with small random perturbation, SIAM Journal on Applied Mathematics 58 (1998) 245–256.
[11] L. Arnold, M.M. Doyle, N. Sri Namachchivaya, Small noise expansion of moment Lyapunov exponents for two-dimensional systems, Dynam Stabil Syst 12 (1997) 187–211.
[12] P. Kozić, G. Janevski, R. Pavlović, Moment Lyapunov exponents and stochastic stability of a double-beam system under compressive axial loading, Int J Solids Struct. 47 (2010) 1435–1442.
[13] Q. Liu, Y. Xu, Y. Li, J. Kurths, X. Liu, Fixed-interval smoothing of an aeroelastic airfoil model with cubic or free-play nonlinearity in incompressible flow, Acta Mechanica Sinica 37 (2021) 1168–1182.
[14] W. Zan, W. Jia, Y. Xu, Reliability of dynamical systems with combined Gaussian and Poisson white noise via path integral method, Probabilistic Engineering Mechanics 68 (2022) 103252.
[15] W. Zan, Y. Xu, R. Metzler, J. Kurths, First-passage problem for stochastic differential equations with combined parametric Gaussian and Lévy white noises via path integral method, Journal of Computational Physics 435 (2021) 110264.
[16] Y. Xu, W. Zan, W. Jia, J. Kurths, Path integral solutions of the governing equation of SDEs excited by Lévy white noise, Journal of Computational Physics 394 (2019) 41–55.
[17] J. Li, Z. Lv, D. Liu, Y. Xu, Stochastic vibration analysis of a nonlinear oscillator with symmetric viscoelastic impact protection under wide-band noise excitations, Meccanica 57 (2022) 1491–1503.
[18] W. Zan, Y. Xu, J. Kurths, Path integral solutions for n-dimensional stochastic differential equations under α-stable Lévy excitation, Theoretical and Applied Mechanics Letters 13 (2023) 100430.
[19] N.A. Saeed, E.M. Awwad, M.A. El-Meligy, E. Abouel Nasr, Sensitivity analysis and vibration control of asymmetric nonlinear rotating shaft system utilizing 4-pole AMBs as an actuator, Eur. J. Mech. Solid. 86 (2021) 104145.
[20] B. Xiao, Y. Li, S. Shi, C. Gao, S. Lu, Analysis of bending-torsional-axial vibration of multi-stage variable-section shaft system, Results Phys 36 (2022) 105460.
[21] F. Yang, Y. Pei, Dynamics and phase-based vibration suppression of rotating flexible shaft with unstressed initial deformation under several parametric excitations, J. Sound Vib. 509 (2021) 116248.
[22] J. Deng, Dynamic stability of a viscoelastic rotating shaft under parametric random excitation, Int. J. Non-Linear Mech. 84 (2016) 56–67.
[23] D. Jordan, P. Smith, Nonlinear Ordinary Differential Equations: An Introduction for Scientists and Engineers, 4th edition, Oxford University Press, 2007.
[24] W. Wedig, Lyapunov exponent of stochastic systems and related bifurcation problem, in: S.T. Ariaratnam, G.I. Schuëller, I. Elishakoff (Eds.), Stochastic Structural Dynamics - Progress in Theory and Applications, Elsevier Applied Science, London, 1988, pp. 315–327.

Part V

From traditional methods to Artificial Intelligence

Modeling and applications of markers in machine learning and technical practice

20

The application of artificial intelligence (AI) in the modeling and solving of mechanical systems represents a significant advance in engineering and scientific disciplines. AI software, particularly those based on machine learning (ML), enables the dynamic adaptation of models to real-world conditions and environmental changes. These systems can accurately estimate unknown parameters of mechanical models based on real measurements and predict future system behavior.

The common problem in mathematical modeling is to estimate the unknown parameters of the model, based on some real-life measurements. Another important scenario is to dynamically adjust some (controllable) model parameters, based on the measurements and due to the change of some global parameters of the environment. In both cases, one can use ML models to obtain such a prediction [1,2]. In this chapter, we introduce some of the fundamental ML models and show examples of their usage, to tackle the above problems. Note that some recent investigations propose ML models that can be used directly to model the solutions to problems from physics and engineering. An example is the so-called physics-informed neural networks. We will not consider these models here, and the reader is referred, for example, to Ref. [3].

Key advantages of AI software in this field include the ability to analyze large datasets quickly and accurately, automatically detect patterns and anomalies, and optimize system performance. For instance, neural networks, which are extensively discussed in this chapter, can be used to model complex mechanical systems such as oscillators and systems with elastic suspensions. These models allow engineers to simulate and analyze systems with multiple degrees of freedom, identify critical failure points, and improve design for better performance and reliability.

20.1 Formulation of the problem

Consider the model f which maps the set of parameters $\mathbf{p} \in \mathbb{R}^k$ to set of the outcomes $\mathbf{y} \in \mathbb{R}^n$. The function f usually requires some cumbersome numerical computation. Assume that we have obtained the measurement vector $\mathbf{m} \in \mathbb{R}^n$ which approximates $\mathbf{y}$. The goal is to provide the estimation of the unknown parameters vector $\mathbf{p}^* \in \mathbb{R}^k$ such that $\mathbf{f}(\mathbf{p}^*) \approx \mathbf{m}$.

The classical approach to this problem is to solve the unconstrained optimization problem

$$\min_{\mathbf{p}\in\mathbb{R}^k} \|\mathbf{f}(\mathbf{p}) - \mathbf{m}\| \tag{20.1}$$

Modeling of Complex Dynamic Systems. DOI: https://doi.org/10.1016/B978-0-443-23942-7.00020-6

See, for example Ref. [4]. This approach usually requires a significant number of evaluations of the function $\mathbf{f}(\cdot)$, which might require extensive computational effort. For example, $\mathbf{f}$ might represent any numerical method considered in the previous chapters. Sometimes, to obtain the single data point on the graph, it required hours, even days of the computation and significant hardware resources.

One way to overcome these difficulties is to train a certain ML model $\mathfrak{M}$ which will take the measurements vector $\mathbf{m}$, and produce the approximation $\widetilde{\mathbf{p}}$ of the optimal solution $\mathbf{p}^*$ of Eq. (20.1). In other words, the alternative to solving the optimization problem (Eq. 20.1) is to obtain the approximate solution by

$$p = \mathfrak{M}(\mathbf{m}) \tag{20.2}$$

where $\mathfrak{M}$ is our (trained) ML model.

The most popular ML models for this and similar purposes are neural networks. In the next section, we give the basic definitions of the neural networks, and briefly describe the training procedure. Then we introduce a toy example of this scenario.

20.2 Neural networks

The artificial neural networks are nowadays one of the most popular ML models, both for their wide range of practical applications and relatively simple design. The structure of a neural network is shown in Fig. 20.1.

The neural network consists of several layers, represented by vectors $\mathbf{z}^{(j)} \in \mathbb{R}^{n_j}$ for $j = 0,1, ...,N$ such that $\mathbf{z}^{(0)}$ is **input**, while $\mathbf{z}^{(N)}$ is the **output** layer. We have that the jth layer contains exactly n_j **neurons**, that is, that the corresponding vector $\mathbf{z}^{(j)}$ has length n_j. These vectors are connected by the following expression:

$$\mathbf{z}^{(j+1)} = \mathbf{f}_{j+1}(\mathbf{W}^{(j)} + \mathbf{z}^{(j)} + \mathbf{b}^{(j)}) \tag{20.3}$$

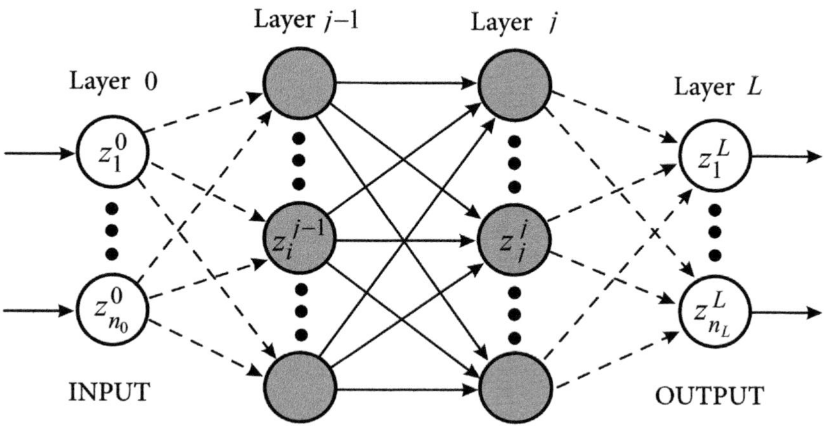

Figure 20.1 Structure of the neural network. A multilayer neural network with the input layer, output layer, and the intermediary hidden layers.

where $\mathbf{W}^{(j)} \quad \mathbb{R}^{n_j \times n_{j+1}}$ and $\mathbf{b}^{(j)} \quad \mathbb{R}^{n_{j+1}}$ for $j = 0, 1, ..., N-1$ are network **parameters**. Function $\mathbf{f}_{j+1}: \mathbb{R}^{n_{j+1}} \rightarrow \mathbb{R}^{n_{j+1}}$ is defined **componentwise**, that is, by $\mathbf{y} = \mathbf{f}_{j+1}(\mathbf{x})$ where $y_i = f_{j+1}(x_i)$ (for $i = 1, 2, ..., n_{j+1}$) where $f_{j+1}: \mathbb{R} \rightarrow \mathbb{R}$ is the **activation** function. The usual choices for the activation function f_j are:

1. **Linear** $f_j(x) = x$;
2. **ReLU**: $f_j(x) = \begin{cases} x & x > 0 \\ 0 & x \leq 0 \end{cases}$;
3. **Sigmoid**: $f_j(x) = (1 + e^{-x})^{-1}$.

If $\mathbf{z}^{(0)} = \mathbf{x}$ is the input vector, then the value $\mathbf{y} = \mathbf{z}^{(N)}$ is the output of the neural network, which we denote by $\mathbf{y} = \mathfrak{M}(\mathbf{x})$.

The main idea is to compute the set of parameters $\mathscr{P} = \{(\mathbf{W}^{(j)}, \mathbf{b}^{(j)}) | j = 0,1, ..., N-1\}$, such that the function $\mathfrak{M}$ which represents the neural network, has certain behavior. In fact, if we have given the set of pairs

$$\mathscr{D} = \{(\mathbf{x}_i, \mathbf{y}_i) \mid i = 1, 2, \ldots, k\}$$

obtained as the response from the certain unknown system (called the **dataset**), the goal is to estimate the parameters $\mathscr{P}$ such that the vector $\mathfrak{M}(\mathbf{x})$ is the approximation of the system response for the input $\mathbf{x}$. The network should be able to provide a certain level of **generalization**, that is, if $\overline{\mathbf{x}} \neq \mathbf{x}_i$ for $i = 1,2, \ldots, k$ is the unknown input point, then the obtained vector $\overline{y} = \mathfrak{M}(\overline{\mathbf{x}})$ should be close to the vector obtained from the system, when the input is $\overline{x}$.

20.3 Training and validation of the neural network

To provide, for a given dataset $\mathscr{D}$, the accurate neural network with a decent generalization property, we need to split $\mathscr{D}$ into two parts: **training** and **validation**. One denotes these two parts by $\mathscr{D}_{\text{train}}$ and $\mathscr{D}_{\text{test}}$. Each sample from $\mathscr{D}$ is in exactly one of these two sets (i.e., $\mathscr{D} = \mathscr{D}_{\text{train}} \cup \mathscr{D}_{\text{test}}$ and the union is disjoint), while$|\mathscr{D}_{\text{train}}|: |\mathscr{D}_{\text{test}}|$ is usually taken to be 80:20. Other options are 70:30, 60:40, etc.

Then, the model parameters are adopted to minimize the total error on the training dataset $\mathscr{D}_{\text{train}}$, and additionally verified on the validation dataset $\mathscr{D}_{\text{test}}$. In other words, the goal is to find optimal values of $\mathscr{P}$ such that the total error on $\mathscr{D}_{\text{train}}$ is minimized, that is,

$$\mathscr{P}^* = \underset{\mathscr{P}}{\operatorname{argmin}} \mathscr{L}(\mathscr{P}, \mathscr{D}_{\text{train}}) + \operatorname{reg}(\mathscr{P}, \lambda), \ \mathscr{L}(\mathscr{P}, \mathscr{D}) = \sum_{(\mathbf{x},\mathbf{y}) \in \mathscr{D}} \operatorname{loss}(y, \mathfrak{M}_{\mathscr{P}}(\mathbf{x})) \tag{20.4}$$

Function loss(·, ·) is known as the **loss** function and represents how well the model $\mathfrak{M}_{\mathscr{P}}$ with parameters, $\mathscr{P}$, approximates the true system response $\mathbf{y}$ at the point $\mathbf{x}$. The Euclidean norm $\operatorname{loss}(\mathbf{y}, \mathbf{y}') = \|\mathbf{y} - \mathbf{y}'\|$ is the most commonly used loss function, but there are also others [5].

The function reg(·, ·) is the **regularization** function and λ are regularization parameters. It is taken into account to prevent the parameters $\mathscr{P}$ adopt the training dataset $\mathscr{D}_{\text{train}}$ too well, and then become unable to provide the required generalization.

This phenomenon is known as the **overfitting**. Despite the regularization, there are as well other techniques how to prevent overfitting (see, e.g., Ref. [5]).

Finally, $\mathscr{L}(\cdot, \cdot)$ is the total objective function which needs to be minimized over the set of all possible parameter values (first argument). For that purpose, unconstrained optimization methods like gradient descent, stochastic gradient descent, etc., are used (see, e.g., Refs. [4,5]).

The total error on the validation set $\mathscr{L}(\mathscr{P}^*, \mathscr{D}_{test})$ is used as the performance measure of the method itself. Since the data points from $\mathscr{D}_{test}$ are not "seen" by the model during the training process (i.e., $\mathscr{P}$ is optimized using only the training set $\mathscr{D}_{train}$), the error on the validation set (i.e., **validation**) error is used to measure the generalization ability of the obtained model $\mathfrak{M}_{\mathscr{P}^*}$.

20.4 Application on the two simple mechanical systems

In Figs. 20.2 and 20.3, the two simplest fundamental models of discrete and continuous mechanical systems are presented, where it is possible to collect response data in two ways:

1. By recording with an ultra high-speed camera,
2. More precisely, using marker sensors placed discretely on the structure or mechanical system.

For the discrete system shown in Fig. 20.2, one can measure the deflection $w(t)$ in the equidistant time intervals, and use that information to obtain the unknown system parameters: mass of the oscillator m and unknown stiffnesses of the springs k_1 and k_2. In other words, we want to use the obtained measured values of $w_j = w(j\Delta t)$ for $j = 0,1, \ldots,n - 1$, to obtain m, k_1 and k_2. In terms of the notation

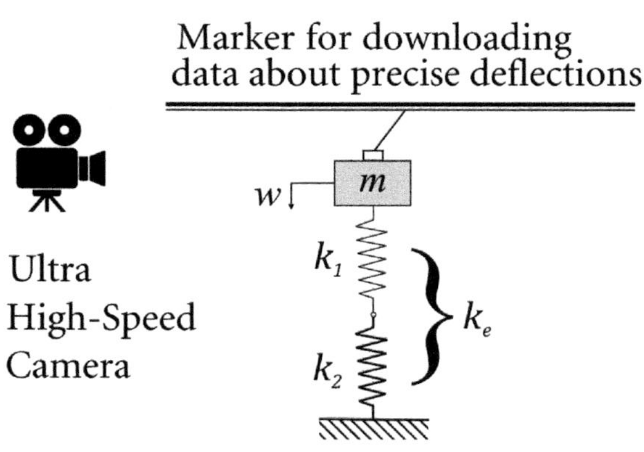

Figure 20.2 Simple discrete mechanical system. This system has a marker placed to track mass displacement, velocity, or acceleration using an ultra high-speed camera.

introduced in previous sections, we have that the input and output vectors **y** and **p** are given by:

$$\mathbf{y} = \begin{bmatrix} w_1 \\ w_2 \\ \vdots \\ w_n \end{bmatrix}, \mathbf{p} = \begin{bmatrix} m \\ k_1 \\ k_2 \end{bmatrix} \tag{20.5}$$

Therefore according to the measurements vector $\mathbf{m} \approx \mathbf{y}$, we obtain the estimations of the unknown parameters $\overline{\mathbf{p}} = \mathfrak{M}(\mathbf{m})$. The system is trained on the generated dataset

$$\mathscr{D} = \{(\mathbf{y}^{(i)}, \mathbf{p}^{(i)}) \mid i = 1, 2, ..., N\} \tag{20.6}$$

using the numerical computation (or analytical solution in this simple case), where N is the total number of training points.

Similarly, in the case of the continuous system, one can form the input vector **y** with the (positions) deflections of all markers in the equidistant time intervals. The unknown parameters vector **p** now consists of:

1. The beam length l, width b, and thickness h;
2. Young's modulus E;
3. Shear modulus of elasticity G;
4. Stiffness of the elastic layer K.

The set of data points $\mathscr{D}$ is generated in the same way as for the discrete system.

It is worth noting that, these simplified models, where the true dependence **f** between **p** and **y** is analytical, can be used to obtain the optimal structure of the model $\mathfrak{M}$ (e.g., if the neural network is used, to obtain the number of layers L, number of nodes per layer n_j, etc.) as well as the optimal number of training points N. The reason for that is the fact that it is much easier to generate required data points $\mathscr{D}$, than for the real model where advanced numerical methods need to be used.

20.5 Application to the design of the advanced mechanical system for wave energy extraction

We describe the complex mechanical system for the extraction of energy from the water waves. The main idea is to use ML models (e.g., neural networks, described in Section 20.3) to predict the wave profile. The entire system is shown in Fig. 20.4.

As it is shown in Fig. 20.4, the system consists of a generator shaft A, which is connected to the floaters B according to the "gedore type" G. This actually means that the useful torque is transferred to the shaft A during the rotation of P in the mathematically negative direction (incidence of waves) while descending to the reference level (after the passage of the water wave), rotation R occurs when the torque is not transmitted to the shaft A. Thus the floaters B are brought to a horizontal position, ready for the encounter of a new water wave with an insignificant loss of energy. Generator shaft A is attached to the foundation (the bottom near the shore). Floaters B are placed symmetrically in relation to the generator shaft A due to the

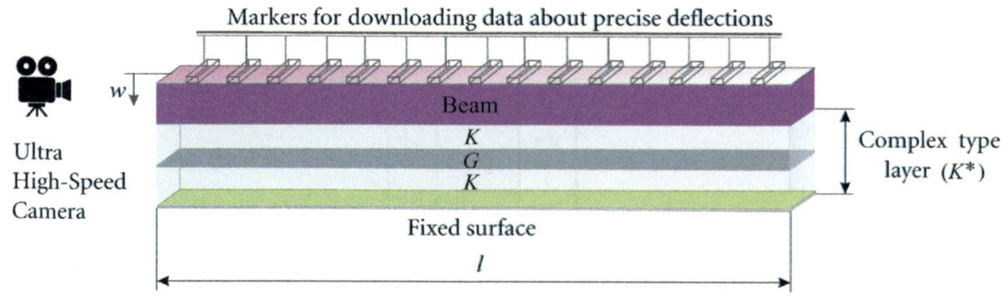

Figure 20.3 Simple continuous mechanical system. This system has markers placed to track beam displacements, velocities, or accelerations at specific points using an ultra-high-speed camera.

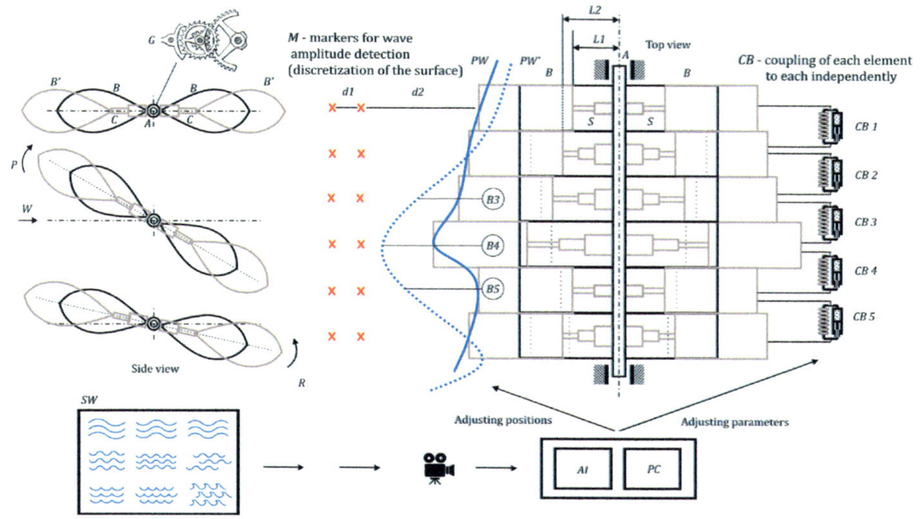

Figure 20.4 The schematic of the complex mechanical system for the energy extraction from water waves. The schematic diagram illustrates the mechanical system of a platform for extracting mechanical energy, including the variable position of the blades, an example of the placement of markers, and the principle of processing input information from them.

stability of the entire mechanical system. However, as the magnitude of the torque transmitted in rotation mode, *P* is conditioned by the distance from the center of gravity of the floaters to the generator shaft *A*. Therefore it is necessary to consider the design conditions according to the usual wave size. The concept of a pair of floaters *B* is designed so that the distance (*L*1–*L*6) between the floater *B* and the shaft *A* can be adjusted pneumatically, which is controlled by low-power servo motors *S* (power for the servo motors can be obtained by solar means). During periodic or stochastic impacts of the water waves, the torque on the generator-type shaft *A* is transmitted intermittently. This fact can cause a significant decrease in the amount of the conserved energy.

In other words, one needs to establish a more uniform torque transmission regime. The idea is that the surface of the incoming waves is prerecorded by placed markers at certain distances d_1 and d_2 (it is left for some time before the actual contact with the floaters B), whereby the peak line PW of the waves could be predicted before the lift of the floaters B (when the useful torque is generated). The continuous line PW represents the line connecting the points of maximum amplitudes of water waves (relative to the reference horizontal position of the generator shaft A) that is closest to the floaters B. The main goal is to record and memorize the surface of the incoming waveform and use it to determine the expected PW line. It will be the domain of possible retrieval of floaters B.

Based on the determined PW line expected in the zone of retrieval of floaters, servo motors S position the floaters (by regulating the lengths of the $L1$–$L6$ floaters from the reference generator shaft A). This is achieved to obtain the maximum torque, at the same moment in time, of all floaters connected to the generator shaft A. It is very important because at that moment the initial torque is obtained which will take the generated mechanical energy to the generator until the next "hit" of a new water wave with a different PW line. At the moment when the amplitude of the floaters begins to decrease, the shaft connected to them using the "Gedore principle" does not drain the energy so that the floaters take the position for the new PW line.

The approximate PW line can be computed using the AI models like neural networks. Denote by $t_k = k\Delta t$ the discrete time, where Δt is the time interval and $k = 0,1,2, \ldots$. Denote by $\boldsymbol{m}^{(k,1)} \quad \mathbb{R}^{n_1}$ and $\boldsymbol{m}^{(k,2)} \quad \mathbb{R}^{n_2}$ the vectors of marker readings at distances d_1 and d_2 , respectively, at the time moments t_k, for $k = 0,1,2, \ldots$. The goal is to predict the current readings $\boldsymbol{m}^{(k,1)}$ at the distance d_1, based on the current readings $\boldsymbol{m}^{(k,2)}$ at the distance $d_2 > d_1$ and historical data $\boldsymbol{m}^{(k-j,1)}$, $\boldsymbol{m}^{(k-j,2)}$ for $j = 1,2, \ldots, m$. In other words, we use the present readings of the second marker set, and past readings of both sets in the time frame $m\Delta t$, to predict current readings $\boldsymbol{m}^{(k,1)}$.

The first model one can consider is the simplest model: a single (feed-forward) neural network. The input and the output vectors at the moment t_k are given by:

$$\boldsymbol{x}^{(k)} = \begin{bmatrix} \boldsymbol{m}^{(k,2)} \\ \boldsymbol{m}^{(k-1,1)} \\ \boldsymbol{m}^{(k-1,2)} \\ \vdots \\ \boldsymbol{m}^{(k-m,1)} \\ \boldsymbol{m}^{(k-m,2)} \end{bmatrix}, \boldsymbol{y}^{(k)} = \boldsymbol{m}^{(k,1)}. \tag{20.7}$$

The goal is to train the model $\overline{\mathbf{y}}^{(k)} = \mathfrak{M}(\boldsymbol{x}^{(k)})$ according to the training dataset:

$$\mathscr{D} = \{(\mathbf{y}^{(i)}, \mathbf{x}^{(i)}) \mid i = m, m+1, \ldots, m+N\}, \tag{20.8}$$

which requires a total $N + m$ sets of marker readings.

The second possibility is to consider the different architecture consisting of two models $\mathfrak{M}_0$ and $\mathfrak{M}_1$. The idea is to process the input at a specific time moment t_{k-j}

separately using the (same) model $\mathfrak{M}_1$ and then to combine the outputs as an input to the final model $\mathfrak{M}_0$. In other words, one computes

$$z^{(k,j)} = \mathfrak{M}_1(\boldsymbol{x}^{(k,j)}), j = 1, 2, \ldots, m, \tag{20.9}$$

$$\bar{\boldsymbol{y}}^{(k)} = \mathfrak{M}_0(\boldsymbol{x}^{(k,0)}), \tag{20.10}$$

where

$$\boldsymbol{x}^{(k,0)} = \begin{bmatrix} \boldsymbol{m}^{(k,2)} \\ z^{(k,1)} \\ \vdots \\ z^{(k,m)} \end{bmatrix}, \boldsymbol{x}^{(k,j)} = \begin{bmatrix} \boldsymbol{m}^{(j,1)} \\ \boldsymbol{m}^{(j,2)} \end{bmatrix}, j = 1, 2, \ldots, m. \tag{20.11}$$

Vectors $z^{(k,j)}$, $j = 1,2, \ldots, m$ are obtained as the transformations of the reading vectors $\boldsymbol{m}^{(j,1)}$ and $\boldsymbol{m}^{(j,2)}$ at the past moment t_{k-j}, by the transformation model $\mathfrak{M}_1$. The purpose of the model $\mathfrak{M}_1$ is to remove the redundances and extract only the relevant information at each past moment t_j, necessary to obtain the final prediction. Transformed values are then joined together with the present readings $\boldsymbol{x}^{(k,2)}$ to form the input vector to the final model $\mathfrak{M}_0$. The training dataset is also given by the Expression (20.8), with the only difference that the components of the input vector $\boldsymbol{x}^{(i)}$ are separately used to simultaneously train the models $\mathfrak{M}_0$ and $\mathfrak{M}_1$.

Note that one can possibly combine the models $\mathfrak{M}_0$ and $\mathfrak{M}_1$ in the other way, or introduce more than two different models, which are trained simultaneously. Such combinations of different models in one architecture are used in modern transformer networks [6].

20.6 Conclusions

The implementation of ML and AI in engineering, particularly in the context of mechanical systems, is a promising field of research. This chapter addresses and illustrates examples, while also examining the dynamics of mechanical systems through the application of AI tools. It demonstrates how measurable parameters from real mechanical systems can be leveraged to predict their dynamics when conditions change. The chapter emphasizes the advantages of using AI tools in complex mechanical systems with multiple degrees of freedom. An advanced technical solution for generating electrical energy from wave power serves as an example of how optimal geometric and material parameters in a mechanical system can be determined, and how the shape of an incoming wave can be predicted to maximize the utilization of mechanical energy. In this scenario, AI software takes precedence by processing the collected data in real time and providing optimal parameters for the mechanical system, which can be adjusted to achieve maximum energy efficiency.

The theoretical foundation and application of AI tools in predicting dynamic behavior are first demonstrated using a simple mechanical system. The same concept is then extended to a more complex model, followed by an application on a novel technical solution from engineering practice. A key conclusion is the potential for using appropriate markers and sensors to collect necessary data from a real mechanical system, processing that data, and obtaining a good approximate solution for the dynamic behavior of both simple and complex models, mechanical systems, or structures. In cases where an analytical solution is impossible to achieve and numerical methods are unstable, the application and processing of data using AI techniques, as described in this chapter, play a crucial role. This not only enhances traditional analytical-numerical analysis techniques but also paves the way for future innovations in engineering.

References

[1] V.M. Nguyen-Thanh, X. Zhuang, T. Rabczuk, A deep energy method for finite deformation hyperelasticity, European Journal of Mechanics/A Solids 80 (2020) 103874.

[2] C. Leake, D. Mortari, Deep theory of functional connections: a new method for estimating the solutions of partial differential equations, Machine Learning and Knowledge Extraction 2 (2020) 3755.

[3] S. Cuomo, V.S. Di Cola, F. Giampaolo, G. Rozza, M. Raissi, F. Piccialli, Scientific machine learning through physics–informed neural networks: where we are and what's next, Journal of Scientific Computing 92 (2022) 88.

[4] L. Grippo, M. Sciandrone, Introduction to Methods for Nonlinear Optimization, Springer, Switzerland, Cham, 2023.

[5] I. Goodfellow, Y. Bengio, A. Courville, Deep Learning, MIT Press, Cambridge, MA, 2016.

[6] U. Kamath, K. Graham, W. Emara, Transformers for Machine Learning: A Deep Dive, Chapman & Hall/CRC, New York, 2022.

Index

Note: Page numbers followed by "*f*" and "*t*" refer to figures and tables, respectively.

D

9780443239427